Specialty Optical Fibers Handbook

Specialty Optical Fibers Handbook

ALEXIS MÉNDEZ
MCH Engineering, LLC, Alameda, California

T. F. MORSE
Photonics Center, Boston University, Boston, Massachusetts

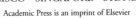

AMSTERDAM • BOSTON • HEIDELBERG • LONDON
NEW YORK • OXFORD • PARIS • SAN DIEGO
SAN FRANCISCO • SINGAPORE • SYDNEY • TOKYO

Academic Press is an imprint of Elsevier

Academic Press in an imprint of Elsevier
30 Corporate Drive, Suite 400, Burlington, MA 01803, USA
525 B Street, Suite 1900, San Diego, California 92101-4495, USA
84 Theobald's Road, London WCIX 8RR, UK

This book is printed on acid-free paper. ⊚

Library of Congress Cataloging-in Publication Data
Application Submitted

British Library Cataloguing in Publication Data
A catalogue record for this book is available from the British Library

ISBN 13: 978-0-12-369406-5
ISBN 10: 0-12-369406-X

For information on all Elsevier Academic Press publications
visit our Web site at www.books.elsevier.com

To my wife Shiva
for her unconditional love, support, and patience
A.M.

"Under the shade of your tresses,
how softly slept my heart,
intoxicated and lovely,
so peaceful and so free..."

RUMI

To Edelgard
for her patience, wisdom, and love.
T.F.M.

Contents

7   Rare Earth-Doped Fibers   **195**
David J. DiGiovanni, Roman Shubochkin, T. F. Morse,
and Borut Lenardic

8   Polarization Maintaining Fibers   **243**
Chris Emslie

Editors

Alexis Méndez   Dr. Alexis Méndez is President and founder of MCH Engineering LLC, a consulting firm specializing in optical fiber sensing technology. Dr. Méndez has over 20 years of experience in optical fiber technology, sensors, and instrumentation. Prior to founding MCH, Dr. Méndez occupied various management positions within the optical communications industry in Silicon Valley. He was the former Group Leader of the Fiber Optic Sensors Lab within ABB Corporate Research (USA), where he led research and development sensor activities for oil and gas, electric utility, and industrial processing applications. He has developed fiber Bragg grating downhole pressure and temperature sensors, fiber optic high voltage and current sensors, and others. He has also conducted research to investigate hydrogen effects on fibers.

Dr. Méndez has written over 45 technical publications, holds four US patents and is the recipient of an R&D 100 award. He is also chairman of the next International Optical Fiber Sensors Conference (OFS-18). Dr. Méndez holds a PhD degree in electrical engineering from Brown University.

T. F. Morse   T. F. Morse received a BA (english literature, 1953) and an MA (history, 1954) from Duke University. He was an International Institute of Education Fellow at Cologne University, Germany in 1954–1955 (history, political science). From 1956–1959, he was employed at Pratt and Whitney Aircraft, East Hartford, Connecticut, during which time he received an ScB (mechanical engineering) from the University of Hartford, and a MSc (mechanical engineering) from the Rensselaer Polytechnic Hartford Graduate Center. Attending graduate school at Northwestern University, he was awarded a PhD (mechanical engineering) in 1961. From 1961–1963, he was a Senior Scientist at ARAP in Princeton, New Jersey where he worked on a variety of theoretical fluid mechanical problems. As an Engineering Professor at Brown University, Providence, Rhode Island (1963–1999), he was the Director of the Laboratory for Lightwave Technology and in 1969–1970, was a Senior Fulbright Research Professor at the Deutsche Versuchs-Anstalt fuer Luft u. Raumforschung. Since 1999, he has been at Boston University as Professor of Electrical and Computing Engineering and Director of the Laboratory for Lightwave Technology. He is the author of over 120 papers and holds five patents. His research interests and areas of expertise are in fiber processing, photonic materials, fiber lasers, and fiber sensors.

List of Contributors

Moshe Ben-David   Dr. Moshe Ben-David received his PhD degree in physics from Tel-Aviv University, Tel-Aviv, Israel in 2003. He has over 10 years experience in developing electro-optical systems for military, telecommunication, entertainment, and medical applications, currently with Glucon Medical. He is the author of over 20 papers, 4 book chapters, and 4 patents. His research fields are: optical fibers and waveguides, laser tissue interaction, optical diagnostics methods in medicine, and light propagation in tissue.

Ryan Bise   Dr. Ryan Bise is a member of the Technical Staff at OFS Laboratories in Murray Hill, New Jersey, formerly the optical fiber research arm of Bell Laboratories, Lucent Technologies. His research focus is on the fabrication and design of microstructured fibers. He received his undergraduate and graduate training in chemistry from UCLA and UC Berkeley, respectively.

Vladimir A. Bogatyrev   Vladimir A. Bogatyrev graduated from the Moscow Physical Technological Institute in 1972. From 1972 to 1982, he investigated high-power neodymium lasers in PN Lebedev Physical Institute RAS. Since 1982, his research interests were moved to the technology of optical fibers and related topics (the fiber drawing process, properties of polymer and hermetic coatings, strength and fatigue of optical fibers). He is currently a Research Fellow of the Fiber Optics Research Center RAS, Moscow, Russia. His main activity is focused on metal-coated fibers (technology of fabrication and investigation of their mechanical and optical properties).

Adrian Carter   Dr. Adrian Carter is the founder and CTO of Nufern. Prior to that, he was an Assistant Professor at the Laboratory for Lightwave Technology at Brown University. Dr. Carter was a Postdoctoral Fellow at the Optical Fibre Technology Center in Sydney, Australia where he focused on the design and fabrication of novel specialty optical fibers, having also been a Research Fellow at the Technische Universitaet Hamburg-Harburg, Germany. He received his PhD in the Department of Physical and Theoretical Chemistry and his BSc in mathematics and chemistry from the University of Sydney, where he is currently also an Honorary Research Associate.

Kai H. Chang   Kai H. Chang has been the Engineering Manager at Heraeus Tenevo USA since 2005. From 1986 to 2005, he worked at Bell Laboratories of

AT&T/Lucent Technologies/OFS in Norcross, Georgia, and he was the Technical Manager of MCVD Technology and a Distinguished Inventor. Kai has worked on optical loss mechanisms in silica fibers and was one of the pioneers that developed zero-OH AllWave® fiber. Kai obtained his PhD in physics from University of Toronto in 1984 and worked at Caltech as a Research Fellow in physics from 1984 to 1986.

Åsa Claesson Åsa Claesson graduated with a MSc in materials science from Uppsala University, Sweden, in 1997 and has since been active in developing fiber based optical components, as well as optical specialty fibers. She has co-authored more than 20 scientific publications, 4 patents, and 2 book chapters. She is presently the manager of Acreo Fiberlab in Sweden.

James P. Clarkin James P. Clarkin is the Vice President of Business Development at Polymicro Technologies, LLC. Mr. Clarkin has 20 years experience in the design, performance, and manufacture of all types of optical fibers. Prior to joining Polymicro in 1998, Jim spent 12 years at Ensign Bickford/ Spectran Corporation as Engineering Manager and Product Line Manager. While at Spectran, Jim led the development and production of their specialty optical fiber and cable product lines. Jim has a BS in chemical engineering, an MS in materials science, and an MBA, all from Rensselaer Polytechnic Institute.

André Croteau André Croteau received an MSc degree in physics from Queen's University in Kingston, Ontario, Canada in 1986. From 1986 to 1988, André was a researcher at the fundamental research laboratory of NEC Corporation in Japan, where he worked on the development of electro-optic thin films. In 1988, André joined INO as a researcher in the specialty optical fiber program where he became the manager in 1998. His main research activities include the development of active rare earth-doped fibers, micro structured fibers, and photosensitive fibers. He has published over 20 papers and received 3 patents.

David J. DiGiovanni David J. DiGiovanni is President of OFS Laboratories, LLC, the central research organization of OFS. David began his career with a postdoctoral position in the Optical Fiber Research Department in Bell Laboratories and has weathered the transition from AT&T to Lucent to OFS in what is essentially the same organization. He has worked on various phenomena related to optical fiber design and fabrication and has made notable contributions to erbium-doped optical fibers for amplifiers, high power amplifiers and lasers, Raman amplification, and optical components. David holds several degrees from Brown University, including a PhD in mechanical engineering. He is a member of IEEE and an OSA Fellow.

Chris Emslie   Dr. Chris Emslie is Managing Director of Fibercore Limited. He began his career in optical fibers in 1982 at Corning's pilot manufacturing facility in Wilmington, North Carolina. He received a PhD from the Optical Fiber Group at the University of Southampton (UK). His thesis focused on the manufacture of low-loss polymer fibers, under the guidance of Professors Alec Gambling and David Payne. Dr. Emslie left Southampton in 1987 to take a commercial role at a fledgling optical components company, York VSOP, and taking charge of its Specialty Fiber business in 1989. This business gradually evolved into Fibercore Limited.

Pierre-Yves Fonjallaz   Pierre-Yves Fonjallaz obtained the MSc degree in physical engineering and the PhD degree (fibre Bragg gratings) from the Swiss Federal Institute of Technology in Lausanne in 1990 and 1995. After a postdoctoral at KTH, he started to work at Acreo AB (Sweden) in 1996. He became manager of the Optical Fibre Components group (2000). He was appointed director of the Kista Photonics Research Centre (KPRC) in 2003, and as such coordinates the collaboration between Acreo and the Royal Institute of Technology (KTH) in photonics. He has been organizing a number of workshops and conferences, such as ECOC in 2004.

Israel Gannot   Professor Israel Gannot received his PhD degree in biomedical engineering from Tel-Aviv University, Tel-Aviv, Israel in 1994. Between 1994 and 1997, he held a National Academy Sciences postdoctoral fellowship. Since 1997, he has been a faculty member at Tel-Aviv University, and since 2005 he has been a professor of biomedical engineering at George Washington University. Professor Gannot is a Fellow of the American Institute of Medical and Biological Engineering. He is the author of over 100 papers, 6 book chapters, and 10 patents. His research fields are: optical fibers and waveguides, laser tissue interaction, optical diagnostics methods in medicine, and biomedical informatics.

Azriel Z. Genack   Azriel Genack is a Distinguished Professor of Physics at Queens College of CUNY, where he has been since 1984. He received his BA degree from Columbia College and his PhD in physics from Columbia University. Following his graduate studies, he was a Postdoctoral Research Associate at the City College of CUNY and then at the IBM Research Laboratory in San Jose. He was a researcher at the Exxon Corporate Research Laboratories from 1977 until 1984. He co-founded Chiral Photonics in 1999. His research centers on the photonics of chiral structures and the statistics of propagation and localization of optical and microwave radiation in random media.

James A. Harrington   James A. Harrington is a Professor of Ceramic Science and Engineering at Rutgers University. Dr. Harrington has over 30 years of research experience in IR materials and fibers and is the inventor of both the hollow sapphire and hollow glass waveguides. He is generally recognized as one of the world's leading experts in this continually evolving field. Prior to joining the Fiber Optic Materials Research Program at Rutgers University in 1989, he was Director of Infrared Fiber Operations for Heraeus LaserSonics, and prior to this role, he was the Program Manager for IR fiber optics at Hughes Research Laboratories in Malibu, California.

Juan Hernández-Cordero   Dr. Juan Hernández-Cordero received his BSc degree in electrical engineering from the National Autonomous University of Mexico (UNAM) in 1992. He was awarded a full scholarship to pursue graduate studies at Brown University, where he earned a Master's (1996) and PhD degrees (1999) from the Division of Engineering. After a year as a Postdoctoral Research Associate at the Laboratory for Lightwave Technology in Boston University, he joined the Materials Research Institute (IIM) of the UNAM, where he has established the Fiber Lasers and Fiber Sensors Laboratory. His fields of interest include optical fiber sensors, fiber lasers, and fiber devices.

Mihai Ibanescu   Mihai Ibanescu received his BS and PhD degrees in physics from the Massachusetts Institute of Technology in 2000 and 2005. From 2005 to 2006, he was a Postdoctoral Associate in the research group of Professor John Joannopoulos at MIT. During 2000–2001, and since 2006, he worked with OmniGuide Inc., in Cambridge, Massachusetts. His main research interests are photonic crystals, photonic band gap fibers, and hollow-core fiber applications.

Anne Claire Jacob Poulin   Anne Claire Jacob Poulin joined INO as a researcher in 2000 after a PhD degree in physics from the Center for Optics, Photonics, and Lasers of Laval University, Quebec, Canada and a MSc degree in physics from the Centre de Physique Moléculaire Optique et Hertzienne, Bordeaux I University, France. She first worked in the communication field with the fabrication of passive optical components with photosensitive fibers and the development of optical fiber amplifiers with specialty-doped fluoride fibers. Her current research interests are on the fabrication and application of photonics devices to sensors systems in the agri-food and biomedical fields.

Steven A. Jacobs   Steven Jacobs is the Systems Engineering Group Leader at OmniGuide Inc., where he leads the development of new medical systems that enable minimally invasive laser surgery based on OmniGuide's photonic-band-gap fiber technology. Prior to that position, he was the Theory and Simulations Group Leader. He received his BS degree from MIT and his PhD degree from the University of Wisconsin, both in physics. Before joining OmniGuide in 2001, Dr. Jacobs was a Distinguished Member of Technical Staff at Bell Laboratories. His professional interests include computational electromagnetics and the use of computational and statistical methods for yield and process improvement.

Steven G. Johnson Steven Johnson received his PhD in 2001 from the Department of Physics at MIT. He is currently an Assistant Professor of Applied Mathematics at MIT, and also consults for OmniGuide Inc. He has written several widely-used, free software packages, including the MPB package to solve for photonic eigenmodes and the FFTW fast Fourier transform library (for which he received the 1999 J. H. Wilkinson Prize). In 2002, Kluwer published his PhD thesis as a book, *Photonic Crystals: The Road from Theory to Practice.* His research interests include the development of new semi-analytical and numerical methods for electromagnetism in high-index-contrast systems.

Jinkee Kim Jinkee Kim received BS and MS degrees in electrical engineering from the Seoul National University and a PhD degree in electrical and computer engineering from the Georgia Institute of Technology. His doctoral research was on integrated optics, 100Gbit/s telecommunication, and digital signal processing. He worked at CREOL in Orlando as a Research Scientist, where his research was focused on photonic control systems for phased arrays. In 1996, he joined Bell Laboratories, Lucent Technologies (now OFS), and works in fiber optics R&D. He has designed and commercialized new optical fibers and holds several US patents.

Victor I. Kopp Victor Kopp is the Director of Research and Development at Chiral Photonics, Inc. He received his PhD degree in laser physics from the Vavilov Optical Institute, St. Petersburg, Russia in 1992. In 1999, working as a Research Associate at Queens College of CUNY, he developed the scientific basis for Chiral Photonics, Inc. with Azriel Genack and became a co-founder of the company. His research interests include wave propagation in periodic media, nonlinear optics, and photonic devices. He is the author and co-author of over 25 papers, as well as over 20 US and international patents on photonic devices, lasers, and fiber gratings.

Charles R. Kurkjian Dr. Kurkjian is currently a visiting scientist in the Materials Science and Engineering Department at Rutgers University, Piscataway, New Jersey. He had previously worked at Bell Laboratories in Murray Hill, New Jersey for 35 years and at Telcordia (formerly Bellcore) in Morristown, New Jersey for five years before retiring and joining Rutgers in 1999. He has worked in a number of areas of inorganic glass research and development. Currently, he is concentrating on the mechanical properties of such glasses, as well as the strength and reliability of silica lightguide fibers.

Paul J. Lemaire Paul J. Lemaire is a Senior Lead Engineer with General Dynamics Advanced Information Systems in Florham Park, New Jersey. He has held technical and management positions at OFS, Lucent Technologies, and Bell Laboratories. His work has been in the areas of optical fiber fabrication, hermetic fibers, fiber design, fiber Bragg gratings, photosensitivity, fiber and component reliability, hydrogen aging, and other topics pertaining to photonics, materials, and reliability. He has numerous publications, presentations, and patents in these areas. He received both his BS and PhD degrees from the Department of Materials Science and Engineering at MIT.

Borut Lenardic Borut Lenardic received a BSc degree in solid state physics from the Faculty of Natural Sciences, University of Ljubljana in 1981 and started his work in fiber optics in 1986 as a Development Engineer in Iskra, Slovenia. Later he worked as a Process Specialist in Cabloptic, Switzerland and Fotona, Slovenia. From 1996 to 2001, he worked as a consultant for Nextrom Oy. In 2001 he founded Optacore, a company dedicated to development of preform and fiber fabrication process, based on furnace-supported CVD, in Ljubljana, Slovenia. Since 2004, he has been developing technology and devices for fabrication of rare earth-doped fibers with emphasis on aerosol and high temperature sublimation processes.

Eric A. Lindholm Eric A. Lindholm received his BSci in ceramic engineering and his BA in english from Rutgers University in 1991. He spent five years at Spectran Communication Fiber Technologies as a Fiber Draw Engineer before becoming a Fiber Development Engineer at OFS Specialty Photonics (formerly Spectran Specialty Optics) in 1996. Eric has since focused on the hermetic carbon deposition process, polymer materials, and fiber draw processes designed to enhance the durability of optical fibers used in adverse environment applications, and characterization of the fibers. He has written and given several technical presentations on related subjects at various conferences.

Robert Lingle, Jr. Robert Lingle, Jr. completed his BS degree in physics from the University of Alabama, a PhD in chemical physics from the Louisiana State University, and held a postdoctoral fellowship at UC Berkeley in ultrafast physical chemistry. He joined the Optical Fiber Division of Lucent Technologies, Bell Laboratories in 1997, where he remained through the transition to OFS. He has conducted research on ultrafast electronic and vibrational processes in solution and at interfaces, sol-gel materials, physics and chemistry for optical materials, optical fiber design, and nonlinear impairments in optical transmission. Dr. Lingle is Director of Fiber Design and Transmission Systems at OFS.

John B. MacChesney John MacChesney is a retired Bell Labs Fellow and former member of Lucent's Photonics Materials Research Department. Dr. MacChesney is best known for his invention of the modified chemical vapor deposition (MCVD) process, for which he received the National Academy of Engineering's Charles Stark Draper Prize. He joined Bell Labs in 1959 and holds more than 100 domestic and foreign patents. Dr. MacChesney was elected to the

National Academy of Engineering in 1985, and has received awards from the American Ceramic Society, the IEEE, the American Physical Society, the Society of Sigma Xi, and others. He holds a BA degree from Bowdoin College and a PhD from Pennsylvania State University.

Walter Margulis Walter Margulis received his PhD from Imperial College, London in 1981. Presently, he works on the fabrication, characterization, and applications of fiber components, design, and fabrication of special fibers for active functions, poling of glass, photosensitivity, and applications of Bragg gratings in optical fibers, optical amplifiers, and passive microwave components. He has co-authored ~185 papers/conference contributions, ~15 patent applications and has supervised over 25 graduate students. He is a Senior Scientist at Acreo AB in Sweden, and a Guest Professor at the Royal Institute of Technology in Stockholm.

M. John Matthewson John Matthewson received his BA, MA, and PhD degrees in physics from Cambridge University. He continued at Cambridge concurrently as the Goldsmiths Junior Research Fellow at Churchill College and as a SRC Postdoctoral Fellow. He later worked at the Cambridge University Computing Service, AT&T Bell Laboratories, and IBM Almaden Research Center. He is now a Professor of Materials Science and Engineering at Rutgers University. His research group studies strength and fatigue of optical materials and modeling of materials processing. He has published over 100 papers, many of them concerning optical fiber reliability, and he has been editor or co-editor of six conference proceedings on the same topic.

Eric Mazur Professor Eric Mazur holds a triple appointment as Harvard College Professor, Gordon McKay Professor of Applied Physics, and Professor of Physics at Harvard University. His area of interest is optical physics. He received a PhD degree in experimental physics at the University of Leiden in the Netherlands. Dr. Mazur is author or co-author of 187 scientific publications and has made important contributions to spectroscopy, light scattering, and studies of electronic and structural events in solids that occur on the femtosecond time scale. In 1988, he was awarded a Presidential Young Investigator Award and is a Fellow and Centennial Lecturer of the American Physical Society.

Alan McCurdy Alan McCurdy graduated with degrees in chemical engineering (BS) and physics (BS) from Carnegie-Mellon University, and applied physics (PhD) from Yale University. He spent nine years on the faculty of the Department of Electrical Engineering at the University of Southern California. His telecommunications work began at Lucent Technologies, then Avaya, and most recently OFS. He has done research on high power, electron-beam driven microwave devices, transmission problems in copper-based enterprise network systems, and statistical and nonlinear problems in optical communications. Dr. McCurdy is currently a Distinguished Member of Technical Staff in the Optical Fiber Design Group at OFS.

Thomas D. Monte Thomas D. Monte received a PhD degree in electrical engineering from the University of Illinois at Chicago in 1996. As Principal Photonics Engineer at KVH Industries, Inc., he has been involved in the research of elliptical core polarization maintaining optical fiber components and sensor assemblies. Between 1983 and 2000, Dr. Monte held various engineering and research positions at Andrew Corporation developing fiber optic devices, microwave waveguide components, and antennas. He holds 11 US patents and several international patents in the fields of microwave and fiber optic components.

Stephen Montgomery Stephen Montgomery is the President of ElectroniCast, a firm specializing in communication network products and services demand forecasting. Stephen is also the Director of the Fiber Optics Components group and the Network Communication Products group at ElectroniCast. He has given numerous presentations and published a number of articles on optical fiber markets, technology, applications, and installations. He is a member of the Editorial Advisory Board of Lightwave magazine and the Advisory Board of the Gigabit Ethernet Conference (GEC). Stephen holds a BA and MBA in Technology Management.

Lars-Erik Nilsson Lars-Erik Nilsson graduated with a degree in chemistry from the Royal Institute of Technology, Stockholm in 1973, and has since been active within industry as well as academia with a main focus on development of optical instruments and components. Lars-Erik has been engaged in design and development of specialty optical fibers and fiber-based components for over 10 years and is presently heading the Optical Fiber Component group at Acreo AB in Sweden. He has co-authored over 12 scientific publications and is the inventor/co-inventor of six patents.

David W. Peckham David W. Peckham received BS and ME degrees in electrical engineering from the University of Florida. He started his career at the Bell Labs Transmission Media Laboratory in 1982 working on optical fiber measurement techniques. Since 1989, he has focused on the design, process development, and commercialization of optical fibers for high capacity transmission systems at Bell Labs, Lucent, and currently, OFS. He received the 2002 OSA Engineering Excellence Award recognizing his contributions in the design and commercialization of fibers enabling high speed, wideband WDM networks. He is currently a Consulting Member of Technical Staff at OFS.

J. Renee Pedrazzani   J. Renee Pedrazzani is a PhD candidate at the Institute of Optics of the University of Rochester, where she is engaged in semiconductor device research at the Molecular Beam Epitaxy Laboratory. She received her BS and MS degrees in electrical engineering from the Virginia Polytechnic Institute and State University, and conducted her MS research at the Fiber and Electro-Optic Research Center. Before beginning her doctoral studies, she worked with optical fiber gratings at Lucent Technologies.

Bryce Samson   Dr. Samson is the Vice President of Business Development at Nufern, having joined Nufern from Corning, where he served as Senior Research Scientist in the areas of doped fibers, fiber amplifiers, and lasers. Prior to that, he worked as a Research Fellow at the University of Southampton, focusing on novel fibers and fiber device physics. He received his PhD in physics from Essex University in the UK and his BS degree in applied physics from Heriott-Watt University in Edinburgh, UK. He is an inventor on several patents in the amplifier and fiber laser field and has been published in scores of industry journals.

Steven R. Schmid   Steven R. Schmid is R&D Manager for DSM Desotech's Fiber Optic Materials Research business. He has also held positions in product management, market development, and business management. He has 30 years experience in the UV coatings industry. Steven earned a BS in chemistry (University of Illiniois), an MS degree in chemistry (University of Houston), and an MBA (IIT). Steven has authored over a dozen papers, been awarded 10 patents, and made several international presentations. He was a co-recipient of an IR100 Award in 1987, and also a co-recipient of DSM's Special Inventor Award in 2001.

Sergei Semjonov   Sergei Semjonov graduated from the Moscow Physical Technological Institute in 1982. In 1997, he received his PhD in Physics from General Physics Institute, Russian Academy of Sciences, Moscow, Russia. He is currently a Deputy Director of the Fiber Optics Research Center RAS, Moscow, Russia. His research interests cover different aspects of modern fiber optics: fabrication of preforms, the fiber drawing process, properties of polymer and hermetic coatings, strength and fatigue of optical fibers, influence of drawing conditions on optical properties of optical fibers, development of rare earth-doped as well as highly Ge- and P-doped fibers, photosensitivity of optical fibers, and microstructured fibers.

Roman Shubochkin   Roman Shubochkin received BS and MS degrees in optical engineering from Moscow Power Engineering Institute, Moscow, Russia in 1987 and 1989, respectively. Between 1989 and 1994, he worked as a Junior Research Scientist in Fiber Optics and Solid State Physics Departments in the General Physics Institute of the Russian Academy of Sciences in Moscow. He received an MS and PhD in electrical engineering from Brown University in 1997 and 2003, respectively. Since 2000, he has been a research associate in the Lightwave

Technology Laboratory at Boston University. Dr. Shubochkin's research interests include the study of new techniques and dopants for fabrication of silica fibers, nanopowders, and glasses.

Bolesh J. Skutnik Dr. Bolesh J. Skutnik has been with CeramOptec Group since 1991. He holds a BS in chemistry/math from Seton Hall University, and an MS and PhD on theoretical physical chemistry from Yale University. Dr. Skutnik has been active in fiber optics since 1979. He is inventor of Hard Plastic Clad Silica optical fibers, as well as author of numerous articles and patents on strength, optical, and radiation behavior of step index fibers.

Cheryl A. Smith Cheryl Smith is a sales engineer for CeramOptec Industries, where she is responsible for the investigation of new applications for specialty fibers. Cheryl has over 20 years of experience in sales and marketing for specialty fiber optics and lasers.

Marin Soljačić Marin Soljacic received his PhD from the physics department at Princeton University in 2000. After that, he was a Pappalardo Fellow in the physics department of MIT. In 2003, he became a Principal Research Scientist at the Research Lab of Electronics at MIT. Since September 2005, he has been an Assistant Professor of Physics at MIT. He is the recipient of the Adolph Lomb medal from the Optical Society of America (2005). His main research interests are in photonic crystals and non-linear optics. He is a co-author of 55 scientific articles and is a co-author of 14 patents.

Kanishka Tankala Dr. Kanishka Tankala has been the VP of Operations at Nufern since 2000, and involved in the development and commercialization of specialty fibers and fiber laser subassemblies. Prior to joining Nufern, he was Technical Manager at Lucent Specialty Fiber and a scientist at SpecTran Corporation, where he developed a wide range of specialty fibers, including rare earth-doped double-clad fibers and polarization maintaining fibers. He received his MS and PhD from Pennsylvania State University in metals science and engineering. He received his BE in metallurgy from the Indian Institute of Science and BSc (Hons) in physics from Delhi University, India.

Burak Temelkuran Burak Temelkuran was born in Turkey, 1971. He received his BS (1994), MS (1996), and PhD (2000) degrees in physics from Bilkent University of Turkey. He received "New Focus Student Award" in 1999. He worked as a Postdoctoral Associate at the MIT Research Laboratory of Electronics and Department of Materials Science and Engineering (2000–2002) where he became a research scientist (2002). He joined Omniguide Inc. in 2003, where he is currently employed as a Senior Optical Physicist. He has been a member of OSA since 1998. His research interests include photonic band gap materials and fibers.

Limin Tong   Dr. Limin Tong received his PhD degree in material science and engineering from Zhejiang University in 1997. After four years of assistant and associate professorship in the Department of Physics in Zhejiang University and another three years as a visiting scholar in the Division of Engineering and Applied Science at Harvard University, he joined the Department of Optical Engineering at Zhejiang University in 2004 and is currently a professor of optical engineering. Dr. Tong's research area includes nanophotonics and fiber optic devices.

Anthony Toussaint   Anthony Toussaint is the Vice President of Research and Development at DSM Desotech. Dr. Toussaint has been with DSM Desotech since 1997, where he has held several positions in R&D working primarily in the development of coatings, inks, and matrix materials for fiber optics. He received a PhD in chemical engineering from University College London, England and an MBA from Northwestern University's Kellogg School of Management.

Liming Wang   Liming Wang received a PhD degree in optics in 1990 from the Chinese Academy of Sciences in China. He then carried on his scientific and engineering career in nonlinear optics and optical materials at the Chinese Academy of Sciences (China, 1991–1993), RIKEN and NIRIN (Japan, 1993–1998), and The University of Chicago (1998–2001). In 2001, he joined KVH Industries, Inc. as a Photonics Engineer to participate in the research and development of new products, including high-speed modulators and components for fiber optic gyroscopes. He is an author and co-author of 60 technical papers in refereed professional journals.

Ori Weisberg   Ori Weisberg is the former Applications Engineering and Systems Engineering Group Leader at OmniGuide Inc., where he worked for six years. He received his BS degree in geophysics from Tel-Aviv University, and an MS degree in planetary science from MIT. He is a co-author of six scientific articles and a co-inventor of eight patents. He currently resides near Tel-Aviv, Israel.

Olaf Ziemann   Professor Olaf Ziemann has been the Scientific Director of the POF-AC at the Nuremberg University of Applied Sciences (FH Nürnberg) since 2001. Dr. Ziemann studied physics at the University of Leipzig and received his doctorate degree at the Technical University of Ilmenau in the field of optical telecommunications engineering. Between 1995 and March 2001, he worked in the research center of the Deutsche Telekom (T-Nova) in the special areas of hybrid access networks and building networks. Since 1996, he has been the chairman of the Information Technology Society-Sub-committee "Polymer Optical Fibers" (ITG-SC 5.4.1).

Preface

The transport of radiation through a flexible, inexpensive conduit has changed our lives in more ways than we can imagine. The outstanding success of this concept is embodied in the millions of miles of telecommunications fiber that have spanned the earth, the seas, and utterly transformed the means by which we communicate. This has all been documented with awe over the past several decades.

However, more and more, optical fibers are making an impact and serious commercial inroads in other fields besides communications, such as in industrial sensing, bio-medical laser delivery systems, military gyro sensors, as well as automotive lighting and control—to name just a few—and spanned applications as diverse as oil well downhole pressure sensors to intra-aortic catheters, to high power lasers that can cut and weld steel. The requirements imposed by the broad variety of these new applications have resulted in the evolution of a new subset of custom-tailored optical fibers commonly known as "specialty fibers," which have their material and structure properties modified to render them with new properties and characteristics.

Specialized fibers are increasingly being used to manipulate the guided light within the fiber and to couple light of different wavelengths into and out of the fiber in telecommunications and sensing applications. The field of specialty optical fibers calls on the expertise and skills of a broad set of different disciplines: materials science, ceramic engineering, optics, electrical engineering, physics, polymer chemistry, and several others.

There are three fundamental aspects that one can engineer to develop a specialty fiber:

- Glass composition
- Waveguide design
- Coatings

Glass composition is one of the most basic fiber parameters and variables used for the design of specialty fibers. Commonly, it is possible to alter the basic glass structure of a fiber—silica based or otherwise—by introducing a number of appropriate dopants that would act as either glass formers, modifiers or actives, thus changing a fiber's basic properties such as refractive index or viscosity or, alternatively, by introducing new properties such as lasing capability, fluorescence, enhanced strain or temperature sensitivity, Brillouin effect coefficient, and many others.

Waveguide design was probably the first design parameter exploited and the one that led the way to define single versus multimode fibers. Nowadays waveguide design is more complex and has resulted in the design of specialty fibers that range from fibers with more than one guiding core to those based on one and two dimensional photonic crystal structures.

A prevalent characteristic and common feature in many popular commercial specialty fibers is their coatings. Due to the diverse set of applications and different environmental conditions to which fibers must be subjected to, one of the most common tailored properties of specialty fibers tends to be their coating. This has resulted in the commercial availability of a number of different coated fibers ranging from high temperature polyimides to hermetic carbon coatings. However, more and more, specialty coatings are being designed with specific sensing or actuation purposes and not solely for environmental or mechanical protection of the fiber. Coatings can enhance fiber sensitivity and selectivity to a number of physical and bio-chemical measurands: i.e., humidity, specific hydrocarbons, biochemical agents, electromagnetic fields, etc.

Although many specialty optical fibers were originally developed for and spun-off from the optical telecommunications industry, their present demand and design are primarily governed by the special needs and particular specifications imposed by fiber optic sensors and photonic components. Hence, as the need for optical fiber sensors and specialized components increases, so too will the demand for specialty fibers. Clear examples of this situation are fiber amplifiers and fiber Bragg gratings. Fiber amplifiers require different dopant compositions and guiding structures, while fiber gratings required photosensitive fibers and mode cladding suppressing designs to facilitate their performance.

However, the field of specialty fibers is not without its hurdles. One of the most problematic issues is the fact that specialty fibers tend to be a niche market and, as the name implies, a specialty item. This means that the volume demands are small when compared to their telecommunications cousins. Development time and cost are significant in the fabrication of a new, custom fiber. The development cost per meter of produced fiber is typically $100–1,000. Therefore, unless the application has a significant market and volume demand, many end-users desist in their attempts to have custom-made fibers. As a result, the overall specialty fiber market tends to be very fragmented and much smaller in size when compared to the regular telecommunications fiber business. At present, the overall worldwide market for all specialty fibers is in excess of $150 million and growing. Nevertheless, one fundamental aim of the specialty fiber industry that must be maintained is to offer custom tailoring.

Although by no means exhaustive, it is the purpose of this volume to provide insight into the many types of specialty fibers that are novel with respect to materials, fiber design, and application. As much as it was practical, we tried to cover as many of the most common and useful specialty fiber types as possible. In a similar way, we strove to provide a balanced and broad coverage of the material by ensuring the participation of top experts in the field and from the leading research groups and commercial specialty fiber manufacturers. Much to our disappointment, and not for lack of effort, it was not possible to obtain any contribution to this volume from Corning.

In order to present a comprehensive overview of the different specialty fiber types, as well as to help support some of their common physics and fundamentals, the first six chapters deal with optical fiber technology fundamentals and market considerations. The driving forces behind fiber development are economic, technological, and scientific, and certainly, without an enormous economic incentive, fibers would not have become so ubiquitous. It is appropriate then, that we begin first with a review of the driving factors that have helped develop and grow the specialty fiber industry, and the market opportunities that may arise from novel technical advances and the diverse commercial applications that seek these fibers, as discussed in Chapter 1. Any such volume as proposed here, that in some sense must be self-contained, should certainly carry the introduction of light-guiding principles and fundamental aspects of fiber design which are described in Chapter 2, as well as an overview of the various fiber fabrication techniques employed (Chapter 3). To address the fiber's protection issues, Chapter 4 discusses coating materials and processes. What are the characteristics and limitations of fiber coatings? These range from standard coatings for telecommunications fibers, to low index polymers for double-clad fibers. Rounding out this group is Chapter 5, which covers some of the newer, more specialized, single-mode fibers for telecommunications applications such as the ultra-low OH fibers, dispersion flattened, and dispersion compensating fibers.

Sensing applications utilizing optical fibers, often times have called for esoteric and non-conventional geometries of single-mode fibers such as dual core, multi-core or exocentric core fibers. Other applications have seen the use of fibers with side-holes, embedded metal electrodes or capillary tube holes. This myriad of different specialty single-mode fibers are discussed in Chapter 6.

Only rare earth–doped elements can lase in an amorphous host, and rare earth–doped fibers appear in a variety of important applications (Chapter 7). The fortuitous concurrence of loss in a silica fiber with the amplification in the 1.5 micron region for erbium, has essentially given us wide-band telecommunications systems that span the globe. The other most striking example of an application of a rare earth–doped fiber laser is that of the high power Yb laser, with over 1 kW of optical power. Polarization maintaining (PM) fibers are more difficult to fabricate than fibers that are circularly symmetric. The highest degree of optical anisotropy is obtained through the insertion of stress rods (PANDA) and anisotropic doping (Bowtie). Such fibers are important for many special application and are the subject of Chapter 8. Since K.O. Hill's discovery of an ability to "write" a grating in glass, UV-induced gratings in optical fibers have become an enabling technology that has allowed the possibility of dense WDM (wavelength division multiplexing) that powers the internet. The photosensitive fibers used in the fabrication of fiber Bragg gratings (FBGs) and other devices are covered in Chapter 9.

We tend to think of optical fibers as having a solid core. However, additional guiding mechanisms appear in a fiber with a hollow core surrounded by layers of material with sharply differentiated refractive indices. Low loss, endlessly single-mode fibers have been demonstrated, and if the hollow core is filled with a gain medium, such as an organic dye, new types of lasing phenomena occur. Such fibers allow for the possibility of transmission in the IR far beyond the $2\,\mu$m cut off of a silica fiber. Chapter 10 describes in detail the fundamentals of hollow-core fibers. In a similar fashion, we also tend to think of the size of a waveguide as being of the order of the wavelength of light it is guiding. However, if the diameter of a guiding structure is significantly smaller than the wavelength, guiding still occurs—albeit lossy—with a large evanescent wave. Such phenomena are not only scientifically of interest, but have significant applications in the measurement of objects much smaller than the wavelength of light. Chapter 11 covers new research work on the analysis and fabrication of sub-wavelength diameter fibers and the so-called silica nanowires. Other singular specialty fiber types, chiral fibers, are studied in Chapter 12. Chiral fibers employ a helical periodicity in the core structure, which provides them with unique polarization as well as wavelength selectivity characteristics.

In general, the vast majority of optical fibers are made of silica because of its extreme optical low loss and amazing physical properties. However, their IR cutoff is near $2\,\mu$m, and for many applications that include spectroscopy, sensing, or laser delivery, transmission at longer wavelengths is essential. Chapter 13 covers the various glass and crystal material candidates—such as fluoride, chalcogenides, and halide glasses—used in mid-IR and IR fibers.

It is well known that fibers must be protected from the environment and, in particular, silica fibers must be protected from moisture attack and OH^- radicals that weaken the fiber. This can lead to additional loss in the core as a consequence of the 1380 nm absorption overtone of OH^-. The need for hermetic coatings—impervious to both water moisture and hydrogen gas—occurs in applications of down-hole logging in oil and gas wells. Here, the elevated temperatures accelerate the diffusion of commonly occurring hydrogen molecules into the glass structure, increasing the optical loss and making the need for fiber protection even more important. One common staple of the specialty fiber portfolio is the carbon-coated fiber (Chapter 14), which has facilitated the use of silica fibers in geophysical and other harsh environments. There are some situations in which an even more rugged protective coating on the fiber is desired, either for military or sensing applications. Metal-coated fibers (Chapter 15) have been developed for this purpose by applying thin layers of low-temperature metals, such as gold, tin, aluminum or copper, and even higher temperature alloys, to the glass surface using dipping or evaporation techniques.

We had noted previously that PM (polarization maintaining) fibers can be formed through stress anisotropies. Any alteration in azimuthal symmetry in the fiber can remove the degeneracy of the two polarization modes. This can be accomplished through geometry, as is the case of a fiber with a D-shape for the cladding, or an elliptical shape for the core (Chapter 16).

One of the first types of commercial specialty fibers, multimode and large-core plastic clad silica (PCS) fibers, are still in wide demand and enjoy their consumption in a variety of applications such as biomedical laser delivery, automotive lighting, LAN networks, and many others. The different glass compositions and new advances in design are covered in Chapter 17.

In photonic devices, there is often a need to include specialty micro-components to ensure that the efficiency of an all-fiber device is maintained. These include tapered fibers that allow the NA of a gain section of the device to be matched with suitable pump sources, or in spreading the output intensity of a high power fiber laser with a predetermined near- and far-field pattern. These devices and their design aspects are described in Chapter 18.

Some of the earliest examples of guiding structures in fibers were provided by inserting higher optical density liquids into the core of a hollow fiber. These liquid-core fibers still have many applications in UV light delivery and spectroscopy (Chapter 19).

There are many situations in which ultra-low loss is not a necessity to fulfill a specific design application, and polymeric optical fibers (POF) can often be employed, as discussed in Chapter 21. This has been a steadily growing and maturing area, and the automotive industry has been particularly interested in this field. Moreover, polymeric fibers with a more efficient graded index have been developed, and losses can now be as low as 40 dB/km.

Although silica fibers can be used as temperature sensors up to approximately 1,100 °C, there is interest in a material that could withstand higher temperatures and harsh, corrosive environments. Sapphire fibers, in short lengths, can meet this demand. They are grown from a pedestal technique as single crystal strands, and the smallest diameter is 150 μm. These fibers fulfill a need in spectroscopy in that they are able to transmit further in the IR than a silica fiber (Chapter 21).

One of the great successes of specialty optical fibers is the advent of the high power Yb fiber laser. Nowadays, fiber lasers are available in the multi-kW range, and single mode operation now exceeds 2,000W CW. The materials processing market for industrial lasers is approximately $2 billion/year, and high power double clad fiber lasers are garnering an ever-increasing share. With 1.5 kW it is possible to cut through thick steel. In cutting applications that utilize plasma technology, carbon dioxide lasers, or YAG lasers, it is expected that with the reduced cost of high power pump diodes, high power fiber laser systems may prove to be a "disruptive" technology. We expect that industrial laser applications will grow significantly over the next few years. The specialty fibers needed in fiber lasers and their industrial applications are covered in detail in Chapter 22.

In biomedical applications, the advantages of fibers are easy to enumerate. A silica fiber is biologically inert, its small size allows it to enter the body as a catheter go to any place in the blood system to remove plaque, unclog arteries, or remove cysts. This is clearly a growing area of applications and one where new developments should be expected. Chapter 23 describes fiber delivery systems, specialty fibers required and applications in the biomedical area.

Our final chapter reports on the mechanical strength and reliability of glass fibers. This knowledge is of considerable importance to the design engineer in applying silica fibers to non-telecommunications applications.

In summary, we have presented what we believe to be a fairly complete overview of the many different types of specialty optical fibers, their uses, and the expected directions in which this field will develop, both with regard to elements of basic science as well as applications of this technology. We have sought to enlist the contributions of individuals who have made their mark in the many disparate areas of this field. We hope that this volume may prove helpful to those who wish to further their knowledge of the basic fundamentals of specialty fibers and how their special properties may provide solutions to real-life applications.

A. Méndez
T.F. Morse

Chapter 1

Specialty Optical Fiber Market Overview

Stephen Montgomery

ElectroniCast Corp., San Mateo, California

1.1 MARKET OVERVIEW

ElectroniCast has studied the potential use and market consumption for a variety of specialty optical fibers. All of the fiber types studied show very impressive historic and future growth potential. A few of the near-term stand-outs—in value potential—include polarization (PZ), ytterbium-doped, dispersion-compensating, and photosensitive fibers. Growth is also foreseen for new more esoteric and highly advanced fibers, such as "holey" fibers (photonic crystal fibers).

The global consumption of selected specialty optical fibers—which is composed of actual product sales and research and development (R&D) production—has been expanding rapidly from $239 million in 2000 to reach a forecasted estimate of $4380 million by 2010 (Fig. 1.1). This growth is driven by the challenges presented by the requirements of greater distances (kilometers/link lengths), optical fiber amplifiers (OFAs), dispersion compensation, attenuation, higher data rates, increased number of wavelengths (DWDM), high-powered fiber lasers, and simply the increase in the number of applications, components, and modules—just to name a few "drivers."

1.1.1 Production Versus Consumption

The actual consumption (use) of specialty optical fibers in internal and external R&D applications and in commercial consumption exceeded the production rates in recent years. This factor has since been corrected, because earlier excess inventory of fiber has been absorbed to more manageable levels.

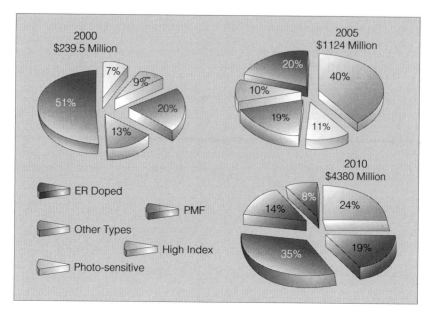

Figure 1.1 Estimated total specialty optical fiber global consumption forecast by type.

At all levels in the fiber optic industry "food chain," there is general agreement that there has been a return to substantial growth, which started in 2004, though not at the dramatic pace seen in North American long-haul/submarine telecommunications in 1999–2000.

The unusually strong fiber optics growth during 1999–2000, and unusually drastic collapse in 2001–2002, was more a function of investment community support than a shift in the basic demand for services. Both venture capital (VC) and mature investment in forward-looking networks (and their supporting equipment and components) collapsed. New networks were frozen in a semi-functional state. Equipment orders were canceled, rippling down through components and devices/parts.

1.1.2 Rapidly Growing Need to Use Fiber Optic Sensors

There is a rapidly growing need to use fiber optic sensors, a major consumer of specialty optical fiber. The wide range of applications (uses) for fiber optic sensors is facilitated by the need for various measurands (types of measurements). Because of the relative average price per unit, Fiber Optic Gyros

(FOGs) sensors, which are used in Military/Aerospace and commercial guidance control applications, is a leading sensor product or function type in terms of consumption value. Additional major fiber optic sensor functions (measurands), which contribute to the use of specialty optic fiber, include but by no means are limited to the following:

- Strain
- Temperature
- Flow
- Pressure
- Gas, liquid
- Acoustic, seismic, vibration
- Detection of objects/sampling
- Magnetic/electric field
- Wavelength monitoring/color

1.1.3 Weapon System Development

For various security, economic, political, and various other reasons, the U.S. weapon system development, production, and deployment is trending toward maximum reliance on the latest possible technologies that may be brought forward to deployment, with reduced dependence on deployment of extensive manpower. The succeeding generations of advanced electronic and optical/photonic capability, however, are advancing much faster than the technologies of the vehicles or shelters that house these advanced systems. This leads, therefore, to the concept that a major vehicle, such as a fighter aircraft, ship, or intercontinental missile, may need to accommodate the insertion of three or four succeeding generations of systems before the basic vehicle or shelter becomes obsolete.

1.1.4 100–1000×Improvements in Performance

Each succeeding military/aerospace system typically achieves a factor of 100- to 1000-fold improvement in performance but must be inserted into the same or smaller volume, with the same or smaller weight. Huge improvements have been made, resulting in smaller bend radius conditions, for example, which make specialty fibers ideal for other applications as well, including medical applications, consuming specialty optical fiber. The use of fiber optics, instead of copper conductors, for high data rate signal transmission and to address EMI/RFI, size/volume, and weight issues, is part of the answer to this challenge. These interconnects, however, must have very high performance reliability under severe environmental conditions or harsh environments. These factors are leading to heavy emphasis on high-reliability specialty optical fiber, fiber optic sensors, high-powered lasers, optic interconnect, which includes optical backplanes, and countless other specialty optical fiber consumers such as passive and active optical components.

1.1.5 High Cost of Functionality

The modest, or relatively small, quantities per year production of most of these military/aerospace systems, combined with the requirement for extreme reliability under severe environments and maximum miniaturization, has previously driven the cost of these components to, typically, 10–100× the cost of functionally similar commercial components. A new emphasis on "commercial off-the-shelf" (COTS) technologies, however, is intended to leverage solutions for commercial applications to reduce these ratios.

1.1.6 Multiple Features in the Same Specialty Fibers

We have observed that extra attention is being paid to very highly doped fibers used to make very short fiber lasers or amplifiers. This family should integrate the double-clad rare-earth doped fibers, which represents a configuration that should allow even larger output power out of fiber lasers and amplifiers. It is forecasted that there will also be more and more rare-earth doped fibers with the PZ-maintaining feature; even PZ-maintaining (PM) double-clad rare-earth doped fibers will become more popular.

Extrapolating further into future possible market opportunities for specialty optical fibers, a probable tendency will be towards combining multiple features in the same specialty fibers. Thus, these multifeature fibers will cost much more to manufacturers but will be sold at a substantially larger price. We expect that the specialty fiber manufacturers' competition will force them to evolve towards more reproducible products that will be able to replace more common products.

1.2 SPECIALTY OPTICAL FIBERS: A FEW SELECTED EXAMPLES

1.2.1 Fluoride Fiber

Fluoride fibers will remain a niche market that will grow at a normal but not exploding pace. ElectroniCast does not expect any breakthrough that would change the market reality concerning these fiber types. These fibers will be used for mid-infrared (MIR) light transmission and to make praseodymium and thulium-doped fiber amplifier and sources. "MIR" optical fibers transmit light between 2 and 5 μm. Specialty optical fibers are often classified and, therefore, marketed by their optical, environmental, and mechanical specifications.

1.2.2 Tellurite Fiber

Tellurite fibers will also remain a niche market. The application is often associated with the wideband fiber amplifier, but other alternative or competitive technology exists. ElectroniCast considers Tellurite optical fibers an R&D fiber. However, Tellurite optical fibers are considered more controllable in terms of optical signal processing versus fluoride fibers, which may provide a market of opportunity, especially because it can be pumped at 980 and 1480 nm. According to our estimates, the worldwide total use (consumption) of Tellurite optical fiber was $200,000 in 2002, increased to $600,000 in 2006, and is forecasted to reach $3.3 million in 2008, as depicted in Fig. 1.2.

1.2.3 Bismuth-Doped Fiber

Bismuth-doped fibers are expected to eventually develop into a very attractive market to produce extended L-band fiber amplifiers. Also, Bismuth-based erbium-doped optical fiber allows for extended L band and C + L band. Although metropolitan area networks (MANs) or "metro" optical networks typically call for relatively shorter link lengths, versus long haul, the emergence of DWDM in this application will open the window of opportunity for the use of OFAs. Bismuth-glass fibers are also thought to provide superior chemical,

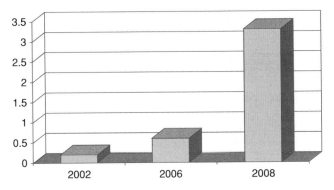

Figure 1.2 Tellurite optical fiber global consumption forecast ($MM).

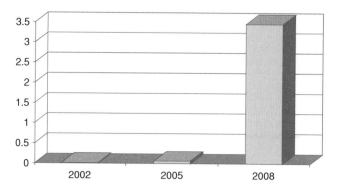

Figure 1.3 Bismuth-doped optical fiber global consumption forecast ($MM).

mechanical, and thermal durability versus competing C + L band technologies, such as tellurite and fluoride specialty optical fibers. As shown in Fig. 1.3, the worldwide total consumption value of bismuth-doped fiber is forecasted to increase from less than $20,000 in 2002 to $3.46 million in 2008.

1.2.4 Polarizing Fiber

PZ fiber is expected to remain a niche market for polarizing applications. Single PZ fiber is used in fiber optic sensors and high-end telecommunication components, such as PZ mode dispersion (PMD) compensators. Figure 1.4 shows the worldwide total use (consumption) of PZ fiber reached $1.1 million in 2004, increased to $2.9 million in 2006, and is forecasted to reach $5.0 million in 2008.

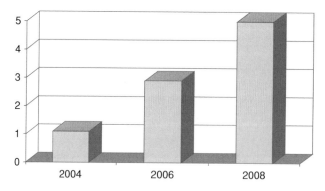

Figure 1.4 Polarization optical fiber global consumption forecast ($MM).

1.2.5 Photonic Crystal Fiber—Holey Fibers

The applications of holey fibers are emerging from R&D. An important application is air-clad fiber for high-power fiber laser pumping. Photonic crystal fibers (PCFs) (holey fibers) are commercial on a "best-effort" basis; however, commercialization is growing mainly from the efforts of selected universities and R&D centers of larger corporations. Current cost estimates range between $100 and $1000/m. We forecast the price to converge towards less than $215/m by year 2008 and less than $100 by year 2012 (Fig. 1.5). Lower prices will be driven by quantity increase, which is driven by commercialization.

PCFs use internal microstructures to guide light through them. Conventional optical fibers, on the other hand, depend on total internal reflection of light in a central core surrounded by a cladding with a lower refractive index. There are at least three types of PCFs:

- Silica solid core: all silica fiber, with a solid core. Cladding of air holes running along the length of the fiber. The guidance mechanism is effectively: total internal reflection. The cladding having the RI of the average of the glass and air in the cladding.
- Hollow core: all silica fiber, with a hollow core—just air in the middle. Cladding of air holes running along the length of the fiber. Guidance mechanism is by a photonic band gap, which is similar to a Bragg effect.
- Liquid crystal–filled PCFs (LC-PCFs): These fibers, which can guide light by the photonic band-gap effect, are thought to be suitable for optical signal processing.

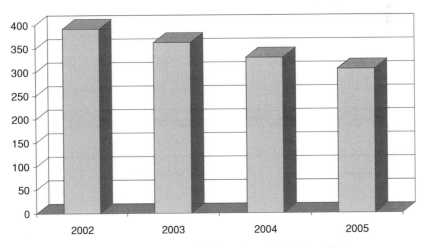

Figure 1.5 Average global selling price of holey fibers ($/m).

The use of holey fibers is beginning to demonstrate commercial potential, emerging from a university-based research-oriented effort. Their commercial potential is seen in certain characteristics such as single-mode operation from the ultraviolet (UV) to infrared (IR) spectral regions, large mode areas with core diameters larger than $20\,\mu$m, highly nonlinear performance with optimized dispersion properties, and numerical aperture (NA) values ranging from arbitrarily low to about 0.9.

1.2.6 Dispersion-Compensating Fiber

Dispersion-compensating fibers (DCFs) will have a substantial market share, with the advent of many new specialty fiber manufacturers fighting for a share of a limited market. This type of fiber will soon be directly available on the market. These fibers will evolve in order to offer a larger bandwidth over which the compensation can be implemented. They will also evolve to meet the 40-Gbps demands.

We believe that more and more multimode specialty optical fibers will become available (photosensitive, rare-earth doped, attenuating) in order for the local area networks (LANs) or access networks to evolve themselves towards all-optical systems, as was the case for long-haul and metropolitan optical networks.

There are several techniques to solve the chromatic dispersion problem; however, DCFs, chirped fiber gratings, etalon material, and electronic chips (electronic dispersion compensator [EDC]) are the most accepted methods. DCF has high levels of dispersion of opposite sign to that of the optical signal carrier fiber (standard single-mode fiber). For example, to compensate for the dispersion over an 80-km span of standard optical fiber, approximately 12–16 km (length) of DCF is linked with the standard fiber into the network. DCF, however, is considered by some vendors and carriers as too large and demonstrates high attenuation and increased optical nonlinear effects. *Filter products*, such as grating-based, or bulk optics such as virtual image phased arrays (VIPAs) and etalon-based chromatic dispersion compensating modules are considered an appropriate solution.

In grating-based products, the grating period is chirped to reflect that slower wavelengths be the faster ones that must travel further into the module before reflections occur. As with other fiber Bragg grating (FBG) devices, an optical circulator is used to segregate the input of the module from the output. As distances increase, transmission rates increase, and as DWDM is used, as well as different types of optical fibers are used, the need to adjust (tunable) the compensating function will increase.

There are three common device techniques to deal with chromatic dispersion in optical networks—the use of DCF; the use of chromatic dispersion-compensating filter modules that are typically FBG or etalon-based or other filter devices; and electronic chips (photonic chips in the future).

In 2004, based on the total global consumption market value of the selected chromatic dispersion-compensating solutions (devices), DCF represented 85% of the market. The filter approach represented 14% and the chip approach represented the balance of 1% in 2004. By 2008, the chips will represent 5.5% of the relative market share and filter solutions increasing to a 49% share. As depicted in Fig. 1.6, the market forecast by 2014 indicates that filter-based modules will represent about 70.1% of the relative market share, as the need for remotely tunable devices drives the market. The chip solution, because of a very low target unit/price, will represent only 9.4% in 2014, with DCF holding the balance, or 20.5%.

Chromatic-dispersion compensation became of concern as deployment of 2.54-Gbps (OC-48) channels ramped up and became urgent as OC-192 (10-Gbps) deployment accelerated, along with the trend to increase spectral width of dense WDM. The growth of the chromatic-dispersion compensator market during 1997–2000 was explosive (*boom*). However, because of the downturn (*bust*) of the high-speed telecommunications market during 2001–2003, this particular product line suffered in terms of major deployment demand. In 2004, the marketplace began to demonstrate recovery, and with global expansion of

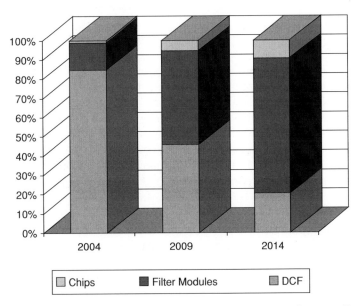

Figure 1.6 Chromatic dispersion compensator market breakdown forecast: dispersion-compensating fiber (DCF) vs. filter-based and chip-based solutions.

high-speed systems, chromatic-dispersion compensation will be required on a larger scale once again.

The big difference between the 1997–2000 chromatic-dispersion compensator marketplace and the marketplace of 2005 is that the filter-based solutions, as well as the alternative electronic chip solutions, have endured a long list of experiences during the 2000–2004 period, such as testbeds/product acceptance, R&D, technology advancements, mergers/acquisitions, and company readiness (e.g., International Standards Organization [ISO] certifications). Therefore, the alternative technologies to the DCF-based product line are much more advanced and are now ready to take on the DCF products in the competitive marketplace.

Early chromatic-dispersion compensators consisted simply of lengths of DCF having a negative dispersion characteristic that, with an appropriate length fiber, canceled the dispersion caused by the preceding transport fiber. These units, however, are expensive to buy and labor-expensive to install; plus, they introduce substantial loss and other problems and a single unit does not completely compensate at all wavelengths. As dense WDM progresses, and as modulation rates move up, it is necessary to de-multiplex the wavelengths and insert, in each wavelength/fiber, a compensator adjusted (variable/tunable) to match the signals on that wavelength at that time. Dispersion compensator technology has evolved, 1997–2000, to include compensators based on FBGs and other optical filter techniques. These cause less loss and have other advantages but still require manual installation at nodes.

Since optical add/drop multiplexers (OADMs) require de-multiplexing of the wavelengths, and later re-multiplexing, along with amplification, they are a natural point for compensator insertion. Manually inserted compensators are not counted in the as-shipped OADM value, because they typically are added separately, after OADM installation. However, for reasons discussed later in this chapter, a trend is developing to use remotely adjustable ("tunable") compensators that are built into the OADM, so these devices are counted as part of OADM component value. The OADM consumption, however, is not the only customer target of the total tunable compensator market. Compensators will be needed at every amplifier node (and regenerator) and there will be demand for amplifiers that will be (are) consolidated with OADMs. However, the physical location of the OADM needs to be convenient to the subscriber base it serves and this may not always coincide with the optimum locations of amplifier nodes. Generally, a subscriber base (typically, a small/medium population city) can be adequately served by one OADM node, which will incorporate four OADMs (as defined for this study): two for bidirectional transport of the active traffic, plus duplicates for the two-fiber protection circuit. Typical long-haul trunk cables, however, range from 48 to 144 fibers, and access loop cables are trending toward 864 fibers or more.

Major R&D efforts have been applied to tunable fiber optic components (compensators, attenuators, laser diodes, photodiodes, and so on) since 1997, and these components have emerged in the commercial market. Their major advantage is elimination of the labor (and overhead) cost of dispatching a craft crew to the OADM and amplifier nodes at each of the remote sites each time there is a need to change the compensation. These "truck rolls" are quite expensive, and there is an increasing shortage of skilled craft persons.

Chromatic-dispersion compensation units are also needed throughout a long-haul link, at OFA nodes, between regeneration points. Improved dispersion-compensation techniques will allow the distance between regeneration points to lengthen, so fewer regenerators per network link will be required and the cost per link will decrease.

1.2.7 High-Index Fiber

Corning Incorporated produces this type of fiber with its patented outside vapor deposition process, providing consistency and uniformity. The fiber uses a dual acrylate system that provides protection from microbend-induced attenuation. The fiber also features good geometry control, high core index of refraction, efficient coupling, and high NA. When used as component pigtails, this fiber allows for efficient coupling within photonic products. It also offers reduced bend attenuation because of its high core index of refraction. Applications include photonic products, fused fiber couplers, and component fiber for erbium-doped fiber amplifiers (EDFAs), couplers, and other DWDM components, laser diode pigtails.

1.2.8 Polarization-Maintaining Fiber

PMF is an optical fiber in which the PZ planes of lightwaves launched into the fiber are maintained during propagation with little or no cross-coupling of optical power between the PZ modes. PZ is the property that describes the orientation, such as time-varying direction and amplitude, of the electric field vector of an electromagnetic wave. Cross sections of PMF range from elliptical to rectangular. As shown in Fig. 1.7, the worldwide consumption of PMF reached $18 million in 2002, increased to $36 million in 2004, and was estimated to reach $66 million in 2005.

If polarized light is launched into a conventional single-mode fiber, the state of PZ will be rapidly lost after a few meters. The PZ state of light traveling through a medium can be influenced by stress within the medium. This can be problematic for ordinary single-mode fiber. Stresses such as bending or twisting

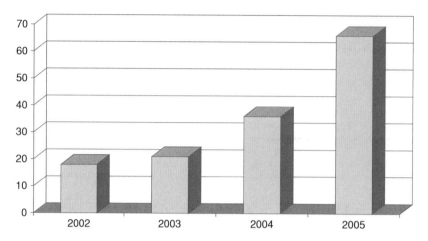

Figure 1.7 Polarization maintaining fiber global consumption ($MM).

will change the PZ state of light traveling through the normal fiber. If the fiber is subjected to any external disturbances such as a change in the fiber's temperature, then the final output PZ will vary with time. This is true for even short lengths of fiber. It is undesirable in many applications that require a constant output PZ from the fiber. To maintain the state of PZ a PZ-preserving fiber (PPF) should be used. PMF is also referred to as *PPF*. The most common type of PMF is the high birefringence type. "Birefringence" refers to a difference in the propagation constant of light traveling through a fiber for two perpendicular PZs. In high birefringence fiber, an asymmetric stress is applied around the core of the fiber that gives slightly different refractive indices to two orthogonal axes. This fast and slow axis will maintain the PZ state launched into the fiber over long distances.

With this type of fiber, it is necessary to align the axes of the fiber with the polarized light. If this is not possible or is inconvenient, then it is possible to use low birefringence fiber to transmit polarized light. In this case, the fiber is made with a much higher degree of symmetry than standard single-mode fiber. The fiber must have perfect geometry and be completely symmetric along the optic axis. In some sensor designs, the fiber itself is used as the sensing element, and in the case of a Faraday sensor, a conductor passing through a coil of fiber will cause a rotation in the polarized light in the coil. An ideal fiber will give a rotation proportional to the current and be insensitive to temperature variations.

PMFs can be used for high-performance transmission laser pigtails, PZ-based modulators, high-data rate communications systems, PZ-sensitive components, and other applications in which the state of PZ of the launched light is to be preserved in the fiber. PMFs are used in special applications, such as in fiber optic sensing and interferometry.

1.2.9 Photosensitive Fiber

FBG is a means to increase optical fiber transmission capacity. It is a technique for building optical filtering functions directly into a piece of optical fiber based on interferometric techniques. Exposing photosensitive fiber to deep UV light through a mask forms regions of higher and lower refractive indices in the fiber core. The number of transmitted channels is limited by the wavelength separation between each Bragg grating. The quality of the Bragg grating depends heavily on the photosensitive fiber used to write the grating.

When the core of a photosensitive optical fiber has been created using dopants, such as germanium, the refractive index of the core (along the length of the fiber) can be modulated. This is done by exposing the core to a pair of interfering beams of UV light.

Photosensitive fibers are used for the writing of long-period grating and FBGs. FBGs have become an enabling technology behind DWDM. EDFA gain-equalizing filters, WDMs, and add-drop multiplexers, created as periodic variations in the refractive index of the core of the fiber itself, have helped in the delivery of increased bandwidth. FBGs can also be used in the fabrication of optical strain and temperature sensors, with quasi-distributed measurements possible using gratings written sequentially into a continuous length of fiber.

1.2.10 Erbium-Doped Fiber

Erbium-doped fibers are used within EDFAs. Since the early use of optical fiber, communication network designers have seen a need for optical amplifiers as an alternative technique for longer distance transmission, along with higher power transmitters, higher sensitivity receivers, and lower loss fiber. EDFAs boost optical signals and eliminate the need for inefficient conversion of optical signals to electrical signals. Erbium-doped fiber is the key component in various applications requiring optical amplification near the 1550-nm wavelength region. Applications include booster amplifiers for long-haul regenerated systems, power amplifiers for terrestrial and cable TV applications, and small-signal amplifiers in optical receivers. New-generation EDFAs include power amplifier, preamplifier, and in-line amplifier for C- and L-bands.

Because of the intrinsic properties of erbium, erbium-doped fibers assist in the regeneration of an optical signal when it passes through the EDFA, a device that amplifies lightwave signals through optical networks. The erbium ions in

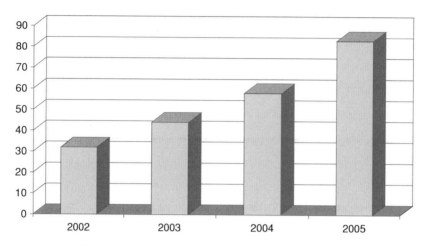

Figure 1.8 Erbium-doped fiber global consumption ($MM).

erbium-doped optical fiber can absorb light at 980 and 1480 nm and re-emit it in the 1550-nm telecommunications band through the process of stimulated or spontaneous emission. This permits one to create an optical amplifier that can restore power to a depleted optical signal in the 1550-nm wavelength band.

The demand for optical amplifiers—and consequently of rare earth-doped fibers—has increased dramatically with the emergence of DWDM networks, as illustrated in Fig. 1.8. Optical amplifiers maintain continuity along the line by amplifying optically the multiple signals simultaneously. They carry the signals to optical demultiplexers, where they are split into the original channels and sent to receivers.

Next-generation amplifiers based on modified doped fiber are touted for amplification over the entire L, S, and C bands. Because erbium-doped amplifiers do not work as well in the other bands, other dopants such as thulium using fluoride or multicomponent silicate options have been explored. Thulium-doped fiber amplifiers (TDFAs) operate similar to EDFAs, the differences lie in the dopant material and pump configuration. Two pumps of the same or different wavelengths are used to achieve optical amplification in TDFAs. TDFA use will be limited because they are seen as less reliable than the silica-based fiber amplifier and not conducive for fusion splicing because of material incompatibilities with the installed base of fibers in the current networks.

Although not as efficient as EDFAs, Pr-doped fiber amplifier (PDFAs) are commercially available for 1300-nm window. For shorter wavelengths, TDFA seems promising. Commercially available PDFAs use fluoride hosts. NTT Laboratory demonstrated a gain flattened Er3+-doped tellurite fiber amplifier

(EDTFA) with a gain of more than 25 dB and noise figure of less than 6 dB over a bandwidth of 50 nm (1560–1610 nm).

The design and manufacturing of fiber-based amplifiers include the use of narrowband-fused couplers with as small as 2 nm and very precise center wavelength alignment. This allows several pump laser modules to be multiplexed, significantly increasing an amplifier's total available power. Fiber-based PZ pump combiners are an improvement over micro-optic–based combiners, which have relatively high insertion loss. With gain-flattening filters, based on FBGs, the amplifier gain curve can be controlled over wide bandwidths. Fiber gratings are periodic refractive index variations in the core of an optical fiber, which cause light to reflect or couple out of the fiber core. FBG filters are used in other applications including lasers for wavelength locking in add/drop for channel selection.

Raman fiber amplifiers offer improved gain over wide bandwidths. High-traffic optical amplification is increasingly required as network capacity increases. Raman fiber amplifiers offer a solution for very broadband gain. However, Raman amplifiers generally require high pumping power to combine a low noise figure with high gain. Possible applications include high-power laser diode chips and fiber-based lasers, claimed to be better in power efficiency and have relatively high power outputs. In Raman amplifiers, the amplifier medium is the fiber itself. The Raman effect is a nonlinear phenomenon, which occurs at high power concentrations in the fiber. These amplifiers use stimulated Raman scattering (SRS) to transfer energy from a higher frequency pump to a lower frequency signal. The combination of Raman amplification with EDFAs offers a very low noise broadband amplifier solution. Raman amplification also allows access to wavelengths where high-performance EDFAs are not available.

Although Raman amplification research dates back before the emergence of EDFAs, the technology did not evolve until high-power pump sources became available in the early 1990s. Both discrete and distributed (gain is spread out over the transmission wavelength) Raman amplifiers are under development and production. Distributed Raman will play a large role in high-speed networks above OC-192 (10 Gbps) in extending system length (reduce spans between amplifiers/electronic regenerators) and offer improved overall performance. Improvements in packaging, Rayleigh scattering, and noise figure are an ongoing effort with Raman amplifier design. Several amplifier manufacturers are offering Raman amplifier products (complete modules and subassemblies).

There is confusion in the terminology of OFAs. The term "optical fiber amplifier" has different meanings in the industry. Typically, ElectroniCast defines the OFA as the final operational assembly, including all electronics required for its operation within the system. The OFA typically consists of the active optical gain block (AOGB), plus dozens of integrated circuits and other electronic components mounted on one or more printed wiring boards.

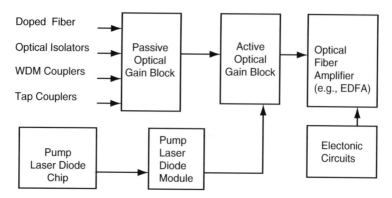

Figure 1.9   Optical fiber amplifier components: value-added progression.

Merchant-market OFAs, as often seen in cable TV and specialty/instrumentation applications, often are further packaged in an enclosure with front-panel controls. The assembly progression of the OFA is outlined in Fig. 1.9. Often, the gain block (active or passive) in the communication industry is referred to as an "optical amplifier." In this explanation, however, these units are called "gain blocks."

The AOGB consists of the passive optical gain block (POGB) plus the required pump laser diode module(s). A unidirectional active gain block may use one, two, or more pump laser diode modules. The pumps may be at the same or different wavelengths and may have other differences.

The POGB typically consists of the doped fiber module plus fiber-pigtailed isolator(s) and couplers fusion spliced into an integral assembly and packaged in an enclosure.

The pump laser diode is assembled from the laser diode chip, plus typically a thermoelectric cooler and a back photodiode, in a semiconductor package with fiber pigtail.

Demand for EDFAs has increased as communications moves toward DWDM technologies, geared for speeds of 10 Gbps and faster. In turn, the demand for engineered fiber as an integral part of these devices has created a good market opportunity for a handful of vendors. Manufacturing companies generally offer a series of different fibers optimized for different applications.

Approximately half the final value of optical fiber amplifiers consists of optical and electronic components used in their fabrication. The balance of the sales price consists of assembly and test labor/overhead cost, general/sales/administrative overhead, and profit. The largest contributor to component value is the pump laser diodes for fiber amplifiers. Numerous other components also are required, however, as illustrated in Fig. 1.10.

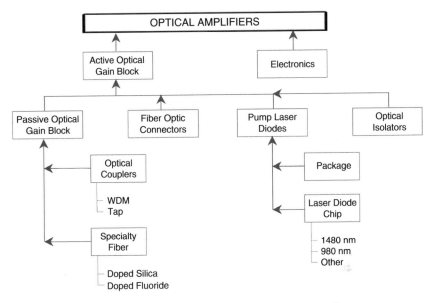

Figure 1.10   Optical fiber amplifier component categories.

1.3 CONCLUSIONS

Optical fibers have evolved from being simply a common transport waveguide for communication applications, to become vital optical components within optical amplifiers, chromatic dispersion compensators, polarizers, sensors, and numerous other devices. A substantial portion of the specialty fiber business comes from fast-growing military/aerospace applications—such as fiber optic gyroscopes, fiber-guided/tethered missiles, and submarine hydrophones, to name a few—as well as oil and gas applications.

The specialty optical fiber market grew from a boutique business in the 1990s to an impressive $239 million market in 2000. Continued dynamic expansion at a rate of more than 30% per year is expected to reach a global market of approximately $4.38 billion in 2010.

Chapter 2

Light-Guiding Fundamentals and Fiber Design

Robert Lingle, Jr., David W. Peckham, Alan McCurdy, and Jinkee Kim

OFS Corporate R&D, Norcross, Georgia

2.1 INTRODUCTION

The purpose of this chapter is to introduce the fundamental concepts of the guidance of light in optical fibers made from dielectric materials such as silica (SiO_2), with an emphasis on optical transmission properties relevant to communications. Some attempt is made to describe fiber properties by explaining how practical measurement issues influence our understanding and application of the theory. This chapter focuses on the properties of fibers intended for single-mode transmission applications. Multimode fibers that reduce system costs in high-speed, short-reach applications are not discussed.

Section 2.2 contains a cursory discussion of the physical structure of a conventional optical fiber used in telecommunications. Section 2.3 introduces the simple step-index—or matched cladding—fiber design and outlines the electromagnetic background of light propagation in a dielectric waveguide. Section 2.4 shows how working definitions of fiber cutoff differ from the theoretical concept of cutoff used in Section 2.3 and describes the criteria used in practice to determine when a fiber is effectively single moded. Section 2.5 describes the important phenomenon of macrobending loss and introduces the depressed cladding concept that is important for the design of fibers with reduced bend sensitivity. Section 2.6 gives a brief discussion of fiber attenuation loss. Some special methods for reducing optical loss are discussed in a subsequent chapter in this volume. Section 2.7 discusses chromatic dispersion, which is the tendency of a fiber to spread an optical pulse in time as it propagates down the fiber. Other chapters in this volume describe fiber designs that tailor the dispersion properties for specific benefits. A different source of dispersion known as *polarization mode dispersion* (PMD) is covered in Section 2.8. The phenomenology and theory of microbending is the subject of Section 2.9, while fiber nonlinearity is briefly introduced in Section 2.10.

2.2 PHYSICAL STRUCTURE OF A TELECOMMUNICATIONS OPTICAL FIBER

Optical fibers are fabricated by first depositing high-purity silica soot, doped with germania (GeO_2) to raise the index of refraction or fluorine (F) to lower the index of refraction, to form a core rod of 1 cm or more in diameter and 1 m or more in length. Fabrication methods [1] include processes known in the industry as "modified chemical vapor deposition" (MCVD) [2, 3], outside vapor deposition (OVD) [4], vapor axial deposition (VAD) [5, 6], and plasma chemical vapor deposition (PCVD) [7]. The MCVD, OVD, and VAD methods involve two steps of deposition and subsequent sintering of oxide soot formed by flame hydrolysis, while the PCVD method produces oxide layers directly in one step. The core rod comprises both the raised index light-guiding core and the portion of the cladding where significant optical power propagates, representing on the order of 10% of the total cross-sectional area of glass.

The core rod plus overcladding glass forms a *preform*. The overclad typically comprises silica of lower purity that may be derived from deposition of flame hydrolysis soot in the form of OVD [4, 8], plasma deposition [9, 10], or sol-gel casting [11, 12]. This material may be either deposited directly onto the core or else formed separately as a tube that is subsequently collapsed onto the core rod [13]. In either case, the preform is drawn down at approximately 2200 °C to a 125-μm diameter optical fiber at speeds greater than 10 m/sec and coated with both a primary and a secondary acrylate ultraviolet (UV)-cured polymer before take-up on a bobbin. The coating serves to preserve strength by protecting the glass surface from particles, to provide some limited protection from environmental moisture, and to provide mechanical protection from stresses that cause microbending losses. The light-guiding core itself comprises the inner 8- to 10-micron diameter of the 125-μm OD glass fiber.

2.3 LINEAR LIGHTWAVE PROPAGATION IN AN OPTICAL FIBER

2.3.1 Electromagnetic Preliminaries

Any treatment of light guiding in a fiber must begin with the Maxwell equations and describe their solution to some degree of mathematical detail.

Many excellent treatments of dielectric waveguides exist [14–19], and the reader would benefit by consulting one or more of these. We draw heavily on Buck's treatment [19].

The Maxwell equations in MKS units can be written as

$$\nabla \times E = -\frac{\partial B}{\partial t} \qquad \nabla \times H = J + \frac{\partial D}{\partial t} \qquad \nabla \bullet D = \rho_{free} \qquad \nabla \bullet B = 0, \quad (2.1)$$

where $D = \varepsilon E$ and $B = \mu H$, where ε and μ are the permittivity and permeability, respectfully, of the medium. In a source-less medium, $J = 0$ and $\rho_{free} = 0$. Using standard manipulations, the wave equations for propagating E and H fields can be derived from the Maxwell equations as

$$\nabla^2 E - \mu\varepsilon\frac{\partial^2 E}{\partial t^2} = 0 \quad \text{and} \quad \nabla^2 H - \mu\varepsilon\frac{\partial^2 H}{\partial t^2} = 0. \quad (2.2)$$

The formulas in Eq. (2.2) are each three-wave equations, one for each vector component of E and H. Assuming time harmonic fields, we may generally write (for a wave propagating in the z direction)

$$E = E_0 \exp[j(\omega t \pm \beta z + \phi)], \quad (2.3)$$

where $\beta = \omega/v = \omega n/c$ is the propagation constant, or the phase shift per length, of a sinusoidal wave measured along the z axis, and $v = 1/\sqrt{\mu\varepsilon}$ is the wave velocity.

The explicit form of the time dependence can be used to simplify the form of the Maxwell equations [17] to

$$\nabla \times E = -j\omega\mu H \quad \nabla \times H = j\omega\varepsilon E \qquad \nabla \bullet \varepsilon E = 0 \quad \nabla \bullet \mu H = 0. \quad (2.4)$$

Defining $k = \omega\sqrt{\mu\varepsilon}$, we can derive the wave equation in phasor form [19] as the vector Helmholtz equations:

$$\nabla^2 E + k^2 E = 0 \quad \text{and} \quad \nabla^2 H + k^2 H = 0. \quad (2.5)$$

The wave number k has units of m^{-1} and is a property of the material layer. The wave vector K points in the direction of energy flow and has magnitude $|K| = k$. The propagation constant β will be used to refer to the rate of accumulation of phase with distance for the electromagnetic wave we seek to calculate. In the case of a plane electromagnetic wave in a uniform, linear, and isotropic medium of dielectric constant, the propagation constant is simply $\beta = k = \omega n/c = nk_0$ where $k_0 = 2\pi/\lambda$ and λ is the wavelength of light *in vacuum*. In a waveguide, however, each region i will be characterized by index n_i, the magnitude of the wave vector in each region i will be $|K_i| = n_i k_0$, and the Helmholtz equations require solution in each region with matching of boundary conditions at the interfaces. We will always choose the direction of propagation in a guide to be along the z direction.

2.3.2 Intuition from the Slab Waveguide

Some intuition can be gained by considering the ray optics picture of a simple slab waveguide. An electromagnetic wave in a guide will spread over at least two regions (core and cladding). The given optical properties of the waveguide structure are characterized by k_i, while the characteristics of the propagating wave for which we seek a solution are described by the propagation constant $\tilde{\beta}$. Figure 2.1 shows a step-index waveguide supporting a guided mode, with index $n_1 > n_2$. The index level of the doped glass is frequently characterized by $\Delta = \frac{n_1^2 - n_2^2}{2n_1^2} \approx \frac{n_1 - n_2}{n_1}$ given in percent. The use of Δ references doped structures in the waveguide core to the cladding index n_2, without regard to the actual value of n_2. In more complex designs, the value of Δ, for additional doped layers adjacent to the central core, are defined analogously, and Δ can be positive or negative. In Fig. 2.1, propagation occurs along the z axis (into the page), and the guiding structure confines light in the x direction. The forward propagation constant β for a guided mode along the z axis is constrained by the relation $n_2 k_0 < \beta < n_1 k_0$, because the mode spreads over both the core and the cladding region.

Figure 2.2 illustrates the geometries of the wave vectors in the two regions of a slab waveguide in the ray optics picture, where rays reflect and transmit according to the Fresnel equations [17, 19] at the interfaces. For waveguides oriented along the z axis, we will write the wave vector in each region as $\mathbf{K}_i = \kappa_i \mathbf{e}_x + 0 \mathbf{e}_y + \beta \mathbf{e}_z$ so that in each region, $\kappa_i^2 + \beta^2 = n_i^2 k_0^2$, $\kappa_i = n_i k_0 \cos \theta_i$, and $\beta = n_i k_0 \sin \theta_i$. Although the guided mode propagates only in the z direction, it is common and convenient to refer to κ and β as the transverse and forward propagation constants, respectively. Intuitively, the propagation constant β must be the same across all regions of a waveguide, while the transverse propagation constants κ_i will differ and may be imaginary. Mathematically, β must be

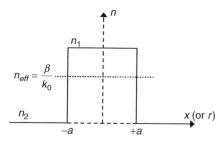

Figure 2.1 A step-index waveguide with mode propagation constant β, where $n_2 k_0 < \beta < n_1 k_0$. The figure can represent a slab waveguide of thickness $2a$ or an optical fiber of core radius $r = a$.

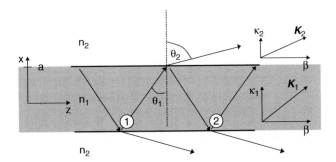

Figure 2.2 Geometry of reflections and propagation in a slab waveguide, where $n_1 > n_2$. Reflections at the interfaces will follow the Fresnel equations for magnitude and phase shifts of electric and magnetic fields, including Snell's law $n_1 \sin \theta_l = n_2 \sin \theta_2$. The propagation constant β in the z-direction must be identical in regions 1 and 2. The case shown corresponds to $\theta_1 \leq \theta_c$. The wave vector K_1 in the core is longer than K_2 in the cladding, because $n_1 > n_2$.

the same across all regions because of the requirement that the tangential components of the fields must be continuous across the interfaces between the regions.

The electric field E_1 in the guiding region 1 assumes the form

$$E_1 \sim \exp(\mp j\kappa_1 x)\exp(-j\beta z), \tag{2.6}$$

where "$-$" corresponds to upward propagation and "$+$" corresponds to downward propagation along the x direction in Fig. 2.2. According to Snell's law, $n_1 \sin \theta_1 = n_2 \sin \theta_2$, total internal reflection will occur when $\theta_1 > \theta_c = \sin^{-1}(n_2/n_1)$. For $\theta_1 = \theta_c$, $\theta_2 \rightarrow 90°$, $\kappa_2 = 0$, and K_2 tilts over to lie along the z axis with $\beta = n_2 k_0$. A guided mode will propagate under the condition of total internal reflection when $\theta_1 > \theta_c$ so that $\beta > n_2 k_0$. In this case, κ_2 becomes imaginary, and we can write $\kappa_2 \rightarrow -j\gamma_2$, where the decay constant γ_2 is a real number so that $\gamma_2 = j\kappa_2 = (\beta^2 - n_2^2 k_0^2)^{1/2}$. Then the electric field E_2 in the cladding region 2 assumes the form

$$E_2 \sim \exp(\mp \gamma_2 x)\exp(-j\beta z). \tag{2.7}$$

To form a propagating mode, the upward traveling waves represented by $\exp(\mp j\kappa_1 x)$ in Eq. (2.6) must be in phase after traversing the waveguide in region 1, including phase shifts caused by reflection from the interfaces, and the total phase shift from (1) to (2) in Fig. 2.2 must be an integral multiple of 2π.

The phase shifts are determined from the Fresnel equations for the transverse electric (TE) and transverse magnetic (TM) cases, where the electric (magnetic) field oscillates entirely in the transverse (xy) plane for TE (TM). *This transverse resonance condition fixes the values of β and γ, leading to one or more guided* *modes that propagate along the z axis, form a standing wave along the x axis in region 1, and decay exponentially in region 2.* The same result can be obtained by solving the wave equation over the regions and matching boundary conditions at the interfaces. The guided mode is said to be cutoff when $\theta_1 \leq \theta_c$, where κ_2 is real and $\beta \leq n_2 k_0$. An unguided wave below cutoff that nevertheless meets the transverse resonance conditions is sometimes known as a "leaky wave."

2.3.3 Optical Fiber: A Cylindrical Waveguide

The ray optics analysis for a cylindrical waveguide such as an optical fiber is complicated by the existence of skew rays in three dimensions, which propagate helically and do not cross the fiber axis. Here, we outline the solution of the field equations to derive the form of the fiber modes in cylindrical coordinates and illustrate their properties. For a step-index optical fiber of the basic form illustrated in Fig. 2.1 with core index n_1 and radius a, we can assume field solutions of the form:

$$E = E_0(r,\phi)\exp(-j\beta z) \qquad \text{and} \qquad H = H_0(r,\phi)\exp(-j\beta z). \qquad (2.8)$$

Again, each vector Helmholtz equation in Eq. (2.5) represents three scalar equations, for a total of six. We can solve for one field component, such as E_z, however, and then use the Maxwell equations to derive the others. Substituting these, the Helmholtz equation for the electric field becomes

$$\nabla_t^2 E_{z1} + (n_1^2 k_0^2 - \beta^2)E_{z1} = 0 \text{ for } r \leq a \qquad (2.9a)$$

and

$$\nabla_t^2 E_{z2} + (n_2^2 k_0^2 - \beta^2)E_{z2} = 0 \text{ for } r \geq a, \qquad (2.9b)$$

where the transverse Laplacian ∇_t^2 includes only the radial and angular derivatives. For notational convenience, we define the transverse propagation constants

$$\beta_{t1}^2 = (n_1^2 k_0^2 - \beta^2) \text{ and } \beta_{t2}^2 = (n_2^2 k_0^2 - \beta^2), \qquad (2.9c)$$

which play the same role as κ_i in the slab waveguide discussion. As in the case of the slab waveguide β_{t2} is imaginary for a guided mode since $\beta > n_2 k_0$.

Writing the solution in the form $E_z = R(r)\Phi(\phi)\exp(-j\beta z)$, and performing the standard separation of variables, one finds that $\Phi(\phi) = \sin(q\phi)$, where q takes integer values and is identified as the azimuthal or angular mode number. For real β_{t1} (in the core region $r \leq a$) the radial solutions are $J_q(\beta_t r)$, the ordinary Bessel functions of the first kind. For imaginary β_{t2} (in the cladding region $r \geq a$), the radial solutions are $K_q(|\beta_t|r)$, the modified Bessel functions that monotonically approach zero for large value of the argument. The normalized transverse propagation and decay constants are defined, respectively, as

$$u = \beta_{t1}a = a(n_1^2 k_0^2 - \beta^2)^{1/2} \text{ and } w = |\beta_{t2}|a = a(\beta^2 - n_2^2 k_0^2)^{1/2}. \tag{2.10a}$$

Then the complete solution can be written as

$$E_z = A J_q(ur/a) \sin(q\phi) \exp(-j\beta z) \text{ for } r \le a, \text{ and} \tag{2.10b}$$
$$E_z = C K_q(wr/a) \sin(q\phi) \exp(-j\beta z) \text{ for } r \ge a. \tag{2.10c}$$

Note that the ϕ and z dependences are identical in the core and cladding, as expected by intuition, with the solutions differing only in the radial dependence. The form of the solution for E_r, E_ϕ, H_r, H_z, and H_ϕ can be derived from the solution for E_z using the Maxwell equations. The solutions are characterized by q, the mode order, or azimuthal mode number, which takes on integer values $q = 0, 1, 2,$ and so on. The eigenvalues are the unique sets of u, w, and β, which match boundary conditions requiring continuity of the tangential field components at $r = a$, thus determining the modes supported by the waveguide. The resulting eigenvalues are numbered by the mode rank, m, or radial mode number, which takes integer values $m = 1, 2, 3,$ and so on. The transverse electric or TE$_{0m}$ set of modes has components $E_z = 0$ (by definition) and E_ϕ, H_z, and $H_r \ne 0$. The transverse magnetic or TM$_{0m}$ set of modes has $H_z = 0$, and E_z, E_r, and $H_\phi \ne 0$. Modes with $q \ne 0$ are labeled EH$_{qm}$ or HE$_{qm}$, which physically correspond to skew rays in the ray optics picture [19].

2.3.4 The Linearly Polarized Mode Set LP$_{lm}$

The problem of solving for the eigenvalues is greatly simplified by the weak guidance approximation $n_1 \approx n_2$ [14]. Cabled telecommunications fibers invariably have values of $\Delta < 1\%$, although dispersion compensating fibers may have values of Δ as high as 2%. This approximation is excellent for fibers with $\Delta < 1\%$, but it may be used with reasonable results for many fiber designs with $\Delta \sim 2\%$. The weak guidance approximation also aids in grouping degenerate modes, which have the same value of β but slightly different field configurations, to form a set of modes referenced by the notation LP$_{lm}$ that are linearly polarized in the transverse plane. These modes are natural modes for describing the fiber, given that communications lasers typically emit linearly polarized light that maintains its polarization in a fiber in the absence of perturbations. The new mode number l is introduced as follows:

$$l = q + 1 \quad \begin{cases} 1 & \text{for TE}_{0m} \text{ or TM}_{0m} \\ q+1 & \text{for EH}_{qm} \\ q-1 & \text{for HE}_{qm} \end{cases}$$

Using the full set of vector field expression for E and H to match boundary condition at $r = a$, and using the weak guidance approximation, the eigenvalue equation to be solved is

$$u \frac{J_{l-1}(u)}{J_l(u)} = -w \frac{K_{l-1}(w)}{K_l(w)}. \tag{2.11}$$

It is also helpful to introduce the normalized spatial frequency V, or V number:

$$V = (u^2 + w^2)^{1/2} = ak_0(n_1^2 - n_2^2)^{1/2} = n_1 ak_0 \sqrt{2\Delta}. \tag{2.12}$$

The definition of V shows that possible values of u and w lie on a circle of radius V, which can be related through Eq. (2.10a) to the fundamental parameters of the waveguide. Larger values of V, due to greater index contrast, shorter wavelength, or larger core, lead to the possibility of more guided modes in the waveguide. A mode is said to be cutoff when it ceases to be confined to the waveguide, that is, when the field in the cladding region 2 ceases to be an evanescent wave. Cutoff, thus, occurs as $\beta \rightarrow n_2 k_0$. Near cutoff, $w \rightarrow 0$ and the evanescent wave extends farther into the cladding. Beyond cutoff, the propagation constant β_{t2} becomes real, the transverse field in the cladding begins to propagate, and the solution becomes a leaky wave rather than a guided mode. The cutoff condition for modes are, thus, found by setting $w = 0$, in which case Eq. (2.11) reduces to $V \frac{J_{l-1}(V)}{J_l(V)} = 0$, showing that the zeros of the Bessel functions give the conditions for mode cutoff.

A few examples are given, referencing Table 3.2 in reference [19]. The LP_{01} mode is simply the HE_{11} mode and is not cutoff at any wavelength. The LP_{11} mode is composed of the TE_{01}, TM_{01}, and HE_{21} modes, and cuts off at a V number of 2.405, the first zero of J_0. The LP_{21} and LP_{02} modes both cut off at $V = 3.832$, the zero of $J_{\pm 1}$. The intensity patterns of the LP_{lm} modes are given by $I_{lm} = E_{lm} E_{lm}^*$ and can be expressed as

$$I_{lm} = I_0 J_l^2 \left(\frac{ur}{a}\right) \cos^2(l\phi) \text{ for } r \leq a \tag{2.13a}$$

$$I_{lm} = I_0 \left(\frac{J_l(u)}{K_l(w)}\right)^2 K_l^2 \left(\frac{wr}{a}\right) \cos^2(l\phi) \text{ for } r \geq a. \tag{2.13b}$$

The physical interpretation of l and m can now be readily understood. The integer value of $m \geq 1$ gives the number of intensity maxima that occur along a radius. A higher value of m means a higher value of u for a given V, resulting in more radial oscillations in the intensity pattern. The value of l is one-half the number of azimuthal maxima in the intensity pattern. Thus, LP_{01}, the fundamental mode, has no azimuthal variation, is maximum on the fiber axis at $r = 0$, and decays monotonically as $r \rightarrow \infty$. The reader is encouraged to consult the references, such as Figure 3.9 in reference [19], for further insight.

2.3.5 Finite Element Analysis for Waveguide Calculations

Index profiles for real fibers may be much more complex than the idealized step-index waveguide of Fig. 2.1. Even matched clad fibers that are step index in principle will depart from an idealized step in various ways, depending on the details of the particular manufacturing process. More complex fiber designs may comprise grading of the refractive index, as well as multiple index layers, including those with index below that of the (nominally) pure silica cladding. The field solutions and propagation constants for LP modes of complex fiber designs, such as those described in Chapter 5, are usually calculated numerically using finite element methods [20, 21]. All subsequent results presented are calculated numerically using FEM methods.

The important optical properties of a fiber that define its utility in a particular application include attenuation, mode field diameter, effective area, cutoff, dispersion, and bending losses. Cutoff, attenuation, and dispersion are described later in this chapter. The radial LP_{01} mode for the step-index fiber is near-Gaussian, approaching cutoff. The fiber mode field diameter (*MFD*) and effective area A_{eff} are defined by

$$MFD^2 = 2\frac{\int_0^\infty |E(r)|^2 r dr}{\int_0^\infty \left|\frac{dE}{dr}(r)\right|^2 r dr} \tag{2.14}$$

and

$$A_{eff} = 2\pi \frac{\left[\int_0^\infty |E(r)|^2 r dr\right]^2}{\int_0^\infty |E(r)|^4 r dr}, \tag{2.15}$$

respectively.

The *MFD* and A_{eff} are inherently wavelength dependent and increase toward longer wavelengths. It is intuitive that longer wavelengths of light will be less confined by the waveguide than shorter wavelengths. From the point of view of physical optics, this is intimately related to the fact that an aperture (the waveguide) of diameter $d = 2a$ diffracts light more strongly as $\lambda \rightarrow 2a$. The wavelength dependence can also be understood from considering the analogy between confinement of light in a region of elevated refractive index and the trapping of a particle in a potential well in mechanics. In particular, the time-independent Schrödinger equation of quantum mechanics is also a scalar Helmholtz equation of the form of one of the field components of Eq. (2.9a). The term

$n_1^2 k_0^2 = (2\pi)^2 n_1^2 / \lambda^2$ corresponds to the potential well depth in the Schrödinger equation. As wavelength λ becomes longer, $n_1^2 k_0^2$ decreases (analogous to a more shallow potential well depth), leading to weaker confinement of light and larger A_{eff} (analogous to a smaller binding energy and spreading of the wave function outside the well). The wave function, the square of which describes the spatial probability distribution for the quantum particle, is analogous to electric field, and the binding energy in a potential well corresponds to values of β^2 for a guided mode.

Figure 2.3 shows the calculated wavelength dependence of MFD and n_{eff} for a step-index fiber of $\Delta = 0.366\%$ and core radius $a = 4.8$ microns, representative of a typical matched clad fiber design with properties compliant to ITU G.652. The *theoretical* cutoff values for this fiber design are 1350 nm for the LP_{11} mode and 840 nm for the LP_{02}. The practical cutoff for the LP_{11} mode will be at a wavelength *shorter* than 1350 nm, as described later. The MFDs at 1310 and 1550 nm for a commercial matched clad fiber are typically specified as $9.2 \pm 0.4\,\mu m$ and $10.4 \pm 0.4\,\mu m$, respectively. It can be seen in Fig. 2.3 that as the MFD increases, then n_{eff} (shown here as the difference in n_{eff} and the cladding index, $n_{eff} - n_2$) decreases as the mode spreads outside the germanium-doped core of diameter 9.6 microns. The effective index of the fundamental mode is ultimately an average of the waveguide refractive indices weighted by the distribution of optical power. The variation in MFD or A_{eff} with wavelength is important in understanding nonlinear effects that impact system performance. The decrease of the effective index with wavelength is closely correlated with tendency for increased macrobending and microbending loss, discussed later in this chapter.

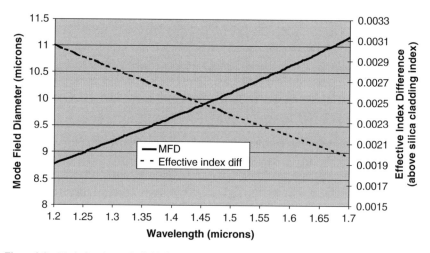

Figure 2.3 Variation in mode field diameter (*MFD*) and ($n_{eff} - n_2$) with wavelength for a step-index fiber with $\Delta = 0.366\%$ and core radius $a = 4.8$ microns.

2.4 WORKING DEFINITIONS OF CUTOFF WAVELENGTH

2.4.1 Introduction

The cutoff wavelength of a single-mode optical fiber is the wavelength above which only a single bound mode, the fundamental LP_{01} mode, propagates. For numerous reasons concerning transmission performance (bandwidth, multipath interference, modal noise, etc.), it is desirable to operate fibers in the regime where only the fundamental mode propagates. (This discussion does not address the intentional use of multimoded fibers for short-reach applications, where as many as 10- to 18-mode groups may be allowed to propagate at the operating wavelength.) In this section, we discuss the theoretical and effective cutoff wavelengths of step-index single-mode fibers.

2.4.2 Theoretical Cutoff Wavelength

It has already been noted that the well-known weakly guiding analysis by Gloge [14] shows that a matched cladding optical fiber supports the propagation of only the fundamental LP_{01} mode when the V number of the waveguide is less than 2.405. Therefore, the theoretical cutoff wavelength for a step-index fiber, λ_c^{th}, is defined as

$$\lambda_c^{th} = \frac{2\pi n_1 a}{2.405} \sqrt{2 \cdot \Delta}, \tag{2.16}$$

where n_1 is the refractive index of the core, a is the core radius, and Δ is the relative index difference between the core and cladding. At wavelengths greater than λ_c^{th}, the transverse propagation constant β_{t2} of the first higher order LP_{11} mode in the cladding region becomes a real number. This changes the solution for the electric field in the cladding from a decaying, evanescent field to an oscillatory, propagating field, thus resulting in radial energy flow (i.e., one that carries energy away from the fiber axis). The bound mode becomes a leaky mode.

2.4.3 Effective Cutoff Wavelengths

Consider the behavior of the LP_{11} at wavelengths shorter than λ_c^{th}. Far below λ_c^{th} the LP_{11} mode is tightly confined within the core region and losses will generally be comparable to those of the fundamental mode. As the wavelength increases, the LP_{11} mode becomes less tightly confined to the core.

The decreasing mode confinement gives rise to excess LP_{11} mode loss when axial imperfections, such as microbends or macrobends, are present. *Microbends* are defined as small-scale random deflections of the fiber axis, small relative to the core size, such as would be present when the fiber is pressed against a rough surface. *Macrobends* are large-scale deflections in the fiber axis, such as loops or the bends associated with placing fiber on a spool. Generally as one approaches, wavelengths 100 nm or so below λ_c^{th}, the LP_{11} mode becomes very loosely confined to the core and very lossy if the fiber axis is not maintained perfectly straight. Even at wavelengths well below λ_c^{th}, the losses on the order of 10 dB/m or more readily occur, so that the LP_{11} does not effectively transmit energy over distances of more than a few meters. This has led to the concept of the *effective cutoff* wavelength, λ_c^{eff}, of a single mode fiber, which is defined phenomenologic-ally as described later in this chapter.

Figure 2.4 shows the power as a function of wavelength that is observed at the output of a short (2 meter) length of fiber at wavelengths that span the fiber's LP_{11} mode cutoff. At the input of the fiber, the launch conditions are such that power P_{01} and P_{11} launched into the LP_{01} and LP_{11} modes, respectively, for all wavelengths. If we assume that the loss of the LP_{01} mode experiences in the 2-m length of fiber is negligible, then the power at the output of the fiber as a function of wavelength is

$$P_{out}(\lambda) = P_{01} + P_{11} \cdot e^{-\alpha_{11}(\lambda) \cdot L}, \tag{2.17}$$

where $\alpha_{11}(\lambda)$ is the LP_{11} attenuation as a function of wavelength. If we consider wavelengths short relative to LP_{11} mode cutoff, then the LP_{11} mode is well

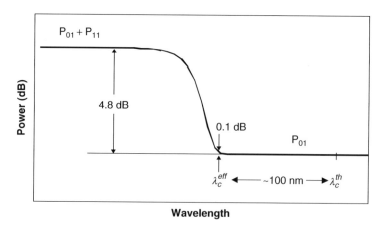

Figure 2.4 Power transmitted through a short length of single-mode fiber near the effective cutoff wavelength.

confined and the excess attenuation is low, so that $P_{out}(\lambda) \sim P_{01} + P_{11}$. Since the LP_{01} mode is a singly degenerate set of modes and the LP_{11} mode is a doubly degenerate set of modes, we assume that $P_{11} = 2*P_{01}$ and then $P_{out}(\lambda) \sim 3*P_{01}$ at short wavelength. At long wavelengths relative to LP_{11} mode cutoff, the attenuation of the LP_{11} mode is very high and $P_{out}(\lambda) \sim P_{01}$. At wavelengths between the two extremes, the power level at the output of the fiber transitions between $3*P_{01}$ and P_{01}, as the fiber transitions from two-moded to single-moded behavior. It is important to note that because the length and layout of the fiber sample will determine the level of excess LP_{01} attenuation, the location of the transition from two-moded to single-moded behavior will also vary with fiber length and layout.

Notice that the wavelength where only LP_{01} power is observed at the output of the fiber is considerably below λ_c^{th}. In other words, at wavelengths considerably below λ_c^{th}, the LP_{11} mode has become effectively cutoff. By convention, the *effective* cutoff wavelength has been defined as the wavelength where $P_{out}(\lambda)$ has risen by 0.1 dB above P_{01}. It can be shown from Eq. (2.17) that the attenuation of the LP_{11} mode at λ_c^{eff} is 19.2 dB.

The *fiber* effective cutoff wavelength has been defined by international standards groups to be measured on a 2-m length of fiber that is deployed in a "nominally" straight configuration except for a single 28-cm diameter loop. This fiber configuration was defined so that fiber manufacturers could easily implement the procedure in their factories on readily available spectral attenuation test benches. Because this factory-friendly measurement configuration may not represent field-deployed conditions, a need arose to relate the *fiber* effective cutoff wavelength to the effective cutoff wavelength of the fiber when it was deployed as a cable section or jumper in an operating transmission system.

Many groups [14, 18, 23, 25] studied how the LP_{11} mode cutoff scales in wavelength as the length and bending configuration of fiber under test is varied. The studies showed that length and bending scaling of cutoff varied significantly across fiber designs. For example, the change in the cutoff wavelength with variation in length of the fiber under test was significantly different for the matched-cladding, depressed-cladding and dispersion shifted fiber designs manufactured during the late 1980s.

In an effort to ensure that fibers are effectively single moded in the various configurations that they are likely to be deployed in, the *cable* effective cutoff wavelength has been defined. The cabled fiber deployment configurations viewed as worst-case scenarios for outside plant, building, and interconnection cables were defined. For outside plant cables, the concern is that a short section of restoration cable, as short as 20 m in length, may be spliced into the transmission path to replace a damaged section of cable. Typically when a telecommunications cable is damaged, for example, by excavation at a construction site near a cable right of way, then the damaged section is removed and a short section of cable is spliced in place to bridge the gap. If the fiber in the short restoration cable is not effectively single moded at the system operating wavelength, then there is the potential for the paired splices to generate excess additive noise, which is referred to as *modal noise*. Much of the energy lost by the incoming LP_{01} mode at the first splice will be coupled into the LP_{11} mode of the restoration fiber. If the transmission loss of the LP_{11} mode in the restoration fiber is low enough, then energy in both the LP_{01} and LP_{11} mode will reach the second splice and will add coherently if the optical path length difference for the two modes is less than the coherence length of the optical source. The energy coupled into LP_{01} of the output fiber of the second splice depends on the electric field shape at the input to the splice, which is a function of the coherent interference of the two modes entering the splice. Because the modal interference at the splice can be time dependent (because of laser wavelength variations, environmental variations that change the optical path length of the restoration fiber, etc.), the LP_{11} mode splice loss of the second splice can be time dependent, which generates *modal noise*. With modal noise in mind, the outside plant cable cutoff wavelength deployment configuration is designed to mimic a 20-m restoration cable and the associated splice closures at its ends. The configuration is defined as a 22-m length of fiber, coiled with minimum bend diameter of 28 cm, with one 75-mm diameter loop at each end of the fiber. Many suppliers of fiber for use in the outside plant specify that the *cable* effective cutoff wavelength of their fiber is 1260 nm or less. Although it depends on the specifics of the fiber design and, therefore, varies considerably, typically the *fiber* effective cutoff wavelength is roughly 100 nm below the theoretical cutoff wavelength for many standard single-mode fibers. Likewise, the *cable* effective cutoff wavelength is typically an additional 60–80 nm below the *fiber* effective cutoff wavelength for typical standard single-mode fibers.

2.5 IMPACT OF PROFILE DESIGN ON MACROBENDING LOSSES

2.5.1 The Depressed Cladding Fiber Design

Historically both matched and *depressed* clad single-mode optical fibers have been widely deployed in telecommunications networks. For example, the original AT&T standard single-mode fiber was a depressed clad fiber with $MFD = 8.8\,\mu m$. A depressed cladding design has an annular ring of $\Delta < 0$, often called a *trench*, between the raised index core and the silica cladding [22–24]. In this case, the effective Δ should be measured between the up-doped core and down-doped trench, so that the depressed clad design has a higher

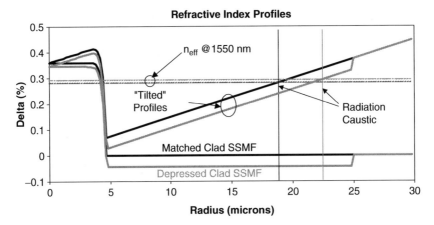

Figure 2.5 The tilted profile model gives the equivalent straight index profile for the bent fiber, shown for both a matched (solid line) and depressed clad profile design. The tilting shows that the effective index (dashed line) for either design drops below the cladding index at some radius, known as the *radiation caustic*. This indicates that a mode in a bent fiber is really a leaky mode, with coupling to radiation modes beyond the radiation caustic.

effective Δ than the matched clad design. Referring to the diagram of Fig. 2.5, the difference between n_{eff} of the LP_{01} mode and the depressed cladding index level is greater than the difference between n_{eff} and the index of the pure silica outer cladding. Therefore, the transverse propagation constant (refer to Eq. [2.9c]) in the depressed cladding region will be a larger imaginary number compared to that in the outer cladding, so that the field decays most rapidly per unit distance in the depressed cladding region. Referring again to the analogy with a particle in a potential energy well, this corresponds to having a repulsive barrier around the central attractive potential well. The depressed cladding region, therefore, decreases the coupling between optical power in the core and optical power in the outer cladding. This reduction in coupling can be used to reduce the sensitivity of a fiber to *macrobending losses*.

Macrobending refers to the loss of power propagating in a guided mode of the fiber when the fiber is held in a curved geometry. In general, macrobending is minimized for waveguides in which optical power is tightly confined to the core of the fiber and when the evanescent wave in the cladding is most rapidly damped. An equivalent condition is to say that the mode in question should have a high effective index n_{eff}. Confinement of the fundamental LP_{01} mode in the core of a step-index fiber can be increased by raising either the core radius a or the index Δ. However, either change decreases the macrobending loss of higher order modes as well, raising the effective cutoff wavelength of the fiber, as described in the previous section. The undesirable increase in cutoff can be mitigated by introducing the depressed cladding feature to the waveguide design. A careful study shows that the matched cladding design can be reoptimized for improved macrobending loss performance using the higher effective Δ to pull the field into the core for a slightly smaller MFD of 8.8 μm, while using a depressed cladding to maintain the cable cutoff less than 1260 nm.

2.5.2 Phenomenology of Macrobending Loss

Macrobending occurs in a large deflection of the fiber axis, where *large* is defined relative to the fiber core diameter, such as that associated with spooling or the presence of loops. The resulting loss consists of the transition loss and pure bending loss [25, 26]. The transition loss occurs at the transitions from straight to bent sections of the fiber and is the result of the mismatch of the field shapes in the straight and the bent fiber. The pure bending loss occurs because of energy radiating in the radial direction along a section of fiber bent at constant radius of curvature. Macrobending is a deterministic problem in a bend at constant radius of curvature, as opposed to the stochastic microbending problem discussed later. Phenomenologically, for small variations around a given profile design, there is a strong and linear correlation between the log of macrobending loss (at fixed radius of curvature R) and the so-called MAC factor, defined as MAC = MFD (μm)/$\lambda c(\mu$m). Either the fiber or cable effective cutoff can be used to calculate the MAC factor.

A rigorous and exact calculation of the macrobending loss of a fiber under constant curvature is computationally very intensive. One simple approximate method results from the realization that by employing a coordinate system transformation, a fiber bent at a constant radius of curvature has equivalent behavior to a straight fiber with index profile that has been altered from that of the bent fiber by a simple linear transformation. The so-called "tilted index profile" model calculates the loss of the equivalent straight fiber with refractive index profile in the plane of the bend as follows [25]:

$$n_s^2(r) = n_o^2(r) + 2n_o^2(o)r/R, \qquad (2.18)$$

where $n_o(r)$ is the index profile of the unperturbed fiber and R is the radius of curvature of the bend.

Figure 2.5 shows graphically the tilted profile macrobending model for a specific bend radius using both matched (black lines) and depressed clad (red lines) designs. The effective indices are indicated by dashed lines using the same color scheme. The depressed cladding design shown here has a trench radius five times the core radius. Far away from the core, at radii more than 19 microns, the effective indices of both profiles are lower than the tilted cladding index level. Thus, the bent fiber supports only a leaky mode instead of a pure guided mode.

Bending loss, thus, occurs by the tunneling of the power from core to the cladding. The point at which the effective index becomes lower than the equivalent straight index (tilted profile) of the bent fiber is the so-called "radiation caustic." Macrobending loss is proportional to the integral of mode power outside the radiation caustic. As the bend radius R decreases, the slope of the tipped index profile increases, and the radiation caustic moves in toward smaller radii. In that case, the fraction of power falling outside the radiation caustic increases, and therefore, the bending loss increases. Clearly for a given bend radius R, a fiber with a higher effective index will be less sensitive to macrobending.

Because of the presence of the depressed index trench, the radiation caustic for the depressed clad fiber is located in this example at approximately 18.8 microns, while that for the matched clad fiber is at about 15.2 microns. This means that in the depressed clad case, the electric field will have decayed to a smaller amplitude when it crosses the cladding index and begins to couple to radiation modes. This is shown quantitatively in Fig. 2.6, where the electric fields for the two cases are plotted on a log scale. The radiation caustics determined from the tilted profile case of Fig. 2.5 are marked to show that the electric field for the depressed clad fiber has decayed by an additional factor of five to six times relative to the matched clad fiber at the point at which power begins to be lost. To continue the analogy with the particle in a potential well, we note that the triangular region between the effective index line and the tilted profile,

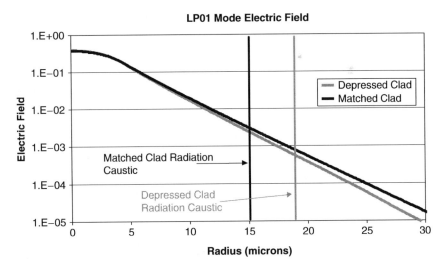

Figure 2.6 The decay of the electric fields associated with matched and depressed clad single-mode fiber designs shown in Fig. 2.5 illustrates the impact of moving the radiation caustic out to larger radii for the depressed clad fiber. The additional decay of the electric field results in less radiative loss beyond the radiation caustic.

between 5 microns and the radiation caustic, represents a tunneling barrier of greater area in the case of the depressed cladding fiber.

Depressed cladding fibers can improve performance in scenarios where low bending losses are important, such as indoor optical wiring, access networks, jumpers, ribbon corner fibers, cables with tight packing densities, and cables intended for very low temperature applications. At the time of this writing, high-quality matched clad fibers are usually specified as having less than 0.05 dB/100 turns for loops of radius R = 25 and 30 mm at 1625 nm and less than 0.05 dB/turn for loop of radius R = 16 mm at 1550 nm. Depressed clad G.652 fibers can give improved performance for loops of this size range, but from the point of view of system performance, losses for modern fibers of either matched or depressed clad designs are rather low in absolute terms for 25 and 30 mm radii. The performance of depressed clad fibers begins to diverge significantly from that of matched clad fibers for radii of R ~ 16 mm or less, where bending losses of commercial matched clad fibers are not currently specified. A single loop of a high-quality G.652-matched clad fiber with radius 10 mm can have bending loss of several decibels. At these tight bending radii, depressed clad fiber may have 5–10 times better macrobending loss performance than matched clad fiber.

2.6 FIBER ATTENUATION LOSS

Optical intensity of light decreases during transmission in a straight fiber because of various absorption and scattering mechanisms. This is represented mathematically in Eq. (2.3) when the longitudinal propagation constant, β, is a complex number. The imaginary part of β is the longitudinal decay constant. The decrease in optical power during transmission is often referred to as "attenuation" or "loss." For modern silica-based fibers, the attenuation within the wavelength range from about 1300 to 1600 nm is dominated by Rayleigh scattering, which results from intrinsic nanoscopic density fluctuations in the glass. Rayleigh scattering loss has wavelength dependence approximately $1/\lambda^4$ [27], as illustrated in the dashed line in Fig. 2.7.

In addition, sources of attenuation in optical fibers result from electronic and vibrational absorption from the silica, intended dopants, and impurities, and possibly from scattering by stress patterns frozen into the core layers during draw. The most commonly used model for the spectral loss, α, in dB/km has been

$$\alpha = A\frac{1}{\lambda^4} + B + C(\lambda), \tag{2.19}$$

where A is the Rayleigh scattering coefficient, B represents the combined wavelength-independent scattering loss mechanisms such as microbending, waveguide imperfections, and other scattering losses, and $C(\lambda)$ represents all other

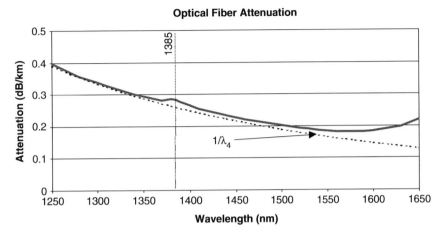

Figure 2.7 Attenuation curve of a very low water peak SSMF wound on a 150-mm diameter bobbin.

wavelength-dependent loss mechanisms such as the OH⁻ absorption peaks. Walker [28] proposed modeling $C(\lambda)$ as

$$C(\lambda) = \alpha_{uv} + \alpha_{IR} + \alpha_{12} + \alpha_{13} + \alpha_{POH} + \alpha_M + \alpha_{XS}, \qquad (2.20)$$

where the UV absorption band edge is modeled by

$$\alpha_{uv} = K_{uv} \cdot w \cdot \exp\left(C_{uv}/\lambda\right), \qquad (2.20a)$$

the infrared absorption band edge is modeled by

$$\alpha_{IR} = K_{IR} \cdot \exp\left(-C_{IR}/\lambda\right), \qquad (2.20b)$$

and the OH⁻ absorption peaks at 1240 and 1383 nm are modeled by the superposition of Gaussian terms

$$\alpha_{12} = \sum_{i=1}^{2} A_{12,i} \exp\left[-(\lambda - \lambda_{12,i})^2 / 2\sigma_{12,i}^2\right] \qquad (2.20c)$$

and

$$\alpha_{13} = \sum_{i=1}^{4} A_{13,i} \exp\left[-(\lambda - \lambda_{13,i})^2 / 2\sigma_{13,i}^2\right], \qquad (2.20d)$$

respectively, where $A_{12,i}$ and $A_{13,i}$ are the individual Gaussian peak amplitudes, $\lambda_{12,i}$, $\lambda_{13,i}$ are the individual Gaussian peak center wavelengths, and $\sigma_{12,i}$ and $\sigma_{13,i}$ are the individual Gaussian peak widths. If phosphorous doping is present in the fiber, the absorption from P-OH, α_{POH}, is

$$\alpha_{POH} = A_{POH} \cdot \exp\left[-(\lambda - \lambda_{POH})^2 / 2\sigma_{POH}^2\right], \qquad (2.20e)$$

where A_{POH} is the peak amplitude, λ_{POH} is the peak center wavelength, and σ_{POH} is the peak width. The macrobending loss, α_M, is modeled by Walker [28] as

$$\alpha_M = A_M \lambda^{-2} \cdot \exp\left[M_1 \lambda^{-1} (2.478 - M_2 \lambda)^3\right], \tag{2.20f}$$

where A_M, M_1, and M_2 are parameters determined by the fiber properties and the bend radius. For high-performance fibers, the excess loss term, α_{XS}, will normally be small and contain measurement error and noise, as well as systematic errors associated with the accuracy of the various terms of the loss model.

2.7 ORIGINS OF CHROMATIC DISPERSION

2.7.1 Introduction

An optical fiber's dispersion is the tendency for the fiber to either broaden or narrow a pulse as it travels along the fiber. The term *chromatic dispersion* is used to refer to the change in pulse shape that results when the velocity of the signal power along the fiber length is a function of optical frequency or wavelength. Chromatic dispersion alters the pulse shape because the signal power has a finite spectral width due to the spectral width of the signal modulation and the spectral width of the laser. The different signal frequencies or wavelengths of the signal travel along the fiber at different velocities, causing a digital pulse to spread in time, or an analog signal to become distorted.

In a single-mode fiber, chromatic dispersion of the fundamental mode is caused by the dispersive properties of the materials that the fiber is made from, referred to as *material dispersion*, as well as by the dispersive properties of the waveguide, referred to as *waveguide dispersion*.

2.7.2 Material Dispersion

As light travels through a material, it is slowed relative to its speed in vacuum, c, by the factor $1/n$, where n is the refractive index, because the electromagnetic wave interacts with the bound electrons in the material. Because the bound electron oscillations have resonance frequencies that are characteristic of the material, the interaction between the electromagnetic wave and the electrons is frequency dependent. This results in a frequency dependent index of refraction which gives rise to chromatic dispersion.

The frequency-dependent relationship for the index of refraction of a material can be obtained from a simple classic model that treats the bound electrons as harmonic oscillators [29]. The well-known Sellmeier equation for the frequency dependence of the index of refraction can be derived from an oscillator model. The Sellmeier equation expressed in terms of optical wavelength is

$$n^2 - 1 = \sum_{j=1}^{M} \frac{\lambda^2 B_j}{\lambda^2 - \lambda_j^2}, \tag{2.21}$$

where M is the number of electron resonances, λ_j are the wavelengths of the electron resonances, and B_j are constants obtained experimentally by the fitting to dispersion measurements.

Values for the coefficients of the Sellmeier equation for fused silica for a three-term fit to experimental data [30, 31] are shown in Table 2.1. Figure 2.8a shows the index of refraction as a function of wavelength for fused silica, germanium-doped silica, and fluorine-doped silica obtained using the Sellmeier coefficients given in Table 2.1.

The group delay per unit length, τ, of a wave propagating along a fiber is given by

$$\tau = \frac{1}{c} \frac{d\beta}{dk_0}, \tag{2.22}$$

where $\beta = n_{eff} k_0$, n_{eff} is the mode effective index, and k_0 is the free space propagation constant. By substituting $\beta = 2\pi n(\lambda)/\lambda$ into Eq. (2.22), we can recast Eq. (2.22) to express the group delay in terms of the index of refraction and its wavelength derivative

$$\tau = \frac{1}{c} \left(n - \lambda \frac{dn}{d\lambda} \right). \tag{2.23}$$

Table 2.1

Sellmeier coefficients for silica, germanium-doped silica, and fluorine-doped silica

Sellmeier coefficient	Undoped silica	Germanium-doped silica (4 mole %)	Fluorine-doped silica (1 mole %)
B_1	0.6968	0.6867	0.6911
λ_1	0.06907	0.07268	0.06840
B_2	0.4082	0.4348	0.4079
λ_2	0.1157	0.1151	0.1162
B_3	0.8908	0.8966	0.8975
λ_3	9.901	10.00	9.896

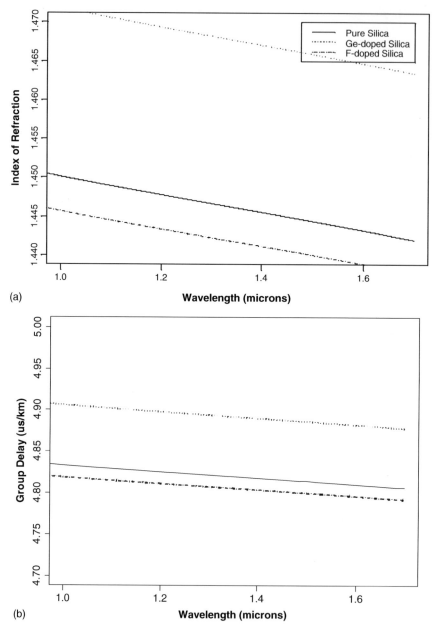

Figure 2.8 (a) Index of refraction as a function of wavelength calculated using Sellmeier coefficients given in Table 2.1. (b) Group delay, normalized by length, as a function of wavelength calculated using Sellmeier coefficients given in Table 2.1.

(*Continued*)

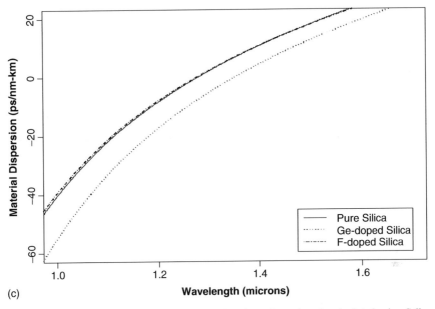

Figure 2.8, cont'd (c) Material dispersion as a function of wavelength calculated using Sellmeier coefficients given in Table 2.1.

We can see clearly from Eq. (2.23) that the group delay is wavelength dependent when the index of refraction varies with wavelength. Figure 2.8b plots as a function of wavelength the length normalized group delay for silica, germanium-doped silica, and fluorine-doped silica calculated from Eq. (2.23) using the index of refraction values plotted in Fig. 2.8a. Pulse distortion results from the dependence of group delay with wavelength because the spectral components of a pulse will experience varying delays as they propagate along a fiber.

The material dispersion of the fiber is given by wavelength derivative of the group delay

$$\frac{d\tau}{d\lambda} = -\frac{\lambda}{c}\frac{d^2n}{d\lambda^2}. \tag{2.24}$$

Figure 2.8c shows the material dispersion curves for silica, germanium-doped silica, and fluorine-doped silica calculated using Eq. (2.24) and the index of refraction curves in Fig. 2.8a. It is important to note that the material dispersion of silica is zero at about 1.270 nm, and it is, therefore, possible to make silica-based fibers with low dispersion in this wavelength region.

2.7.3 Waveguide Dispersion

The chromatic dispersion of a step-index single-mode fiber resulting from waveguide effects is referred to as *waveguide dispersion*. Although not explicitly discussed in Section 2.1, solutions to the Helmholtz equation for the electric fields in the cylindrical waveguide and the associated eigenvalues, u, w, and β, are wavelength dependent. Recall that β and β_t must obey the relationship $\beta_{t1} = (n_1^2 k_0^2 - \beta^2)^{1/2}$. Therefore, even if n_1 is assumed to be constant (i.e., there is no material dispersion), the wavelength dependence of $k_0 = 2\pi/\lambda$ results in variation of β and β_t with wavelength. By recasting Eq. (2.22) for group delay in terms of the wavelength derivative of β,

$$\tau = -\frac{\lambda^2}{2\pi c}\frac{d\beta}{d\lambda}, \tag{2.25}$$

we see that the wavelength dependence of β results in variation in group delay with wavelength.

Waveguide dispersion in a single-mode fiber can be understood through the following heuristic discussion of how the fundamental mode longitudinal propagation constant, β, changes with wavelength. We consider two bounding cases: first, where the optical wavelength is large relative to the core diameter, and second, where it is small relative to the core diameter.

For the case of very long wavelength, the fundamental mode is very loosely confined to the core and the ratio of the power carried in the cladding to the total power approaches unity [14]. At very long wavelength, the longitudinal propagation constant of the fundamental mode, β, approaches that of a plane wave propagating in the cladding, kn_{clad}. The group delay will approach that of a plane wave propagating in the cladding, β approaches its lower limit for a bound mode, kn_{clad}, and the group velocity is maximized. The group delay asymptotically approaches the silica curve in Fig. 2.8b at long wavelength.

Now considering the case of very short wavelength, the fundamental mode is very tightly confined to the core and the ratio of power carried in the cladding to the total power approaches zero. In this case, the longitudinal propagation constant, β, approaches that of a plane wave propagating in the core material, kn_{core}. In this case as β approaches its upper limit, kn_{core} and the group velocity is minimized. Because for most step-index fibers the core material is germanium-doped silica, for the very short wavelength case the group delay approaches that of the curve for germanium-doped silica plotted in Fig. 2.8b.

The finite element method was used to solve the scalar Helmholtz equation and determine the dispersion properties of the fundamental mode for a step-index, single-mode fibers. Figure 2.9 shows the material dispersion, waveguide dispersion, and total dispersion for a typical first-generation matched cladding

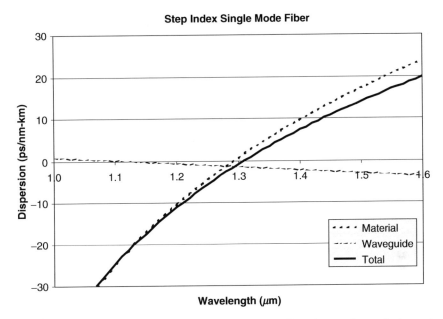

Figure 2.9 The contribution of material and waveguide dispersion to the total observed dispersion of a matched clad fiber design. The zero crossing is located at 1.31 μm.

single-mode fiber, with 4.65 microns core radius and core $\Delta = 0.34\%$. The material dispersion zero is at 1.28 microns. The core dimensions of the fiber have been chosen to provide the necessary waveguide dispersion so that the zero crossing of the total dispersion (material + waveguide) is located at the 1.31-micron local attenuation minimum. With this design choice, the dispersion zero and local attenuation minimum are collocated at 1.3 μm. Figure 2.10 shows the dispersion curves for step-index fibers with core radii ranging from 2.0 to 5.0 microns. As the core radius decreases, the magnitude of the waveguide dispersion curve is seen to increase in absolute value, while the change in the material dispersion is small. The location of the zero of the total dispersion curve, therefore, is shifted from around 1.3 μm at the larger core radii to around 1.55 μm at the smaller radii. This set of design choices results in the total dispersion zero to be collocated with the absolute loss minimum at 1.55 μm.

It was recognized that the variation of the waveguide dispersion with wavelength could be tailored to provide total dispersion curves with flattened shapes, multiple zero crossings, or reduced dispersion slope in the 1.55 μm transmission window [23, 32] by proper design of multilayer index of refraction profiles. These profiles typically have a central core region with the highest delta, a surrounding

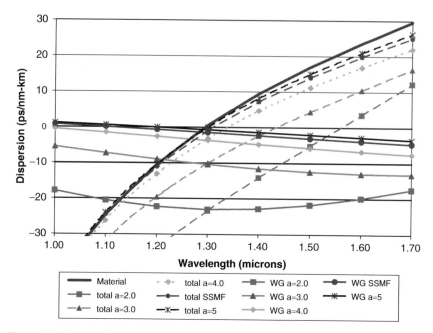

Figure 2.10 Calculations of the material, waveguide and total dispersion of fibers with core radii ranging from 2.0 to 5.0 microns using finite element solution of scalar Helmholtz equation. The dispersion zero of the total dispersion curves shift from ~1.3 to ~1.55 μm as the core radius decreases from 5.0 to 2.0 microns.

region with index reduced close to or below the cladding level, and a third concentric layer with raised index, typically at a level between the first two layers. Computational techniques, such as FEM, are required to obtain quantitative values for the properties of these complicated waveguide structures. However, the previous heuristic discussion can be applied to gain insight into how the multilayer waveguides behave.

Consider a fiber with central core surrounded by a fluorine-doped depressed-index trench and then a raised index ring. At very short wavelength, the electric field is tightly confined within the central layer and the overall group delay and dispersion approaches that of a plane wave propagating within this region. As wavelength increases, the mode starts to extend more into the trench region. Because of the depressed index of refraction of the trench, a plane wave traveling in this region has faster velocity and lower group delay relative to one traveling within the core. As the mode extends into the trench, the group delay and dispersion properties start to tend toward those of the trench region. The larger the contrast in the group velocity between the central region and the trench region, the greater is the ability to tailor the group delay and dispersion of the waveguide. The presence of the third raised index layer adds an additional guiding layer to the waveguide and as wavelength increases further, when overall structure is properly designed the energy will spread across the entire three layer structure and remain confined. As the mode spreads out over the entire structure, the fraction of energy contained within the trench region peaks and eventually decreases as the ring layer provides guidance. With appropriate choices of waveguide dimensions, the waveguide can be designed so that as the mode grows with wavelength and extends outward into the trench and ring regions, the group delay and the magnitude and shape of the waveguide dispersion curve of the mode can be tailored to provide total dispersion multiple zeros, or flattened shape. Multilayer index profiles with extreme contrast in index between the core and trench can provide very high values of waveguide dispersion (e.g., -150 ps/nm-km, for use as dispersion compensation devices [33].

2.8 POLARIZATION MODE DISPERSION

2.8.1 Overview

In the early 1990s, the deleterious effects of PMD were first reported in transmission of analog signals over Hybrid Fiber Coax (HFC) networks [34, 35]. The transmission distances in these networks were not long (typically <50 km), but the analog signals (unlike digital ones) were very sensitive to small levels of impairment that can be generated by unwanted dispersion. Here, it was found that chromatic dispersion, which is deterministic and can be compensated, was not the only impairment, but that PMD, coupled with source laser frequency chirp, was also playing a critical role. Since that time, fiber PMD has been improved far beyond the limits where this analog limitation is observed (for systems with small source frequency chirp). However, digital signal transmission rates increased, and PMD was found to be a limiting factor in long-haul transmission. Modern state-of-the art optical fiber for transmission is capable of carrying signal line rates of 10–40 Gbps over distances of thousands of kilometers without serious degradation due to PMD. Dispersion compensating fibers and some other specialty fibers pose a PMD concern because of the difficulty in manufacturing a small fiber core with an optically uniform circular cross-section. In this case, careful manufacturing controls and measurement are required to produce a reliable product. A fair amount of work has been devoted to PMD compensation. This generally is possible when the PMD is due to a fixed component (such as a LiNb modulator). For the fiber generated PMD, the compensation problem is much more difficult, because dynamic adjustments must be made on a per-channel basis (for each transmission wavelength).

This section will not attempt to cover the full spectrum of PMD topics. The intent is rather to give a basic understanding of the phenomenon and a few references for the interested reader. A more comprehensive review can be found in reference [36]. The requirements imposed by high-speed digital transmission are addressed in Chapter 5.

2.8.2 Background

The first concept to be understood is that of lightwave polarization, because PMD originates when different light polarizations travel at different average speeds in the optical fiber. The polarization of light (or any electromagnetic wave) is a description of how the electric field vector of the wave varies in time at a fixed position in the fiber. In general, the tip of the electric field vector traces out an elliptical shape in a plane transverse to the fiber axis as time evolves. This elliptical polarization can, in special cases, degenerate to a line or circle (linear or circular polarization). Details can be found in most elementary texts on electromagnetic fields or optics [37]. The speed that light will travel in an optical fiber is a function of the effective index of refraction. This effective index can be a function of polarization (leading to PMD) and wavelength (leading to chromatic dispersion). This phenomenon is familiar from the optical concept of birefringence (typically due to index of refraction variation with lightwave electric field orientation in a crystalline material).

Confusion often arises over the difference between *differential group delay* (DGD) and PMD. It has been shown mathematically that for any fixed wavelength and length of optical fiber, there are two *principle states of polarization* (PSP) [38]. These states possess, among other properties, the maximal and minimal transit time through the fiber of any input state of polarization. The difference in transit time between these states, over a given length of fiber, is the DGD. The DGD is generally a function of wavelength. In a uniform birefringent material, there are fast and slow axes corresponding to the orientation of the two PSPs. In this case, the PSPs are independent of wavelength and remain invariant along the direction of light propagation (because the fiber cross-section does not vary). Light that is input in any polarization other than the PSPs suffers a periodic evolution in polarization state along the propagation direction. This spatial period depends on the wavelength and birefringence strength and is called the *beat length*. The DGD builds up through the material in a linear fashion, with a net delay that is directly proportional to the propagation distance in the material. In the case of an optical fiber, the birefringence is generally not fixed but fluctuates along the fiber length because of variations in core transverse geometry or core mechanical stress (see reference [39] for an overview of the many physical reasons for this birefringence). In this case, the PSPs vary along the fiber length and are wavelength dependent. For any fixed set of fiber terminals (input and output positions along the fiber), there is a unique set of PSPs. Because of this, the DGD (at a fixed wavelength) builds up along the fiber in a complicated way, instead of simple linear growth. An example is given by the simulation results shown in Fig. 2.11. Here, four erratic curves show the buildup of DGD in a group of fibers with slightly different distribution of birefringence along their length. Note that the DGD is almost as likely to decrease as to increase as the lightwave propagates down the fiber.

The fiber length over which the polarization "randomizes" is termed the field *correlation length*. This correlation length can be understood in terms of a fictitious experiment: launch a lightwave along a local PSP into each of an ensemble of equivalent fibers and determine the average position in these fibers where the optical power in the polarization mode orthogonal to that launched is within $1/e^2$ of the launch power. Note that the lightwave and the birefringence may randomize independently and have different correlation lengths. The PMD is defined as the average (linear or rms) of the DGD over all realizations of the system (which may vary in time, wavelength, or birefringence pattern in the fiber). The net PMD (or DGD) obtained after transit through a length of fiber is called the *PMD value* (or DGD value) with units typically in picoseconds. The rate of increase in PMD (or DGD) along a length of fiber is known

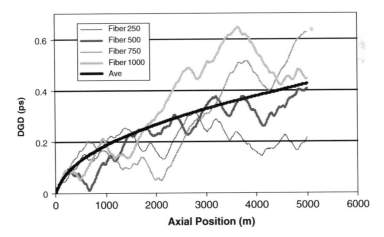

Figure 2.11 Monte Carlo simulation of differential group delay (DGD) evolution in 5 km length of optical fiber. Ten thousand fibers, with varying random component of birefringence are used. Beat length = 10 m, wave-plate length = 1 m. Several individual cases are shown, with the average of the 10,000 simulations shown by black bold line (behavior of Eq. [2.32] is noted).

as the *PMD* (or *DGD*) coefficient with typical units of picosecond per kilometer (for very short fibers) or picosecond/km$^{1/2}$ for long fibers. Because any polarized lightsource consists of a range of wavelengths, after propagation through a length of fiber a range of polarizations will be present, one for each wavelength. In this way, a source will depolarize because of its own finite line width (the PMD required for depolarization is roughly equal to the source coherence time).

2.8.3 Modeling and Simulation

The evolution of the lightwave polarization is most conveniently studied in terms of Stokes vectors in Poincaré space [36]. The three-dimensional normalized Stokes vector, $\bar{s} = (s_1, s_2, s_3)$, consists of combinations of the time-harmonic transverse electric field components for the lightwave:

$$s_1 = \frac{|E_x|^2 - |E_y|^2}{|E_x|^2 + |E_y|^2}, \quad s_2 = \frac{2\mathrm{Re}\{E_x E_y^*\}}{|E_x|^2 + |E_y|^2}, \quad s_3 = \frac{2\mathrm{Im}\{E_x E_y^*\}}{|E_x|^2 + |E_y|^2} \qquad (2.26)$$

and polarization-dependent loss (PDL), which describes the case in which the PSPs have unequal attenuation, is neglected. The polarization at any point in the fiber can be determined from the launch polarization by the Muller Matrix (which, in the absence of PDL, is just a rotation matrix) $\overline{\overline{R}}(z,\omega)$, where ω is the angular frequency of the lightwave and z is the distance traveled along the fiber from the launch point

$$\bar{s}(z,\omega) = \overline{\overline{R}}(z,\omega)\bar{s}(0,\omega). \qquad (2.27)$$

The evolution of $\bar{s}$ in frequency and axial position can be found by differentiating Eq. (2.27) with respect to ω and z, respectively [40]. Eliminating the launch polarization from the resulting equations gives

$$\frac{\partial}{\partial\omega}\bar{s}(z,\omega) = \left[\frac{\partial}{\partial\omega}\overline{\overline{R}}(z,\omega)\right]\overline{\overline{R}}^{-1}(z,\omega)\bar{s}(z,\omega) \equiv \overline{\Omega}(z,\omega) \times \bar{s}(z,\omega), \qquad (2.28a)$$

$$\frac{\partial}{\partial z}\bar{s}(z,\omega) = \left[\frac{\partial}{\partial z}\overline{\overline{R}}(z,\omega)\right]\overline{\overline{R}}^{-1}(z,\omega)\bar{s}(z,\omega) \equiv \overline{W}(z,\omega) \times \bar{s}(z,\omega), \qquad (2.28b)$$

where the operator identities for the polarization dispersion vector (PDV), $\overline{\Omega}$, and the material birefringence, $\overline{W}$, are

$$\overline{\Omega}(z,\omega)\times \equiv \left[\frac{\partial}{\partial\omega}\overline{\overline{R}}(z,\omega)\right]\overline{\overline{R}}^{-1}(z,\omega) \qquad (2.29a)$$

$$\overline{W}(z,\omega)\times \equiv \left[\frac{\partial}{\partial z}\overline{\overline{R}}(z,\omega)\right]\overline{\overline{R}}^{-1}(z,\omega). \qquad (2.29b)$$

The cross-product model works because with no PDL, any change in $\bar{s}$ can only result in a change in orientation (magnitude is fixed at unity). Thus, $\bar{s}$ and its derivatives are perpendicular. A dynamical equation for the PDV can be found from Eqs. (2.28a and 2.28b) [41]:

$$\frac{\partial}{\partial z}\overline{\Omega}(z,\omega) = \frac{\partial}{\partial \omega}\overline{W}(z,\omega) + \overline{W}(z,\omega) \times \overline{\Omega}(z,\omega). \qquad (2.30)$$

Under the assumption of a suitably narrow source spectral width, the PDV can be written as a power series in frequency:

$$\overline{\Omega}(z,\omega) \approx \overline{\Omega}(z,\omega_0) + (\omega - \omega_0)\frac{\partial}{\partial \omega}\overline{\Omega}(z,\omega_0) + \dots, \qquad (2.31)$$

where the terms on the right-hand side of Eq. (2.31) are the first and second-order PDVs. Using Eq. (2.31) in Eq. (2.30), one can obtain equations for each PDV order. In particular, $|\overline{\Omega}(z,\omega_0)|$ is the first order DGD and $\frac{\overline{\Omega}(z,\omega_0)}{|\overline{\Omega}(z,\omega_0)|}$ is the slow PSP at position z. Early models considered the fiber to be a stack of wave plates with Gaussian distributed birefringence components and fixed axial length [41]. This model requires a physical average over optical wavelength or random birefringence to obtain the PMD. Another approach is to directly form stochastic differential equations that can be solved for expectation values of the polarization dispersion vector [40, 42]. The stochastic equation approach is the most commonly used today. Useful concatenation rules for the PDV can be found in Annex A of [IEC 61282-3]. From a variety of models, the PMD as a function of fiber length is found [41, 43, 44]:

$$PMD = \sqrt{2}h\frac{\partial \Delta}{\partial \omega}\sqrt{e^{-z/h} + z/h - 1}, \qquad (2.32)$$

where the PMD is defined as the rms value of the magnitude of the stochastic PDV, h is the fiber correlation length, and Δ is the expectation value of the birefringence magnitude. This equation shows the expected linear growth of PMD for short fiber lengths leading to growth of PMD depending on the square root of the fiber length at long lengths (when the DGD due to short segments of fiber add statistically to previous segments).

2.8.4 Control of PMD in Fiber Manufacturing

Asymmetric stresses or noncircular core geometries are difficult to completely eliminate in optical fiber manufacturing. Even worse, such non-uniformities in fiber cross-section may tend to be rather uniform along the fiber length. This causes the resulting birefringence to be deterministic on possibly kilometer-length scales. The resulting PMD can be unacceptably large. In addition, the fiber becomes very sensitive to environmental conditions (temperature, cabling stresses, or movement). Thus, it is very hard to predict the PMD performance for the end-user.

The origin of noncircular core geometry is usually related to non-uniform materials or processes in the preform manufacture. Fibers made with the MCVD, OVD, or PCVD methods [1] require the collapse of a hollow glass cylinder at some point in manufacture of the core. Careful control of the collapse process [45] must be maintained to avoid introduction of excess ovality. The MCVD and PCVD processes, most capable for fabrication of complex index profiles, use starting tubes that must be specified to be highly circular and of uniform wall thickness to avoid resulting geometrical imperfections in the resulting fiber.

A source of both geometrical and stress non-uniformity is the presence of trapped vapor bubbles in the preform, which translate to airlines in the optical fiber. Various methods have been devised to screen performs for bubbles and fiber for airlines.

In spite of careful process control, there is inevitably some non-uniformity to the fiber cross-section, in either the geometry or the stress profile. The only known practical solution to this problem is to spin the fiber during the draw process [46, 47]. The most effective way to spin the fiber is to use a device to rotate the fiber just below the draw furnace. This causes the fiber to rotate, as a rigid body, with the molten glass at the preform tip accommodating the deformation (and relieving most stresses). As the glass transitions to a solid, the spin is "frozen-in" with little residual stress. The transmission effect is to average out the azimuthal non-uniformities experienced by the lightwave. For spinning to be effective, it must suitably randomize the deterministic birefringence and do so on a length scale less than a beat length (otherwise the lightwave polarization will tend to follow the spin as in the case of a weakly twisted polarization maintaining fiber). Optimal spin parameters have been studied theoretically [48, 49] (beyond the notion of making several fiber spins per beat length), although in practice little improvement is found because of random birefringence effects and production uncertainties. Spinning can be accompanied by mechanical twist. This term is used to describe the case where the fiber is twisted after the glass has solidified. Here, the twist is accompanied by an elastic stress, which results in a circular birefringence [50], because of the stress-optic effect and unwanted DGD [51]. Generally, a small amount of mechanical twist lowers DGD in an unspun fiber, while the DGD in a well-spun fiber is always increased by mechanical twist. The sensitivity of fibers to twist is relatively independent of fiber type. Mechanical twist levels, as small as one twist per meter, can noticeably degrade the PMD performance of modern fibers. Twist control can be maintained by ensuring proper functioning of the spin device and alignment of the draw/spin/takeup process. Some process monitoring is required to maintain low twist production.

2.8.5 Measurement of PMD

PMD measurement is complicated by several factors including the statistical nature of the phenomenon and its sensitivity to the surrounding environment (temperature or mechanical stress changes). It is useful then to describe measurement of spun and unspun fiber, the fiber environment during the measurement, and the way the DGD averaging is accomplished. The standardized reference test method is based on Jones Matrix Eigenanalysis (JME). Test sets based on this method measure DGD over a limited wavelength range. Other test methods include those based on interferometry and wavelength scanning (see list of useful standards at the end of this section for information on these methods).

From the foregoing discussion, it is clear that the level of mechanical twist in the fiber must be accurately assessed before making claims of the intrinsic fiber PMD. In addition, spool-based measurements (even large diameter or collapsible spools) are unreliable on unspun fiber because inadvertent mode coupling occurs and has an enormous effect on the PMD (PMD generally appears much lower than it really is). Figure 2.12 shows some measurements that indicate that this PMD difference can be as large as an order of magnitude. The spooled geometry provides several fiber stresses, which lead to mode coupling. These are bending, lateral stress (caused by winding tension), and fiber crossovers. These effects occur on a scale length of tens of centimeters, much smaller than the correlation length of a typical unspun fiber. Because of this, mode mixing occurs and the fiber PMD is lowered. When the fiber is unspooled (e.g., during the cabling operation), it returns to its former, weakly coupled, state and the PMD once

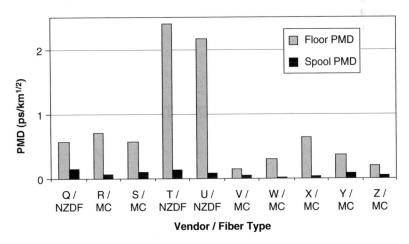

Figure 2.12 On- and off-spool measurements on unspun fiber from different manufacturers.

again is large. Modern well-spun fiber can have very low PMD, often well below $0.02\,\text{ps/km}^{1/2}$ [52]. Accurate measure of PMD for this fiber is difficult because spool effects usually raise the PMD (because of fiber bending and tension), and instrument bandwidths are generally insufficient to reduce the measurement uncertainty to an acceptable value [53]. Figure 2.13 shows the PMD difference obtained on a group of non-zero dispersion-shifted fibers (NZDFs) when measured on a typical shipping spool (elevated PMD) and on a large diameter collapsible spool with manual disturbance. Fortunately, spun fiber is less susceptible to small random external effects, so some approximate methods can be employed. One is a loose winding of fiber on a large (generally >30-cm) diameter spool. Depending on fiber design, the resulting fiber crossovers may or may not mask the true fiber PMD. Other methods include spreading the fiber on a large flat surface (usually >10 m diameter) or measuring the fiber in a quiescent cable design such as a low fiber count central core. With 20-km lengths of fiber (or cable), one can lower fundamental statistical measurement uncertainties to below 30%. To reduce this uncertainty further, manual disturbances have been used to randomize the birefringence and obtain more independent DGD samples.

Because no two fibers (or cables) from a given manufacturer are exactly alike, either among a production run or when comparing the same fiber in the factory and the field, there is doubt as to the actual quality of the product the end-user receives. One standardized means of specification is to use the *link design value* (LDV or PMD_Q). This statistical specification creates virtual links by randomly

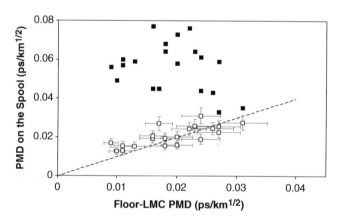

Figure 2.13 Large-diameter collapsible spool PMD measurement (open squares) on NZDF fiber compared to measurement on the floor (reference line dashed). The same fibers measured on a 160-mm diameter shipping spool under 35 g winding tension (solid squares).

selecting and concatenating M fibers from the production fiber PMD distribution, then determining the probability of a maximum PMD being exceeded. A typical specification may state that the LDV for a given fiber product is $0.04 \, \text{ps/km}^{1/2}$ for $M = 20$ and $Q = 10^{-3}$. This says that when 20 fibers are randomly concatenated to form links, only 0.1% (Q^*100) of these links will have PMD above $0.04 \, \text{ps/km}^{1/2}$. Clearly, a specification that requires more fibers to be used (larger M) or higher fraction of fibers exceeding the specification (high Q) is weaker than a low M, low Q specification. The computation of LDV is often done using the Monte Carlo technique, but for reasonably smooth distributions (with single maxima), an analytical method due to Jacobs [54, 55] provides accurate results. The concatenation rule for PMD coefficients (note that this does not work for DGD) is

$$PMD_{TOTAL} = \sqrt{\frac{\sum\limits_{k=1}^{M} PMD_k^2 L_k}{\sum\limits_{k=1}^{M} L_k}}, \tag{2.33}$$

where PMD_k and L_k are the PMD coefficient and length of the kth fiber in the concatenation and M is the total number of fibers.

2.8.6 Fiber-to-Cable-to-Field PMD Mapping

Ensuring good PMD performance in cable requires adequate process control in fiber and cable production, as well as sufficient measurement to sample production populations. A particularly useful measurement is one that follows fibers through the cabling process and installation, thereby producing a mapping function that, on an individual fiber basis, compares changes in PMD due to cabling and installation. Figure 2.14 shows a typical mapping between uncabled and cabled fiber for NZDF fiber in a central core cable. Manual disturbance is very helpful in reducing uncertainties in these measurements. Figure 2.15 shows the results of a blind test run on a cable with two identical tubes of 12 unique fibers. In Fig. 2.15a, the PMD for the corresponding fibers in each tube do not appear related. After 10 manual disturbances, Fig. 2.15b shows that the measurement uncertainty has been reduced to the point where the identical fiber pairs are obvious. In field tests and in fiber production, manual disturbance of the sample is usually not possible. In these cases, a large number of samples are used (with the longest lengths possible) to estimate the PMD mapping. Although this type of measurement runs the danger of confusing DGD and PMD, the results have been quite successful. An example is the data obtained for a 40 Gb/sec network built by MCI [52].

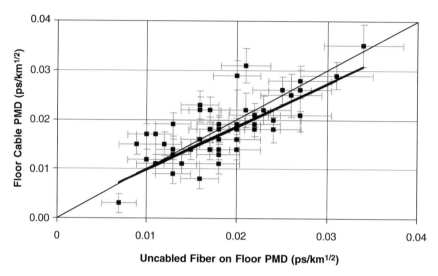

Figure 2.14   Mapping between uncabled NZDF fiber and same fiber in central core cable.

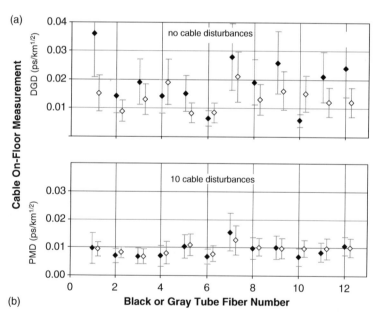

Figure 2.15   Effect of manual disturbance on PMD measurement. Two identical cabled tubes (of 12 unique fibers each) are compared (a) without and (b) with manual disturbance. Solid (open) points are fibers in black (gray) tube.

Useful International Electrotechnical Commission (IEC) documents on PMD system and measurement include the following:

IEC 60793-1-48: Optical fibres—Part 1–48: Measurement methods and test procedures—Polarization mode dispersion

IEC 61280-4-4: Fibre optic communication subsystem basic test procedures—Part 4–4: Cable plants and links—Polarization mode dispersion measurement for installed links

IEC 61290-11-1: Optical amplifier test methods—Part 11–1: Polarization mode dispersion—Jones matrix eigenanalysis method (JME)

IEC 61290-11-2: Optical amplifiers—Test methods—Part 11–2: Polarization mode dispersion parameter—Poincaré sphere analysis method

IEC/TR 61282-3: Fibre optic communication system design guides—Part 3: Calculation of polarization mode dispersion

IEC/TR 61282-5: Optical amplifiers—Part 5: Polarization mode dispersion parameter—General information

IEC/TR 61282-9: Fibre optic communication system design guides—Part 9: Guidance on polarization mode dispersion measurements and theory.

2.9 MICROBENDING LOSS

2.9.1 Microbending

Fibers often exhibit excess loss when they are spooled or cabled as the result of small deflections of the fiber axis that are of random amplitude and are randomly distributed along the fiber. The loss induced in optical fiber by these small random bends and stress in the fiber axis is called *microbending loss.*

Figure 2.16 cartoons the impact of a single microbend, at which, analogous to a splice, power can be coupled from the fundamental mode into higher order leaky modes. Because external forces are transmitted to the glass fiber through the polymer coating material, the coating material properties and dimensions, as well as external factors, such as temperature and humidity, affect the microbending sensitivity of a fiber. Further, microbending sensitivity is also affected by coating irregularities such as variations in coating dimensions, the presence of particles such as those in the pigments of color coatings, and inhomogeneities in the properties of the coating materials that vary along the fiber axis. Coating surface slickness can also affect the mechanical state into which a fiber relaxes after spooling or within a cable structure, thereby affecting microbending loss.

The fiber axis perturbations that cause microbending loss are random in magnitude and are randomly distributed along the fiber. The perturbations are, therefore, modeled by a stochastic process characterized by broadband

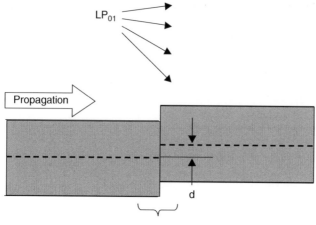

Deflection in the fiber axis (d)

Figure 2.16 Model of the core of a fiber in the vicinity of a highly exaggerated microbend. Power carried in the fundamental mode before the microbend is coupled into the fundamental as well as higher order modes at the microbend, similar to the case of a non-ideal splice.

and randomly phased spatial frequency components. The statistics of fiber deformation may be unknown, in contrast to the limiting case of a deterministic long-period fiber grating. The statistics of microbends can only be measured indirectly. Profilometry can be used to measure the roughness of the inner surface of a cable core tube. Fourier analysis of the roughness profile of the extruded polymer surface might yield spatial frequency content in a range around 500 microns. However, the properties of the fiber as a stiff beam serve to impose a low pass filter on the spatial frequency content, with fairly sharp cutoff characteristics varying as the fourth power [56].

Mode coupling results when microbends occur, transferring power from guided modes to radiation modes. Many approximate analyses of microbending loss have been proposed [25, 57–60]. One of the simplest metrics used for parameterizing the sensitivity of fibers to microbending loss is the MAC factor defined in the earlier discussion of macrobending [60]. Theory and experiment predict decreasing microbending sensitivity with decreasing MFD and increasing cutoff, and thus, decreasing MAC factor. When using MAC factor for the microbending sensitivity analysis, care has to be taken to compare fibers with similar refractive index profiles.

Analyzing microbend loss more rigorously requires time-intensive calculations of the modal coupling coefficients between the guided and unguided modes. Unguided modes can be represented as cladding, leaky, or radiation modes [61]. Marcuse [57] predicted the microbend loss of single mode fibers as a function of wavelength based on mode coupling theory between guided modes and cladding mode using analytic expressions for the LP modes. A stochastic model was developed for the random bends assuming a Gaussian-shaped auto-correlation function with rms perturbation amplitude, σ, and correlation length, L_c. The magnitude and wavelength dependence of the predicted microbending loss was found to be strongly dependent on the value of the correlation length of the bends.

In general, we follow Marcuse's approach except that we use finite element techniques [20] to find the solutions for the fiber modes. In Marcuse's derivation, the fundamental mode field shape was approximated as Gaussian, while a number of simplifying assumptions were used in approximating the cladding mode solutions. Bjarklev [58] improved the accuracy of the microbending calculations by using the more accurate approach of solving for the cladding modes of a coated fiber surrounded by air. Here, we also use the more accurate approach of coupling to the set of leaky modes, rather than cladding modes.

Microbend loss following coupled mode theory [57] can be represented as

$$2\alpha_m = \sum_{p=1}^{\infty} C_{1p}^2 \Phi(\beta_{01} - \beta_{1p}), \qquad (2.34)$$

where C_{1p} is the coupling coefficient between the fundamental guided mode and p-th cladding mode and Φ is the power spectrum of the axis deformation function. β_{01} is the propagation constant of the guided mode, and β_{1p} is the propagation constant of p-th leaky mode. The coupling coefficient is calculated as

$$C_{1p}^2 = \frac{k^2}{2} \frac{\left(\int\limits_0^\infty \frac{dn}{dr} E_{01} E_{1p} r dr\right)^2}{\int\limits_0^\infty E_{01}^2 r dr \int\limits_0^\infty E_{1p}^2 r dr}, \qquad (2.35)$$

where E_{01} and E_{1p} are electric fields of guided mode and leaky mode, respectively. The power spectrum of the Gaussian deformation is

$$\Phi(\beta_{01} - \beta_{1p}) = \sqrt{\pi}\sigma^2 L_c \cdot \exp\left\{-\left[\frac{1}{2}(\beta_{01} - \beta_{1p})L_c\right]^2\right\}, \qquad (2.36)$$

where σ is the rms deviation of the distortion function, and L_c is the correlation length. The physical significance of the correlation length L_c can be understood through a discussion of the spectral analysis of the fiber axis deformation $f(z)$. The spectral analysis of aperiodic signals is often accomplished by taking the Fourier transform of the auto-correlation function of the signal. The auto-correlation function of a periodic signal is also periodic with the same frequency spectrum as the original signal. In contrast, the auto-correlation function of a

aperiodic signal, with randomly varying amplitude and phase, will decay monotonically from the maximum of one to zero for large shift. The environmental
roughness impressed on an optical fiber, resulting in deformation of the fiber
axis, will be a function of this type. The power spectrum of such an auto-
correlation will reflect the length scales present in the environmental roughness
and decay to zero at higher frequencies.

For simplicity in an analytic formulation, Marcuse assumed that the
auto-correlation of the unspecified function $f(z)$ is Gaussian of the form
$R(u) = \sigma^2 \exp\left[-\left(\frac{u}{L_c}\right)^2\right]$, where σ is the rms deviation of the distortion function
$f(z)$ and L_c is its correlation length. Therefore, the Fourier transform must be a
Gaussian function of spatial frequency, leading to the simple expression of Eq.
(2.36). The argument to the power spectrum of the perturbations is the difference
in propagation constants $(\beta_{01} - \beta_{1p})$, of the modes that are exchanging power.
The strength of the coupling is proportional to power spectrum evaluated
at $\beta_{01} - \beta_{1p}$ and the overlap integral of the electric fields of the coupled modes,
as shown in Eq. (2.35). The correlation length is, thus, a measure of the shortest
length scale represented in the random perturbation function of the fiber axis.

As an example of the dependence of microbending loss on fiber profile when
the statistics of the fiber axis perturbations are known, we describe microbend
loss for typical matched clad (MC) and depressed clad (DC) fiber designs. Figure
2.17 shows calculated microbend loss for the MC and DC fibers as a function of
wavelength for the cases where correlation length, L_c, is 50, 450, and 800 μm.

The microbending loss is largest for small L_c, $= 50 \mu$m, and is slowly decreasing as wavelength increases. The large magnitude microbending loss occurs for
short correlation length because the broad width of the perturbation power
spectrum results in coupling to several leaky modes. The variation with wavelength is low because the wavelength-dependent coupling of individual modes is
averaged across several modes. However, in practice, perturbations on this
length scale are only weakly transmitted to the fiber core because of the strong
spatial filtering from the coating and the stiffness of the fiber. For a more
physically relevant correlation length of 450 μm, the microbend loss is seen to
increase as the wavelength increases, as is typical of microbend added loss
spectra, as observed in real fibers in cables or modules. In this regime, the
depressed cladding design shows lower microbend sensitivity than standard
matched clad fiber. This is also observed in practice. The calculated increase
with wavelength in this range is also reasonable. For a longer correlation length
of 800 μm, the trend with wavelength is similar to the 450-μm case, but with
an overall reduction in magnitude. The nearest leaky mode, to which microbending may power out of the fundamental LP_{01} mode, is usually LP_{11}. As
$L_c \rightarrow \infty$, $2\pi/\Lambda$ becomes too small to couple even these two most closely spaced
modes, and microbending loss becomes negligible.

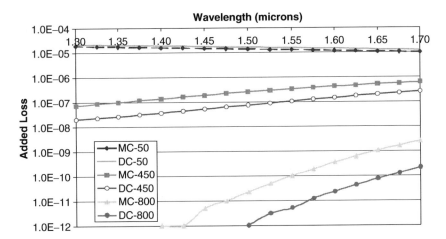

Figure 2.17 Calculated microbend loss for the standard matched clad fiber (MC) and the depressed-clad fiber (DC) for three value of correlation length L_c as a function of wavelength.

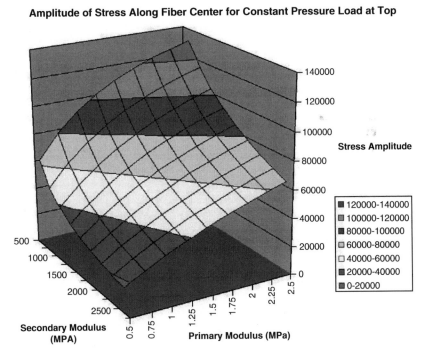

Figure 2.18 Stress amplitude at the fiber core versus modulus of the primary and secondary (finite element calculation). (Figure courtesy Dr. Harry Garner of OFS, Norcross, GA.)

The dual coating is used to protect the fiber and to reduce the stress at the fiber when external force exists to the fiber due to cabling or spooling. Yang [62] describes the impact of coating properties on microbending loss: Microbending losses decrease with increasing thickness, Young's modulus, and Poisson's ratio of the primary coating. Similarly, changes in refractive index in the glass fiber decrease with the increasing Young's modulus and Poisson's ratio of the secondary coating.

Figure 2.18 shows a calculation of the stress at the fiber axis as a function of primary modulus and secondary modulus following Yang's derivation [62]. As the modulus of the secondary coating increases, the stress decreases. However, as the modulus of the primary coating increases, the stress increases. The combination of soft primary and hard secondary are desirable for best microbending performance. Since the microbend loss is proportional to the stress, we can design coating geometry and materials to reduce the microbend loss. However, other coating performance metrics must also be kept in balance, so there are limits to the improvement available in microbending performance by tailoring coating properties.

2.10 FIBER NONLINEARITIES

2.10.1 Overview

Highly focused coherent laser light, propagating with low loss through optical fiber over long distances (kilometers), is an ideal breeding ground for nonlinear interaction with the glass material. Although nonlinear effects were found in early optical transmission work with analog signal delivery (CATV, etc.), much attention lately has been given to resolution of nonlinear problems in long-haul optical communications and high-power operation in specialty fibers. In particular, new fiber types have been developed to overcome nonlinear impairments. As fiber design introduced dispersion-shifted fibers (DSFs) in the early 1990s, to overcome chromatic dispersion impairments, it was soon found that multiple lightwaves, with different wavelengths, were able to efficiently interact through a four-wave mixing (FWM) process since the coupling waves were well matched in phase and group velocity. This led to the development of NZDFs that struck a balance between the high chromatic dispersion of standard single-mode fiber and the very low dispersion, at operating wavelengths, of DSFs. With the advent of high-power erbium-doped fiber amplifiers (EDFAs) and high-power laser diodes, many nonlinear issues arose because of the long distance between signal regeneration points and the multiple optical wavelengths that could simultaneously be used. In particular, stimulated Brillouin scattering became apparent (at 5–10 dBm levels with laser line widths <5 MHz). This required new features in transmitters to broaden the effective source line width. Self- and cross-phase modulation issues were also noted. Generally, these problems increased with small effective area fibers (such as those often used in specialty applications). In the late 1990s, Raman amplification received renewed attention because of potential noise improvements due to its distributed nature. This amplifier was based on stimulated Raman scattering of a signal wavelength by a high-powered laser pump in a transmission fiber medium. Here, we outline some of the general principles of these interactions and provide sufficient references to start the interested reader on a course of further study.

2.10.2 Background

The optical transmission nonlinearities occur as a result of interaction of the traveling lightwave with the doped glass dielectric medium. The wave equation for lightwave transmission in an optical fiber can be written as

$$\nabla^2 E - \frac{1}{c^2}\frac{\partial^2 E}{\partial t^2} = \mu_0 \frac{\partial^2 P}{\partial t^2}, \text{ where } P \cong \varepsilon_0 \left\{ \chi^{(1)} E + \chi^{(3)} \vdots EEE \right\}. \tag{2.37}$$

Here, the nonlinear polarization due to the dielectric is written in terms of the tensor third-order susceptibility, because the second-order effect is negligible due to molecular symmetry of silica. Time retardation, which may be important in some nonlinear interactions, is not included (see reference [63] for this correction). Because the nonlinearity in fibers is weak, a perturbation approximation is normally used in solving Eq. (2.37). That is, the full solution is approximated as a slight modification of the solution with $\chi^{(3)} = 0$. Changing to a time harmonic representation of the fields, an index of refraction can be related to the susceptibilities:

$$n = n_0 + n_2|E|^2,$$

where

$$n_0 = 1 + \frac{1}{2}\text{Re}\{\chi^{(1)}\} \text{ and } n_2 = \frac{3}{8n_0}\text{Re}\{\chi^{(3)}_{xxxx}\}. \tag{2.38}$$

Here, $\chi^{(3)}_{xxxx}$ is the component of the susceptibility tensor aligned with the wave polarization and $\text{Re}\{z\}$ indicates the real part of z. In proceeding to derive propagation equations from Eq. (2.37), it is found that the nonlinear effects are proportional to n_2/A_{eff}.

Because nonlinear effects depend on the lightwave intensity, proportional to $|E|^2$, these effects are reduced as the lightwave propagates into the fiber because the intensity drops by conventional fiber attenuation. Thus, the effective length, L_{eff}, of fiber over which the nonlinearity is important depends both on the physical fiber length, L, and the optical power attenuation coefficient, α:

$$L_{eff} = \frac{1}{\alpha}\left[1 - e^{-\alpha L}\right]. \tag{2.39}$$

Clearly for high attenuation, α, the $L_{eff} \sim 1/\alpha$ and for small attenuation $L_{eff} \sim L$, and the effective length equals the physical fiber length.

The formalism based on the nonlinear index is appropriate for describing the intensity-dependent phase change seen in self-phase modulation (SPM), cross-phase modulation (XPM) and FWM. Modifications are needed to describe scattering phenomena such as stimulated Brillouin scattering (SBS) and stimulated Raman scattering (SRS). These involve the material dynamics of acoustic waves and molecular vibrations. Parametric processes, such as FWM and parametric gain, are caused by mixing of lightwaves of different frequencies through the nonlinear fiber index. Phase matching is required for these effects to become significant. The following provides a rapid overview of these nonlinearities.

SPM occurs when varying signal intensity changes the phase through the signal pulse, causing new frequencies to develop. This effect can be encapsulated in the "nonlinear phase":

$$\Phi(z) = \int k\,dz, \quad \Phi_{NL} = \frac{2\pi n_2 P L_{eff}}{\lambda A_{eff}}, \tag{2.40}$$

which is obtained by integration of the optical wavenumber, k, ($k = n\omega/c$) over the fiber length in which nonlinear effects have consequence (L_{eff}). Here, the nonlinear relation for n, given by Eq. (2.2), is used and P and λ are the optical power and wavelength, respectively. The time derivative of the optical phase results in a frequency, so Φ_{NL} results in frequency changes on amplitude modulated signals (where $dP/dt \neq 0$). This gives spectral broadening and pulse distortion. For a signal pulse, the leading edge is up-shifted in wavelength and the trailing edge is downshifted. Fiber chromatic dispersion can act on this wavelength spectrum to cause pulse broadening.

XPM occurs when varying signal intensity on one channel causes phase change in other channels. Again, Eq. (2.40) applies, but in this case the power, P, is that of the interfering channel [64]. The frequency broadening is twice that of the SPM effect because of two field combinations that can occur in Eq. (2.37) for the two distinct signals. Despite that there can be many channels simultaneously interfering in a WDM system due to XPM, chromatic dispersion prevents continual coincidence of these channels as the signals propagate down the fiber. In addition, as pulses pass through each other (because of their different group velocities), the XPM frequency shifts due to the leading and trailing edges will average to zero in the ideal case of a lossless fiber.

Stimulated scattering comprises another class of nonlinear fiber effects. An early analysis, which is still highly relevant, is the paper by Smith [65]. This class includes scattering via the Raman and Brillouin interactions. Both occur spontaneously when light scatters off of vibrational modes of the glass, exchanging one vibrational quanta of energy, thus shifting the frequency of the lightwave. The scattered waves become large when the interaction becomes stimulated, that is, where the pump and scattered waves self-consistently generate vibrations or acoustic waves.

SBS gives backward radiation in fibers [66], a low-intensity threshold, and is narrow band (40 MHz). The interaction consists of a forward traveling, high-power "pump" lightwave, a co-propagating acoustic wave, and a backscattered frequency-downshifted lightwave. Using the dispersion relations for the three waves, and requiring momentum conservation, we find the relation between the acoustic and lightwave frequencies and wave numbers:

$$\omega_A = \frac{2n_1 \nu_A}{c} \omega_P, \text{ and } k_A \cong 2k_P, \tag{2.41}$$

where the ω_A, ν_A, and k_A are the frequency, velocity, and wave number of the acoustic wave, while ω_P and k_P refer to the pump lightwave. The backscattered lightwave is frequency downshifted by about 11 GHz for a 1550-nm pump. The acoustic wave is formed by density variations created by electrostriction [67]. Because the electrostriction causes high density everywhere there is high light intensity, it is clear that the wave number of the acoustic wave will be twice that of the pump wave. The interaction line width is set by the decay time of the acoustic wave, leading to line widths of less than 40 MHz. The frequency spectra of the backscattered lightwave may consist of multiple peaks due to the presence of various radial acoustic modes in the fiber. Figure 2.19 shows an NZDF fiber with this type of structure, along with some theoretical predictions [68]. The peak gain is given by

$$g_B = \frac{2\pi n^7 p_{12}^2}{c\lambda_p^2 \rho_0 \nu_A \Delta \nu_B}, \tag{2.42}$$

where p_{12} is a stress-optic coefficient, λ_p is the pump wavelength, ρ_0 is the material density, and $\Delta \nu_B$ is the width of the dominant Brillouin spectral peak. This equation holds for pump line widths much less than the $\Delta \nu_B$. The gain drops for broader pump line widths [69]. When the gain is higher than the fiber loss, the backscatter signal builds up along the length of the fiber toward the lightwave source. The effect is stronger for longer fibers and comes at the expense of optical power which would otherwise be transmitted.

SBS has largely receded into the background as an impairment for systems deployed over cabled communications fiber because phase dithering of the transmitter can almost always raise the SBS threshold high enough for practical applications. SBS remains relevant for high-power device applications such as high-power amplifiers or fiber lasers.

SRS gives forward or backward scattered waves, has a higher intensity threshold than SBS, and is wideband ($\sim$12 THz). The spontaneous Raman

Light-Guiding Fundamentals and Fiber Design

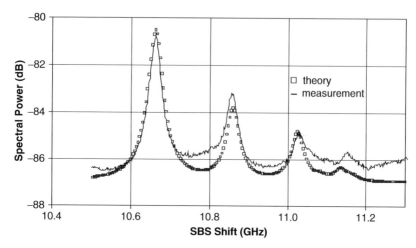

Figure 2.19 SBS frequency spectrum of an NZDF fiber with multiple radial modes.

interaction consists of an input optical signal generating up- and down-frequency shifted lightwaves (anti-Stokes and Stokes waves) as the waves exchange energy due to transitions between molecular vibration levels. The new frequencies are sum and difference frequencies from the original lightwave and the vibrational transition frequencies. For a large pump signal, the anti-Stokes wave is suppressed and the Stokes wave grows. For maximal gain, the polarizations of signal and pump must be aligned. SRS gain factors for a number of host materials are given in references [63] and [67], which can be scaled to the signal wavelength of interest.

The primary current interest in SRS in the telecommunications world is its application for Raman amplification. The advantages are several. First, the optical signal-to-noise ratio is improved when amplification is distributed throughout the transmission fiber, so the signal is not allowed to attenuate to a very low level. Second, because the Raman gain spectrum (Fig. 2.20) is approximately 13 THz below the pump frequency, one can obtain gain at any wavelength if appropriate pump wavelength and power are available [70]. The bandwidth of a Raman amplifier can be expanded by multiplexing pumps at optimally chosen wavelengths. Transmission fibers that have been designed to enhance Raman performance [71, 72] are discussed in Chapter 5.

Figure 2.20 shows the measured Raman gain curve of a matched clad standard single-mode fiber when pumped at 1453 nm. The Raman gain efficiency C_R, measured consistently with IEC Technical Report 62324, can be used to calculate the on–off gain by the relationship $G_{on-off} = \exp[C_R P_{pump} L_{eff}]$, where P_{pump} is the Raman pump power. This relation holds in the regimen where the signal power is

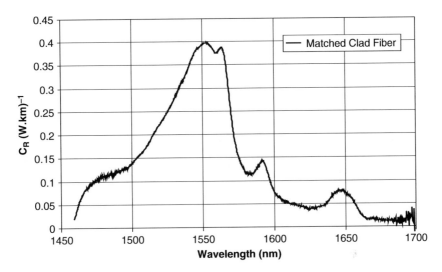

Figure 2.20 Raman gain curve of a matched-clad, standard single-mode fiber, plotted versus wavelength. The pump laser was set at 1453 nm, resulting in Stokes Raman gain peaking at 13.4 THz to the red of the pump wavelength, which is approximately 100 nm in this wavelength range. (Figure courtesy Bera Pálsdóttir of OFS, Brøndby, Denmark.)

low compared to the Raman pump power P_{pump}, so depletion of the pump power is negligible. Multiple copies of the gain shape in Fig. 2.20 can be superposed to form flat gain across a band by multiplexing Raman pumps at appropriately chosen wavelengths and powers. The output from a single Raman pump laser may be several hundred milliwatts; the total power for multiplexed pumps may approach 500 mW or higher, depending on design and eye safety targets.

REFERENCES

[1] Li, T., ed. 1985. *Optical Fiber Communications*, Vol. 1, *Fiber Fabrication*. Academic Press, Orlando.

[2] MacChesney, J. B. et al. 1974. A new technique for the preparation of low-loss and graded-index optical fibers *Proc. IEEE* 62:1280–1281, 1974.

[3] Nagel, S. R. et al. 1985. Modified chemical vapor deposition. In: *Optical Fiber Communications* (T. Y. Li, ed.), Vol. 1, *Fiber Fabrication*. Academic Press, Orlando.

[4] Morrow, A. J. et al. 1985. Outside vapor deposition. In: *Optical Fiber Communications* (T. Y. Li, ed.), Vol. 1, *Fiber Fabrication*. Academic Press, Orlando.

[5] Izawa, T. et al. 1977. Continuous fabrication of high silica fiber perform. *Integrated Optics and Optical Communication*, Cl-1:375, 1977.

[6] Niizeki, N. et al. 1985. Vapor-phase axial deposition method. In: *Optical Fiber Communications* (T. Y. Li, ed.), Vol. 1, *Fiber Fabrication*. Academic Press, Orlando.

[7] Geittner, P. et al. 1976. Low-loss optical fibers prepared by plasma-activated chemical vapor deposition. *Appl. Phys. Lett.* 28(11).

[8] Glodis, P. F., et al. 1994. The application of synthetic silica tubing for large preform manufacture using MCVD. In: *Proceedings of the IWCS International Wire Cable Symposium*, Inc., Eatontown, NJ.

[9] Dorn, R. and C. Le Sergent, 1988, Preform technologies for optical fibers. *Electrical Communication* 62.

[10] Fleming, J. W. et al. 1989. Preform overcladding using plasma fusion of sol-gel powder. *Proceedings of the European Conference on Optical Communication.*

[11] MacChesney, J. B. et al. 1997. Optical fibers using sol-gel silica overcladding tubes. *Electron. Lett.* 33:1573.

[12] Trevor, D. J. 2005. Fabrication of large near net shapes of fiber optic quality silica. In: *Handbook of Sol-Gel Science and Technology V. III: Applications of Sol-Gel Technology* (S. Sakka, ed.), pp. 27–65. Kluwer Academic, Boston.

[13] Chang, K. H. et al. 2005. Next generation fiber manufacturing for the highest performing conventional single mode fiber. In: *Optical Fiber Communication Conference, Technical Digest.* Optical Society of America, Washington, DC. Paper JWA5.

[14] Gloge, D. 1971. Weakly guiding fibers. *Applied Optics* 10:2252–2258.

[15] Cherin, A. H. 1983. *An Introduction to Optical Fibers.* McGraw-Hill, New York.

[16] Snyder, A. W. and J. D. Love. 1983. *Optical Waveguide Theory.* Chapman and Hall, London.

[17] Haus, H. A. 1984. *Waves and Fields in Optoelectronics.* Prentice Hall, Englewood Cliffs, NJ.

[18] Marcuse, D. 1991. *Theory of Dielectric Optical Waveguides*, 2nd edition. Academic Press, Boston.

[19] Buck, J. A. 1995. *Fundamentals of Optical Fibers.* Wiley Interscience, New York.

[20] Lenahan, T. A. 1983. Calculation of modes in an optical fiber using the finite element method and EISPACK. *Bell System Technical J.* 62:2663.

[21] Kawano, K. and T. Kitoh. 2001. *Introduction to Optical Waveguide Analysis.* John Wiley & Sons, New York.

[22] Ainslie, B. J. et al. 1982. The design and fabrication of monomode optical fiber. *IEEE J. Quantum Electr.* QE-18:514.

[23] Cohen, L. G. et al. 1982. Radiating leaky-mode losses in single-mode lightguides with depressed-index claddings. *IEEE J. Quantum Electr.* QE-18:1467.

[24] Lazay, P. D. and A. D. Pearson. 1982. Developments in single-mode fiber design, materials, and performance at Bell Laboratories. *IEEE J. Quantum Electronics* QE-18:504.

[25] Petermann, K. 1976. Theory of microbending loss in monomode fibres with arbitrary refractive index profile. *AEU, Bd.* 30(9):337–342.

[26] Povlsen, J. and S. Andreasen. 1986. Analysis on splice, microbending, macrobending, and Rayleigh losses in GeO_2-doped dispersion-shifted single-mode fibers. *IEEE J. Lightwave Technmol.* LT-4:706–710.

[27] Jenkins, F. A., and H. E. White. 1957. *Fundamentals of Optics.* McGarw-Hill, New York.

[28] Walker, S. S. 1986. Rapid modeling and estimation of total spectral loss in optical fibers. *J. Lightwave Technol.* 4:1125.

[29] Marcuse, D. 1982. *Light Transmission Optics.* Van Nostrand Reinhold, New York.

[30] Fleming, J. W. 1976. Material and mode dispersion in GeO_2, B_2O_3, SiO_2 glasses,. *J. Am. Ceramics Soc.* 59(11-12):503–507.

[31] Fleming, J. W. 1978. Material dispersion in lightguide glasses. *Electr. Lett.* 14(11):326–328.

[32] Bhagavatula, V. A. (1987). Low dispersion, low-loss single-mode optical waveguide. U.S. Patent 4,715,679, Dec. 29, 1987.

[33] Vengsarkar, A. M. et al. 1995. Article comprising a dispersion-compensating optical waveguide. U.S. Patent 5,448,674.

[34] Poole, C. D. and T. E. Darcie. 1993. Distortion related to polarization mode dispersion in analog lightwave systems. *J. Lightwave Technol.* 11:1749.

[35] Refi, J. J. et al. 1993. Polarization mode dispersion and its effect on optical transmission. NCTA Show, June 7, 1993.

[36] Kogelnik, H. et al. 2002. Polarization mode dispersion. In: *Optical Fiber Telecommunications IVB* (I. P. Kaminow and T. L. Li, eds.). Academic, San Diego.

[37] Born, M. and E. Wolf. 1986. *Principles of Optics*, Chapter 1. Pergamon, New York.

[38] Poole, C. D. and J. Nagel. 1997. Polarization effects in lightwave systems. In: *Optical Fiber Telecommunications IIIA* (I. P. Kaminow and T. L. Koch, eds.). Academic Press, San Diego.

[39] Rashleigh, S. C. 1983. Origins and control of polarization effects in single-mode fibers. *J. Lightwave Technol.* LT-1:312–331, 1983.

[40] Foschini, G. J. and C. D. Poole. 1991. Statistical theory of polarization dispersion in single mode fibers. *IEEE J. Lightwave Technol.* 9:1439.

[41] Poole, C. D. et al. 1991. Dynamical equation for polarization dispersion. *Opt. Lett.* 16:372.

[42] Poole, C. D. 1988. Statistical treatment of polarization dispersion in single-mode fiber. *Opt. Lett.* 13:687.

[43] Gisin, N. et al. 1991. Polarization mode dispersion of short and long single-mode fibers. *IEEE J. Lightwave Technol.* 9:821.

[44] Wei, P. K. A. and C. R. Menyuk. 1996. Polarization mode dispersion, decorrelation, and diffusion in optical fibers with randomly varying birefringence. *IEEE J. Lightwave Technol.* 14:148.

[45] Geyling, F. T. et al. 1983. The viscous collapse of thick-walled tubes. *J. Applied Mechanics* 50:303.

[46] Barlow, A. J. et al. 1981. Birefringence and polarization mode dispersion in spun single-mode fibers. *Appl. Opt.* 20:2962.

[47] Hart, A. C. et al. 1994. Method of making a fiber having low polarization mode dispersion due to a permanent spin. U.S. patent 5,298,047, March 29, 1994.

[48] Schuh, R. E. et al. 1998. Polarization mode dispersion in spun fibers with different linear birefringence and spinning parameters. *IEEE J. Lightwave Technol.* 16:1583.

[49] Pizzinat, A. et al. 2003. Influence of the model for random birefringence on the differential group delay of periodically spun fibers. *IEEE J. Lightwave Technol.* 15:819.

[50] Ulrich, R. and A. Simon. 1979. Polarization optics of twisted single-mode fibers. *Appl. Opt.* 18:2241.

[51] Schuh, R. E. et al. 1995. Theoretical analysis and measurement of effects of fibre twist on polarization mode dispersion of optical fibers. *Electron. Lett.* 31:1772.

[52] McCurdy, A. H. et al. 2005. Control of polarization mode dispersion when building high bit rate optical transmission systems. In: *The National Fiber Optic Engineers Conference.* Optical Society of America, Washington, DC. Paper NThC1.

[53] Gisin, N. et al. 1996. How accurately can one measure a statistical quantity like polarization mode dispersion? *IEEE Phot. Tech. Lett.* 12:1671.

[54] Jacobs, S. A. et al. 1997. Statistical estimation of the PMD coefficients for system design. *Electron. Lett.* 33:619.

[55] IEC 61282-3 standard, "Fibre optic communication system design guides—Part 3: Calculation of polarization mode dispersion." International Electrotechnical Commission (IEC) standard, 2002.

[56] Gloge, D. 1975. Optical-fiber packaging and its influence on fiber straightness and loss. *Bell System Techn. J.* 54:245.

[57] Marcuse, D. 1984. Microdeformation losses of single-mode fibers. *Applied Opt.* 23: 1082–1091.

[58] Bjarklev, A. 1986. Microdeformation losses of single-mode fibers with step-index profiles. *JLT* LT-4(3):341–346.

[59] Petermann, K. and R. Kuhne. 1986. Upper and lower limits for the microbending loss in arbitrary single-mode fibers. *J. Lightwave Technol.* 4:2.

[60] Unger, C. and W. Stocklein. 1994. Investigation of the microbending sensitivity of fibers. *J. Lightwave Technol.* 12:591.

[61] Marcuse, D. 1976. Microbending losses of single-mode, step-index, and multimode, parabolic-index fibers. *Bell Syst. Tech. J.* 55:937–955.

[62] Yang, Y. et al. 2002. Elasto-optics in double-coated optical fibers induced by axial strain and hydrostatic pressure. *Applied Opt.* 41:1989–1994.

[63] Agrawal, G. P. 1995. *Nonlinear Fiber Optics.* Academic Press, San Diego.

[64] Islam, M. N. et al. 1987. Cross phase modulation in optical fibers. *Opt. Lett.* 12:625.

[65] Smith, R. G. 1972. Optical power handling capacity of low loss optical fibers as determined by stimulated Raman and Brillouin scattering. *Applied Optics* 11:2489.

[66] Ippen, E. P. and R. H. Stolen. 1972. Stimulated Brillouin scattering in optical fibers. *Applied Phys. Lett.* 21:539.

[67] Boyd, R. W. 2003. *Nonlinear Optics*, Chapter 9. Academic Press, San Diego.

[68] McCurdy, A. H. 2005. Modeling of stimulated Brillouin scattering in optical fibers with arbitrary radial index profile. *J. Lightwave Technol.* 23:3509.

[69] Lichtman, E. and A. A. Friesem. 1987. Stimulated Brillouin scattering excited by a multi-mode laser in single-mode optical fibers. *Opt. Commun.* 65:544.

[70] Bromage, J. 2003. Raman amplification for fiber communication systems. In: *Optical Fiber Communication Conference, Technical Digest.* Optical Society of America, Washington, DC. Tutorial Tu C1.

[71] Zhu, B. et al. 2001. 3.08 Tb/s (77 × 42.7 Gbps) Transmission over 1200 km of NZDF with 100-km spans using C- and L-band distributed Raman amplification. In: *Optical Fiber Communication Conference, Technical Digest.* Optical Society of America, Washington, DC. Paper PD23.

[72] Pálsdóttir, B. and C. Larsen. 2005. Raman gain efficiency measured on 16 Mm of Raman optimized NZDF fiber. In: *Optical Fiber Communication Conference, Technical Digest.* Optical Society of America, Washington, DC. Paper OThF7.

Chapter 3

Overview of Materials and Fabrication Technologies

John B. MacChesney,[1] Ryan Bise,[1] and Alexis Méndez[2]

[1]*OFS Bell Labs, Murray Hill, New Jersey*
[2]*MCH Engineering LLC, Alameda, California*

3.1 DOUBLE-CRUCIBLE TECHNIQUE

The first attempt at producing high-purity glass, the so-called "double-crucible technique," proceeded along the lines of conventional glass melting but used specially prepared constituents [1, 2]. Soda-lime-silicate and sodium-borosilicate glasses were made from materials purified to parts-per-billion (ppb) levels of transition metal impurities by ion exchange, electrolysis, recrystallization, or solvent extraction. These starting glasses were melted, fined, drawn to cane, and fed into an ingenious continuous casting system composed of concentric platinum crucibles, shown in Fig. 3.1. A thin stream of core glass flowed from the upper crucible, passed through the reservoir of cladding glass, and was concentrically surrounded by the cladding as it flowed through the orifice of the lower thimble. The time and temperature of core-cladding contact in the cladding reservoir were controlled, enabling diffusion to produce the index gradient needed to minimize intermodal dispersion.

Despite its elegance, overwhelming problems beset this method from the start. First, contamination during processing raised the impurity level from the ppb level in the constituents to parts-per-million (ppm) levels in the fiber. Many attempts were made to eliminate this contamination and a partial solution was achieved by using the oxygen partial pressure of the atmosphere during processing to control the redox conditions within the molten glass. Absorption by iron and copper, the two principal contaminants, could thus be minimized by altering their valence state. Iron could be oxidized primarily to the Fe^{3+} state and copper retained the monovalent state by processing in a controlled oxygen atmosphere.

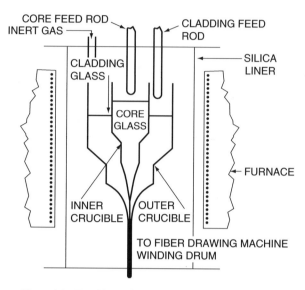

Figure 3.1   Double-crucible method for fiber fabrication.

Thus, strong absorptions by Fe^{2+} and Cu^{2+} at near-infrared wavelengths were diminished. Fibers adequate for commercial systems of the time were made this way. Losses as low as 5 dB/km were achieved at $0.9\,\mu m$, but the lower losses offered by the 1.3- to $1.5\text{-}\mu m$ spectral window were unattainable using this technique. Fundamental electronic vibrations and severe OH^- contamination were intrinsic to the starting materials and could not be appreciably lowered by improved processing. The method was stillborn as it was introduced to the market due to the advent of a superior technology.

3.2 VAPOR-DEPOSITION TECHNIQUES

The double-crucible technique was short-lived because vapor-deposition techniques soon appeared, which were capable of lower losses from the visible into the infrared. These techniques appeared in the early 1970s and may be categorized as either inside or outside processes. Both use oxidation of silicon tetrachloride vapor to produce submicron amorphous silica particles. Other chloride vapors such as germanium tetrachloride and phosphorus oxychloride are used as sources of dopants in the silica.

Outside deposition uses flame hydrolysis whereby chloride vapors pass through a propane-oxygen or hydrogen-oxygen flame to produce a "soot" of SiO_2 particles. The particles partially sinter as they collect on a mandrel. The inside process uses these same reactants together with oxygen, but the reaction occurs inside a silica tube in the absence of hydrogen. The high temperatures needed for reacting halide vapors with oxygen are provided by an oxygen-hydrogen burner, which traverses along the tube as it rotates on a glass working lathe. The reactions produce particles by oxidation rather than hydrolysis. These particles are deposited on the inside wall of the tube downstream of the torch and are sintered to form a vitreous layer as the torch moves past the deposit.

3.3 OUTSIDE VAPOR DEPOSITION

Two versions of outside processes have been developed. These are the "outside vapor deposition" (OVD) [3] process developed by Coming Glass Works and the "vertical axial deposition" (VAD) [4] version developed by a consortium of Japanese cable makers and Nippon Telephone and Telegraph Corporation. The OVD process is one of the most common techniques used for optical fiber fabrication. A schematic of the steps involved in using the OVD technique is shown in Fig. 3.2 [5]. During the soot-deposition step of the OVD process, silica and doped silica particles are generated in a methane/oxygen flame via hydrolysis reaction. Silica soot preforms are formed by multilayer deposition of vapors and particles on a rotating cylindrical target rod by traversing the soot-containing flame along the axis of the cylindrical target.

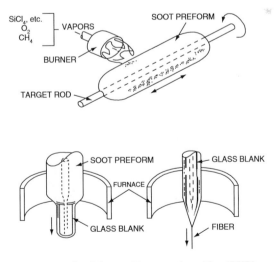

Figure 3.2 Schematic of the outside vapor deposition (OVD) process.

During the soot-deposition process, dopants are added to certain segments of the silica-based preform to modify the refractive index of these soot layers. Typical dopants used in the OVD process include germania, boron, fluorine, phosphorus, and titania. The most widely used dopant is germania (GeO_2), which is often used in the central core region of the fiber. Germania doping of the silica preform is achieved by the introduction of a dopant precursor (e.g., $GeCl_4$) to the deposition burner. The precursor undergoes oxidation in the flame, resulting in the formation of GeO_2 particles and GeO vapors. This is in contrast to the case of SiO_2 in the flame, which is predominantly in particulate form. The relative rates of the germania particle and vapor deposition, as well as the morphology of the resulting deposit (i.e., amorphous or crystalline germania), are controlled by the chosen process conditions.

Following the soot-deposition step, the porous soot preform is treated with a drying agent (e.g., chlorine) for removal of water and metal impurities [6–8]. Preform drying isotopically performed at temperatures between 950 and 1250 °C, where the diffusion rates and reaction kinetics of the drying reaction with impurities (e.g., Fe reaction with chlorine) are fast and the preform is not significantly densified. The reaction of the drying agent with the hydroxyl species in the soot may be written as [9]

$$SiOH + Cl_2 \leftrightarrow SiOCl + HCl \qquad (3.1)$$

While the hydroxyl content of the soot preform may be reduced to some degree by merely heating the preform to a high temperature [10], the remaining OH content is still not sufficient to enable acceptable absorption contribution to attenuation for telecom-based applications.

Following the drying process, the porous soot preforms are sintered into glass blanks at temperatures ranging from 1200 to 1600 °C [6, 7]. For silica-based soot preforms, surface energy–driven viscous flow is the dominant mechanism of sintering. A "unit cell" model was provided by Mackenzie and Shuttleworth [11] to predict the rate of increase of density during surface energy–driven viscous flow sintering of granular solids.

The final step of the OVD process involves drawing the sintered glass blanks into 125-μm diameter optical fibers for use in telecommunication systems [12, 13]. During the fiber draw process, the glass preforms are heated (typically in an inert atmosphere) to temperatures above the softening point of the glass (2000–2200 °C for silica-based fibers), followed by drawing into fibers by applying axial tension to the samples. Precision control of the fiber draw conditions (i.e., draw speed, tension, furnace temperature, fiber diameter, etc.) is of great importance, as this plays a large role in both the physical and chemical characteristics of the glass fiber and the overall optical system performance of the waveguides [14, 15].

3.4 VERTICAL AXIAL DEPOSITION

The VAD process is a variant on the OVD method, where the core and clad glasses may be deposited either simultaneously or separately [16–21]. The VAD process also forms a cylindrical body using soot, but deposition occurs end-on, as shown in Fig. 3.3. Here, a porous soot cylinder is formed without a hole by depositing the core and cladding simultaneously using two torches. When complete, the body is sintered under conditions similar to those used for OVD. A fundamental difference between the two processes is that while the composition profile of the OVD preform is determined by changing the composition of each layer, the VAD profile depends on subtle control of the gaseous constituents in the flame and the shape and temperature distribution across the face of the growing soot boule.

Critical to the development of VAD was the design of a torch composed of up to 10 concentric silica tubes. Typically, reactant vapors pass through one or more of the central passages where they are protected from premature reaction by a

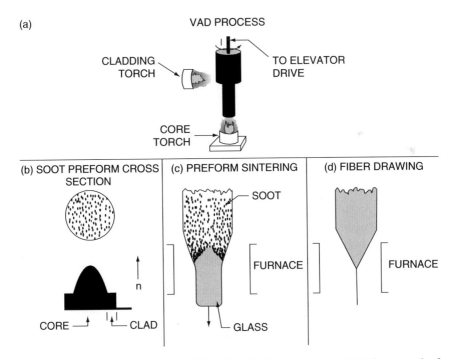

Figure 3.3 Vertical axial deposition (VAD) fiber fabrication process. (a) End-on growth of boule; (b) profile of soot preform after removal of mandrel; (c) preform sintering; (d) fiber drawing.

ring of inert shield gas. The outer series of tubes alternate between hydrogen and oxygen to compose the flame. By manipulation of gas flows, the temperature and particle distribution in the flame can be controlled to determine the surface temperature distribution and the shape of the boule.

In spite of this rather fragile control of composition, VAD had one significant advantage over first-generation OVD. Recall that at this time, transmission systems were using graded index multimode fiber. The high refractive index differences—between core and cladding required by such fiber—were obtained with heavy core doping. This produced a large mismatch in thermal expansion between core and cladding and caused cracking of consolidated OVD preforms at the inner surface as the preform cooled below the glass transition temperature. Because VAD preforms do not have a central hole, they can better withstand thermal stress.

The major challenge to VAD was how to create an optimized index profile to minimize mode dispersion. Initially it was thought that control of the GeO_2 distribution across the boule required several $GeCl_4$ sources, each of different composition. However, it was found that such grading could be accomplished by control of the boule surface temperature distribution. Eventually, process development focused critically on the shape of the growth face and the temperature profile across it. Figure 3.4 [22] shows GeO_2 incorporation into silica as a function of the temperature of the boule end-face. Below 400 °C, GeO_2 is lost by vaporization of discrete crystalline particles when the boule is sintered at high temperature.

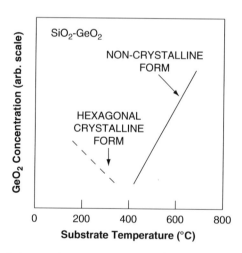

Figure 3.4 Relation between substrate temperature and GeO_2 concentration in the vertical axial deposition (VAD) process.

3.5 DIRECT NANOPARTICLE DEPOSITION

The Direct Nanoparticle Deposition (DND) technology, developed and commercialized by Liekki Corporation (Finland), is a new fiber manufacturing process, ideally suited for the demanding needs of advanced high-power fiber laser applications. DND provides the flexibility to engineer the glass matrix into which the rare-earth ions are dissolved [23]. This capability makes it possible to increase the rare-earth (RE) concentration without sacrificing the fiber performance with effects resulting from too high local RE ion concentration [24, 25]. Furthermore, DND technology inherently provides radial control of dopants, resulting in excellent flatness of the refractive index profile, a key attribute in large, low numerical aperture cores. This ability also facilitates the fabrication of advanced waveguiding structures in which the gain provided by rare-earth ion is independently controlled.

The DND process is based on the combustion of gaseous and atomized liquid raw materials in an atmospheric oxy-hydrogen flame. The flexibility in how raw materials are fed to the process gives the freedom of incorporating materials with very different vapor pressures. The glass is doped in-flame where the glass particles are formed, thus the clustering tendency is low. The DND process makes it possible to mix the refractive index effecting materials (e.g., alumina, germanium, phosphorous, etc.) with other doping materials (e.g., Yb, Er, Nd, etc.) already during the deposition of the glass particles. This improves the homogeneity of the glass composition prior to the sintering phase. Various parameters affect the formation process, for example, the vapor pressure of the metals, temperature of the flame, gas velocities, droplet route through the flame and the Gibbs free energy of the raw materials. The qualitative results obtained from the particle size distribution measurement shows a single-peaked particle-size distribution, which indicates that the particles are formed through evaporation-condensation process. Rapid quenching and a short residence time produce small particles with narrow size distribution.

Fiber manufacturing using the DND process can be described as a special form of OVD where nano-size particles of the typical glass former materials and the gain dopants are deposited simultaneously onto a target alumina rod mounted on a rotating glass lathe to form the fiber's core and cladding regions. The glass modifiers and RE dopant elements are fed into the process—in liquid or vapor phase—directly into the reaction zone through independent and controlled channels of a specially designed flame burner (Fig. 3.5), while a $SiCl_4$ gas bubbler is used as a source for the silica glass former. The atomized particles leaving the burner are uniform in composition and can be controlled in size from 10 to 100 nm. The rapid particle formation and deposition process allows extremely high doping levels with reduced clustering and photo darkening, as

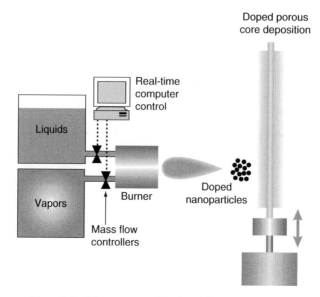

Figure 3.5   Direct nanoparticle deposition process (DND).

well as a higher fiber damage threshold. After the glass formation and doping stages, the alumina mandrel is gently removed from the deposited outside preform and handling tubes are attached to it. The preform is then inserted into a furnace where the first step is drying and cleaning. Finally, the porous glass is sintered into a solid clear core fiber preform.

In contrast to the multistep modified chemical vapor deposition (MCVD) solution doping process (soaking and diffusion)—which produces core absorption values up to 1200 dB/m at 976 nm—the DND process can reach 2000 dB/m and beyond. Furthermore, as a result of this new preform fabrication process, core-to-clad ratios of up to 0.5 are attainable with DND, whereas MCVD-produced fibers are limited to values of 0.16. In addition, the DND process is fast, efficient, and particularly well suited for producing large-mode-area double-clad (DC) fibers with a large core-to-clad ratio (e.g., a highly ytterbium-doped DC fiber with 20-μm core and 125-μm cladding diameters). The DND process is applicable to single-mode, DC, and DC-PM fiber, as well as to more complex fiber designs, such as fibers with rectangular (or other noncircular) cores and claddings, multicore fibers, or coupled multiple waveguiding-element fibers.

With the control and flexibility DND provides, it is possible to integrate new functionality into the active fiber. The DND is directly applicable to complicated fiber designs such as polarization maintaining (PM) PANDA-fibers [26], all-glass coated fibers, multilayered core and cladding designs for spectral filtering and reduction non-linear effects as well as other advanced fiber designs. The combination of these features will ultimately result in low-cost, fully functionally integrated, all-glass active fibers for high-power fiber lasers and amplifiers. Such truly monolithic (un-spliced) structures provide a common platform for a range of fiber applications and the necessary means to reduce the cost of fiber lasers dramatically.

3.6 MODIFIED CHEMICAL VAPOR DEPOSITION

Inside processes, such as MCVD, had a different origin. Following the tradition of the electronics industry, chemical vapor deposition (CVD) techniques were used to produce doped silica layers inside silica substrate tubes [27]. As in CVD, the concentration of reactants was very low to inhibit gas phase reaction in favor of a heterogeneous wall reaction that produced a vitreous particle-free deposit on the tube wall. The tube was collapsed to a rod and relatively low loss fiber obtained. However, deposition rates were impractically low and attempts to increase them always produced silica particles that deposited on the tube wall and resulted in excess loss. The solution was to exactly reverse the CVD practice: intentionally produce a gas phase reaction by increasing the reactant flows by more than 10 times. Submicron particles were, thus, produced that deposited on the tube wall and were fused into clear pore-free glass as the torch traversed along the tube. MCVD was, thus, developed [28] as the process shown in Fig. 3.6. High-purity gas mixtures are injected

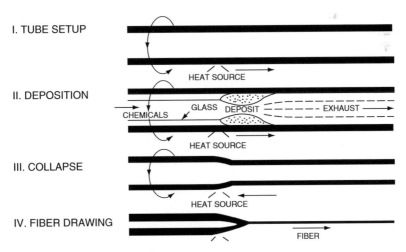

Figure 3.6 The modified chemical vapor deposition (MCVD) process consists of deposition of glass layers inside a silica tube, collapse of tube to a solid rod, and drawing of preform into fiber.

into a rotating tube, which is mounted in a glass working lathe and heated by a traversing oxy-hydrogen torch. Homogeneous gas phase reaction occurs in the hot zone created by the torch to produce amorphous particles, which deposit downstream of the hot zone. The heat from the moving torch sinters this deposit to form a pure glass layer. Typical torch temperatures are sufficiently high to sinter the deposited material, but not so high as to deform the substrate tube. The torch is traversed repeatedly to build up, layer by layer, the core or cladding. Composition of the individual layers is varied between traversals to build the desired fiber index structure. Typically, 30 to 100 layers are deposited to make either single-mode or graded index multimode fiber.

3.6.1 Chemical Equilibria: Dopant Incorporation

After the initial demonstration of feasibility, fundamental investigations established the knowledge required to create a commercial process. For instance, it was necessary to better understand the chemistry of the MCVD process in order to control the incorporation of GeO_2 and limit hydroxyl impurities. In addition, to increase fabrication efficiency, it was necessary to understand the mechanism by which particles deposit on the substrate tube, as well as the manner in which the silica particles are sintered into pore-free glass. Although process development preceded quantitative understanding, optimization of the commercial process required this knowledge.

The chemistry of $SiCl_4$ and $GeCl_4$ oxidation was investigated by infrared spectroscopy [29]. Samples of effluent gases from typical MCVD reactions demonstrated that as the maximum hot-zone temperature reached $1300\,°K$, $SiCl_4$ began to oxidize to Si_2OCl_6 (Fig. 3.7). Up to $1450\,°K$, the amount of oxychloride increases to a maximum, whereas at higher temperatures the $SiCl_4$, Si_2OCl_6, and $POCl_3$ contents decrease until their concentration in the effluent is insignificant above about $1750\,°K$. Above this temperature, all reactants are converted to oxides.

The behavior of $GeCl_4$ is different. Its concentration in the effluent gas stream decreases between 1500 and $1700\,°K$, but above $1700\,°K$ remains approximately 50% of its original value. It is clear that the majority of the initial germanium is unreacted and escapes in the effluent.

These results indicate that at low temperatures $(T < 1600\,°K)$, the extent of the reaction for $SiCl_4$, $GeCl_4$, and $POCl_3$ is controlled by reaction kinetics, while at higher temperatures thermodynamic equilibria become dominant. It is clear from rate studies that the residence times in the hot zone are sufficient to produce equilibrium above $1700\,°K$. The $SiCl_4$ and $GeCl_4$ concentrations at high temperatures are strongly influenced by the equilibria:

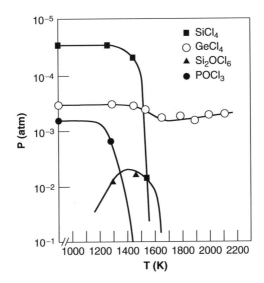

Figure 3.7 Modified chemical vapor deposition (MCVD) effluent composition as a function of hot-zone temperature. Starting reactants: 0.5 g/min SiCl₄, 0.05 g/min GeCl₄, 0.016 g/min POCl₃, 1540 cm³/min O₂.

$$\text{SiCl4 (g)} + \text{O}_2 \text{ (g)} \rightarrow \text{SiO}_2 \text{ (s)} + 2\text{Cl}_2 \text{ (g)} \tag{3.2}$$

and

$$\text{GeCl}_4 \text{ (g)} + \text{O}_2 \text{ (g)} \rightarrow \text{GeO}_2 \text{ (s)} + 2\text{Cl}_2 \text{ (g)} \tag{3.3}$$

Equilibrium constants for these reactions may be written

$$K_{\text{SiO}_2} = (a_{\text{SiO}_2})(P_{\text{Cl}_2})^2 / (P_{\text{SiCl}_4})(P_{\text{O}_2}) \tag{3.4}$$

$$K_{\text{GeO}_2} = (a_{\text{GeO}_2})(P_{\text{Cl}_2})^2 / (P_{\text{GeCl}_4})(P_{\text{O}_2}), \tag{3.5}$$

where P_i are the partial pressures of gaseous species and a_i represents the chemical activities of the solid species. The activities can be approximated by $\gamma_i x_i$, where x_i is the mole fraction of the particular species in the solid and γ_i is the activity coefficient. An activity coefficient of unity implies an ideal solution obeying Raoult's law. The equilibrium constants for these reactions have been determined as a function of temperature and indicate that Eq. (3.2) strongly favors the formation of SiO₂ at high temperature, as verified by the experiments described earlier. Oxidation of GeCl₄ by Eq. (3.3), on the other hand, is incomplete because the equilibrium constant, K_{GeO_2}, is less than unity at temperatures higher than 1400°K. This means that only a fraction of the germanium starting composition will be present as GeO₂. The presence of significant Cl₂ concentration resulting from the complete oxidation of $SiCl_4$ shifts the equilibrium further toward $GeCl_4$ by the law of mass action. Low oxygen partial pressure has the same effect.

3.6.2 Purification from Hydroxyl Contamination

A second important aspect of MCVD chemistry is the incorporation of the impurity OH^- [30] because reduction of OH^- in optical fibers to ppb levels is essential for realization of low attenuation in the 1.3- to 1.55 μm region. Hydrogen species originate from three sources: diffusion of OH^- from the substrate tube during processing, impurities in the starting reagents and carrier oxygen gas, and contamination from leaks in the chemical delivery system.

The OH^- level in the fiber is controlled by the reaction

$$H_2O + Cl_2 \rightarrow 2HCl + {}^1\!/_2 O_2 \tag{3.6}$$

with equilibrium constant

$$K_{OH} = (P_{HCl})2(P_{O_2})^{1/2}/(P_{H_2O})(P_{Cl_2}). \tag{3.7}$$

The concentration of OH^- incorporated into the glass, c_{SiOH} [3], is described by

$$C_{SiOH} = (P_{H_2O\,initial})(P_{Cl_2})^{1/2}/(P_{O_2})^{1/4}. \tag{3.8}$$

During deposition in MCVD, Cl_2 is typically present in the 3–10% range because of oxidation of the chloride reactants. This is sufficient to reduce OH^- by a factor of about 4000. However, chlorine is typically not present during collapse and significant amounts of OH^- can be incorporated by diffusion of torch byproducts through the silica tube. Figure 3.8 shows the dependence of the SiOH concentration in the resultant glass as a function of typical P_{O_2} and P_{Cl_2} concentrations used during MCVD deposition and collapse with 10 ppm H_2O in the starting gas. Figure 3.8 also shows typical consolidation of the VAD and OVD soot processes.

3.6.3 Thermophoresis

Turning now from the reaction equilibria, we consider the mechanism of deposition of particles on the tube walls. The SiO_2 particles produced by vapor phase reaction have diameters in the range 0.02–0.1 μm and are, thus, entrained in the gas flow. Without the imposition of a temperature gradient, they would remain in the gas stream and exit from the tube end. However, temperature gradients in the gas stream produced by the traveling torch give rise to the phenomenon of thermophoresis [31]. Here, particles residing in a thermal

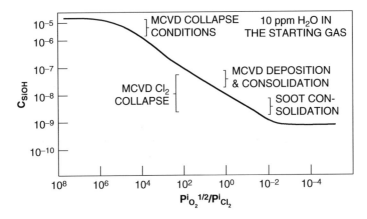

Figure 3.8 Typical incorporation of OH⁻ during processing stages of modified chemical vapor deposition (MCVD), for 10 ppm H_2O in chemical precursors.

gradient are bombarded by energetic gas molecules from the hot region and less energetic molecules from the cool region. A net momentum transfer forces the particle toward the cooler region. Within an MCVD substrate tube, because the wall is cooler than the center of the gas downstream of the torch, particles are driven toward the wall where they deposit. The MCVD process is shown schematically in Fig. 3.9 in terms of (1) heat transfer in the hot zone, (2) reaction, (3) particle formation, (4) particle deposition beyond the hot zone where the tube wall becomes cool relative to the gas stream, and (5) consolidation of previously deposited particles in the hot zones as the torch traverses to the right.

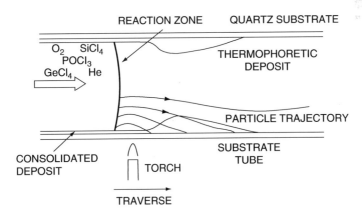

Figure 3.9 Particle formation and thermophoretic deposition in modified chemical vapor deposition (MCVD).

A mathematical model for thermophoretic deposition [32], experimentally verified, concluded that deposition efficiency (ratio of SiO_2 equivalent entering tube to that contained in exhaust) may be expressed as $e = 0.8(1 - T_e/T_{rxn})$, where T_{rxn} is the gas reaction temperature and T_e is the temperature downstream of the torch at which the gas and the tube wall equilibrate. Typically, T_e is about $400\,°C$ and T_{rxn} about $2000\,°C$, giving an efficiency value on the order of 60%. Note that the efficiency is not a function of the maximum tube temperature.

Examination of the process of consolidation of the soot layer on the inner surface of the silica tube revealed the mechanism to be viscous sintering [33]. By this mechanism, the rate of consolidation is proportional to the sintering time and surface tension and inversely proportional to the void size, initial soot density, and glass viscosity.

3.7 PLASMA CHEMICAL VAPOR DEPOSITION

A second inside process, plasma chemical vapor deposition (PCVD) [34], is similar to MCVD in that it uses the same reactants inside a silica substrate tube that is collapsed after deposition and drawn into a fiber. The primary difference between the two methods is that the oxidation of reactants in the tube is initiated by a non-isothermal microwave plasma inside the tube rather than by heating the exterior of the tube, as shown in Fig. 3.10. In the PCVD process, most of the deposition is in vapor form, instead of soot form, as is seen via the MCVD technique.

The generation of the plasma requires a reactant vapor pressure of only a few torr. A microwave cavity—operating at 2.45 GHz, which produces the plasma heat source—traverses along the substrate tube and promotes chemical reaction. During the PCVD step, the vapors diffuse to the wall of the tube and undergo heterogeneous reaction to form deposit [35, 36]. Soot formation, while possible,

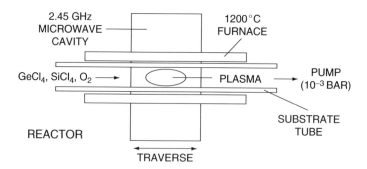

Figure 3.10   Schematic representation of plasma chemical vapor deposition (PCVD) process.

is not desired during the PCVD deposition process, as this can result in glass defects (bubbles, etc.). Furthermore, the reaction and deposition of both GeO_2 and SiO_2 is much more efficient than in MCVD, approaching 100%. However, this leads to a high level of water retention in the glass, which results in higher optical losses. Typical methods to address the hydroxyl issues include the use of fluorine-based dopants (C_2F_6, SF_6, etc.), along with the reactants during the PCVD step [37].

Another advantage—especially for multimode preforms—is that because the plasma involves no latent heat, it can be traversed very rapidly to produce hundreds of layers. The resulting deposit, thus, has a very smooth and precise index profile essential for minimizing intermodal dispersion.

3.8 SOL-GEL PROCESSES

Optical fiber drawing technology and each of the processes associated with preform fabrication—OVD, VAD, MCVD, and PCVD—have been developed to the point at which they can easily yield both multimode and single-mode fiber in long lengths, with losses limited only by the intrinsic properties of fused silica, their principal constituent. In their initial form, each produced preform typically yields only about 10 km of fiber. The quantity of silica glass necessary to make up the core and primary cladding of any conventional single-mode or multimode optical fibers is small, totaling less than 20% of the fiber's eventual mass. The remainder is composed of the substrate tube and overclad cylinder. To reduce the cost of the latter (conventionally made by flame deposition and machined to desired dimensions), and to increase the yield of drawn fibers from each preform, an "overcladding" technique based on sol-gel technology was developed and introduced commercially into fiber production [38]. With overcladding, after a preform core rod with an oversized core region is fabricated by conventional means, the outer diameter is built up to attain the proper proportions and increase the effective glass mass content to facilitate multi-kilometer drawing of fiber. The outer diameter is increased either by jacketing the preform with a second silica tube or by using it as a bait rod for subsequent OVD or VAD soot deposition. This procedure increases the length of fiber produced from a vapor-deposited preform, yielding up to a hundred kilometers of fiber.

3.8.1 Alkoxide Sol-Gel Processing

Suitable overcladding material may be prepared by sol-gel and powder form-ing techniques. In one instance, a chemical precursor, typically a silicon alkoxide such as $Si(OC_2H_5)_4$, is reacted with water in the presence of ethanol and an acid catalyst. The "sol" is cast into cylindrical molds and poly-condensation of the resulting silanol groups produces a filamentary siloxane gel network. The gel body is dried and consolidated to form silica glass as films or bulk bodies [39]. Alternatively, commercial colloidal powders obtained from flame hydrolysis, commonly known as "fumed silica," are formed into bodies by mechanical compaction [40], centrifugation [41], or casting/gelation [42]. In this last approach, the silica particles (generally 0.05–0.5 μm) are dispersed in water to form a sol. Control of pH or addition of surface active agents is used to promote electrostatic and steric stabilization to inhibit interparticle attractive forces, which cause agglomeration. The dispersed colloid containing up to 60 weight percent silica is cast after the stabilizing forces are dissipated. Gelation by van der Waals forces soon follows to produce a semirigid body. After drying, the porous silica body can be sintered to glass much like the soot boules formed in the OVD and VAD processes.

Drying of the gel body is accompanied by large stresses due to shrinkage and capillary forces, which generally cause the body to fracture. To date, bodies as large as overcladding tubes have not been successfully fabricated using this approach. However, there has been much effort to fabricate all-gel preforms because the chemical precursors are available in high purity and the refractive index of the glass may be altered by adding dopant alkoxides. Fibers with a raised index core, doped with alkoxides such as $Ge(OC_2H_5)_4$, have not yielded fiber with losses comparable to that produced by the vapor technique because the germanium dioxide either dissolves in the liquor or precipitates in some crystallized form. The usual product of consolidation is bubbled silica, whose index is raised only marginally.

The alkoxide route has achieved its best success in fabricating fibers with a silica core and lowered index cladding [43]. Hydrolysis and polycondensation of $Si(OC_2H_5)_3$ F lowers the index by incorporating fluorine. The sol is cast, gelled, and dried to yield a porous silica body with a surface area of 200–650 m^2/g. Such high surface area allows consolidation at low temperatures in a fluorine containing atmosphere. The result is a down-doped tube ($\Delta = -0.62\%$), which is collapsed with a stream of oxygen flowing down the center. This removes the fluorine from the inner tube wall and produces a core region with a higher refractive index than the fluorine-doped cladding. Losses as low as 0.4 dB/km have been reported for such fiber.

3.8.2 Colloidal Sol-Gel Processing

The colloidal approach has achieved more success in producing large bodies for overcladding. The colloidal process implemented for this purpose differed from that of the more publicized tetraethyl orthosilicate (TEOS) process. This process produces low-density gels (volume fraction of silica) and limits the size of monolithic bodies produced. For commercial fiber production, bodies weighing more than 10 kg are needed and are produced by higher silica loading of the gel. This was achieved by using nano-dimensioned silica particles produced commercially by flame hydrolysis of silica tetrachloride. These are marketed by various names; one commonly used is OX-50 (Degussa AG, Frankfurt, Germany). It is dispersed in water with loading of 50wt% or more and can be gelled to yield large cylinders strong enough to survive processing to net-shape and large glass bodies of waveguide-quality silica. Briefly, the colloidal silica is dispersed in water containing quaternary ammonium hydrochloride. This produces a high pH (>12) and a negative and repulsive surface charge on the particles. The resulting fluid sol is centrifuged. Gelling results from addition of an ester—typically methylformate—which lowers pH and dissipates the surface charge. After a few hours in the mold, the cylinder is removed and carefully dried to prevent warping or stress-cracking. Finally, the cylinder is suspended in a sintering/purification furnace, where it is "burned out" to remove organics and purified in a chlorine containing atmosphere, and then consolidated into glass at about 1500 °C in a helium gas atmosphere. A detailed discussion of this process is provided by Trevor [44].

For fiber production using this method, extraordinary care must be exercised in the equipment design; high precision molds and driers are required to achieve uniform dried bodies without warping and cracking. Likewise, purification requires similar care: Water and transition metal impurities are removed in inert, oxidizing, and chlorine containing atmospheres. Purification from transition metals may be enhanced by use of a low oxygen partial pressure atmosphere, as indicated by the reaction:

$$Fe_2O_3 + 2Cl_2 \rightarrow 2FeCl_2 + 3/2\ O_2 \qquad (3.9)$$

By firing in an atmosphere protected from air intrusion, oxygen partial pressures can be in the range of 10^{-6} atm. Thus, at temperatures between 600 and 1000 °C, iron and other impurities are effectively removed [45, 46]. This was demonstrated by intentionally contaminating a gel body with 1wt% hematite. After a two-step dehydration and consolidation treatment, the residual iron content was only 40 ppb.

Finally, an additional equilibration in thioryl chloride is used to remove minute quantities of zirconia present in less than the parts/trillion (ppt). These, if present, cause low strength breaks in the hundreds of kilometers drawn from the gel-silica preforms.

The process of overcladding with gel-derived material may be accomplished using two strategies, as shown in Fig. 3.11.

On the left side, in the "rod-in-tube" process, an overcladding tube is formed from gel and then consolidated directly onto a core rod [47]. Tubes for the

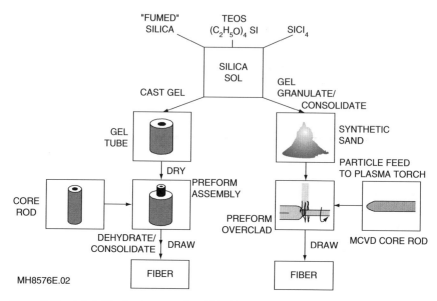

Figure 3.11 A hybrid sol-gel strategy in which (right side) gel is cast into tubes and used to overclad a core rod. And, alternatively (left side), gel is granulated and then fusion-sprayed onto a preform to accomplish overcladding.

rod-in-tube process are formed by dispersion, milling, casting, and gelation of colloidal silica. After removal from the mold and air-drying, they are placed over a core rod and the assembly is dehydrated, consolidated, and drawn into fiber. A satisfactory interface, free of bubbles and other defects, must be obtained between the core rod and the gel-derived overcladding tube. By proper cleaning of the core rod and appropriate consolidation conditions, the loss of the eventual fiber can be as low as that of the original core rod.

On the right side of the graph, instead of casting a tube, the wet gel is granulated into particles, which are fed through an oxygen plasma torch to deposit glass droplets onto the core rod [48]. Because these particles are 100 μm in diameter, they deposit on the rod by impaction, rather than by weak thermophoretic forces. Deposition efficiency is, thus, quite high.

3.9 SOL-GEL MICROSTRUCTURE FIBER FABRICATION

Vapor phase synthetic silica processes have led the way toward extremely low-loss index-guided transmission optical fiber. Although these methods have demonstrated of high purity and precise control over the index structure, these methods are generally constrained to cylindrically symmetric structures and modest levels of index contrast ($\Delta n \leq 0.03$). Microstructured optical fibers consist of an array of holes, which extended longitudinally along the z-axis of the fiber. The large index ($\Delta n = 0.45$), combined with the ability to pattern structures with dimensions similar to the wavelength of light, yields novel waveguide properties such as photonic band-gap guidance [49–51], such as endlessly single-mode behavior [52], low bend loss [53], high birefringence [54], high nonlinearities [55], and dispersion control [56]. Increased sophistication in fiber fabrication has yielded losses as low as 0.28 dB/km for index-guided structures [57] and 1.7 dB/km for hollow core structures [58], making these fibers suitable for both transmission and device applications. Several methods have been adopted for fabrication of microstructured fibers such as the stacking and drawing of glass capillaries [59], extrusion of soft glasses [60], preform drilling [61], and sol-gel casting [62, 63]. Here, we describe the sol-gel casting method for the fabrication of microstructured fiber, developed originally at Bell Laboratories and continued at OFS Laboratories [62, 64, 65].

The sol-gel casting process, originally developed for the fabrication of large precision overcladding tubes for optical fiber preforms [66, 67], has been adapted for the fabrication of microstructured fiber designs. The sol-gel process offers advantages in the fabrication of microstructured fiber including low-cost starting materials, high purity, design flexibility, reusable mold and mandrel elements, and the ability to scale up to large bodies for the generation of low-cost long lengths of microstructured optical fiber. Figure 3.12 outlines the steps involved in the fabrication of an exemplary sol-gel microstructured preform. A mold

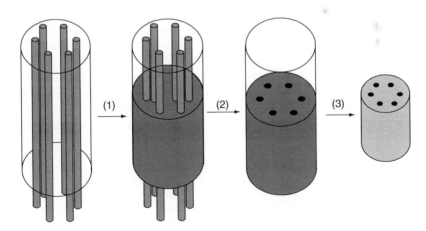

Figure 3.12 Fabrication processing of microstructured preforms using sol-gel casting. (1) Casting and gelation, (2) mandrel removal, and (3) drying, purification, and sintering of gel body.

containing an array of mandrel elements is assembled. The mandrels are individually tensioned to ensure uniformity along the length of the mold. The mold is subsequently filled with colloidal silica dispersed at high pH with an average particle size of 40 nm.

The pH is lowered by the addition of an organic ester, causing the sol to gel. At the wet gel stage, the mandrel elements are removed, leaving air columns within the gel body. The gel body is then dried, purified thermochemically to remove organic and transition metal contaminants, and sintered into vitreous silica. Because contaminants are removed in the dried gel body, the sol-gel process is reasonably insensitive to contaminants introduced during mold assembly, unlike stack-and-draw, or drilling methods, which may potentially introduce contamination at the glass stage. The sintered preform is then available for draw. If required, additional glass processing steps such as etching, overcladding, or stretching may be performed on the microstructured preform. The air holes are pressurized during draw to obtain the desired size and air-fill fraction in the fiber, and the hole size is monitored during draw using online measurements and offline measurements with an optical microscope.

Images of several microstructured fibers fabricated using the sol-gel process are displayed in Fig. 3.13, demonstrating the wide range of fiber designs afforded by the casting process. Unlike the stack-and-draw technique, the sol-gel process allows the hole size, position, and shape to be adjusted independently in non–closest-packed structures such as circular arrays. Furthermore, the sol-gel technique can generate structures consisting of several hundreds of holes required for low confinement losses in both index-guided [68] and photonic band-gap fibers [58], which are expensive and challenging to fabricate by methods such as preform drilling. To date, continuous lengths of more than 10 km of sol-gel microstructured fibers have been drawn with variations in hole size of approximately less than 2% over kilometer length with optical losses of roughly 1 dB/km at 1550 nm and OH peak absorption loss at 1385 nm of 1.5 dB/km, values that are competitive with microstructured fibers produced from high-quality VAD capillaries. Continued refinements in powder source, processing, and glass surface finish will continue to lower the optical loss. Preform dimensions can be readily scaled to preform sizes of more than 200 km with reusable molds and mold assemblies, providing a route toward large-scale production of low-cost microstructured fiber.

3.10 FIBER DRAWING

Typical preforms produced by any of the previously described methods are about a meter in length and between 2.0 and 7.5 cm in diameter. These preforms are drawn into 125-μm diameter fiber by holding the preform vertically and

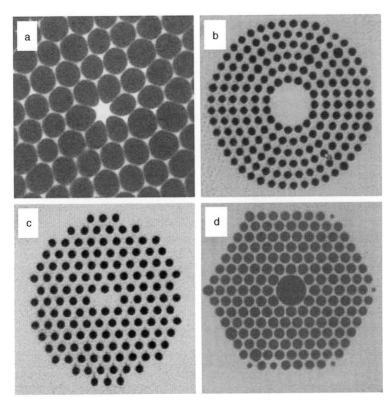

Figure 3.13 Examples of microstructured optical fibers fabricated using the sol-gel casting method. Examples include (a) small-core high delta fiber, (b) circular single-mode, (c) birefringent fiber, and (d) air-core fiber.

heating the end of the preform above the glass softening temperature until a glob of glass falls from the end. This forms a neck-down region, which provides transition to a small-diameter filament. Uniform traction on this filament results in a continuous length of fiber. Before this fiber contacts a solid surface, a polymer coating is applied to protect the fiber from abrasion and preserve the intrinsic strength of the pristine silica. The fiber is then wound on a drum.

Although the basic principles of fiber drawing were established before the advent of optical fiber technology, stringent fiber requirements necessitated improvements in process control and understanding of the effects of draw conditions on optical performance. Fiber is now drawn without inducing excess loss while maintaining high strength and dimensional precision and uniformity [69].

The essential components of a draw tower, shown schematically in Fig. 3.14, are a preform feed mechanism, a furnace capable of 1950–2200 °C, a diameter

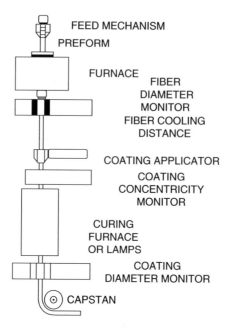

Figure 3.14 Schematic of a fiber draw tower.

monitor, a polymer coating applicator, a coating curing unit, a traction capstan, and a take-up unit. The furnace is typically either a graphite resistance type or an inductively coupled radiofrequency zirconia furnace. The former requires an inert atmosphere to prevent oxidation of the graphite element. The zirconia furnace may be operated in air but must be held above 1600 °C, even when not in use, because the volume change associated with the crystallographic transition of zirconia at this temperature can cause stress-induced fracture. The advantage of this furnace is that it generally has less contamination by particles emitted from the heater element.

The uniformity of the fiber diameter depends on control of the preform feed rate, the preform temperature, and the pulling tension. Over long lengths of fiber (>100 cm), diameter variations can result from changes in preform diameter and drifts in furnace temperature and the speeds of the feed and capstan motors. Diameter variations with shorter length period arise from perturbations in the temperature of the neck-down region caused by thermal fluctuations. These may be minimized by control of convective currents and nonuniform gas flows inside the furnace, as well as acoustical and mechanical vibrations. The diameter monitor positioned below the furnace typically provides feedback to the capstan, adjusting the draw tension to maintain constant fiber diameter.

ACKNOWLEDGMENTS

The authors would like to thank David DiGiovanni, Michael Murtagh and Mikko Söderlund for valuable contributions to this chapter. R. B. would like to thank the following people for assistance in areas of design, modeling, sol-gel, and preform processing and fiber draw: Dennis Trevor, John Alonzo, Tom Stockert, John Fini, Frank Dimarcello, Eric Monberg, Jim Fleming, and George Zydzik.

REFERENCES

[1] Pearson, A. D. and W. G. French. 1972. Low loss glass fibers for optical transmission. *Bell Laboratories Record* 50:103–106.

[2] Beals, K. J. and C. R. Day. 1980. A review of glass fibers for optical communication. *Phys. Chem. Glasses* 21:5–19.

[3] Keck, D. B. et al. 1973. U.S. Patent no. 3,737,292.

[4] Izawa, T. and N. Inagaki. 1980. Materials and processes of optical fiber fabricating. *Proc. IEEE*, pp. 1184–1187.

[5] Morrow, A. J. et al. 1985. Outside vapor deposition. In: *Optical Fiber Communications* (T. Ki, ed.). Academic Press, Orlando, FL.

[6] Scherer, G. W. 1977. Sintering of low density glasses: I. Theory. *J. Am. Ceramic Soc.* 60(5-6):236–239.

[7] Scherer, G. W. 1979. Sintering inhomogeneous glasses: Application to optical waveguides. *J. Non-Crystalline Solids* 34:239–256.

[8] Sakaguchi, S. 1994. Consolidation of GeO_2 soot body prepared by flame hydrolysis reaction. *J. Non-Crystalline Solids* 171:228–235.

[9] Wood, D. L. et al. 1987. Germanium chemistry in the MCVD process for optical fiber fabrication. *J. Lightwave Technol.* LT-5:277–285.

[10] Moulson, A. J. and J. P. Roberts. 1960. Water in silica glass. *Trans. Br. Ceramic. Soc.* 59:388.

[11] Mackenzie, J. K. and R. Shuttleworth. 1949. A phenomenological theory of sintering. *Proc. Phys. Soc.* 62(Section B):833–852.

[12] Yin, Z. and Y. Jaluria. 1998. Thermal transport and flow in high-speed optical fiber drawing. *Trans ASME* 120:916–930.

[13] Lee, S. H. and Y. Jaluria. 1997. Simulation of the transport processes in the neck down region of a furnace drawn optical fiber. *Int. J. Heat Mass Transfer* 40(4):843–856.

[14] Hanafusa, H. et al. 1985. Formation mechanism of drawing induced E'-centers in silica optical fibers. *J. Appl. Phys.* 58:1356.

[15] Thomas, G. A. et al. 2000. Towards the clarity limit in optical fiber. *Nature* 404:262–264.

[16] Sudo, S. et al. 1981. Refractive index control techniques in the vapor-phase axial deposition method. *Trans. IECE Japan* E64:536.

[17] Izawa, T. and S. Sudo. 1987. *Optical Fibers: Materials and Fabrication.*. KTK Scientific Publishers/D. Beidel Publishing Co., Boston, MA..

[18] Nizeki, N. et al. 1985. Vapor-phase axial deposition process. In: *Optical Fiber Communications, Vol. I Fiber Fabrication*, (T. Li, Ed.), pp. 97–177, Academic Press, Inc., Orlando, FL. 1985.

[19] Kawachi, M. et al. 1980. Deposition properties of SiO_2-GeO_2 particles in the flame hydrolysis reaction for optical fiber fabrication. *Jpn. J. Appl. Phys.* 19(2):L69.

[20] Potkay, E. et al. 1988. Characterization of soot from multimode vapor phase axial deposition optical fiber preforms. *J. Lightwave Tech.* 6:1338.

[21] Sudo, S. 1982. *Studies of the Vapor-Phase Axial Deposition Method for Optical Fiber Fabrication*, Ph. D. Dissertation, Tokyo University, 1982.

[22] Edahiro, T. et al. 1980. Deposition properties of high-silica particles in the flame hydrolysis reaction for optical fiber fabrication. *Jpn. J. Appl. Phys.* 19:2047–2054.

[23] Tammela, S. et al., "Potential of nanoparticle technologies for next generation erbium-doped fibers", OFC 2004 Technical digest, FB5 (2004).

[24] Tammela, S. et al., "Direct nanoparticle Deposition Process for Manufacturing Very Short High Gain Er-doped Silica Glass Fiber", ECOC 2002 Proceedings, paper 9.4.1, 2002.

[25] Koponen, J, et al., "Measuring photodarkening from Yb-doped fibers," in Proceedings of CLEO/Europe '05, CP2-2-THU (2005).

[26] Söderlund, M.J. et al.,"Design considerations for large-mode-area polarization maintaining double clad fibers" Proc. SPIE 5987, 99 (2005).

[27] MacChesney, J. B. et al. 1973. Low-loss silica core-borosilicate clad fiber optical waveguide. *Appl. Phys. Lett.* Vol. 23, No. 6, pp. 340–341.

[28] MacChesney, J. B. et al. 1974. Preparation of low loss optical fibers using simultaneous vapor phase deposition and fusion. *Xth mt. Congress on Glass, Kyoto, Japan*, pp. 6–40.

[29] Wook, D. L. et al. 1987. The germanium chemistry in the MCVD process for optical fiber fabrication. *J. Lightwave Techl.* LT-5:277–283.

[30] Walker, K. L. et al. 1981. Reduction of hydroxyl contamination in optical fiber preforms. *Tech. Digest 3rd mt. Conf. on Integ. Optics and Opt. Fiber Comm., San Francisco, CA*, pp. 86–88.

[31] Simplins, P. G. et al. 1979. Thermophoresis: The mass transfer mechanism in modified chemical vapor deposition. *J. Appl. Phys.* 50:5676–5681.

[32] Walker, K. L. et al. 1980. Thermophoretic deposition of small particles in modified chemical vapor deposition process. *J. Am. Cer. Soc.* 63:96–102.

[33] Walker, K. L. et al. 1980. Consolidation of particulate layers in the fabrication of optical fibers preforms. *J. Am. Cer. Soc.* 63:92–96.

[34] Kuppers, D. and H. Lydtin. 1980. Preparation of optical waveguides with the aid of plasma activated chemical vapor deposition at low pressures. *Topics Current Chem.* 89:109.

[35] Bauch, H. et al. 1987. Chemical vapor deposition in microwave produced plasmas for fiber preforms. *J. Opt. Comm.* 8:130–135.

[36] Weling, F. 1985. A model for the plasma-activated chemical vapor deposition process. *J. Appl. Phys.* 57(9):4441–4446.

[37] Rau, H. et al. 1984. Incorporation of OH in PCVD optical fibers and its reduction by fluorine doping. *Mat. Res. Bull.* 19:1621–1628.

[38] MacChesney, J. B. et al. 1998. Optical fibers by a hybrid process using sol-gel silica overcladding tubes. *J. Non-Crystalline Solids* 263:232–238.

[39] Zarzycki, J. 1985. The gel-glass process. In: *Glass: Current Issues* (A. F. Wright and J. Dupois, eds.), pp. 203–231. Martinez-Nijoff, Boston.

[40] Dorn, K. et al. 1987. Glass from mechanically shaped preforms. *Glastch Ber.* 66:79–32.

[41] Buchmann, P. et al. 1988. Preparation of quartz tubes by centrifugational deposition of silica particles. *Proc. 14th European Conference on Optical Communications, Brighton, UK*.

[42] Shibata, S. and T. Kitagawa. 1986. Fabrication of SiO_2–GeO_2 glass by the sol-gel method. *J. Appl. Phys.* 25:L323–L324.

[43] Shibata, S. et al. 1987. Wholly synthesized fluorine-doped silica optical fibers by the sol-gel method. *Tech. Digest 13th European Conf. on Optical Communication, Helsinki, Findland.*

[44] Trevor, D. J. 2005. Fabrication of large near net shapes of fiber optic quality silica. In: *Handbook of Sol-gel Science and Technology* (S. Sakka, ed.), pp. 27–65. Kluwer Academic, Boston.

[45] MacChesney, J. B. et al. 1987. Influence of dehydration/sintering conditions on the distribution of impurities in sol-gel derived silica glass. *Mat. Res. Bull.* 22:1209–1216.

[46] Clasen, K. 1988. Preparation of glass and ceramics by sintering colloidal particles deposited from the gas phase. *Glast Ber.* 61:119–126.

[47] MacChesney, J. B. et al. 1987. Hybridized sol-gel process for optical fibers. *Elec. Lett.* 23:1005–1007.

[48] Fleming, J. W. 1987. Sol-gel techniques for lightwave applications. *Tech. Conf. on Optical Fiber Comm., Reno, Nevada*, Paper MH-1.

[49] Bise, R. T. et al. 2002. Tunable photonic band gap fiber. Paper presented at the Optical Fiber Communication Conference, Anaheim.

[50] Smith, C. M. et al. 2003. Low-loss hollow-core silica/air photonic band gap fibre. *Nature* 424:657.

[51] Cregan, R. F. et al. 1999. Single-mode photonic band gap guidance of light in air. *Science* 285(5433):1537–1539.

[52] Birks, T. A. et al. 1997. Endlessly single-mode photonic crystal fiber. *Opt. Lett.* 22(13): 961–963.

[53] Hasegawa, T. et al. 2003. Bend-insensitive single-mode holey fibre with SMF-compatibility for optical wiring applications. Paper presented at the European Conference on Optical Communications, Rimini.

[54] Saitoh, K. and M. Koshiba. 2002. Photonic bandgap fibers with high birefringence. *IEEE Photonics. Technol. Lett.* 14(9):1291–1293.

[55] Ranka, J. K. and R. S. Windeler. 2000. Nonlinear interactions in air-silica microstructure optical fibers. *Opt. Photonics News* 11(8):20–25.

[56] Ferrando, A. et al. 2000. Nearly zero ultraflattened dispersion in photonic crystal fibers. *Opt. Lett.* 25(11):790–792.

[57] Tajima, K. et al. 2003. Low water peak photonic crystal fibers. Paper presented at the European Conference on Optical Communication, Rimini, Italy.

[58] Mangan, B. J., et al. 2004. Low loss (1.7 dB/km) hollow core photonic bandgap fiber. Paper presented at the Optical Fiber Communication, Los Angeles.

[59] Knight, J. C. et al. 1996. All-silica single-mode optical fiber with photonic crystal cladding. *Opt. Lett.* 21(19):1547–1549.

[60] Monro, T. 2002. High nonlinearity extruded single-mode optical fiber. Paper presented at the Optical Fiber Communications Conference, Anaheim, CA, 2002.

[61] Hasegawa, T. et al. 2001. Hole-assisted lightguide fiber for large anomalous dispersion and low optical loss. *Opt. Exp.* 9(13):681–686.

[62] Bise, R. et al. 2002. Impact of preform fabrication and fiber draw on the optical properties of microstructured fibers. Paper presented at the International Wire and Cable Symposium, Orlando.

[63] De Hazan, Y. et al. 2002. U.S. patent no. 6,467,312 B1 (Oct. 22, 2002).

[64] Bise, R. T. et al. 2004. Holey fibers. Paper presented at the IEEE Lasers and Electro-Optics Society, Puerto Rice.

[65] Bise, R. T. and D. J. Trevor. 2005. Solgel-derived microstructured fibers: Fabrication and characterization. Paper presented at the Optical Fiber Conference and National Fiber Optical Engineering Conference, Anaheim.

[66] MacChesney, J. B. et al. 1997. Optical fibers using sol-gel silica overcladding tubes. *Electron. Lett.* 33(18):1573–1574.

[67] MacChesney, J. B. et al. 1998. Optical fibers by a hybrid process using sol-gel silica overcladding tubes. *J. Non-Cryst. Solids* 226(3):232–238.

[68] White, T. P. et al. Confinement losses in microstructured optical fibers. *Opt. Lett.* 26(21):1660–1662.

[69] DiMarcello, F. V. et al. Fiber drawing and strength properties. In: *Optical Fiber Communications* (T. Yi, ed.), Vol. 1, Fiber Fabrication. Academic Press, Orlando, FL 1985.

Chapter 4

Optical Fiber Coatings

Steven R. Schmid and Anthony F. Toussaint

DSM Desotech, Elgin, Illinois

4.1 INTRODUCTION

It could be said that the modern era of fiber optics began in 1966, with the publication of the paper "Dielectric-Fibre Surface Waveguides for Optical Frequencies" by Dr. C. K. Kao and G. A. Hockham of Standard Telecommunications Laboratories Ltd. (STL) [1]. This paper discussed the theory and potential use of optical fiber for communications. Dr. Kao believed that fiber loss could be reduced below 20 dB/km by eliminating metal impurities in the glass. Such attenuation would allow 1% of the light entering 1 km of this type of fiber to successfully reach the other end.

During the same period, Corning Glass Works researchers, R. Mauer, D. Keck, and P. Schultz, were experimenting with high-purity fused silica [2]. They employed titanium as a dopant to produce a higher refractive index (RI) (1.466) fiber core, compared with the surrounding lower RI (1.4584) silica cladding. They were able to draw fibers, at 1900 °C, to a diameter of 100 μm. In 1970, "Method of Producing Optical Waveguide Fibers" was filed with the U.S. Patent Office and issued 3 years later as U.S. 3,711,262.

Before 1966, fibers had losses of roughly 1000 dB/km [2]. The Corning team effort reduced this loss to only 16 dB/km, thus demonstrating the feasibility of optical fibers as a communication medium. Such fibers were said to be capable of carrying 65,000 times more information than copper wire [3].

This chapter focuses on the development, science, and performance of protective coatings for optical fiber. The widespread deployment of high-bandwidth optical fiber would not have been feasible without the immediate protection afforded by ultraviolet (UV)-curable coatings. Applied on fiber draw towers, at speeds approaching 60 mph, these coatings enable fibers to survive the rigors of proof testing, cabling, and installation. UV-curable coatings have also proven to be durable in the field, with millions of kilometers of fiber already in operation for more than 20 years.

4.2 EARLY HISTORY OF COATINGS FOR OPTICAL FIBER

Although the inherent strength of pure silica [4] is known to be nearly 14,000 N/mm^2, it soon became apparent, during the development of optical fibers, that some type of protective coating was required to shield the fibers from abrasion to preserve their strength. The combination of moisture and stress causes microscopic flaws in the glass to propagate, resulting in fiber failure. Without protective coatings, optical fibers would never have been a practical alternative to copper for telecommunications.

According to Stevens and Keough [5], the prime requirements for optical fiber coatings were protection against microbending and static fatigue. This necessitated that cured coatings "be concentric about the fiber, be continuous over the length of application, be of constant thickness, be abrasion resistant and moisture retardant." Suitable liquid coatings, consequently, had to be of a workable and stable viscosity (minimally 6 months) [6], adhere to glass, have a relatively low surface tension, be free of particle contamination, have minimal hydrogen generation ($10^{-2} - 10^{-4} \mu l/g$) [6, 7], and have fast cure speeds. In addition, it was expected that cured coatings would have stable modulus and adhesion properties over the 25-year operating life of the installed cable.

In the early 1970s, J. E. Ritter, Jr. [8], explored the effectiveness of polymeric coatings in preventing the degradation of abraded soda-lime glass in the presence of moisture. His results indicated "acrylic, epoxy, and silicone coatings all significantly increased the short-term strength of abraded glass." He believed that the coatings functioned as a diffusion barrier and limited the availability of water at the glass surface.

Wang et al. [9] and Wei [10] issued additional reports on the contributions of polymeric coatings towards providing resistance to moisture, in the early and mid-1980s. Wang et al. [9] reported on the accelerating effects of water on the deterioration of fiber strength through fatigue and aging. Wei [10] further described how the combination of moisture and stress initiates crack propagation and ultimately reduces fiber strength and provided examples of how this was retarded by polymeric coatings.

Early coating materials used in the protection of optical fiber included two package systems, blocked urethanes, solvent-based lacquers, silicone rubbers, and UV radiation–curable epoxy acrylates [11, 12]. By the early 1980s, UV-curable coatings became the popular choice for protection of optical fibers, largely because of their rapid cure response and easily tailored properties [6, 13, 14, 15].

In 1974, H. N. Vazirani of Bell Telephone Laboratories developed one of the first UV-curable coatings for optical fiber. U.S. Patent 4,099,837 [13] describes that the "polymer coating comprises the polymerization product of a prepolymer mixture resulting from reacting acrylic acid with a 0.4 to 1.0 weight ratio mixture of aliphatic diglycidyl ether to aromatic diglycidyl ether, and further characterized in that polymerization product contains a UV sensitizer in order to cure said prepolymer mixture with ultraviolet light." Additional claims cited 1,4-butanediol diglycidyl ether as the aliphatic ether and the diglycidyl ether of halogenated bisphenol A as the aromatic ether. The use of a silane or titanate coupling agent was also claimed.

The fiber coating, described earlier, was applied and cured as the fiber was drawn at a speed of 25 m/min. Physical properties of the polymer coating were reported as a Young's modulus of 6000 psi and an elongation at break of approximately 20%. The resulting coated fiber was evaluated [11, 16] and was found to have tensile strengths greater than 500,000 psi in 1-km gauge lengths and long-term strength retention properties in moist environments [17, 18]. Schonhorn et al. [19] also demonstrated that the interaction between the coating and the glass interface determines the ability of the coating to prevent degradation of fiber strength and that this is enhanced by the inclusion of silane coupling agents within the coating.

4.3 EVOLUTION OF OPTICAL FIBERS AND PROTECTIVE COATINGS

Attenuation losses associated with optical fibers continued to decline during the 1970s: 4 dB/km (1975), 0.5 dB/km (1976), and 0.2 dB/km (1979). The latter value corresponds to 63% of a light signal reaching the end of a 10-km long fiber [20]. In 1982, Corning achieved an attenuation of 0.16 dB/km, on single-mode fibers transmitting at 1550 nm, representing a 100-fold improvement over the 1970 transmission of Mauer's first low-loss fiber [21]. It has been reported that if the ocean had the same transparency as the glass in such low-loss fibers, one could see to the bottom of the Mariana Trench in the Pacific Ocean, a little more than 6 miles below sea level [22].

4.3.1 Coating Contributions to Microbending Minimization

Concurrent with the aforementioned improvement in fiber transmission properties, fiber coatings evolved from single-layer to dual-layer systems. In the early 1980s, the outer diameter for dual-layer systems was standardized to between

245 and 250 microns, while the outer diameter for the inner coating, contacting the glass, ranged from 190 to 210 microns. The dual-layer coating system was designed to enhance protection for fibers against microbending-induced attenuation. This phenomenon is caused by microscopic departures from straightness in the waveguide axis [23]. Varying causes of microbending include longitudinal shrinkage of the fiber coating, poor drawing or cable manufacturing methods, or stresses imposed during cable installation [24].

D. Gloge [25] first reported that microbending, losses could be reduced by shielding the fiber from outside forces by using a soft inner coating, having a modulus of 14,000 psi (~100 MPa), and an outer shell of a material having a modulus of 140,000 psi (~1000 MPa). The inner primary coating is designed to act as a shock absorber, under the tougher outer layer, to minimize attenuation caused by microbending. It has a very low crosslink density and current primary coatings typically have a modulus between 0.5 and 3.0 MPa. It must adhere to the glass, yet strip cleanly from the glass, to facilitate splicing and connecting.

The outer primary coating, sometimes called the *secondary coating*, protects the primary coating against mechanical damage and acts as a barrier to lateral forces. It also serves as a barrier to moisture. It is a hard coating, having a high modulus and Tg, to facilitate good handling and durability. It is generally fast curing, for ease of processing, and has good chemical resistance to solvents, cable filling gels, and moisture. The surface properties of the secondary coating must be carefully controlled to allow good adhesion of the ink used in color identification while allowing for good winding onto takeup spools. A schematic diagram of an optical fiber is shown in Fig. 4.1.

During the development of dual-layer coating systems, it was important to consider the modulus of the inner coating not only at room temperature but also at colder temperatures to which fibers could reasonably be exposed [12]. Viscoelastic coating materials are known to increase in modulus as temperature drops (i.e., they become stiffer). If these coatings also adhere tightly to glass, they can impose forces on the fiber that will produce microbending-induced signal attenuation.

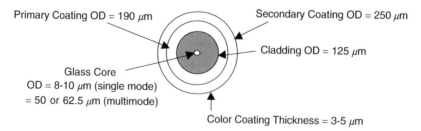

Figure 4.1   Schematic of coated fiber cross-section.

Kuzushita et al. [26] reported on the low-temperature modulus properties of nine coatings and correlated them with the added attenuation observed at $-30\,^{\circ}\mathrm{C}$ for fibers coated with four resin types: polyester-polyol–type urethane acrylate, polyether-polyol–type urethane acrylate, polybutadiene acrylate, and UV-curable silicone.

The modulus of coating films at cold temperatures was initially determined by Instron tensile strength testing of films in temperature-controlled chambers. This process proved fairly time consuming, because of the time required for temperature equilibration between runs.

The development of dynamic mechanical analyzers gave rise to the exceedingly more efficient dynamic mechanical analysis (DMA) of UV-cured films. This nondestructive test allows for temperature sweeps at a chosen frequency to define a material's modulus as a function of temperature. The slope of a coating's modulus curve as it changes from the glassy phase to the rubbery phase can, in part, determine the suitability of coatings for use at low temperatures. Sarkar et al. [27] shared examples of DMA curves for soft RTV silicone and UV-cured urethane acrylate primary coatings and showed how they compared with the attenuation properties of fibers on which they were applied. This reference also illustrates improved temperature-induced attenuation at 1300 nm for the UV-cured urethane acrylate coating, which exhibited a significantly lower modulus profile at lower temperatures.

4.3.2 Glass Fiber Fracture Mechanics and Coating Contributions to Fiber Strength Retention

Fiber fatigue is an important mechanical property of optical fibers. Pristine silica fibers have strengths of approximately 7 GPa at ambient condition. However, fibers can experience fatigue when subjected to lower stresses for long periods. Fiber fatigue is thought to occur by crack growth of existing flaws on the glass surface due to interaction between the Si-O bond and the moisture in the environment when the fiber is subjected to stress.

The coating is thought to contribute to fiber fatigue in that basic compounds present in the composition can accelerate glass corrosion. Conversely, acidic components have been shown to improve fatigue resistance. Skutnik et al. [28] found that coatings with strong adhesion facilitated a greater n-value fatigue parameter for coated fibers. Supporting this is the observation that both n-value and adhesion generally increase with time after draw.

When measuring fatigue, it is typical to measure either static fatigue or dynamic fatigue. In static fatigue, a constant stress is applied to the fiber, and the time to failure is measured. In dynamic fatigue, the strength of the fiber is measured as a function of the applied stress rate.

There are two techniques for measuring dynamic fatigue: tension and two-point bending. In the tensile test, a fiber is gripped at each end and pulled in tension until it breaks. In the two-point bending test, fibers are bent between two faceplates that move toward each other at a controlled rate until the fibers break [29–32].

Gulati [33] published one of the earliest papers that discussed test methods for measuring the tensile and bending strengths of optical fibers. He also calculated the required proof stress level to ensure fiber durability of at least 20 years when subjected to a known value of service stress.

Michalske and Bunker [34–38] published a number of studies on the fracture mechanics of glass and glass fibers. Helfinstine [39] published a very thorough review on delayed failure or sub-critical crack growth in glass.

Wang and Zupko [17] found fiber strength retention to be a function of fiber wetting and adhesion by protective polymeric coatings. Fiber strength decreased when application viscosity was increased and wetting of the fiber was consequently decreased.

Sakaguchi et al. [40] demonstrated that silane treatment stabilizes glass surfaces against water. The group evaluated the behavior of untreated, silicone-treated, and silane-treated glass in water at different temperatures. Infrared spectroscopy was used to monitor the interaction of water molecules with glass silanol groups.

Schlef et al. [41] evaluated the performance of a number of UV-curable primary and secondary coatings via static fatigue, dynamic tensile testing, and proof testing and concluded that these coatings "retain the initial strength and fatigue resistance of optical fibers."

Dunn and Smith [42] performed a variety of abrasion and static fatigue tests, demonstrating that the use of hard UV-cured secondary coatings yielded fibers with improved strength and handling characteristics, compared with silicone single coatings.

4.3.3 Durability of Fiber Optic Coatings

Long-term durability of protective coatings was considered to be of considerable importance because fibers installed into outside plant networks were expected to have a minimum service lifetime of 25 years [43–45]. D. R. Young [46] provided "the first report on a long-term, long length static fatigue test in an outdoor, in-ground trough environment." He reported that UV-cured acrylate composite protective coatings were shown to provide excellent protection against a variety of environments: 65 °C air, high and low pH solutions, and temperature/humidity cycling of 150 days 65 °C/98% relative humidity (RH) to 10 °C/4% RH. Weibull plots, which describe the probability of failure at a given stress for a given length, were provided for fibers exposed to temperature/humidity cycling and immersed in petroleum jelly, similar to what is used as loose tube filling compound.

In 1984, O. R. Cutler [47] reported on the results of his durability testing on UV-curable coating films. In his study, Cutler [47] exposed 75 μm thick films of a variety of commercial fiber coatings to temperatures of 38, 54, 88, 125, and 175 °C, for up to 1 year. He developed Arrhenius plots that showed, based on the time required to double the coating's modulus, operating lifetimes of primary and secondary coatings could extend beyond 100 years at room temperature. Several coatings had similar calculated service lifetime when aged continuously at 54 °C (130 °F).

Nevins and Taylor [43] issued their report on the effect of a variety of environmental conditions on the "three key characteristics of fiber optic waveguides which may be effected by environmental conditions: strength, attenuation and resistance to losses caused by microbending." In addition to the conditions reported earlier by Young [46], the research team of Nevin and Taylor mentioned exposure to seawater, fungi, and abrasives. They also cited an experiment, Procustes, designed to correlate accelerated aging tests with actual long-term aging.

Simoff et al. [48] studied the aging of a polyether urethane acrylate primary coating, both in films and on fiber, and correlated the changes in physical properties of the films with the stripping force required to remove the coating from the fiber.

In 1993, Chawla et al. [49] reported on fiber optic coating durability, as was measured by weight changes and shifts in DMA modulus profiles. DMA provides insight into a material's durability following exposure to a wide variety of environments such as hydrolytic, thermo-oxidative, and chemical exposure. Comparison of DMAs before and after exposure to these conditions allows one to monitor changes in the material's glass transition temperature profile and the material's equilibrium modulus. The equilibrium modulus region of the DMA curve is observed at the modulus plateau reached in the rubbery phase. This modulus can be related to the crosslink density of a cured coating's network through the equation

$$\rho_0 = E_0/6kT, \tag{4.1}$$

where ρ_0 represents crosslink density, k is the Boltzmann constant, and T is the temperature, in degrees-Kelvin.

Decreases in equilibrium modulus indicate a reduction of a coating network's crosslink density through chain scission, and hence a weaker coating. Conversely, increases in the equilibrium modulus can signify embrittlement of the coating network through crosslinking. Chawla et al. [49] demonstrated excellent durability of several primary coatings after 1 year of aging at 125 °C.

The results of additional durability studies [44, 45, 50–56] have been published and offer a deeper understanding of this subject.

4.4 CABLING OF OPTICAL FIBERS

Early 250-μm outer diameter coated fibers were further protected from the potential hazards of the cabling process by placement in buffer tubes. The tubes had an inside diameter many times larger than the diameters of the enclosed fibers. Fiber length was designed to slightly exceed the length of the buffer tubes, to introduce fiber slack. This slack was necessary to prevent tensile loading, which could compromise the optical transmission of the fiber. The buffer tube also contained water-blocking filling compound. The filling compound not only blocked water from reaching the fiber, but also provided a medium in which the fibers could freely move past each other during thermal expansions and contractions. Protective fiber coatings had to be resistant to changes in material properties induced by such filling compounds, which often comprised fumed silica dispersed in mineral oil.

An alternative to the "loose-tube" cable design was the "direct-strand buffer" construction. An example consisted of a fiber that was coated with a silicone coating to 400-μm outer diameter and then over extruded with a nylon jacket to 900-μm outer diameter. Fibers of this type were wound about a strength member that had a lower expansion coefficient than that of the tight-buffer coating materials.

In tight-buffered or up-jacketed fiber, 250-μm fiber is overcoated with a tough thermoplastic extrusion, such as a Nylon 12, polyethylene or PVC. These thicker fibers provide improved handling and mitigate the use of a cabling gel. Tight-buffered cables have been used in premise applications for more than 2 decades. Though widely used, thermoplastic extrusions are limited by slow line speeds, 100–200 m/min, and high scrap rates.

UV materials have successfully been employed as a cushioning layer between flame-retardant thermoplastic extrusion and the fiber. Construction of these types of fibers involves up-jacketing a standard 250-μm fiber to 400–500 μm with a UV coating and then extruding with a colored thermoplastic to bring the final thickness to 900 μm.

Another approach to tight buffering using UV-curable materials involves up-jacketing a colored 250-μm fiber with a clear UV resin directly to 500 or 900 μm. The development of flame-retardant UV cure coatings has been reported [57, 58]. Montgomery et al. [59] described the use of a tight-buffer coating containing a pigmented flame retardant, thus giving performance typically observed by thermoplastics. These UV-curable up-jacketed fibers can be processed on a modified ink-coloring machine, giving much faster processing speeds, up to 600 m/min, than thermoplastics [60].

During the early to mid-1980s, Japan began using UV-curable matrix materials to bond fibers in four-fiber ribbon arrays. AT&T had earlier pioneered 12-fiber Adhesive Sandwich Layer Ribbons, in which fibers were packaged between two plastic tapes coated with pressure-sensitive adhesive. UV-curable matrix materials for ribbon application became more popular globally, with the growing demand for fiber, in the 1990s.

Prior to the cabling of fiber, in either loose-tube or ribbon configuration, the vast majority is colored with a UV-curable ink, in an off-line operation. UV-curable inks are designed to provide high-quality color on fiber for good identification and are typically applied at 3- to 5-μm thickness. The UV-cured ink layer has a hard and slick surface finish to allow suitability in both loose tubes and ribbons. It is also designed to have excellent adhesion to the secondary coating and to provide excellent resistance to cabling gel. Colors are designed to be bright and distinct, giving good color retention over time.

4.5 SPECIALTY COATINGS

Low RI coatings are used for cladding of plastic optical fiber or glass cores. They are by nature relatively low-modulus materials and are, therefore, often protected by overlaying a standard secondary coating. The lower RI coating is preferred for fiber used in laser and amplifier applications. The low RI coating increases the numerical aperture (NA) value, allowing for higher power inputs and thus higher powered lasers. These coatings typically have an RI less than 1.41. For application as a single coat, a modulus of approximately 120–200 MPa is desirable; however, when used as the primary layer in a dual-coating system, a modulus of 15–50 MPa may be used.

4.6 BASICS OF OPTICAL FIBER CHEMISTRY

In the late 1970s and early 1980s, UV-curable coatings designed for the protection of optical fiber transitioned from single coatings based on epoxy acrylate chemistry to dual-layer coatings based on urethane acrylate chemistry. These coatings are composed of one or more urethane acrylate oligomers [12, 61], diluent monomer(s), photoinitiator(s), and various additives.

4.6.1 Oligomers

Urethane acrylate oligomers are based on stoichiometric combinations of di-isocyanates (DICs), polyols, and some type of hydroxy-functional terminating species containing a UV-reactive terminus (A). Urethanes are known for their toughness and flexibility, a combination that adds value to the performance and protective nature of the coatings in which they are contained (Fig. 4.2).

Depending on the properties desired, different types of polyols are chosen. Typically, oligomers are made using the types of polyols shown in Figs. 4.3–4.5:

A – DIC – polyol – DIC – A

Figure 4.2   Typical urethane acrylate structure.

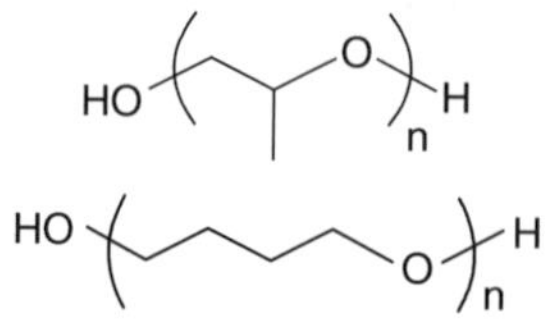

Figure 4.3   (a) Polyether–polypropylene glycol (PPG). (b) Polyether-polytetramethylene glycol (PTMG).

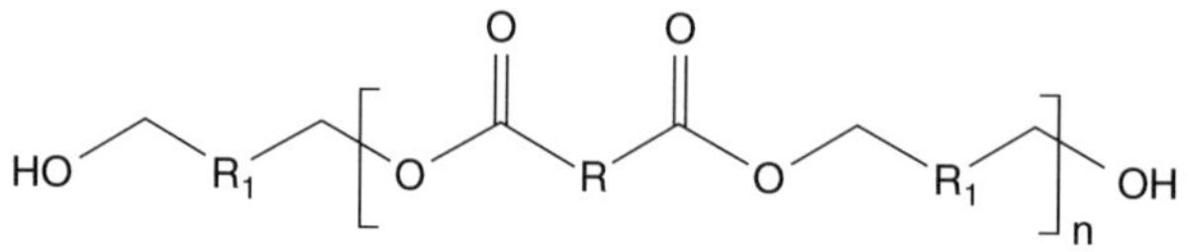

Figure 4.4   Polyester

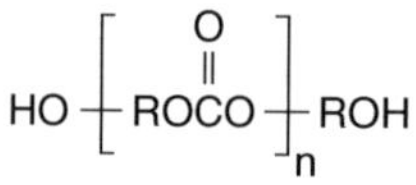

Figure 4.5   Polycarbonate.

Urethane acrylate oligomers employed in UV-curable optical fiber coatings generally range between 1000 and 10,000 number average molecular weight. Higher molecular weights tend to be exceedingly viscous. The viscosity can be reduced by either heat and/or addition of diluent monomer. Typical viscosities of coatings supplied to producers of optical fiber are less than 10,000 mPa·s. The viscosity of coatings, as they are applied to fiber, is typically 2000 mPa·s or less. In general, high-molecular-weight urethane acrylate oligomers are used in soft low-modulus coatings, while low-molecular-weight oligomers are employed to produce hard high-modulus coatings.

Acrylated epoxies are another type of oligomer commonly used in optical fiber coatings. Epoxy acrylates are tough fast-curing materials that have good chemical resistance. They tend to be used in secondary coatings (Fig. 4.6).

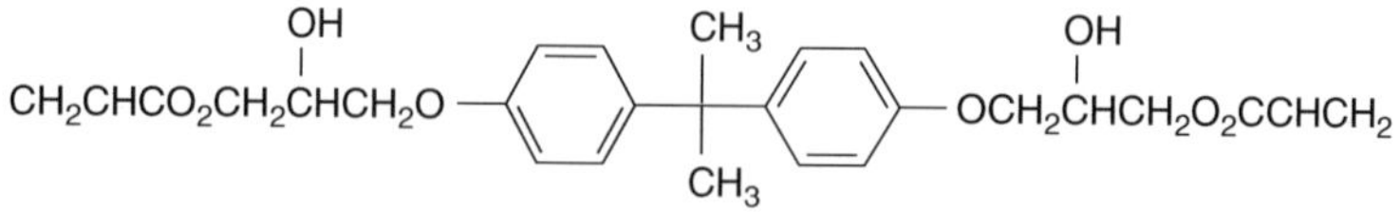

Figure 4.6   Bisphenol Alepoxy diacrylate.

4.6.2 Monomers

Diluent monomers can be monofunctional or multifunctional with respect to UV-reactive terminal groups. Monofunctional diluent monomers are generally more efficient in their diluency; however, they are only capable of reacting linearly and will not add to the crosslinking of polymer networks. Monofunctional monomers can be aliphatic (e.g., isodecyl acrylate [IDA]), aromatic (e.g., phenoxyethyl acrylate [PEA]), or alicyclic (e.g., isobornyl acrylate [IBOA]). In general, monofunctional monomers find their greatest utility in soft primary coatings, which are designed to have very low modulus. Monofunctional monomers generally give good flexibility and low shrinkage.

Multifunctional monomers are most often employed in secondary coatings, which have higher modulus and crosslink density relative to primary coatings. Multifunctional monomers are added for fast cure speed, increased crosslinking, and tensile strength [61].

Monomers possess a variety of useful properties beyond their acrylate functionality. These property differences provide skilled formulators the latitude to design coatings with optimized performance with regards to viscosity, cure speed, tensile properties, glass transition temperature profile, oleophobic/hydrophobic balance, adhesion, and long-term durability in various environments.

4.6.3 Photoinitiators

Photoinitiators, as their name implies, initiate the photopolymerization process by absorbing light. In UV-curable acrylate systems, they form radicals by cleavage (Norrish type I) or hydrogen abstraction (Norrish type II). The effectiveness and efficiency of photoinitiators is largely governed by the their absorption spectra and extinction coefficients. It is important to appropriately match the absorption spectrum of a photoinitiator with the emission spectra of the lamps employed for coating cure.

4.6.4 Adhesion Promoters

Primary coatings based exclusively on oligomers, monomers, and photoinitiators generally do not provide adequate adhesion to the glass fiber. This is especially true in humid environments. It is, therefore, necessary to incorporate an adhesion promoter or glass-coupling agent. Typically, adhesion promoters have an organic functional group, which bonds, or associates, with the coating. The adhesion promoter bonds to the glass surface through hydrolysis and condensation reactions.

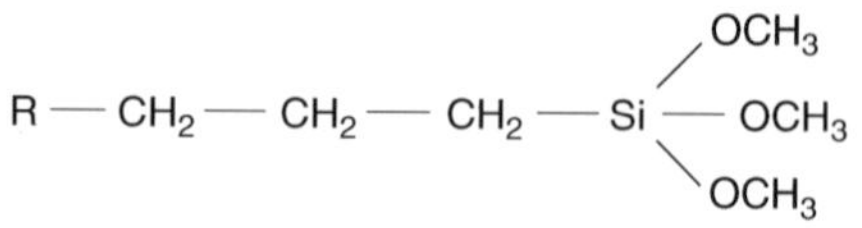

$$R — CH_2 — CH_2 — CH_2 — Si \overset{\displaystyle OCH_3}{\underset{\displaystyle OCH_3}{— OCH_3}}$$

Figure 4.7 Organopropyl trimethoxysilane.

Alkoxy silanes are commonly employed as adhesion promoters in fiber optic coatings.

The organofunctional end of the adhesion promoter covalently links with the oligomer or other radiation-curable moieties (Fig. 4.7). The glass-coupling moiety is generally inorganic in nature and bonds with the glass surface. Types of organic functionality include amino, epoxy, vinyl, methacryloxy, isocyanato, mercapto, polysulfide, and ureido. The inorganic end of the coupling agent must have hydrolyzable groups, which leave silanol groups that can condense with the surface silanol groups of the glass fiber surface. Methoxy and ethoxy groups are typically employed as hydrolyzable groups of silane coupling agents.

Several types of non-silane adhesion promoters include chromium, orthosilicate, inorganic ester, titanium, and zirconium systems.

4.6.5 Other Additives

Other additives in fiber optic coatings include antioxidants and stabilizers and slip additives. Antioxidants and stabilizers are added to coatings to maintain the desired viscosity of the liquid composition and impart resistance to cured films against degradation by light and oxidation. Hindered phenols are widely used antioxidants in fiber optic coatings, whereas hindered amines provide stability against light. Additives such as silicones are added to the coatings to modify the surface properties.

The standard components of a fiber optic coating are shown in Table 4.1, together with typical levels at which they are present, and their contribution to the coating.

As mentioned, acrylate systems cure by free radical photopolymerization. Photopolymerization occurs in several stages, as shown in the following subsections.

4.6.5.1 Initiation

Initiation involves the absorption of light, which, via several intramolecular energy transfers, yields primary radicals. This is followed by the addition of the resulting primary radicals in a Michael type of reaction to a suitable double bond, thereby resulting in propagating radicals [63] (Fig. 4.8).

Table 4.1

Typical fiber coating composition

Component	Contribution	% concentration
Oligomer acrylate	Controls final cured film properties, flexibility, chemical resistance	30–70
Reactive diluent or monomer	Reduces coating viscosity and contributes to coating cure speed	20–60
Photoinitiator	Absorbs ultraviolet light and initiates polymerization	<5
Additives (inhibitors, adhesion promoters, slip additives)	Stabilizes liquid coating, enhance cured film adhesion to glass, and reduce surface friction	<2

4.6.5.2 Propagation

As propagation reactions, all additions after the primary addition are counted and treated in a similar manner (Fig. 4.9).

4.6.5.3 Termination

Bimolecular termination: the standard that is employed in all classic descriptions of radical polymerization.

Primary radical termination: the propagating radicals are terminated by initiator radicals (Fig. 4.10).

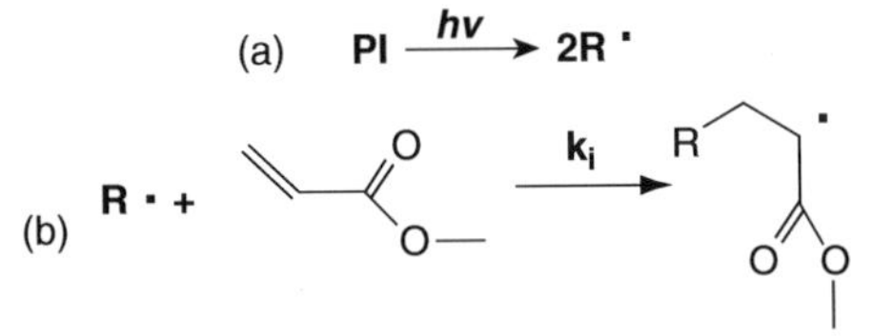

Figure 4.8 (a) Photoinitiation fragmentation. (b) Free radical initiation.

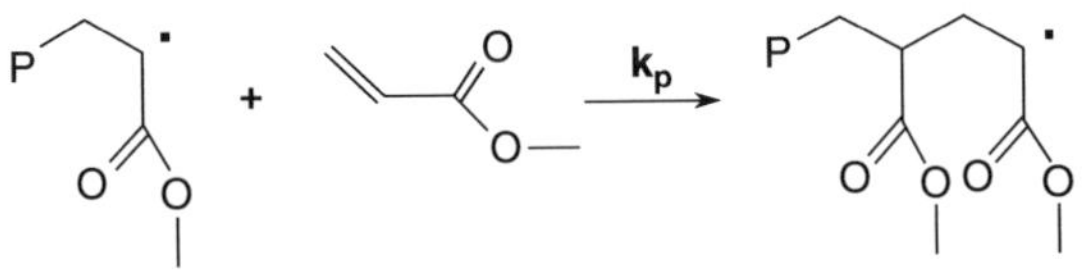

Figure 4.9 Chain propagation.

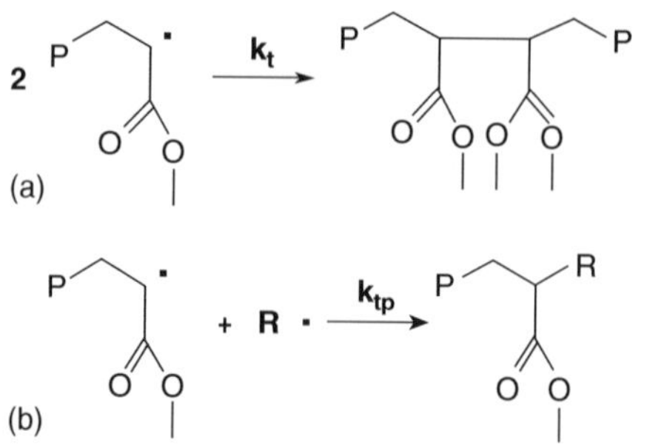

Figure 4.10 (a) Bimolecular termination. (b) Primary radical termination.

In addition to acrylate systems, free radical polymerization may also employ thiol-ene chemistry. An alternative curing mechanism to free radical polymerization is cationic polymerization.

In thiol-ene chemistry [64], a thiol (mercaptan) radical reacts with an olefin in the presence of UV light. While thiol-ene systems have been evaluated for their potential use in optical fibers, to our knowledge, they have never been used commercially.

Cationic polymerization involves the acid polymerization of an epoxy or vinyl ether group. Typically diaryliodium or triarylsulfonium salts are used as photoinitiators. Epoxies are fairly slow curing; however, vinyl ethers cure extremely fast. Advantages of cationic cure include an absence of oxygen inhibition and good through-cure. However, cationic cure suffers from moisture inhibition, limited raw material selection, and higher cost. With cationic polymerization, curing continues after irradiation.

Some hybrid systems such as acrylate free radical and cationic cure have also been employed in fiber optic coatings, but these are very rare [65].

4.7 APPLICATION OF COATINGS ON THE DRAW TOWER

Optical fibers are drawn from glass preforms at temperatures approximating 2000 °C. The majority of tower systems employ the wet-on-dry application method, wherein fibers pass through a primary coating applicator, followed by one or more UV radiation–emitting lamps, and then through a secondary coating applicator, again followed by one or more UV radiation–emitting lamps. In an alternative wet-on-wet application protocol [66, 67], fibers are passed through both the primary and the secondary coating before UV exposure.

It is imperative to apply concentric coating layers, to prevent damage to the fiber during the drawing operation, and to maximize fiber strength and micro-bending resistance. Nonconcentric coating application puts differential stress on the fiber when cured. An unevenly coated fiber will experience non-uniform forces during periods of coating expansion and contraction and is susceptible to greater attenuation of light signals passing though the fiber.

The earliest coating applicators used were of open cup geometry with either rigid or flexible tips [68]. Some of these included a temperature-controlled cylinder, which contained fitted split conical dies [23, 69] that produced "constant coating thickness over a wide range of coating viscosities and drawing speeds." Tapered dies aided concentric application of each of the coatings to the fiber [42, 70].

At drawing speeds greater than 1.5 m/sec, open cup applicators tended to have problems with coating meniscus collapse and bubble entrainment. Paek and Schroeder [71] studied a number of factors that were critical to increasing operating speed. These factors included fiber cooling rate, applicator efficiency, and available UV energy. They noted that shear force in the coating die needed to be minimized to avoid beading on the fiber due to non-Newtonian flow effects of the coating. The conical dies, used in open-cup applicators, were calculated to produce a shear rate of 10,000 sec^{-1}, at a speed of 1 m/sec.

Paek and Schroeder [72] were among the first to publish results on the behavior of forced cooling of bare glass fibers, during draw, to prevent deterioration of the coating. This doubled the allowable drawing speed from 1 to 2 m/sec. Paek increased the height and diameter of the coating cup to increase the "wetting time and minimize the air-trapping effect in the interface between the coating and the fiber." These modifications facilitated a coating speed of approximately 5 m/sec.

Kimura et al. [73] reported on the effect of fiber temperature, forced convection of liquid resin, and viscosity of coating. They found that the fiber temperature had to be maintained at less than 50 °C when applying a viscous coating, with a standard open-cup die. A fiber temperature of up to 120 °C could be tolerated when a forced convection of liquid was brought about by the use of a rotating rotor within the die. Draw speeds could be increased when the viscosity of coating was decreased from 2500 to 1200 mPa·s.

Geyling et al. [74] reported on their development of a pressurized applicator, wherein the coating was injected radially from a cylindrical plenum and filled a thin bore at the applicator. The thinner coating channel stabilized the coating application and facilitated application speed of approximately 2.5–3.0 m/sec. Pressurized die construction evolved during the early 1980s and became the standard coating application tool, because of the improvement it offered in coating circulation and meniscus stability.

Kassahun and Viriyayuthakorn [75] described the improvements offered by pressurized coating application, 40–80 N/cm^2, towards bubble-free coatings on fiber.

Paek and Schroeder [76] reported substantial improvement in high-speed drawing up to 12 m/sec. Fibers were determined to have uniform 50-micron coating thickness and concentricity along a 50-km length. The fibers survived a proof test stress level of 0.7 GN/m^2 (100,000 psi), and most survived stress levels of twice that level. Draw speed rates of 0.7 and 10 m/sec were found to produce single-mode fibers with identical optical properties.

Draw tower coating speeds ranged from 1 to 2 m/sec in the 1979–1981 time-frame [71]. Improvements in optical glass technology began to yield larger preforms, which significantly improved production efficiency. As a result, fiber prices became more competitive, fiber consumption jumped, and the demand for faster drawing technology was created.

In 1985, Sakaguchi and Kimura [77] published one of the first reports on pressurized coatings being applied to fiber at speeds up to 20 m/sec.

Paek and Schroeder [71] found that for an open-cup applicator, the coating thickness remains constant as the draw speed increases. They found, however, that with a pressurized applicator, thickness depends heavily on the draw speed and decreases as the speed increases [78].

The two primary controls for fiber coating application consist of temperature and pressure. UV-curable coatings have a nonlinear decrease in viscosity with increasing temperature. Viscosities drop precipitously during a temperature increase from room temperature to 45 °C.

Increasing temperature further to 55 and 65 °C produces diminishing decreases in viscosity (Fig. 4.11).

It is important to have knowledge of the rheological properties of optical fiber coating resins at high shear rates. Steeman et al. [79] described this relationship using the well-known time–temperature–superposition technique to avoid limitations of traditional direct measurement steady state methods, such as viscous heating in capillary viscometers and torque limitations in rotational viscometers.

A typical viscosity versus shear rate plot of a fiber optic coating is shown in Fig. 4.12. Fiber coatings exhibit shear thinning behavior, whereby the viscosity is Newtonian at low shear and decreases with increasing shear rate. The shear thinning behavior is generally controlled by the oligomer in the coating. Fiber optic coatings exhibit viscoelastic behavior.

4.7.1 Coating Cure Speed Measurement Techniques

During the mid- to late 1980s, coating cure speeds were increased through design improvements. Many of these improvements came from the availability of

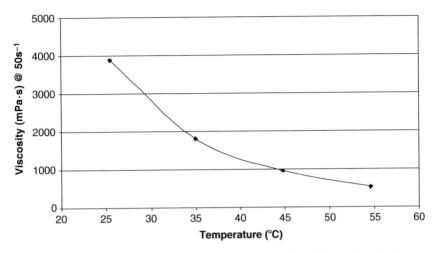

Figure 4.11 Typical viscosity versus temperature curve for a fiber optic coating.

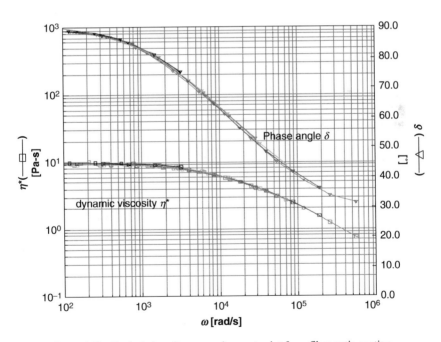

Figure 4.12 Typical viscosity versus shear rate plot for a fiber optic coating.

new coating raw materials. As cure speeds improved, so did the test methods employed for their measurement.

An early method monitored the development of modulus in free coating films as a function of UV dose, or more appropriately termed *UV energy density.* Typical curves showed a rapid rise in modulus at low UV energy density and then a slow rise after further exposure. Some coatings did not reach their ultimate modulus until exposed to an energy density of $2\,J/cm^2$ or greater; $3.5\,J/cm^2$ was chosen as the standard energy density for determination of a cured film's physical properties: tensile strength, elongation at break, and modulus at 2.5% elongation. It is typical to measure the cure speed in terms of the dosage required to cure the coating to 95% of its ultimate modulus, as shown in Fig. 4.13.

As demand for faster curing coatings grew, and with the advent of faster curing monomers and more efficient photoinitiators, UV energy densities of $1\,J/cm^2$ or less were required to achieve fully developed physical properties. Around this time, another analytical method became popular for the characterization of cure speed. Fourier transform infrared (FTIR) spectroscopy had long been used to identify chemical species and to follow chemical reactions. In the 1970s and 1980s, a number of papers were published [80–90], which described how this technique could be used to follow the disappearance of absorbance peaks corresponding to acrylate unsaturation.

Julian and Millon [91] reported on how FTIR could be used as a quality-control test for monitoring UV-curable coating cure speed. However, they cautioned that UV doses required for property development were different from those required to cause disappearance of acrylate unsaturation, suggesting that

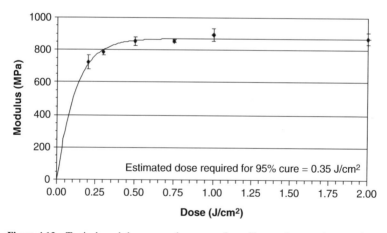

Figure 4.13 Typical modulus versus dose curve for a fiber optic secondary coating.

"loss of acrylate unsaturation is only one of a number of mechanisms which lead to the development of a properly cured coating film."

While the FTIR cure speed measurement method related more to the inherent chemical crosslinking taking place on UV exposure, the development of real-time DMA resulted in a more time-efficient means to evaluate the property development of a coating.

Coating cure is dependent on the peak intensity, emission spectrum, temperature, and geometry of the lamps to which it is exposed. It is also dependent on the coating's absorption of incident UV radiation. It has not been possible to exactly measure the environment encountered by the coating as it is exposed to UV energy on the draw tower. However, it is possible to measure both the reacted acrylate unsaturation (RAU) on cured fiber coatings and the *in situ* modulus of coatings cured on fiber. Therefore, some approximations can be made for the conditions that produced these types of properties.

Other cure measurement analytical tools include monitoring of glass transition temperature development, dielectric relaxation, and differential scanning photocalorimetry.

4.7.2 Cured Properties of Coatings on Fiber

Hussain [23] employed a torsional pendulum technique comprising small coated optical fiber specimens to determine the suitability of coatings and their processing parameters. Using this technique, he was able to compare the shear modulus properties of different primary coatings as cured on fiber.

Frantz et al. [92] evaluated a number of techniques for determining the extent of cure on optical fibers. These techniques included infrared spectroscopy (FTIR), differential photochemistry (DPC), solvent evaporation rate analysis (ERA), solvent extraction, water/solvent soak, dielectric measurement, hydrogen generation, weight loss, oxidation onset, abrasion resistance, coating strip force, and coating pullout. FTIR, DPC, ERA, and solvent extraction were found to have the greatest sensitivity and reproducibility for measuring the degree of cure of a given coating on fiber.

In situ modulus is a useful test for measuring the modulus of the coating on the fiber. It has been observed that the *in situ* modulus could be substantially lower than the flat film modulus even at high cure degree. Small variation of the *in situ* RAU can result in large differences in *in situ* modulus. Even with the same degree of cure, large variations of polymer networks may be formed, largely because of variations in the intensity of UV light and the resultant heat from the drawing process [93].

The shear modulus of primary coatings can be measured *in situ* by pullout test using TEM or DMA [93–95] (Fig. 4.14).

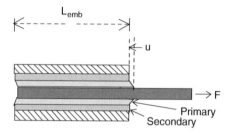

Figure 4.14 Illustration of pullout testing for determination of shear modulus.

Traditional techniques involved the use of an adhesive to immobilize the fiber on a substrate cylinder. A new method using mechanical clamping [93] has the advantage of easy sample preparation and very small data scattering. Steeman et al. [93] measured the primary modulus on a DMA by applying an oscillating shearing stress to the coating.

The *in situ* secondary modulus can be measured by applying an oscillating tensile stress to a stripped coating tube on a DMA. The coating tube is prepared by separating the glass fiber from the coating, in liquid N_2. Because the modulus of the primary coating is several orders of magnitude lower than that of the secondary coating, it can be assumed that its effects are negligible, so the measured Young modulus of the coating tube can be assumed to come from the secondary coating.

Strip force provides an indication of the ease of stripping the coating from the glass. The coating must have adequate adhesion to the glass while being mechanically strippable for splicing and connectorization. There is little or no correlation between strip force and primary coating adhesion. The modulus, deformation, and failure mode of the secondary coating are the dominating factors that influence strip force [96].

Pullout force measures the force required to pull a glass fiber out from the coating layers. It gives a measure of the adhesion of the primary coating to the glass. The sample is glued to a support, prepared by gluing the fiber onto a support tab. The application area, the uniformity of the glue, and drying time can significantly affect the results. A mechanical clamping method has been tested, but the measured force inevitably is a complex result of adhesion force and friction force.

The FTIR/ATR technique can be used to measure cure of both the primary and the secondary coating on fiber. In terms of %RAU, the cure of the inner surface of primary coating is measured by cutting through the coating layers to the glass, removing the glass fiber, and laying the coating flat on a diamond attenuated total reflectance (ATR) crystal, thus exposing the primary. The degree of cure of the secondary coating can be measured by directly measuring the %RAU on the fiber surface.

4.7.3 Test Methods for UV-Curable Liquids and UV-Cured Films

Julian [97] reviewed many techniques used to characterize the quality of radiation-curable coatings. For UV-curable coatings, it is also important to characterize the quality of the raw materials that are used to formulate the coating.

For coatings to be of commercial utility, they must have sufficient liquid stability in a variety of containers (plastic bottles and lined steel drums) and in heated application dies and coating reservoirs. Pasternack et al. [98] mentioned that changes in properties could be "evaluated by accelerated aging of the liquid coating at temperatures high enough to cause changes within reasonable times but still below thermal degradation levels." A typical temperature for such testing is 54 °C (130 °F).

Chawla et al. [99] published a very thorough study on the factors that affect measurement of the mechanical properties (tensile strength, elongation, and modulus) used to characterize UV-cured films of optical fiber coatings. The study led to a number of improvements in equipment (more sensitive load cells and more accurate gauges for specimen measurement) and environmental controls (temperature, relative humidity, and cured film preparation parameters). These improvements were important to both characterizing the properties of new coatings and providing accurate and precise measurements of batch-to-batch variation in commercial coating production.

The modulus of primary coatings is typically calculated by taking the slope of the stress–strain curve from the origin to 2.5% strain (i.e., the segment modulus technique). For secondary coatings, secant modulus is typically measured and calculated from a tangent to the single point of the stress–strain curve at 2.5% strain. Chawla et al. [99] found that there was no difference in which approach was used to measure the modulus of secondary coatings, but the segment modulus was found to give greater accuracy for primary coatings.

For optical fiber coatings, it is customary to characterize the mechanical properties of a coating by running a temperature sweep on a dynamic mechanical analyzer. A coating film is subjected to a controlled oscillating strain at a fixed amplitude and frequency, whereby the elastic modulus E', the viscous modulus, E'', and tan δ are measured. Information regarding crosslink density and glass transition can be obtained from the plot. There are a variety of methods for measuring the Tg of a coating. The peak tan δ is often referred to as the "glass transition point," but because the glass transition is a region rather than a single point, it is customary to quote the temperatures at which $E' = 1000$ MPa and $E' = 100$ MPa. Typically, polymers exhibit glassy behavior at low temperatures and rubbery behavior at high temperatures. The equilibrium modulus, measured at the minimum of the E′ curve, is proportional to a coating's crosslink density. An example of DMA curves for a fiber optic primary coating is shown in Fig. 4.15.

As mentioned earlier in this chapter, DMA curves are useful to characterize the aging behavior of a film by inspecting the changes between spectra taken before and after environmental exposures.

Szum [100] observed that coatings with glass transition temperatures above room temperature, notably secondary coatings, exhibited changes in physical properties over time. DMA analysis showed that these changes in properties were due to relaxation of the polymer network, rather than chemical changes. In general, it was observed that a high Tg crosslinked polymer's properties were dependent on heat history. If a UV-cured film was quenched (i.e., quickly cooled after cure), it was found that the modulus of the film tended to increase over time, through densification, as the cured coating relaxed.

Julian [97] reviewed a number of characterization methods for UV-curable liquids and cured films. These included viscosity measurement, cleanliness (freedom from particulate contamination), molecular weight determination, cure speed, mechanical property measurement (tensile properties and DMA properties), and thermal stability.

Julian cited three goals for equipment utilization, as mentioned in the paper: to provide a thorough characterization of materials, both raw materials and finished coating formulation; to perform analyses at reasonable cost; and to improve product consistency.

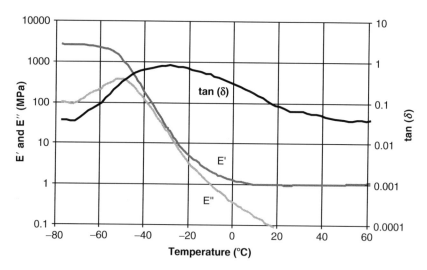

Figure 4.15 Typical DMA curves for a fiber optic primary coating.

4.7.4 Coating Adhesion

Coating adhesion was thought to be a prerequisite for retention of fiber strength. This is why many optical fiber coatings in direct contact with glass have contained some type of adhesion-promoting agent since the time of Vazarani's patent [13]. Silane coupling agents have typically filled this role. Many of these silane molecules contain three alkoxy silane groups on one end of the molecule and an organofunctional terminus of some type at the opposite end. The organofunctional group is selected to react with the coating network and, thus, be tied to the glass through the alkoxy silane groups. Silane coupling agents improve the adhesion of coatings in wet and dry environments. In 1982, Plueddemann [101] produced an excellent reference on the chemistry of silane coupling agents.

Overton and Taylor [102] stated that reliable adhesion of the primary to glass is necessary to protect against coating delamination. Coating delamination decreases the fiber's robustness and degrade its performance. They reported that adhesion is achieved though intermolecular forces, achieved by intimate contact between coating and glass, and chemical bonding afforded by silane coupling agents.

The adhesion to glass for some early fiber primary coatings was very high and made coating removal for splicing and terminations difficult. Chemical removal was often the preferred method to avoid damaging fibers in the field. As cable fiber counts began to increase, outside craft personnel spent more and more time making fiber connections. Amos et al. [103] reported on a primary coating with improved lower stripping force and its suitability for mechanical removal without damage to the underlying fiber.

Overton and Taylor [102] described a coating-removal test that resolved the problem of measuring the adhesive strength between the primary coating and the glass. The pullout test revealed that the temperature dependence of the pullout force is similar to that of the primary coating's modulus. Continued decline in pullout strength at temperatures much greater than the coating Tg was attributed to the relaxation of hoop stress in the secondary coating.

4.8 SUMMARY

Many aspects of the protective coatings have an impact on the final optical fiber performance, including the following:

Primary coating feature	Performance impact
Composition	Fiber strength/fatigue, long-term durability
Reactivity	Drawing speed
Refractive index	Concentricity monitoring

Viscosity	Ease of processing
Tg	Low-temperature microbending loss
Modulus	Buffering (protection against microbending); strippability (residue left on fiber)
Adhesion	Delamination resistance, strippability

Secondary coating	**Performance impact**
Composition	Long-term durability
Modulus	Shielding (protection against lateral load), strippability, abrasion resistance, handling ability
Surface properties	Ink adhesion, fiber friction
Crosslink density	Chemical resistance

ACKNOWLEDGMENTS

We would like to thank Myron Bezdicek, Huimin Cao, Johan Janssen, Ad Abel, and Mark Tilley for contributions to this chapter.

REFERENCES

[1] Kao, C. K. 1966. Dielectric-fibre surface waveguides for optical frequencies. *Proc. IEEE* 113(7):1151.
[2] An interview with Robert D. Maurer. 2002. *OE Reports SPIE. The International Society for Optical Engineering* No. 198 (June).
[3] http://inventors.about.com/gi/dynamic/offsite.htm?site=http://web.mit.edu/invent/iow/MaurKeckSchul.html Inventor of the Week Fiberoptic communications Massachusetts Institute of Technology MIT School of Engineering (May 1997).
[4] Dormeus, R. H. 1973. *Glass Science*, p. 284. John Wiley & Sons, New York.
[5] Stevens, J. M. and A. Keough. 1978. *The Application of UV Coatings to Glass Optical Fiber.* FC78-551. The Association for Finishing Processes of SME (September).
[6] Kar, G. and J. Toler. 1988. Polymeric coatings for optical fibers. In: *Proceedings of ANTEC '88*, pp. 372–376.
[7] Honjo, M. and S. Masuda. Required properties of UV curable resin for optical fiber. In: *Proceedings of Radiation Curing Asia, Tokyo, Japan*, pp. 169–172.
[8] Ritter, J. E., Jr. 1973. Stress corrosion susceptibility of polymeric-coated soda-lime glass. *J. Am. Chem. Soc.* 56(7).
[9] Wang, T. T. et al. 1979. *J. Applied Polymer Sci.* 23:887.
[10] Wei, T. 1986. The effects of polymer coatings on the strength and fatigue properties of optical fibers. In: *Proceedings of American Ceramics Society, Chicago, IL* (April).
[11] *The Western Electric Engineer* (Winter 1980):57.
[12] Ansel, R. E. and J. J. Stanton. 1980. An overview of ultraviolet light (UV) curing systems used as optical wave guide coatings. Presented at the Conference on Physics of Fiber Optics, American Ceramic Society (April).

[13] Vazarani, H. N. (Inventor). 1978. Bell Telephone Laboratories, Incorporated (Assignee), U.S. Patent 4,099,837, filed May 26, 1976; issued July 11, 1978.

[14] Eccleston, D. J. 1983. Environmental testing of coated optical fibers. *Wire Technol.*

[15] Geyling, F. T. and T. J. Louzon. (Inventors). 1981. Bell Telephone Laboratories, Incorporated (Assignee), U.S. Patent 4,374,161, filed April, 24, 1981; granted.

[16] Vazarani, H. N. et al. 1977. Coatings and plastics preprints. *Am. Chem. Soc.* 37(2):253.

[17] Wang, T. T. and H. M. Zupko. 1978. Long-term mechanical behaviour of optical fibers coated with a UV-curable epoxy acrylate. *J. Materials Sci.* 13:2241–2248.

[18] Schonhorn, H. et al. 1976. *Appl. Phys. Lett.* 29:712.

[19] Schonhorn, H. et al. 1979. *J. Applied Polymer Sci.* 23:75.

[20] Davis, C. Fiber optic technology and its role in the information revolution Christopher. Available at: www.ece.umd.edu/~davis/optfib.html.

[21] Duke, D. A. 1983. *A History of Optical Communications.* Corning Special Report SR-7 (April).

[22] Gunderson, L. C. and D. B. Keck. 1983. Optical fibers: Where light outperforms electrons. *Technol. Rev.* 86(4)(May/June).

[23] Hussain, A. 1983. Optimization of UV-curing for optical fibers using a torsional pendulum technique. Paper presented at: FC83-257 Society of Manufacturing Engineers RADCURE 1983 (May).

[24] The Photonics Dictionary, Photonics.com: Optical, Laser and Fiber Optics Resource. 1996–2005. Laurin Publishing.

[25] Gloge, D. 1975. *Bell System Tech. J.* 54(2):245.

[26] Kuzushita, H. et al. 1987. Study on transmission characteristics of UV curable resin-coated optical fibers at low temperatures. In: *Proceedings of the 36th International Wire & Cable Symposium* p. 217. Arlington, VA. Nov. 17–19, 1987.

[27] Sarkar, A. et al. 1987. High performance UV-cured optical fiber primary coating. *Fiber Integrated Optics* 6(2).

[28] Skutnik, B. J. et al. 1986. Coating adhesion effects on fiber strength and fatigue properties. In: *Materials Research Symposia Proceedings, 88: Optical Fiber Materials and Properties, Dec. 3–5, 1986, Boston*, pp. 27–34.

[29] Rondinella, V. V. and M. J. Matthewson. 1993. Effect of loading mode and coating on dynamic fatigue of optical fiber in two-point bending. *J. Am Chem. Soc.* 76(1):139–144.

[30] Griffeon, W. et al. 1990. Two-point bending apparatus, fracturing optical fibres at different speeds in one run; measurements in standard vacuum environment. In: *Proceedings of the 39th International Wire & Cable Symposium/Focus, 1990*, pp. 368–372. Reno, NV. Nov. 13–15, 1990.

[31] Matthewson, M. J. 1986. Strength measurement of optical fibers by bending. *J. Am. Ceramic Soc.* 69(11):815.

[32] Rondinella, V. V. 1993. Effect of loading mode and coating on dynamic fatigue of optical fiber in two-point bending. *J. Am. Ceramic Soc.* 76(1)139.

[33] Gulati, S. T. 1977. Strength and static fatigue of optical fibers. In: *Proceedings of the First International Telecommunications Exp. (Atlanta, GA) Oct. 9–15, 1977*, pp. 702–711.

[34] Michalske, T. A. and B. C. Bunker. 1987. The fracturing of glass. *Sci. Am.* 255(12): 122–129.

[35] Michalske, T. A. et al. 1991. Fatigue mechanisms in high-strength silica-glass fibers. *J. Am. Ceramic Soc.* 74(8):1993–1996.

[36] Michalske, T. A. and B. C. Bunker. 1993. A chemical kinetics model for glass fracture. *J. Am. Ceramic Soc.* 76(10):2613–2618.

[37] Michalske, T. A. and B. C. Bunker. 1987. Steric effects in stress corrosion fracture of glass. *J. Am. Ceramic Soc.* 70(10):780–784.

[38] Bunker, B. C. and T. A. Michalske. 1986. Effect of stress corrosion on glass fracture. In: *Fracture Mechanics of Ceramics* (Bradt et al., eds.), Vol. 8, pp. 391–411. Plenum Press.

[39] Helfinstine, J. D. 1983. Delayed failure or subcritical crack growth in glass. The American Scientific Glassblowers Society, #58-3756. In: *Proceedings of the 28th Symposium on the Art of Glassblowing.*

[40] Sakaguchi, S. et al. *Influence of Glass-Plastic Interface on Optical Fiber Strength.* OQE79-138.

[41] Schlef, C. L. et al. 1982. UV cured resin coating for optical fiber/cable. *J. Radiation Curing* (April):11.

[42] Dunn, P. L. and D. C. Smith. 1982. Double coating of optical fibres. *Plastics Telecommunications III* (October):14-1-8.

[43] Nevins, R. C. and D. H. Taylor. 1985. *Testing for a Long Service Life, TELEPHONY* (Feb. 11, 1985).

[44] Kurki, J. et al. 1989. Reliability and environmental performance of cabled single-mode optical fibers. In: *Proceedings of 38th International Wire and Cable Symposium 1989*, pp. 380–389. Atlanta, GA. Nov. 14–16, 1989.

[45] Wang, T. T. et al. 1977. UV-cured epoxy acrylate coatings of optical fiber III: Effect of environment on long term strength. *J. Radiat. Curing* (Oct):22–24.

[46] Young, Jr., D. R. 1981. Environmental effects on acrylate coated optical fibers. In: *Proceedings of the 30th International Wire & Cable Symposium*, p. 51.

[47] Cutler, O.R. 1984. Therm al and aging characteristics of UV curable optical fiber coatings. In: *Proceedings of Radcure 1984 International Conference on Radiation Curing.*

[48] Simoff, D. et al. 1988. Thermo-oxidative aging of a primary lightguide coating in films and dual-coated fibers. In: *ANTEC'88 Proceedings*, pp. 386–389.

[49] Chawla, C. P. et al. 1993. Aspects of optical fiber durability. In: *Proceedings of International Symposium On Fiber Optic Networks And Video Communications, Berlin, Germany (April 1993).*

[50] Simoff, D. A. et al. 1988. Thermo-oxidative aging of a primary lightguide coating in films and dual coated fibers. In: *Proceedings of ANTEC'88 (1988)*, pp. 586–589.

[51] Chawla, C. P. and D. M. Szum. 1991. Effect of chemical environments of UV curable optical fiber coatings. In: *Proceedings of the 40th International Wire & Cable Symposium/ Focus*, pp. 141–148. St. Louis, MO. Nov. 18–21, 1991.

[52] Bishop, T. E. et al. 1992. Aspects of thermo-oxidative and hydrolytic degradation in optical fiber cable matrix materials. In: *Proceedings of the 41st International Wire & Cable Symposium/Focus*, pp. 442–446. St. Louis, MO. Nov. 18–21, 1991.

[53] Bishop, T. E. and D. M. Szum. 1991. A new test method for measuring hydrolytic stability of UV curable coatings. In: *Proceedings of Radtech Europe 1991*, pp. 124–132.

[54] Heyword, L. P. et al. 1985. Effects of antioxidants on the thermo-oxidative stabilities of ultraviolet-cured coatings. In: *Polymer Stabilization and Degradation* (P. P. Klemchuk, ed.), pp. 299–311. American Chemical Society.

[55] Chan, M. G. et al. 1987. The stabilization of UV curable coatings for optical fibers. In: *Proceeding of Plastics in Telecommunications IV*, pp. 14-1 to 14-9.

[56] Spencer, J. F. 1989. Environmental testing of optical fibre at OWA. In: *Proceedings of Australian Conference on Optical Fibre Technology*, pp. 105–108.

[57] Bishop, T. E. 1994. UV curable halogen-free flame retardant compositions. In: *Proceedings of RadTech'94, May 1–5, 1994, Orlando, FL.*

[58] Guo, W. 1992. Flame-retardant modification of UV-curable resins with monomers containing bromine and phosphorous. *J. Polymer Sci. Part A Polymer Chem* 30:819–827.

[59] Montgomery, E. I. et al. UV-curable buffer resins vs. thermoplastics: A closer look at new flame retardant, UV-curable materials in tight buffered cables. In: *Proceedings of the 52nd International Wire & Cable Symposium/Focus.*

[60] Chase, D. et al. 2001. UV cure fiber optic buffering resins. In: *Proceedings of the 50th International Wire & Cable Symposium/Focus*, pp. 529–531.

[61] Pachuta, J. M. 1982. Fiber optics: Lighting the way in communications. *Radiat. Curing* 16(February).

[62] Mori, M. et al. 1998. Enhancement of tensile strength of low Young's modulus UV curable urethane acrylates. In: *Proceedings of the International UV/EB Processing Conference, April 19–22, 1998, Chicago, IL.*

[63] Turro, N. J. et al. 2000. *Angew. Chem. Int. Ed. Engl.* 39:4436.

[64] Jacobine, A. F. Thiol-ene polymers. In: *Radiation Curing in Polymer Science and Technology* (J. P. Fouassier and J. F. Rabek, eds.), Vol. III, *Polymerisation Mechanisms*, Chapter 7.

[65] Bahadur, M. et al. 2003. Dual cationic and free radical UV curable optical fiber coating system. In: *Proceedings of the 52nd International Wire & Cable Symposium*. Philadelphia, PA. Nov. 17–20, 2003.

[66] Taylor, C. R. (Inventor). 1984. AT&T Bell Laboratories (Assignee), U.S. Patent 4,474,830, filed December 29, 1982; granted October 2, 1984.

[67] Perry, G. and J. Kurki. 1987. High-speed optical fiber drawing system. *Wire Technol.* July.

[68] Hart, A. C. et al. 1977. An improved fabrication technique for applying coatings to optical fiber waveguides. Presented at the Conference on Optical Fiber Transmission II Williamsburg, VA.

[69] France, P. W. et al. 1979. *Fiber Integrated Optics* 2(3-4):2674.

[70] Ohls, J. W. (Inventor). 1981. Corning Glass Works (Assignee), U.S. Patent 4,246,299, filed January 7, 1979; granted January 20, 1981.

[71] Paek, U. C. and C. M. Schroeder. 1981. High speed coating of optical fibers with UV curable materials at a rate of greater than 5 m/sec. *Applied Optics* 20(23):4028–4034.

[72] Paek, U. C. and C. M. Schroeder. 1979. Forced convective cooling of optical fibers in high-speed coating. *J. Appl. Phys.* 50(10):6144–6148.

[73] Kimura, T. et al. Coating technique for high speed drawing.

[74] Geyling, F. T. et al. Pressure coating of optical fibers.

[75] Kassahun, B. and M. Viriyayuthakorn. 1984. International Publication no. WO84/01227. March 29, 1984; priority date September 15, 1982 U.S.

[76] Paek, U. C. and C. M. Schroeder. 1984. Optical fibres coated with a UV-curable material at a speed of 12 m/s. *Electronics Lett.* 20(7):304–305.

[77] Sakaguchi, S. and T. Kimura. 1985. A 1200 m/min. Speed drawing of optical fibers with pressurized coatings. In: *Proceedings of the Conference on Optical Fiber Communications and Optical Fiber Sensors, San Diego, CA, 11–13 February 1985.*

[78] Paek, U. C. and C. M. Schroeder. 1984. High strength in a long length for fibers coated at a speed of 5 m/s. *J. Lightwave Technol.* LT-2(4).

[79] Steeman, P. et al. 2004. Rheological properties of optical fiber coating resins at high shear rates. In: *Proceedings of the 53rd International Wire & Cable Symposium/Focus.* Philadelphia, PA. Nov. 17–20, 2003.

[80] Chambers, S. et al. 1986. *J. Polymer Comm.* 27:209.

[81] Collins, G. L. and J. R. Costanza. 1979. Reactions of UV curable resin formulations and neat multifunctional acrylates II. *J. Coatings Technol.* 51(648):57.

[82] Collins, G. L. and J. R. Costanza. 1976. Reactions of UV curable resin formulations and neat multifunctional acrylates I. *J. Coatings Technol.* 48(618):48.

[83] Decker, C. 1984. UV curing of acrylate coatings by laser beams. *J. Coatings Technol.* 56(724):29.

[84] Decker, C. 1983. J. Polymer Sci. *J. Polymer Chem Ed.* 21:2451.

[85] Doorkian, G. A. and G. A. Lee. 1977. *J. Rad. Curing* 4:2.

[86] Imaura, M. et al. 1974. *J. Appl. Polymer Sci.* 18:3445.

[87] LieBerman, R. A. and F. S. Stowe. 1985. *J. Radiat. Curing* 12:16.

[88] LieBerman, R. A. 1984. *J. Radiat. Curing* 11:22.

[89] Plews, G. and R. Phillips. 1979. Some factors affecting cure of UV curing inks and varnishes. *J. Coatings Technol.* 51(648):69.

[90] Van Neerbos, A. 1978. *J. Oil Colour Chemists Assoc.* 61:241.

[91] Julian, J. M. and A. M. Millon. 1978. Quality control testing of UV curable coatings using FTIR. *J. Coatings Technol.* 60(765).

[92] Frantz, R. A. et al. 1991. Evaluation of techniques for determining the extent of cure of optical fiber coatings. In: *Proceedings of the 40th International Wire & Cable Symposium/ Focus*, pp. 134–140.

[93] Steeman, P. et al. 2003. Mechanical analysis of the in-situ primary coating modulus test for optical fibers. In: *Proceedings of the 52nd International Wire & Cable Symposium/Focus*.

[94] Taylor, C. R. 1985. In-situ mechanical measurements of optical fiber coatings. In: *Proceedings of the Conference on Optical Fiber Communication, San Diego, CA, 1985.*

[95] Katsuta, T. et al. 2000. In-situ measurement of primary coating modulus on optical fiber by pull-out-modulus technique. In: *Proceedings of the 49th International Wire & Cable Symposium/Focus*.

[96] Mann, J. D. et al. 1998. Mechanical stripping of lightguide coatings: Coating morphology. In: *Proceedings of the National Fiber Optic Engineers Conference*, 271.

[97] Julian, J. 1985. Instrumental techniques for analyzing radiation curable coatings. *Modern Paint Coatings* Feb:50.

[98] Pasternack, G. et al. 1986. Property measurement of radiation curable coatings based on acrylate chemistry. In: *Proceedings of Radcure '86 Conference, Association Finishing Processes of SME*. Baltimore, MD. Sept. 8–11, 1986.

[99] Chawla, C. et al. 1994. Development of improved methods for the measurement of mechanical properties of UV cured optical fiber coatings. In: *Proceedings of Radtech 1994*, p. 275. Orlando, FL. May 1–5, 1994.

[100] Szum, D. Polymer relaxation effects on physical properties of UV curable materials. In: *Proceedings of Radtech, 1994*, p. 220. Orlando, FL. May 1–5, 1994.

[101] Plueddemann, E. P. 1982. *Silane Coupling Agents*. Plenum Press, New York.

[102] Overton, B. and C. Taylor. 1988. The temperature dependence of dual coated lightguide pullout measurements. *ANTEC'88*, pp. 392–394.

[103] Amos, L. G. et al. 1981. Improvements in optical fiber coatings. In: *Plastics in Telecommunications III*.

Chapter 5

Single-Mode Fibers for Communications

Robert Lingle, Jr., David W. Peckham, Kai H. Chang, and Alan McCurdy

OFS Corporate R&D, Norcross, Georgia

5.1 INTRODUCTION

System performance can be maximized and total system cost savings can be realized by choosing an optical fiber design optimized for a particular system application. The cabled optical fiber that forms the backbone of the physical layer is one part of an optical transmission line that also comprises amplifiers and dispersion compensation modules (DCMs). The designs of the amplifier, DCM, and cabled transmission fiber are not mutually independent, and an integrated view of the transmission line design is necessary to optimize performance and drive cost out of the total system.

A digital transmission system relies on the ability of a receiver to discriminate whether a transmitted bit is a "1" or "0." The ability to do this can be degraded by multiple factors. First, the signal-to-noise ratio (SNR) can be degraded in either the optical or the electrical domain. Insufficient optical SNR (OSNR) can be due to noise in the transmitter, optical attenuation in the fiber span, the injection of noise by an amplifier, or insufficient receiver sensitivity. Second, intersymbol interference (ISI) occurs as adjacent bits (symbols) spread into one another and overlap due to dispersion in the fiber. ISI can cause a "0" to be corrupted by a neighboring "1." ISI arises from both chromatic and polarization mode dispersion. In a purely linear system, ISI can be fully compensated, at least theoretically, by a dispersion compensation device.

Nonlinearity in the optical transmission line can permanently distort the shape of an optical pulse. The combination of nonlinearity with dispersion leads to uncorrectable ISI penalties. Impairments can limit the reach, bit rate, or capacity of a system, whereas devices that mitigate impairments add cost and complexity to a system.

123

In choosing optical fiber design targets to maximize performance in specific application segments such as metro, long-haul, or ultralong haul, a clear understanding of how fiber properties affect current and future system performance must be combined with an understanding of the system cost impact. The technologies used to amplify signals, compensate dispersion, and route wavelengths will interact with the fiber properties to affect the cost–performance balance. The constraints due to practical considerations of cabling (microbending and macrobending sensitivity), limits on cutoff wavelength, the availability of ultrapure materials, and cost of fabrication are also important in developing a new commercial product.

This chapter explores several key network segments for which optimized optical fibers have been designed and commercialized. This chapter focuses on the application-specific transmission fibers that are deployed in optical cables in the outside plant as the primary infrastructure of a communications system. Fibers for dispersion compensation or nonlinear signal processing are deployed in modules in a terminal or hut and are the subject of other chapters in this volume. First, an overview of the system limitations and penalties that drive optical fiber design is presented. A summary of International Telecommunication Union (ITU) fiber standards follows. Fibers optimized for lowest loss in shorter reach applications are discussed next. Then we present design principles for optimizing an optical transmission line for high bit rate, high-capacity dense wavelength diversion multiplexing (DWDM) applications. Finally, the waveguide design and properties of nonzero dispersion-shifted fibers (NZDFs) are discussed in detail.

5.2 SYSTEM IMPAIRMENTS INFLUENCING FIBER DESIGN

5.2.1 Limitations from Optical Signal-to-Noise Ratio

Two cases are considered in which optical transmission may be limited by the degradation of OSNR. The trivial example is a low-cost nonamplified link that is simply loss limited. A receiver has a specified sensitivity in dBm (logarithmic unit of power), and the power budget specifies the maximum loss allowed to achieve the desired bit error ratio (BER) at the data rate. If we assume a 10-Gbps PIN diode receiver sensitivity of -18 dBm and a launch power in the range of a few dBm, then a loss budget of approximately 20 dB is available for fiber, splice, and connector loss. Carriers often budget for a worst-case cabled fiber loss of approximately 0.25 dB/km at 1550 nm.[1] As an example, allowing 2 dB for

---

[1] The loss of good-quality modern fiber on spool is typically in the range of 0.19 dB/km at 1550 nm. The loss may change somewhat in cable, depending on cable design and quality or upon installation if the cable is left in a condition of high mechanical stress.

splices and connectors results in a calculated span length of 72 km. For many years, almost a third of the optical transmission window between 1260 and 1625 nm was unusable because of the presence of a large "water peak" at 1383 nm with loss as high as 2 dB/km. Today, zero water peak (ZWP) fibers, described in detail later, open previously unusable spectrum for low-cost coarse wavelength division multiplexing (CWDM).

The more interesting case is that of multiple spans, in which noise accumulates in an amplifier chain. An expression that approximates the OSNR for a link comprising N_{span} spans of loss L_{span} in which each amplifier has noise figure, NF, is given in [1] as

$$OSNR = 58 + P_{ch} - NF - L_{span} - 10 \log 10(N_{span}), \qquad (5.1)$$

where all units are in dB or dBm as appropriate and P_{ch} is the span launch power. The required OSNR scales inversely as the data rate increases, increasing by 6 dB from 10 to 40 Gbps. This is true because for a fixed received power, the number of photons per bit decreases by the same factor that the bit rate increases. All else being equal, a requirement of 6 dB higher OSNR would decrease the reach of a 40-Gbps system by a factor of four relative to a 10-Gbps link. Equation (5.1) shows that the most effective method for increasing OSNR in an amplified span is to reduce the loss by spacing amplifiers more closely, because the OSNR improves linearly with span loss but only rises logarithmically with the number of amplifiers. However, this option is available only to the submarine system designer, in which repeater spacing is flexible. Distributed Raman amplification is a key-enabling technology for 40-Gb transmission, improving OSNR by about 3 dB, as well as forward error correction and advanced modulation formats. Transmission fiber designs that improve the efficiency and cost-effectiveness of Raman amplification are described in the following subsections.

5.2.2 Limitations from Intersymbol Interference

ISI occurs when pulses broaden out of their assigned bit slots because of some form of dispersion in system components. Chromatic dispersion (see Chapter 2) arises from the wavelength dependence of the propagation constant in the fiber and is proportional to fiber length and laser line width. The requirements on chromatic dispersion become very severe for higher bit rates, increasing as the square of bit rate. PMD arises from the small usually randomized birefringence of the fiber and grows with the square root of fiber length. Systems operating at 2.5 Gbps or less permit the use of low-cost directly modulated lasers with wider line widths arising from chirp. Systems running standard NRZ modulation at 10 Gbps over more than 10–20 km of G.652 standard single-mode fiber (SSMF) rely on externally modulated, CW sources with very narrow intrinsic line widths.

As in the case of OSNR, we may distinguish two cases of dispersion limitation: one applicable to low-cost systems, in which it is desirable to avoid in-line dispersion compensation altogether, and the other relevant to long-reach systems based on concatenated amplified spans.

The simplest ISI limit occurs for uncompensated transmission, in which the dispersion tolerance of a "receiver"—often defined as that dispersion in ps/nm resulting in a 1-dB power penalty—limits the reach of link. For SSMF, with a 1550-nm dispersion of approximately 17 ps/nm-km, that limit is usually reached at 60–80 km. At the time of this writing, efforts are underway to write standards for electronically *equalized* receivers [2] that extend uncompensated reach with G.652 SSMF from 80 to 120 km. However, optical fiber designs with one-fourth the dispersion of SSMF can increase the reach of uncompensated links by a factor of four.

Maintaining tight dispersion tolerances over many channels in an amplified DWDM system comprising many spans once posed a significant challenge for transmission line design. This obstacle was overcome by the milestone development of the slope-matched DCM, which is capable of tightly compensating dispersion across the entire C- or L-band in one broadband device [3, 4]. For the purpose of this chapter, it is critical to note that transmission fibers can be designed so they co-optimize the design of compensating fiber in the DCM for optimum broadband compensation.

PMD is a special challenge because it is an inherently statistical quantity that depends sensitively on the mechanical state of the fiber. The single number known as the PMD coefficient used to characterize an optical fiber is intended to describe the (normally) maxwellian distribution of differential group delay (DGD) values that a fiber can assume. A single fiber displays its full range of DGD values as its mechanical stress state changes through all possible configurations, usually varying with temperature cycles or mechanical vibrations (e.g., for a cable laid alongside a railroad track). The link design value (LDV), defined in Chapter 2, is used to identify the worst-case PMD that a link comprising many cabled fiber segments will experience. As shown later in this chapter, there is a small but nonzero probability that DGD in a system will fluctuate through very high values that can halt transmission for short periods each year.

5.2.3 Limitations from Nonlinearity

Nonlinear impairments can broaden the frequency content and temporal evolution of a single pulse, cause crosstalk between pulses in adjacent channels, and cause two adjacent channels to produce noise in a third channel. Nonlinear impairments are most pronounced in very long systems in which small nonlinear products are able to accumulate. Cross-phase modulation (XPM), as an *inter*channel impairment, is usually the dominant nonlinearity in 10-Gbps DWDM systems. Four-wave mixing (FWM) can be essentially eliminated as an impairment between channels in 10-Gbps systems as long as the fiber dispersion across the band is greater than approximately 2 ps/nm-km [5]. Self-phase modulation (SPM), XPM, or FWM between temporally adjacent bits at the same wavelength (i.e., *intra*-channel nonlinearities) tend to dominate in 40-Gbps systems. The OSNR improves linearly (on a decibel scale) as transmitted power increases; however, the penalty due to nonlinear impairments increases with transmitted power. An optical transmission system is often designed to operate in the sweet spot that maximizes OSNR but avoids significant nonlinear distortions. It will be shown that the same transmission fiber design strategy that allows co-optimization of broadband dispersion compensation also reduces system nonlinear penalties.

5.2.4 Limitations from Amplifier Technology

The advent of the erbium-doped fiber amplifier (EDFA) [6] in the early 1990s revolutionized the economics of optical transmission in several ways. Not only did it make expensive optical-to-electronic-to-optical (OEO) conversion, also known as *regeneration*, necessary only at the endpoint of a link, but it also favored adding as many WDM wavelengths as practical within the amplifier bandwidth. The EDFA is most efficient in the C-band but has been successfully extended as a practical L-band amplifier as well. Although the C- and L-bands together provide tremendous capacity with DWDM technology at 40 Gbps, it is prudent to consider future expansion needs because optical fiber cable is rated for a lifetime more than 20 years.

Raman amplification not only is an enabling technology for 40 Gbps but can also provide gain where other technologies are not available. The gain spectrum for stimulated Raman scattering is based on the phonon (vibrational) spectrum of the silica glass, not on the electronic spectrum of dopants in the glass. Essentially the peak of the Raman gain curve will lie approximately 13 THz lower than the pump frequency, which is approximately 100 nm to the red of the pump wavelength in the telecommunications bands. This means, for example, that Raman pumps can be placed at 1410 nm to provide gain for S-band channels near 1505 nm [7].

5.2.5 Can Fiber Design Be Used to Optimize a Transmission System?

The question addressed in this chapter is How can optical fiber design facilitate the mitigation of these impairments to enable higher capacity, more cost-effective transmission? To answer that question, one must address three issues:

1. The impact of fiber of properties on transmission impairments
2. The fiber properties that can be manipulated
3. The design tradeoffs that must be considered

The first question regarding how fiber properties affect transmission impairments is more complex than commonly supposed. The cabled fiber is often referred to as the "transmission fiber" (Tx fiber) to distinguish it from the dispersion compensation fiber (DCF). The Tx fiber is only one part of a transmission line, but the Tx fiber properties critically affect the design and performance of the other components of the transmission line. It is the system OSNR, PMD, and nonlinearity that must be optimized, not the properties of the individual components. Figure 5.1 illustrates the basic elements of the optical transmission line. The loss of the fiber and connectors must be offset by either EDFA, Raman amplification, or a combination of both. For distances greater than approximately 80 km, the chromatic dispersion of the fiber must be compensated to eliminate ISI. For older fibers, or even modern fibers manufactured without "spinning," some form of PMD compensation may be required. The key point here is that the optical properties of the Tx fiber *directly* impact the requirements for efficiency, linearity, loss, and PMD of the other elements of the transmission line.

The second issue concerns which fiber properties can be effectively manipulated by the fiber design. The primary fiber property available for modification is the dispersion curve, which describes how an optical pulse of a given wavelength and spectral width will broaden in time as it propagates down the fiber. Altering the dispersion curve necessarily involves altering the effective area (A_{eff}) of the fiber as well, which also affects fiber cutoff wavelength.

Finally, it is critical to understand that design tradeoffs are inevitable. The Maxwell equations dictate that key transmission properties of an optical fiber cannot be varied independently. In general, adding waveguide dispersion to the

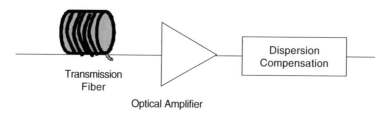

Figure 5.1 The basic elements of an optical transmission line, the designs of which are highly interdependent.

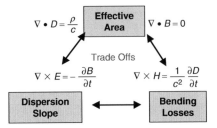

Figure 5.2 The Maxwell equations constrain the design of nonzero dispersion-shifted optical fiber so that tradeoffs must be made. It is not possible to independently choose the dispersion slope, bending losses, and effective area of a fiber, unless the fiber cutoff is allowed to rise above the operating wavelength. Bending losses are fixed by restrictions on allowed cable and handling losses. It is generally not possible to retain a desirable low slope while retaining a large effective area.

fiber to modify the dispersion curve will reduce the mode field diameter (MFD) and A_{eff}. System design issues dictate that the ideal fiber will have a lower dispersion at 1550 nm than standard fiber (see Chapter 2), as well as a lower dispersion slope (see later discussion). The cartoon in Fig. 5.2 illustrates the fundamental problem. If we assume that the cabled cutoff must be maintained below 1260 nm to permit 1310-nm applications, then manipulation of the dispersion curve to reduce the dispersion and dispersion slope must be balanced against the need to maintain a reasonable A_{eff} while keeping bending losses low. A_{eff} must be kept moderate to high to limit nonlinearities. Low macrobending and microbending losses are critical for good cable performance in the field. Once we choose any two of the attributes in Fig. 5.2, the third is chosen for us by the Maxwell equations. Intelligent decisions on these tradeoffs are the key factor in fiber design and require close coupling of expertise in fiber design and processing with understanding of optical transmission systems.

5.3 OVERVIEW OF ITU STANDARDS FIBER CATEGORIES

The ITU is an agency of the United Nations and has, over the last 20 years or so, categorized single-mode fiber to assist suppliers and their customers in providing optical fiber designed to meet specific telecom applications. The development of ITU recommendations is currently occurring in ITU-T Study Group 15 (ITU-T is the telecommunications standards branch of ITU). It is to be noted that many fiber properties are actually specified in cable, because that is generally the way the fiber is used, and many fiber properties can change once the fiber has been cabled. Other statistically based properties are specified on a link basis. It should be emphasized that ITU standards are developed in a multiparty process that sometimes favors weakening requirements in order to gain broad consensus. For example, it will noted that ITU requirements on PMD are relatively loose compared to the system requirements to be described in Section IV, Table 5.1. Although ITU standards play a vital role in educating and informing the broader fiber optic community, they should be viewed as *necessary but not sufficient conditions* for an optical fiber to enable a particular application.

The most fundamental way to group optical fiber designs is by the characteristics of the dispersion curve. Figure 5.3 shows a selection of historically significant dispersion curves labeled by their ITU categories. Characteristics of the dispersion curve that may be significant are the point where the dispersion is zero, called the zero-dispersion wavelength (ZDW) or λ_0, the value of dispersion in the transmission band, and the slope of dispersion across a band. The significance of these quantities is described in detail in Sections IV and V.

The original standard single-mode fiber is specified in ITU recommendation G.652. This fiber has a *MFD* in the range 8.6–9.5 microns, a maximum cable cutoff wavelength of 1260 nm and ZDW in the range 1300–1324 nm. This fiber typically has a chromatic dispersion of 17 ps/nm-km at 1550 nm (which can be excessive for dispersion sensitive applications). Although there are several cat-

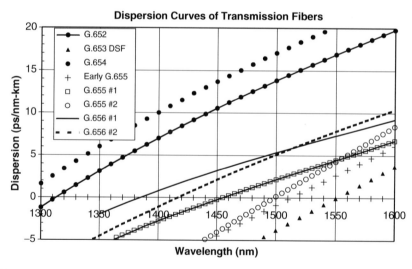

Figure 5.3 Illustration of the International Telecommunication Union (ITU) fiber categories using chromatic dispersion curves representative of various fiber designs. The chromatic dispersion properties of fibers manufactured according to a particular fiber design will vary within stated tolerances determined by the design sensitivity and manufacturing process limits.

egories within G.652, the most modern is G.652D fiber that has low attenuation in cable (maximum of 0.3 dB/km at 1550 nm), good PMD performance (better than $0.2\,ps/km^{1/2}$ link design value), and low water peak (LWP) attenuation at 1383 nm. The LWP designation requires that loss at 1383 nm be less than the maximum loss at other wavelengths from 1310 to 1625 nm. Based on 1310-nm specifications, LWP generally means 1383 nm loss $<0.35\,dB/km$. However, some manufacturers have virtually eliminated the water peak by careful processing [8]. Because the E-band is available on account of the LWP loss, the G.652D fiber is optimized for full spectrum use at data rates up to 10 Gbps (STM-64). It is important to note that some manufacturers routinely offer G.652 fibers with a PMD specification three to four times tighter than required by the standard. Low PMD is critical for future upgrades to a fiber network.

The dispersion-shifted fiber (DSF) was developed in the late 1980s to support transmission in the low-loss 1550-nm window. These systems transmitted a single channel in the vicinity of the loss minimum at 1550 nm where the ZDW was located. This allowed increasing the loss limited transmission distance. DSF was standardized in G.653 with a ZDW in the range 1500–1600 nm. Today, this fiber has limited application because of nonlinearities that occur between WDM optical channels when they are close to the ZDW. The embedded base of DSF in some countries has been upgraded by moving to L-band systems for DWDM.

Cutoff-shifted fiber was standardized in G.654 to provide lower loss and allow higher optical power for transmission over long distances. This fiber has a larger *MFD* than G.652 (some categories go as high as 13 micron), a cutoff wavelength as high as 1530 nm, with low attenuation limits (0.22 dB/km at 1550 nm) and a tight PMD specification (as low as $0.2\,ps/km^{1/2}$). The chromatic dispersion is specified at 1550 nm and is similar in size to that of G.652 fiber. This fiber has been applied in submarine systems [9], in combination with a cabled inverse dispersion fiber [10], as well as long unamplified links.

As high bit rate transmission systems were designed using wideband EDFAs and the low fiber loss in the C-band, it became clear that the high chromatic dispersion of G.652 fiber at 1550 nm would limit transmission capacity because of dispersion-related signal impairment, and that the zero dispersion of DSF near 1550 nm would result in signal impairment related to nonlinear propagation effects. A better solution was found by shifting the ZDW away from the C-band.

A balance of fiber properties for long-haul applications is found in the NZDFs that are standardized in G.655. These fibers support transmission rates of 40 Gbps (STM-256) over long distances. The key features are low (but nonzero) chromatic dispersion in the C-band and low PMD ($<0.2\,ps/km^{1/2}$). Cable cutoff is held below 1450 nm. A classification is now being developed in ITU that will group G.655 fibers according to "low" or "medium" dispersion. In addition, limitations to the dispersion over the entire 1460- to 1625-nm range will be specified. The G.655 upper limit on PMD should be contrasted with the tighter and more realistic requirements discussed in Section IV, because it could be argued that PMD is the most critical specification for upgradeability to 40-Gbps operation.[2]

A final fiber category is the CWDM/DWDM optimized fibers outlined in G.656. The original purpose of this category was to have low dispersion from 1460 to 1625 nm to decrease ISI that limits uncompensated CWDM transmission. However, the requirements evolved substantially during consideration and debate. In the final analysis, G.656 can be considered to be the wideband Raman-enabled fiber standard. These fibers have specified performance over the 1460- to 1625-nm wavelength range (similar to the newer G.655 table). In fact, fibers in the proposed medium dispersion category of G.655 also fulfill the requirements for G.656. The key feature of the G.656 fiber is that there is a minimum chromatic dispersion of 2 ps/nm-km at 1460 nm, which enables good performance (free from nonlinearities) for signal channels, as well as Raman pumps at short wavelengths. Some G.655 fibers are lacking in this regard. Fibers according to G.656 fibers allow highest performance with optical channels spaced over a wide band at 40 Gbps and over long distances.

5.4 OPTICAL FIBERS FOR REDUCED ATTENUATION

For systems in which span loss limits performance in unamplified systems, the fiber loss (in cable) is the key enabling performance parameter. Two categories of fiber for such applications have been commercially important: ZWP peak and pure silica core fibers. In contrast to fibers with novel dispersion properties controlled by a special profile design, loss reduction is primarily accomplished by improving the materials chemistry and physics of the silica optical materials and carefully controlling fiber processing.

Since the beginning of silica optical fiber development, the (previously) ubiquitous loss peak centered around 1383 nm was recognized as the vibrational second-overtone absorption of the hydroxyl group OH. This OH or "water" peak causes increased optical loss from about 1360 to 1460 nm, a wavelength region now designated as the E-band by standards bodies. Because of the high OH loss in typical commercial single-mode fibers, most telecommunication systems until recently avoided the E-band and instead used the two windows at either side of the OH peak, namely, the O- and C-bands centered at 1310 and 1550 nm, respectively. Although the benefits of completely eliminating the OH

---

[2] An NZDF fiber should have PMD LDV less than or equal to 0.04 ps/km$^{1/2}$ to be considered generally ready for 40-Gbps transmission.

loss in commercial fibers had been obvious for more than 2 decades, the enormous technical and commercial challenges of realizing a viable zero-OH manufacturing fostered a belief in its impracticality. Today ZWP performance ($L_{1383nm} < 0.31$ dB/km) can be guaranteed at essentially no additional cost to a carrier.

5.4.1 Pure Silica Core Fiber

The design of pure silica core fiber seeks to minimize Rayleigh scattering losses due to nanoscopic fluctuations in the index of refraction (determined by either the glass density or the chemical composition). Average loss of 0.168 dB/km is commercially obtained at 1550 nm, while the loss minimum near 1580 nm is a few thousandths lower still. This is achieved by eliminating germania as the dopant to create the waveguide. Instead, the cladding is heavily fluorinated to reduce its index of refraction, creating a profile very similar to the matched or depressed cladding profiles shown in Fig. 2.5 of Chapter 2, with the exception that all index of refraction values are shifted down by approximately 0.35%.

However, the properties of silica core fiber—and the index profile itself—are extremely sensitive to the drawing conditions. Because the high-temperature viscosity of pure silica is much higher than that of doped silica (whether doped with Ge or F), most of the tension during draw is carried by the core. In the extreme case, the stresses and strains that remain in the final drawn fiber can alter the refractive index profile to the degree that the core index is reduced close to that of the F-doped cladding [11]. Furthermore, the high viscosity of pure silica leads to a relatively more rapid quenching of the glass nanostructure as the fiber rapidly cools. This results in less time for annealing of density fluctuations and thus greater Rayleigh scattering. As a result, silica core fiber is typically drawn at speeds of 1–2 m/sec to achieve the loss performance noted earlier.

This slow draw speed, combined with a complex preform fabrication process, makes silica core fiber traditionally very expensive compared to the highest quality G.652D fiber. As a result, the application of silica core fiber has been limited primarily to unamplified (or "unrepeatered") submarine systems of several hundred kilometers in length, where the lower loss is a requirement for system operation.

5.4.2 Zero Water Peak Fiber

A more practical and commercially viable low-loss fiber design has been one that eliminates the water peak loss from standard single-mode fiber, whether of matched or depressed cladding profile design. Such fiber meets and exceeds the

ITU G.652D standard and has become the *de facto* standard for high-quality fiber in many applications spaces. The water peak loss at 1383 nm has historically raised loss across approximately 30% of the available spectrum in the low-loss window between 1260 and 1625 nm. For many years, the water peak specification limit in standard fibers was as high as 2 dB/km for several major fiber suppliers. The elimination of water peak loss opened up spectrum for application of low-cost 16-channel CWDM, as well as removed a major limitation to the efficiency of Raman amplifiers. It was in fact the realization of such zero water peak (ZWP) fiber or, in less demanding applications, a LWP fiber, that provided the key enabling component and the impetus for the serious work on CWDM components. In the medium- and short-distance local and access networks, low-cost CWDM with ZWP/LWP single-mode fiber is ideally suited to aggregate traffic from the rapidly growing number of FTTx broadband users and deliver it to the core network in a cost-effective way.

It should be noted that although the focus of ZWP/LWP fiber development and applications has been primarily for G.652 fibers with a zero dispersion near 1310 nm, aspects of the zero-OH technology described here can also benefit fibers of other designs and for different applications. For example, LWP NZDFs have increased bandwidth for DWDM applications and improved Raman pumping efficiency at 1450 nm because of the lower fiber loss at that wavelength. Water peak reduction is now a common feature of many fibers, including G.655 and G.656 NZDF fibers. This section focuses on the requirements necessary to manufacture and specify ZWP G.652 fibers of the basic profile shapes described in Chapter 2, whether of the matched or depressed clad profile design.

While it has always been clear that to achieve LWP or ZWP fibers, one has to keep the OH contamination in silica to an extremely low level (i.e., to <0.1 ppb of OH in the core of ZWP fiber for <0.005 dB/km of added loss at 1383 nm, for instance), the biggest challenge was to devise a practical, high-yield, and low-cost manufacturing process. To put this challenge in perspective, commercial single-mode fibers are still sold with a 1383-nm loss as high as 2 dB/km; this value is equivalent to about 40 ppb of OH in the fiber core. An optical preform comprises a core made from high-quality deposited glass, lower purity overcladding glass, and an interface between them. The water peak loss at 1383 nm can be broken down into components as

$$L_{total} = L_{Rayleigh} + L_{deposit} + L_{tube} + L_{interface} + L_{cladding} + L_{aging}, \qquad (5.2)$$

where $L_{Rayleigh}$ represents the background Rayleigh scattering level at 1383 nm, which is independent of OH contamination. This value is approximately 0.26 dB/km. $L_{deposit}$ represents the contribution from light propagating in the deposited core material, which is the highest purity glass, where the optical power is maximum. This glass may be formed with the VAD, MCVD, OVD, or PCVD methods (see Chapter 3), all of which have shown the capacity to support the LWP performance level (typically $L_{total} < 0.35$ dB/km for G.652C/D fibers or $L_{total} < 0.38$ dB/km for G.655 or G.656 fibers).

However, the VAD core method has proven most adept at making ZWP fiber in large preforms. The VAD or OVD methods allow for an explicit dehydration step during glass deposition where silica in a porous soot form can be exposed to Cl_2 at temperatures above 800 °C to eliminate OH before sintering. VAD cores are formed as solid rods, while OVD soot is deposited onto a mandrel and requires an additional step, to collapse the glass annulus after sintering, during which ultimate dryness must be maintained to avoid contaminating the center-line of the preform with OH. The PCVD and MCVD processes deposit glass inside a substrate tube, whose additional contribution to 1383-nm loss is represented above by L_{tube}, referring to Fig. 5.4. Even a high-quality substrate tube can have OH concentration as high as 200 ppb, although higher purity is available at a cost premium. Substrate tube glass is approximately 9–10 microns from the fiber centerline for MCVD or PCVD fiber drawn from large-diameter preforms, so the optical power is lower. Nevertheless, L_{tube} can be on the order of 0.05 dB/km. Like OVD, PCVD and MCVD also form an annular glass body requiring a very high temperature collapse process during which the centerline of the preform must be protected from OH contamination. Neither PCVD nor MCVD allows for a separate dehydration step. The PCVD process is noted

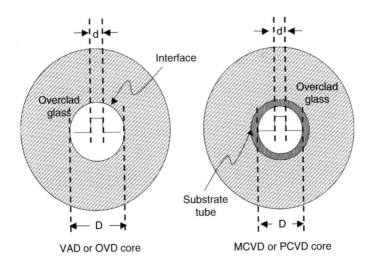

Figure 5.4 Cross-section, not to scale, of an optical fiber preform. Left figure is appropriate for a monolithic core rod made with OVD or VAD. A typical *D/d* ratio would be in the range of 3–5. Right figure shows the case in which core material is deposited inside a substrate tube in the MCVD or PCVD processes. In this case, *D/d* may be in the range of 2.5 or less.

for its ability to incorporate a wide range of feedstock materials, including any impurities.

The contribution to OH loss from the interface between the higher purity glass core rod (regardless of core fabrication technology) and the generally less expensive cladding glass is represented by $L_{interface}$. For rod-in-tube technology, this interface is formed by collapsing one glass surface onto another at high temperature. The opportunity is thus present to contaminate the surfaces and diffuse moisture into the core. In the case of soot overcladding, this interface is the result of depositing high-density soot on glass, yielding an interface that is more difficult to dehydrate than bulk soot. In the case of plasma overcladding, molten quartz particles are sprayed onto a glass core. $L_{cladding}$ is generally negligible because the total power in the cladding is relatively small, on the order of 0.01–0.1% depending on fiber design and D/d.

It is important to note that it is always possible to minimize $L_{interface}$ and $L_{cladding}$ by raising the ratio D/d, shown in Fig. 5.4, where D is the diameter of the core body (before overcladding) and d is the diameter of the GeO_2-doped region of the waveguide. A large D/d ratio means a higher percentage of the total glass is formed in the high-purity core deposition step, pushing the interface and the cladding glass farther away from the fiber centerline where the optical power is maximum. Applying this tactic to make ZWP or LWP fibers generally requires making a smaller less economical preform. This is necessarily true since practical issues of fluid flow, heat transfer, or even legacy machine design limit the total amount of core deposit possible. Clearly smaller preforms or large D/d cores are only cost effective for more expensive specialty fiber products.

However, eliminating OH contamination is only part of the challenge. As a result of the manufacture and study of ZWP fiber, an important new hydrogen aging loss mechanism was discovered [8] where the OH peak in ZWP/LWP fibers can grow by as much as a few tenths of dB/km or more when certain reactive atomic defects were present in the fibers to rapidly react with a trace amount of molecular hydrogen at ambient temperature. Because of the ubiquitous nature of these reactive defects in silica fibers and the inevitable presence of small amounts of molecular hydrogen in fiber installations, a solution to the hydrogen aging loss problem must be found for ZWP/LWP fibers so the loss, particularly at the OH peak, remains permanently low throughout their service lifetime. Thus, the following sections also include a discussion on the solution to the hydrogen aging loss problem and the development of the standards for hydrogen aging test.

Although the elimination or reduction of the OH loss is the key to ZWP/LWP fibers and low-cost CWDM applications, other fiber performance parameters are also important. In particular, to support future migration to 40 Gbps and use of the L-band, high-performance in fiber PMD and macrobending, respectively, are required. But perhaps even more important than technical performance, the cost for manufacturing ZWP/LWP fibers must be low because it is the critical driver for any mass fiber deployment toward the end-users. These different aspects of ZWP/LWP fiber performance, manufacturing process, and cost are discussed in the following subsections.

5.4.2.1 A Cost-Effective Fabrication Technology for ZWP Fiber

ZWP fiber was first developed by Lucent Technologies, Bell Laboratories in 1997 as the commercial AllWave ZWP fiber using the Rod-in-Tube (RIT) process [12]. A method for solving the technical and economic challenges of fabricating ZWP fiber, which has been described in detail in the literature [13], is the Rod-in-Cylinder (RIC) and Overclad-during-Draw (ODD) manufacturing process. This fabrication process increases the preform size to more than 5000 fiber km and achieves low cost and ZWP fiber performance exceeding ITU G.652D standards.

Essentially, the RIC-ODD process [14–16] entails a totally mechanical RIC assembly and drawing this large assembly directly into fiber with the ODD process. The RIC assembly is constructed by inserting a VAD core rod and an optional thin-walled first overclad tube into a large, hollow overclad cylinder of up to 170 mm outside diameter (OD), 60-mm inside diameter (ID) and 3 m in length to form a preform capable of yielding more than 5000 km of fiber. The bottom end of the cylinder is machined into a conical taper and a hole is drilled through its walls so that a quartz plug-and-pin assembly can be inserted. The plug-and-pin is used to hold up the VAD core rod and the first overclad tube inside the cylinder. The conical taper of cylinder bottom facilitates the initial seal and glass drop during the RIC-ODD draw process. The cylinder also has a quartz handle attached to the top end for handling by robotic manipulator. When this completely mechanical RIC assembly is lowered into the draw furnace and a vacuum is applied through the hollow handle at the top, the overclad cylinder and the first overclad tube collapse onto the core rod and form a seal at the tapered end of the cylinder. The plug-and-pin and rest of cylinder taper will then be melted and dropped off to begin the high-speed fiber draw as the RIC assembly is further lowered into the furnace and the vacuum-assisted ODD collapse continues.

5.4.2.2 Maintaining ZWP Fiber Performance

To achieve ZWP performance, the VAD core soot body is dehydrated with chlorine prior to consolidation into a core rod. The VAD core rod is then etched with a hydrogen-free (i.e., "dry") plasma torch to remove its surface OH. The first overclad tube and the overclad cylinder, both made by the OVD process with typically 300 ppb OH or less, are also lightly etched by acids to remove surface contaminants and OH. From FTIR and fiber loss analyses, it has been found that less than 2 ppm OH remained at the interfaces (which had a width of about 1 μm in a 125-μm OD fiber). Such a small amount of interface OH will not adversely affect the fiber 1383-nm loss to a significant degree as long as the interfaces are placed at a sufficiently large distance from the fiber core (e.g., a clad-to-core ratio $D/d > 3$). The median 1383- and 1550-nm losses for fiber produced by the RIC-ODD process achieved world-class levels of 0.276 and 0.187 dB/km, respectively. The essentially zero-OH loss (i.e., no observable OH peak) at 1383 nm in particular far exceeds the requirements for G.652C/D fibers.

It should be noted that a process like RIC-ODD has some inherent advantages for low 1383-nm loss: (1) The VAD core bodies have no open holes or center lines like those appearing in the MCVD, PCVD, and OVD processes so it is relatively easy to keep OH completely out of core bodies with a separate dehydration and consolidation process, (2) the RIT or RIC overclad material, made by a separate OVD process, can also be completely dehydrated, (3) the surface OH on the core rods and overclad tubes or cylinders can be easily removed by plasma or acid etch, and (4) it is fairly easy to keep the interfaces in RIT or RIC dry with a vacuum ODD process or with a more aggressive dehydration or etching procedure utilizing Cl- or F-containing gases during overclad. In contrast, the complete dehydration of soot-on-glass interfaces for the alternative OVD and VAD overclad processes can be difficult, especially for large preforms. For MCVD and PCVD processes where the glass layers are typically deposited without forming soot layers first, a complete dehydration of consolidated glass layers can also be difficult. However, even in these cases, it is possible to greatly reduce the 1383-nm OH loss peak to the LWP level through vigilant practice of eliminating any leakage and avoiding any moisture contamination throughout the chemical delivery systems.

For ultralow fiber PMD, the patented draw process [17] must be implemented to impart frozen-in spins in the core of the fiber. The result is a typical low-mode-coupled PMD $\leq 0.02\,\mathrm{ps}/\sqrt{\mathrm{km}}$ for RIC-ODD fiber, which again far surpasses the G652C/D requirements. Cylinder surface preparation and cleaning procedures result in no interface problems such as airlines or bubbles and achieve a fiber break rate of less than 5/fMm at 100 kpsi (0.7 GPa) proof test. The RIC-ODD fiber performance in core eccentricity, clad non-circularity, and fiber curl is excellent, supporting excellent typical splice loss of less than 0.02 dB. The G.652D depressed cladding fiber design described in Chapter 3 can also be implemented with a similar large preform, ZWP, low PMD process to yield a bend-insensitive fiber well suited to FTTH access networks.

5.4.2.3 Hydrogen Aging Losses

The optical loss in fibers can degrade with time due to the chemical reaction between the inevitable atomic defects in fibers and the trace amount of molecular hydrogen normally present in or around optical cables. This hydrogen aging loss is a particularly important problem for the ZWP/LWP fibers because all known hydrogen reactions with silica fibers will result in at least the loss increase at the 1383-nm OH peak (other additional hydrogen aging loss components are possible, depending on the defects and reaction types; see later discussion). During the development of ZWP fiber, it was discovered that a very common but previously unknown silica defect is extremely reactive and can cause significant hydrogen aging loss at the OH peak upon brief exposure to trace amount of hydrogen, even at room temperature. We discuss this and two other types of hydrogen aging losses relevant to ZWP/LWP fibers, as well as the countermeasures against them.

Basically, there are three types of hydrogen aging losses that must be avoided in Ge-doped silica fibers to ensure reliability in optical transmission over the service lifetime of 25 years or more. These hydrogen aging losses are caused by different types of atomic defects or impurities present in the silica fibers. The severity of hydrogen aging loss degradation is entirely dependent on the fiber manufacturing process and the purity of the silica material used.

The first two types of hydrogen aging losses [8] involve two species of extremely reactive silica defects: (1) a pair of nonbridging oxygen hole centers (NBOHCs, $\equiv$Si-O• •O-Si$\equiv$) and (2) the peroxy radical plus Si E' center ($\equiv$Si-O-O• •Si$\equiv$), where • denotes an unpaired electron at the broken chemical bond. These two types of silica defects (involving Si and O atoms only and no Ge) even at room temperature can react almost instantaneously with trace amounts of hydrogen and cause significant loss increases of up to a few tenths of dB/km or more. The two hydrogen reaction mechanisms can be described as follows:

$$\equiv\text{Si-O• •O-Si}\equiv + \text{H}_2 \rightarrow \equiv\text{Si-O-H} + \text{H-O-Si}\equiv$$

$$\textbf{NBOHCs} \qquad\qquad\qquad \textbf{1383 nm} \qquad\qquad\qquad\qquad (5.3)$$

$$\equiv\text{Si-O-O• •Si}\equiv + \text{H}_2 \rightarrow \equiv\text{Si-O-O-H} + \text{H-Si}\equiv \rightarrow \equiv\text{Si-O-O-Si}\equiv + \text{H}_2$$

$$\textbf{peroxy radical} + \textbf{Si E}' \qquad\qquad \textbf{1383} + \textbf{1530 nm} \qquad\qquad\qquad (5.4)$$

Hydrogen reaction (5.3) with NBOHCs results in an OH peak at 1383 nm. Hydrogen reaction (5.4) with peroxy radical and Si E' center results in an OH peak at 1383 nm and a "SiH" peak at 1530 nm, which are metastable and can decay at room temperature, although a significant fraction of the loss increases will remain after a few months. Room-temperature hydrogen tests [18–20] have shown that the two hydrogen reactions above typically reach saturation in less than 4 days in 0.01 atmospheres of hydrogen and this implies that the partial pressure of hydrogen in the cable installation needs to be much less than 4 ppm over the 25-year lifetime to avoid these two hydrogen aging losses: an impractical solution as the measured hydrogen partial pressure in cable installation is on the order of 400 ppm or more. So, to reduce the risk of hydrogen aging loss, it is very important to minimize the above silica defects in the fiber manufacturing process. This can be achieved by adjusting the oxidation and reduction conditions in dehydration and consolidation in the preform making process as well as optimizing the fiber draw. In addition, it is possible to passivate [21] any remaining reactive silica defects in fiber by treating the drawn fiber spools with a small amount of deuterium at ambient temperature [22]. Deuterium treatment can completely eliminate these particularly egregious hydrogen aging losses due to the two reactive silica defects.

The deuterium reactions work in a similar way as the hydrogen reactions (5.3) and (5.4):

$$\equiv\text{Si-O}\bullet \; \bullet\text{O-Si}\equiv + D_2 \rightarrow \equiv\text{Si-O-D} + \text{D-O-Si}\equiv$$

NBOHCs   **1900 nm**   (5.5)

$$\equiv\text{Si-O-O}\bullet \; \bullet\text{Si}\equiv + D_2 \rightarrow \equiv\text{Si-O-O-D} + \text{D-Si}\equiv \rightarrow \equiv\text{Si-O-O-Si}\equiv + D_2$$

peroxy radical + Si E′   **1900 + 2100 nm**   (5.6)

But the OD and SiD absorption losses are now harmless because they occur at much longer wavelengths (>1625 nm) and are completely outside the normal operating wavelength windows. Furthermore, the reactive silica defects after being passivated by the deuterium reaction are no longer capable of causing additional hydrogen aging loss in the field.

Though less reactive, the third type of hydrogen aging loss that is of concern is when there is alkali (Na, Li, K, etc.) contamination in the Ge-doped silica fiber [23, 24]. Alkali contamination can be as low as a fraction of parts per million

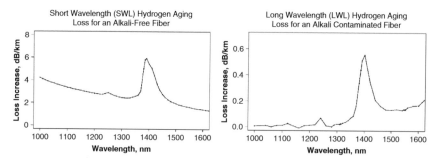

Figure 5.5 An example of short wavelength (SWL) hydrogen aging loss (left) and long wavelength (LWL) hydrogen aging loss (right).

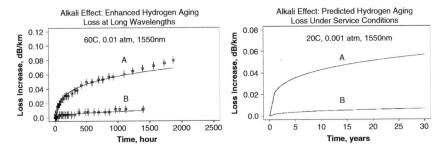

Figure 5.6 Left: The 1550-nm hydrogen aging loss versus time is much higher for Fiber A with Na + Li = 1.35 ppma than that of Fiber B with Na + Li < 0.3 ppma in a 60 °C, 0.01 atm H_2 test. The solid curves are from model calculations. Right: Model predicted hydrogen aging losses at 20 °C, 0.001 atm H_2 service conditions again show the alkali contaminated Fiber A has a significant risk of loss degradation over its lifetime.

atomic (ppma) and still results in significant hydrogen aging loss over time. This is because the activation energy for hydrogen reaction is greatly reduced when the normal high-activation energy Ge defects (which are inevitably generated in the fiber manufacturing process) interact with alkali impurities. The hydrogen reaction can be described as follows:

$$Na^+ \qquad\qquad Na \qquad Na$$
$$\equiv Si\text{-}O\bullet\ \bullet O\text{-}Ge\equiv + H_2 \rightarrow Si\text{-}O\text{-}H + H\text{-}O\text{-}Ge$$
$$Na^+ \qquad\qquad \textbf{1383 nm} + \textbf{Long } \lambda \textbf{ loss} \qquad (5.7)$$

Alkali contamination can arise from the use of natural quartz material, insufficient purification, or contamination in preform processing. When there is alkali contamination in the hydrogen aging, loss has an OH peak as well as a "Long Wavelength Loss" that increases with wavelength beyond 1360 nm (Fig. 5.5). We have developed an accurate quantitative model based on our extensive hydrogen studies and it predicts a hydrogen aging loss in the range of 0.02–0.04 dB/km from 1360 to 1625 nm (which includes the OH peak) for a Ge-doped fiber with 1 ppma alkali contamination after 25 years under typical cable operating conditions of 20 °C and 400 ppm H_2 (Fig. 5.6).

5.5 OPTICAL FIBER DESIGN PRINCIPLES FOR WIDEBAND AND HIGH BIT RATE TRANSMISSION

The design of transmission fibers to optimize performance of high-capacity, long-reach DWDM systems requires careful consideration of how fiber design affects the balance between OSNR, dispersion compensation, and management of nonlinearities across the entire system. Three critical aspects need to be considered. First, precise dispersion compensation across the amplifier band-width is crucial to permit cost-effective 40-Gbps transmission. This requires excellent chromatic dispersion compensation and minimizing PMD. Second, the contributions to nonlinearity from each element of the transmission line must be considered, and the nonlinearity of the composite transmission line should be minimized. Third, Raman amplification is a key enabling technology for improving OSNR for high-bit-rate ultralong haul transmission.

5.5.1 Precise Dispersion Compensation

Pulse spreading due to the nonzero spectral width of the laser and dispersion of the Tx fiber must be compensated in systems longer than a few spans of SSMF (dispersion $\sim 17 \, \text{ps/nm-km}$ at 1550 nm). With the NRZ modulation format, the allowable dispersion (in ps/nm) corresponding to a 1-dB power penalty is approximately given by $100 \, \text{Gbps-nm}^{-1}/B^2$, where B is the bit rate (in Gbps) [5]. The squared dependence on bit rate is due to the fact that the transmitter line width is approximately equal to the modulation bandwidth, while the temporal width of the bit is inversely proportional to B. Thus, a DWDM system running at 40 Gbps over the full C-band is limited to approximately 60 ps/nm of accu-mulated residual dispersion across all wavelengths from 1530 to 1565 nm, com-pared to a limit of approximately 1000 ps/nm for a 10-Gbps system. The requirement of 60 ps/nm over 1000 km means that the transmission fiber and the DCF must be sufficiently matched to give a net 0.06 ps/nm-km residual dispersion when paired together. Failure to meet this limit would necessitate the use of expensive per-channel compensation at the terminal. This section shows that the ability to design a wideband DCM to meet stricter limits on residual dispersion is closely related to the design of the Tx fiber.

5.5.2 Dispersion Compensation Fiber Technology

The development and commercialization of DCF in robust packaging has been one of the two most critical enabling technologies for the deployment of cost-effective DWDM transmission over amplified spans. DCMs are passive fiber-based devices located at the amplifier, and often at terminals as well, that are capable of compensating the dispersion of the transmission fiber span over a range of wavelengths [25]. In addition to compensating dispersion itself, the periodic use of per-span, in-line dispersion compensation at amplifier sites is a key element in dispersion management techniques to control nonlinearities such as cross-phase modulation and other nonlinearities at 10 Gbps and beyond. Key work on the design and fabrication of DCF with negative dispersion and negative dispersion slope began in the early 1990s [26–29]. Successful large-scale manufacturing was in place by 1998 [30].

Transmission fibers used in terrestrial applications have positive dispersion and positive dispersion slopes, as shown in Fig. 5.3. DCFs have high negative dispersions in the range of -100 to -250 ps/nm-km, as well as negative slopes. To achieve the negative dispersion and dispersion slope, a large amount of waveguide dispersion must be added by strongly confining the mode in a narrow and high Δ Ge-doped core, surrounded by a low Δ deeply F-doped trench, with an additional ring of positive Δ Ge-doped silica surrounding that. DCF index profiles may include central cores with Δ as high as 2%, compared to 0.3–0.6% for transmission fibers. As the wavelength grows longer, the optical mode progressively spreads out of the core, having more power in the trench and ring. The effective index is, thus, forced to change rapidly with wavelength, and both the dispersion and the slope can be designed to be negative [31]. One consequence of building-in the large waveguide dispersion is a necessarily small A_{eff} in the range of 15–21 μm^2. Because of the high Δ, which results in a softer glass at processing temperatures, the core ovality of a DCF is generally more difficult to control than that of a Tx fiber, and the stress asymmetry resulting from core ovality is also higher, resulting in somewhat higher PMD values. However, good process control and the use of spinning [17] allows modern DCFs to have PMD values of 0.1 ps/rt(km) or lower [25]. The large Ge doping in the core increases Rayleigh scattering and the large change in material properties at core–trench interface elevates the loss of DCF to a typical range of approximately 0.5 dB/km. Typical slope-compensating module insertion losses and PMD are noted in Section 5.6.3.

Because of the higher loss and PMD, as well as the desire to minimize the size of the fiber bobbin, it is desirable to maximize the magnitude of the negative dispersion as practical of the fiber to minimize the length of DCM. As a consequence, the dispersion divided by the fiber loss is often used as a figure of merit for DCM. Well-designed DCF should have a figure of merit larger than 200.

5.5.3 Full-Band Dispersion Compensation

To guarantee that a DCM precisely compensates the dispersion across the entire C- or L-band, it is necessary for the relative dispersion slope (*RDS*) of the Tx fiber and the DCF to be equal. Let us approximate the dispersion of an optical fiber as a linear function over a wavelength band,

$$D(\lambda) = D(\lambda_c) + (\lambda - \lambda_c)D'(\lambda_c), \qquad (5.8)$$

where the center of the band is λ_c and $D' = dD/d\lambda$ is the dispersion slope. Then

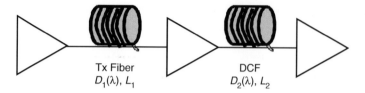

Figure 5.7 Typical compensated span in an amplified system for a long-haul DWDM link with erbium-doped fiber amplifier (EDFA) technology, in which the dispersion compensation fiber (DCF) in module is placed in a two-stage EDFA.

$$RDS = \frac{D'(\lambda_c)}{D(\lambda_c)}. \tag{5.9}$$

The *RDS* should be considered a fundamental parameter and figure of merit for a telecommunication optical fiber.

With reference to Fig. 5.7, consider that the transmission fiber is characterized by length L_1 and dispersion $D_1(\lambda)$, while the DCF is characterized by length L_2 and dispersion $D_2(\lambda)$. Then the condition for 100% dispersion compensation across the band of interest is that

$$L_1 D_1(\lambda) + L_2 D_2(\lambda) = 0 \tag{5.10}$$

must be satisfied for every value of λ. Substituting Eq. (5.8) into Eq. (5.10) and regrouping terms yields the following expression:

$$[L_1 D_1(\lambda_c) + L_2 D_2(\lambda_c)] = (\lambda_c - \lambda)[L_1 D_1'(\lambda_c) + L_2 D_2'(\lambda_c)]. \tag{5.11}$$

In order for this expression to be true for any λ within the band, the two expressions in brackets must both be identically zero. Equating the two expressions in brackets yields

$$\frac{D_1'}{D_1} = \frac{D_2'}{D_2} \text{ or, using different notation, } RDS_1 = RDS_2. \tag{5.12}$$

This is the first requirement for precise wideband dispersion compensation. Matching the *RDS* of the Tx fiber and the DCF to achieve wideband compensation is often referred to in the literature as "slope compensation" or "slope matching." DCMs, which offer 100% slope-matching across a band, are then referred to as "dispersion slope compensating modules" (DSCMs).

5.5.4 Requirement for Low Residual Dispersion

The second requirement for precise wideband dispersion compensation is that the Tx and DCFs have a low *RDS*. The assumption in Eq. (5.9) that the fiber dispersion curve $D(\lambda)$ is linear is more valid for the Tx fiber than for the DCF. In reality, $D_2(\lambda)$ may have significant quadratic and cubic terms in the Taylor expansion. For 40-Gbps transmission line design, it is useful to define the relative dispersion curvature $RDC = D''/D$, which can also be matched between Tx fiber and DCF. The term *residual dispersion* refers to the net dispersion of a compensated span or a concatenated series of compensated spans as illustrated in Fig. 5.7. If the RDS values of the Tx fiber and DCF have been matched, then the residual dispersion will be essentially parabolic across a band. In cases in which the RDC can also been matched, then the residual dispersion takes on a cubic shape [25]. It is this residual dispersion over the target transmission distance that must be compared to the dispersion tolerance of the receiver.

It can be shown [25, 31, 32] that the available bandwidth (over which low residual dispersions are achieved) for a Tx fiber and a slope-matched DCF varies inversely with RDS. Essentially, DCFs with lower residual dispersions tend to have lower curvature $D''(\lambda)$. Empirically, the optical properties of DCF with low RDS values also tend to be more robust to manufacturing variation.

Thus, the RDS of the Tx fiber becomes a critical parameter in determining how well the dispersion of the link can be compensated. From Eq. (5.9), it is obvious that higher dispersion and lower slope facilitate the design of a matching dispersion compensation solution. As a rule of thumb, it is relatively easier to design and manufacture DCMs with tight residual dispersion tolerances for Tx fibers with RDS values well below $0.01\,\text{nm}^{-1}$, while residual dispersions will tend to be high and PMD and insertion loss penalties greater for fibers with RDS values significantly above $0.01\,\text{nm}^{-1}$. Referring to Fig. 5.1, fibers with high dispersion ($\sim 17-20\,\text{ps/nm-km}$) near 1550 nm, as well as those with medium dispersion ($\sim 7-8\,\text{ps/nm-km}$) and lower dispersion slopes, will fall into the former category. A fiber with a lower value of dispersion ($\sim 4-5\,\text{ps/nm-km}$) at 1550 nm will fall into the latter category if its slope is high. A lower dispersion Tx fiber with low slope can still have good dispersion slope matching with a DCF, while maintaining lower module loss and PMD, if the RDS is not higher than $0.01\,\text{nm}^{-1}$.

5.5.5 Factors Affecting Nonlinearity

Nonlinearity can arise in optical fiber transmission from the cabled Tx fiber itself, the amplifier, or in the DCM. The fundamentals of fiber nonlinearity and the manifestations in communication links are well summarized in the literature [5, 33, and annual OFC and ECOC Technical Digests]. For a Tx fiber, the important parameters affecting nonlinearity are the nonlinear index n_2, the effective area A_{eff}, the zero dispersion wavelength λ_0, and the level of local dispersion. The value of n_2 varies linearly with the overlap integral of optical power and the Ge-doping profile in the core of the waveguide [34]. It is intuitively obvious that nonlinear interactions scale inversely with A_{eff}. The nonlinear coefficient $\gamma \propto n_2/A_{eff}$ combines these two parameters in a useful ratio.

It has been well understood since the early 1990s that DWDM transmission requires nonzero dispersion along the signal path to suppress FWM between channels, thus requiring that λ_0 be well away from the transmission band. Moving away from λ_0, FWM products between channels are suppressed approximately as $1/D^2$ [5], requiring about 2 ps/nm-km across the signal bands in case of 10 Gbps. This consideration guided the development of the G.655 NZDFs in Fig. 5.3. XPM between channels is left as the dominant source of nonlinearity for 10-Gbps transmission. The local dispersion of the Tx fiber does play a role in suppressing XPM in conjunction with a properly designed dispersion map. In dispersion management [5], the optimum dispersion map prescribes the levels of pre-, post-, and in-line optical dispersion compensation so that the pulse collisions that generate non-linear interactions are minimized, on average, near the beginning of each span where optical power is maximum and within the DCM. Similarly, the level of local dispersion in the Tx fiber also affects the magnitude of penalties due to SPM, XPM, and FWM between pulses within the same channel that are the dominant source of nonlinear penalties at 40 Gbps. A somewhat higher local dispersion offered by the G.656 fibers can be useful, with the proper dispersion map, to further suppress these effects in the more demanding applications. It can be shown from a simple analysis of dispersion curves shown in Fig. 5.3 that λ_0 is inversely proportional to the fiber's RDS. Thus, a lower RDS value will correlate with a greater bandwidth over which nonlinearity is suppressed.

Another contribution to nonlinearity arises from the DCM, which contains kilometers of fiber with larger values of γ (on the order of $5\,\text{W}^{-1}\,\text{km}^{-1}$) compared to Tx fiber (on the order of $1.5\,\text{W}^{-1}\,\text{km}^{-1}$). The length of DCF necessary to compensate the Tx fiber scales with the length of the span (typically 60–100 km) and the dispersion of the fiber in the operating band. Full-slope compensation of fibers with a high RDS generally also requires longer lengths of DCF in the DCM. Significantly, G.652 fiber has far more dispersion in the C- and L-bands than is necessary or useful for suppression FWM or other nonlinear effects, leading to excessive loss, PMD, and nonlinearity in the DCM. For example, in the case of full C-band compensation of an 80-km span of SSMF (17 ps/nm-km at 1550 nm), approximately 35% of the nonlinear phase shift will occur in the DCM [25]. For the case of a low-dispersion (4.5 ps/nm-km at 1550 nm), low-slope (0.045 ps/nm²-km at 1550 nm) fiber such as TrueWave RS, less than 10% of the nonlinear phase shift for the 80-km span occurs in the DCM for full C-band compensation [25]. A comparison of the impact of fiber dispersion on module loss and PMD is made in Table 5.2.

5.5.6 Impairments Affecting Raman Amplification

The advent of Raman amplification has drawn attention to the importance of FWM effects between the high-power Raman pumps and between the DWDM signals and high-power Raman pumps. Essentially, it is desirable to place λ_0 at a shorter wavelength than any signal or pump wavelength used in the system. For example, studies of Raman amplification led to the observation of FWM products between counter-propagating high-power pumps at 1429, 1447, and 1465 nm and C- and L-band signals when λ_0 of the Tx fiber is very close to 1500 nm. Significant noise peaks due to FWM were observed near 1528, 1548, and 1568 nm [35], causing OSNR degradation. These deleterious features did not appear for a Tx fiber with $\lambda_0 < 1405$ nm to the short wavelength side of the Raman pumps. Again, a low *RDS* value correlates with a greater bandwidth over which FWM impairments are suppressed.

5.5.7 Systems Implications of Tx Fiber PMD

In system work, one needs to account for both the PMD of each component of the transmission line shown in Fig. 5.1. A general rule of thumb is to allow half the system PMD for the fiber and half for the components. Although the components may have a relatively fixed DGD, components in the middle of a system span appear to have a statistically varying DGD because of the fiber that surrounds them. The total pulse spreading resulting from PMD is usually allocated no more than one-tenth of the bit period. Hence, for a 40-Gbps transmission system, the total pulse spread due to PMD could be as high as 2.5 ps. For a system of 1000 km length, the fiber PMD coefficient could be no higher than $0.04\,\text{ps/km}^{1/2}$. A more sophisticated analysis estimates the power penalty due to DGD for a given modulation format at a fixed line rate (power penalty is the increase in signal power required to overcome the impairment). For example, an NRZ system roughly suffers an eye-closure penalty of [36]

$$\text{Penalty (dB)} = 15[\text{DGD}^* \text{bit rate}]^2 \qquad (5.13)$$

so for a 1-dB penalty, $\text{DGD}_{max} = 0.26/(\text{bit rate})$. This $\text{DGD} = \text{DGD}_{max}$ value is taken at the limit at which the system fails. Because of the statistical nature of DGD, the actual instantaneous value of DGD could be far above the PMD value for the same system (since PMD is the average of the DGD). In fact the maxwellian tail of the DGD distribution extends to infinity (though with very small probability density). The system designer uses the DGD_{max} to find the probability of instantaneous DGD causing the system to fail. Knowing an acceptable outage probability, the overall PMD is adjusted downward to guarantee that DGD_{max} is far enough out on the maxwellian tail so that system

Table 5.1

Optical fiber PMD requirements for future telecommunications needs

Speed (Gbps)	Format	DGD_{max} (ps)	System PMD (ps)	Fiber PMD (ps)	System length (km)	Fiber PMD coefficient $(ps/km^{1/2})$
10	NRZ	26	7.5	5.3	100	0.53
10	NRZ	26	7.5	5.3	4000	0.08
40	NRZ	6.5	1.9	1.3	100	0.13
40	NRZ	6.5	1.9	1.3	2000	0.03
40	RZ	9.1	2.6	1.9	2000	0.04
80	RZ	4.6	1.3	0.9	1000	0.03

failure probability is kept within the design limits. It can be shown that the outage probability of 10^{-6} (or 30 seconds per year) requires a PMD value that is less than one-third of DGD_{max}. Some examples of PMD fiber requirements based on this analysis are shown in Table 5.1 [36].

5.5.8 Summary of Design Principles

Based on the previous discussion, a Tx fiber with reduced dispersion (relative to G.652 fiber) and low *RDS* offers system advantages in wideband dispersion compensation, reduced attenuation, PMD, and nonlinear penalties for the DCM module, avoidance of nonlinear impairments in distributed Raman amplification, as well as high efficiency for Raman gain. The Tx fiber PMD must also be controlled to ultralow levels per Table 5.1 to take advantage of a fiber designed to these principles.

5.6 DESIGN OF NONZERO DISPERSION FIBERS

The design process for NZDFs can be summarized by the following simplified steps:

1. Determine an acceptable compromise between the relevant fiber transmission properties that will support the intended applications. This primarily requires determining a mutually attainable set of values for the *MFD*, dispersion and dispersion slope within the transmission band, and macrobending and microbending loss sensitivity. This is normally accomplished using waveguide modeling tools.

2. Determine the radial shape of the refractive index profile that yields the desired transmission properties.

3. Ensure that the index profile can be robustly manufactured with the available preform fabrication process.

Because the physics of optical fiber transmission and the state of the art of glass processing technology will not always allow the simultaneous achievement of the targets and constraints, the process by nature is a compromise.

It would be impractical to implement this process without simulation tools that enable a "virtual" search of the fiber design parameter space for acceptable solutions. Typically, finite element techniques are employed to solve the Helmholtz equation for a given waveguide structure and determine values for the pertinent fiber properties. Optimization techniques can be applied to the design search to locate solutions that best fit the design targets while maintaining the appropriate index profile constraints to ensure that the solutions obtained are realizable. However, the uncertainty introduced by the imperfect nature of the practical physical models of optical fiber behavior will, in general, make the process an iterative process of modeling, prototyping, and performance testing, and subsequently repeating the process loop to "tune" a design to an acceptable optimum. In particular, it is challenging to make highly accurate absolute predictions of bending losses and cutoff from first principles.

5.6.1 Fiber Transmission Parameter Tradeoffs

A complicating factor of the fiber design problem is that there are several coupled problems that must be simultaneously solved when searching for an optimum set of fiber transmission parameters. We would like to design fibers that minimize the impairments resulting from nonlinear propagation effects, but we may simultaneously desire that the fibers be robust to cabling and environmental changes while providing efficient low-noise distributed-Raman gain. This combination of desirable fiber properties puts conflicting demands on the size of the *MFD*. Further, we would like flexibility to engineer the fiber dispersion curve (zero dispersion wavelength and dispersion slope) so that we (1) maintain dispersion within an optimum range across the entire transmission band to help mitigate some nonlinear impairments by braking phase matching conditions, but (2) we also want to have low *RDS*, to match the available dispersion compensating fiber technology so that linear dispersion can be optimally compensated over a broad bandwidth. However, obtaining low-dispersion slope tends to preclude achieving simultaneously both large *MFD* and insensitivity to microbending loss.

In general, to minimize the transmission performance degradation caused by nonlinear propagation effects, large *MFD* and A_{eff} are desirable. However, the macrobending and microbending loss sensitivities have strong dependencies on *MFD* and tend to favor small *MFD*. Raman gain efficiency is inversely proportional to A_{eff} and the square of *MFD*, so small *MFD* will increase Raman gain efficiency and lower the Raman pump power required for a desired value of gain. The dependency of nonlinear impairments, Raman gain efficiency, and bending-induced loss sensitivity on *MFD* requires finding the appropriate balance of these properties that support the performance requirements of the intended application.

As discussed in Chapter 2, the chromatic dispersion of a silica-based fiber is a function of the dispersion of the silica and the dispersive effects associated with propagation within the waveguide (i.e., waveguide dispersion). In the wavelength region where a fiber is single moded, the waveguide dispersion of the LP_{01} mode is normally negative and the material dispersion is normally positive, as was shown in Fig. 2.8 (see Chapter 2). The waveguide dispersion effects result from the mode becoming less confined to the high index of refraction core region and spreading out into the lower index of refraction, faster phase velocity, cladding material as wavelength increases (see Fig. 2.10, Chapter 2). As wavelength increases, the LP_{01} mode β decreases, the group velocity increases and waveguide dispersion becomes more negative. Engineering the shape of the dispersion curve is primarily a process of designing a multilayer index profile shape so that the LP_{01} mode transitions from regions of high-index, slow-phase velocity to regions of lower index faster phase velocity so that the group velocity of the mode and, therefore, the dispersion are controlled. Superimposed on the dispersion curve engineering is the need to maintain the *MFD* at the signal wavelength, the bending sensitivity and single-mode operation at desired values.

5.6.2 Realizability, Manufacturability, and Scalability

When engineering a particular waveguide design, one needs to consider that the index profile shape must be realizable using the available manufacturing processing. Typical considerations are the availability of dopants and related glass processing to achieve the required index delta of the various waveguide regions, the ability of the process to fabricate the required shape of the index layers, the ability of the process to achieve the required tolerances on the index levels and shapes, as well as the tolerances on radial dimensions. Most modern commercialized transmission fibers are doped with germanium to raise the refractive index, to relative deltas (Δ, see Chapter 2) up to about 1%; while fluorine doping is used to depress the refractive index, to Δ down to about -0.3%.

The realizability of a design will depend on the fabrication process available for manufacturing. The VAD process is well suited for fabricating the step-index core shapes typically used for SSMF designs. However, the VAD method has been used to fabricate the more intricate waveguide shapes required for more

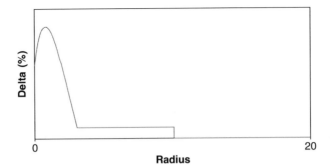

Figure 5.8 Simple core-pedestal index profile shape used for moderate A_{eff} and dispersion slope nonzero dispersion-shifted fiber (NZDF) often made with the VAD process.

advanced designs such as G.653 and G.655 fibers. The first-generation VAD G.653 fibers were made with relatively simple waveguide shapes, as shown in Fig. 5.8. The core torch is carefully controlled to fabricate a relatively high index, alpha profile shape central core region, with alpha typically less than about 10. A cladding torch with GeCl4 vapor is used to form the raised index ring region adjacent to the core. This relatively simple waveguide shape has been used for G.653 and G.655 fibers with moderate A_{eff} ($\sim$50–55 μm^2) and moderate dispersion slope ($\sim$0.07–0.08 ps/nm^2-km).

More complicated radial waveguide shapes are generally required to allow further enhancement in the A_{eff} or further tailoring of the dispersion curve. Figure 5.9 shows an index profile shape that provides an increase in the effective area while maintaining the bend sensitivity near that of SSMF. The dispersion slopes associated with this family of designs can vary from moderate to high,

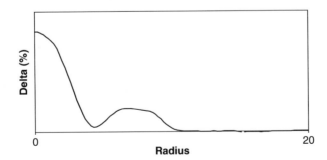

Figure 5.9 Core-ring index profile shape used for large A_{eff} nonzero dispersion-shifted fiber (NZDF).

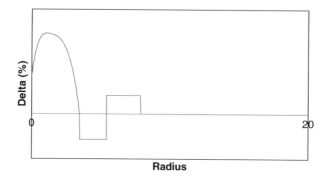

Figure 5.10   Three-layer index profile with core, trench, and ring layers used for low dispersion slope nonzero dispersion-shifted fiber (NZDF) with moderate A_{eff}.

ranging from 0.052 to 0.09 ps/nm²-km. The three-layer core structure includes a high-index, low-alpha, central core, surrounded by a narrow annular region with index that is nearly matched to the cladding index. This structure is then surrounded by a second annular region with index raised above the cladding index. Fabrication techniques that build the index structure by radially layering multiple deposition steps, such as the OVD, MCVD, or PCVD methods, have been successfully used for fibers of these designs. It is also possible to use the VAD method when multiple deposition torches (as many as five torches have been used with some success) are employed.

Figure 5.10 shows the index profile shape generally used in tailoring the dispersion curve to provide low dispersion slope. A large contrast between the phase velocities of the material in the central region and the first annular region is required. To enhance this contrast, the index of the first annular region is reduced below that of the cladding by doping with fluorine. In general, fibers with low dispersion slope (<0.05 ps/nm²-km) must have moderate values for $A_{eff}(<60\,\mu m^2)$ in order to have bending loss properties that allow placement in high fiber count, low size cable structures. The MCVD and PCVD processes are more amenable to low-cost fabrication of depressed index layers than are the OVD and VAD processes. However, multiple VAD or OVD deposition, dehydration, doping, and sintering steps may be employed to fabricate annular fluorine-doped regions embedded within germanium or pure silica regions.

5.6.3 Low-Dispersion NZDFs

The family of G.655 fibers with low dispersion was co-developed as research progressed on 10-Gbps DWDM systems in the mid-1990s. This class of fiber attempts to reduce the excess dispersion of G.652 SSMF in the C- and L-bands

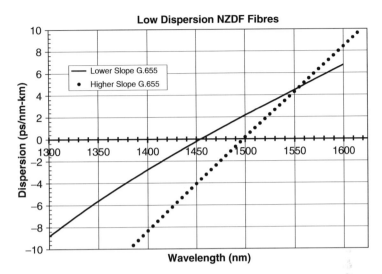

Figure 5.11 These two fiber designs manage the tradeoffs noted in Fig. 5.2 differently. One maximizes A_{eff}, and the other minimizes the dispersion slope. The larger A_{eff} fiber offers the advantage of launching approximately 1.0–1.5 dB more power into the span at a fixed level of Tx span nonlinearity, in cases in which additional optical signal-to-noise ration (OSNR) is required. The tradeoff is that a higher slope fiber has a larger relative dispersion slope (*RDS*) and is generally less well suited for broadband compensation, incurring greater penalties and more residual dispersion for a slope-matched dispersion compensation module (DCM).

while maintaining sufficient dispersion at the lower end of the C-band to fully suppress FWM. Removing the excess dispersion reduces the length of fiber in the DCM and thus reduces DCM loss, PMD, and nonlinear penalties. Two widely deployed commercial fibers make up this category, as shown in Fig. 5.11. They have approximately one-quarter of the dispersion of G.652 SSMF, so both necessarily have lower values of A_{eff} as well. One fiber has a larger $A_{eff}(\sim 72\,\mu m^2)$ but higher dispersion slope ($\sim 0.09\,ps/nm^2$-km) because of the tradeoffs discussed previously. The other fiber has a moderate $A_{eff}(\sim 54\,\mu m^2)$ but a low slope of $0.045\,ps/nm^2$-km and maintains cabled cutoff less than 1260 nm. The lower slope fiber requires the F-doped trench profile design, as shown in Fig. 5.10. Both fibers are available with a standard LWP specification at 1383 nm. In fact, 16-channel × 10-Gbps CWDM transmission has been demonstrated over the low-slope G.655 fiber [37].

These fibers offer a direct benefit over G.652 SSMF in the area of uncompensated metro transmission. Because the dispersion at 1550 nm is four times lower than at SSMF, it is possible to transmit four times the distance without compensation; distances of 320 km have been demonstrated at 10 Gbps by

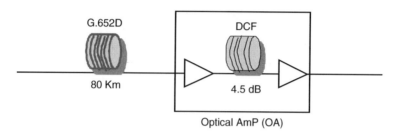

Figure 5.12 A typical amplified and compensated span includes a dispersion compensation module (DCM) at the midstage access point of an erbium-doped fiber amplifier (EDFA). Use of a low-dispersion G.655 fiber allows elimination of the DCM and simplification of the EDFA by eliminating the second stage for transmission up to 320 km.

systems vendors and more than 400 km at 2.5 Gbps. This not only offers the cost benefit of eliminating the DCM but also enables the use of simpler cheaper EDFAs. The added loss of a good metro DCM, perhaps 3.5 dB for an 80-km span of G.652 SSMF, is usually inserted at a midstage access point in a two-stage EDFA, as shown in Fig. 5.12. Thus, by removing the DCM, a simpler, lower cost amplifier can be employed as well. In this way, the total system cost can be reduced by 8% for a 2000 km, 10G network using low-dispersion G.655 fiber [38].

Laboratory results with optical duo-binary modulation and a maximum likelihood sequence estimation (MLSE) equalizer at the receiver indicate that low-dispersion NZDFs can support 10-Gbps transmission over 1200 km without in-line compensation using simplified EDFAs. In one demonstration, DCF at the transmitter and duo-binary modulation were used in conjunction with an MLSE receiver to transmit 50-GHz spaced channels over 1200 km of the high-slope G.655 fiber in Fig. 5.11 [39]. In a different approach, chirp-managed lasers with duo-binary modulation were used in conjunction with an MLSE receiver to transmit 50-GHz spaced channels over 1200 km of TrueWave RS, which is the low-slope NZDF in Fig. 5.11 [40].

The low-dispersion fibers of Fig. 5.11 have been extensively deployed in regional and long-haul networks around the world starting in the late 1990s. Transmission experiments using TrueWave RS have demonstrated that the lower dispersion across the C-band remains sufficient to suppress nonlinearities for high-bit-rate DWDM applications. For example, 40- by 40-Gbps channels were transmitted over 5- by 100-km spans of TrueWave RS in the C-band using a slope-matched DCM and hybrid Raman/EDFA amplification [41]. An early demonstration of a commercialized all-Raman transmission system demonstrated 64- by 40-Gbps channels over 1600 km of TrueWave RS between 1554 and 1608 nm, with 50- by 40-Gbps channels over a slightly shorter distance for the high-slope G.655 fiber [42]. This system avoids the lower C-band to circumvent difficulties associated with (1) compensating the lower C-band for

Table 5.2

Impact of Tx fiber properties on DCM properties[a]

Transmission fiber attributes			Slope-matched DCM properties		
ITU category	Tx fiber dispersion @1550 nm	RDS (nm^{-1})	Residual dispersion (ps/nm-km)	Typical loss (dB) 80-km module	Typical PMD (ps) 80-km module
G.652	17	0.0035	±0.10	6.0	0.33
G.656	7.3	0.0055	±0.05	4.5	0.25
G.655	4.5	0.01	±0.15	2.8	0.15
G655	4.2	0.02	±0.20	3.5	0.36

[a] The residual dispersion, loss, and PMD for 100% slope-matched dispersion compensation modules (DCMs) in the C-band matched to G.652 SSMF, a G.656 NZDF, and two G.655 NZDFs. Values for dispersion, loss, and *RDS* refer to 1550 nm. The residual dispersion is across the entire C-band. (Values from reference [43].)

fibers with high *RDS* and (2) applying Raman amplification in fibers with λ_0 too close to the signals.

The beneficial impact on the DCM properties from reducing Tx fiber dispersion is shown in Table 5.2. To illustrate the principles, values are drawn from a family of commercial slope-matched DCMs intended for precise compensation applications, on the market at the time of writing [43]. Comparing the 17-ps/nm-km value for G.652 Tx fiber in the first line of Table 5.2 with the 4.5-ps/nm-km value for G.655 fiber in the second line shows the impact of reducing the length of DCF in the module on both loss and PMD. The Tx fiber with $RDS \sim 0.01\,nm^{-1}$ does suffer a modest penalty in residual dispersion for reasons described in Section 5.5.4. The Tx fiber with $RDS \sim 0.02\,nm^{-1}$ suffers a larger penalty in residual dispersion and loses the benefit of lower dispersion with regard to reducing PMD in the DCF.

5.6.4 Medium-Dispersion NZDFs

The fibers that most closely meet the design principles for enabling precise dispersion compensation, reducing total system nonlinearity, and promoting efficient use of Raman amplification are the "medium-dispersion" NZDFs (MDFs) that meet both the G.656 and the G.655 ITU specification. Figure 5.13 shows dispersion curves for two commercially deployed examples. To reduce the *RDS* and maintain a low λ_0 while balancing the dispersion slope with effective area as required by Fig. 5.2, the 1550-nm dispersion should be in the range of 7–8 ps/nm-km. By reducing the dispersion slope in the C-band to 0.04 ps/nm-km, the TrueWave *REACH* LWP fiber [44–47] shown in the solid

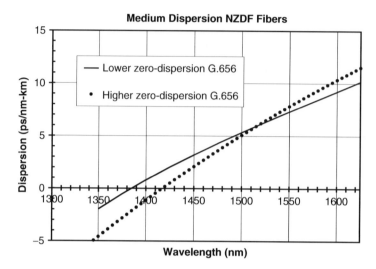

Figure 5.13 Dispersion curves for two G.656 "medium dispersion fibers." The fiber shown by the solid line has a 1550-nm dispersion, relative dispersion slope (*RDS*), and A_{eff} of 7.3 ps/nm-km, 0.0055 nm^{-1}, and 55 μm^2, respectively. The fiber shown by the dashed line has values 8 ps/nm-km, 0.0065 nm^{-1}, and 63 μm^2, respectively.

line achieves a nominal $\lambda_0 \sim 1385$ nm. This represents a significant reduction of the excess dispersion of G.652 SSMF while making the entire S-, C-, and L-bands available for signals and Raman pumps [7, 48], by keeping $\lambda_0 < 1385$ nm. The combination of medium dispersion and low *RDS* gives this class of fibers a moderate A_{eff} that enables efficient, cost-effective application of Raman gain. Numerous record-setting experiments for high-capacity, long-distance product experiments used transmission lines based on MDFs [48–51].

From Table 5.2, it is clear that reducing the excess dispersion of the G.652 SSMF while maintaining a low *RDS* allows the corresponding DCM to have better wideband residual dispersion, lower insertion loss, and lower PMD. The nonlinear performance of the DCM is also improved. It has been shown that the low *RDS* ~ 0.0055 nm^{-1} of the TrueWave *REACH* fiber supports the design of an industrialized, full C+L band slope-matched DCM with ± 0.15 ps/nm-km residual dispersion [43, 52]. This device was the basis for a demonstration of 80 channels of 40-Gbps transmission over the C- and L-bands of TrueWave *REACH* using a single wideband DCM plus all Raman amplification [50].

To enable wideband and high-bit-rate systems, it is important for the fiber design to aid the system designer in getting the maximum benefit from Raman amplification. The combination of low *RDS* and medium dispersion leads to a moderate A_{eff} and enhanced Raman gain efficiency. For this higher Raman gain to be used effectively in system design, the Raman gain coefficient must have a tight manufacturing distribution. This, in turn, requires good manufacturing control over A_{eff} as well as water peak loss in the Raman pump region. If the water peak loss is reduced from 0.5 to 0.35 dB/km at 1383 nm, then the attenuation at 1410 nm is correspondingly reduced from about 0.335 to 0.285 dB/km. Figure 5.14 shows both the measured manufacturing distribution of the Raman gain efficiency C_R of 16,000 km of TrueWave *REACH* LWP fiber [47], as well as Raman gain curves for TrueWave *REACH* compared to G.652 SSMF fiber. The tight distribution of Raman gain around the mean supports practical systems engineering requirements.

The higher peak Raman gain efficiency of the G.656 fibers compared to SSMF makes it possible to use less pump power to achieve a targeted Raman gain. Although Raman amplifiers have proven reliability, prudence is still recommended with high-power transmission. The use of lower pump power can improve thermal management in equipment design and avoid power levels that raise the possibility of connector end-face damage or burning of the coating at incidental fiber bends. Alternatively, a higher Raman efficiency allows significantly higher Raman gain for a fixed pump power level. This situation arises where the 500-mW IEC Class IIIb laser safety limit is of concern. Consider the situation where two co-propagating pumps and three counter-propagating pumps are used, but the total power including signals is constrained to 500 mW at each end of the span. A span of the G.656 fiber with $A_{eff} = 55\,\mu m^2$ has 50% higher peak Raman gain efficiency and supports more than 16 dB of flat Raman gain, whereas the SSMF supports just less than 12 dB with similar gain flatness. For a typical 100 km span with 21 dB total attenuation, the optimum balance of EDFA amplified spontaneous emission noise and Raman multipath interference noise performance occurs with about 16 dB of Raman gain.

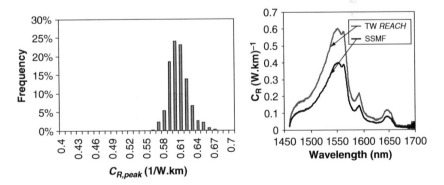

Figure 5.14 Manufacturing distribution of Raman gain for 16,000 km of TrueWave *REACH* (left) and Raman gain curves for G.656 TrueWave REACH versus G.652 SSMF (right). TrueWave REACH has an $A_{eff} = 55\,\mu m^2$, compared to $85\,\mu m^2$ for SSMF. (Data courtesy Bera Pálsdóttir, OFS Fitel Denmark, Brøndby.)

The ability to employ high levels of Raman gain with cost and power efficiency without noise from nonlinear impairments makes possible very high capacity transmission in the traditional C- and L-bands, as well as novel architectures for lower cost systems in demanding applications. In the area of high capacity-distance product transmission, 160- × 42.7-Gbps channels spaced at 50 GHz have been transmitted over 3200 km of TrueWave *REACH* using the single C+L DCM [52], for a spectral efficiency of 0.8 bps/Hz and capacity-distance product of 20 Pbps*km [53]. It has been shown experimentally that the moderate A_{eff} of the NZDF Tx fibers, responsible for the larger Raman gain, presents only a small XPM penalty across a wide range of launch powers when mixing 10 and 40G channels in a system with industrial margins [54].

It is also possible to use high levels of co-pumped Raman gain to accomplish a doubling of the span length from about 100 to 200 km, sometimes known as "hut skipping," for ultra long haul 10G systems. Adding large Raman co-gain allows one to reduce the launch power and thus the average signal power over the first ~40 km of the span where the power is highest. This reduces the impact of Kerr non-linearities and thus reduces the required OSNR to attain the targeted bit error ratio [55], such that 40 channels × 10.66 Gbps transmission over 12- × 200-km spans of TrueWave *REACH* fiber was accomplished with 14.5 dB of co-pumped gain, 22 dB of counter-pumped gain, plus EDFA. A related challenge is the case of un-repeated transmission over a very long, single span to a remote area with amplifiers and DCM only at the terminals. In this case, 20- × 10-Gbps channels were transmitted over 300 km (or 64 dB of loss) on TrueWave *REACH* fiber, using 25 dB of counter-gain, 12.5 dB of co-gain [56], plus EDFA.

A study [57] on the TrueWave *REACH* fiber (chromatic dispersion curve shown as the G.656 #1 in Fig. 5.3) has demonstrated PMD performance that will enable transmission at the highest bit rates (40–80 Gbps) shown in Table 5.1. This work involved measurements of PMD from the fiber manufacturing process, through the cabling process and into field installation. The statistical performance of links formed from fiber at each stage in the process was evaluated so that link design values of installed cabled fiber could be inferred from measurements on uncabled factory fiber samples. This demonstration verifies that a commercially available G.656 fiber has low enough PMD to handle next-generation transmission networks.

5.7 A NEW PARADIGM IN TRANSMISSION LINE DESIGN

Instead of placing the DCF in a module, it is possible to design a compensation fiber that can be cabled. In the former case, loss and nonlinearity are added to the span by the DCM without transporting the signal any closer to the end terminal; in the latter case, the loss and nonlinearity associated with dispersion compensation also contribute to system reach. Submarine systems have traditionally been built from spans in which both positive and negative dispersion fibers were cabled. A span comprising a positive and a negative dispersion fiber with equal *RDS*—both fibers being cabled—is known as a dispersion-managed fiber (DMF) span. DMF spans were first deployed in a 10-Gbps trans-Pacific submarine fiber link [9], requiring extremely precise dispersion compensation over 96 channels. The positive dispersion fiber was a super large area (SLA) fiber with dispersion +20 ps/nm-km and $A_{eff} = 107\,\mu m^2$, while the negative dispersion fiber was an inverse dispersion fiber (IDF) with dispersion approximately -40 ps/nm-km and $A_{eff} = 31\,\mu m^2$ [10], representative of the UltraWave fibers. The accumulation of nonlinear penalties can also be minimized by employing DMF spans. The first two-thirds of each span is an SLA fiber, to accommodate a high launch power, while the greatly attenuated signal traverses the IDF fiber with smaller A_{eff} in the final third of the span. Submarine experiments have also demonstrated Raman amplification with DMF spans [58].

DMF spans represent an optimized transmission line for ultralong haul terrestrial routes as well as submarine systems. In the past few years, exceedingly high capacity–distance product transmission experiments have been conducted with these fibers using both hybrid Raman-EDFA [59] and all-Raman [60, 61] amplification. In proposed terrestrial applications adapting the aforementioned submarine fibers, the first third of the span is SLA, the middle third is IDF, and the final third is again SLA. This symmetrical arrangement allows launch from either direction as well as helping to reduce nonlinearity when employing Raman amplification. The highest power levels occur at launch in the first SLA segment (perhaps including co-pumped Raman gain), while the second highest powers occur in the final SLA segment due to counter-pumped Raman gain. It can be shown that the exact position of the middle third of IDF fiber is not critical to amplifier noise figure improvement. DMF spans have also been demonstrated using the NZDF fiber labeled as G.656 #2 in Fig. 5.3 with a slope-matched negative dispersion fiber having -16 ps/nm-km at 1550 nm [62, 63].

In addition to enabling ultralong haul transmission over the widest possible bandwidth, DMF spans may also prove an enabling technology for all-optical networking. The dispersion of a properly designed DMF span is inherently compensated across the entire C- and L-bands, providing complete flexibility in wavelength routing. Further consideration of technical aspects of DMF spans can be found in reference [25].

REFERENCES

[1] Zyskind, J. L. et al. 1997. Erbium-doped fiber amplifiers. In: *Optical Fiber Telecommunications IIIB* (I. Kaminov and T. Koch, eds.). Academic Press, San Diego.

[2] Ghiasi, A. et al. 2006. Experimental results of EDC based receivers for 2400 ps/nm at 10.7 Gb/s for emerging telecom standards. In: *Optical Fiber Communications Conference, Technical Digest*. Optical Society of America, Washington, DC. Paper OTuE3.

[3] Grüner-Nielsen, L. et al. 1999. Design and manufacture of dispersion compensating fibre for simultaneous compensation of dispersion and dispersion slope. In: *Optical Fiber Communications Conference, Technical Digest*. Optical Society of America, Washington, DC. Paper WM13.

[4] Grüner-Nielsen, L. et al. 2000. New dispersion compensating fibres for simultaneous compensation of dispersion and dispersion slope of non-zero dispersion shifted fibres in the C or L band. In: *Optical Fiber Communications Conference, Technical Digest*. Optical Society of America, Washington, DC. Paper TuG6.

[5] Forghieri, F. et al. 1997. Fiber nonlinearities and their impact on transmission systems. In: *Optical Fiber Telecommunications IIIA* (I. Kaminov and T. Koch, eds.). Academic Press, San Diego.

[6] Desurvire, E. 2002. *Erbium-Doped Fiber Amplifiers*. John Wiley & Sons, Hoboken, NJ.

[7] Bromage, J. et al. 2001. S-band all Raman amplifier for 40 × 10 Gbps transmission over 6 × 100 km of non-zero dispersion fiber. In: *Optical Fiber Communications Conference, Technical Digest*. Optical Society of America, Washington, DC. Paper PD4.

[8] Chang, K. H. et al. 1999. New hydrogen aging loss mechanism in the 1400 nm window. In: *Optical Fiber Communications Conference, Technical Digest*. Optical Society of America, Washington, DC. Paper PD22.

[9] Bakhshi, B. et al. 2003. Terabit/s field trial over the first installed dispersion-flattened transpacific system. In: *Optical Fiber Communications Conference, Technical Digest*. Optical Society of America, Washington, DC. Paper PD27.

[10] Knudsen, S. N. et al. 2000. Optimization of dispersion compensating fibres for cabled long-haul applications. *Electron. Lett.* V36:2067.

[11] Ohashi, M. et al. 1992. Optical loss property of silica-based single-mode fibers. *J. Lightwave Technol.* v10:539.

[12] Chang, K. H. et al. 2000. Method of making a fiber having low loss at 1385 nm by cladding a VAD preform with a $D/d < 7.5$. U.S. Patent 6131415. Oct. 17, 2000.

[13] Chang, K. H. 2005. Hydrogen aging loss. In: *Optical Fiber Communication Conference, Technical Digest*. Optical Society of America, Washington, DC. Paper OThQ3.

[14] Dong, X. et al. 2002. Collapsing a multitube assembly and subsequent optical fiber drawing in the same furnace. U.S. Patent 6460378B1. Oct. 8, 2002.

[15] Fabian, H. and T. J. Miller. Method for producing an optical fiber. U.S. Patent 7, 028508B2. Apr. 18, 2006.

[16] Fletcher, J. P., III et al. 2004. Rod-in-tube optical fiber preform and method. U.S. patent application 2004/0107735A1.

[17] Hart, A. C. et al. 1994. Method of making a fiber having low polarization mode dispersion due to a permanent spin. U.S. Patent 5298047. March 29, 1994.

[18] Shimizu, M. et al. 2001. Hydrogen aging tests for optical fibers. In: *Proceedings of the 50th International Wire and Cable Symposium, Inc. (IWCS)*, pp. 219–223. Eatontown, NJ.

[19] Ohkubo, F. et al. 2002. Stability of low water peak SMF against hydrogen aging. In: *Proceedings of the 51st International Wire and Cable Symposium, Inc. (IWCS)*, pp. 461–465. Eatontown, NJ.

[20] Jason, J. et al. 2003. Hydrogen aging performance of various fiber types under different aging conditions. In: *Proceedings of the 52nd International Wire and Cable Symposium, Inc. (IWCS)*, pp. 75–81. Eatontown, NJ.

[21] Lemaire, P. J. and K. L. Walker. 2002. Glass optical waveguides passivated against hydrogen-induced loss increases. U.S. Patent 6499318B1. Dec. 31, 2002.

[22] Chang, K. H. et al. 2004. Method of making an optical fiber using preform dehydration in an environment of chlorine-containing gas, fluorine-containing gases and carbon monoxide. U.S. Patent 6776012B2. Aug. 17, 2004.

[23] Chang, K. H. et al. 2005. Next generation fiber manufacturing for the highest performing conventional single mode fiber, In: *Optical Fiber Communication Conference, Technical Digest*. Optical Society of America, Washington, DC. Paper JWA5.

[24] Ogai, M. et al. 1987. Behavior of alkali impurities and their adverse effect on germanium doped silica fibers. *J. Lightwave Technol.* LT-5(9):1214–1218.

[25] Grüner-Nielsen, L. et al. 2005. Dispersion compensating fibers. *J. Lightwave Technol.* 23:3566.

[26] Dugan, J. M. et al. 1992. All-optical, fiber-based 1550 nm dispersion compensation in a 10 Gbps, 150 km transmission experiment over 1310 nm optimized fiber. In: *Optical Fiber Communications Conference, Technical Digest*. Optical Society of America, Washington, DC. Paper PD14.

[27] Antos, A. J. and D. K. Smith. 1994. Design and fabrication of dispersion compensating fiber based on the LP01 mode. *J. Lightwave Technol.* 12:1739.

[28] Vengsarkar, A. M. and W. A. Reed. 1993. Dispersion compensating single-mode fibers: Efficient designs for first- and second-order compensation. *Opt. Lett.* **18**:924.

[29] Vengsrkar, A. M. and D. W. Peckham. 1995. Recent progress in dispersion compensating fibers. Invited paper at IOOC 95, Hong Kong, June 26–30, 1995.

[30] Grüner-Nielsen, L. et al. 1998. Large volume manufacturing of dispersion compensating fibres. *Optical Fiber Communications Conference, Technical Digest*. Optical Society of America, Washington, DC. Paper TuD5.

[31] Kristensen, P. 2004. Design of dispersion compensating fiber. In: *Proceedings of the European Conference on Optical Communication*. Stockholm, Sweden. Paper 3.3.1.

[32] Rathje, J. 2002. Relationship between relative dispersion slope of a transmission fiber and the usable bandwidth after dispersion compensating. In: *Proceedings of the European Conference on Optical Communication*. Copenhagen, Denmark. Paper 1.23.

[33] Agrawal, G. P. 1995. *Nonlinear Fiber Optics*. Academic Press, San Diego.

[34] Philen, D. L. et al. 2000. Measurement of the non-linear index of refraction, N_2, for various fiber types. In: *Optical Fiber Communications Conference, Technical Digest*. Optical Society of America, Washington, DC.

[35] Leng, L. et al. 2003. Experimental investigation of the impact of NZDF zero-dispersion wavelength on broadband transmission in Raman-enhanced systems. In: *Optical Fiber Communications Conference, Technical Digest*. Optical Society of America, Washington, DC. Paper WE4.

[36] Judy, A. et al. 2003. Fiber PMD—room for improvement. In: *Proceedings of the 19th National Fiber Optic Engineers Conference*, p. 1208. Telcordia Technologies, Piscataway, NJ.

[37] Thiele, H. J. et al. 2004. 160-Gb/s CWDM capacity upgrade using 2.5-Gb/s rated uncooled directly modulated lasers. *Photonics Tech. Lett.* 16:2389.

[38] Georges Kechichian. 2004. Reducing DWDM cost. Presented at the Optical Fiber Communications Conference 2004. Workshop MD-8. Los Angeles, CA.

[39] Downie, J. D. et al. 2006. Flexible 10.7 Gbps DWDM transmission over up to 1200 km with optical in-line or post-compensation of dispersion using MSLE-EDC. In: *Optical Fiber Communications Conference, Technical Digest 2006*. Optical Society of America, Anaheim, CA. Paper JThB5.

[40] Chandrasekhar, S. et al. 2006. Chirp-managed laser and MLSE-Rx enables transmission over 1200 km at 1550 nm in a DWDM environment in NZDSF at 10-Gbps without any optical dispersion compensation. *Photonics Technol. Lett.* 18:1560.

[41] Leng, L. et al. 2002. 1.6 Tb/s (40 × 40 Gb/s) transmission over 500 km of nonzero dispersion fiber with 100-km amplified spans compensated by extra-high-slope dispersion-compensating fiber. In: *Optical Fiber Communications Conference, Technical Digest*. Optical Society of America, Washington, DC. Paper ThX2.

[42] Banerjee, S. et al. 2003. Long-haul 64 × 40 Gbit/s DWDM transmission over commercial fibre types with large operating margins. *Elec. Lett.* 39:92.

[43] OFS Fitel Denmark, RightWave DCM literature. Available at: www.ofs.dk.

[44] Zhu, B. et al. 2000. 800 Gbps NRZ transmission over 3200 km of TrueWave fiber with 100-km amplified spans employing distributed Raman amplification. In: *Proceedings of European Conference on Optical Communication*. Munich, Germany. Paper 2.2.3.

[45] Kalish, D. A. et al. 2005. Method for the manufacture of optical fibers, improved optical fibers, and improved Raman fiber amplifier communication systems. U.S. Patents 6,904,217 (June 7, 2005) and 6,952,517 (October 4, 2005).

[46] Geisler, T. et al. 2005. Large volume fiber and cable results for low slope NZDF for 40 Gbps. *National Fiber Optics Engineers Conference, Technical Digest*. Paper NWE4.

[47] Pálsdóttir, B. and C. Larsen. 2005. Raman gain efficiency measured on 16 Mm of Raman optimized NZDF fiber. In: *Optical Fiber Communication Conference, Technical Digest*. Optical Society of America, Washington, DC. Paper OThF7.

[48] Zhu, B. et al. 2001. 3.08 Tb/s (77 × 42.7 Gbps) transmission over 1200 km of non-zero dispersion-shifted fiber with 100-km spans using C- and L-band distributed Raman amplification. In: *Optical Fiber Communications Conference, Technical Digest*. Optical Society of American, Washington, DC. Paper PD23.

[49] Bigo, S. et al. 2001. Transmission of 125 WDM channels at 42.7 Gb/s (5 Tb/s capacity) over 12 × 100 km of Teralight Ultra fiber. In: *Proceedings of the European Conference on Optical Communication*. Amsterdam, The Netherlands. Paper PD.M.1.1.

[50] Zhu, B. et al. 2002. 3.2 Tb/s (80 × 42.7 Gb/s) transmission over 20 × 100 km of non-zero dispersion fiber with simultaneous C + L-band dispersion compensation. In: *Optical Fiber Communications Conference, Technical Digest*. Optical Society of America, Washington, DC. Paper FC8.

[51] Zhu, B. et al. 2004. High spectral density long-haul 40-Gb/s transmission using CSRZ-DPSK format. *J. Lightwave Technol.* 22:208.

[52] Grüner-Nielsen, L. et al. 2002. Module for simultaneous C + L band dispersion compensation and Raman amplification. In: *Optical Fiber Communications Conference 2002, Technical Digest*. Optical Society of America, Washington, DC. Paper TuJ6.

[53] Zhu, B. et al. 2003. 6.4 Tb/s (150 × 42.7 Gb/s) transmission with 0.8 bit/s/Hz spectral efficiency over 32 × 100 km of fiber using CSRZ-DPSK format. *Optical Fiber Communications Conference, Technical Digest*. Optical Society of America, Washington, D.C. Paper PD19.

[54] Agarwal, A. 2003. Ultralong-haul transmission of 40-Gb/s RZ-DPSK in a 10/40 G hybrid system over 2500 km of NZ-DSF. *Phot. Tech. Lett.* 15:1779.

[55] Bromage, J. et al. 2004. WDM transmission over multiple long spans with bidirectional Raman pumping. *J. Lightwave Technol.* 22:225.

[56] Du, M. et al. 2005. Unrepeatered transmission over 300 km non-zero dispersion-shifted fiber with bi-directionally pumped Raman amplification. In: *Proceedings of European Conference on Optical Communication*. Glasgow, Scotland. Paper MO 4.2.6.

[57] McCurdy, A. H. et al. 2005. Control of polarization mode dispersion when building high bit rate optical transmission systems. The National Fiber Optic Engineers Conference. Optical Society of America, Washington, D.C. Paper NThC1.

[58] Tsuritani, T. et al. 2004. 70 GHz-spaced 40×42.7 Gbit/s transmission over 9400 km using prefiltered CS-RZ DPSK signal, all-Raman repeaters and symmetrically dispersion-managed fiber span. *J. Lightwave Technol.* 22:215.

[59] Zhu, B. et al. 2002. Transmission of 3.2 Tb/s (80×42.7 Gb/s) over 52×100 km of Ultra-Wave fiber with 100-km dispersion-managed spans using RZ-DPSK format. In: *Proceedings of the European Conference on Optical Communications, 2002, Copenhagen, Denmark.* Paper PD2.4.

[60] Rasmussen, C. 2004. DWDM 40 G transmission over trans-Pacific distance (10,000 km) using CSRZ-DPSK, enhanced FEC and all-Raman amplified 100 km UltraWave fiber spans. *J. Lightwave Technol.* 22:203.

[61] Sauer, M. et al. 2003. 1.6 Tbit/s transmission over 2160 km of field-deployed dispersion-managed fibre without per channel dispersion compensation. *Elec. Lett.* 39:728.

[62] du Mouza, L. et al. 2001. 1.28 Tbit/s (32×40 Gbit/s) WDM transmission over 2400 km of TeraLight/Reverse TeraLight fibres using distributed all-Raman amplification. *Elec. Lett.* 37:1300.

[63] Schuh, K. 2002. 4×160 Gbit/s DWDM / OTDM transmission over 3×80 km TeraLight-Reverse TeraLight fibre. Proceedings of the European Conference on Optical Fiber Communication. Copenhagen, Denmark. Paper 2.2.1.

Chapter 6

Specialty Single-Mode Fibers

Lars-Erik Nilsson, Åsa Claesson, Walter Margulis, and Pierre-Yves Fonjallaz

Acreo AB, Kista, Sweden

6.1 INTRODUCTION

The last two decades of the twentieth century saw an immense increase in the number of applications in which optical fibers of different kinds were used. The explosive development of fiber optics for communication was a major driving factor behind this progress, providing the necessary tools and enabling technologies. Contrary to popular belief, however, the development of the optical fiber was not initially driven by the needs of the communications sector but has a much longer history. For instance, the "controlled" guiding of light in a transparent water jet was first described in 1841 [1]. The technology subsequently found use for the spectacular illumination of water fountains at the great exhibitions of the late nineteenth century. A short notice in *The Lancet* [2] issue of 1889 described the use of a glass light guide for purposes of medical examination, and the 1920s and 1930s saw the development of the concept of fiber bundles for image transfer in medical and other applications. These were the driving forces behind such important inventions as, for example, the glass clad fiber, which gained commercial success in medical endoscopes as well as faceplates for image intensifier devices in the early 1960s. Up to this time, optical fibers had been of the multimode type. The single-mode optical fiber was discovered and described about the same time [3]. With the invention of the laser and the techniques for making low-loss fibers in the early 1970s [4], made available the basic prerequisites for efficient fiber optic communication. The enormous inherent technical and economical advantages of fiber optic communication spurred a huge global research and development (R&D) effort into commercializing such systems and large-scale

165

deployment started in the late 1970s—at first with multimode fibers, but from the beginning of the 1980s onwards with increasing deployment of single-mode fibers to meet the requirements of high-speed long-haul communications.

With an in-depth understanding of the special properties of single-mode fibers (e.g., see Chapter 2) and the tools available for manufacturing them, as described in Chapter 3, the field was open for extending their application beyond the one of pure light transport. Specialty single-mode fibers have, therefore, been developed to incorporate several functionalities (i.e., multifunctional fiber), and today, they find important uses in a diverse range of applications such as optical amplifiers and lasers, sensors, signal restoration, and optical filtering.

To illustrate this diversity, we devote this chapter to components based on some unusual multifunctional, specialty single-mode fibers, namely macrohole fibers, multicore fibers, fibers with internal electrodes, and fibers for high-temperature–resistant fiber Bragg gratings. This set of examples is by no means exhaustive, but it is a broad set of the most common specialty SM fibers used in the past.

6.2 MACROHOLE FIBER

Macrohole fibers belong to the group of microstructured fibers, which encompass a wide variety of fibers with air holes or other structures extending in the axial direction. There are several classes of microstructured fibers, and the terminology is not standardized. Here, we choose to categorize the fibers by their physical appearance and use the terms macrohole and microhole fibers, where the macroholes typically are several times larger than the wavelength of light. Microhole fibers, commonly called "holey" or "photonic crystal fibers" (PCFs), are treated in detail in Chapter 10 of this handbook. A good review can also be found in reference [5].

While silica PCFs are typically fabricated by the stack-and-draw method [5], which is a quite labor-intensive and complex procedure, the manufacturing of a macrohole fiber can be significantly less demanding. In a simple case, it involves three steps: manufacturing of the starting preform, preparation of the required holes and structures in the preform, and drawing of the fiber. The preform manufacturing does not differ from other fiber types, as discussed in other chapters of this handbook, and will, therefore, not be further discussed here.

The holes are machined into the preform using high-precision diamond drills, laser ablation, or ultrasonic tools. Hole sizes and positions depend on the desired structure of the final fiber, a typical hole is 3 mm in diameter and machined into a 25-mm diameter preform. Features such as grooves or flats are machined onto the preform. The processed preform is carefully cleaned from any residual debris and contamination and can be further stretched and sleeved if necessary to meet the geometrical requirements. Finally, the preform is mounted in the fiber drawing tower, and drawn under accurate control of the drawing conditions. The holes are carefully pressurized using dry, inert gas during drawing, to maintain the desired geometry of the fiber [6]. The manufacturing procedures allow for flexibility in the fiber design, in terms of shape, size and position of holes, material composition, and size of the fiber core. As will be discussed later in this chapter, the use of two or more cores may also be advantageous. A choice of protective fiber coatings can be applied in the drawing, typically acrylates or polyimides.

Figure 6.1 shows two examples of macrohole fibers manufactured using the described technique. The fiber to the left was designed for use in all-fiber electro-optic devices, as described elsewhere in this chapter. The core has a high numerical aperture, to allow for tight confinement of the mode, and the positions of the holes are chosen so that one hole is further away from the core than the other. To prepare such components, the holes are metal-filled in a post-processing procedure described later in this chapter. The second image depicts a one-hole fiber, with an additional groove machined on the outside of the fiber. The groove allows for simple alignment of the fiber in post-processing procedures.

There are a large number of proposed and published component applications for macrohole fibers [6–19]. Although these fibers in some cases compete with PCFs, the macrohole fibers have two unique properties:

- The relatively large size of the holes enables efficient introduction of various materials into the holes. These materials can be used for interaction with the cladding modes or evanescent field of the guided mode of the light or to perform active functions.
- The larger structures make the fibers easier to manufacture, compared to PCFs, and they are, therefore, attractive to cost-sensitive applications. However, unique guiding properties, such as those enabled by photonic band-gap structures, are not possible to implement in these fibers.

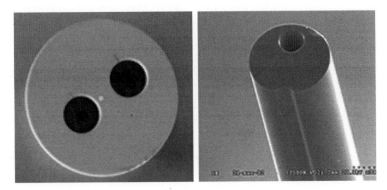

Figure 6.1 Two examples of macrohole fibers: side-hole (left) and off-center hole (right) for poling purposes. (Courtesy Acreo AB.)

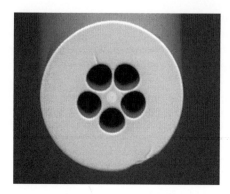

Figure 6.2   A penta hole fiber, used as an active element in a fiber laser [11].

Functions performed using macrostructured fibers include supercontinuum generation in tapered hole fibers [7], dispersion management [8], fibers with decreased bend loss for compact optical fiber wiring [9], and fibers for polarimetric sensing [10] and lasers [11]; this latter one is shown in Fig. 6.2.

The introduction of materials in the holes adds the attractive possibility to manipulate the guiding properties of the fiber. This can be achieved by interaction of the guided mode with actively controllable materials in the holes or by using the inserted materials for other active functions, such as a metal electrode to implement electro-optic control of the fiber. The holes in macrostructured fibers can be filled with different materials with relative ease. The materials are introduced into the holes by pressure. Short sections can be filled using capillary forces only. In many cases, the force achievable by mere vacuum pumping (<1 bar) is sufficient to fill a long enough section of the fiber. If a larger force is needed, materials are dispensed from a pressurized vessel. In this manner, fibers can be filled with gases and liquids. Solids, such as metals or polymers, are introduced to the holes in a molten or a precursor state and post-solidified. The filling procedure is described in some detail later in this chapter.

6.2.1 Microfluidic Devices

Eggleton et al. [12] introduced a class of hybrid devices using macrostructured fibers with movable plugs of fluids. The fiber, commonly called the "grapefruit" fiber, has six large air holes forming a circular inner cladding with a diameter of approximately $34 \mu m$, around a single-mode germanium-doped core. The holes are placed far from the single-mode core so there is no significant interaction of the fundamental mode with the material in the holes in the unaltered state. The interaction is achieved either by tapering the entire structure so that the guided mode expands into the holes or by using long period gratings (LPGs) to couple light between core and cladding. A liquid plug is inserted into the holes, and the holes are sealed by splicing to a standard fiber on both ends. The structure is hence an air-hole fiber with a short section of the holes filled with a liquid plug. On-fiber heaters are used to thermally control the air-filled sections of the hole fiber. When heated, the air expands and displaces the fluid plug. In this manner, the position of the fluid plug along the hole-fiber device can be controlled.

The tapered devices are formed by heating and stretching a section of the fiber to an outer diameter of approximately 50 μm, taking care that the holes do not collapse, and that the transition is adiabatic. In the tapered section, the mode is guided by total internal refraction on the holes. When the liquid plug, in this case methylene iodide, which has an index much higher than silica, is positioned in the tapered section, the mode leaks into the liquid and is attenuated. A thermally controlled variable optical attenuator (VOA) with an extinction of 45 dB is demonstrated [12]. By selectively filling only one of the holes with a movable liquid plug, an adjustable polarizer is described in reference [12]. A more simple VOA using a solid UV-cured acrylate in the tapered section was demonstrated in reference [13], by thermally controlling the refractive index of the polymer. In reference [14], a series of low-index fluid plugs are inserted into the holes, acting as a LPG in the tapered section. By thermally expanding air in the holes, the fluid LPG is compressed, changing the period of the LPG, hence, changing the resonance peak of the LPG filter.

In an untapered device, the interaction between core and cladding is performed by inscribed LPGs in the core of the fiber. The loss spectrum of the LPG is determined by the propagation properties of the core and cladding modes. By displacing the high- or low-index liquid plug over the LPG, the filter properties of the LPG are altered [12]. A similar approach, using a solid UV-curable polymer, is described in reference [15].

6.3 FIBERS WITH INTERNAL ELECTRODES

The explosive growth of the field of microstructured fibers has been accompanied by the development of devices based on the insertion of various materials in the holes running parallel to the fiber core as described earlier. Besides applications with liquids and gases, discussed in detail elsewhere in this handbook, new applications of fibers with internal electrodes are emerging. With a long electrode running parallel to the fiber length, one can subject the core of the fiber to a very strong electric field, because the isolation capability of silica is excellent (typical practical breakdown field $>3 \times 10^8$ V/m). Continuous electrodes of tens up to hundreds of meters in length have been reported.

Applications of fibers with electrodes include active control of the refractive index through the electro-optical effect, control of the fiber birefringence through the passage of current in the electrode, and "poling," a process after which the fiber gains an effective second order nonlinearity. Lithography of electrodes in D-shaped and twin-hole fibers has also been reported to produce a periodic structure for quasi-phase matching. The sections to follow review these devices.

6.3.1 Electrode Incorporation

In the early 1980s the PANDA fiber was invented [16], where a pair of stress elements were inserted in the cladding on opposite sides of the core introducing birefringence, which results in polarization maintenance. Soon after, a similar fiber was designed where the space occupied by the stress elements was instead left open, resulting in a "side-hole" [17] or "hollow-section" fiber [18], which found applications in pressure sensing. Based on such geometry, in 1986 Luksun Li et al. [19] described a technique to fabricate long "internal electrodes." In that pioneering work, they used a fiber with two holes. Aided by a few atmospheres of overpressure, they pumped an indium/gallium alloy—a liquid metal—into the holes. Metal-filled pieces as long as 30 m were reported, and the authors also mentioned preliminary attempts to directly draw a preform incorporating a metal. Such fibers were then used for Kerr modulation, although many practical aspects of that experiment are unavailable. The same group also showed that a metal-filled fiber (BiSn alloy, in this case) could be used to polarize light [20], because the optical loss for the TE and TM polarizations can differ by more than 40 dB for a few centimeters-long device. This early work was, however, discontinued and the techniques presented were left unexploited for many years.

For almost a decade, the use of fibers with electrodes laid dormant, until the discovery in 1991 of the possibility of inducing second-order nonlinearity in silica glasses by thermal poling [21]. By subjecting a silica disk—and as later shown—an optical fiber [22, 23] to a high-voltage bias ($\sim$4 kV) at a temperature in the neighborhood of 280 °C, the displacement of cations led to the creation of a permanent strong electric field distribution in the sample. This paved the way to making fiber components such as Pockel's cells and frequency doublers. The second-order nonlinearity induced by poling is in general confined to a thin layer of approximately 10-μm thickness adjacent to the anode electrode [21], and therefore, the core of the fiber has ideally to be inside that region (i.e., very close to the electrode). Furthermore, the need for a high-voltage bias during poling and the relatively weak nonlinearity induced favored poling of long pieces of fiber with the electrodes inside the glass to prevent electrical breakdown of the material. The internal electrode configuration was the natural choice. An important improvement to the arrangement was demonstrated by researchers at Sydney University, who by side-polishing the fiber could gain access to the metal and still be able to splice the active fiber from both ends [24, 25]. A simple but time-consuming technique of manually inserting a thin metal wire into the holes was used by the various groups working in the field [26–28]. Wire insertion requires skill, is time consuming (i.e., costly), and is not appropriate for fabrication of long devices (tens of centimeters or more). Furthermore, the position of the wire in the holes varies along the fiber and from device to device, leading to a non-uniform field distribution, uncertainty in the performance, reproducibility problems, and impedance variation along the device.

Nevertheless, long pieces of fiber have been manufactured with an electrode inserted during drawing [29]. A 200-m long piece of fiber with one internal electrode became available. For poling, the need for an outer electrode behaving as ground led to the interesting development of electrically conductive coatings. Two types of coatings were reported, one involving polyimide [30] and the other one carbon-loaded acrylate [31], with better performance in terms of conductivity and uniformity. Metal-filled capillaries for applications in medical microprobing have also been developed, and silica-isolated micrometer-size electrodes are available [32].

The technique of pumping liquid metal into a ready drawn fiber with holes was redeveloped in 2002 [33], and a schematic of the arrangement used is shown in Fig. 6.3. One end of the fiber with holes is inserted into the molten metal and the other end is kept free in the atmosphere. The liquid metal is contained in a small crucible in a sealed cell that is pressurized, forcing the melt into the fiber and filling the holes up to where the fiber is cold (and the metal solidifies). The entire cross-section of the hole is filled with metal, improving reproducibility. Up to 14 pieces of fiber are filled at a time, so the fabrication becomes quicker and cheaper.

Various types of alloys have been used for fiber components. A euthetic alloy of Bi (43%) and Sn (57%) melts at 137 °C, which allows the standard acrylate coating to be intact while filling the holes. The euthetic Au (80%) and Sn (20%) melts at 282 °C and can be inserted in liquid form at 300 °C and used as a solid electrode for poling at 260 °C. The time to fill the fiber depends on hole size, type of alloy, temperature, and pressure. Typically, a 1-m long fiber device is filled in less than 1 minute. The technique is compatible with 125-μm fibers (as shown in the insert of Fig. 6.3) and the hole size ranges from about 20 to 40 μm. The internal electrodes are accessed by side-polishing in approximately 1 minute, and the fibers are polished with the primary coating still on. Electrical contact is made with a thin wire or with conductive epoxy. Splicing the fibers with holes to standard telecom fibers can give losses of approximately 0.1 dB, but the end to be spliced needs to be free of metal. This is achieved by initially leaving approximately a 20-cm long piece of the fiber outside the oven, as shown in Fig. 6.3. By inserting half this length into the oven and removing the supply of metal, the

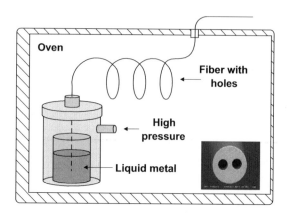

Figure 6.3 Schematic of the filling method used to prepare electrodes. (Insert shows a fiber cross-section made for poling.)

molten metal column can be displaced about 10 cm further along the fiber, freeing also the other end from metal.

In some cases, it is advantageous to use thin conductive films on the surface of the holes [34], rather than a solid alloy electrode, because the stress introduced by a thin film is small, and periodic electrodes for quasi-phase matching can be easily fabricated. Silver film electrodes were deposited on the inside of a two-hole fiber by mixing a silver nitrate solution with a reducer flowing in the holes [35]. To limit the size of the particles, and to prevent clogging of the approximately 30-μm holes, the two solutions were mixed at low temperature ($\sim 8\,^{\circ}$C) to reduce the reaction rate and inserted by high pressure. The films were usually conducting for fiber lengths more than 2 m, and devices with electrodes longer than 3 m were produced. Such hole-coated fibers could be spliced to standard fibers without further processing. The thickness of the Ag layer was measured to be in the range approximately $0.1-1.0\,\mu$m. Periodic internal electrodes could be fabricated by point-by-point side-exposure to 0.53-μm radiation through the acrylate coating, causing laser ablation. The fiber was translated and the top electrode ablated, after which the beam and lens were translated by 40 μm along the fiber and a new period recorded. Only the surface of the hole closest to the core of the fiber was exposed to the ablating beam, so that the continuity of the electrode was not jeopardized, the film remaining continuous on the back part of the hole. A typical pattern produced is shown in Fig. 6.4, with openings approximately 15 μm in diameter and center-to-center separation of approximately 40 μm, close to the beat length of 1064 and 532 nm in standard telecom fibers. In applications demanding periods of a few microns only, side exposure of a photoresist with lower power light through the acrylate coating and conventional lithography inside the fiber should be possible.

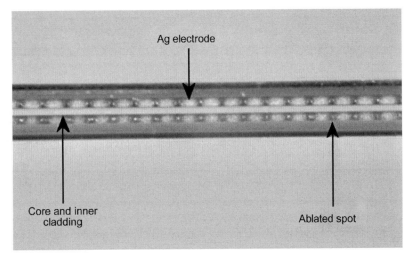

Figure 6.4 Example of periodic structure recorded inside a twin-hole fiber by ablation lithography [35].

6.3.2 Applications

A component for tunable polarization control based on a metal-filled fiber has been developed [36, 37], exploiting the tight physical contact between the metal alloy and the surface of the fiber hole. Current is run through the alloy causing heating through Ohmic dissipation. Because of thermal expansion, the heated alloy exerts pressure on the glass, which strains the core asymmetrically, leading to birefringence and a change in the polarization state of the signal in the fiber. Complete coverage of the Pointcare Sphere can be accomplished with two metal filled fiber components spliced at 45 °C [37].

Most applications of internal electrode fibers, however, are related to their use as electro-optical modulators. In this case, a voltage is applied between the electrodes, causing an electric field to be established across the core and a change in the refractive index and, thus, optical path. A Mach-Zehnder interferometer comprising an active fiber in one arm and a passive fiber in the other is sufficient means to provide for switching, transforming phase into amplitude modulation. When the phase changes by π radians, the interference of the signals in the two arms changes from constructive to destructive and light is switched from one output fiber to the other (2×2 switching). Interferometers were built exploiting the Kerr effect in 1-m long pieces of twin-hole fiber prepared with BiSn electrodes. Figure 6.5 illustrates the quadratic dependence of the Kerr switch when driven by voltages as high as 4 kV. The switching voltage was a few hundred

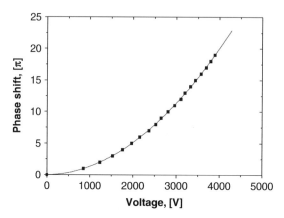

Figure 6.5 Quadratic dependence of the phase shift on the voltage. The data are taken where the phase changes by π [48].

volts and the phase excursion becomes a more rapidly varying function of the applied voltage as the voltage is increased. For example, with a 3.8-kV DC bias, the required voltage for a π phase shift is approximately 100 V.

If the fiber is first poled at approximately 280 °C for a few minutes with high voltage applied between the two electrodes, a large permanent electric field can be recorded across the core [22, 23] and the refractive index gains a linear dependence on the applied voltage. After poling, tens of volts are sufficient for full switching in the interferometer [38, 39]. Although this is still an order of magnitude larger than for LiNbO$_3$ modulators, all-fiber devices exhibit potentially lower loss and higher optical power handling capability. One such switch has been used for video transmission (i.e., as a modulator) and as a 2×2 switch for protection of a fiber network operating at 10 Gbps without degradation of the signal quality [40].

Fiber interferometers are long devices (typically 1-m long arm length) whose transmission function exhibits a sinusoidal wavelength dependence if the optical paths are unequal. The application of a control voltage signal to the active fiber results in electro-optical tuning of the sinusoidal spectral response. Therefore, electro-optical filtering can be accomplished in an unbalanced Mach-Zehnder interferometer incorporating a poled fiber in one of the arms. This has been exploited in the construction of a stepwise tunable ring fiber laser [41]. The ring incorporated an Er-doped fiber, a circulator with a sampled[1] fiber Bragg grating and an unbalanced Mach-Zehnder interferometer. The laser could be made to

---

[1] A number of Bragg gratings with different resonance wavelengths are recorded on top of each other.

operate at any of the 16 wavelengths of the sampled grating. The interferometer is tuned to transmit one of the sampled grating's wavelengths by the control voltage and the laser will oscillate at this frequency since it exhibits the lowest loss. Both mode-locking [42] and Q-switching operation have also been accomplished with poled fibers with internal electrodes.

Internal electrode fibers have also been exploited for wavelength conversion and in particular frequency doubling. D-fibers were poled with a film electrode defined by lithography on the flat surface of the fiber and an internal wire electrode [43, 44]. Up to 20% conversion efficiency was accomplished with high-power femtosecond optical pulses. Furthermore, quasi–phase-matched fibers have been constructed by periodically illuminating a continuously poled fiber with UV light selectively erasing the permanent field [45, 46]. The period was carefully adjusted for phase matching. The highest conversion efficiency demonstrated was 2.5% achieved in an 11.5-cm long device with pump peak power of only 108 W. Higher conversion efficiency is, however, expected with an improved laser source. By bending the periodically poled fiber, it was possible to achieve 27-nm tunability [47].

The electrode deposition techniques described here also opens a number of opportunities for steering active media inside the fiber, such as liquid crystals and magnetic powders. Components based on fibers with internal electrodes are versatile and can be used to perform a number of functions, such as optical switching, wavelength conversion, and active polarization control. The most attractive feature of fiber components is that they to a large extent inherit the characteristics of standard telecom fibers in terms of low loss, ease of splicing, and competitive price. It is, therefore, likely that we will see a growth in the number of applications of fibers with internal electrodes.

6.4 MULTICORE FIBERS AND COMPONENTS

An optical fiber is generally conceived as consisting of a light-guiding core concentrically positioned in a surrounding cladding structure. The concept of embedding two or more cores in a common cladding structure was, however, launched quite early in the history of single-mode fibers [49, 50]. Although multicore fibers have since attracted considerable attention and a large variety of applications and designs have been proposed, the commercial success of such fibers has been quite limited. The proposed applications for multi-core fibers span over lasers and amplifiers, transport fibers for broadband communications, passive and active fiber optic components such as filters, multiplexers, and so on, and various kinds of sensors. The ultimate multicore fiber is, of course, the image fiber used in endoscopes and other such devices. It may contain tens of thousands of individual cores in a fiber of a diameter of less than 2 mm. These cores are, however, generally not single mode but support a limited number of propagating modes of visible wavelengths and are not covered in this chapter (see Chapter 23 of this handbook for more information).

Multi-core fibers can be divided into two categories. In the first category, the fibers are designed for a controlled coupling of the guided modes between the cores, whereas for the second category, the design is such that the modes in the different cores are practically decoupled.

6.4.1 Coupled Cores

The coupling of energy between the modes guided in the various cores in multiwaveguide structures, including fibers, has been extensively analyzed theoretically (see references [51–55]). The most common fiber is the twin-core[2] fiber where two identical single-mode cores, waveguide 1 (WG1) and waveguide 2 (WG2) in Fig. 6.6, are symmetrically located in the fiber cladding. Illuminating one of the cores will equally excite two transversal modes in the fiber, one symmetrical and one asymmetrical, as indicated in Fig. 6.6, with slightly different propagation constants.

The intensity distribution in the fiber structure is the result of the summation of the two modes: Energy is periodically transferred back and forth between WG1 and WG2 as a result of the beating between the two modes. The first core will be completely depleted of its energy after a certain length. This energy is now

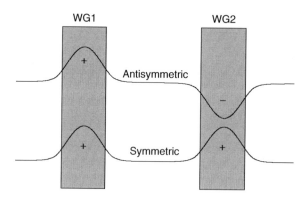

Figure 6.6 Symmetrical and asymmetrical modes excited in a twin-core fiber.

---

[2] For fibers with two cores the expressions "twin-core" and "dual core" are commonly used. Herein, a "twin-core" fiber is a fiber with two identical cores, whereas the cores of a "dual-core" fiber can be unequal.

guided in the second core and will start to transfer back to the original core. The process continues *ad infinitum* as light travels down the fiber. The fiber length necessary for completion of one cycle is called the *beat length* and depends on core separation, core geometry, refractive index, and wavelength of the light. Figure 6.7 shows a BeamProp simulation of the coupling of light, with a beat length of 32 mm, between two cores 16 μm apart in a fiber. The coupling of energy between cores is, of course, extensively used in various fiber coupler schemes, a subject that is not further treated here (refer to the extensive literature on the subject [54, 56–57]). The use of coupled multicore fibers has been proposed for many other applications and some typical examples are presented in the following subsection.

6.4.1.1 Optical Amplifiers

The possibility to "split" light from one core and propagate it for a certain distance in a parallel core has been exploited in active fibers for use in amplifiers and lasers. The use of a twin-core erbium-doped fiber for channel gain equalization was proposed early by Zervas et al. [58, 59]. The pump and WDM signals are launched into one of the cores. Since the beat length is wavelength dependent by approximately λ^{-3} [58], the various WDM channels will travel partly different

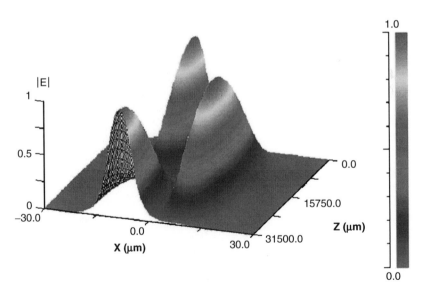

Figure 6.7 Simulation of coupling in a twin-core fiber with a 16-μm core separation. Beat length is 32 mm. (Courtesy Acreo AB.)

roads down the fiber and thus interact and saturate different subsets of erbium ions. The gains of the different channels are, thus, spatially decoupled, resulting in inhomogeneous broadening. If the power in one channel increases with respect to the others, the corresponding gain will ultimately decrease, leading towards spectral gain equalization. A slightly different approach was proposed by Lu and Chu [60, 61], in which case only one of the cores was doped with erbium ions while the other remains passive. Since the cores are different, with one core being amplifying, there is a nonreciprocity in the coupling between them. For the C-band amplifier, the pump and signal are launched into the erbium-doped core. The fiber is designed so that at the peak of the erbium gain (1533 nm) the coupling from the active to the passive core is at maximum while the coupling from the passive core is at minimum and subsequently excess energy at 1533 nm is wasted. The net result is an overall flattening of the gain in the C-band. In a similar manner, a flattened L-band amplifier can be constructed. In this case, the pump and signal are coupled into the passive core. The coupling coefficient from the passive to the active core increases with wavelength, thus compensating for the decrease in the erbium gain.

For Raman amplifiers, Kakkar et al. [62] suggested in a theoretical paper an asymmetrical twin-core fiber with a high NA central core close to a low NA core with a larger radius. Both cores are single mode and have the same propagation constant at a selected phase-matching wavelength. Signal and pump are launched into the central core. By appropriate fiber design, the pump and signals will be confined to the central core at wavelengths much shorter than the phase-matching wavelength, resulting in a high pump and signal overlap. As the signal wavelength approaches the phase-matching wavelength, a larger part of the signal's power will be confined in the second core, decreasing the pump and signal overlap and increasing in the effective mode area (A_{eff}). As the fiber's effective Raman gain coefficient (γ_{eff}) is related to the material's Raman gain coefficient (g_{eff}) and the effective area as $\gamma_{eff} = g_{eff}/A_{eff}$, an increase in g_{eff} can be compensated for by an increase in A_{eff}.

6.4.1.2 Fiber Lasers

Winful and Walton [63] proposed the use a of a twin-core fiber for passive mode locking of a fiber laser. The fiber is similar to the ones proposed by Lu and Chu [60, 61] but, in this case, with the active core enclosed in a cavity consisting of a high reflector and an output coupler. The length of the cavity is half a beat length in the absence of amplification. A low-intensity pulse will completely couple into the passive core where it is lost. However, the central portion of the high-intensity pulse will induce changes in the active cores index, de-tuning the coupler, while the wings of the pulse will couple to the passive core. The system was expected to operate in the pulsed mode because high-intensity pulses minimize losses. A similar approach for a ring laser was proposed by Oh et al. [64] and the concept was further elaborated by Martí-Panameño et al. [65].

Graydon et al. [66] practically demonstrated a triple-frequency erbium ring laser where the individual gains of the different lasing wavelengths are partially decoupled from the others due to the inhomogeneous broadening introduced by the twin-core design.

Other practical implementations of lasers were demonstrated by Kaňka et al. [67] and Peterka et al. [68] who used erbium-doped dual-core fibers for line narrowing and wavelength stabilization as well as high-speed pulse generation.

Wrage et al. [69] presented a multicore fiber laser array for high power. Eighteen single-mode Nd-doped cores were equally distributed close to the annulus of a multimode fiber carrying the pump modes. The bad beam quality, usually associated with laser arrays, caused by the different beams interacting incoherently, was overcome by phase locking the 18 resonators in a Talbot cavity. Yanming et al. [70] presented an alternative solution based on an isometric concept with up to 19 equidistant Yb-doped cores. These cores couple to each other via the cladding through the evanescent fields facilitating in-phase oscillations for all cores. These modes combine to supermodes with 80% of the energy in the lowest order (in-phase) mode with a Gaussian-like appearance [70].

6.4.1.3 Miscellaneous Applications

Ortega and Dong [71] demonstrated a tuning procedure for adjusting the coupling wavelength of a twin-core fiber. By tapering the fiber, the propagation constants for the two cores can be tuned to coincide at a predetermined wavelength. At wavelengths on both sides of the coupling wavelength, the mismatch in propagation constants is sufficiently large to suppress coupling and the fiber constitutes an optical filter with a high temperature and mechanical stability. Another way to tune the coupling was demonstrated by Atkins et al. [72]. Ge-doped cores are inherently photosensitive, that is, the refractive index changes in response to exposure to UV light. By illuminating one of the cores while monitoring the coupling, the index is trimmed until a maximum response is obtained.

An et al. [73] have reported the use of a LPG–assisted twin-core fiber for an add–drop filter. An LPG is written into one of the cores and the photosensitive cladding surrounding the core. The mismatch of the cores prevents coupling while the LPG at the coupling wavelength ensures that the overlap of the two core modes for this wavelength is large enough for efficient coupling to occur. Finally, Chu and Wu [74] used the large nonlinear effect in a symmetrical erbium-doped twin-core fiber to demonstrate optical switching.

Coupled twin-core fibers have also been proposed for sensing applications. An early, but perhaps not so practical, attempt was published in 1983 by Meltz et al. [75], who demonstrated the temperature-dependent coupling between the cores in a twin-core fiber.

6.4.2 Uncoupled Cores

If the difference in propagation constants of a multicore fiber or the distances between the cores are sufficiently large, the coupling of energy between the waveguides will be so weak that effectively no coupling occurs over the length of the fiber. The use of such fiber for transmission purposes for cost-effective access networks was proposed early [50, 76] and was later continued by, for example, Le Noane et al. [77] at CNET in the mid-1990s and Rosinski [78]. The technique, however, does not seem to have been commercially implemented on a larger scale.

The two uncoupled cores of a dual-core fiber can constitute the two arms of a Mach-Zehnder interferometer. Compared to an ordinary Mach-Zehnder interferometer, in which the two light paths are separated in two physically different fibers, this approach facilitates a much higher stability because common mode disturbances, such as temperature drift and external vibrations, will have a similar effect on the two cores and, thus, be effectively canceled. The coupling of light into the two arms and the subsequent recombination can quite easily be achieved by tapering a small part of the fiber, thus creating, for example, an in-fiber 3-dB coupler. By manipulating with different means the refractive index difference between the cores, the transmission through the interferometer can be controlled for various purposes.

6.4.2.1 Switching and Multiplexing

Nayar et al. [79] demonstrated all-optical switching in a 200-m nonlinear Mach-Zehnder interferometer in 1991. The signal is added to the input port (Fig. 6.8) and split 30:70 by the coupler into each of the two interferometer arms. The two propagating beams are recombined by the output coupler and the energy is partitioned between output 1 and output 2 depending on the phase difference. By suitable phase-shifting means (not shown), the energy can be directed to only one of the ports. Increasing the intensity will induce an unequal index change in the two cores through the optical Kerr effect. When the corresponding phase shift between the propagating beams is π, the energy will have been switched to the other output.

In an implementation of an optical add–drop multiplexer, Yvernault et al. [80] inscribed a Bragg grating in each of the two arms of a twin-core fiber Mach-Zehnder interferometer (Fig. 6.9).

Figure 6.8 A twin-core Mach-Zehnder interferometer with integral couplers. (Used, with permission, from reference [79].)

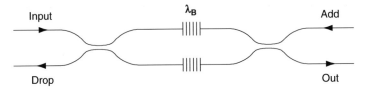

Figure 6.9 A twin-core add–drop multiplexer (Used, with permission, from reference [80].)

The gratings were phase matched through UV exposure. The unit constitutes a four-port device: "In" and "Drop" ports on one side of the interferometer and "Out" and "Add" ports on the other. Wavelengths present at the In port are equally split by the integral 3-dB coupler to the two arms. Signals at the Bragg wavelength are reflected in both arms and interfere constructively at the Drop port, while other signals are transmitted through and interfere constructively at the Out port where they appear unattenuated. Similarly, a signal at the Bragg wavelength present at the Add port will appear at the Out port. The unit can be wavelength tuned through Ohmic heating caused by passing an electrical current through a thin metal layer deposited on the fiber surface [81]. The same group also proposed a variable optical attenuator based on a thermally tuned twin-core Mach-Zehnder interferometer [82].

6.4.2.2 Fiber Sensors

Uncoupled multicore fibers have been proposed for several sensing applications in a large variety of configurations and some examples are given here. Noda et al. [83] already demonstrated how the twisting angle of a fiber could be measured in a Mach-Zehnder–like twin-core interferometer by monitoring the far-field interference fringes created by the two emerging beams. Tanak et al. [84] constructed a system for quench detection of superconducting magnets using a similar setup. The small temperature increase of a superconducting magnet mockup was detected by observing the far-field interferogram from a double-core fiber monitoring the cooled magnet before the temperature rise could cause loss of superconductivity.

Khotiaintsev et al. [85] demonstrated the use of a twin-core fiber with an integrated coupler as a small-sized probe for invasive laser Doppler anemometry. Parasitic phase modulation due to external disturbances was reported to be greatly attenuated because of the integrated design. Wosinski et al. [86] in 1994 proposed the use of a dual-core fiber with different strain and temperature dependencies for the two cores for monitoring strain in, for example, composites. Changes in temperature and/or strain introduce refractive index changes that can be detected with, for example, fiber Bragg gratings inscribed in the cores. Strain and temperature are then easily obtained from two independent equations.

The sensor examples so far cited employ dual-core fibers. Gander et al. [87] have demonstrated how the far-field interference pattern from a four-core fiber can be used for the simultaneous measurement of bending about two orthogonal axes. The strain caused by bending the fiber will introduce different amounts of phase shifts for the propagating light in the four cores depending of the orientation of the fiber relative to the bend direction. By inscribing gratings in the cores, similar information can be obtained from the changes in reflection spectra upon bending the fiber [88].

Bulut and Inci [89] used a four-core fiber for creating a stable light pattern illuminating an object for three-dimensional Fourier transform profilometry. The interference pattern, created by the interference of the wave fronts emitted from the four-fiber cores, is projected onto an object. The resulting deformed pattern contains information about the surface's topography, which can be retrieved through Fourier analysis.

An unusual application of multicore fiber was presented by Watson et al. [90]. Through laser ablation, the material around each core of a four-core fiber was removed to create pits about 15 μm deep. The pits were covered by thin reflective films so each pit finally constituted a pressure-sensitive Fabry-Perot interferometer and the whole assembly an ultra-miniature four-channel pressure sensor.

6.4.3 Manufacturing Multicore Fibers

There are basically two approaches to manufacture an optical fiber with several cores. Dorosz and Romaniuk [91] describe a multicrucible technique.

This method is schematically illustrated in Fig. 6.10. An outer crucible contains the molten glass that will eventually constitute the cladding of the fiber to be (A), while an inner crucible contains a glass melt that will make up the cores (B). The core glass flows through the nozzles at the bottom of the inner crucible into the cladding glass and the combined glass streams jointly exit the outer crucible through the bottom nozzle without intermixing of the streams. The fiber is then drawn and coated in the normal fashion as described in

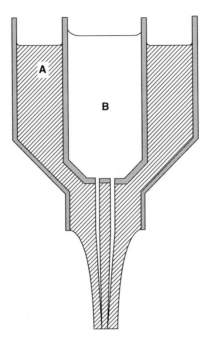

Figure 6.10 Schematic view of a double crucible for making a dual-core fiber with cladding (A) and core (B) glass melts. (Used, with permission, from reference [91].)

Chapter 3 of this handbook. The multicrucible technique is best suited for soft glasses, silica, or nonsilica based, with low process temperatures. Very complex fiber structures can be manufactured with a large range of different glass compositions. The attenuation of these fibers is, however, quite high, 0.1–1 dB/m. This is mostly because of impurities in the raw materials used for making the bulk glass.

For glasses with higher processing temperatures ($\geq 2000\,^\circ$C) such as the silica glasses today used in most optical fibers, the multicrucible technique is less suitable. Instead such fibers, or more correctly preforms, are manufactured by assembling the core rods in a cladding structure by various methods and then fusing the assembly into a homogenous preform from which the fiber is drawn in the normal fashion. The core rods can be manufactured with, say, standard modified chemical vapor deposition (MCVD) or outside vapor deposition (OVD) techniques, as described in Chapter 4. In this case, the core, doped with, say, Ge, Er, Al, and so on, must be separated from the surrounding cladding structure, which can be done with, for example, etching with hydrofluoric acid. Cores can also be manufactured from bulk glass if the higher attenuation usually associated with such glass can be tolerated.

Figure 6.11 A multicore preform with holes drilled to accept core rods (in front). (Courtesy Acreo AB.)

The cladding can be in the form of a solid rod into which are drilled holes at the appropriate places to house the core rods. After thorough cleaning and drying, the rods are inserted into the holes and the assembly is fused at high temperature into a solid preform rod. Stretching and overcladding techniques, as mentioned earlier, can also be used where appropriate. Figure 6.11 depicts a cladding rod with holes drilled into which core rods (front) will be inserted.

An alternative technique for making multicore preforms is to machine from ordinary MCVD or OVD preforms the core and some of the surrounding cladding, which subsequently is inserted into properly shaped grooves in the multicore preform, as illustrated in Fig. 6.12.

From a core preform (Fig. 6.12a), the core and part of the cladding is removed by sawing, yielding a core piece (Fig. 6.12b). A matching groove is machined out of a preform, leaving the central core intact (Fig. 6.12c) and the core piece is inserted into the groove. The assembly is inserted into a silica tube to keep it together (Fig. 6.12d). Upon heating, the tube collapses on the rod and the whole assemble fuses into one solid dual-core preform (Fig. 6.12e) subsequently drawn into a fiber with standard methods.

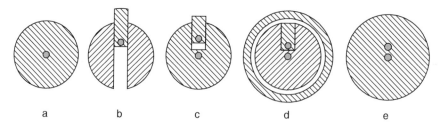

Figure 6.12 Different stages in the making of a dual-core preform.

6.5 FIBERS FOR HIGH-TEMPERATURE–RESISTANT GRATINGS

Fibers with a high photosensitivity, for an efficient fabrication of fiber Bragg gratings, are presented in Chapter 9. Here, specialty fibers developed for obtaining high-temperature–resistant fiber gratings are reviewed. The chosen grating fabrication techniques have a determinant role and will, therefore, also be reviewed. High-temperature–resistant fiber gratings find an obvious application for fiber-sensing measurements at high temperatures and in particular the measurement of the high temperatures themselves. Optical fiber thermometry is especially advantageous in hostile environments, such as in the presence of corrosive chemicals, high electric, magnetic, or strong radiofrequency fields. Examples are the measurement of temperature and pressure in oil wells where the temperature can exceed several hundreds degrees centigrade, measurement of the temperatures reached in engines or electrical transformers, and the measurement of temperatures in industrial processes involving highly exothermic chemical reactions. Another situation where high-temperature–resistant gratings are required is when the fiber containing the grating needs to be metallized for subsequent welding or embedded in a metallic structure. Fiber gratings known to be high-temperature–resistant have been found successfully stable in the presence of ionizing radiations [92]. One hypothesis, which has not yet (November 2005) been tested or confirmed as far as we know, is that high-temperature–resistant fiber gratings also exhibit a higher stability when subject to very large optical fields, as is the case in high-power fiber lasers.

Although the photosensitivity of optical fibers is usually defined as a permanent refractive index change following an exposure to light, fiber gratings are not stable at elevated temperatures and can always be totally erased when a sufficiently high temperature is applied. Even for room temperature use, a grating needs to be annealed at an elevated temperature to stabilize its refractive index modulation. Afterwards, at temperatures lower than the annealing temperature, the grating generally does not experience substantial thermal decay. The thermal annealing operates an accelerated aging process, which removes the lesser stable contributions to the index change [93]. The thermal stability depends not only on the type of fiber used and on the fabrication conditions (irradiation wavelength, intensity and dose applied), but also on presensitization techniques. The latter techniques help in getting quicker index changes but, not so surprisingly, give rise to significantly lower stability. The strength of gratings in hydrogen-loaded fibers, for example, is already strongly reduced at temperatures of only about 100 °C [94–96].

The first fiber gratings, which appeared to have good stability at elevated temperatures, were the so-called "type II gratings." The formation of these gratings, first obtained in 1992, is caused by a highly increased nonlinear absorption when applying a large UV intensity single pulse. They usually appear at the first core–cladding interface as a corrugation of the glass due to remelting. These gratings have shown to be stable at temperatures as high as 800 °C for a period of 24 hours [97]. However, type II gratings are difficult to manufacture in a controlled manner and exhibit huge loss due to coupling to cladding modes. Another type of grating with increased temperature resistance are the so-called "type IIa gratings." These gratings are obtained by high-dose irradiation at 193 nm in high germanium concentration fibers (never obtained in hydrogen-loaded fibers). They are stable up to about 300 °C [98].

Let us now review the effect of the fiber composition on the thermal stability of fiber gratings. Most fiber gratings are obtained in fibers containing a substantial concentration of germanium. However, strong gratings could be obtained in 1996 in a germanium-free fiber based on a nitrogen-doped silica core. These gratings when written with 193-nm light and of type IIa, exhibited a good thermal stability up to temperatures of about 600 °C [99]. Later and still in the same type of fibers, gratings with a post-exposure increasing strength at elevated temperatures were reported [100]. This new phenomenon was associated to the thermodiffusion of nitrogen in the UV-irradiated regions of the fiber. These gratings after proper annealing were shown to be stable up to 900 °C. Gratings written with 248-nm light in tin-doped silica fibers also exhibit very high thermal stability [101]. No major degradation can be observed at 850 °C over a few hours and extrapolations from experimental data indicate that gratings operating at 500 K for 10 years will retain more than 99% of their initial strength. Tin-doped silica fibers are low loss at 1.55 μm and exhibit large photosensitivity, but they usually have a large numerical aperture. Finally, an antimony-germanium (Sb-Ge) co-doped fiber with high-temperature sustainability has been developed [102]. Gratings written in this fiber with 248-nm light have been shown to be quite stable at temperatures up to 500 °C and the decrease as a function of temperature above that temperature is not very steep (still a significant reflectivity at 950 °C).

The underlying mechanisms of the photosensitivity are the creation or modification of defects and structural modifications of the glass matrix itself either by relaxation of internal stresses or by compaction, which is a natural tendency of amorphous solids. The highest temperatures for thermal stability of induced defects are between 100 and 600 °C. Structural changes are more stable and the highest temperatures in these cases are between 400 and 1000 °C. The next mechanism of even higher thermal stability is that which is limited by viscous flow or diffusion of dopants [103]. In 1996, a method was proposed to create high-temperature stable fiber gratings by periodically modifying the concentration of fluorine in the core of the exposed fiber [104]. This type of grating is usually called *chemical composition gratings* (CCGs). They are manufactured by writing a type I grating in a hydrogen-loaded fiber containing a fluorine-doped core. Exposure to UV light induces a periodic variation of OH bondings in the core due to a photo-induced reaction between molecular hydrogen and the glass matrix. When the fiber is heated at a temperature about 900 to 1100 °C, the original type I grating totally disappears, as expected, and a new one appears after several tens of minutes and remains stable after development. It is believed that a chemical reaction takes place at these elevated temperatures between the OH bondings and the fluorine atoms, leading to the formation of HF molecules. The latter have a high mobility and rapidly diffuse out of the core region, leading to a periodic extraction of fluorine. With simple diffusion models, it has been shown that the stability of these gratings at an elevated temperature only depends on the diffusion process between the dark regions (low exposure to UV light) with larger concentration of fluorine and the bright regions having experienced a depletion of fluorine. These CCGs have an excellent stability at temperatures below approximately 800 °C, as illustrated by Fig. 6.13.

A comparison between different fiber grating types, in particular types I and IIa and CCGs, has been realized in 2002 and clearly shows the significantly higher thermal stability of the CCGs. A similar type of CCG, with similar

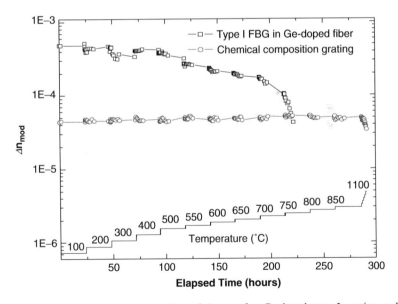

Figure 6.13 Comparison of the variation of Δn_{mod} of a Ge-doped type I grating and a "chemical composition grating" with elapsed time over various temperature ranges. The gratings were maintained for around 24 hours at each temperature before going to the next higher temperature [105].

stability behavior, has been obtained in erbium-doped fibers, which did not contain any fluorine. A new type of process that produces gratings with even higher temperature stability has been discovered. The gratings are obtained with large exposure doses and, therefore, large concentration of OH bondings. It is believed that the CCG development leads to the formation of molecular water and, hence, periodic extraction of oxygen in the irradiated cores in that case. The strength of these gratings was basically constant at more than 1100 °C for 60 minutes [105].

A totally new fabrication technique, with a huge potential, was demonstrated in 2003. It apparently works in any type of fiber and without any pre-photo-sensitization [106]. It consists of irradiating the fibers with ultrashort pulses from near-infrared (IR) lasers. The refractive index changes occur through a multi-photon absorption (most probably four or five photons) and the spatial locations where they appear can, hence, be chosen by properly focusing the writing laser beam. Gratings fabricated in these conditions are highly thermally stable. Here, also type I and type II gratings can be distinguished depending on the intensity of the IR laser beam, and the latter ones, obtained above a certain ionization threshold, are shown to be stable up to the glass transition temperatures [107]. In 2004, a grating fabricated with this technique in a multimode crystalline sapphire fiber was reported and it was shown that no reduction in the grating reflectivity or any hysteresis in the Bragg resonance was detectable up to 1500 °C [108].

6.6 SUMMARY

The single-mode optical fiber has been shown to have great potential beyond the one of mere light transport for use in a diverse range of applications. The combination of several technologies in one physical fiber allows the construction of multifunctional entities with the inherent advantages of small size, light weight, and high stability. This chapter has demonstrated a large number of components and solutions based on multifunctional single-mode fibers. Given the rapid development in material sciences and of manufacturing technologies, the future is likely to continue to bring forward new exciting fiber-based solutions to new and old problems.

REFERENCES

[1] Hecht, J. 1999. *City of Light*. Oxford University Press, New York.
[2] Our own correspondent, Vienna. 1889. A new method for illuminating internal organs. *Lancet* Jan 5:52.
[3] Snitzer, E. and H. Osterberg. 1961. Observed dielectric waveguide modes in the visible spectrum. *J. Opt. Soc. Am.* 51(5):499–505.

[4] MacChesney, J. B. et al. 1973. Low-loss silica core-borosilicate clad fiber optical wave-guide. *Appl. Phys. Lett.* 23(6):340–341.

[5] Bjarklev, A. et al. 2003. *Photonic Crystal Fibres*. Kluwer Academic Publishers. Boston, MA.

[6] Yoshida, K. and T. Morikawa. 1996. Fabrication and characterization of side-hole single-mode optical fibers. *Opt. Fiber Technol.* 2(3):285–290.

[7] Chanda, J. K. et al. 2001. Adiabatic coupling in tapered air–silica microstructured optical fiber. *IEEE Photon Technol. Lett.* 13(1):52–54.

[8] Hasegawa, T. et al. 2001. Hole-assisted lightguide fiber for large anomalous dispersion and low optical loss. *Optics Express* 9(13):681–686.

[9] Nakajima, K. et al. 2003. Hole-assisted fiber design for small bending and splice losses. *IEEE Photon Technol. Lett.* 15(12):1737–1739.

[10] Rogers, A. J. et al. 2005. Distributed measurement of fluid pressure via optical-fibre backscatter polarimetry. *OFS17, Proc. SPIE* 5855:230–233.

[11] Glas, P. et al. 1995. Large-mode-area, Nd-doped single-transverse-mode dual wavelength microstructured fiber laser. *Optics Express* 13:7884–7892.

[12] Kerbage, C. and B. J. Eggleton. 2004. Manipulating light by microfluidic motion in microstructured optical fibers. *Opt. Fiber Technol.* 10:133–149.

[13] Kerbage, C. et al. 2001. Integrated all-fiber variable attenuator based on hybrid micro-structure fiber. *Appl. Phys. Lett.* 79(19):3191–3193.

[14] Kerbage, C. and B. J. Eggleton. 2003. Tunable microfluidic optical fiber gratings. *Appl. Phys. Lett.* 82(9):1338–1340.

[15] Abramov, A. A. et al. 1999. Widely tunable long-period fibre gratings. *Electron. Lett.* 35(1):81–82.

[16] Hosaka, T. et al. 1981. Low-loss single polarization fibers with asymmetrical strain birefringence. *Electron. Lett.* 17:530–531.

[17] Xie, H. M. et al. 1986. Side-hole fiber for fiber-optic pressure sensing. *Opt. Lett.* 11: 333–335.

[18] Dakin, J. and B. Culshaw. 1988. *Optical Fiber Sensors: Principles and Components*. Vol. 1. Artech House. Norwood, MA.

[19] Li, L. et al. 1986. An all-fiber electro-optic Kerr modulator. *Digest IEE Colloquium of Adv. Fib. Devices, IEE, London* 10–13.

[20] Li, L. et al. 1986. *Electron. Lett.* 22:1020.

[21] Myers, R. A. et al. 1991. Large second order nonlinearity in poled fused silica. *Opt. Lett.* 16(22):1732–1734.

[22] Kazansky, P. G. et al. 1994. High second order nonlinearity in poled silicate fibers. *Opt. Lett.* 19(10):701–703.

[23] Long, X. C. et al. 1996. A poled electrooptic fiber. *IEEE Photon Technol. Lett.* 8(2): 227–229.

[24] Wong, D. et al. 1999. Frozen-in electrical field in thermally poled fibers. *Opt. Fiber Technol.* 5(2):235–241.

[25] Long, X. C. and S. R. J. Brueck. 1997. Large-signal phase retardation with a poled electrooptic fiber. *Phot. Technol. Lett.* 9(6):767–769.

[26] Kazansky, P. G. et al. 1994. Optical fiber electrets: Observation of electro-acousto-optic transduction. *Electron. Lett.* 30(17):1436–1437.

[27] Fujiwara, T. et al. 1995. Large electrooptic modulation in a thermally poled germanosili-cate fiber. *IEEE Photon Technol. Lett.* 7(10):1177–1179.

[28] Quiquempois, Y. et al. 1998. Study of organized chi(2) susceptibility in germanosilicate optical fibers. *Optical Materials* 9:361–367.

[29] Lee, K. et al. 2004. A 200-m long fiber with internal electrode and poling demonstration. Presented at the European Conference on Optical Communication ECOC 2004, Stockholm. Paper Tu 4.3.5.

[30] Lee, K. et al. 2005. Application of a conductive polyimide coating in thermal poling of long lengths of fibre. Presented at European Conference on Optical Communication ECOC 2005, Glasgow. Paper Tu 4.6.2.

[31] Lee, K. et al. A conductive fibre coating for poling arbitrary fibre lengths. In: Conference *Proceedings of the OSA Topical Meeting on Bragg Gratings, Poling and Photosensitivity (BGPP 05)*. Paper Mo17.30. Sydney, Australia.

[32] Thomas RECORDING GmbH, Winchester Strasse 8, Europaviertel, D-3539 Giessen, Germany.

[33] Fokine, M. et al. 2002. Integrated fiber Mach-Zehnder for electrooptical switching. *Opt. Lett.* 27(18):1643–1645.

[34] Rabii, C. D. et al. 1999. Processing and characterization of silver films used to fabricate hollow glass waveguides. *Applied Opt.* 38(21):4486–4493.

[35] Myren, N. and W. Margulis. 2004. In-fiber lithography. *J. Opt. Soc. Am. JOSA B* 21(12):2085–2088.

[36] Claesson, A. et al. 2003. Internal electrode fiber polarization controller. In: *Conference Proceedings of the OFC 2003*, p. 39. Paper MF35. Anaheim, CA.

[37] Tarasenko, O. et al. 2005. All-fibre polarisation control. Presented at the Australian Conference on Optical Fibre Technology ACOFT (2005), Sydney. Paper Mo 17.15.

[38] Long, X. C. et al. 2000. A high-speed poled all-fiber switch. Bragg gratings, photosensitivity, and poling in glass waveguides. *OSA Trends Optics Photonics Series* 33:355.

[39] Margulis, W. and N. Myrén. 2005. Recent progress on poled glass and devices. Presented at the Optical Fiber Conference OFC05, Anaheim. Paper OThQ1.

[40] Li, J. et al. 2005. Systems measurements of 2x2 poled fiber switch. *IEEE Photon Technol. Lett.* 17(12): 2571–2573.

[41] Myrén, N. and W. Margulis. 2005. All-fiber electro-optic tuning of fiber laser. Presented at the European Conference on Optical Communication ECOC, Glasgow. Paper Mo 3.4.4.

[42] Myrén, N. and W. Margulis. 2005. All-fiber electrooptical mode-locking and tuning. *IEEE Photon Technol. Lett.* 17(10):2047–2049.

[43] Pruneri, V. et al. 1998. Efficient frequency doubling of 1.5 μm femtosecond laser pulses in quasi-phase-matched optical fibers. *Appl. Phys. Lett.* 72(9):1007–1009.

[44] Pruneri, V. et al. 1999. Greater than 20%-efficient frequency doubling of 1532-nm nanosecond pulses in quasi-phase-matched germanosilicate optical fibers. *Opt. Lett.* 24(4):208–210.

[45] Bonfrate, G. et al. 2000. Periodic UV erasure of the nonlinearity for quasi-phase-matching in optical fibers. Conference on Lasers and Electro-Optics (CLEO 2000), 7–12 May, 73.

[46] Corbari, C. et al. 2005. All-fiber frequency conversion in long periodically poled silica fibers. Presented at the Optical Fiber Conference OFC 2005, Anaheim. Paper OFB3.

[47] Canagasabey, A. et al. 2005. Tuneable second harmonic generation in periodically poled fibers. Presented at the Optical Fiber Conference OFC 2005, Anaheim. Paper OThQ3.

[48] Fokine, M. et al. 2004. A fibre-based Kerr switch and modulator. Presented at the European Conference on Optical Communications ECOC 04, Stockholm (2004). Paper Tu 4.3.3.

[49] Meltz, G. and E. Snitzer. 1980. Thermal and dispersive characteristics of multicore fibers. Int. URSI Symp. on Electromagn. Waves, NBS Special Publications (1980), 314B/1-B/2.

[50] Inao, S. et al. 1979. High density multicore-fiber cable. In: *Conference Proceedings of the 28th International Wire & Cable Symposium*. Cherry Hill, NJ.

[51] Kishi, N. et al. 1989. Modal and coupling-field analysis of optical fibers with linearly distributed multiple cores. *J. Lightwave Technol.* 7(6):902–905.

[52] Dong, L. et al. 1994. Intermodal coupling by periodic microbending in dual-core fibers—comparison of experiment and theory. *J. Lightwave Technol.* 12(1):24–27.

[53] Burns, W. K. and A. F. Milton. 1988. Waveguide transitions and junctions. In: *Guided Wave Optoelectronics* (T. Tamir, ed.), Vol. 26, *Springer Series in Electronics and Photonics*, pp. 89–144. Springer, Berlin, Heidelberg.

[54] Kashima, N. 1995. Passive optical components for optical fiber transmission, 225 ff. Artech House, Boston.

[55] Snyder, A. W. and J. D. Love. 1982. *Optical Waveguide Theory*. Chapman and Hall, London.

[56] Tekippe, V. J. 1999. Production, performance, and reliability of fused couplers. *Proc. SPIE* 3666:56–61.

[57] Chaudhuri, P. R. et al. 2001. Understanding coupling mechanism in fused fiber coupler-based components: Role of core- and cladding modes. *Proc. SPIE* 4417:403–408.

[58] Zervas, M. Z. et al. 1993. Advanced erbium-doped fibre amplifiers: Channel equalizers. *Proc. SPIE* 2073, Fiber Lasers and Amplifiers V:76–84.

[59] Poulsen, C. V. et al. 1997. Passive spectral gain control of a four channel WDM link employing twincore erbium-doped amplifiers. *Proc. OFC'97* 82–83.

[60] Lu, Y. B. and P. L. Chu. 2000. Gain flattening by using dual-core fiber in erbium-doped amplifier. *IEEE Photon Technol. Lett.* 12(12):1616–1617.

[61] Lu, Y. B. and P. L. Chu. 2002. Gain flattened L-band erbium-doped fibre amplifier using dual-core fibre. *Proc. OFC'02* 631–632.

[62] Kakkar, C. and K. Thyagarajan. 2005. Broadband, lossless, dispersion-compensating, asymmetrical twin-core fiber design with flat-gain Raman amplification. *Appl. Opt.* 44(12):396–401.

[63] Winful, H. G. and D. T. Walton. 1992. Passive mode locking through nonlinear coupling in a dual-core fiber laser. *Opt. Lett.* 17(23):1688–1690.

[64] Oh, Y. et al. 1995. Robust operation of dual-core fibre ring laser. *J. Opt. Soc. Am. JOSA B* 12(12):2502–2507.

[65] Martí-Panameño, E. et al. 2001. Self-mode-locking action in a dual-core ring fiber laser. *Opt. Comm.* 194:409–414.

[66] Graydon, O. et al. 1996. Triple frequency of an Er-doped twincore fiber loop laser. *IEEE Photon Technol. Lett.* 8(6):63–65.

[67] Kaňka, J. et al. 2000. Line-narrowing and wavelength stabilization in a tuneable Er-Yb fibre ring laser with and Er twin-core fibre. In: *Proceedings of the SPIE*, Vol. 4016, *Photonics, Devices and Systems*, pp. 315–319. Prague, Czech Republic.

[68] Peterka, P. et al. 2003. Generation of high repetition rate pulse trains in fiber laser through a twin-core fiber. In: *Proc. SPIE*, Vol. 5036, *Photonics, Devices and Systems II*, pp. 376–381. Bellingham, WA. SPIE—The International Society for Optical Engineering.

[69] Wrage, M. et al. 2001. Structured mirrors supporting phase-locking and beam-shaping of a multicore fiber laser. In: *Conference Proceedings of the CLEO*, pp. 364–365. Baltimore, MD.

[70] Yanming, H. et al. 2004. Fundamental mode operation of a 19-core phase-locked Yb-doped fiber amplifier. *Optics Express* 12(25):6230–6240.

[71] Ortega, B. and L. Dong. 1999. Accurate tuning of mismatched twin-core fiber filter. *Opt. Lett.* 23(16):1277–1279.

[72] Atkins, G. R. et al. 1994. UV tuning of coupling in twin-core optical fibres. *Electron. Lett.* 30(25):2165–2166.

[73] An, H. et al. 2004. Long-period-grating-assisted optical add-drop filter based on mismatched twin-core photosensitive-cladding fiber. *Opt. Lett.* 29(4):343–345.

[74] Chu, P. L. and B. Wu. 1992. Optical switching in twin-core erbium-doped fibers. *Opt. Lett.* 17(4):255–257.

[75] Meltz, G. et al. 1983. Cross-talk fiber-optic temperature sensor. *Appl. Opt.* 22(3): 464–477.

[76] Romaniuk, R. S. 1985. Broadband buses based on multicore fibres. In: *Proceedings of SPIE*, Vol. 585, *Fiber Optic Broadband Networks*, pp. 260–267. Bellingham, WA. SPIE—The International Society for Optical Engineering.

[77] Le Noane, G. et al. 1996. Towards FTTH networks based on new passive, active devices and installation techniques. In: *Proceedings of the European Conference on Networks and Optical Communications*, pp. 274–277. IOS Press, USA.

[78] Rosinski, B. et al. 1999. Multichannel transmission of a multicore fiber coupled with a vertical-cavity surface-emitting laser. *J. Lightwave Technol.* 17(5):807–810.

[79] Nayar, B. K. et al. 1991. All-optical switching in 200-m twin-core fiber nonlinear Mach-Zhender interferometer. *Opt. Lett.* 16(6):408–410.

[80] Yvernault, P. et al. 2001. Passive athermal Mach-Zehnder interferometer twin-core fiber optical add/drop multiplexer. In: *Proceedings of the 27th European Conference on Optical Communication*, Vol. 6, pp. 88–89. Copenhagen, Denmark, IEEE.

[81] Gauden, D. et al. 2004. Tunable Mach-Zehnder–based add-drop mulitplexer. *Electron. Lett.* 40(21):1374–1375.

[82] Gauden, G. et al. 2004. Variable optical attenuator based on thermally tuned Mach-Zehnder interferometer within a twin-core fiber. *Opt. Comm.* 231:213–216.

[83] Noda, K. et al. 1984. Twisting angle sensing using dual core fiber. In: *Proceedings of the 2nd International Conference on Optical Fiber Sensors*. Sttutgard, Germany, SPIE.

[84] Tanak, T. et al. 1990. Quench detection of superconducting magnets at very low temperature using optical fibers. In: *Proceedings of the 7th Optical Fibre Sensors Conference*, pp. 399–402. Sydney, Australia.

[85] Khotiaintsev, S. N. et al. 1996. Laser Doppler velicometer miniature differential probe for biomedical applications. *Proc. SPIE*, 2928:158–164.

[86] Wosinski, L. et al. 1994. Experimental system considerations for distributed sensing with grating Fabry-Perot interferometers. *Proc. SPIE* 2360:146–149.

[87] Gander, M. J. et al. 2002. Two-axis bend measurement using multicore optical fibre. *Opt. Comm.* 182:115–121.

[88] MacPherson, W. N. et al. 2004. Pitch and roll sensing using fibre Bragg gratings in multicore fibre. *Meas. Sci. Technol.* 15:1642–1646.

[89] Bulut, K. and M. N. Inci. 2005. Three-dimensional optical profilometry using a four-core optical fibre. *Optics Laser Technol.* 37:463–469.

[90] Watson, S. et al. 2004. Laser-machined fibres as ultra-miniature pressure sensors. *Proc. SPIE* 5502:112–115.

[91] Dorosz, J. and R. Romaniuk. 1999. Current developments of multi-crucible technology of tailored optical fibers. *Proc. SPIE* 3731:32–58.

[92] Fernandez A. et al. 2004. Behaviour of chemical composition gratings in a very hard mixed gamma neutron irradiation field. In: *Proceedings of the Radiation Effects on Components and Systems Conference, RADECS2004*, pp. 97–100.

[93] Erdogan, T. et al. 1994. Decay of ultraviolet-induced fiber Bragg gratings. *J. Appl. Phys.* 76:73–80.

[94] Patrick, H. et al. 1995. Annealing of Bragg gratings in hydrogen-loaded optical fiber. *J. Appl. Phys.* 78:2940–2945.

[95] Grubsky V. et al. 1999. Effect of molecular water on thermal stability of gratings in hydrogen-loaded optical fibers. In: *Proceedings of OFC'99, ThD2*, pp. 53–55. San Diego, CA. OSA—Optical Society of America.

[96] Kannan, S. et al. 1996. Thermal reliability of Bragg gratings written in hydrogen-sensitized fibers. In: *Proceedings of OFC'96, TuO4*. San Jose, CA. OSA.

[97] Archambault, J. L. et al. 1993. 100% reflectivity Bragg reflectors produced in optical fibres by single excimer laser pulses. *Electron. Lett.* 29(5):453–455.

[98] L. Dong et al. "Thermal decay of fiber Bragg gratings of positive and negative index changes formed at 193 nm in a boron-co-doped germanosicilate fiber", Appl. Opt. 36(31) (1997) 8222-8226.

[99] Dianov, E. M. et al. 1997. Grating formation in a germanium free silicon oxynitride fibre. *Electron. Lett.* 33(3):236–238.

[100] Butov, O. V. et al. 2002. Ultra-thermo–resistant Bragg gratings written in nitrogen-doped silica fibres. *Electron. Lett.* 38(11):523–525.

[101] Brambilla, G. et al. 2002. Fiber Bragg gratings with enhanced thermal stability. *Appl. Phys. Lett.* 80(18):3259–3261.

[102] Shen, Y. et al. 2004. High-temperature sustainability of strong fiber Bragg gratings written into a Sb-Ge-codoped photosensitive fiber. Decay mechanisms involved during annealing. *Opt. Lett.* 29(6):554–556.

[103] Fokine, M. 2002. Photosensitivity, chemical composition gratings, and optical fiber based components [Doctoral Thesis], Royal Institute of Technology, Sweden.

[104] Fokine, M. et al. 1996. High temperature resistant Bragg gratings fabricated in silica optical fibres. *Proc. ACOFT'96, PD2.*

[105] Pal, S. et al. 2003. Characteristics of potential fibre Bragg grating sensor-based devices at elevated temperatures. *Meas. Sci. Technol.* 14:1131–1136.

[106] Mihailov, S. J. 2003. Fiber Bragg gratings made with a phase mask and 800-nm femtosecond radiation. *Opt. Lett.* 28(12):995–997.

[107] Smelser, C. W. et al. 2005. Type I-IR and type II-IR fiber Bragg grating formation with an ultrafast infrared source and a phase mask. In: *Proceedings of BGPP' 2005*. Sydney, Australia, OSA.

[108] Grobnic, D. et al. 2004. Sapphire fiber Bragg grating sensor made using femtosecond laser radiation for ultrahigh temperature applications. *IEEE Photon Technol. Lett.* 16(11)2505–2507.

Chapter 7

Rare Earth-Doped Fibers

David J. DiGiovanni,[1] Roman Shubochkin,[2] T. F. Morse,[2] and Borut Lenardic[3]

[1] *OFS Bell Labs, Murray Hill, NJ*
[2] *Laboratory for Lighthouse Technology, Boston University, Boston, MA*
[3] *Optacore d.o.o., Ljubljana, Slovenia*

7.1 INTRODUCTION

Rare earth (RE) doping of optical fibers dates back to the 1960s and was one of the forces driving development of guided wave optical fibers. The goal was to exploit the long path length provided by wave-guiding media to improve operation of Nd, Er, and Er/Yb fiber lasers. Then, as now, the fiber consisted of regions with raised refractive index to guide light and some distribution of RE ions that interacted with this light. Very simply, the goal remains to exploit the optical activity of the RE elements to create a laser or amplifier. From its inception to the late 1980s, such fibers were primarily a research platform for study of various optical phenomena, which, though interesting, did not mature into commercial products. There were no compelling applications or needs that were filled by fiber lasers. This changed very quickly and very dramatically with the discovery of the erbium-doped fiber amplifier (EDFA) because a critical need for optical amplification arose, and because the EDFA was able to fulfill that need exceptionally well.

The success of the erbium-doped fiber spawned an industry driven to improve and surpass its performance. This included the search for alternative dopants and hosts, as well as improved fiber designs. Naturally, these efforts morphed into other uses and applications for amplifier fiber technology, the most active one being the quest for very high power fiber lasers and amplifiers for industrial materials processing applications. RE-doped silica fibers play an important role in a variety of modern technologies. Fiber lasers and amplifiers using such fibers are extensively used in basic and applied research, medicine, and military applications. They are replacing gas and solid state devices in industrial materials processing and are used in a variety of fiber optic sensors.

Many of these applications take advantage of unique properties of silica-based glasses, such as excellent optical transmission from ultraviolet (UV) to near-infrared wavelengths (with some of the lowest losses possible), isotropy of optical properties of the glass, excellent refractive index homogeneity with a low nonlinear refractive index, small strain birefringence, a very low coefficient of thermal expansion, very high thermal stability, very high chemical and environmental durability, high mechanical strength, and resistance to radiation. Although the field is by no means new, quickly developing military and commercial applications, particularly involving high-power lasers and amplifiers, put new and demanding requirements on quality and long-term reliability of such fibers, as well as stimulate further research and development in the field of RE fiber fabrication.

This chapter discusses two main categories of RE-doped fibers: fibers for telecommunications applications and fibers for high-power source applications. The former comprises Er and Er/Yb fibers while the latter covers a broader range of RE elements (Er, Er/Yb, Yb, Tm, Ho, etc.) and fiber architectures. A brief discussion of the general optical activity of RE dopants is followed by discussion of fiber fabrication. A more thorough treatment of fiber fabrication technology can be found in Chapter 3, but the section here focuses on unique aspects of incorporating RE ions into glass. This section is followed by details of the design and operation of the most important types of RE-doped fiber. The discussion is limited to glass fibers and is dominated by silica, which by far has been the most successful. Though discussed briefly here, a more comprehensive review of waveguide design can be found in Chapter 2, while RE fiber design for specific industrial applications can be found in Chapter 22.

7.2 MOTIVATION

Why put RE elements in an optical fiber? Because RE ions in glass are optically active, meaning they absorb light at one wavelength and emit light at another [1]. This can be useful in creating a laser, a broadband source, or a signal amplifier at the emission wavelength. Figure 7.1 illustrates this concept using the well-known example of an erbium-doped amplifier. Figure 7.1 shows the absorption spectrum of Er in a silica fiber, while the inset shows the bottom three levels of the energy level diagram. When light at a wavelength near 980 or 1480 nm is incident upon Er ions in the fiber, the photons are absorbed, causing excitation into higher energy levels. These relax to the lowest excited level whereupon final relaxation back to the ground state is accompanied by emission at a wavelength around 1530 nm. This light is quickly reabsorbed by ions in the ground state, but if a population inversion exists (number of ions in upper state exceeds that in ground state), emission will exceed absorption and gain about 1530 nm results.

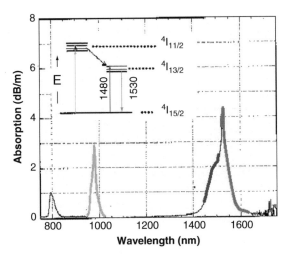

Figure 7.1 Absorption spectrum of Er in silica. Inset shows the lowest energy levels of the Er ion, with indication of absorption at 980 and 1480 nm and emission at 1530 nm.

The spectra and optical behavior are the result of a quirk in the electronic levels of the REs, which include the lanthanide and actinide series (atomic numbers 57–103). Using the classic description of an atomic nucleus surrounded by shells of electrons, for most of the periodic chart, as the atomic number increases, the electron shells are filled with progressively increasing radius. However, beginning with lanthanum (atomic number 57), the situation changes. Lanthanum has filled 5s and 5p shells and addition of another electron occupies the 4f shell, which has a smaller radius than the 5s and 5p shells. The filled outer 5s and 5p shells of the RE elements effectively shield the unfilled bands and cause the ions to exhibit atomic-like properties. Although many RE ions can exist in a divalent state, they are most commonly incorporated in glasses as trivalent ions in which two electrons are absent from the 6s shell and one from the 4f shell. The resulting electron structure is similar to xenon with only partially filled electron orbital f^{N-1}. Normally, elements doped into a solid give up an electron to the matrix because the outermost wave functions are delocalized. For RE elements, shielding results in relatively well-defined energy levels, which are exhibited as narrow absorption and emission bands.

The incompletely filled 4f energy levels are composed of states that are spread in energy due to spin–spin and spin–orbit coupling. These levels are spread still further when the ions interact with a host material such as a crystal or solid because the degeneracy of levels is lifted somewhat by slight asymmetry of the environment. This Stark splitting gives rise to multiple transitions, which comprise the spectra of the REs. In Er, for example, the ground state has eight

Stark levels and the first excited state has seven, so the emission around 1530 nm results from an ensemble of 56 individual transitions. Although transitions within the 4f shell are strictly forbidden in an electric dipole because the initial and final states have the same parity, asymmetry from the local environment causes some admixture of higher lying states with opposite parity and strongly influences the oscillator strength. The local environmental asymmetry present in glasses causes crystal field splitting of the energy levels and results in spectra that are much broader than found in a crystalline host, but considerably more narrow than that exhibited by other elements, such as transition metals in glass. This is evident in the spectrum of Fig. 7.1 and is true of all of the lanthanide series. See reference [2] for an excellent review of the electronic properties of RE ions.

7.3 HOST GLASSES FOR RARE EARTH IONS

The highly polymerized structure of pure silica glass does not allow easy accommodation of RE ions even at low concentrations. Trivalent RE ions do not substitute for silicon in the glass network easily, if at all [3], and at the same time they need six to eight oxygen ions for their coordination. Silica has very few nonbridging oxygen ions that can provide such coordination and RE ions are forced to cluster together to share those few oxygen ions that are present in the network. Such clustering can lead to enhanced energy transfers between f^{N-1} energy levels of neighboring RE ions, concentration self-quenching of luminescence even at low doping, and rapid phase separation at higher doping levels. For example, the formation of Nd—O—Nd bonds was directly observed in silica glass doped with 2400 ppm of Nd_2O_3 by the extended x-ray absorption fine structure (EXAFS) spectroscopy with Nd—Nd and Nd—O distances similar to those found in crystalline Nd_2O_3 [4]. Such close pairing can result in luminescence self-quenching due to very efficient cooperative energy transfers, which cause losses of ion excitations. The signature of concentration quenching is the decrease of luminescence lifetime with increasing RE concentration.

To increase RE solubility in silica and decrease the negative effects of clustering, co-doping is most often used in all fabrication methods. Co-dopants also provide the index modification needed for creating a wave-guiding structure and are often used to alter spectroscopic properties, such as increasing the emission bandwidth or shifting undesirable excited state absorption features. A special category of co-dopants consists of other lanthanides such as lanthanum and lutetium that tend to cluster together with optically active ions shielding them from one another, increasing the distance between neighboring optically active ions, and therefore decreasing efficiency of cooperative energy transfers [5–7]. This can allow higher active RE concentration or improved device performance.

The most popular solubilizer is aluminum. Aluminum can be incorporated into the silica network either in tetrahedral coordination as a network former or in octahedral coordination as a network modifier. The four coordinated aluminum shares nonbridging oxygen ions with RE ions, therefore reducing RE ion clustering [8]. As an illustration, direct probing of neodymium-doped aluminosilicate glasses using EXAFS spectroscopy showed a breakup of Nd—O—Nd linkages and the formation of Nd—O—Si/Al bonds at their expense. From the solution chemistry point of view, alumina dissolves well in silica, while RE oxides dissolve well in alumina. Therefore, alumina forms a solvation shell around RE ions, allowing them to become soluble in a silica network [9]. Arai et al. [9] showed that a molar ratio Al/Nd of 10 was enough to make a neodymium-doped glass suitable for laser applications [9], as confirmed by Magic Angle Spinning (MAS), Nuclear Magnetic Resonance (NMR), and Electron Paramagnetic Resonance (EPR) measurements [10], while a similar impact for Er has been confirmed [11].

Phosphorus is another popular co-dopant, and in fact, phosphate glasses provide an excellent matrix into which high concentrations of RE ions can be readily incorporated. High gain coefficient phosphate glass fiber amplifiers with high gain per unit length were demonstrated [12], although their poor compatibility with high silica telecommunication fibers and low mechanical and thermal stability are still known issues. When co-doped into silica, phosphorus plays a role similar to aluminum and incorporates in two tetrahedral configurations P_2O_5 and P_2O_4, with the second configuration attracting charge compensation cations, such as RE ions. A solvation shell is formed around RE ions, allowing their accommodation within silica network without clustering. A molar ratio P/Nd of 15 or more was shown to be sufficient for laser applications [9]. Similar to use of Al, addition of P can be used to solubilize the RE or alter the spectral properties. Phosphorus is essential for efficient operation of Er/Yb–co-doped systems, as is discussed in detail later in this chapter.

Other less common dopants that were used for co-doping RE-doped fibers include alkali and alkaline earth metals [13, 14]. Quite often, a combination of several dopants both from the first and second groups are used in order to obtain desired wave-guiding and spectral properties.

The primary goals in altering the host glass for RE ions for photonic applications is to tailor the absorption and emission spectra, influence excited state properties, and improve the glass-forming characteristics. Unfortunately, the atomic structure, which confers unique optical properties on RE ions, also conspires to make these properties relatively immune to changes in local glass environment. A huge literature has been developed on hosts for RE ions for solid state lasers and other optical and nonoptical applications, but relatively little of this has been (or can be) applied to optical fiber. Many glass systems are not appropriate because of difficulty in creating the required index structure to produce wave guiding or because of optical attenuation, which would be excessive in a fiber device that has a path length of meters. The dominant host has been silica, but the range of compositions is small because of glass-forming limitations. Even use of modest dopant levels is problematic because most elements cause excess optical attenuation and must be avoided, whereas the remaining few have only modest influence. Some, such as phosphorus as a co-dopant for Er, cause undesirable changes in the spectrum, such as spectral narrowing.

More radical modification requires use of other hosts, such as phosphates, borates, tellurites, and fluorophosphates, as well as nonoxygen hosts such as fluorides, sulfides, and other chalcogenides. For these, there are several benefits. The dominant influence is a profoundly altered phonon energy spectrum, which reduces nonradiative excited state decay and allows radiative emission, which would be quenched in silica. In addition, other host glasses allow lasing transitions that occur beyond the 2.0-μm absorption edge of silica. These glasses are mainly nonoxides and heavy oxide materials like TeO_2. Glasses such as phosphates are useful because the dopant concentration can be very high. Finally, some hosts are chosen because the high polarizability enhances Stark splitting and creates a more desirable spectrum. The range of possibilities is too broad for this discussion, but nonsilica materials are discussed briefly in the following sections.

7.4 FABRICATION OF RARE EARTH-DOPED FIBERS

Before discussing details of the manner in which RE ions are incorporated into optical fiber, the basic methods for fiber fabrication are reviewed. A more comprehensive discussion can be found in Chapter 3, but the discussion here focuses on the issues that are salient for RE incorporation. Several comprehensive reviews on the subject of RE fiber fabrication are available from the early years of development [15–17].

7.4.1 Overview of Optical Fiber Fabrication

Fabrication of low-loss optical fiber requires very pure precursor materials and a method to produce the cylindrically symmetrical preform from which the fiber is drawn. To achieve the current level of transparency for optical communications transmission fiber, the concentration of impurities such as iron (which gives window glass its characteristic green color) must be reduced to less than one part per billion (ppb). Hydroxyl (OH^-) impurities from water contamination create a characteristic absorption band at 1.39 μm and are typically reduced to less than 10 ppb. Amazingly, such high purity is readily achieved in large-scale manufacturing using several vapor-phase processes practiced globally. All possess the similar characteristic that silicon-containing vapor (such as $SiCl_4$) is reacted with oxygen to form small particles of silica (SiO_2), as first reported by Maurer [5] in 1973. Such vapor-deposition processes have been refined to comprise two categories: outside vapor deposition (OVD) and internal deposition processes such as modified chemical vapor deposition (MCVD) reported by MacChesney et al. in 1974 [6].

In MCVD, the first process for commercial fabrication, this reaction occurs inside a silica tube, as shown schematically in Fig. 7.2. Vapor of silicon tetrachloride is delivered into a rotating silica tube and reacted with oxygen at approximately 1300 °C to create submicron silica particles. The composition of the particles is determined by the composition of the gas phase, so the addition of $GeCl_4$, for example, will produce Ge-doped particles. These particles are driven to the cooler glass wall by thermophoresis, where they deposit on the inner surface of the tube. The reaction hot zone is produced using an external torch, such as an oxy-hydrogen burner. As this burner traverses down the tube, the deposition zone also translates. As the torch passes over the previously deposited "soot," the particles are sintered to form clear glass. Because the glass is formed at high temperature in the presence of chlorine, the primary impurities (transition metals and water) remain as volatile chlorides and exit the tube. This allows purification by many orders of magnitude and is a critical aspect of the process. If the precursors contain contaminants such as hydrogen, each layer must be purified before being fully consolidated.

To build the desired wave-guiding structure in the glass, multiple layers are deposited, each of appropriate composition to either raise (e.g., using germanium) or depress (using fluorine) the refractive index relative to silica. The tube is collapsed to a solid rod by surface tension when heated to approximately 2300 °C. The result is a cylinder of glass with a doped region at the center.

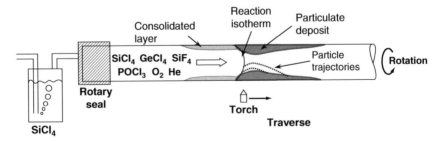

Figure 7.2 Schematic of modified chemical vapor deposition (MCVD) showing vapor-phase precursors, reaction at hot zone, creation and deposition of particles, and sintering of particulate soot into solid glass inside a rotating glass tube.

Alternative techniques for making this rod entail creation of the silica particles in a flame through hydrolysis rather than oxidation. In OVD, the flame traverses along a mandrel and the silica particles are collected as porous soot. Multiple passes of the torch build up the soot (and the index structure) radially. After the mandrel is removed, the soot is purified in a chlorine-containing atmosphere in a furnace and sintered to clear glass. In vapor-phase axial deposition (VAD), the chemistry is similar, but the flame is directed at the end of a rotating rod. As the rod is gradually withdrawn, a long soot boule is grown axially. Multiple torches are used to form the Ge-doped core and the pure silica cladding. Because porous boules created by outside processes are purified before sintering, the glass precursors may contain hydrogen or other impurities. This allows a wider range of starting chemicals to be used, such as organometallic precursors instead of $SiCl_4$.

For all of these processes, after the core rods are formed, they are overclad or jacketed to build up the outer diameter and then drawn into fiber. These aspects of the process are unchanged when producing RE-doped fibers, so the reader is referred to Chapter 3 for a more complete discussion. Several books discussing these processes are available [18, 19].

7.4.2 Incorporation of Rare Earth Elements

A key feature of the preceding discussion is that the glass precursors are vapors formed from high vapor pressure compounds such as the chlorides. In addition, the use of vapor methods precludes transport of transition metal ions and other contaminants to the reaction zone during deposition and allows for the routine fabrication of fiber approaching the theoretical minimum loss in silica. However, these same properties prohibit incorporation of many potentially useful dopants and only several elements (such as Si, Ge, P, B, Ti) can be conveniently incorporated from liquid phase through vapor deposition at relatively high levels. Because the desired RE elements do not possess suitable high vapor pressure compounds, a host of alternative doping methods have been developed for use with OVD, VAD, and MCVD [20, 21]. The most widely used techniques are discussed here. Several other methods exist such as rod-in-tube [22, 23], seed fiber [24], and various vapor-delivery techniques [20, 25, 26], but none of these is practiced on a wide scale.

7.4.2.1 Vapor-Phase Methods

All-vapor delivery avoids contamination of the system, accelerates the glass formation process, and allows superior control of the refractive index profile. Vapor delivery may be accomplished by heating RE chlorides to around 1000 °C

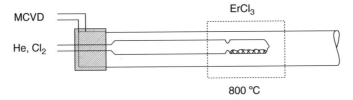

Figure 7.3 Apparatus for delivery of RE chlorides in modified chemical vapor deposition (MCVD). Note that RE is protected from oxidation at high temperature.

to generate sufficient vapor [27]. Such an apparatus is shown in Fig. 7.3 and consists of an "injector" containing a suitable RE halide or nitride situated inside a rotating MCVD silica substrate tube and heated by an external furnace. Because the vapor pressure of the RE precursor increases with temperature exponentially, however, such a process is difficult to control. Note that the understanding of early work [25, 28, 29] using RE halides situated on the inside surface of the substrate tube was flawed because the halides quickly react with oxygen at elevated temperature to form refractory materials and are not delivered as halides.

While heated RE halides produce enough vapor to obtain modest glass doping levels, their vapor pressure is still quite low (well under 1 Torr at 600 °C). However, it is well known that aluminum chloride readily complexes with many metal chlorides, forming a high vapor pressure complex. Thus, by passing aluminum chloride over erbium chloride, the complex $Al_xErCl_{3(x+1)}$ will be generated [30, 31], increasing the delivery rate by seven orders of magnitude at 600 °C and reducing the variation of vapor pressure with temperature. Using this technique, many core layers may be deposited and dehydration is not required. However, it becomes complicated if multiple dopants are desired, such as for Er/Yb–doped fibers, which is discussed later in this chapter.

The surface plasma chemical vapor deposition (SPCVD) technique is a modification of the MCVD technique that allows a lower deposition temperature. In a further development, vapor phase SPCVD deposition of erbium and aluminum from heated anhydrous salts allowed fabrication of erbium-doped fibers with 5700 ppm weight of erbium [32]. An amplifier made from this fiber demonstrated a 0.5-dB/cm gain when pumped at 980 nm. It was argued that the lower deposition temperature that the SPCVD provides decreases RE clustering, leading to the higher gain.

7.4.2.1.1 Chelates

Vapor phase delivery is also possible with certain metal organic compounds of RE metals heated to around 200 °C. Such methods were used for the fabrication of optical waveguides [33] and fiber optic preforms using both OVD [34] and

MCVD [35]. The precursor materials are typically β-diketonate complexes, which are fluorinated to increase volatility. To aid processing of high Nd^{3+} and Yb^{3+} MCVD preforms, unfluorinated β-diketonate compounds have been used [36]. In these processes, a carrier gas of helium or argon is used to transport the vaporized material to the OVD or MCVD preform being processed. Although this requires heated delivery lines and careful metering of gas flows, the process is readily controlled and is suitable for multiple RE dopants. With any metal organic delivery, hydrogen is present in the precursor material, so the deposited glass must be dehydrated in the same manner as for solution doping.

7.4.2.2 Nonvapor Methods

The first neodymium-doped optical fiber was fabricated by the "rod-in-tube" method from a multicomponent silicate glass rod overclad with a soda-lime silicate tube [37]. The same approach was used to fabricate RE-doped fibers with core glass that was compatible with a pure silica overcladding tube [23]. Similar to these methods, the powder-in-tube technique allowed fabrication of ultrahigh RE concentration silica fibers (54 wt% of Tb_2O_3) for Faraday isolator applications [38]. The technique allows a single draw directly to the final fiber with a small core diameter, so a much higher thermal expansion mismatch can be handled than would ever be possible in regular preform fabrication methods.

Vapor methods that were described earlier offer significant advantages over these "bulk glass" techniques, which suffer from poor dimensional control and high background attenuation. In addition, they may require significant modifications of existing standard equipment. However, multiple co-dopants can be easily incorporated and they allow high levels of RE and other dopants.

Various nonvapor-delivery methods were developed to circumvent limitations of both of the vapor phase approaches earlier bulk glass methods. Contamination from transition metal ions and hydroxyl complexes can generally be avoided by a careful choice of high-quality dopants and an appropriate drying/sintering treatment [39].

7.4.2.2.1 Solution Doping

The simplest and most widely used method for doping optical preforms with alternative dopants is solution doping or molecular stuffing [25, 26, 40]. In this process, as applied to MCVD, the torch temperature during glass deposition is reduced so that sintering of the soot is incomplete and the glass layer left behind the torch remains porous. The tube is then soaked in a liquid that contains the desired dopant to permeate the porous glass. As the body is dried, the dopants are adsorbed or precipitate in the pores of the glass and become incorporated in the glass when it is finally sintered. Additional doped layers can be built up sequentially. Conventional dopants such as germanium, phosphorus, and fluorine are incorporated from the gas phase while other elements such as $AlCl_3$ and low-vapor pressure precursors such as RE chlorides or nitrates are dissolved in the liquid. Virtually any solvent and any soluble precursor free of impurities may be used. A wide range of elements have been successfully incorporated. Because the porous glass soot is sintered at high temperature ($>2100\,°C$ if necessary), crystallization of the glass is avoided. This allows high concentration of dopants such as Al to be incorporated readily without crystallization.

Although solution doping is easy to implement, control of dopant concentration, which depends on surface adsorption, solute concentration, soot density, and dehydration conditions, is difficult. For example, the soot density is controlled by glass viscosity, which varies exponentially with temperature and the concentration of dopants such as Ge. In addition, to achieve uniform doping, the solution must be drained and dried evenly, avoiding a buildup of material at the bottom or on one side of the body due to gravitational settling. Dehydration, which is most effectively accomplished by treating the body in an atmosphere containing Cl_2 at $800\,°C$ for about an hour [41], may alter the composition by volatilizing the dopants as chlorides. Finally, it is difficult to build up large or complex cores because each layer requires considerable time to deposit, cool, soak, dry, dehydrate, and then sinter each layer, with the probability of forming a defect in the soot increasing with each pass.

The soot processes, VAD and OVD, lend themselves readily to solution doping because the glass boule is already in the form of a porous body. However, there are several critical limitations that complicate the process. Because the porous boule is typically many millimeters in diameter, soluble dopants migrate to the surface as the solvent evaporates. This creates a pronounced radial gradient in dopant concentration. The effect of this gradient can be reduced by removing the outer regions of the cylinder. Alternatively, the solute can be immobilized using hypercritical drying, freeze-drying, or precipitation. Because dehydration and sintering of the boules occurs below about $1400\,°C$, there is a strong tendency for the doped bodies to devitrify before densification is complete. This can limit aluminum concentration to only a few percent to avoid formation of mullite. As is shown later in this chapter, Er-doped fiber has optimum spectral properties when the Al concentration is greater than 10 mol%. This level is difficult to achieve with outside processes.

The solvents most commonly used are water and alcohols. Water is often preferred because it provides higher solubility limits and lower viscosity at high concentration, which is important for faster adsorption [42]. Various acids can also be used as a solvent and may provide an added advantage of lower pH, which can potentially enhance dopant adsorption [42]. For example, phosphoric acid was used to prepare fiber RE concentration up to 2.8 mol% while serving as a precursor for phosphorus [43, 44].

Despite the relative simplicity of water and alcohol solution doping, questions of glass homogeneity and the influence of process variables on glass quality remain and little published experimental data exist to clarify the issue. It is possible that evaporation of the solvent leaves a crystalline residue of hydrated RE salts in the pores of the frit, which form clusters or microcrystallites of refractory RE oxide upon heating and sintering [45, 46]. Even if the halide residue is washed away with acetone or other solvents (see reference [42], p. 47), the ion adsorption preferentially occurs inside the pores of the frit, so it is possible that areas of highly doped glass are formed after these pores collapse during sintering. This doping technique would then be inherently prone to some initial non-uniformity, with the frit nanostructure and pore size playing an important role.

Adsorption during solution doping can be of a physical or chemical nature, with indirect experimental evidence supporting the prevalence of the latter (see reference [42], p. 45). For example, efficiency of adsorption strongly depends on the type of metal ion. While some ions such as lanthanides are incorporated quite easily, others like transition metal ions are hardly adsorbed at all. There exists incorporation interdependence between ions of different species, with aluminum, for example, enhancing the adsorption efficiency of the RE ions. The ion size does not seem to play a role, as would be expected if only physical trapping was taking place. Solution acidity is also quite important with the pH range over which maximum adsorption occurs, strongly depending on the cation in question (see reference [43], p. 62). Surface cation sorption onto silica is often explained by an ion-exchange mechanism, with silica acting as an ion exchanger of the weakly acid type (see p. 60 in reference [43] and [46]).

Assuming that chemical bonding and ion-exchange reactions are mostly responsible for cation adsorption during solution doping, a linear dependence between solution strength and cation adsorption efficiency implies the existence of equilibrium of the exchange reaction between "free" cations in the solution and cations bonded to the frit. The equilibrium is determined by both relative concentration of cations and the bond strength between them and the frit. Saturation of the cation incorporation at higher solution strengths may suggest a decrease of available bonding sites (see reference [42], p. 29).

7.4.2.3 Aerosol Doping

The limited availability of volatile glass reagents led to the development of liquid aerosol methods that do not rely on the vapor pressure of the precursors. Bulk glasses, including silicates, were made by flame hydrolysis of the vapors of glass precursors while being co-doped with various metal ions through aerosol delivery of nebulized solutions to the reaction zone [47]. The same approach using flame hydrolysis of chloride vapors and aerosolized aqueous solutions of metal salts was used for fabrication of passive optical silica fibers [48, 49]. It was observed that in bulk glasses, because of poor mixing of glass precursor vapors and aerosol, the presence of large aerosol particles in the aerosol stream lead to non-uniformity and solid inclusions of the final glass. For the fiber fabrication, a specially designed multiconduit burner was used to improve the mixing of incoming reagent streams and to increase the dopant content of fiber preforms.

In these techniques, vapors of liquid glass-forming chlorides, such as $SiCl_4$, $GeCl_4$, and $POCl_3$, were used as glass-forming precursors. This, in addition to the observed inhomogeneity due to the poor mixing with aerosol, may also hinder efficient incorporation of metal oxides because of the production of high amounts of chlorine as a byproduct of the oxidation reaction.

These problems are avoided using an aerosol-doping technique that combines all glass components in one solution without significant amount of chlorine. It can be applied to both MCVD and OVD [50]. For the MCVD setup, an ultrasonic nebulizer filled with a solution produces fine ($4-10\,\mu m$) aerosol mist that is delivered by a carrier gas to the reaction hot zone. In Fig. 7.4, we see the results of nebulization from a 1.6-MHz transducer. The nebulization rate from a single transducer can be as high as 2 g/min, which is more than adequate for core doping of single-mode preforms. Similar to the arrangement shown in Fig. 7.3, an injection tube carries the aerosol from the nebulizer into the substrate tube. To maintain constant spacing between the injector exit and the hot zone, the tube is moved axially along with the burner. This produces a more uniform deposit.

As with other nonvapor methods, some post-deposition processing is required before sintering. Depending on the nature of solution, post-processing steps may include removal of residual carbon from the soot and drying. Because this is

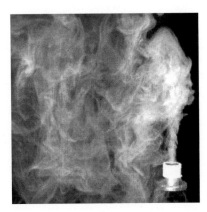

Figure 7.4 Aerosol generated by ultrasonic nebulizer.

carried out with a substrate tube on the glass MCVD lathe, processing is trivial and far less complex than the drying process necessitated by OVD fabrication. Dry and low-loss fibers with high RE concentrations of 10 wt% and higher can be routinely fabricated.

Unlike solution and sol-gel doping, aerosol doping does not require removal and subsequent remounting of the substrate tube on the lathe. Other additional steps like soaking, draining, and drying for solution doping are also unnecessary. At the same time, the deposition efficiency depends on the physical and chemical properties of a particular solution, whereas axial uniformity requires control of process parameters, just as in vapor deposition.

The same technique was modified for use in the OVD. While the described high-frequency nebulizer can be successfully used in both MCVD and OVD setups, in the OVD it can be replaced with a lower frequency ultrasonic atomizer that produced larger (up to $70\,\mu$m) aerosol droplets, providing higher delivery rates of material. Because the solution is fed directly into the nozzle and immediately atomized at the nozzle tip, the requirements on solution viscosity and solvent evaporation rates are greatly relaxed.

Two basic kinds of solutions are normally used: inorganic and organic. Inorganic solutions with metal halides and organic solutions with metal alkoxides and salts are similar to the solutions used for solution doping and the sol-gel method, respectively. Tetraethyl-orthosilicate (TEOS) is most often used as a silica precursor in organic solutions. It is normally mixed with organic solvents used to dissolve various metal organic compounds, including precursors for aluminum, REs, phosphorus, boron, germanium, or any other chemical elements and their combinations. This makes the process extremely accommodating as to which chemical element can be doped in the glass because almost any element can be successfully put in a solution with the appropriate combination of a solute and solvent. For example, TEOS can be combined with other precursors for glass-forming oxides, such as germanium and boron ethoxides, for fabrication of photosensitive fibers from one solution. Fibers co-doped with both boron and germanium oxides were successfully fabricated with concentrations estimated to be more than 10 mol% of each oxide [51]. Simultaneous co-doping of such glasses with RE ions is also possible.

It is also to be noted that many organic precursors do not have distinct boiling points, so when they enter the hot zone in either MCVD or OVD, they decompose. Thus, the oxides are formed from partially dissociated precursors. This, along with initial homogeneity at the molecular level in the precursor liquid, leads to a more homogeneous glass structure and tends to minimize clustering.

There are several important general constraints on choice of solution chemistry. In addition to mutual solubility and miscibility, solutions should have low viscosity and surface tension to enable efficient mist formation. The material transport of chemicals from the solution to the reaction zone should predominantly proceed through the aerosol and not the vapor phase, so the vapor pressures of the components should be low. In the aerosol process, evaporation of components with relatively high vapor pressure leads to changes in solution composition over time and may affect nebulization and deposition efficiencies, as well as glass composition. Low mutual reactivity is also quite important, especially if such reactions in the solution lead to precipitations or increased viscosity.

7.4.2.4 Nanoparticle Doping

A process very similar to the aerosol approach as applied to the outside process OVD has been termed *direct nanoparticle deposition* (DND). It, too, is based on the combustion of gaseous and atomized liquid raw materials in an atmospheric oxy-hydrogen flame. As shown in Fig. 7.5, the reagents include both the standard refractive index changing materials such as aluminum and germanium, as well as other dopants such as the REs. Oxidation and hydrolysis in the flame produce doped glass particles, which are collected on a target and then sintered using conventional methods. The particle formation process depends on parameters such as the vapor pressure of the metals, the temperature and

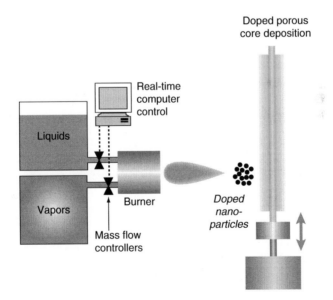

Figure 7.5 Direct nanoparticle deposition process, showing source chemicals, formation of nanometer-sized particles in an oxy-hydrogen flame and collection on a target to form a porous glass body.

temperature distribution in the flame, gas velocities, droplet route through the flame, and the Gibbs free energy of the raw materials. Particle size measurements show a single-peaked particle-size distribution, which indicates that the particles are formed through evaporation–condensation process with a narrow size distribution, which can be adjusted between 10 and 100 nm. The large number of layers deposited onto the target rod allows accurate and independent control of the radial index difference and active dopant profiles in the fiber core.

7.4.3 Summary of Rare Earth-Doped Fabrication Techniques

A review of RE-doped fiber fabrication methods was published in the early days of RE fiber commercialization [16]. This work included a chart showing a comparison of the various methods. In Table 7.1, we present a somewhat updated version of this chart.

7.5 ERBIUM-DOPED FIBER

Before the invention of EDFAs, the effects of optical loss in a transmission system were compensated every few tens of kilometers by an electronic repeater. At each repeater, the signal was detected, electronically filtered, and retransmitted by a new laser diode. This architecture is expensive and restricted to a single

Table 7.1

Comparison of different techniques for the fabrication of RE-doped silica fibers

	Intrinsic loss	Co-dopant concentration control	RE-dopant profile control	Alternate hosts	Deposition rate	High RE doping levels
MCVD						
Vapor phase	+	−	+	−	+	+
Solution	−	+	−	+	−	+
Chelate vapors	+	−	+	−	−	+
Sol-gel	−	+	−	+	−	+
Aerosol	−	+	+	+	−	+
OVD/VAD						
Vapor phase	+	−	−	−	−	−
Liquid phase	−	+	−	+	−	+
Aerosol	−	+	+	+	+	+

channel at a single wavelength. Network architecture changed dramatically when single-mode EDFAs for direct optical amplification of 1.5-μm signals were simultaneously developed by AT&T Bell Laboratories [52] and the University of Southampton [53] in 1987. The efforts directed at development of techniques for producing high-quality RE fibers suddenly bore fruit as Er amplifiers emerged to fill a critical need. Optical amplifiers are a tremendous advantage because they directly amplify the optical signal, support multiple wavelengths, and are independent of bit rate. By 1989, the first EDFA pumped by an efficient semiconductor laser had been demonstrated [54], and in 1994, the first undersea optical amplifiers were deployed by AT&T Submarine Systems in a communication system between Florida and St. Thomas. By 1995, the first trans-Atlantic communication system using optical amplifiers was deployed, followed soon by trans-Pacific systems and the ubiquitous deployment of EDFAs in terrestrial communication systems. The development of EDFAs enabled an explosion in the capacity of fiber optic communication systems. When the first optically amplified systems were deployed, the aggregate data rate per fiber was only a few gigabits per second. Current communications systems now transmit hundreds of gigabits per second while many laboratory demonstrations have exceeded 10 terabits/sec (Tbps).

The need for facile amplification of optical signals arose as transmission bandwidth began to outstrip the performance of electronics and erbium-doped amplifiers filled this niche. This occurred just as the industry was migrating from signal wavelength of 1310 to 1550 nm to exploit the lower fiber attenuation that resulted from improved fiber manufacturing. It is a wonderful coincidence of nature that erbium fits this wavelength window and performs with simplicity and efficiency.

7.5.1 Principles of Operation

As discussed earlier in Section 7.2, and illustrated in Fig. 7.1, erbium-doped silica can be pumped around 980 or 1480 nm to excite the Er ions and produce emission about 1530 nm. Emission can be stimulated by injecting a signal around 1530 nm, allowing creation of a traveling wave amplifier, as illustrated in Fig. 7.6. A simple optical amplifier consists of a signal source, a short length (5–20 m) of Er-doped fiber, a pump diode, a multiplexer for combining a signal and the pump, and optical isolators to prevent undesirable effects from backward-scattered light. Because the system is all optical, amplification is independent of bit rate and can operate over many channels simultaneously. Because the signal is continually reabsorbed by the ground state of the Er ion, the spectral shape of emission depends critically on the details of the population inversion of the Er ions, as illustrated by the emission spectra shown in Fig. 7.7. Note that the

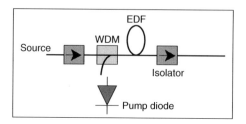

Figure 7.6 Traveling wave amplifier consisting of short length of erbium-doped fiber (EDF), wavelength division multiplexer (WDM) to combiner signal light with the pump light from the pump diode and optical isolators to prevent unwanted feedback to the EDF.

spectra nicely cover the lowest loss region of silica around 1550 nm, shown in Fig. 7.1. The broad emission of Er readily extends from 1530 to 1565 nm and defines the so-called "amplifier C-band." Because the emission spectrum extends to long wavelength, slight changes in amplifier architecture and the population inversion can extend the spectrum into the L-band, from 1565 to about 1615 nm. The number of channels supported over this bandwidth obviously depends on channel spacing, but more than 80 channels at 10 Gbps can be readily accommodated in a single fiber. Tighter channel spacing can increase this even further.

The excellent performance of EDFA is a direct consequence of the energy level diagram. Because the transitions are strictly dipole forbidden but allowed due to slight crystal field asymmetries caused by the local environment of Er in a solid, the excited state lifetime is quite long, around 10 ms. In addition, because there are no states below $^4I_{13/2}$, the only way for the ion to relax back to the

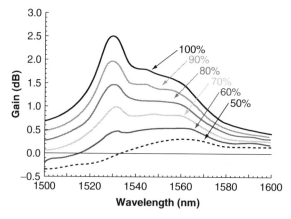

Figure 7.7 Emission spectrum of Er-doped fiber at various levels of population inversion. Note that positive emission continues beyond 1600 nm.

ground state is either radiatively by emission of photon or nonradiatively by coupling energy into the vibrational modes of the glass network. However, since the average phonon energy of silica is about $1100 \, \text{cm}^{-1}$, bridging the energy gap requires five phonons. Because the probability of decay decreases exponentially with the number of phonons, the nonradiative lifetime is very long. As a consequence, Er is an excellent energy storage medium and optical conversion efficiencies very close to theoretical maximum are readily achieved [55].

The optical gain experienced by the signal can be represented as

$$G = 4.343 \Gamma L N_0 [\sigma_e(\lambda) - (\sigma_a(\lambda) n_2 - \sigma_a(\lambda))], \tag{7.1}$$

where gain is expressed in decibels, Γ is the spatial overlap between optical field and dopant distribution, L is fiber length, N_0 is the concentration of RE, σ_e and σ_a are the emission and absorption cross-sections, and n_2 is the population of excited RE ions. This simple equation illustrates the impact of several fiber parameters, which make up the discussion in this section:

- The overlap integral, which is determined by waveguide design
- The absorption and emission cross-sections, which are determined by glass composition
- Excited state population, which is determined by the first two bullets plus device architecture.

The goal of fiber design is to optimize the transfer of energy from pump to signal and to control the spectral characteristics of gain, including both the overall bandwidth and the flatness of the spectrum. A critical aspect for telecommunications applications is generation of noise due to spontaneous emission, but this is highly dependent on the specifics of amplifier architecture and is beyond this discussion. An excellent discussion of noise issues is contained in reference [56].

7.5.2 Fiber Design Issues

The energy level diagram in Fig. 7.1 illustrates the three-level nature of the Er system. That is, the ground state is also the terminal emission level, which means that the signal is constantly reabsorbed as it transits the fiber and full population inversion ($n_2 = 0$) cannot be achieved. Without pump light to excite the ions, the fiber is highly absorbing, with about 5–10 dB/m absorption given typical dopant levels. As pump power increases, however, the population of ions in the ground state, and hence the absorption, decreases. At the threshold power for a particular wavelength, the fiber is transparent ($G = 0$), with just as many photons absorbed as emitted. This competition between absorption and emission implies that high optical intensity will more fully invert the erbium population and increase gain [57].

In the early days of amplifier development, the limited available pump power (tens of milliwatts) put a premium on the gain slope efficiency, expressed as dB of gain per mW of pump power. Achieving high slope efficiency is readily accomplished by maximizing the pump intensity seen by the Er. The most effective route is to increase the index of the core (expressed as Δ, the difference between core and cladding refractive index normalized by the core index) to reduce the mode field diameter. For example, Δ of 2%, compared to 0.35% for standard fiber, are typical and are realized simply by increasing the germanium concentration. In addition, for maximum pump efficiency, the cutoff wavelengths are tuned to minimize mode-field diameter. Optimum cutoff wavelengths occur at 800 and 900 nm for pumping at 980 and 1480 nm, respectively, and are roughly independent of Δ [55]. Further increase in slope efficiency is provided by confining Er doping to the central part of the core, as illustrated in Fig. 7.8. Such high Δ-confined dopant cores are useful for application with limited pump power, such as remotely pumped or battery-operated amplifiers, because slope efficiency can exceed 12 dB/mW.

However, there are several impairments of such designs. High Δ cores have an anomalous scattering loss [58], which varies [59] as Δ^2 as Δ exceeds about 1.2%. This loss is due to scattering from thermodynamic fluctuations in dopant concentration and irregularities at the core–clad interface, and although it can be reduced by grading the index profile of the core and drawing the fiber at lower temperature [60], it cannot be eliminated. Although a typical required length of erbium-doped fiber is only about 20 m, losses can approach 0.01 dB/m, causing a noticeable penalty in performance. The impact of higher background loss is exacerbated by the longer fiber length required for high Δ designs. Because the overlap between pump light and doped region is reduced by a factor of more than two for such fibers, the fiber length must be proportionally increased.

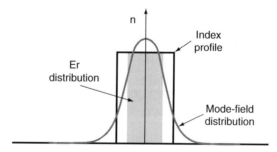

Figure 7.8   Index profile and resulting mode-field distribution. Localizing the Er distribution to the center of the core (Er shown as shaded region) allows interaction with only the highest optical intensity.

A second concern raised by the high-delta small-core fiber is increased loss when spliced to a standard fiber. Because of the small mode-field diameter of erbium-doped fibers, typically less than $3\,\mu m$, losses of more than 3 dB may be incurred upon butt coupling to dispersion-shifted fibers. However, the mode-field diameter can be altered during fusion splicing by extending the duration of the arc to tens of seconds. The extended exposure time at high temperature allows significant diffusion of the highly doped core, reducing the effective index and increasing the core diameter. The mode-field diameter of the standard fiber will also increase, but at a slower rate. For each pair of fibers, there is an optimum splicing schedule to minimize splice loss [61]. Splice losses less than 0.1 dB are readily achieved.

Finally, although high Δ fiber is efficient at low power, at higher power several excited state interactions come into play and create additional loss mechanisms. This is illustrated in Fig. 7.9, which shows gain as function of pump power for a range of core Δ. At high pump intensity, the rate of non-radiative decay from the $^4I_{11/2}$ state to the $^4I_{13/2}$ state becomes a limiting factor, and excited state absorption from $^4I_{11/2}$ promotes the Er ion to higher energy levels. These levels relax nonradiatively back to the $^4I_{11/2}$, resulting in loss of one photon. In addition, since Er is never homogeneously distributed in the glass, energy can migrate from ion to ion until it finds a suitable trap to decay nonradiatively. The higher pumping intensity increases the probability that this can occur. As can be seen in Fig. 7.9, these effects occur even for modest pumping powers, so optimum fiber design depends on the operating power level.

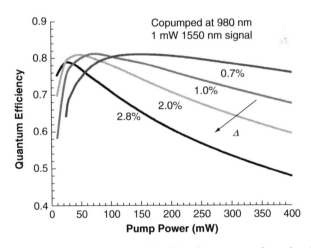

Figure 7.9 Quantum amplifier efficiency as a function of pump power for various levels of core Δ. Note that high Δ cores are efficient at low power but degrade quickly as pump power increases.

As available pump power has increased over time, the requirement on reduced mode-field diameter has relaxed, so Δ is typically in the range of about 1%, although communications applications with higher optical output signal power use fiber with even lower Δ. These designs do not require confined dopant distribution to improve efficiency. On the contrary, the distribution of Er may be reduced at the center of the core or extended into the cladding to manipulate the interaction between pump and signal intensities to control the population inversion. As illustrated in Fig. 7.7, since the emission spectrum is highly dependent on the population inversion, such schemes can be used to tailor the emission spectrum.

7.5.3 Fiber Composition Issues

The amazing initial success of optical amplifiers inspired considerable effort to further improve fiber performance. The primary goals were to broaden the emission bandwidth to support even more signal channels, and to flatten the gain spectrum so that each channel behaves similarly. Despite heroic efforts, however, the technology remains dominated by the first materials explored: Al-doped silica, with Ge used to produce the desired index profile. The atomic structure of Er, which produces the long excited state lifetime, and well-defined absorption and emission spectra also conspire to make these features relatively immune to changes in local glass environment. Many elements of the Periodic Table have been explored, with most eliminated because they cause excess optical attenuation in silica. The remaining few such as alkalis, alkaline earths, and P, Sb and Bi have only modest influence, and many co-dopants, such as phosphorus and the alkalis, narrow rather than broaden the spectrum. Despite much work, the dominant tweaks beyond waveguide design discussed in the previous section remain refinement of the Er and Al concentrations.

The most critical feature of doping Er into silica is the detrimental effect of clustering of Er ions. In particular, RE ions have low solubility in silica, which means they tend to cluster and even crystallize at concentrations of fractions of 1%. Close physical proximity of the ions allows energy exchange and several excited state and quenching phenomena, which invariably reduce amplifier performance. The different phenomena come into play under different amplifier operating conditions, but in all cases, the driving design direction is to reduce Er concentration and increase the length of fiber needed in the amplifier. The choice of ion concentration requires a balance between the detrimental clustering effects and the benefits of reduced fiber length, including less impact from background loss and lower fiber cost and package size.

As discussed in Section 7.3, the high field strength of RE ions results in low solubility in silica and a tendency for the ions to cluster together, even at concentrations of less than 1% [62]. Even before outright crystallization occurs, these clusters exhibit several long-range optical interactions that cause loss of energy [63]. The most serious effect is a cooperative upconversion process called "pair-induced quenching" in which pairs of ions in the $^4I_{13/2}$ state combine their energy to promote one ion to a higher level while the other goes to the ground state [64]. Decay back to the $^4I_{13/2}$ state is nonradiative and results in loss of one photon. This mechanism is unsaturable and behaves as a strongly wavelength-dependent attenuation. Energy may also be lost through concentration quenching in which excited state energy migrates from ion to ion until it finds a suitable trap to decay nonradiatively. A higher fraction of clustered ions increases the probability for such decay. Aluminum co-doping is the silver bullet used to homogenize the glass to inhibit scattering losses due to crystallization and inefficiency due to concentration quenching, and as discussed in Section 7.3, only a small amount is necessary to improve clustering effects. However, because the ion–ion interaction occurs over long distance (up to 3 unit cells in the silica matrix), quenching will occur at some concentration even in a homogeneous glass. The onset of quenching typically occurs at a concentration of Er of about 500 ppm.

An additional significant benefit of aluminum doping is the modification of the emission spectrum, as shown in Fig. 7.10. Aluminum flattens and broadens the emission spectrum around 1550 nm, thereby decreasing the sensitivity to signal wavelength, and increases the absorption at the pumping wavelength of 1.48 μm. The flatter spectrum is very beneficial for high channel count operation. In this case, very high concentration of Al is most desirable.

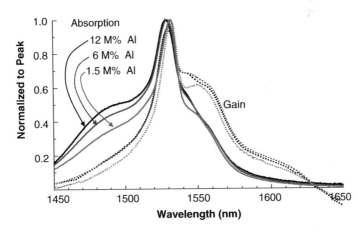

Figure 7.10 Absorption and emission spectra for Al-doped silica.

Although aluminum doping is beneficial, only limited modification occurs in a silica host. More radical modification requires use of other hosts, such as phosphates, borates, tellurites, and fluorophosphates, as well as nonoxygen hosts such as fluorides, sulfides, and other chalcogenides. The first of these to be explored in earnest were fluorides. These glasses, of the base composition ZrF_4-BaF_2-LaF_3-AlF_3-NaF (ZBLAN), were discovered in the late 1970s and investigated [65] very actively because they offer the potential of very low attenuation (<0.001 dB/km at $2.6\,\mu$m) because of reduced Rayleigh scattering [66] and a low vibrational energy spectrum, which pushes the phonon attenuation edge to long wavelength. To date, this potential has not been realized because of contamination issues and scattering resulting from the low threshold for crystallization during fiber processing. Contamination is the result of batch melting methods used for preform fabrication, which contrast sharply with gas-phase methods used for silica fiber processes, which are free from contact with solid containers and provide intrinsic purification during glass formation. The lowest losses achieved are in the range of 2 dB/km, which is inadequate for long-distance transmission but suitable for fiber device applications such as amplifiers, which require only tens of meters of fiber length.

As a host for Er, since RE elements may simply substitute for some of the LaF_3 in the base glass network, solubility is not an issue as it is for silicates. Fluorides produce an inherently broader and flatter emission spectrum because of details of crystal field splitting [67], but the change is only modest and has been insufficient to justify the added cost and complexity of using fluoride fibers. However, the low phonon energy of fluorides provides an advantage over silica for other RE systems. The vibrational frequency of the glass network is in the range 600–400 cm^{-1} (compared to 1100 cm^{-1} for silica), and because the probability of phonon relaxation from an excited state decreases exponentially with the number of phonons required to bridge the energy gap, radiative emission is possible for transitions that would otherwise be quenched. For example, optical amplification can be achieved in Pr at 1310 nm and Tm at 1460 nm. These are discussed briefly in the next section.

Because the emission spectrum of REs is dictated by Stark splitting of the intraband transitions, the most effective way to manipulate the spectrum is to increase the polarizability of the local environment. To this end, heavy elements such as tellurium, antimony, and bismuth are effective, although these are not compatible with standard glass-formation methods. In particular, chlorine is typically used to remove hydroxyl ions (OH^{-1}) during silica fabrication, but this also removes these elements and makes fabrication of low-loss silica fibers difficult. In addition, because the allowable dopant concentration must remain low to maintain glass stability, the contribution from these elements is limited. Therefore, it is more effective to abandon use of high silica glass and explore other glass hosts with high polarizability.

Erbium doping of tellurite glasses (TeO_2) fully leverages such an environment, and wideband operation extending almost 80 nm in a single amplifier has been demonstrated [68]. Although the tellurites are good glass formers and optical amplification is reasonably efficient, the added cost and complication of using such glasses has inhibited further development. One of the more effective approaches has been to produce a multicomponent antimony silicate glass (MCS) using batch processing to prepare the low-temperature base glasses and a crucible approach to draw the fiber. As shown in Fig. 7.11, the spectrum is broader than achieved in typical Al-doped silica. Output power exceeding 20 dBm with greater than 80% quantum efficiency has been demonstrated. Interestingly, the bandwidth is also extended beyond 1610 nm, offering the possibility for adding channels to the L-band, where fiber attenuation is still reasonably low. This occurs because of reduced excited state absorption from the $^4I_{13/2}$ state [69]. Despite superior spectral properties, this technology lies dormant because current applications do not justify the added complexity of this approach.

7.5.4 Short Wavelength Amplifiers

The low-loss transmission window of silica fiber shown in Fig. 7.1 is considerably wider than the spectrum covered by C- and L-band optical amplifiers. This suggests the opportunity for broadening the Er-gain spectrum to lower

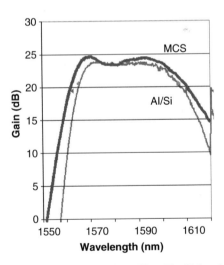

Figure 7.11 Gain spectrum of an antimony silicate fiber. The high polarizability provided by Sb broadens the spectrum considerably.

wavelength or using other RE elements. The Er emission spectrum in Fig. 7.7 certainly shows the potential for gain far shorter than 1500 nm, but competing emission at the gain peak near 1530 nm renders such amplifiers inefficient and noisy. A clever solution to quench the amplified spontaneous emission from this peak is to create a waveguide design that cuts off the fundamental mode beyond about 1500 nm [70]. An index profile such as shown in Fig. 7.12 uses a deep trench of carefully controlled width to make the loss of the fundamental mode very sensitive to bending. This can create a sharp increase in attenuation with wavelength so the fiber operates like a short-pass filter. By tuning the design and bend diameter, the loss edge can be placed to extinguish unwanted 1530-nm emission with little impact on gain at shorter wavelengths. Such an approach has produced efficient low-noise amplifiers operating in the so-called "S-band" from about 1488 to 1508 nm. This allows amplification of 27 additional transmission channels, and long-haul transmission over 100 km has been demonstrated [71].

RE elements other than Er can provide amplification in the S-band. In particular, Tm shows efficient gain in a four-level scheme, as shown in Fig. 7.13, as long as low-phonon energy glasses such as fluorides are used [72]. Although the terminal $^3F_4$ state for 1470-nm emission has relatively long lifetime, multiwavelength pumping can be used to manipulate the excited state populations to achieve efficient gain. The figure shows upconversion pumping at 1100 nm to empty the $^3F_4$ level. Amplifiers based on such pumping schemes are well studied and have demonstrated excellent performance.

If amplification around the 1550-nm transmission window is readily accomplished, what about amplification at 1310 nm, a wavelength that is still being actively deployed? Several possibilities exist for such amplifiers, the dominant ones being Nd and Pr. In silica, emission from Nd is poor because of competition from excited state absorption. However, low-phonon energy glasses such as fluorides and fluoroberyllates have better performance [73]. Unfortunately, however, the peak emission is at slightly longer wavelength, and performance at 1310 nm is rather poor for all-glass hosts [74].

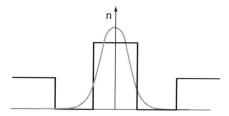

Figure 7.12 Index profile to cut off the fundamental mode using bend loss. The cladding beyond the trench causes "tunneling" of higher wavelength light when the fiber is bent. This design operates as a short-pass filter whose loss edge can be tuned by varying the bend diameter.

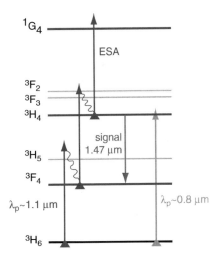

Figure 7.13 Energy level diagram of Tm in a fluoride host. Signal amplification at 1470 nm can be achieved with care.

Significantly better performance can be achieved using a Pr-doped fluoride glass [75]. The energy level diagram in Fig. 7.14 illustrates the four-level nature of the transition responsible for emission around 1300 nm and the close spacing of the intermediate levels. In silica, the $^1G_4$ state would rapidly decay nonradiatively down this ladder, with no optical emission, but the low phonon energy of the fluoride host inhibits nonradiative decay, lengthens the $^1G_4$ lifetime to about 0.1 ms, and provides relatively efficient gain. Performance of such amplifiers has been reasonably good, with predictions of high small-signal gain exceeding 40 dB, excellent saturated output power, and quantum conversion efficiency greater than 30% [76]. Although Pr amplifiers were available commercially for a brief period, the applications for 1310-nm transmission are extremely price sensitive and could not justify use of such an amplifier.

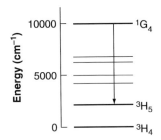

Figure 7.14 Energy level diagram of Pr^{3+}.

7.6 THE CO-DOPED Er/Yb SYSTEM

Erbium-doped amplifiers require pumping in narrow wavelength bands around 980 or 1480 nm. To extend pumping into the 1060- to 1100-nm band and allow use of higher power pumps such as solid state lasers made from crystals of Nd:YAG or Nd:YLF [77], Er may be "sensitized" by addition of Yb. Pump wavelength from about 900 to 1100 nm may be absorbed by Yb^{3+}, which can transfer the energy to erbium if the glass contains both dopants. Absorption of Yb^{3+} peaks around 975 nm and populates the $^2F_{5/2}$ level, as shown in Fig. 7.15. Energy may transfer to the $^4I_{11/2}$ band of erbium, from which nonradiative relaxation populates the $^4I_{13/2}$ state. Emission and signal amplification is then quite similar to standard Er-doped fiber. Such energy transfer was first demonstrated in an alkali silicate host [78] and later in a phosphate glass. The efficiency of this sensitization process depends on several factors. The ions must be in close physical proximity for energy transfer to occur, meaning that the dopant concentrations of Er and Yb must be considerably higher than in standard erbium-doped fibers. In particular, Er concentration is typically approximately 2000 ppm. Note that the pair-induced quenching process described in previous sections results in strong unsaturable absorption peaked around 1530 nm, which effectively shifts the gain spectrum to longer wavelength than in erbium-doped fiber by about 10 nm.

To facilitate energy transfer, the lifetime of the Yb^{3+} $^2F_{5/2}$ level must be sufficiently long to avoid spontaneous Yb emission and the lifetime of the Er^{3+} $^4I_{11/2}$ state must be short enough to inhibit back-transfer. Both effects occur in phosphate glasses and phosphorus-doped silica, allowing remarkable efficiency [79]. Typical fibers have about 10 mol% P_2O_5. Addition of Al is found to be detrimental, presumably because it associates more closely with the RE ions and diminishes the influence of phosphorous.

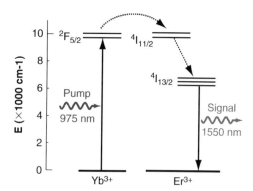

Figure 7.15   Energy level diagram showing transfer from the excited state of Yb to Er.

Whereas the Er system is well understood and can be simulated with excellent accuracy, the Er/Yb system is more complicated and modeling of amplifier performance is more uncertain. Simulation involves treatment of more energy levels and the details of energy transfer, but the higher RE concentration also brings more ion–ion interactions into play. At the same time, because the goal of the Er/Yb system is to allow operation at higher power, additional excited state phenomena begin to dominate. Despite this complexity, current models now have adequate accuracy for most high-power applications.

There is limited use for Er/Yb fiber lasers or amplifiers in single-mode operation in telecommunications because the efficiency and operating bandwidth are inferior to erbium-doped fibers. However, the high pump absorption, around 915–975 nm, allows pumping with very high power diodes, and utility has shifted to architectures that can leverage these broad area pumps, which have single emitter power levels exceeding 6 W. This is the topic of the next section.

7.7 DOUBLE-CLAD FIBER

In the fibers discussed earlier in this chapter, pump light must be launched into the single-mode core, placing a significant limitation on diode power because of the amount of current coupled into the active area of the diodes and restrictions on facet intensity to minimize damage. A clever solution to these limits is to break the constraint on single-mode diode pumping. By surrounding the fiber cladding with a second cladding that has a lower index than the inner cladding, the entire inner cladding region becomes a waveguide [80], as shown schematically in Fig. 7.16. Pump light introduced into this region propagates along the fiber, occasionally crossing the core where it is absorbed by, and

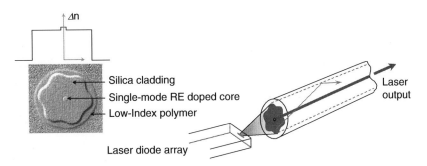

Figure 7.16 Schematic of the double-clad fiber concept in which broad-area diode pump light is coupled into the inner cladding of the fiber. This pumps the rare earth core, allowing gain in a diffraction-limited waveguide.

transfers energy to, an optically active RE dopant. This allows high-power strongly multimode pump light to be efficiently converted to a diffraction-limited output beam, with an enormous increase in optical brightness. Such a design, called a "double-clad" or "cladding-pumped" device, was first demonstrated as a 1060-nm neodymium laser pumped by a collection of 807-nm broad-area diodes [80].

To collect the greatest amount of pump light, the second cladding should have as low an index as possible. This is best achieved using a low index polymer cladding directly on a silica fiber, for which the numerical aperture (NA) is around 0.45. With a modest facet intensity of only $5\,mW/\mu m^2$, more than 500 W could theoretically be coupled into a 180-μm diameter fiber with NA = 0.45. Applications of these devices are described more fully in Chapters 22 and 23, while the discussion here focuses on fiber design issues.

In a double-clad laser, multimode pump power must cross the core to be effective. In circular fiber, this is inefficient because much of the light follows helical paths and is not absorbed. To prevent this, the fiber is made noncircular so that fewer helical modes are supported and all rays eventually cross the core. An alternative and more accurate viewpoint is that the cladding waveguide supports more modes that have nonzero power at the centerline, so a larger fraction of the supported modes experience absorption. Because only the core absorbs pump light, the effective absorption rate of the fiber, α_{clad}, is roughly equal to the core absorption times the ratio of core area to cladding area, $\alpha_{clad} = \alpha_{core} \cdot A_{core}/A_{clad}$. This relation assumes that the cross-section of the fiber is uniformly illuminated and that all available optical modes are excited. Because the core absorption at the pump wavelength is typically several decibels per centimeters for common dopants (Yb, Tm, Nd), laser lengths are typically less than 100 m and in many cases only a few meters. Note that the large pumping area results in relatively low pump intensity compared with the signal intensity, which is guided in the central core. This poses limits on fiber and amplifier design if there is significant ground state signal absorption. For example, Yb-doped double-clad lasers operating at 980 nm where there is high absorption require very short fiber lengths and high pump intensity to achieve population inversion because Yb at 980 nm is a three-level system. Operation in regimens with minimal ground state absorption, such as Yb at 1060–1120 nm or Nd at 1060 nm, is much more efficient and has greater design flexibility, because in this wavelength regimen, they are quasi-four level systems.

The basic double-clad structure may consist of a laser if a resonant cavity is present, such as provided by fiber Bragg gratings as shown in Fig. 7.17a, or a signal amplifier if a suitable pump-signal multiplexer is provided. In either case, pump light from high-power broad-area diodes must be coupled into the cladding of the active fiber. Multiplexing may be accomplished using lenses in conventional free-space optics, whereas an all-fiber architecture may use a

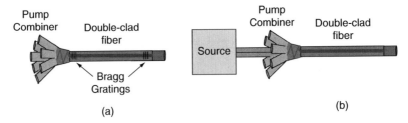

Figure 7.17 (a) Double-clad fiber laser using fiber Bragg gratings to form the resonant laser cavity and (b) a traveling wave signal amplifier.

fused-fiber pump combiner, as shown in the figures, V-groove pump injection [81], or side-pumping in which the pump and signal fibers lie side-by-side in a common low-index polymer coating [82]. Each method has its advantages and proponents. The amplifier architecture can have one or more stages and an array of interstage components depending on the required operation.

The pump combiners and double-clad fibers should preserve pump brightness as efficiently as possible, where brightness is approximated as the product of fiber area and the square of NA, $A_{clad} \cdot NA^2$. This puts a premium on the NA of the inner pump cladding. For silica-based fiber, materials constraints limit the NA if both the inner and the outer claddings are glass. Higher NA requires use of a low-index polymer to define the outer cladding in addition to protecting the glass surface from mechanical damage. Fluorinated acrylate and silicone polymers replace the standard fiber coating, achieving NA from 0.45 to 0.60 with no compromise in mechanical properties. To pump these fibers, broad-area diodes are commonly pigtailed with fibers with NA approximately 0.22, allowing reduction in cladding area of a factor of more than 4 in multiplexing to the gain fiber. This means that in the fused bundle approach, the pigtails of 19 broad-area diodes with 0.22 NA pigtails can be bundled together, fused, and tapered to an output diameter only about 1.6 times the diameter of an individual pigtail fiber. At the fiber exit, the beam is typically diffraction limited and the fiber can be spliced to other fibers using conventional techniques.

Double-clad fiber lasers and amplifiers based on the aforementioned concept are increasing in performance and continue to penetrate wider markets and applications [83]. Fiber lasers, in particular, offer several very attractive benefits over more conventional lasers, the most notable being better heat dissipation and relative immunity to detrimental wave-guiding effects. Fiber lasers can be compact, rugged, reliable, and efficient and have low cost of ownership. Pushing fiber technology to higher performance in terms of average and peak power will allow greater penetration into existing markets and enable new directions. This requires new developments in fiber and component design if fibers are to replace established solid state technology or create new niche applications. These issues are explored in the following subsection.

7.7.1 Limitations of Fiber Lasers

Design of a fiber laser capable of generating high pulse energies with high average and peak power requires careful attention to limitations from the extractable energy of the gain medium and nonlinear limits of the fiber. The saturation energy of the gain medium is a key parameter for determining how much energy can be stored in an amplifier and is given by [84]

$$E_{sat} = \frac{h\nu_s \, A_{eff}}{(\sigma_{es} + \sigma_{as})\Gamma_s},$$ (7.2)

where σ_{es} and σ_{as} are the emission and absorption cross-sections at the signal wavelength, $h\nu_s$ is signal energy at frequency ν_s, A_{eff} is area of the active doped region, and Γ_s is signal overlap with the active dopant.

Although long fiber length is beneficial in managing heat dissipation and increasing pump absorption, long length becomes an impairment because of nonlinear interactions. Two deleterious nonlinear effects of concern are stimulated Brillouin scattering (SBS) and stimulated Raman scattering (SRS). Both rob power from the signal and can cause catastrophic damage if allowed to build uncontrollably. For SRS, the threshold for peak power P_{th} before onset of serious Raman scattering in passive fibers is given by [85]

$$P_{th} = \frac{16 A_{eff}}{g_R L},$$ (7.3)

where A_{eff} is the effective mode area of the fiber, g_R is the Raman gain coefficient, and L is the fiber length. For example, for a fiber with 25-μm core diameter, $P_{th} \cdot L \sim 70\,\mathrm{kWm}$. Because typical fiber lengths exceed 5 m, this indicates peak powers of about 20 kW before Raman scattering becomes severe.

SBS arises from interaction of the signal with longitudinal acoustic modes of the fiber, causing part of the signal to be reflected backwards. Similar to the case of SRS, the threshold condition for SBS can be written as [85]

$$P_{th} = \frac{21 A_{eff}}{g_B L}\left(1 + \frac{\Delta\nu}{\Delta\nu_{SiO_2}}\right),$$ (7.4)

where g_B is the Brillouin gain coefficient, $\Delta\nu$ is the bandwidth of the signal, and $\Delta\nu_{SiO2}$ is the Brillouin bandwidth of the fiber ($\sim$50 MHz for silica). If the signal has bandwidth comparable to $\Delta\nu_{SiO_2}$, then for a fiber with 25-μm core diameter, $P_{th} \cdot L \sim 350\,\mathrm{Wm}$. This is obviously a severe constraint and mitigation is necessary for narrow line-width lasers. Although narrow line width is important for many applications in spectroscopy and frequency conversion, for materials processing, the line width is less important and the threshold can be increased considerably.

7.7.2 Methods to Improve Performance

7.7.2.1 Core Design

For both SBS and SRS impairments, as well as other limits like four-wave mixing and self-phase modulation, mitigation is possible by increasing the modal area and decreasing the fiber length. To increase mode area, the core diameter can be increased while the core index is decreased to maintain single-mode operation. For example, a core with $\Delta n = 0.0012$ will remain single mode at 1060 nm up to core diameter of about 20 μm. At such a low core index, the fiber becomes sensitive to external perturbations, which can cause excessive bend loss. This is generally considered the lower limit in index before severe problems with fiber handling and fixturing set in [86].

For larger core diameter, the fiber will support multiple modes. In some cases, this is acceptable, but generally single-mode performance is preferable, especially where the output beam requires diffraction-limited focus, a high level of collimation, or where the beam is to be used for frequency conversion. In such cases, because the higher order modes are inherently more sensitive to bending than the fundamental mode, they can be stripped by coiling the fiber without incurring excessive loss of the fundamental mode [87]. To enhance the "leakiness" of the higher order modes, cladding features such as holes or raised index rings can be used. If the effective index of an unwanted core mode is matched to that of a cladding mode supported by the added features, energy can mix between the regions and cause high mode extinction [88]. This method was successfully applied [89] to a fiber with effective mode area of about 1400 μm^2.

However, as core diameter increases, the difference in propagation constant or effective index between modes decreases and mode mixing becomes severe. Because the number of modes and the power transfer rate between modes vary quadratically with core diameter, there is a practical limit in core size dictated by excessive power transfer, as is well known from studies of multimode transmission fiber. A practical limit is about 25 μm if the refractive index contrast between core and cladding are kept as low as permitted by bend loss. Thus, there is a practical limit to increasing core diameter while maintaining effectively single-mode operation. Use of microstructured fiber to increase the mode area has been discussed, but the mode diameters are not significantly larger than has been achieved in solid fibers with comparable effective refractive indices [90].

As core diameter increases, the mode-field diameter also increases, but at a slower rate, as shown in Fig. 7.18. Around 15 μm, the mode-field and core

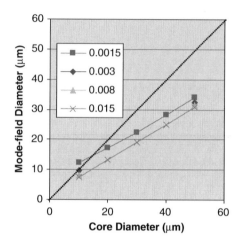

Figure 7.18 Calculation of mode-field diameter as a function of core diameter for different core index differences, Δn. The dotted line shows where core and mode-field diameters are equal.

diameters are approximately equal, but for a 40-μm core, the mode field occupies only about half the core area. The effect is relatively insensitive to core index. This poor overlap between the mode and the core can generate excessive amplified spontaneous emission if the entire radius of the core contains RE dopant. However, as explained for erbium-doped fibers in Fig. 7.8, confinement of the RE dopant can improve efficiency, in some cases by up to 10 percentage points [91].

A further problem with large mode-area designs is the well-known but often overlooked property of bend-induced mode distortion. From a mode propagation point of view, bending of a fiber is analogous to tilting the refractive index profile, transforming a step index, for example, into a profile. This is illustrated in Fig. 7.19, which shows the mode-field diameters for bend (solid lines) and unbent (dotted lines) configurations. Because the mode is increasingly sensitive to features at large radius, bend distortion increases with core diameter. As seen in the figure, the 30-μm core experiences little change on bending, whereas the 50-μm core shows almost 50% reduction with even a modest 20-cm diameter bend. This distortion has profound implications for packaging of large mode-area fibers. This effect can be avoided using non–step-index profiles. For example, a parabolic core profile is immune to such distortion [92] (a tilted parabola is still a parabola).

Modification of the core index profile can also be used to resolve many of the other issues discussed earlier. For example, step-index designs produce Gaussian mode profiles that do not fully fill the core and have high intensity at the center. By adding a high index ring to the periphery of the core, as shown in Fig. 7.20,

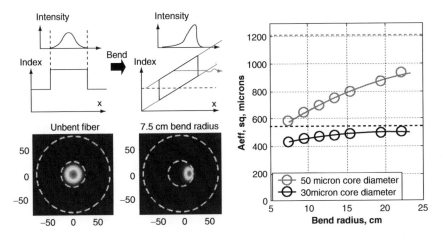

Figure 7.19 Simulation of bend-induced distortion, illustrating the effective "tilt" of the index profile and resulting decrease in mode-field area. In the plot on the right, the dotted lines represent the unbent effective area. Note that the 30-μm core shows little distortion.

the mode is drawn outward and flattened [93]. This increases the effective mode area, reduces the peak intensity, and provides better overlap between the mode and the core. Such mode-flattened designs achieve a five times increase in nonlinear threshold. Despite this improvement, several additional problems arise. While the fundamental mode has better overlap with the core, the higher order modes, such as LP_{11} and LP_{02} are stabilized by the high index ring and become

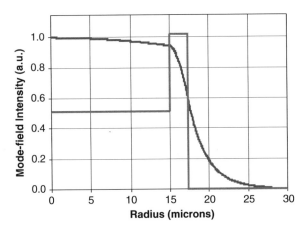

Figure 7.20 Mode flattened index profile and resulting mode-field distribution. Note that the peak intensity is reduced and the mode more fully fills the core.

more difficult to remove by bending. These features also exacerbate the distortion caused by bending because the ring becomes a more dominant feature on one side of the profile. As a consequence, mode-flattened designs have the greatest mode distortion sensitivity of any other design. In addition, because the mode profile is highly non-Gaussian, coupling to a conventional step-index fiber can result in exceedingly high loss and is often undesirable for the output beam. Although this is problematic, solutions exist and are discussed in the following subsections.

7.7.2.2 Microstructure Fiber

The design constraints discussed in the previous section can be relieved by adding air holes to the cladding. First used in 1974 [94], holes add a degree of freedom in waveguide design for controlling cladding index and mode propagation. This can open the design space for achieving higher performance with larger core area, shorter fiber length, better mode purity, and potentially more robust device operation. Microstructure fiber has been used to demonstrate an assortment of waveguide phenomena, including supercontinuum generation in small-core, extremely high index waveguides [95], endlessly single-mode operation [96] in which higher order modes are suppressed, extremely bend-insensitive fiber designs [97], microfluidics [98], and even true photonic band-gap operation [99]. Applied to RE-doped gain fibers, the primary design goal has been to increase the effective area beyond levels practical with solid fibers.

The addition of holes to the cladding reduces the average refractive index of the cladding and can render the cladding index wavelength dependent. For suitable structures in which the ratio of hole diameter to hole spacing is less than about 0.4, the cladding index approaches that of the core index and prohibits propagation of any higher order modes. In essence, the fiber can become single mode at all wavelengths, a feature not possible in solid fibers. This phenomenon was also thought to allow operation at arbitrary mode-field diameter, offering breakthrough performance for fiber amplifiers [100]. However, for large core size, the effect of the holey cladding is simply to reduce the average index and the fiber behavior is very similar to a conventional low NA waveguide. Thus, addition of holes in this design regimen does not provide relief from bend sensitivity or mode-coupling issues plaguing solid fiber. To overcome these issues, the fibers can be made very rigid to inhibit microbending and held straight to avoid bend loss. Such rodlike fibers have achieved very high levels of performance, but their practicality remains questionable because of packaging and free-space alignment issues [101].

A second design regimen using holes can be defined based on higher order mode suppression. If the holes are large and few, higher order modes experience significantly greater bend loss than the fundamental mode, allowing large core

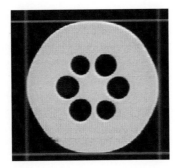

Figure 7.21 Photograph of a microstructure fiber that provides single-mode operation by stripping out higher order modes.

operation with an effectively higher core index. This reduces the bend loss of the fundamental mode and inhibits mode coupling. Fibers based on this principle offer a practical improvement over conventional fiber, with core area around $1417\,\mu m^2$ and excellent immunity to bend loss having been achieved [102]. However, just as for conventional large core solid fibers, this design still suffers from severe bend-induced mode distortion described earlier [92].

7.7.2.3 Concentration-Induced Problems

As discussed earlier, in addition to increasing the effective area of the signal, high-power devices can avoid nonlinear limitations by using shorter fiber lengths. This is readily achieved by increasing the concentration of RE dopants. Unfortunately, just as erbium-doped fibers suffer from pair-induced quenching if the ion concentration is too high (see Section 7.3), other RE ions show deleterious concentration-induced problems as well. For example, Tm is known to photo-darken in an aluminum-doped host when exposed to 1064-nm light [103]. The rate of photo-darkening depends on the light intensity to the power of 4.7. Also, in neodymium-doped silica fibers, if the Nd concentration exceeds about 7 wt%, phase separation occurs, resulting in high scattering losses [104]. Even for relatively low concentrations, however, an excited Nd ion can transfer part of its energy to an ion in the ground state, placing both ions into an intermediate energy level from which nonradiative decay occurs [105, 106]. This cross-relaxation process limits the concentration of Nd in a silica host to only a few hundred parts per million.

In Yb-doped fibers, the workhorse of high-power fiber lasers, because the energy levels in a dielectric host consist of only two manifolds, there are few possibilities for quenching phenomena. However, when a highly Yb-doped fiber is exposed to intense pump radiation, the signal degrades over time. This photo-darkening is likely due to formation of a color center in the glass and appears as a strong absorption at visible wavelengths, accompanied by a strong absorption around 975 nm [107]. Although the mechanism for photo-darkening has not been fully resolved, a number of studies have probed various aspects of the problem. Similar to other RE ions at high concentration, excited state decay of Yb can be decomposed into an initial rapid decay, followed by a long fluorescence lifetime [108]. The latter is associated with isolated Yb^{3+} ions in the glass matrix, with lifetimes from 0.8 ms in an aluminum-rich host to about 1.2 ms in a phosphorus-rich host. The fast component with lifetime of less than $200\,\mu s$ appears to be independent of host composition and is likely the result of a nonradiative transition. The fast component is not found in fibers with low Yb concentration. Although Yb exhibits cooperative luminescence in which two excited ions combine energy to produce emission from a virtual excited state, the fast component of lifetime indicates that the behavior of Yb is more complex than expected from a simple ion embedded in a dielectric host.

When exposed to only 150 mW of pump light at 975 nm, a single-mode core with 8-μm diameter can exhibit a rapid reduction in transmitted power, even with fiber length as short as 30 cm. Note that this optical intensity is equivalent to about 100 W of pump launched into a double-clad fiber with a 200-μm outer diameter. The time trace of 975-nm transmission through two fibers of different hosts and roughly the same Yb concentration is shown in Fig. 7.22a. Note that pump throughput normalized to the initial power, T_0, initially decays very rapidly and appears to saturate. This behavior suggests a kinetic process that depletes a population of precursors and creates some form of defect, perhaps a color

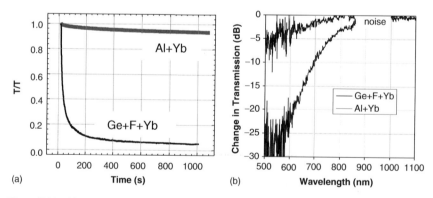

(a)   (b)

Figure 7.22 (a) Transmission of 975-nm light through a single-mode Yb-doped core, illustrating photodarkening. Yb concentration in both fibers produces about 150 dB/m core absorption at 915 nm. (b) Change in the absorption spectrum for 30-cm fiber length.

center. Based on the behavior of similar phenomena such as decay of UV-written Bragg gratings [109], it is likely that the population of precursors has a distribution of activation energies, with higher optical intensity reaching deeper into the distribution to form more color centers. For the Ge + F composition shown in the figure, the activation energy for defect formation apparently is relatively low.

Although the decay of 975-nm transmission is accompanied by strong absorption in the visible (Fig. 7.22b), the correlation in attenuation between both spectral regions is very sensitive to host composition. Note that the Ge + F–doped fiber has relatively little change in visible attenuation despite catastrophic attenuation at 975 nm. Because of this, it can be misleading to use the change in visible attenuation as a metric for photo-darkening. This is unfortunate because the visible spectrum is quite easy to measure or monitor, but it does indicate that the color center is very sensitive to glass host, unlike the usual spectra of typical RE ions, which behave like isolated trivalent ions embedded in a glass host. Note also in Fig. 7.22b that the tail of the absorption feature in the visible does not extend into the pump or signal wavelengths and, thus, is not responsible for the degradation in amplifier or laser performance. This suggests some phenomenon related to the Yb-excited state.

The dependence of photo-darkening on optical intensity has been studied using several methods. Early work indicated a strong unsaturable absorption associated with the 975-nm absorption peak in Yb [107]. Also, because photo-darkening appears to result from the incidence of pump light rather than signal light at wavelengths beyond about 1020 nm, the $^2F_{5/2}$ state must be involved in defect formation. By exposing sets of fibers to varying intensity of 920-nm light at constant temperature and using a numerical model to calculate the Yb population inversion, the change in attenuation at 633 nm is found to vary with population inversion to the seventh power [110]. Thus, CW fiber lasers with limited population inversion are much less susceptible to degradation than pulsed lasers, which experience high inversion between pulses. As described earlier, the degree of photo-darkening also increases with increasing Yb concentration.

To further probe the defect state, photo-darkened fibers were heated to 500 °C in air and found to recover completely. On the other hand, unexposed fibers heated in H_2 gas showed growth of a strong absorption band in the visible, which was not altered by exposure to intense 975-nm light and which could not be annealed by heating in the absence of H_2 [111]. The spectral changes induced by reaction with H_2 were different from those induced photolytically, but one would expect a difference caused by the presence of H^+ near the defect site. The change in attenuation shown in Fig. 7.23 indicates several emission peaks. The location of these peaks does not agree with those found from cooperative luminescence [108] or other contaminant RE ions and is likely due to fluorescence from the color center.

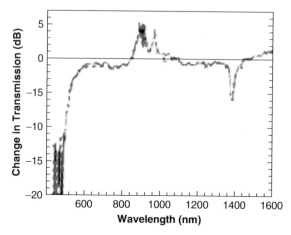

Figure 7.23 Change in attenuation spectrum of Yb-Al fiber after photodarkening. Note the narrow fluorescence features at 480 and 510 nm.

These observations, and the ones described earlier, are consistent with a model in which Yb^{3+} is reduced to Yb^{2+}. The reduced state is responsible for the strong broad absorption band in the visible, which is very sensitive to glass host composition (including the presence of H^+). It is also responsible for the fluorescence of photo-darkened fiber. Yb^{2+} has been studied in several types of fluoride crystals in which it exhibits similar spectral absorption and emission features [112]. Based on this model, Yb^{3+} is reduced to form a metastable exciton composed of Yb^{2+} and a trapped hole. Multiphoton excitation for this transformation is likely aided by the presence of nearby excited Yb^{3+} ions, which contribute energy. As illustrated in Fig. 7.24, during operation of a fiber amplifier, energy transfer from excited Yb^{3+} ions to Yb^{2+} ions results in loss of a pump photon through nonradiative decay at the defect. The formation of color centers and the degradation of the amplifier, thus, increase with Yb concentration because of the formation of pairs or clusters, which enable migration of energy from the excited state. Note that the pair shown on the right in Fig. 7.24 will appear spectroscopically as an unsaturable absorption feature composed of the superposition of the two spectral absorptions. This model explains all the observed features of photo-darkening and indicates why reduction of Yb concentration has been effective in reducing the effect. It also suggests a possible route to eliminating the effect if one can create a glass host in which holes are not available to stabilize the formation of Yb^{2+}. To date, it does not appear that this problem has been solved because all known fibers have exhibited similar levels of photo-darkening despite the attempts at the usual glass modifications discussed in Section 7.3.

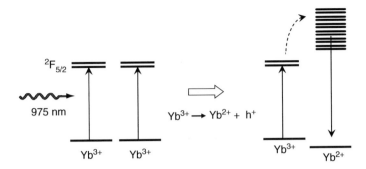

Figure 7.24 Proposed model for photodarkening.

7.7.2.4 Cladding Design

Given the limitations in design of the fiber core waveguide and composition, an additional means of decreasing fiber length is to decrease the pump area by increasing the NA of the pump waveguide. Recall that for double-clad designs, conservation of brightness dictates that the product of pump NA times the diameter of the pump region cannot exceed that of the pump diodes or the fiber pigtail that delivers the light. Typical low-index polymers used as the pump cladding allow NA of about 0.45, while all-glass structures have an even lower NA, typically about 0.22, because of the limited amount of fluorine that can be incorporated into silica to reduce the refractive index. An alternative approach is to surround the pump region with air in a microstructured fiber [113], as shown in the photo in Fig. 7.25. In such an air-clad structure, the NA can be increased to 0.6–0.8, allowing reduction in fiber length by almost a factor of 2. An additional advantage of this design is that the outer glass region can be made

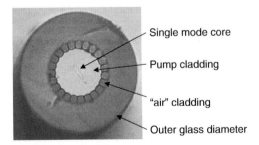

Single mode core

Pump cladding

"air" cladding

Outer glass diameter

Figure 7.25 Cross-section of an air-clad optical fiber. The pump is guided within an inner cladding defined by a ring of air holes. This construction allows higher pump NA and elimination of low-index polymer.

quite thick because it is not part of the wave-guiding structure. This reduces external mechanical perturbations that can cause deleterious mode mixing.

Air-clad designs offer the additional benefit of higher pump intensity. This can be beneficial for lasers or amplifiers that operate as three-level laser systems. Because such systems have non-negligible ground state absorption at the signal wavelength, to achieve significant gain, the population inversion must be maintained as high as possible. Air-clad designs are useful and even imperative for obtaining emission at moderate pump powers from Yb at 980 nm, Nd at 940 nm, or even Er at 1530 nm.

7.7.2.5 Device Assembly

In most cases, it is essential that the large mode area signal be preserved in the components and through the fusion splices of the device. The key issue is development of components and assembly methods that preserve the purity of the signal mode with low signal and pump attenuation. For high-power lasers, however, the fiber designs discussed earlier depart radically from traditional core index profiles and sizes. Large area fibers support propagation of many optical modes and should have maximum possible mode area and optimum overlap between the signal mode-field and the gain material. Although the conventional techniques for fusion splicing are used, it can be quite difficult to understand the mode content after the splice and verify that the region can reliably handle tens of watts of power.

To maintain signal integrity, mode transformation technology allows facile coupling between conventional step-index single-mode fiber with Gaussian mode shape and large mode area fiber with non-Gaussian mode shape. This allows the gain fiber design to be optimized for peak performance independent of concerns for signal launch and output beam quality. With suitable mode transformation methods, the construction shown in Fig. 7.26 is most desirable. In this example, the output beam is near Gaussian, has very high mode purity, and a mode-field diameter of about 15μm at 1550 nm. Depending on the specific fiber designs, the mode transformers can have losses less than 0.1 dB. Similar results have been achieved with output mode-field diameter of 28μm at 1080 nm using different pigtail fiber [114].

Much of the published literature on double-clad devices uses free-space coupling into and out of the fibers. Although this is acceptable for laboratory demonstrations or fiber and amplifier performance, it is generally unsuitable for robust commercial applications. A fused approach such as discussed earlier will be essential for this technology to have significant commercial impact. As fiber designs become increasingly exotic, this aspect becomes quite a challenge and novel methods for mode transformation and multiplexing of pump and signal will be required.

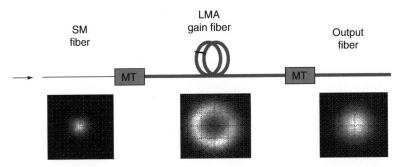

Figure 7.26 Mode transformer (MT) converts desired Gaussian mode shape into and out of non-Gaussian gain fiber. Near field images were measured at 1300 nm. The index profile of the gain fiber is shown in Fig. 7.8.

7.8 CONCLUSION

In this chapter, we have discussed the fabrication, design, and characteristics of RE-doped optical fibers. Almost all applications for such fibers are in fiber-based amplifiers or lasers and the demands of improved performance and reliability continue to drive the evolution of many types of fiber for many different applications. A wealth of publications have been devoted to the erbium-doped optical amplifier, which has become ubiquitous and essential for optical communications worldwide. Although applications for other types of RE fiber are not nearly as widespread or significant, the rate of progress and adoption has been breathtaking over the last few years. RE lasers and amplifiers have begun to penetrate many industries, from materials processing (welding, drilling, cutting, micromachining) to medical (imaging, surgery, therapeutics) to military (weapons, imaging, ranging). We can expect both performance and commercial applications to grow rapidly in the next few years.

REFERENCES

[1] Urquhart, P. 1988. Review of rare earth-doped fiber lasers and amplifiers. *IEEE Proc.* 135(6):385–407.

[2] Miniscalco, W. J. 2001. Optical and electronic properties of rare earth ions in glasses. In: *Rare Earth Fiber Lasers and Amplifiers* (M. Digonnet, ed), p. 17. CRC Press, Orlando, Florida.

[3] Stone, B. T. and K. L. Bray. 1996. Fluorescence properties of Er^{3+}-doped sol-gel glasses. *J. Noncrystalline Solids* 197:136–144.

[4] Sen, S. 2000. Atomic environment of high-field strength Nd and Al cations as dopants and major components in silicate glasses: A Nd L-$_{III}$-edge and Al K-edge X-ray absorption spectroscopic study. *J. Noncrystalline Solids* 261:226–236.

[5] Nakazawa, M. and Y. Kimura. 1991. Lanthanum codoped erbium fiber amplifier. *Electron. Lett.* 27:1065–1067.

[6] Myslinski, P. et al. 1999. Performance of high-concentration erbium-doped fiber amplifiers. *IEEE Photonics Technol. Lett.* 11:973–975.

[7] Samson, B. N. et al. 1998. 1.2 dB/cm gain in erbium:lutecium co-doped Al/P silica fibre. *Electron. Lett.* 34:111–113.

[8] Lee, L. L. and D. S. Tsai. 1994. Ion clustering and crystallization of sol gel-derived erbium silicate glass. *J. Materials Sci. Lett.* 13:615–617.

[9] Arai, K. et al. 1986. Aluminum or phosphorus codoping effects on fluorescence and structural properties of neodymium-doped silica glass. *J. Appl. Phys.* 59(10):3430–3436.

[10] Sen, S. and J. F. Stebbins. 1995. Structural role of Nd^{3+} and Al^{3+} cations in SiO_2 glass—A ^{29}Si MAS-NMR spin-lattice relaxation, ^{27}Al NMR and EPR study. *J. Noncrystalline Solids* 188:54–62.

[11] Craig-Ryan, S. P. et al. 1990. Optical study of low concentration Er^{3+} fibers for efficient power amplifiers. *ECOC'90 Proc.* 1:571–574.

[12] Lange, M. R. et al. 2003. High gain coefficient phosphate glass fiber amplifier. In: *Technical Proceedings of the NFOEC, Orlando, FL, Sep. 8–10, 2003*, p. 126.

[13] MacChesney, J. B. and J. R. Simpson. Multiconstituent optical fiber. Patent No. US4666247, AT&T, Bell Laboratories, Murray Hill, NJ, 1987.

[14] Belov, A. V. et al. 2000. Erbium-doped fibers based on cesium-silicate glasses. *Optical Fiber Technol.* 6:61–67.

[15] Simpson, J. R. 1993. Rare earth-doped fiber fabrication: Techniques and physical properties. In: *Rare Earth-Doped Fiber Lasers and Amplifiers* (M. J. F. Digonnet, ed.), Vol. 37, *Optical Engineering*, p. 659. Marcel Dekker, New York.

[16] Simpson, J. R. 1989. Fabrication of rare earth-doped glass fibers. In: *Proceedings of SPIE: Fiber Laser Sources and Amplifiers*, Vol. 1171, pp. 2–7. SPIE—The International Society for Optical Engineering, Bellingham, Washington.

[17] Ainslie, B. J. 1991. A review of the fabrication and properties of erbium-doped fibers for optical amplifiers. *J. Lightwave Technol.* 9:220–227.

[18] Li, T. 1985. Fiber fabrication. In: *Optical Fiber Communications Series*, Vol. 1, p. 363. Academic Press, Inc., Orlando, FL.

[19] Izawa, T. and S. Sudo. 1987. *Optical Fibers: Materials and Fabrication*, Vol. 1. KTK Scientific Publishers, Tokyo, Boston.

[20] Ainslie, B. J. 1991. A review of the fabrication and properties of erbium doped fibers for optical amplifiers. *J. Lightwave Technol.* 9(20):220–227.

[21] DiGiovanni, D. J. 1990. Fabrication of rare earth-doped optical fiber. In: *SPIE*, Vol. 1373, Paper 01, *Fiber Laser Sources and Amplifiers II*. SPIE—The International Society for Optical Engineering, Bellingham, Washington.

[22] Yamashita, T. T. 1989. Nd^{3+} and Er^{3+} doped phosphate glasses for fiber lasers. In: *SPIE*, Vol. 1171, pp. 291–297, *Fiber Laser Sources and Amplifiers*. SPIE—The International Society for Optical Engineering, Bellingham, Washington.

[23] Snitzer, E and R. Tumminelli. 1989. SiO_2 clad fibers with selectively volatilized soft glass cores. *Opt. Lett.* 14(14):757–759.

[24] Simpson, J. R. et al. 1990. A distributed erbium fiber amplifier. In: *Conference on Optical Fiber Communication, Technical Digest.* Paper PD19-1. Optical Society of America, Washington, DC.

[25] Poole S. et al. 1986. Fabrication and characterization of low-loss optical fibers containing rare earth ions. *J. Lightwave Technol.* LT-4(7):870–876.

[26] Gozen, T. 1988. Development of high Nd^{3+} content VAD singlemode fiber by molecular stuffing technique. In: *Conference on Optical Fiber Communication*, Paper WQ1. *Technical Digest.* Optical Society of America, Washington, DC.

[27] U.S. Patent #4,66,247.

[28] Poole, S. B. et al. 1985. Fabrication of low-loss optical fibres containing rare-earth ions. *Electronics Lett.* 21:737–738.

[29] Ainslie, B. J. et al. 1987. The fabrication and optical properties of Nd^{3+} in silica-based optical fibers. *Materials Lett.* 5:143–146.

[30] Dewing, E. W. 1970. Gaseous complexes formed between trichlorides $AlCl_3$ and $FeCl_3$ and dichlorides. *Metallurgical Trans.* 1:2169–2174.

[31] Øye, H. A. and D. M. Gruen. 1969. Neodymium chloride–aluminum chloride vapor complexes. *J. Am. Chem. Soc.* 91:2229.

[32] Dianov, E. M. et al. 1995. Efficient amplification in erbium-doped high-concentration fibers fabricated by reduced-pressure plasma CVD. In: *Proceedings of OFC, San Diego, CA, USA, Feb. 26–March 3, 1995*, pp. 174–175. Paper WP5. OSA—Optical Society of America, Washington, D.C.

[33] Miller, S. B. et al. 1985. Process for making sintered glasses and ceramics. Patent No. US4501602. Corning Glass Works, USA, p. 10.

[34] Bocko, P. L. 1989. Rare earth-doped optical fibers by the outside vapor deposition process. In: *Proceedings of OFC: Summaries of Papers. Houston, TX, USA, Feb. 6–9, 1989*, p. 20. OSA—Optical Society of America, Washington, D.C.

[35] Blair, R. G. 1986. Method for introducing dopants in optical fiber preforms. Patent No. WO8607348, Hughes Aircraft Company, p. 24.

[36] Tumminelli, R. P. et al. 1990. Fabrication of high concentration rare earth-doped optical fibers using chelates. *J. Lightwave Technol.* 8(11):1680–1683.

[37] Snitzer, E. 1961. Optical maser action of Nd^{+3} in a barium crown glass. *Phys. Rev. Lett.* 7:444–446.

[38] Ballato, J. and E. Snitzer. 1995. Fabrication of fibers with high rare-earth concentrations for Faraday isolator applications. *Applied Optics* 34:6848–6854.

[39] MacChesney, J. B. et al. 1987. Influence of dehydration/sintering conditions on the distribution of impurities in sol-gel derived silica glass. *Materials Res. Bull.* 22:1209–1216.

[40] Stone, J. and C. A. Burrus. 1973. Nd^{3+} Doped SiO_2 lasers in end pumped fiber geometry. *Appl. Phys. Lett.* 23:388–389.

[41] Townsend, J. E. et al. 1987. Solution-doping technique for fabrication of rare earth-doped optical fibers. *Electron. Lett.* 23(7):329–331.

[42] Townsend, J. E. 1990. The development of optical fibers doped with rare earth ions, Ph.D. Thesis, University of Southampton, p. 265.

[43] Carter, A. L. G. 1994. High phosphate content silicate optical fibres fabricated via flash-condensation, Ph.D. Thesis, University of Sydney, p. 326.

[44] Carter, A. L. G. et al. 1992. Flash-condensation technique for the fabrication of high-phosphorus-content rare earth-doped fibers. *Electronics Lett.* 28:2009–2011.

[45] DiGiovanni, D. J. and J. B. MacChesney. 1992. Sol-gel doping of optical fiber preform. Patent No. US5123940, AT&T Corp., Murray Hill, NJ, USA, p. 16.

[46] Wu, F. Q. et al. 1993. Low-loss rare earth-doped single-mode fiber by sol-gel method. *Materials Res. Bull.* 28:637–644.

[47] Randall, E. N. 1975. Method of producing glass in a flame. Patent No. US3883336, Corning Glass Works, Corning, NY, USA, p. 6.

[48] Takahashi, S. et al. 1983. Apparatus for producing multi-component glass fiber preform. Patent No. US4388098, NT&T (Japan), Fujikura Ltd. (Japan), p. 10.

[49] Takahashi, S. et al. 1982. Method for producing multi-component glass fiber preform. Patent No. US4336049, NT&T (Japan), Fujikura Ltd. (Japan), p. 10.

[50] Morse, T. F. et al. 1989. Aerosol doping technique for MCVD and OVD. In: *Proceedings of SPIE: Fiber Laser Sources and Amplifiers, Boston, MA, Sep. 6–8, 1989*, Vol. 1171, pp. 72–79. SPIE—The International Society for Optical Engineering, Bellingham, Washington.

[51] Morse, T. F. et al. 2002. Aerosol techniques for photosensitive glasses. In: *Proceedings of Summer School on Photosensitivity in Optical Waveguides and Glasses, St. Petersburg, Russia, June 17–21, 2002*, Paper FB1. Technical proceedings.

[52] Desurvire, E. et al. 1987. High-gain erbium-doped traveling-wave fiber amplifier. *Opt. Lett.* 12:888.

[53] Mears, R. J. et al. 1987. Low-noise erbium-doped fibre amplifier operation at 1.54 μm. *Elect. Lett.* 23:1026.

[54] Nakazawa, M. et al. 1989. Efficient Er^3-doped optical fiber amplifier pumped by a 1.48 μm InGaAsP laser diode. *Appl. Phys. Lett.* 54:295.

[55] Pederson, B. et al. 1991. The design of erbium-doped fiber amplifiers. *J. Lightwave Technol.* 9(9):1105–1112.

[56] Becker, P. C. et al. 1999. *Erbium-Doped Fiber Amplifiers*. Academic Press, New York.

[57] Armitage, J. R. 1988. Three level fiber laser amplifier. *Appl. Opt.* 27:4831.

[58] Sudo, S. and H. Itoh. 1990. Efficient non-linear optical fibres and their applications. *Optical Quantum Electron.* 22:187–212.

[59] Davey, S. T. et al. 1989. The fabrication of low loss high NA silica fibres for Raman amplification. In: *SPIE*, Vol. 1171, pp. 181–191, *Fiber Laser Sources and Amplifiers*. SPIE—The International Society for Optical Engineering, Bellingham, Washington.

[60] Lines, M. E. et al. 1999. Explanation of anomalous loss in high delta singlemode fibers. *Electron. Lett.* 35(12):1009–1010.

[61] Yablon, A. D. 2005. Optical fiber fusion splicing. In: *Optical Sciences*, Vol. 103. Springer, Heidelberg, Germany.

[62] Snitzer, E. 1966. Glass lasers. *Appl. Opt.* 5(10):1487–1499.

[63] Miniscalco, W. J. 1991. Erbium doped glasses for fiber amplifiers at 1500 nm. *J. Lightwave Technol.* 9(2):234–250.

[64] Delevaque, E. et al. 1993. Modeling of pair-induced quenching in erbium-doped silicate fibers. *IEEE Photon. Tech. Lett.* 5(1):73–75.

[65] Poulain, M. et al. 1977. New fluoride glasses. *Mater. Res. Bull.* 12:131.

[66] Lines, M. E. 1988. Theoretical limits of low optic loss in multicomponent halide glass materials. *J. Non-Cryst. Solids* 103:265.

[67] Clesca, B. et al. 1994. Gain flatness comparison between erbium doped fluoride and silica fiber amplifiers with wavelength-multiplexed signals. *IEEE Photon. Tech. Lett.* 6(4):509–512.

[68] Yamada, M. et al. 1998. Gain-flattened tellurite-based EDFA with a flat amplification bandwidth of 76 nm. *Photon Tech. Lett.* 10(9):1244–1246.

[69] Goforth, D. E. et al. 2000. Ultra-wide band erbium amplifiers using a multi-component silicate fiber. *OAA*.

[70] Arbore, M. A. et al. 2002. 34dB gain at 1500nm in S-band EDFA with distributed ASE suppression. In: *Conference Proceedings of ECOC 2002. 28th European Conference on Optical Communications*, IEEE Copenhagen, Denmark.

[71] Arbore, M. et al. 2003. S-band erbium-doped fiber amplifiers for WDM transmission between 1488 and 1508 nm. Presented at OFC 2003, 23–28 March 2003. Paper WK2.

[72] Komukai, T. et al. 1995. Upconversion pumped thulium-doped fluoride fiber amplifier and laser operating at 1.47 μm. *IEEE J. Quantum Electron.* 31:1880–1889.

[73] Zemon, S. et al. 1992. Nd^{3+} doped fluoroberyllate glasses for fiber amplifiers at 1300nm. In: *Optical Amplifiers and Their Applications. Technical Digest Series*, Vol. 17. Optical Society of America, Washington, DC. Paper WB3.

[74] Dakss, M. L. and W. J. Miniscalco. 1990. Fundamental limits on Nd^{3+}-doped fiber amplifier performance at 1.3 μm. *IEEE Photon. Tech. Lett.* 2(9):650–652.

[75] Ohishi, Y. et al. 1991. A high gain, high output saturation power Pr^{3+} doped fluoride fiber amplifier operating at 1.3 μm. *IEEE PTL* 3(8):715–717.

[76] Pederson, B. et al. 1992. Neodymium and praseodymium doped fiber power amplifiers. In: *Optical Amplifiers and Their Applications. Technical Digest Series*, Vol. 17. Optical Society of America, Washington, DC. Paper WB4.

[77] Grubb, S. G. et al. 1992. High output power Er^{3+}/Yb^{3+} codoped optical amplifiers pumped by diode-pumped Nd^{3+} lasers. In: *Optical Amplifiers and Their Applications. Technical Digest Series*, Vol. 17. Optical Society of America, Washington, DC. Paper FD1-1.

[78] Snitzer, E. and R. Woodcock. 1965. Yb^{3+}-Er^{3+} glass laser. *Appl. Phys. Lett.* 6(3):45–46.

[79] Townsend, J. E. et al. 1991. Yb^{3+s} ensitised Er^{3+} doped silica optical fibre with ultrahigh transfer efficiency. *Elect. Lett.* 27(21):1958–1959.

[80] Po, H. et al. 1989. Double clad high brightness Nd^{3+} fiber laser pumped by GaAlAs phased array. In: *Conference on Optical Fiber Communication, OFC'89 Technical Digest*, Vol. 5, pp. 220–223. Optical Society of America, Washington, DC. Paper PD7-1.

[81] Goldberg, L. 1998. Method and apparatus for side pumping and optical fiber. U.S. Patent 5 854 865, Dec. 29, 1998.

[82] Yla-Jarkko, K. H. et al. 2003. Low-noise intelligent cladding-pumped L-band EDFA. *IEEE Photon. Tech. Lett.* 15(7).

[83] Kincade, K. and S. G. Anderson. 2004. Review and forecast of the laser markets, part 1: Nondiode lasers. *Laser Focus World* January:75–90.

[84] Renaud, C. C. et al. 2001. *IEEE JQE* 37(2):199–206. Characteristics of Q-switched cladding-pumped ytterbium-doped fiber lasers with different high-energy fiber designs.

[85] Agrawal, G. P. 1995. *Nonlinear Fiber Optics*. Academic Press, San Diego.

[86] Fermann, M. 1998. Singlemode excitation of multimode fibers with ultrashort pulses. *Optics Lett.* 23(1):52–54.

[87] Koplow, J. P. et al. 2000. Singlemode operation of a coiled multimode fiber amplifier. *Optics Lett.* 25(7):442–444.

[88] Fini, J. 2005. Design of solid and microstructure fibers for suppression of higher-order modes. *Op. Express* 13(9):3477–3490.

[89] Wong, W. S. et al. 2005. Robust single-mode propagation in optical fibers with record effective areas. CLEO paper CPDB10.

[90] Limpert, J. et al. 2004. Low nonlinearity single-transverse mode ytterbium doped photonic crystal fiber amplifier. *Op. Express* 12(7):1313–1319.

[91] Oh, J. M. et al. 2006. Increased amplifier efficiency by matching the area of the doped fiber region with the fundamental fiber mode. To appear OFC'06, Anaheim, CA.

[92] Fini, J. M. 2006. Bend-resistant design of conventional and microstructure fibers with very large mode area. *Opt. Express* 14:69–81.

[93] Dawson, J. W. et al. 2004. Large flattened mode optical fiber for reduction of non-linear effects in optical fiber lasers. In: *Proceedings of SPIE*. 5335:132.

[94] Kaiser, P. and H. W. Astle. 1974. Low loss single-material fibres made from pure fused silica. *Bell Syst. Technol. J.* 53:1021–1039.

[95] Ranka, J. K. and R. S. Windeler. 2000. Nonlinear interactions in air-silica microstructure optical fibers. *Optics Photonics News* 11(8):20–25.

[96] Birks, T. A. et al. 1997. Endlessly single-mode photonic crystal fiber. *Optics Lett.* 22(13):961–963.

[97] Hasegawa, T. et al. 2003. Bend-insensitive single-mode holey fibre with SMF-compatibility for optical wiring applications. Presented at the ECOC, Rimini, Italy.

[98] Eggleton, B. J. et al. 2001. Microstructured optical fiber devices. *Optics Express* 9(13): 698–713.

[99] Venkataraman, N. et al. 2002. Low loss (13 dB/km) air core photonic band gap fibre. Presented at the 28th European Conference on Optical Communication, ECOC'02, Copenhagen, DK.

[100] Birks, T. A. et al. 1998. Single-mode photonic crystal fiber with an indefinitely large core. Presented at the Summaries of Papers at the Conference on Lasers and Electro-Optics. Conference Edition. 1998 Technical Digest Series, Vol. 6 (IEEE Cat. No. 98CH36178).

[101] Limpert, J. et al. 2006. High-power ultrafast fiber laser systems. *IEEE J. Sel. Topics Quant. Electron.* 12(2).

[102] Wong, W. S. et al. 2005. Robust single-mode propagation in optical fibers with record effective areas. Paper presented at the Conference on Lasers and Electro-Optics. Paper CPDB10.

[103] Boer, M. M. et al. 1993. Highly nonlinear near-resonant photodarkening in a thulium-doped aluminosilicate glass fiber. *Optics Lett.* 18(10):799–801.

[104] Ainslie, B. J. et al. 1989. Optical and structural analysis of neodymium-doped silica-based fibre. *Materials Lett.* 8(6,7):204–208.

[105] Arai, K. et al. 1986. Aluminum or phosphorus co-doping effects on the fluorescence and structural properties of neodymium-doped silica glass. *J. Applied Physics* 59(10): 3430–3436.

[106] Miniscalco, W. J. 1991. Erbium-doped glasses for fiber amplifiers at 1500 nm. *J. Lightwave Technol.* 9(2):234–250.

[107] Paschotta, R. et al. 1997. Lifetime quenching in Yb-doped fibres. *Opt. Comm.* 136: 375–378.

[108] Choi, Y. G. et al. 2002. Spectral evolution of cooperative luminescence in an Yb3+ -doped silica optical fiber. *Chem. Phys. Lett.* 364:200–205.

[109] Kannan, S. et al. 1997. Thermal stability analysis of UV-induced fiber Bragg gratings. *J. Lightwave Technol.* 15(8):1478–1483.

[110] Koponen, J. et al. 2006. Photodarkening rate in ytterbium doped silica fibers. *Photonics West.*

[111] Jasapara, J. et al. Effect of heat and H_2 gas on the photodarkening of Yb^{3+} fibers. Paper presented at Conference on Lasers and ElectroOptics. Long Beach, CA. Paper CTuQ.

[112] Kaczmarek, S. M. et al. 2005. Optical study of Yb^{3+}/Yb^{2+} conversion in CaF_2 crystals. *J. Phys. Condens. Matter* 17:3771–3776.

[113] U.S. patent 5,907,652. Article comprising an air-clad optical fiber.

[114] DiGiovanni, D. J. 2006. Progress in all-fiber components. Presented at SPIE, San Jose, CA. Paper 6103.

Chapter 8

Polarization Maintaining Fibers

Chris Emslie

Fibercore Ltd., Hampshire, England

The purpose of this chapter is to provide a practical, technical introduction to the field of polarization maintaining (PM) fiber that will equip the reader with the basic knowledge and understanding necessary to use or specify this category of specialty fiber.

The chapter begins by explaining how PM fibers work and provides brief examples of their various applications in sensing, medicine, and telecommunications, before describing the main fiber designs and the fabrication techniques used to produce them. The relatively new technology of "holey" or "microstructure" fibers is addressed briefly, as is one of PM fiber's more exotic cousins, polarizing fiber. The second half of the chapter moves into the more practical areas of performance criteria, measurements, and environmental effects. Wherever possible, Web-based review article–type references have been used, in recognition of the way this book is likely to be used.

To the novice, this chapter provides a clear framework to assist the acquisition and development of knowledge gained from practical experience, and to the more experienced individual, it provides insight to stimulate deeper understanding and more effective problem-solving when using PM fibers.

8.1 WHAT IS A POLARIZATION MAINTAINING FIBER?

"Polarization maintaining," "PM," "polarization preserving," "HiBi," or even occasionally "polarization retaining fiber" are all different names to describe the same thing—any optical fiber that will faithfully preserve and transmit

243

the polarization state of the light that is launched into it, even when subjected to environmental perturbations.

8.2 WHY USE PM FIBERS?—APPLICATIONS

PM fiber is used in any application that requires the transmission and delivery of polarized light.

8.2.1 Interferometry

The applications of PM fibers cover a broad spectrum encompassing telecommunications, medicine, and sensors. Most typically, these applications are interferometric and take advantage of the fiber's ability to prevent signal fade by ensuring that the light traveling in the signal and reference arms of the interferometer always recombines with the same state of polarization, ensuring optical constructive interference, as depicted in Fig. 8.1. If conventional single-mode fiber were used, the polarization state of the light traveling within each arm would vary independently with time, causing the recombined signal to fade between a maximum and zero as the relative polarization state of the two waveforms varied over 360 degrees. In broad terms, this basic principle is relevant to the interferometric techniques used in all three of the main application areas.

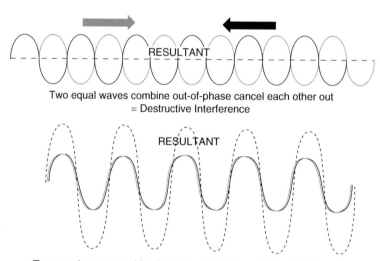

Two equal waves combine out-of-phase cancel each other out = Destructive Interference

Two equal waves combine in-phase generate a resultant of twice the amplitude but the same wavelength = Constructive Interference

Figure 8.1 Interfering waveforms.

8.2.2 The Fiber Optic Gyroscope

The interferometric fiber optic sensor that has achieved the most commercial success is the fiber optic gyroscope (FOG) [1]. In essence, a FOG is a rotation and rotation-rate sensor that comprises up to three coils of PM fiber, one for each degree of freedom required (in aircraft terms: roll, pitch, and yaw). Light is launched into both ends of each coil simultaneously and recombined at a detector. If the coil experiences a rotation, the light will undergo a Doppler-shift (the Sagnac effect [2]) with the result that the counter-propagating beams will recombine out of phase, creating interference that may be analyzed to determine the degree and rate of rotation.

The basic design of a FOG ably illustrates the key benefit of using fiber as an intrinsic optical sensing element; the fiber's ability to guide light enables exceptionally long path lengths to be confined within small physical volumes. These long path lengths magnify relatively weak optical effects, enabling the manufacturing of very compact high-precision sensors. Typical FOG coils contain between about 100 and 5000 m of PM fiber, depending on the desired performance, and are capable of challenging the precision of the very best spinning-mass or ring-laser gyros. It is interesting to note that the original experiments that demonstrated the basic principle by which the FOG operates were performed in the early 1920s using free-space optics, deployed over an area of several square kilometers [3]; in stark contrast, the same measurements could be performed today using a FOG that would not be much bigger than a teacup.

8.2.3 Coherent Communications

PM fibers have also been used in telecommunications since the earliest days of specialty fiber technology. Coherent communications [4], or "Cocomms," used

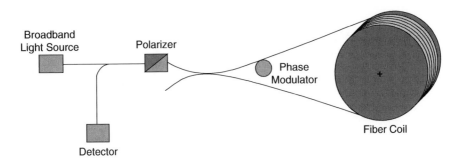

Figure 8.2 A simplified open-loop fiber optic gyroscope.

interferometric detection techniques that could improve both receiver sensitivity and selectivity by around two orders of magnitude, when compared with the conventional state of the art of the 1980s. Unlike conventional direct detection, where the optical signal is converted directly into a demodulated electrical output, a coherent receiver first adds a locally generated optical wave and then detects the combination. The resultant signal carries all of the information of the original yet is at a frequency low enough to allow further signal processing to be performed using conventional electronics. The technique is described as "coherent" because some degree of coherence between the signal and locally generated waveforms is essential; if the phase relationship between the two waves were to vary with time, then the data would become corrupted. The development of commercial coherent communications systems for broad-area networks effectively ceased in the early 1990s, as the advent of the erbium-doped fiber amplifier (EDFA) [5], combined with dense wavelength division multiplexing (DWDM) presented a simpler and more versatile solution to high-bandwidth repeaterless transmission. However, Cocomms has survived as a specialized technology for use in applications that require vast amounts of data to be processed in real time, particularly to enable antenna remoting in military phased-array radar [6] deployments.

8.2.4 Integrated Optics

Signal processing in interferometric sensors and transmission or detection in both conventional and coherent communications both use another significant technology that is enabled by PM fiber: integrated optics (IO) [7]. Today, IO is

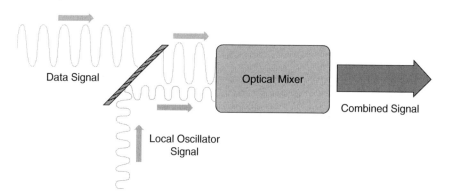

Figure 8.3 A coherent communications receiver.

most often encountered in the lithium niobate (LiNbO₃) modulators used in telecommunications transmitters. A typical modulator consists of a lithium niobate chip into which titania-doped waveguides, flanked by gold electrodes, have been diffused. A PM fiber pigtail delivers a stable polarization state, aligned to the birefringent axes of the chip. The device functions because of the Pockel effect [8]; in other words, when a voltage is applied to the electrodes, the refractive index of the substrate is changed in proportion to that voltage. The resulting change in effective optical path length can be used to generate interference that, depending on the precise design of the titania-doped waveguides, may be manipulated to provide modulation of phase, frequency, or amplitude or even to switch optical power between channels.

8.2.5 Laser Doppler Anemometry and Velocimetry

Whereas some interferometers may be constructed entirely of PM fiber, through the use of fused or polished all-fiber components, in many instances, the main function performed by the PM fiber is merely to provide a flexible delivery system that enables fragile optics and signal processing electronics to be isolated from the delivery point. Examples of this sort of application are laser Doppler velocimetry (LDV) and laser Doppler anemometry (LDA) [9]—noncontact techniques for flow velocity measurement, for example, air-flow

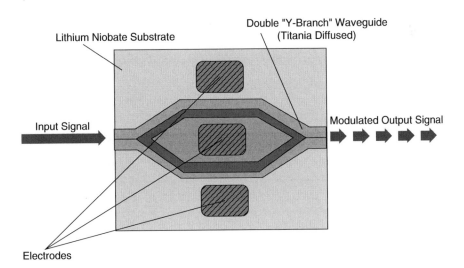

Figure 8.4 An integrated optic modulator.

in wind tunnels or even blood flow in veins and arteries.[1] In LDA and LDV, flow velocity is determined by measuring the Doppler shift of the light scattered from the fluid.

To make a measurement, linearly polarized light from a laser source is split into two equal components and transmitted to the measurement site through two identical lengths of PM fiber. At the output of these fibers, lenses focus the two beams down to a small spot within the moving fluid. At this point, the two beams converge to form interference fringes. Small particles within the fluid scatter light from each beam at slightly different Doppler frequencies because of their motion relative to the two beam directions. Some of this scattered light is then collected by a large-core multimode fiber and transmitted to a photodetector where the two frequencies combine to form a temporal beat frequency. This beat frequency is linearly related to the difference in the Doppler frequencies created from each laser beam and, therefore, to the particles' velocities.

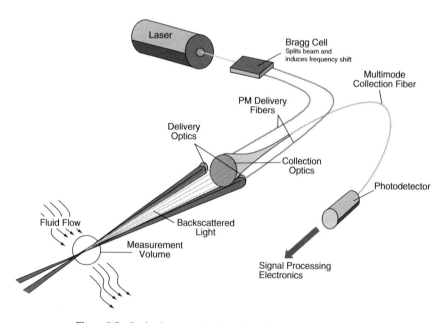

Figure 8.5   Basic elements of a laser doppler anemometry system.

---

[1] The terms *velocimetry* and *anemometry* are not strictly interchangeable. *Velocimetry* refers to the velocity measurement of solid surfaces/liquids, whereas *anemometry* refers to vapors/gasses. *Velocimetry* is also usually interpreted to encompass "vibrometry," the science of vibration measurement.

8.2.6 EDFA Pump Combiners, Reflection-Suppression Schemes, Current Sensing, and Optical Coherence Tomography

The use of PM fiber to enable the flexible and remote delivery of polarized light extends to a variety of other applications across the full spectrum of industries. Developments in telecommunications systems' architectures over the last 5 years have demanded ever increasing power outputs from EDFAs that, in some designs, have been achieved through the polarization multiplexing of 980- or 1480-nm pump diodes. Similarly, pump diodes have also been pigtailed in PM fiber to enable polarization-based schemes for the suppression of back-reflection. In sensing, the Faraday effect current sensor [10] is experiencing something of a renaissance. As a polarimetric device, the current sensor relies on the delivery of a stable and known polarization state to the sensor head, and typically, this is achieved via PM fiber. Finally, in medicine, coronary heart disease patients suffering from the condition known as "total occlusion," where a blood vessel is totally blocked, are being treated with the assistance of a special catheter, or "guidewire," that uses PM fiber. The PM fiber enables the surgeon to differentiate between the vessel wall and the blockage itself through the technique of optical coherence reflectometry (OCR [11]), thereby facilitating its safe removal.

8.3 HOW DO PM FIBERS WORK?

Under laboratory conditions, polarization maintenance may be demonstrated in virtually any single-mode fiber—provided that it is kept short enough, straight enough, and isolated from any form of environmental perturbation. The problems tend to occur when it becomes necessary to use the fiber in more practical situations.

The fundamental (TEM_{00}) mode that propagates within a single-mode fiber is actually a degenerate combination of two orthogonally polarized modes. In a conventional telecommunications-type fiber, these two components have the same propagation constant (i.e., they travel at the same velocity). This property makes it very easy for optical energy to transfer, or "cross-couple," from one of these modes to the other, if it encounters any sort of perturbation within the fiber. These perturbations may be intrinsic, caused by microscopic geometric variations within the core or residual thermal stress, locked-in by the fiber fabrication process, or extrinsic, induced by the environment in which the fiber has been deployed. Extrinsic perturbations are usually mechanical stress-related phenomena caused by microbending or macrobending, typically in combination with the effects of the thermal behavior of the fiber coating (buffer) material.

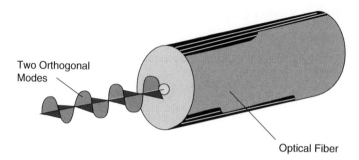

Two Orthogonal
Modes

Optical Fiber

Figure 8.6 Two modes of a single-mode optical fiber.

In essence, these are the same phenomena that generate polarization mode dispersion (PMD), for the simple reason that birefringence and PMD are essentially one and the same thing.[2]

In the real world, in which fibers cannot reasonably be protected from environmental stress (temperature fluctuations, etc.), it is necessary to use a purpose-designed PM fiber. PM fibers are engineered in such a way that the two orthogonally polarized modes are forced to travel at different velocities (i.e., with different propagation constants). This difference in velocities makes it very difficult for optical energy to cross-couple, with the result that the polarization state of the transmitted light is preserved.

This difference is created through the introduction of anisotropy within the core of the fiber, either geometric, by making the core elliptical, or, more typically, through the application of a controlled uniaxial stress. These two designs are described as *form birefringent* or *stress birefringent*.

8.4 PM FIBER TYPES: STRESS AND FORM BIREFRINGENT

8.4.1 Stress-Birefringent Fibers: Bowtie, PANDA, and Elliptical Jacket

The vast majority of PM fibers used today have one of the three basic stress-birefringent geometries: bowtie, PANDA, and elliptical jacket. All three designs function in the same way; the cores are flanked by areas of high-expansion glass

---

[2] This is not to suggest that modern telecoms fibers with very low PMD values are highly birefringent or PM. In fact, they are the very opposite: designed with very low values of intrinsic birefringence to make them effectively transparent to the polarization state of the transmitted light. These fibers are still adversely affected by environmentally induced birefringence.

that shrink-back more than the surrounding silica, as the fiber is drawn, and freeze the core in tension. This tension induces birefringence (i.e., it creates two different indices of refraction: a higher index parallel and a lower index perpendicular to the direction of the applied stress). In essence, the phenomenon is very similar to that which creates visible interference fringes when transparent plastics are stressed except that, in a PM fiber, the effect is highly controlled and its magnitude is at least an order of magnitude lower than may be achieved in an organic glass (polymer/plastic).

In any of these designs, when polarized light is launched along the "slow axis," it is forced to travel at a lower velocity than if it had been launched along the "fast axis" and vice-versa. The cross-coupling of light from one axis to the other, therefore, becomes very difficult because it would require a perturbation capable of making a significant change in the velocity of the transmitted light. The greater the applied stress, the greater the difference in propagation constant (light velocity) between the two axes and the higher the birefringence. PM ability is enhanced because a larger perturbation is, therefore, needed to generate cross-coupling from one axis to the other. It should be clarified at this stage that polarization maintenance is not a loss mechanism, as is sometimes believed; under the vast majority of practical circumstances, there is no measurable difference in attenuation between the fast and slow axes. However, to this day, most applications use the slow axis exclusively. Although the origins of this practice are thought to lie in the theoretically superior resistance to bend-induced loss provided by the slight increase in numerical aperture (NA) on this axis, the principal benefit has been to provide a useful degree of standardization.

Each of the three designs is capable of generating sufficient birefringence for even the most demanding of applications, so the precise choice of fiber is typically determined by other criteria, ranging from handling characteristics to history. In telecommunications applications, the fiber of choice is usually the

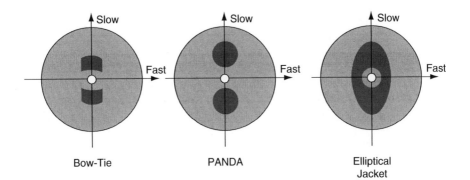

Figure 8.7 Cross-sections of bowtie, elliptical jacket, and PANDA geometries.

PANDA design, invented by Nippon Telegraph and Telephone in the early 1980s [12], then developed and commercialized in the United States and Europe throughout the 1990s. In essence, PANDA is a telecommunications fiber, modified through the insertion of stress rods (usually referred to as *stress applying parts* [SAPs]) to provide PM properties. The fiber was conceived in support of the large volume of Japanese work in the area of coherent communications that had been driven by that country's unique geography (i.e., a series of islands interlinked by stretches of ocean that could be 100 km or more in width). This topography challenged the direct detection technologies of the day, but provided an ideal test case for the improved receiver sensitivity of Cocomms. Fiber attenuation and mode-field diameter (MFD) were well matched to those of single-mode telecommunications fibers so that, when Cocomms was overtaken by the EDFA in the late 1980s, it was a natural extension to continue using the fiber in conjunction with the IO modulators that successfully made the transition between the two technologies.

To this day, bowtie fibers are most typically encountered in sensor applications, and indeed, the majority of FOGs worldwide use a fiber of this design. It should, therefore, come as no surprise that the fiber was conceived in the early 1980s as a sensor fiber, developed at the University of Southampton Optical Fiber Group [13] in support of the FOG program at British Aerospace. Without the constraints of a telecommunications fiber design, bowtie was introduced with a high NA to provide increased resistance to the bend-induced loss that could arise in small-diameter sensor coils, and the elliptical core generated by its fabrication process was accepted readily for sensor use. In extremis, the bowtie design can be shown capable of creating more birefringence than any other stressed design [14, 15], simply because it is based on two opposing wedges, the simplest and most efficient means of applying stress to a point. However, in all but the most exotic of applications, the fundamental design is implemented in a suboptimum condition in the interests of manufacturing yield and consequent cost. It is interesting to note that although redesigned variants of both PANDA and bowtie have subsequently been introduced to address the sensor and telecommunications markets, respectively, more than 20 years later, both designs continue to dominate the applications for which they were originally conceived. The bowtie fiber has been developed and commercialized by Fibercore Limited, a spinout company from the University of Southampton Optical Fiber Group since the fiber's invention in 1983.

The elliptical jacket fiber also had its origins in Japan, but this time with Hitachi. In common with PANDA, the original interest was in a fiber suitable for coherent communications use. However, the fiber was manufactured in the United States initially by a company called Eotech that was subsequently acquired by 3M, who is probably best known for the commercialization of this product as a sensor fiber throughout the United States and Europe. Whereas the elliptical jacket fiber is capable of similar levels of birefringence to the PANDA and bowtie designs, this performance is achieved at the expense of handling characteristics. As can be seen from the Fig. 8.7, this fiber is unique in that the SAP extends all around the core, generating a significant amount of parasitic stress that serves to reduce the fiber's birefringence. The oversized stress member necessitated by this characteristic can compromise performance in the reduced-diameter fibers used in many sensor applications, for the simple reason that there can be insufficient room to locate a SAP of the necessary dimensions. Furthermore, levels of parasitic stress can be considerable, leading to uneven fracture when cleaved and reducing fusion splice yields.

8.4.2 Elliptical Core, Form-Birefringent Fiber

In conventional single-mode fibers, core ellipticity is undesirable simply because it creates birefringence (a.k.a. PMD) that reduces performance in high data-rate systems. In a form-birefringent fiber, this effect is taken to the extreme.

Form-birefringent fibers, with their simple design of a highly elliptical core, combined with a very high NA and small mode, predate stress-birefringent fibers and may be traced back to Hitachi in the late 1970s. However, the high levels of attenuation and lack of compatibility, created by their highly germania-doped cores and consequent small modes, made them unsuitable for telecommunications use. However, they did find some application in sensors as the "E-core" fiber [16] manufactured by Andrew Corporation and sold throughout the 1980s, and the original Hitachi designs were developed by Corning as its PMF38 fiber throughout the latter part of the 1990s. However, despite best efforts to demonstrate how the basic incompatibility caused by the small mode and high loss could be overcome, and very attractive pricing, the product failed to gain market

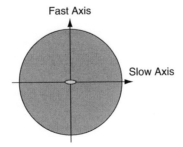

Figure 8.8 Cross-section of an elliptical core, form—birefringent fiber.

acceptance. Today, form-birefringent fibers, tracing their lineage back to the original Andrew E-core fiber, are still used in KVH's range of FOGs.

8.4.3 Microstructure ("Holey") Fibers

Although the original form-birefringent technology may have been sidelined to a significant degree, it is interesting to note that in the future, form-birefringent designs, in the guise of microstructure fibers [17] (also referred to as "holey" or sometimes "photonic crystal" or "photonic band-gap" fibers), may prove to offer the ultimate in terms of PM performance. Although there has been much debate about the precise mechanism by which these fibers work, it is now generally accepted that the vast majority of examples function because of the index difference created by the microstructure of air-gaps within the cladding, making PM variants effectively form birefringent. By virtue of the unitary index of the air that makes up the bulk of the optical cladding, a huge degree of anisotropy may be generated, with correspondingly huge levels of birefringence, up to an order of magnitude greater than has been achieved from conventional stress-birefringent designs [18, 19].

8.4.4 Polarizing Fiber

A type of very highly birefringent fiber that has been known since the earliest years of the technology but remains a rather rare and exotic variant of PM fiber, is polarizing fiber, also known as "zing" [20, 21] or simply "PZ." Zing fibers take advantage of the fact that light polarized along the slow axis is guided slightly

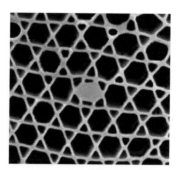

Figure 8.9 A microstructure PM fiber. (Reproduced, with permission of Southampton University Opto-Electronics Research Centre Optical Fiber Group.)

more strongly than that polarized along the fast axis and will, therefore, be less sensitive to bend-induced optical loss. For this reason, a bend diameter may be found at which the fast axis is attenuated very strongly and only the slow axis propagates.

With the correct choice of fiber length, bend diameter, cutoff, and transmission wavelengths, a significant degree of polarizing behavior may be demonstrated in virtually any standard PM fiber. However, in practical sensor applications, it is typically advisable to use a purpose-designed polarizing fiber to ensure environmentally robust performance over a sufficiently broad operating window. Zing fibers are typically designed with short cutoff wavelengths (λ_c) relative to the intended polarizing window, together with low NAs. These design characteristics weaken the strength of optical guidance and, therefore, "encourage" both the fast and the slow mode to radiate. To ensure that the slow mode remains guided for an operating window extending perhaps 50 nm beyond the wavelength at which the fast mode is lost, the birefringence of the fiber is also increased by about 40% over values usually encountered in PM fibers. As a refinement to this basic design, structures may also be incorporated within the inner cladding to preferentially attenuate the fast mode.

Although there is no fundamental reason that both form- and stress-birefringent designs may be used to create polarizing fibers, and indeed, microstructure

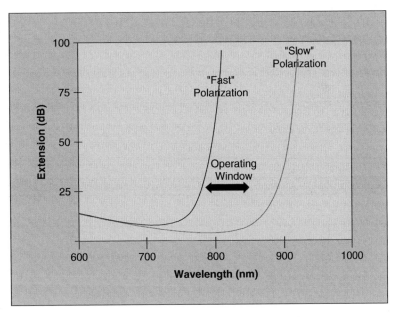

Figure 8.10 Operating window of a polarizing fiber.

designs may be highly suited due to the extreme levels of birefringence that have been demonstrated, to date only stress-birefringent configurations have been used commercially, primarily the zing bowtie fiber introduced by York VSOP in the mid 1980s and the PZ elliptical jacket fiber commercialized by 3M around 1990.

8.5 PM FIBER FABRICATION METHODS

In common with their more conventional counterparts, the manufacturing process for PM fibers comprises two parts: preform fabrication and fiber drawing. The fiber drawing process is practically identical for all fiber types, both PM and non-PM. For this reason, only the preform fabrication stage is addressed in this section. A more comprehensive treatment of conventional fiber fabrication processes may be found in Chapter 3, as well as in any one of a number of standard fiber optic textbooks [22].

8.5.1 Bowtie Fibers

Bowtie preforms are fabricated on a lathe [23] using inside vapor-phase oxidation (IVPO) with stress members created by the process of gas-phase etching [24]. First of all, a ring of boron-doped silica (effectively boric oxide B_2O_3 mixed with silica SiO_2) is deposited within a high-purity, synthetic-silica substrate tube by the oxidation of boron tribromide (BBr_3) in combination with silicon tetrachloride ($SiCl_4$).

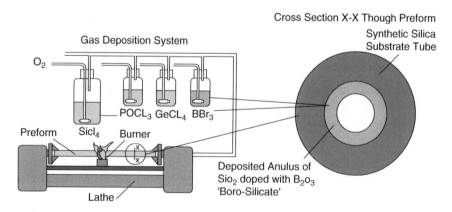

Figure 8.11 Deposition of the boron-doped ring in a MCVD process.

When a sufficiently thick layer has been created, the rotation of the lathe is stopped to allow two diametrically opposed sections to be etched away. The material is etched by passing a suitable etchant (typically sulfur hexafluoride SF_6) through the center of the tube and activating it by means of a narrow-zone etching burner. The final shape of the bowtie SAPs may be controlled and stress levels optimized by varying the arc through which the etching burner is rotated.

After the completion of the etching stage, rotation is recommenced and inner-cladding, followed by core layers, are deposited. The inner-cladding is typically fused silica containing a suitable viscosity modifier (usually phosphorus pent-oxide P_2O_5, synthesized by the oxidation of phosphoryl chloride $POCl_3$, and index matched using fluorine). The core is germanosilicate (a mixture of germania GeO_2 and SiO_2), created through the oxidation of a combination of $SiCl_4$ and $GeCl_4$.

Preform fabrication is completed with a controlled collapse process. An overpressure of dry nitrogen is introduced to balance the surface-tension forces that gradually overpower the viscosity of the material as the process temperature is increased to almost 2000 °C. The characteristic bowtie shape is created as the central hole is collapsed to form a solid cylindrical preform. When this preform is

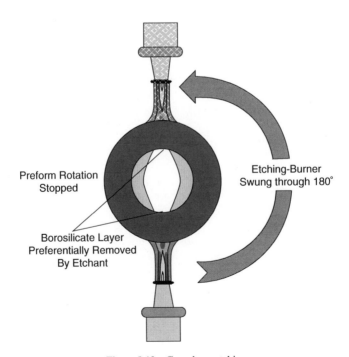

Preform Rotation
Stopped

Etching-Burner
Swung through 180°

Borosilicate Layer
Preferentially Removed
By Etchant

Figure 8.12 Gas-phase etching.

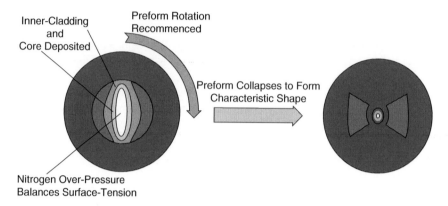

Inner-Cladding
and
Core Deposited

Preform Rotation
Recommenced

Preform Collapses to Form
Characteristic Shape

Nitrogen Over-Pressure
Balances Surface-Tension

Figure 8.13   Core deposition and controlled collapse.

drawn, surface tension forces also ensure that the geometry of the preform is faithfully reproduced in the fiber.

8.5.2 PANDA Fiber

As benefits its origins as a modified telecommunications fiber, the starting point for a PANDA fiber is a circularly symmetric "telecoms-type" preform, into which two diametrically opposed holes have been drilled ultrasonically. The PANDA preform is then completed by the insertion of a boron-doped stress rod into each hole. These stress rods may be fabricated either by vapor deposition or by a sol-gel process [25]. When drawn, the low melt viscosity of these stress rods relative to the surrounding silica ensures that the boron-doped material entirely fills the holes to form the characteristic PANDA shape.

Whereas the fabrication process for PANDA may appear relatively straight-forward, the drawing process must be well controlled to prevent distortion of both the stress rods and the fiber itself and achieve the exceptional geometric precision for which the design is renowned. Similarly, despite various proposals for alternative PANDA fabrication methods, the tight machining tolerances and tool stability necessary to drill side-holes of significant depth have continued to make the manufacture of very high-yielding preforms technically challenging.

8.5.3 Elliptical Jacket Fiber

The fabrication method for elliptical jacket preforms combines processes from both bowtie and PANDA manufacture: boric oxide deposition by MCVD and ultrasonic machining. The initial fabrication steps are identical to

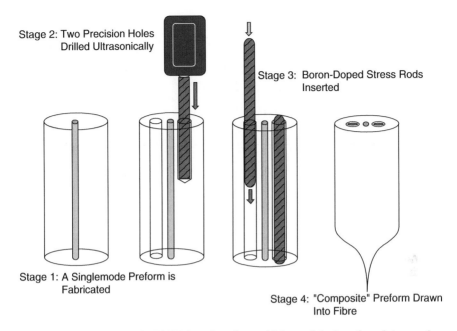

Stage 2: Two Precision Holes
Drilled Ultrasonically

Stage 3: Boron-Doped Stress Rods
Inserted

Stage 1: A Singlemode Preform is
Fabricated

Stage 4: "Composite" Preform Drawn
Into Fibre

Figure 8.14 Fabrication of a PANDA preform by machining and the insertion of stress rods.

those for a bowtie preform, except that the gas-phase etching is omitted, leaving an unbroken circular annulus of boron-doped material surrounding the core.

After this circularly symmetric preform has been completed, its symmetry is broken by the machining of two flats, one on each side of the core. Although it is possible to use conventional mechanical grinding techniques, the associated force and vibration would make the highly stressed preform vulnerable to fracture. For this reason, a more benign ultrasonic machining process is typically used.

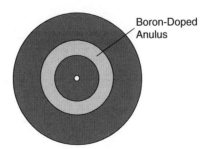

Boron-Doped
Anulus

Figure 8.15 Stage 1 of elliptical jacket fabrication: MCVD preform with a boron-doped ring (c.f. Figure 8.11).

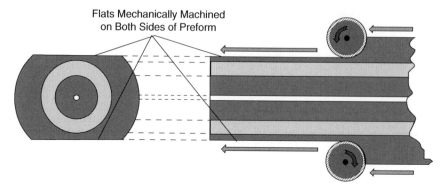

Figure 8.16   Stage 2 of elliptical jacket fabrication: preform machining.

When drawn, the high surface tension forces within the molten glass force the preform to circularize, thereby creating the characteristic elliptical jacket shape.

8.5.4 Elliptical Core, Form-Birefringent Fiber

Some degree of ellipticity may be induced in the core of any single-mode fiber if process controls are inadequate. For example, in MCVD, if the surface tension forces within the collapsing preform are not accurately balanced by pressurizing the substrate during the final stages of the fabrication process, then the preform will "pull flat" to some degree and create an elliptical core. To fabricate a core that is deliberately elliptical, it is necessary to take this phenomenon to an extreme by also removing the circularizing effect of the rotation of the lathe,

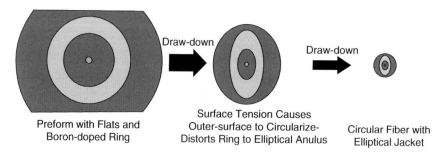

Preform with Flats and
Boron-doped Ring

Surface Tension Causes
Outer-surface to Circularize-
Distorts Ring to Elliptical Anulus

Circular Fiber with
Elliptical Jacket

NB: Draw-down is not to scale

Figure 8.17   Stage 3 of elliptical jacket fabrication: flattened preform circularizes during fiber drawing.

depositing the core (by MCVD/IVPO) within a stationary substrate tube, then collapsing it under vacuum to create the characteristic high-ratio ellipse. The resulting cladding ellipticity is low because of the very small amount of material movement actually taking place, relative to the total volume of the preform, and any residual cladding ellipticity can either be removed by including an additional high-temperature "rounding" step during preform fabrication or even during the fiber drawing process itself.

8.5.5 Microstructure ("Holey") Fibers

The fabrication technique for PM microstructure fibers bears little resemblance to that of any other PM fiber because it does not rely on direct chemical vapor deposition to create the preform. Microstructure fiber performs are fabricated by building a close-packed arrangement of silica tubes around a central silica rod that replicates the desired fiber structure. Precision-machined jigs are used to facilitate this process and the completed preform is typically held together with platinum wire during drawing. Exceptional precision is essential during the assembly of the preform, together with fine control of all drawing

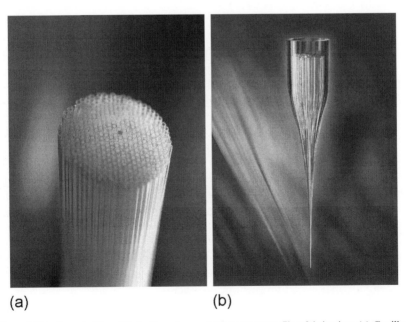

(a)   **(b)**

Figure 8.18 Aspect of two fabrication stages of microstructure fiber fabrication: (a) Capillary stack, before insertion into substrate tube, and (b) microstructure preform showing draw-down. (Reproduced, with permission, of Crystal-Fibre A/S, Denmark.)

conditions to ensure that viscous forces do not distort the fiber during formation. Please note that chemical vapor deposition may still be used to fabricate the high-purity, fused silica components that make up the preform.

8.6 KEY PERFORMANCE PARAMETERS

This section introduces the important optical and physical parameters that influence a PM fiber's performance. It is intended to act as a practical guide to specifying a PM fiber for any specific application. Explanations are provided where these parameters differ from those of more conventional fibers, particularly in cases in which these differences should raise practical concerns, and in doing so, some popularly held myths are dispelled.

8.6.1 Attenuation (α)

Attenuation values for PM fibers are typically higher than those encountered in other single-mode fibers for three reasons: transmission wavelength, the proximity of the SAPs, and core dopants.

Many PM fibers are designed to operate at wavelengths outside the second (1310 nm) or third (1550 nm) telecommunications windows. At short wavelengths, the components of attenuation contributed by both Rayleigh Scattering (λ^{-4}) [26] and electronic transitions (e^x) [27] are still substantial resulting in attenuation values as high as 50 dB/km in the blue (488 nm), falling to approximately 30 dB/km at 514 nm, around 12 dB/km at 633 nm and roughly 4 dB/km in the first telecoms window around 850 nm. These relatively high values are no reflection of the quality of the fiber itself, merely a practical illustration of the fundamental physics that underpins the spectral attenuation cure of the silicate glasses involved.

Even at 1550 nm, the losses of sensor-optimized PM designs tend to be significantly higher than the 0.2-dB/km theoretical limited that has become the norm for use in telecommunications, typically around 1.5 dB/km. Once again, the explanation here does not lie with the quality of the fiber, but in the influence of boric oxide (B_2O_3) used to increase the expansion coefficient of the SAPs. Boron has a very strong vibrational absorption overtone centered close to 1550 nm. The 2000 °C temperatures—and above—prevalent during both the final stages of preform fabrication and throughout the fiber drawing process cause this boron to diffuse into the inner-cladding. If the SAPs have been located very close to the core (say, within 4 μm or so) to generate additional birefringence, this diffusion will bring the strong absorption of the boron into contact with the edge of the optical field, resulting in a higher value of fiber attenuation.

8.6.2 Numerical Aperture (NA)

To cope with the small-diameter bends often encountered in sensor coils, many PM fibers are designed with higher NAs than generally found in telecommunications fibers. Typical NA values of 0.16 or 0.20 are achieved by incorporating higher levels of germania (GeO_2) within the fiber core. Whereas germania, like silica, is a transparent amorphous glass, it is also susceptible to the formation of defects when subjected to the high ($\sim$1800–2200 °C) temperatures of both preform fabrication and fiber drawing. These defects are created when thermal energy knocks electrons from the material lattice to create empty sites, the presence of which intensifies the contribution from the electronic absorption edge (or Urbach edge [27]) to the attenuation of the fiber. This contribution can be significant, particularly at wavelengths in the ultraviolet and visible spectrum, where it may reach several tens of dB/km in extremis. Although careful attention to fabrication conditions, and in particular, by drawing the fiber at the lowest viable temperature, can minimize the creation of these defects, it is difficult to eliminate them entirely. For this reason, the tolerances on attenuation for PM fibers designed for the blue and green (488 nm/514 nm) or the red (633 nm/680 nm) tend to be significantly broader than those designed for the infrared, particularly for fiber designs that preclude drawing at reduced temperatures.

Another unique aspect of PM fibers is that birefringence creates a lower and a higher value of NA for the fast and slow axes, respectively. In normal PM fiber designs, this difference is perhaps 2.5% with the result that there is no practical difference between the two axes in terms of resistance to bend-induced loss. However, if taken to extremes, for example, in a polarizing fiber, a bending condition may be found for which the slow axis continues to guide, yet the fast axis does not. In this case, the guidance of the slow axis is also assisted, to a small extent, by the presence of the low refractive index regions formed by the SAPs.

Fast and slow axes are clearly visible in the aforementioned refractive index profiles, identified as sections through the relevant axes. The presence of the SAPs is indicated by the low index regions that flank the core in the slow axis view. NAs for PM fibers are typically determined directly from the fiber refractive index profile using the equation

$$NA = \sqrt{(n_1^2 - n_2^2)} \approx \sqrt{2n_2 \delta n}, \qquad (8.1)$$

where

n_1 core refractive index
n_2 cladding refractive index
δ_n refractive index difference between core and cladding.

Please note that for the slow axis, the value of n_2 is that of the inner-cladding only—not that of the SAP. It should also be borne in mind that this method of

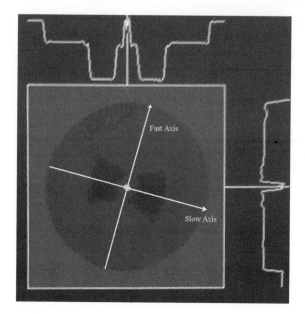

Figure 8.19   Refractive index profiles of the two axes of a PM fiber.

measuring NA directly from the fiber refractive index profile will typically generate a value higher than that derived from taking the sine of the half-angle of the output cone due to the influence of cutoff wavelength.

8.6.3 Is There a Connection Between Polarization Maintenance and Attenuation?

One question that is often asked is What is the difference in attenuation between the fast and slow axes? The answer, in all but the most extreme applications, is Nothing. Theoretically, the slow axis is more strongly guiding and has been both the transmission axis of choice in the majority of applications and the axis most often "keyed" in PM connectors, but the truth is that you can use either axis, to equally good effect, most of the time. The attenuation of a PM fiber is certainly influenced by the presence of the SAPs, but polarization maintenance is fundamentally not a loss mechanism.

8.6.4 Cutoff Wavelength (λ_c)

To address their many and varied applications, PM fibers have been available across a broad range of cutoff wavelengths since their commercial introduction in the early 1980s. Most manufacturers offer standard specifications optimized for 488–532 nm in the blue and green, 633–680 nm in the red, 780–850 nm in the first telecommunications window, 940–1100 nm for YAG, YLF, and semiconductor EDFA pump lasers, 1310–1550 nm in the second and third telecommunications windows, and 1530–1610 nm for the so-called S-, C-, and L-bands.

All of the basic rules that apply to conventional telecommunications fibers also apply to PM fibers. Progressively fewer modes are supported as the cutoff wavelength is approached, reducing to only a single mode when cutoff is reached. The fiber will then continue to support this one remaining mode, usually for at least 200 nm beyond cutoff. This single-mode waveband will increase if the fiber follows a relatively straight physical path and gradually reduce as the fiber is deployed in smaller and smaller diameter coils. This waveband is also extended by the increased NAs of many "bend-insensitive" PM fiber designs, with an NA of around 0.2 capable of providing practical single-mode transmission over perhaps 500 nm or more.

A theoretical complication introduced by PM designs is that the cutoff wavelength is actually fractionally shorter for the fast axis than it is for the slow, because of the difference in index difference (δn) created by the birefringence. However, in practice, this difference amounts to only a few nanometers (in fact, it is typically within the limits of measurement accuracy)—unless, of course, it has been deliberately accentuated through a polarizing fiber design deployed at a relatively small diameter.

8.6.5 Mode-Field Diameter (MFD)

The MFD, or sometimes "spot size," is simply the diameter of the optical field within the fiber. Since the distribution of energy within the fundamental mode (TEM_{00}) is Gaussian in shape, this diameter may be defined in a number of different ways including $1/e^2$, $1/e$, and full width half maximum (FWHM) (i.e., the diameter at which the optical intensity falls to $1/e^2$ of its peak value, etc.). Most typically, it is the $1/e^2$ value that is quoted in product specifications. The MFD is always larger than the actual "physical" core diameter, defined at the point at which the refractive index falls to that of the cladding, because the tail of the Gaussian penetrates into the inner cladding, typically by up to 1 micron.

The higher NAs and shorter cutoff wavelengths used in many PM designs generate MFDs that are significantly smaller than those of more conventional single-mode fibers. For example, MFDs of typical PM fibers may vary from as little as 2.75 μm, for a fiber designed for 488 nm, up to around 8.0 μm for a 1550-nm fiber; these figures compare with 8–10 μm for standard telecommunications fibers. Although users are often concerned about handling difficulties that may result from the smaller MFDs of many PM fibers, and it is true that alignment for connectorization or fusion splicing is far more critical in these designs, it should be noted that it is not possible to design a fiber with a larger MFD without reducing its resistance to bend-induced loss, perhaps significantly. For this reason, the MFD of a custom-designed fiber is often determined by the relative importance of bend-induced or splice/connector loss within the intended application.

Another aspect of MFD in which some PM fibers also differ is mode shape. Both form-birefringent fibers and most bowtie[3] geometry stress-birefringent fibers have elliptical cores that generate elliptical modes. In the case of the form-birefringent fiber, this shape is intrinsic to the fundamental mechanism by which the fiber functions. However, in bowtie fibers, the ellipse is created by the flow of the SAP and inner-cladding materials during the collapse phase of preform fabrication. The shape is not the result of applied stress, as is sometimes believed, or even a design factor intended to enhance PM performance by creating additional form birefringence; in fact the very small amount of form birefringence generated in this way actually works against the primary stress birefringence within the fiber, but to an insignificant degree.

One interesting phenomenon associated with elliptical modes is that the shape reverses from the near to the far field. In the near field, the mode shape matches that of the core, whereas in the far field, the major and minor axes of the ellipse correspond to the "slow" and "fast" axes, respectively.

This phenomenon has frequently led users to believe that the mode of these fibers is indeed circular for the simple reason that the human eye is relatively insensitive to ellipticity and the mode will appear circular if projected onto a screen positioned close to the midpoint of its transition.

Users accustomed to the perfectly circular modes of conventional fibers are sometimes resistant to the idea of an elliptical mode, particularly in telecommunications. However, in the vast majority of applications, mode shape has absolutely no impact on performance, and in fact, elliptical modes can even be beneficial when pig-tailing to devices that also have noncircular modes, including planar waveguides and some semiconductor lasers. The only true exceptions to this rule that are generally encountered are in LDA and in the fabrication of fiber-based depolarizers. In LDA, an elliptical mode will generate different particle transit times in the two orthogonal directions, thereby creating potential problems in data analysis, and in depolarizers, the insertion loss of the 45-degree splice will be increased because of the incomplete overlap of the elliptical modes.

---

[3] Bowtie fibers designed for use in sensors typically have an elliptical mode shape; however, those designed for telecommunications are manufactured using a modified process that produces a circular mode.

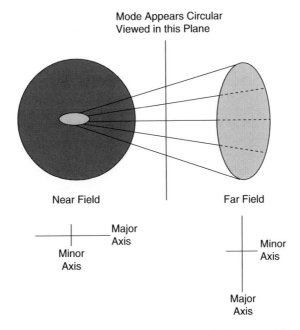

Figure 8.20 Mode shape of an elliptical core fiber in the near and far fields.

8.6.6 Beat Length (L_p)

Beat length is arguably the parameter that best represents the fundamental ability of a PM fiber to preserve polarization. Beat length is particularly useful because, unlike H-parameter or extinction ratio (ER), it is independent of fiber length or the way in which the fiber has been deployed. H-parameter and ER will both be affected adversely by high winding tensions, small coil diameters, point intrusions, or multilayered winds, whereas beat length will remain constant. For this reason, beat length is an invaluable tool that enables you to compare the PM properties of different fibers; put quite simply, the shorter the beat length, the better the performance.

When light is launched into a PM fiber with a linear component along each of its two birefringent axes, the difference in velocities of these two components causes the resultant polarization state to vary along the length of the fiber. The beat length is the distance over which this polarization rotates through 360 degrees. The greater the birefringence within the fiber, the greater the difference between the two velocities and the shorter the beat length.

Beat length is related to birefringence (B) by the following equation:

$$L_p = \frac{\lambda}{B,} \tag{8.2}$$

where λ is the wavelength at which the beat length is measured, and birefringence (B) is the difference between the refractive indices of the fast and slow axes ($n_{slow} - n_{fast}$).

From this equation, it may be seen that beat length varies with wavelength in a linear fashion. For example, a beat length of 1.3 mm measured at 633 nm would increase to approximately 2.7 mm at 1310 nm and 3.2 mm at 1550 nm.

The most direct way of measuring beat length is to launch light into the fiber at 45 degrees to its birefringent axes so that equal amounts of optical power are coupled into each axis. Rayleigh scattering [26] within the fiber causes beats to appear as alternating regions of light and dark as the two modes interfere as they move in and out of phase along the length of the fiber. In this instance, the beat length is merely the distance between two successive light or dark regions. In many cases, the simplest thing to do is to use red light (typically 633 or 680 nm) in the measurement so that these beats may be observed with the naked eye and measured on a millimeter scale. Alternatively, an infrared diode laser may be used (say, at 830 nm), but in this case, a scanning photodiode is required to measure the beats. This simple and direct technique does have limitations, primarily that it is relatively coarse because it relies on an operator counting the number of beats within a relatively short distance and is, therefore, susceptible to rounding errors caused by partial beats, but its precision is sufficient for the vast majority of applications. Furthermore, because it is possible to engineer launch conditions so that a single mode at 633 nm may propagate for a few meters in virtually any fiber, irrespective of cutoff wavelength, the technique is exceptionally versatile.

If a more precise measurement of beat length is required, then this may be achieved by, once again, exciting both axes of the fiber equally but this time capturing the output on a spectrum analyser.

Interference between the two modes may be seen in the fringe pattern and the beat length calculated in the following manner:

$$L_p = \frac{\text{Fringe Width}}{\text{Central Wavelength} \times \text{Fiber Length under Test}} \tag{8.3}$$

The wavelength dependence of beat length is also clearly indicated in the steady increase in fringe width visible throughout the spectrum.[4] Whilst this technique is

---

[4] The characteristic spectral shape is due to the use of an ASE source based on erbium-doped fiber, similar to that which is rapidly becoming an industry standard in high-accuracy fiber optic gyroscopes

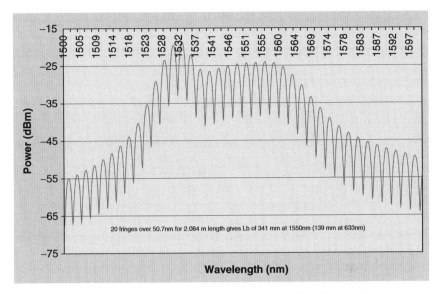

Figure 8.21 Beat-length measurement using an optical spectrum analyzer.

more accurate than direct visual measurement, its suitability is limited by the wavelength range of available spectrum analyzers, together with that of suitable polarizers and sources. For this reason, the technique is usually restricted to measurement at 1550 nm (see Fig. 8.21).

8.6.7 Extinction Ratio (ER)

Whereas beat length provides an invaluable means by which the PM performance of different PM fibers may be compared, the influence of environment is so significant that absolute performance may only be determined experimentally—by deploying the fiber in a way that is representative of the intended application and measuring the purity of the output polarization.

One of the most frequently used measures of the PM performance of a PM fiber is the ER. If a "perfect" PM fiber were used under ideal conditions, all transmitted optical power would remain within the same birefringent axis that it was originally launched into (i.e., light polarized along the fast axis would remain entirely in the fast axis and vice-versa). In practice, this does not happen. Instead, some energy will always transfer into the orthogonal axis, causing a degraded state of polarization to emerge from the delivery end of the fiber. The

ER is merely the ratio of this "unwanted" power, to the amount of power remaining in the launch axis, typically expressed in decibels:

$$ER = 10 \log_{10} (P_u/P_w), \tag{8.4}$$

where

P_u = optical power transferred to the unwanted axis
P_w = optical power remaining in the launch (wanted) axis.

In other words, an ER of -30 dB indicates a ratio of wanted to unwanted optical power of 1000:1, -20 dB is 100:1, -10 dB is 10:1, and so on.

8.6.8 H-Parameter

Another frequently used measure of PM fiber's performance is the so-called *H-parameter*. The H-parameter is simply an ER, expressed as a decimal (not to be confused with *decibel*) per unit length of fiber. For example, an ER of -30 dB, achieved over 1000 m of fiber is the same as an H-parameter of

$$\frac{1}{1000 \times 1000} \text{ per meter}$$

Typically written as simply 10^{-6}. Similarly, an ER of -20 dB achieved over 100 m is the same as an H-parameter of

$$\frac{1}{100 \times 100} \text{ per meter} \quad \text{or } 10^{-4}$$

8.6.9 Effect of Test Conditions and Environment on Polarization Maintaining Performance

One of the most frequently asked questions about PM fibers is What is the H-parameter (or ER)? Despite its apparent simplicity, this question is perhaps one of the most difficult to answer satisfactorily. The fact that H-parameter is expressed normalized to a unit length may lead one to believe that PM performance and length have a linear relationship and indeed, straight-line graphs of PM performance against length appeared on a number of fiber data sheets published during the late 1980s. However, in practice, PM performance is dictated by a combination of test conditions, deployment conditions, packaging, and beat length. Put quite simply, under the "right" conditions, ERs of -30 dB may be achieved over perhaps 5 km of fiber (an H-parameter of 2×10^{-7}), whereas, under the "wrong" conditions, even the best PM fiber could struggle to preserve any polarization at all, even over a few meters.

The fundamental mechanism by which most PM fibers work is birefringence created by uniaxial stress generated within the fiber structure; the greater the stress, the better the fiber in terms of PM performance. It should, therefore, come as no surprise that this internal stress may be reduced, along with the fiber's ability to preserve polarization, by any externally induced stress that the fiber may be subjected to. By far the most common and the most damaging sources of externally induced stress are point defects (essentially microbends) that occur when the fiber is inadvertently tensioned over any small rigid object. In a sensor coil, these point defects may occur where irregularities within the wind cause fibers to cross over each other or where leads cross un-radioed corners within the package or even where air or gas bubbles are entrapped within potting compounds. This last type of point defect is also frequently encountered in PM connector manufacture or in device pig-tailing where the fiber must be cemented into a ferrule of V groove. The physical size of these defects is typically on the same order as the fiber diameter, or smaller. Differential thermal expansion and changes in the Young modulus of different materials used within these systems also conspire to exacerbate the effect.

The damaging effects of these point defects may be reduced through careful attention to both the fiber design itself and that of its packaging. Fibers with a shorter beat length (in the range 1.0–1.3 mm at 633 nm) have greater levels of internal stress and are, therefore, intrinsically more resistant to the effects of externally generated stress than those with longer beat lengths. However, because of the very high levels of localized stress that can be generated by point defects, it is also essential that the coating package (buffer) be capable of isolating the fiber from their penetration over the full operating temperature range of the sensor coil or component. To achieve this isolation, it is necessary to use a primary coating that is very soft and that remains soft throughout the desired temperature range. Primary buffer materials are typically urethane acrylates that, in common with all polymer glasses, get significantly harder at temperatures below 0 °C. In this context, "soft" should be interpreted as maintaining a Young modulus of less than 100 MPa for the majority of the required operating temperature range and less than 1000 MPa for the entire range. At room temperature, these primary materials have a typical modulus in the region of 1 MPa and, for this reason, must be protected with a secondary coating that is significantly harder to impart strength and abrasion resistance. Young's modulus should also be taken into account when selecting ferrule cements, with softer urethane-based materials being more suitable for use in conjunction with PM fibers than the harder yet more popular epoxies.

In general, the environments and packaging in which PM fibers designed for sensor applications tend to be used are significantly more challenging than those encountered in telecommunications. In sensors, longer lengths of fiber tend to be packaged in smaller volumes and used over much wider temperature ranges.

For example, it is not unknown for 200 m of PM fiber to be coiled to a diameter of 20 mm and expected to perform down to -55 °C, whereas even in telecommunications components, bends smaller than 40 mm are rarely encountered and room-temperature operation is often the norm. The more aggressive deployment environments of sensors should always be reflected in the choice of coating materials for a PM fiber, and for this reason, care should be taken when specifying fibers for these applications.

It should also be noted that the fundamental PM properties of stress-birefringent fibers are also temperature dependent. This phenomenon may be explained intuitively by reference to the mechanism by which the birefringence is generated—the mismatch in thermal expansion coefficients between the boron-doped SAPs and the silica that makes up the bulk of the fiber structure. At lower temperatures, this mismatch is accentuated, with a corresponding reduction in beat length. Conversely, at elevated temperatures, the beat length increases, with birefringence disappearing completely at around 600 °C [28]. In theory, this effect helps to provide some compensation for the hardening of buffer materials at low temperatures. In practice, the effect is negligible for the vast majority of applications. However, there are some sensor applications in which it may be advantageous to use a PM with reduced temperature sensitivity. For example, situations in which it is difficult to differentiate the effects of temperature from those of the desired measurand. In these cases, the use of a form-birefringent fiber should be considered because the lack of reliance on thermal stress within these designs reduces their temperature coefficients to practically zero.

When winding sensor coils, avoidance of crossovers and the maintenance of a constant and low tension (typically a few grams) are essential if good values of H-parameter are required. When these points are considered, it becomes clear why values of ER or H-parameter measured on a standard shipping spool provide little or no indication of fundamental PM performance. The main considerations when spooling fiber for shipment are speed and robustness, to minimize cost and ensure that the fiber stays firmly on the reel during transit. To optimize PM performance, you need a low-tension precision wind.

One area of PM fiber measurement often overlooked is the fundamental limitations of the test equipment used. Any commercially available PM fiber should be able to demonstrate an ER in excess of -35 dB, if tested under ideal conditions and using suitable equipment. Something that may easily be forgotten is that PM and polarizing fiber are very different things: PM fiber is designed merely to *maintain* the state of polarization launched into it, and only a polarizing fiber can improve it. What this means is that in practice, any polarization measurement performed on a PM fiber will be limited by the ER of the worst component, so if the source, or polarizer used at launch is capable of only -20 dB, then that is the best result that could be expected from the fiber, even if it may be theoretically capable of −35 dB or better. When assembling equipment to make these measurements, you should consider that most ER or H-parameter rigs put together from nonoptimized components will struggle to measure better than about −25 dB. Achieving −30 dB or higher requires either a great deal of luck or the hand selection of top-quality components.

8.7 MECHANICAL AND LIFETIME PROPERTIES

8.7.1 Strength Paradox I: Fragile Preforms Make Exceptionally Strong Fibers

In all stress-birefringent fibers, it is the preform fabrication stage that provides the greatest challenge to process yields—particularly for the bowtie and elliptical jacket fibers. As stressed structures formed from brittle materials, both bowtie and elliptical jacket preforms are highly vulnerable to brittle fracture. Process controls have evolved to minimize this risk, through optimization of fiber geometries and avoidance of mechanical or thermal shock, so very few preforms are lost in this way today. Although there is some danger that the stress rods themselves may fracture within a PANDA preform, because the rods are mechanically decoupled from the surrounding material until the preform enters the hot zone of the drawing furnace, immediately before fiber formation, this risk is actually very small.

Paradoxically, although the fiber preforms themselves are very fragile, there is absolutely no danger that stress-birefringent fibers will shatter in use, as is sometimes feared. Although both the preform and the fiber are stressed brittle structures and, therefore, potentially unstable, the fibers themselves cannot shatter because their small dimensions hold insufficient surface energy to create a new crack surface. A stressed fiber's inability to shatter is, in fact, a perfect illustration of Griffith's Theory of Brittle Fracture [29].

In practice, PM fibers can exceed the tensile strength requirements set out in Telcordia GR-20-CORE by typical margins of 30% or more with dynamic stress corrosion parameters (n_d) in the low to mid 20s against a requirement of 18 [30]. Advances in materials, particularly the availability of ultrahigh purity synthetic silica substrate tubes, combined with state-of-the art fabrication practices, including drawing in a Class 100 clean-room environment, mean that the Telcordia requirements for tensile strength have become extremely conservative and any fiber of quality manufacture should be expected to exceed them. High though they are, the tensile strength values of stress-birefringent PM fibers are usually a few percentage points below those of standard telecommunications-grade single-mode fibers (see Fig. 8.22). This observation may be explained intuitively by reference to the fiber cross-sections; in PM fibers, a significant fraction of the

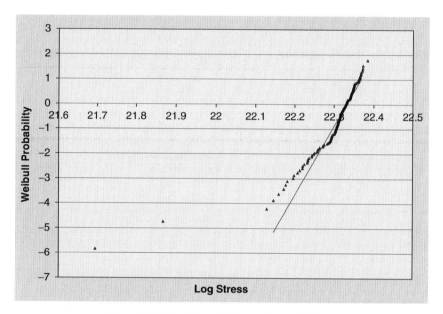

Figure 8.22 Typical weibull curves for a PM fiber.

cross-sectional area is composed of boro-silicate, a material with a tensile strength approximately 20% lower than that of pure silica.

In all practical applications, fiber tensile strength is relatively unimportant provided that it exceeds the GR-20-CORE minimum limits of 3.14 GPa/455 kpsi (unaged) and 2.76 GPa/400 kpsi (aged) [30] because tensile failure (i.e., pulling a fiber too hard) is very rarely encountered. Lifetime, as determined by a fiber's resistance to static fatigue, is of far greater consequence, particularly for PM fibers that are used in sensors or components, because they are usually packaged in far smaller volumes and, therefore, are under far greater strain than would be encountered in typical telecoms installations. Fibers subjected to a constant strain, typically induced by bending or coiling, will eventually succumb to static fatigue because strain causes intrinsic microflaws (cracks) located at the fiber surface to grow. When these cracks have grown beyond a critical limit, they will quickly propagate across the fiber and cause it to fracture. In these circumstances, the stressed nature of PM fibers may actually increase fiber lifetime values because SAPs that place the core in tension also place the fiber surface in compression, thereby resisting crack growth. Hard evidence of this strengthening phenomenon is difficult to find because of the highly statistical nature of fiber lifetime data. However, it is my experience—from reviewing reliability data drawn from several million meters of PM fiber destined for FOG applications—that values of n_d are typically in the upper range of what might be expected of a more conventional single-mode fiber.

8.7.2 Strength Paradox II: Thin Fibers Can Be Stronger Than Thicker Ones

Review of the PM fiber data sheets from any manufacturer will reveal that many specifications are available in both 125-μm and 80-μm variants. A 125-μm glass diameter is a standard that was originally developed for the telecommunications industry and that has a number of advantages, primarily it is big enough to be seen and handled easily while being small enough to enable multifiber cable designs of acceptable dimensions and, being a *standard* means that a wide variety of compatible components are available. However, once again, the more aggressive service environments often encountered by PM fibers should always be considered before making a final choice on fiber diameter. Put very simply, if a PM fiber is to be used in a small-diameter sensor coil, an 80-μm glass diameter will offer a significant increase in lifetime because of its enhanced resistance to mechanical failure through static fatigue.

Static fatigue is a phenomenon by which optical fibers can fracture spontaneously if subjected to bending stress or an invariant tensile load. The stress induces intrinsic and microscopic flaws located on the fiber surface to grow, causing fracture as soon as any flaw grows beyond the critical limit for the material. Fibers are made more resistant to static fatigue, typically by applying one or more of two methods: increased proof test level and reduced glass diameter. Proof test involves straining the fiber to destruction to screen out intrinsic flaws; the higher the proof test level, the smaller the size of intrinsic flaw that will remain in the surviving fiber and the lower the probability that the fiber will fail under static fatigue. The telecommunications industry standard for proof test is 1% strain. Deploy such a fiber in a 20-mm diameter FOG coil and it could fracture within months. When analyzing the reliability data for any fiber, you should always consider that lifetime predictions represent the valid design constraint of the very worst case of what *could* happen—not what *will* happen. If a fiber is proof tested to 1% strain ($\sim$100 kpsi), the resulting lifetime estimate will assume that the fiber still contains a flaw only fractionally below the size necessary to have caused spontaneous mechanical failure during that proof test. In truth, the fiber has only been tested to around one-sixth of its tensile limit, so this flaw *probably* does not exist. In other words, observations that "none of the 10 units made in the laboratory 5 years ago have broken yet," is not a valid reason to discount these lifetime predictions when progressing to high-volume production.

A more direct way to enhance lifetime is to limit bending stress levels by reducing the outside diameter of the fiber itself. A typical FOG fiber has a

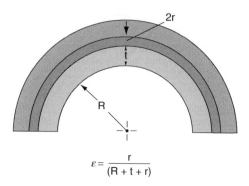

$$\varepsilon = \frac{r}{(R + t + r)}$$

Figure 8.23   Relationship between fiber bend radius and strain.

diameter of 80 μm—a little more than two-thirds of that of a standard telecommunications fiber (125 μm). When bent, the induced stress within these fibers is around 40% lower than that of the larger fiber (Fig. 8.23), slowing growth of intrinsic flaws and boosting lifetimes from mere months to 20 years or more.

When used in the FOG or other intrinsic fiber sensors, reduced-diameter fibers have the added benefit of enabling more fiber to be coiled within a given volume, enhancing the sensitivity that may be achieved from a specific form factor.

In conclusion, if you intend to use a PM fiber in a small package or coil, due consideration should be given to the choice of both glass diameter and proof-test level in any reliability-critical application.

REFERENCES

[1] LeFevre, H. 1993. *The Fiber Optic Gyroscope.* Artech, Norwood.
[2] Sagnac, G. 1913. *Comptes Redus de l'Academie des Sciences (Paris)* 157:708–710, 1410–1413.
[3] Available at www.mathpages.com/rr/s2-07.htm.
[4] Betti, S. et al. 1995. *Coherent Optical Communications Systems.* Wiley-Interscience. John Wiley and Sons, Inc. New York, NY.
[5] Mears, R. J. et al. 1987. U.K. Patent GB 2180392B, Fibre-optic lasers and amplifiers. Published March 25, 1987.
[6] Fenn, A.J. *et al.* The development of phased-array radar technology. 2000. *MIT Lincoln Lab. J.* 12(2).
[7] Hunsperger, R. G. 1995. *Integrated Optics—Theory and Technology,* 4th edition. Springer-Verlag Telos.
[8] Available at http://en.wikipedia.org/wiki/Pockels_effect.
[9] Measurement principles of LDA. Available at http://www.dantecdynamics.com/LDA/Princip/Index.html. Dantee Dynamics, AYS. Denmark.

[10] Edwards, H. O. et al. 1989. Optimal design of optical fibers for electric current measurement. *Appl. Opt.* 28(11):1989.

[11] Schmitt, J. M. 1999. Optical coherence tomography (OCT): A review. *IEEE Selected Topics Quantum Electronics* 5(4):1205–1215.

[12] Hosaka, T. et al. 1981. Low-loss, single polarisation fibres with asymmetrical strain birefringence. *Electr. Lett.* 17(15).

[13] Birch, R. D. et al. 1982. Fabrication of polarisation—maintaining fibres using gas-phase etching. *Electr. Lett.* 18(24):1036–1038.

[14] Varnham, M. P. et al. 1983. Analytic solution for the birefringence produced by thermal stress in polarization-maintaining optical fibers. *J. Lightwave Technol.* LT-1(2).

[15] Noda, J. et al. 1986. Polarisation-maintaining fibers and their applications. *J. Lightwave Technol.* LT-4(8).

[16] Dyott, R. B. et al. 1979. Preservation of polarization in optical-fibre waveguides with elliptical cores. *Electr. Lett.* 15:380–382.

[17] Bennett, P. J. et al. 1999. Toward practical holey fiber technology, fabrication, splicing, modeling and characterization. *Optics Lett.* 24(17):1203–1205.

[18] Suzuki, K. et al. 2001. Optical properties of a low-loss, polarization-maintaining, photonic crystal fiber. *Optics Express* 9(13).

[19] Libori, S. B. et al. 2001. Highly birefringent, index-guiding photonic crystal fibers. *IEEE Photonics Technol. Lett.* 13:588–590.

[20] Varnham, M. P. et al. 1983. Single-polarisation operation of highly birefringent bow-tie optical fibres. *Electr. Lett.* 19(7):246–247.

[21] Messerly, M. J. et al. 1991. A broad-band single polarization optical fiber. *J. Lightwave, Technol.* 9(7).

[22] Midwinter, J. E. 1979. *Optical Fibres for Transmission,* pp. 178–183. John Wiley & Sons.

[23] S.G. Controls. Available at www.newtonhall.co.uk/mcvd.htm.

[24] Birch, R. D. and Payne, D. N. 1986. Method of making optical fibres and optical fibre preforms. U.K. Patent GB 2122599B.

[25] Simax Technologies, Inc. Irvine, CA, USA. Available at www.simaxtech.com.

[26] Lord Rayleigh. 1891. *Philosophical Magazine* 12:81.

[27] Urbach, F. 1953. *Physics Rev.* 92:1324.

[28] Ourmazd, A. et al. 1983. Thermal properties of highly birefringent optical fibers and preforms. *Applied Optics* 22:2374–2379.

[29] Griffith, A. A. 1920. *Philos. Trans. R. Soc. London* 221A:160.

[30] Telcordia GR-20-CORE, *Generic Requirements for Optical Fiber and Optical Fiber Cable,* Issue 2, July 1998.

Chapter 9

Photosensitive Fibers

André Croteau and Anne Claire Jacob Poulin

INO, Sainte-Foy, Québec, Canada

9.1 INTRODUCTION

Photosensitivity is another amazing feature of most specialty optical fibers (SOFs). The photosensitivity phenomenon is different from photo-darkening and radiation-darkening, which induces excess losses. There is then no added background losses due to the fiber Bragg gratings (FBGs) inscribed in the fiber's core besides filtered wavelengths.

Photosensitivity of a medium is defined as its capacity to have its refractive index permanently changed by a modification of its physical or chemical properties through light exposure. Photosensitivity is a complex phenomenon because of the diversity of both parameters and effects that are observed. Fiber composition, fabrication process, operation wavelength, and even light source are all different parameters that can have a significant influence on the photosensitive properties of a fiber. Effect of interest in this chapter is the photosensitivity of Ge-doped silica fiber for which the core refractive index can be permanently modified by an ultraviolet (UV) irradiation [1].

In 1978, photosensitivity was first observed by Hill et al. [1, 2] at the Communication Research Centre (CRC) in Canada. To improve the performance of his tunable fiber Raman laser, Ken Hill ordered two spools of high numerical aperture (HNA) single-mode SOF from Bell Northern Research (BNR). These fibers turned out to be inadequate for use in the fiber Raman laser but accidentally led to the discovery of the photosensitivity. The experiment consisted of injecting light from a single-frequency Argon laser (514 nm) into the core of a doped silica fiber. Hill observed that a fraction of the input power was being reflected by the fiber itself and this phenomenon was attributed to the formation of a permanent index grating permanently inscribed in the fiber core by the light propagating in the fiber.

Progress in optical fiber photosensitivity research was initially slow, probably because the scientific community perceived the phenomenon as a peculiarity of the BNR fibers. An important development was realized 11 years later by Meltz et al. [3], who reported the formation of FBGs by a transverse holographic method. It consisted of illuminating the core from the fiber's side with the interference pattern of two beams of coherent UV radiation in the 244-nm germania oxygen-vacancy defect band. By using this holographic method, the periodic gratings could then be photo-imprinted (to resonate at wavelengths in the infrared [IR] region of the spectrum) to operate as Bragg reflectors at any wavelength longer than the writing wavelength, rather than be resonant near the writing wavelength. The work of Meltz et al. motivated the optical fiber communication community because of the increased possibility that grating-based devices could be fabricated in the IR region, which is their main spectral region of interest. Moreover, side-written gratings can be produced in a few minutes (in contrast with relief gratings), with low loss, and without any other modification of the fiber.

With this new technology and its potential applications, Meltz et al. generated a huge amount of interest, resulting in worldwide research into the fabrication and applications of side-photo-imprinted fiber gratings. In less than 5 years, progress has been extremely rapid on many fronts. One key point of the technology development is the use of phase masks for writing FBGs instead of using the transverse holographic method, with first a static writing through the mask [4] and next with the translation of the UV beam through the mask into the fiber core [5]. The phase mask method simplified the fabrication of FBGs and permitted them to get longer gratings. Using a phase mask, Hill wrote the first fiber gratings in the spring of 1992 [4]. Currently, the phase mask method is still the principal method for manufacturing fiber gratings.

With the emergence of multiple applications based on FBGs, it appeared necessary to enhance the photosensitivity of optical fibers. A lot of work has been performed in order to increase the intrinsic photosensitivity of Ge-doped fibers by modifying its chemical composition with the addition of constituents like boron [6], phosphorous [7, 8], aluminum [7], cerium [9], and others [10]. New methods relying on postfabrication techniques, such as flame-brushing and hydrogen-loading, have also been developed to enhance the Ge-doped fibers' photosensitivity. At this time, the molecular hydrogen diffusion method, which has been developed at AT&T Bell Labs in 1993 [11], is one of the most widely used methods to enhance the two-photon absorption mechanism involved in the 244- to 248-nm writing.

Photosensitivity in Ge-doped silica fibers is now a well-known and mature subject. Several mechanisms contribute to and explain this phenomenon: changes of color centers [12], volume [13], or stress [14]. The discovery of fiber photosensitivity led to the realization of optical components based on FBGs or long-period gratings [15]. Reflectors for fiber lasers [16], wavelength division filters [16], gain-flattening filters for fiber amplifiers [16], chirped FBGs for pulse compression and dispersion compensation [16], short- [17–19] or long-period gratings [20–22] for temperature, strain, and pressure sensors are all examples showing that photosensitivity revolutionized both the optical fiber communication and the optical fiber sensor field. Before the advent of photosensitive fibers and FBGs, these components were made out of bulk optics.

9.2 DESIGN AND FABRICATION

Since the discovery of photosensitivity in optical fibers, many designs driven by the desire to extend the photosensitive fiber's limited performances have been proposed. First, the enhancement of the photosensitivity of standard single-mode fibers (SSMFs) was demonstrated to be efficiently realized by hydrogen (H_2) loading. The HNA fiber design, which requires a high Ge concentration in the core, turned out to be intrinsically photosensitive; moreover, it offsets the FBGs cladding modes in transmission. This type of fiber led Hill's discovery of the photosensitivity phenomenon. The cladding mode suppression (CMS) design further enhanced the FBG bandwidth in transmission. The rare earth–doped (RED) photosensitive fiber design facilitated the building of short-cavity fiber lasers. The polarization maintaining (PM) photosensitive fiber was designed for applications requiring a photosensitive fiber that maintains the polarization plane of the transmitted light in the fiber. Furthermore, the photosensitivity of soft glass fluoride fibers also facilitated the building of short-cavity fiber lasers. Polymeric fiber, for which FBGs can be tuned over a wider range than glass FBGs, was also investigated for its sensor applications. In each case, the fiber design consisted of simulating the optimal cross-section, in terms of refractive index profile (RIP), geometry, and chemical composition, to achieve the requested fiber performances. The photosensitive chemical elements are nonmetal in groups IIIA, IVA, and VA of the periodic table, combined with oxygen to form oxide glasses (Table 9.1). The different photosensitive fiber designs are treated separately in subsequent sections in this chapter.

Photosensitive fibers like all other SOFs are fabricated by drawing optical preforms. The fibers are then coated on-line by standard UV-cured acrylate. The know-how rests in preform design and fabrication. The silica fiber preforms are fabricated by different chemical vapor-deposition (CVD) processes including modified CVD (MCVD), plasma CVD (PCVD), outside vapor deposition (OVD), and vertical axial deposition (VAD). Soft-glass fluoride fibers are fabricated by glass melting and preform assembly called the "rod-in-tube" method.

Table 9.1

Periodic table highlighting the photosensitive elements and silica glass dopants

Group	IA	IIA												IIIA	IVA	VA	VIA	VIIA	O
	Li	Be												B	C	N	O	F	Ne
	Na	Mg			Transition elements									Al	Si	P	S	Cl	Ar
	K	Ca	Sc	Ti	V	Cr	Mn	Fe	Co	Ni	Cu	Zn		Ga	Ge	As	Se	Br	Kr
	Rb	Sr	Y	Zr	Nb	Mo	Tc	Ru	Rh	Pd	Ag	Cd		In	Sn	Sb	Te	I	Xe
	Cs	Ba	La	Hf	Ta	W	Re	Os	Ir	Pt	Au	Hg		Tl	Pb	Bi	Po	At	Rn
	Fr	Ra	Ac																

Rare Earth elements

Ce	Pr	Nd	Pm	Sm	Eu	Gd	Tb	Dy	Ho	Er	Tm	Yb	Lu	
Th	Pa	U	Np	Pu	Am	Cm	Bk	Cf	Es	Fm	Md	No	Lr	Actinide elements

■ Photosensitive elements
▨ Silica glass dopants

The photosensitivity level of silica fibers is critically dependent on several MCVD preform fabrication parameters, in particular the collapsing procedure, the atmospheric conditions during collapsing, and the concentration of the co-dopants in the fiber core [23]. In the industry, preform collapsing is commonly performed under reduced atmosphere to favor the germanium-related oxygen-deficient centers (GODCs). Another MCVD approach developed by Dianov et al. [24] is the deposition of white soot layers and their sintering under a reduced atmosphere.

9.3 STANDARD NUMERICAL APERTURE FIBERS

Over the years, a standard numerical aperture (NA) of about 0.12 was established for SSMFs used by the optical communication industry. The germanium dioxide (GeO_2) concentration necessary to obtain such an NA is about 4 mol% and the fiber RIP is of the step-index type with a Δn (defined by $n_{core} - n_{clad}$) of about 0.005 (see Chapter 5 on SSMF for communications). The core dopant choice, RIP, and NA of the SSMF are the result of minimization of the fiber losses at 1310 and 1550 nm, but the enhancement of the photosensitivity was not considered when optimizing these fiber parameters. The development of a highly photosensitive fiber having the same optical characteristics as the SSMF was realized by boron doping of germano-silicate fibers [6], as well as antimony doping [25] and tin doping [26]. The design, fabrication particularities, characteristics, and applications of these types are described in the following subsections.

9.3.1 Standard Single-Mode Fibers

Since the introduction of FBG as a commercial product in 1995, the use of FBG has increased in the telecommunications and the sensors fields [27], with the development of new components built with SSMFs. Indeed, the filtering capabilities (small insertion loss, low transmission loss, high wavelength selectivity) of FBGs make them an ideal candidate as well for add–drop filters and dispersion compensators than for temperature or strain sensors. Most of the FBG applications require gratings with high reflectivity, so fiber photosensitivity is an important factor in the development of specific FBGs. For example dense wavelength division multiplexers (DWDMs) require highly photosensitive fibers for efficient inscription of several gratings. Moreover, the performances of a system depend on some FBG characteristics like bandwidth. FBGs for WDM filters have very narrow spectral widths, whereas FBGs for dispersion compensators have wider bandwidths.

Intrinsically, the SSMF photosensitivity is too small for most of these applications. Typically, without any sensitizations, UV-induced index changes are limited to about 3×10^{-5} [28]. However, it is possible to overcome this problem by enhancing the photosensitivity of SSMF through hydrogen (H_2) loading [11] and then increase refractive index modulation to the order of 10^{-3}. Similar results have been obtained using other MCVD and VAD SSMFs. The mechanism is, therefore, not dependent on fiber or preform processing, but on the interactions between GeO_2 and H_2 molecules, coupled with the UV exposure conditions. Various sensitization methods have been developed including UV hypersensitization [29], OH-flooding [30], and under strain [31]. From a practical point of view, the H_2 loading method remains the best technological solution for writing low-loss, strong short-period gratings with a superior Bragg wavelength stability [30].

9.3.2 Boron-Doped Germano-Silicate Fibers

Fiber photosensitivity can be enhanced by increasing the germanium concentration. Boron co-doping is then used to lower the RIP of the Ge-enriched fibers, so a larger amount of GeO_2 can be incorporated without increasing the NA [6]. Boron co-doped fiber has an excellent photosensitive response. Konstantaki et al. [32] studied the effects of Ge concentration, boron co-doping, and hydrogenation on FBG characteristics. They showed that an unloaded boron co-doped fiber increased the photosensitivity only slightly. In contrast, hydrogenation with the presence of boron in the fiber core significantly enhances the photo-induced increase rate of the refractive index and results in FBGs with wider bandwidths. The modulation depth of the induced refractive index change is stronger and results in FBGs with high reflectivity. Different photosensitivity-enhancement mechanisms observed individually due to boron co-doping and hydrogenation contribute additively to the final photosensitivity in hydrogenated boron co-doped fibers.

Typical B_2O_3 and GeO_2 concentrations are 8 and 12 mol%. INO's fiber model PS-SMF-30 corresponds to this type of fiber (Table 9.2). The photo-induced refractive index change is about 3×10^{-3}.

This type of fiber is mainly fabricated using the MCVD technique. First, a standard P_2O_5 and F co-doped silica cladding is deposited. Afterwards, $SiCl_4$, BCl_3, and $GeCl_4$ vapors carried by O_2 gas are used as core precursors for the MCVD process. The collapse is under nitrogen (N_2) to favor the GODCs. A typical preform chemical composition profile obtained by electron probe microanalysis is shown in Fig. 9.1. For SSMF, the NA or Δn is the same for the preform and fiber (i.e., there is no change in the Δn during the drawing process). This is not the case for boron-doped germano-silicate fibers. The Δn is lowered during the drawing process because of some stress effect, which means that the Δn of the preform must then be higher to compensate for the Δn reduction during the drawing. To illustrate this fact, a typical preform RIP having a maximum Δn of 0.012 is shown in Fig. 9.2 and the corresponding fiber RIP with a maximum Δn of 0.009 is shown in Fig. 9.3. Hence, the Δn was lowered by 0.003 during the drawing. Another difference that can be observed from Fig. 9.3 is that the index dip observed in the PS-SMF-30 is wider and deeper than the one typically observed for SSMF. As can be seen on Fig. 9.4, the background loss is also higher for this fiber compared to the SSMF, especially in the telecommunication window of 1.55 μm due to

Table 9.2

Boron-doped germano-silicate fiber data

Fiber model:	PS-SMF-30
Core diameter:	$6.2 \pm 0.5\ \mu m$
Core dopants:	B_2O_3 and GeO_2
B_2O_3 concentration:	8 mol%
GeO_2 concentration:	12 mol%
Numerical aperture:	0.16 ± 0.02
LP_{11} cutoff wavelength:	1250 ± 50 nm
Mode-field diameter @ 1550 nm:	$9 \pm 1\ \mu m$
Attenuation @ 1550 nm:	<65 dB/km
Cladding diameter:	$125 \pm 1\ \mu m$
Coating diameter:	$250 \pm 2\ \mu m$
Screen proof tested:	>100 kpsi
Splice loss with a SSMF:	<0.2 dB

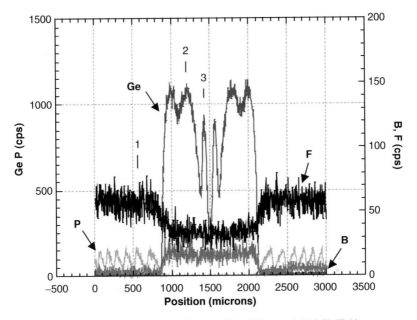

Figure 9.1 Preform composition profile of fiber model PS-SMF-30.

the higher phonon energy of the boron dopant. However, it is worth noting that the higher background loss does not limit the applications because of the short length required per component. All other characteristics are similar to SMF.

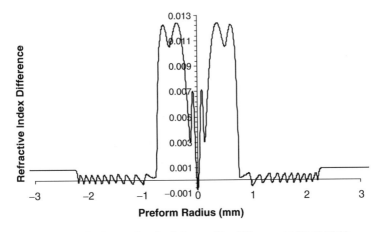

Figure 9.2 Preform refractive index profile of fiber model PS-SMF-30.

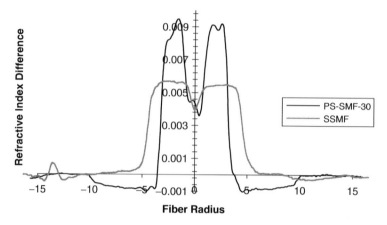

Figure 9.3   Refractive index profile fiber model PS-SMF-30 and SSMF.

9.3.3 Antimony-Doped Fibers

A standard NA antimony (Sb) doped silica optical fiber was developed by Oh et al. [25]. The RIP and the index difference (Δn) between the core and cladding were similar to the SSMF. In D_2-loaded samples, they observed UV photosensitivity with an initial refractive index growth rate six times higher than in SSMF. The fiber core diameter was 8.5 μm and the Sb_2O_3 concentration was 3.5 mol%. The attenuation at 1550 nm was 700 dB/km, which is very high, and it was attributed mainly to the OH absorption induced in the wet sol-gel process. Antimony is a promising element, but the background losses problem would need to be solved.

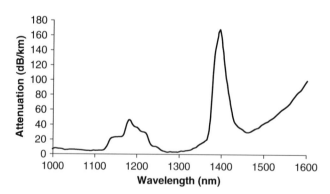

Figure 9.4   Attenuation spectrum of a typical fiber model PS-SMF-30 boron-doped germanosilicate fiber.

9.3.4 Tin-Doped Fibers

A standard NA tin (Sn)-doped silica optical fiber was developed by Brambilla et al. [26]. The fiber preform was fabricated by the MCVD process by depositing a soot rich in SnO_2 at 1300 °C and then sintered at 1800 °C. During the collapse process, the temperature was kept under 2200 °C to avoid massive SnO_2 volatilization. The core NA was about 0.1 with a large index dip and the cutoff wavelength was about 1300 nm. The fiber exhibited a moderate photosensitivity with an index modulation of up to 2.5×10^{-4}.

9.4 HIGH NUMERICAL APERTURE FIBERS

The HNA refers to silica optical fibers with an NA from about 0.2 to 0.4 maximum. This design, requiring a high Ge concentration in the core, turned out to be intrinsically photosensitive, which historically played an important role in Hill's discovery of the photosensitivity phenomenon. Because of the HNA, the fiber needs a smaller core diameter than with the SSMF to be single mode in the operating wavelength between 1310 and 1610 nm. The number of modes supported by an optical fiber is described by the fiber parameter V, given by

$$V = \frac{2\pi a}{\lambda} NA,$$ (9.1)

where a is the core radius and λ is the wavelength. For a given design, the fiber is single mode when the V parameter is smaller than 2.405 (i.e., for wavelengths longer than the LP_{11} cutoff wavelength [λ_c] of the fiber). For optimal performance, λ_c should be lower than the operating wavelength by about 50 nm. Depending on the NA, the core diameter will range from 2 to 7 μm. The higher the NA, the smaller the core diameter has to be for the fiber to remain single mode. The mode-field diameter (MFD) will also be smaller for the HNA fiber than for the SSMF, leading to a higher splice loss to SSMF due to mode mismatch. Typically, the splice losses between an HNA fiber and an SSMF are < 0.2 dB, whereas they are less than 0.05 dB between two SSMFs.

Another interesting feature of HNA fibers is that they push the cladding mode coupling loss on the short wavelength's side from the Bragg wavelength in transmission [33, 34]. When the grating first starts to grow, wave vector matching requirements lead to a gap between the Bragg wavelength and the onset of the cladding mode coupling. The width of this initial offset is given by

$$\Delta\lambda_{off} = \lambda_B - \lambda_L = \Lambda(n_{eff} - n_{clad}),$$ (9.2)

where λ_B is the Bragg wavelength grating defined by $\lambda_B = 2n_{eff}\Lambda$, λ_L is the longest wavelength of the cladding mode, n_{clad} is the cladding index, Λ is the grating pitch, and n_{eff} is the effective index of the propagating mode.

n_{eff} can be approximated by

$$n_{eff} \approx n_{core} + \Delta n_{UV}/2, \tag{9.3}$$

where n_{core} is the refractive index of the core and Δn_{UV} is the change in the core index caused by the UV radiation, and, such that considering the Δn, $\Delta\lambda_{off}$ is now expressed as

$$\Delta\lambda_{off} \approx \Lambda(\Delta n + \Delta n_{UV}/2). \tag{9.4}$$

Therefore, $\Delta\lambda_{off}$ becomes larger as Δn is increased. A larger $\Delta\lambda_{off}$ causes the excess loss region, due to the cladding modes, to be shifted far from the Bragg reflection wavelength. For a moderate NA of 0.25, the $\Delta\lambda_{off}$ is 4.5 nm and for an ultrahigh NA of 0.40, the $\Delta\lambda_{off}$ is 12 nm. This feature is very important in developing DWDM components.

Furthermore, other elements were found to enhance the photosensitivity of HNA fibers including tin (Sn) [35] and indium (In) [36]. The design, fabrication particularities, characteristics, and applications of these three types are described in the following subsections.

9.4.1 Heavily Ge-Doped Silica Optical Fibers

Intrinsically, the photosensitivity of heavily Ge-doped silica fibers is sufficient for most FBG applications. INO's fiber model PS-HNA-40 is a photosensitive step-index HNA fiber, whose physical characteristics are listed in Table 9.3. The photo-induced refractive index change is 1×10^{-3}.

HNA fibers are fabricated by drawing silica optical preforms obtained by the CVD processes, including MCVD, PCVD, OVD, or VAD. The preform fabrication by MCVD requires special care during the core deposition, precollapse, and collapse. Only $SiCl_4$ and $GeCl_4$ with an O_2 carrier gas are used as core precursors. Compared to SSMF fabrication, the $GeCl_4$ flow must be increased and deposition temperature lowered to favor the Ge deposition. The precollapse and collapse steps are performed under nitrogen (N_2) atmosphere to favor the GODCs. On the preform, the maximum Δn measured was 0.022. The RIP of the corresponding fiber is shown in Fig. 9.5. Note that this fiber has a maximum Δn of 0.021, which indicates that contrary to the standard NA case, the Δn difference between the preform and the fiber in the HNA scenario is negligible. The background loss is higher than an SSMF but lower than the boron co-doped germano-silicate fiber (Fig. 9.6). All other characteristics are similar to SSMF.

Table 9.3

High numerical aperture silica optical fiber data

Fiber model:	PS-HNA-40
Core diameter:	5.6 ± 0.5 μm
Core dopant:	GeO_2
GeO_2 concentration:	12 mol%
Numerical aperture:	0.20 ± 0.02
LP_{11} cutoff wavelength:	1450 ± 50 nm
Mode-field diameter:	6.5 ± 1.0 μm
Attenuation @ 1550 nm:	<3 dB/km
Cladding mode offset:	5 nm
Cladding diameter:	125 ± 1 μm
Coating diameter:	250 ± 2 μm
Screen-proof tested:	>100 kpsi
Splice loss with a SSMF:	<0.25 dB

HNA fiber is intrinsically photosensitive and its primary purpose is to offset radiation and cladding modes in the transmission spectra of FBGs. Its low birefringence is also an attractive specification in DWDM applications.

9.4.2 Tin-Doped Germano-Silicate Fibers

An HNA tin (Sn)-doped germano-silicate optical fiber was developed by Dong et al. [35]. The preform was fabricated by introducing $SnCl_4$ vapor in the MCVD process. The required extra $SnCl_4$ bubbler needs to be heated at 39 °C because of the low vapor pressure of $SnCl_4$. Nitrogen instead of oxygen was used as the

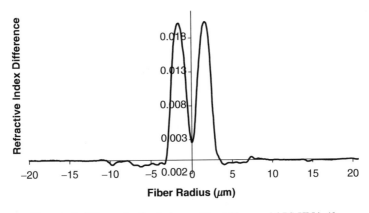

Figure 9.5 Fiber refractive index profile of fiber model PS-HNA-40.

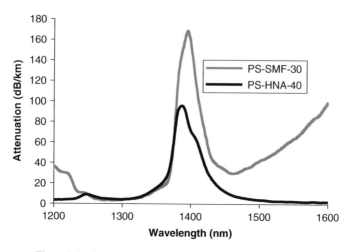

Figure 9.6   Attenuation spectra PS-HNA-40 and PS-SMF-30.

carrier gas for SnCl$_4$. Two SnO$_2$-GeO$_2$-SiO$_2$ porous soot layers were deposited at about 1250 °C and they were sintered into a transparent glass at about 1600 °C. The preform was then collapsed into a solid rod in the conventional manner. The core diameter of the resulting fiber was 4.8 μm, the NA was in the 0.20 range, and the cutoff wavelength was about 1250 nm. Background losses at 1300 and 1550 nm were 3 and 2 dB/km, respectively, which are much lower than those with boron-doped germano-silicate fibers (25 dB/km and 115 dB/km at the same wavelengths). The photo-induced refractive index change was 3 × 10^{-3}, which is three times larger than that for heavily Ge-doped silica fibers but about the same as boron-doped germano-silicate fibers. Sn is an interesting photosensitive element but requires an extra SnCl$_4$ bubbler for the preform fabrication, which is not common among fiber manufacturers.

9.4.3 Indium-Doped Germano-Silicate Fibers

An HNA indium (In)-doped germano-silicate optical fiber was developed by Shen et al. [36]. The In was chosen for its large cation size (80 pm) compared to that of Sn (71 pm). Higher temperature sustainability is, thus, expected for the FBG written in these fibers. The preform was fabricated by the MCVD process followed by the solution doping technique. The solution was prepared from In$_2$O$_3$ powder instead of commonly used salts. The resulting preform core was brown instead of transparent as usual. The preform was drawn into a 125-μm fiber. The core diameter was 5 μm, the core NA was about 0.21, and the calculated cutoff wavelength was 1375 nm. The In_2O_3 concentration was about 0.05 mol%, and the GeO_2 was about 9 mol%. The fiber exhibited a refractive index modulation of 3.2×10^{-4}. Annealed FBGs in this fiber type survived a high temperature of 900 °C for 24 hours and even 1000 °C for more than 2 hours. Unfortunately, the background loss was not reported by the authors, but the brown color of the core suggests it might be quite high. This specification needs improvement to classify indium as a good photosensitive element.

There are numerous applications for HNA fibers in areas such as filters, narrow-band reflectors for fiber lasers, optical strain/temperature sensors, and modal couplers. HNA fibers are also perfectly suitable to adapt the cladding mode offset in order to optimize the channel spacing in telecommunications applications.

9.5 CLADDING MODE SUPPRESSION

The CMS design enhances the FBGs bandwidth in transmission. Add–drop filters or multiplexer–demultiplexer components can be fabricated at potentially low cost using this technology. However, an FBG written in an SSMF induces coupling to the cladding modes in addition to coupling to the fundamental counter propagating mode LP_{01}. This coupling to higher order cladding modes induces several resonance dips in the transmission spectrum on the short-wavelength side of the Bragg reflection. The performance of such devices can be improved by using appropriately designed fibers. Several methods have been proposed to reduce this effect, but they do not give optimal performances. The first design was an HNA fiber (see Section 9.4) that allows shifting of the cladding modes to a shorter wavelength. A second design is the depressed cladding fiber [37], where the cladding mode coupling loss is reduced, but improvement is modest and the field distribution is different from an SSMF. The third design, proposed by Delevaque et al. [38], consists of a photosensitive cladding to obtain a uniform photosensitivity over the spatial extent of the guided orthogonal mode that ensures a negligible coupling. This last design is still considered the best technical solution. The core of the CMS design can have a standard NA or an HNA.

The standard NA PS-RMS-28 [39] fiber was fabricated by the MCVD process and the design consists of a matched photosensitive cladding made of F and Ge co-doped silica and a Ge-doped silica core. The core refractive index had a profile similar to that of SSMF and F has been incorporated into the cladding in sufficient concentration to lower the index of refraction to the silica level. The PS-RMS-50 design consists of an HNA photosensitive fiber with the cladding mode suppression feature. The matched photosensitive cladding is made of B and Ge co-doped silica and the silica core is doped with Ge. The advantages of the PS-RMS-28 over PS-RMS-50 include an NA, MFD, and attenuation within the tolerances of the SSMF. The average splice loss of a PS-RMS-28 to an SSMF is about 0.03 dB at 1550 nm. Typical specifications for SSMF, PS-RMS-28, and PS-RMS-50 fibers are compared in Table 9.4.

To compare the cladding mode losses between different fibers, the FBG photo-inscription method is used to study the photosensitivity of the fibers. All of the gratings presented here have been imprinted using a CW frequency-doubled argon laser at 244 nm with a standard holographic phase mask. The PS-RMS-50 fiber is intrinsically photosensitive and does not require H_2 loading. A 30-dB grating has been photo induced in this fiber and in an H_2-loaded SSMF. The exposure time was 2 minutes for the H_2-loaded SSMF and 1 minute for the PS-RMS-50, meaning that this fiber is twice as photosensitive, even without H_2 loading. The cladding mode coupling losses in the PS-RMS-50 are below 0.1 dB, compared to almost 1.5 dB for the SSMF fiber. This substantial improvement is obtained by making the fiber cladding photosensitive.

Limitations of the PS-RMS-50 are the unwanted propagation losses, which are as high as 50 dB/km at 1550 nm. These high losses are due to the high concentration of B in the cladding, which is necessary to counteract the increase of the refractive index attributable to the high Ge content. To overcome these limitations, a new type of fiber (PS-RMS-28) has been designed. The goal was to reduce the transmission losses while preserving the same low cladding mode losses. The low Ge concentration of the new fiber required hydrogen loading to make it more photosensitive. The cladding modes coupling losses measured on an H_2-loaded PS-RMS-28 fiber were lower than 0.1 dB. For comparison, the transmission spectrum of the gratings imprinted in the SSMF and PS-RMS-28 is shown in Fig. 9.7.

The first strong cladding mode coupling loss observed on the short-wavelength side of the Bragg peak in Fig. 9.7 often referred as the "ghost-dip" [40] is likely due to coupling to the LP_{11} mode. Because of the photosensitive cladding around the core and lateral exposure, the UV light is preferentially absorbed on one side, inducing an asymmetry in the RIP, as illustrated in Fig. 9.8. It has been

Table 9.4

Comparison table of SSMF, PS-RMS-28 and PS-RMS-50 fibers specifications

Fiber code	NA	Photosensitive cladding thickness (μm)	Ge in clad/core	MFD (μm) at 1550 nm	Cutoff wavelength (nm)	Att. (dB/km) at 1550 nm	Average splice loss (dB)
SSMF	0.14	N/A	N/A	10.4 ± 0.8	1260 ± 70	0.2	N/A
PS-RMS-28	0.14	10	0.92	10.3	1390	0.22	0.03
PS-RMS-50	0.24	2.2	N/A	5.9	1280	50	<0.1

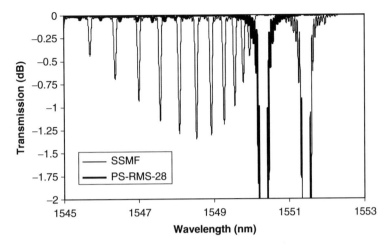

Figure 9.7 Transmission spectra of Bragg gratings written in an SSMF and a PS-RMS-28.

shown by Poulsen et al. [40] that in fibers having a highly UV-sensitive core, the transverse UV absorption allows coupling to asymmetric cladding modes, even in the absence of a blaze angle during grating writing. Their model emphasizes that the index asymmetry, due to the UV light absorption, can explain the strength of the first cladding mode dip that cannot be explained by an unintentionally induced blaze angle.

9.6 RARE EARTH-DOPED PHOTOSENSITIVE FIBERS

The RED photosensitive fiber is an important component for the development and fabrication of fiber laser cavities. Unfortunately, most RED fibers are not photosensitive because of the replacement of the Ge by Al and/or P necessary to reduce the effect of quenching and lifetime shortening [41]. It is well known, however, that the presence of P bleaches the absorption band centered on 240 nm and, thus, reduces the photo-induced index change. The first reported photosensitive RED fiber was by Bilodeau et al. [42] on an Er/Ge–doped fiber. The Er concentration of that fiber was too low for fiber laser and amplifier applications but showed some weak photosensitivity with a Δn_{uv} of about 3×10^{-5}. A higher Er concentration without quenching requires an alumino-silicate glass host, which reduces the photosensitivity. Since then, many photosensitive RED fibers have been reported [43] (Table 9.5). Different rare earth elements, fiber designs, fabrication particularities, characteristics, and applications are described in the following subsections.

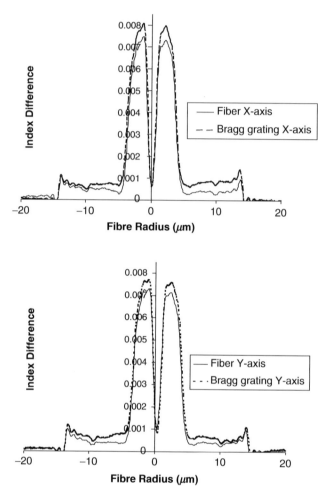

Figure 9.8 Refractive index profile of a PS-RMS-28 fiber before and after Bragg grating writing in the X (top) and Y (bottom) orthogonal axes. The asymmetry is along the X-axis.

9.6.1 Germano-Alumino-Silicate Glass Host Core

The RED Ge/Al/Si glass host core fiber is intrinsically photosensitive as long as the GeO_2 concentration is 20 mol% or more. Below that and depending on the application, the fiber might need to be H_2 loaded before the FBG fabrication. The Ge/Al/Si core design was applied to the Er-doped silica fiber. The GeO_2 concentration was 20 mol%, which makes this fiber intrinsically photosensitive.

Table 9.5

Rare earth-doped silica photosensitive fiber

Dopants	B$_2$O$_3$ (mol%)	GeO$_2$ (mol%)	P$_2$O$_5$ (mol%)	Fiber design	H$_2$ loading	UV source wavelength (nm)	Induced Δn	Reference/ fiber model
Er/Ge/Al	0	20	0		Yes	244 CW	1.5×10^{-3}	Er 304[a]
Nd/Al/Ge/P	0	8	<0.9		Yes	244 CW	3.0×10^{-3}	Nd 100[a]
Er/Yb/P	0	6	0	Confined core	Yes	244 CW	1.6×10^{-3}	EY 701[a] Ref [71]
Er/Yb/P B/Ge	20	7	0	Photos. clad	No	248 pulsed	5×10^{-4}	Reference [72]
Yb/Al/Ge				Photos. clad				
Ge/P	0	4/14	0/0.7		Yes	244 CW	3.3×10^{-3}	Yb 708[a]
Er/Yb/P				Confined core				
B/Ge clad	8	6/4	0	Photos. clad	Yes	244 CW	1.8×10^{-4}	EY 305[a]
Tm/Sb/Al	0	0	0		Yes	244 CW	1×10^{-4}	Reference [73]

[a] Δn values are given for UV-inscribed FBG after complete annealing

Table 9.6

Er-doped silica optical fiber data

Fiber model:	Er 304
Core host material:	Germano-alumino-silicate
Core diameter:	$4.0 \pm 0.5 \ \mu m$
Numerical aperture:	0.20 ± 0.02
LP_{11} cutoff wavelength:	900 ± 50 nm
Mode-field diameter @ 1550 nm:	$6 \pm 1 \ \mu m$
Absorption peak @ 976 nm:	3 dB/m
Absorption peak @ 1538 nm:	4 dB/m
Background loss @ 1200 nm:	<5 dB/km
Cladding diameter:	$125 \pm 1 \ \mu m$
Coating diameter:	$250 \pm 5 \ \mu m$
Screen proof tested:	>100 kpsi
Splice loss with a HI1060:	<0.20 dB

This single clad design was developed for fiber laser operating at 1550 nm. This fiber corresponds to fiber model Er 304 (Table 9.6).

The Ge/Al/P/Si core design was also applied to the Nd-doped silica fiber. The P_2O_5 concentration being lower than 1 mol% did not reduce the fiber photosensitivity. The GeO_2 concentration was only 8 mol%, which is not sufficient for the fiber to be intrinsically photosensitive. This fiber then needed to be H_2 loaded to become photosensitive. The single clad fiber was designed for the development of fiber lasers operating at 1064 nm. This fiber corresponds to fiber model Nd 100 (Table 9.7).

The RED photosensitive fibers are fabricated by drawing silica optical preforms obtained by a combination of MCVD process and solution doping technique.

Table 9.7

Nd doped silica optical fiber data

Fiber model:	Nd 100
Core host material:	Germano-alumino-phospho-silicate
Core diameter:	$4.0 \pm 0.5 \ \mu m$
Numerical aperture:	0.17 ± 0.02
LP_{11} cutoff wavelength:	800 ± 50 nm
Absorption peak @ 807 nm:	2 dB/m
Background loss @ 1060 nm:	<10 dB/km
Cladding diameter:	$125 \pm 1 \ \mu m$
Coating diameter:	$250 \pm 5 \ \mu m$
Screen proof tested:	>100 kpsi
Splice loss with a HI1060:	<0.10 dB

9.6.2 Confined Core

The confined core design was developed to achieve efficient Er/Yb–co-doped photosensitive fibers [44]. Actually, the Er and Yb ions are confined in the guiding core. The photosensitivity is obtained from the Ge-doped ring surrounding the Er/Yb–co-doped phospho-silicate glass host core necessary to make an efficient energy transfer between the Yb and Er ions and to avoid ion quenching. The confined core ring design was applied to the all-silica double-clad hexagonal Er/Yb co-doped single-mode photosensitive fiber. The all-silica composition and geometry allow this fiber to be fusion-spliced with fiber pigtailed laser diodes. Moreover, it ensures long-term reliability. The double-clad (DC) design allows the coupling of high pump power lasers to the active fiber core where the hexagonal shape enables efficient mode mixing. This all-silica guiding structure has the advantage of producing a more robust fiber, with better resistance to heating at the pump launching surface than the double-clad fibers, which use a low-index polymer coating to form the outer cladding. This fiber corresponds to fiber model EY 701 (Table 9.8). The confined core design could be applied to all RED fibers.

The mother preform of this fiber was fabricated by a combination of MCVD process and solution-doping technique. The Ge-doped ring core was deposited

Table 9.8

Er/Yb co-doped and confined DC SM all-silica optical fiber data

Fiber model:	EY 701
Er/Yb core host material:	Phospho-silicate
Core ring host material:	Germano-silicate
Core diameter:	$6.0 \pm 0.5\ \mu m$
Confinement factor:	0.8 ± 0.1
Numerical aperture:	0.18 ± 0.02
Theoretical LP_{11} cutoff wavelength:	1400 ± 50 nm
Peak absorption @ 976 nm:	>1.2 dB/m
Peak absorption @ 1538 nm:	>0.06 dB/m
Estimated background loss @ 1550 nm:	<75 dB/km
First clad material:	Silica
First clad shape:	Hexagonal
First clad distance between parallel planes:	$123\ \mu m$
First clad NA:	0.264
Second clad material:	F-doped silica
Second clad diameter:	$140 \pm 1\ \mu m$
Coating diameter:	$262 \pm 2\ \mu m$
Screen-proof tested:	>100 kpsi
Splice loss with a SSMF:	<0.1 dB

by MCVD and the Er/Yb–doped phospho-silicate core was fabricated by the solution-doping technique. The mother preform was then sleeved to enlarge its diameter and meet the targeted LP_{11} cutoff wavelengths. For the double-clad design, the preform was then shaped into a hexagon. Meanwhile, the silica core of a Fluosil preform was drilled. The NA of the Fluosil preform was 0.264. The final double-clad preform was obtained by fusing the drilled Fluosil preform onto the hexagonal mother preform. After drawing, the resulting all-silica fiber was round with a hexagonal first clad and a round core (Fig. 9.9). This double-clad fiber was characterized in terms of refractive index profile, core diameter, LP_{11} cutoff wavelength, effective NA, spectral attenuation and absorption, photosensitivity, and splicing losses. The RIP of the fiber was measured by the refracted near-field technique (Fig. 9.10). From this fiber index profile, the measured core diameter at 50% of maximum Δn is 5.6 $\propto$m, the calculated effective core NA is 0.17, and the calculated LP_{11} cutoff wavelength is 1250 nm. The estimated background losses at 1550 nm is 210 dB/km and the Er/Yb absorption of the multimode pump guide at 976 nm is 1.2 dB/m. To measure the photosensitivity of the fiber, it was hydrogen loaded for 10 days at 1500 psi. A 10-mm-long FBG was then written into that fiber and into hydrogen-loaded SSMF for comparison. The photosensitivity results presented in Fig. 9.11 clearly show that this all-silica Er/Yb fiber was indeed photosensitive, which indicates that it would be suitable for FBG writing to realize a cladding pumped fiber laser cavity. The single-mode splice loss has been measured between this all-silica Er/Yb fiber and an SSMF. The

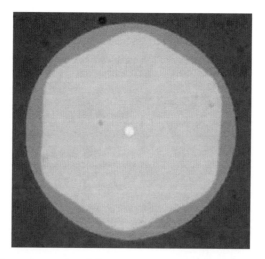

Figure 9.9   Photomicrography of fiber model EY 701. The outer diameter is 125 μm.

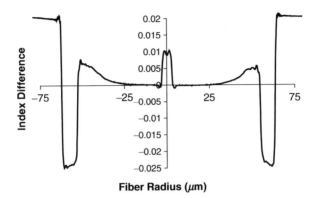

Figure 9.10   Refractive index profile of fiber model EY 701.

multimode splice loss was measured between this all-silica Er/Yb fiber and a 125 μm all-silica fiber pigtail having a 113 μm core diameter and a 0.242 NA. The results of the splice losses measurements at a 1310-nm wavelength are given in Table 9.9. The splicing losses to an SSMF were 0.09 dB, and they were 0.07 dB to a multimode fiber pigtailed laser diode pump. Gratings of more than 20 dB were obtained on a hydrogen-loaded Er/Yb fiber compared to 16 dB for a hydrogen-loaded SSMF. Both gratings were written under the same conditions.

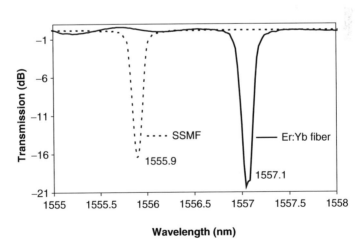

Figure 9.11   Photosensitivity results of fiber model EY 701 compared to the SSMF.

Table 9.9

Experimental splice loss for the all-silica Er/Yb fiber
with a SSMF and with a multimode pigtail fiber

Fiber type	Splice loss (dB/splice)
SSMF	0.09 ± 0.01
Multi-mode fiber	0.07 ± 0.01

9.6.3 Photosensitive-Clad

The photosensitive clad design was developed by Dong et al. [45] to achieve efficient Er/Yb–co-doped silica fibers. The fiber is single clad with a highly photosensitive B/Ge–doped silica clad surrounding the single-mode Er/Yb–doped phospho-silicate core. B and Ge were chosen because both elements increase the photosensitivity. Furthermore, B lowers the refractive index of silica, whereas Ge increases it. By using the right concentrations of both elements, it is possible to obtain a B/Ge clad having the same refractive index as the pure silica clad. The Er/Yb–co-doped core is not affected by the B/Ge–co-doped silica cladding. The highly photosensitive cladding allows strong (>99%) 10-mm-long gratings to be easily achieved in these fibers, despite the reduced overlap between the grating and the guided optical field. The high absorption at 980 nm permits efficient pump absorption over a short fiber length.

The photosensitive clad design was also applied to the all-silica double-clad, single-mode, Yb-doped photosensitive optical fiber. This fiber corresponds to fiber model Yb 708, whose physical characteristics are presented in Table 9.10. The photosensitive clad design could be applied to all RED fibers.

9.6.4 Confined Core and Photosensitive Clad

The confined core and photosensitive clad design is a combination of the two aforementioned designs. It was developed to achieve efficient Er/Yb–co-doped photosensitive fibers. The central core is single mode over a 1500-nm operating wavelength. The ring around the Er/Yb confined core is co-doped with Ge, making the core photosensitive. This feature allows fabrication of FBGs in the ring, which will make fabrication of a fiber laser cavity possible. This fiber corresponds to fiber model EY 305 (Table 9.11). This design could be applied to all RED fibers.

Table 9.10

Yb-doped DC SM all-silica optical fiber data

Fiber model:	Yb 708
Core host material:	Germano-alumino-silicate
Core diameter:	$7.0 \pm 0.5 \ \mu m$
Numerical aperture:	0.11 ± 0.02
Theoretical LP_{11} cutoff wavelength:	1000 ± 50 nm
Peak absorption @ 976 nm:	8 dB/m
Background loss @ 1030 nm:	<5 dB/km
First clad material:	Germano-phospho-silicate
First clad shape:	Round
First clad diameter:	$32 \ \mu m$
First clad NA:	0.19
Second clad material:	Silica
Second clad diameter:	$125 \pm 1 \ \mu m$
Coating diameter:	$250 \pm 15 \ \mu m$
Screen-proof tested:	>100 kpsi

9.6.5 Antimony-Doped Alumino-Silicate

A germanium-free antimony (Sb) and thulium co-doped alumino-silicate optical fiber was designed and developed by Sahu et al. [46]. The fiber was fabricated by the MCVD process, along with the solution-doping technique. The fiber, being not intrinsically photosensitive, was H_2 loaded for 2 weeks

Table 9.11

Er/Yb co-doped and confined SM silica optical fiber data

Fiber model:	EY 305
Er/Yb core host material:	Phospho-silicate
Core ring host material:	Germano-silicate
Core diameter:	$6.0 \pm 0.5 \ \mu m$
Confinement factor:	0.8 ± 0.1
Numerical aperture:	0.17 ± 0.02
LP_{11} cutoff wavelength:	1250 ± 50 nm
Mode-field diameter @ 1550 nm:	$7 \pm 1 \ \mu m$
Absorption peak @ 976 nm:	>380 dB/m
Absorption peak @ 1538 nm:	>18 dB/m
Estimated background loss @ 1550 nm:	<40 dB/km
Cladding ring dopants:	Ge & B
Cladding diameter:	$125 \pm 1 \ \mu m$
Coating diameter:	$250 \pm 2 \ \mu m$
Screen-proof tested:	>100 kpsi
Splice loss with a SSMF:	<0.1 dB

before the FBG fabrication. Afterwards, the fiber was annealed at 100 °C for 24 hours to outgas any residual hydrogen in the loaded sample and to stabilize the index modulation at room temperature. The FBGs were fabricated using an illuminating beam at a 244-nm wavelength, having an intensity of $300 \, \text{W}/\text{cm}^2$. The resulting index modulation achieved in these conditions was 1×10^{-3}. The fiber had a 120-μm diameter, an NA of 0.16, a cutoff wavelength of 1500 nm, and a Tm concentration of about 1000 ppm. These preliminary results suggest that Sb might become a promising photosensitive element for the development of RED photosensitive fibers, which are much needed for the fiber laser industry.

9.7 POLARIZATION MAINTAINING

The polarization maintaining (PM) photosensitive fiber design [44] was developed for applications requiring a photosensitive fiber that maintains the polarization plane of the transmitted light in the fiber. PM fibers are used for various applications because they can be designed as well to optimize the performances of optical fiber sensors than to get optimal specifications for telecommunication components. The requirements for PM photosensitive fibers include good photosensitivity, low crosstalk, low transmission loss, and good mechanical bending characteristics. The PM design selected for the photosensitive fiber is a PANDA type. Many designs of PM fibers have been developed (see Chapter 8). INO's fiber model PS-PM-60 corresponds to an HNA photosensitive fiber with a PANDA design (Table 9.12).

Table 9.12

High numerical aperture photosensitive polarization maintaining silica optical fiber data

Fiber model:	PS-PM-60
Core diameter:	$4.0 \pm 0.5 \, \mu$m
Numerical aperture:	0.24 ± 0.2
LP_{11} cutoff wavelength:	1500 ± 50 nm
Mode-field diameter @ 1550 nm:	$6 \pm 1 \, \mu$m
Attenuation @ 1550 nm:	<5 dB/km
Birefringence @ 1550 nm:	$<3 \times 10^{-4}$
Beat length @ 1550 nm:	<5 mm
Cross talk @ 1550 nm:	<-40 dB on 4 m
Cladding diameter:	$125 \pm 2 \, \mu$m
Coating diameter:	$250 \pm 2 \, \mu$m
Screen-proof tested:	>100 kpsi
Splice loss with a PM 1550 SSMF:	<0.35 dB

The photosensitive core is fabricated by the MCVD process, as described in Section 9.4. The PANDA feature is obtained by drilling two holes on both sides of the core and inserting stress-applying part (SAP) rods in the holes. The SAPs are also manufactured by the MCVD process. The SAPs have a larger thermal expansion coefficient and a lower vitrification temperature than the other parts of the preform. Hence, a residual stress remains in the fiber after the drawing procedure and the presence of this stress across the fiber core region induces a large birefringence in the fiber. The RIP for the X-scan perpendicular to the SAPs and Y-scan along the SAPs is given in Fig. 9.12. The background loss is affected by the insertion of the SAPs, as shown in Fig. 9.13. For this fiber, the crosstalk is −44 dB on a 4-m long fiber, which is sufficient for most applications. The beat length is 4.9 mm at 1550 nm, corresponding to a birefringence of 3.2×10^{-4}.

9.8 OTHER PHOTOSENSITIVE FIBER TYPES

This section covers other photosensitive fiber types including polymer optical fibers (POFs), cerium-doped fluoride glass fibers, and heavily P-doped silica fibers. The fabrication processes and applications are varied and are presented in the following subsection.

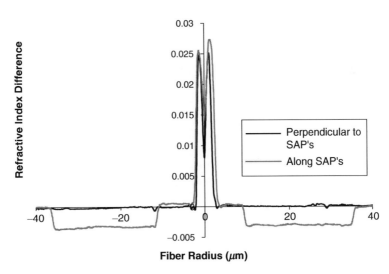

Figure 9.12 Fiber model PS-PM-60 RIP along and perpendicular to the stress-applying parts (SAPs).

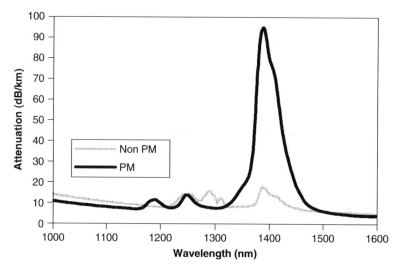

Figure 9.13 Spectral attenuation of non–polarization maintaining (PM) and PM-photosensitive fibers.

9.8.1 Polymer Optical Fibers

Silica fiber photosensitivity finds applications in many fields such as optical signal processing, communications, and sensors. Many types of silica FBGs like chirped gratings, apodized gratings, or long-period gratings have been extensively studied over the past 2 decades. The technology is now completely mature and relatively inexpensive.

The POF was developed at about the same time as the silica fiber by the U.S. company Dupont in 1968. During the following years, POF was neglected in front of silica fiber. POF technology was mostly developed in Japan with a new fabrication process [45] to get a reduction of the attenuation and increase the bandwidth for telecommunications applications. Zubia et al. [46] present in their review the historical evolution of the most important landmarks related to POF from its first fabrication to the year 2000.

Photosensitivity of polymethylmethacrylate (PMMA) was discovered 35 years ago with the work of Tomlinson et al. [47] who showed that the refractive index of PMMA could increase after a 325- or 365-nm UV irradiation. After subsequent work by Kaminov et al. [48], Welker et al. [49], and Peng et al. [50], it was demonstrated that a significant photosensitivity could be induced in POFs not only with doping materials but also with non-doped basic material that could give significant photosensitive effects at UV wavelengths, so that POF grating may be possible to fabricate. The first POF grating was written with a few-moded POF and a writing wavelength of 325 nm [53]. After the development of single-mode POF by Koike [51], 1-cm-long gratings were written with a reflection bandwidth reduced to approximately 1 nm for a maximum reflectivity of about 80% [52].

The photorefractive effect involved in polymer fiber gratings is the photo-polymerization of the fiber. This phenomenon is different from the photosensitive phenomenon observed in silica fibers. Indeed, the incident UV light launched in the fiber causes polymerization of the unreacted monomer, which increases the polymer density and hence the refractive index of the fiber. Photo-polymerization in PMMA POF is induced at a wavelength of 325 nm. The method to write FBG can be the same as for silica fiber with the use of a phase mask. An example of the growth dynamics of a polymer-based FBG is presented in Fig. 9.14 [53]. A detailed description of the fabrication of POF Bragg grating is given in the POF reference manuscript by Peng and Chu [54].

Both types I and II can be observed for polymer FBGs. Indeed, there is a distinctive threshold in UV exposure during the grating growing. This threshold separates the two grating formation stages. An example of their specificities is summarized in Table 9.13 [55]. Note that researchers have found that because of the thermal stress induced by UV irradiation, polymer FBG growth is a writing-power–dependent process [56].

The photosensitivity in CYTOP POF is smaller than in PMMA POF. The CYTOP fiber, composed of perfluorinated polymer, has been developed by Keio

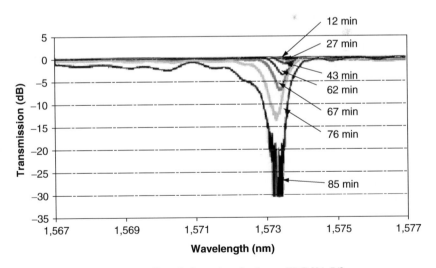

Figure 9.14   Growth dynamics of polymer FBG [56, 74].

Table 9.13

Types I and II specifications

Fiber type	Dimensions	Index diff. (nc-ncl)	Single mode	UV exposure
PMMA	6/125 (μm)	8.6×10^{-3}	1150 nm	325 nm, 6 mJ

FBG type	Time exposure	Index modulation	Index change
Type I	62 min	2.5×10^{-4}	Negative
Type II	113 min	2×10^{-3}	None

Source: Used, with permission, from reference [59].

University and Asahi Glass Company in Japan. Its photosensitive properties have been investigated by Tanio et al. [57]. It is possible to write FBGs in CYTOP fiber. Although the index modulation is smaller (0.6 times the PMMA), gratings are much more thermally stable than with PMMA fiber [58]. The interest in using perfluorinated polymer is its low attenuation of 0.3 dB/km at 1550 nm, which is close to the silica fiber attenuation, and is of interest for short-distance communications.

Strain and temperature have been extensively studied for silica-based FBGs because of the number of applications based on these properties. Despite their high polymer fiber attenuation, PMMA-based FBGs have interesting potential applications because their grating lengths are short ($\sim$1 cm long). Indeed, it has been demonstrated that the Young modulus of polymer fibers is about 30 times less than that of the silica, so that mechanical tunability of polymer-based FBGs can be 30 times larger than for a silica-based FBG. Moreover, because the thermal strain coefficient in polymer fibers is higher compared with the silica fiber, the thermal tuning range of polymer Bragg gratings can be significantly wider. Table 9.14 summarizes mechanical and thermal properties of silica and PMMA materials. A great difference between silica-based and PMMA-based FBGs has been pointed out by Liu et al. [59] when they demonstrated that both strain and thermal sensitivities of types I and II PMMA-based FBGs are similar.

Applications of PMMA-based FBGs are various. Because of their wide tuning range, polymer FBGs have interesting applications in the sensors field, with simultaneous strain and temperature sensors. Figure 9.15 shows a simple sensor constructed by combining two FBGs, one made with a silica fiber, and the other with a polymer fiber. The setup uses a broadband source. To sense the temperature change and a strain change independently but simultaneously, the Bragg wavelength shift due to each grating is recorded. These quantities can be expressed in terms of the temperature change and the strain change. With one grating made of silica and the other made of polymer, it is possible to separate

Table 9.14

Comparison of FBG tunability properties

FBG material	Young's modulus	Wavelength strain coefficient $(d\lambda/\mu\varepsilon)$	Thermal expansion coefficient	Wavelength temperature coeff. $d\lambda/dT$
PMMA	$2.5 \times 10^{+9}$ N/m^2	1.71 pm/$\mu\varepsilon$	$5 \times 10^{-5}/°$C	-148 pm/$°$C
Silica	$7.25 \times 10^{+10}$N/m^2	1.15 pm/$\mu\varepsilon$	$5.5 \times 10^{-7}/°$C	$+13.7$ pm/$°$C

the contribution of temperature and strain and sense both changes more accurately [59].

Optical add–drop multiplexer is an important device in WDM communication and sensing. Its basic configuration consists of an FBG inserted between two broadband optical circulators. With a silica-based FBG, the device is not easily tunable, but by replacing the silica-based FBG with a polymer-based FBG, the wavelength is then tunable (mechanically or thermally) over 20 nm [56]. Another configuration is the Mach-Zehnder interferometric configuration with two identical FBGs. Tunability of the device can be obtained by replacing the silica FBGs with polymer FBGs [57].

Dispersion compensator is also a very important device in communication systems. An interesting means to get a higher tunability would be with the use of a polymer-based chirped Bragg grating.

Bragg gratings in POF are more tunable than in silica fibers. Indeed, a temperature step of about 50 degrees will give a 10-nm variation of the POF FBG wavelength compared with 1 nm for silica FBG. Variations of the same order of magnitude can be obtained by stretching the POF FBG. FBGs with UV photosensitive cores in PMMA-based POFs are still under development [60]. FBGs with polymer fiber are interesting candidates to be used as medical sensors because of their biocompatibility properties, but no commercial products are available at this time.

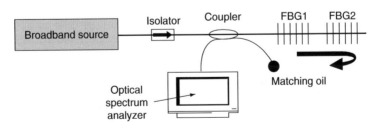

Figure 9.15 Sensor with a combination of polymer and silica fiber gratings [56, 63].

9.8.2 Fluoride Glass

Fluoride (or nonoxide) glasses were discovered in 1975 by Marcel and Michel Poulain [61]. At that time, one of the most important companies making fluoride fibers was "Le Verre Fluoré" [62]. The principal interest of these glasses is their transmission spectra because they are transparent to light from 300 to 6500 nm. ZBLAN is the most common composition (ZrFM4-BaF2-LaF3-AlF3-NaF), but other compositions permit to reach different mechanical or optical properties. Fluoride fibers can be used in a variety of scientific and industrial applications such as infrared and near-infrared fiber spectroscopy, ultrafast fiber pyrometry, astronomical interferometry, or telecommunications with fluoride glass optical amplifier modules. Lasers can be developed with fluoride fibers for the visible spectrum and for infrared wavelengths. Fluoride fibers can also be interesting for medicine and surgery, more particularly for Er/YAG lasers. IRphotonic has developed a special fluoride fiber with low moisture suitable for Er/YAG transmission [63].

Until now, research on the photosensitivity of fluoride glasses was not very extended, with only a few studies on this topic and no available commercial product. The first studies on the photosensitivity of fluoride glasses or fibers were done with trivalent rare earth ion cerium (Ce^{3+})-doped glasses [64, 65]. They present the influence of the cerium concentration on the photosensitivity of the fluoride fibers with 1550 nm Bragg grating inscription with a 246-nm UV light exposure and prove that enhancement of the Ce^{3+} concentration contributes to increase the photo-induced refractive index change up to 4×10^{-4} in a single-mode Ce^{3+}-doped fluoride fiber [68]. Other studies have been performed on RED fluorozirconate (FZ) glasses with permanent holographic gratings written at 248 nm in bulk samples of fluoride glasses doped with Tb^{3+}, Pr^{3+}, Tm^{3+}, or Ce^{3+}, with refractive index changes at less than 1×10^{-5} for cerium and lower for the others [66]. Other work has been performed by Zeller et al. [67], who report on undoped fluoride glass slides exposed to pulsed 193 nm UV irradiation. Photosensitivity of fluoroaluminate (FA), FZ, and fluorozircoaluminate (FZA) glasses is compared. FA and FZ glasses provide small index changes (2.0×10^{-6} for FA and 2.6×10^{-6} for FZ), but the index change up to 1.75×10^{-4} has been evaluated in FZA glass at 1550 nm. Commercial products are not yet suitable, but the photosensitivity of fluoride fibers is still under development for research applications.

9.8.3 Heavily P-Doped Silica Fibers

The low-loss heavily P_2O_5-doped silica fiber is a good candidate for the development of an efficient high-power Raman fiber laser (RFL) used as a pump source for Raman fiber amplifiers (RFAs) or erbium-doped fiber amplifiers (EDFAs) [68]. The RFL gain block consists of a P-doped fiber as the gain medium; the laser cavity is obtained by FBG inscribed in Ge-doped silica fibers. The two FBGs are then spliced to the gain medium fiber. The splice losses could be reduced or even avoided by inscribing the FBGs directly into the heavily P-doped silica fibers. The number of cavities depends on the Stokes shift. For a Ge-doped silica fiber, the Stokes shift is limited to 13.2 THz. For example, five laser cavities are needed to shift the Yb-doped fiber laser pump emitting at 1060 nm to the pump wavelength of 1480 nm for EDFAs. Using a heavily P-doped silica fiber, the number of cavities is reduced to only two, thanks to the Stokes shift of 39.9 THz. This shift is three times as large as the Ge-doped silica fiber. In only one step or one pair of FBGs, a pumped P-doped silica fiber at 1060 nm will convert the pump light to 1240 nm. This wavelength is suitable to pump RFAs for the 1310-nm window. Although the P-doped silica fiber is very promising, it is not intrinsically photosensitive and needs to be treated to become so [69, 70]. Once applied to the P-doped silica fiber, the standard hydrogen-loading process has many disadvantages such as significant OH formation, hydrogen out diffusion issues, and a contribution to component instability from an undesired index change. The OH formation is due to the hygroscopic nature of P, and consequently, it increases the background losses, reducing the efficiency of Raman converters. A proposed solution is the photolytic hypersensitization [73] or two-stage exposure process [74]. The first stage consists of irradiating the hydrogen-loaded P-doped fiber, which is the photolytic hypersensitization, with a low-exposure dose homogeneous beam. Then, the remaining free hydrogen is removed from the fiber by leaving it at room temperature for many days. The second stage consists of writing the FBGs as usual. The FBGs were written directly through a phase mask at 193 nm from an ArF source. The 193-nm source wavelength was found to be the most efficient writing wavelength with the lowest induced OH. The fiber was fabricated by the MCVD process. The P_2O_5 concentration was 15 mol%. Other specifications are given in Table 9.15.

9.9 CONCLUSIONS

This chapter reviewed the design, fabrication, special features, and applications of several photosensitive SOFs. Naturally, most of SOFs are not intrinsically photosensitive and need some special design and fabrication care to become so. The emphasis was made on the following three types: HNA, CMS, and RED photosensitive fibers. They are the most commonly used photosensitive SOFs in the optical communication industry for applications such as laser diode wavelength stabilization, band-stop or bandpass filters, and reflectors for single-frequency fiber lasers. The RED photosensitive fiber is also used in the material

Table 9.15

Heavily P_2O_5-doped Raman silica fiber data

Fiber type:	Heavily P-doped Raman silica fiber
Core diameter:	4.0 ± 0.5 μm
P_2O_5 concentration:	15 mol%
Numerical aperture:	0.18 ± 0.02
LP_{11} cutoff wavelength:	850 ± 50 nm
Mode-field diameter @ 1550 nm:	8 ± 1 μm
Attenuation @ 1550 nm:	<2 dB/km
Cladding diameter:	125 ± 2 μm
Coating diameter:	250 ± 2 μm
Screen-proof tested:	>100 kpsi

processing industry as the main optical component of compact high-power fiber lasers. For an economic reason, hydrogen loading of the very low price standard single-mode fiber is even more widely used in the optical communication industry and in the sensor field. This subject was treated briefly because it is well covered in the literature [11, 30]. Photosensitive SOFs treated in this chapter were all made by the MCVD process. The rare earth, aluminum, or high phosphorous content when needed was incorporated by the solution-doping technique. Other CVD processes are used but were not discussed here. Upon future needs, improved photosensitive fibers and new photosensitive fiber types will be developed using new materials, fabrication processes, and designs.

ACKNOWLEDGMENTS

The authors acknowledge Claude Paré and Antoine Proulx for their comments.

REFERENCES

[1] Hill, K. O. et al. 1978. Photosensitivity in optical fiber waveguides: Application to reflection filter fabrication. *Appl. Phys. Lett.* 32:647–649.
[2] Hill, K. O. 2000. Photosensitivity in optical fiber waveguides: From discovery to commercialization. *IEEE J.* 6:1186–1189.
[3] Meltz, G. et al. 1989. Formation of Bragg gratings in optical fibers by the transverse holographic method. *Opt. Lett.* 14:823–825.
[4] Hill, K. O. et al. 1993. Bragg gratings fabricated in monomode photosensitive optical fiber with UV exposure through a phase mask. *Appl. Phys. Lett.* 62:1035–1037.
[5] Martin, J. et al. 1994. Novel writing technique of long and highly reflective in-fiber gratings. *Elec. Lett.* 30:811–812.

[6] Williams, D. L. et al. 1993. Enhanced UV photosensitivity in boron codoped germanosilicate fibers. *Elec. Lett.* 29:45–47.

[7] Archambault, J. L. et al. 1994. High reflectivity photorefractive Bragg gratings in germania-free optical fibers. In: *Conference Proceedings of CLEO'94, Paper CWK3,* p. 242. Anaheim, CA. OSA—The Optical Society of America.

[8] Zagorulko, K. A. et al. 2004. Fabrication of fiber Bragg gratings with 267 nm femtosecond radiation. *Opt. Express* 12:5993–6001.

[9] Taunay, T. et al. 1997. Ultraviolet-enhanced photosensitivity in cerium-doped aluminosilicate fibers and glasses through high-pressure hydrogen loading. *J. Opt. Soc. Am. B* 14:1–14.

[10] Taunay, T. 1997. Photosensitization of terbium doped alumino-silicate fibers through high pressure H_2 loading. *Optics Comm.* 133:454–462.

[11] Lemaire, P. J. et al. 1993. High-pressure H_2 loading as a technique for achieving ultrahigh UV photosensitivity and thermal sensitivity in GeO_2 doped optical fibers. *Elec. Lett.* 29:1191–1193.

[12] Hand, D. P. et al. 1990. Photoinduced refractive index changes in germanosilicate fibers. *Opt. Lett.* 15:102–104.

[13] Poumellec, B. et al. 1995. UV induced densification during Bragg grating inscription in Ge: SiO_2 preforms: Interferometric microscopy investigations. *Opt. Mat.* 4:404–409.

[14] Limberger, H. G. et al. 1996. Compaction and photo elastic induced index changes in fiber Bragg gratings. *Appl. Phys. Lett.* 68:3069–3071.

[15] Kashyap, R. 1999. *Fiber Bragg Gratings.* Academic Press. San Diego, CA.

[16] Fiber Bragg grating components from Teraxion. Available at http://www.teraxion.com/.

[17] Frazao, O. et al. 2004. Simultaneous measurement of strain and temperature using a Bragg grating structure written in germanosilicate fibres. *J. Opt. A Pure Appl. Opt.* 6:553–556.

[18] Liang, W. et al. 2005. Highly sensitive fibre Bragg grating refractive index sensors. *App. Phy. Lett.* 86:151122-1–151122-3.

[19] Sun, A. et al. 2005. Study of simultaneous measurement of temperature and pressure using double fiber Bragg gratings with polymer package. *Opt. Eng.* 44:034402-1.

[20] Bhatia, V. 1999. Applications of long-period gratings to single and multi-parameter sensing. *Opt. Exp.* 4:457–466.

[21] Khaliq, S. et al. 2002. Enhanced sensitivity fibre optic long period grating temperature sensor. *Meas. Sci. Tech.* 13:792–795.

[22] James, S. W. et al. 2003. Optical fibre long-period grating sensors: Characteristics and application. *Meas. Sci. Tech.* 14(5):R49–R61.

[23] Dong, L. et al. 1994. Ultraviolet absorption in modified chemical vapor deposition preforms. *J. Opt. Soc. Am. B* 11:2106–2111.

[24] Dianov, E. M. et al. 2000. Photorefractive effect and photoinduced $\chi^{(2)}$ grating formation in germanosilicate core fibers co-doped with nitrogen in MCVD process. *SPIE* 4083: 144–152.

[25] Oh, K. et al. 2002. Ultraviolet photosensitive response in an antimony-doped optical fiber. *Opt. Lett.* 27:488–490.

[26] Brambilla, G. and V. Pruneri. 2001. Enhanced photo refractivity in tin-doped silica optical fibers (review). *IEEE J.* 7(3):403–408.

[27] Othonos, A. and K. Kali. 1999. *Fiber Bragg Gratings, Fundamentals and Applications in Telecommunications and Sensing.* Artech House, Boston.

[28] Williams, D. L. et al. 1992. Enhanced photosensitivity in germania doped silica fibers for future optical networks. In: *Proc. ECOC'92,* pp. 425–428. IEEE—The Institute of Electrical and Electronic Engineers.

[29] Canagasabey, A. et al. 2003. 355 nm hyper sensitization of optical fibers. *Opt. Lett.* 28:1108–1110.

[30] Lancry, M. et al. 2005. Comparing the properties of various sensitization methods in H_2-loaded, UV hyper sensitized or OH-flooded standard germano-silicate fibers. *Opt. Exp.* 13:4037–4043.

[31] Salik, E. et al. 2000. Increase of photosensitivity in Ge-doped fibers under strain. *Opt. Lett.* 25:1147.

[32] Konstantaki, M. et al. 2005. Effect of Ge concentration, boron co-doping, and hydrogenation on fiber Bragg grating characteristics. *Microwave Opt. Tech. Lett.* 44:148–152.

[33] Mizrahi, V. et al. 1993. Optical properties of photosensitive fiber phase gratings. *JLT* 11:1513–1517.

[34] Komukai, T. et al. 1995. Efficient fiber gratings formed on high NA dispersion-shifted fibers. Conference Proceedings of *ECOC'95*, Paper Mo.A.3.3, pp. 31–34. IEEE. Brussels, Belgium.

[35] Dong, L. et al. 1995. Enhanced photosensitivity in tin-codoped germanosilicate optical fibers. *IEEE Phot. Tech. Lett.* 7:1048–1050.

[36] Shen, Y. et al. 2004. Photosensitive indium doped germano silica fiber for strong FBGs with high temperature sustainability. *IEEE Phot. Tech. Lett.* 16:1319–1321.

[37] Hewlett, S. J. et al. 1995. Cladding-mode coupling characteristics of Bragg gratings in depressed cladding fiber. *Elect. Lett.* 31:820–822.

[38] Delevaque, E. et al. 1995. Optical fiber design for strong gratings photo-imprinting with radiation mode suppression. *OFC 1995, Techni. Digest,* PD5-2.

[39] Croteau, A. et al. 2003. Influence of H_2 loading on the performance of a cladding-mode suppression photosensitive fiber. *Photonics North'03, SPIE* 5260:266–271.

[40] Poulsen, T. et al. 1998. Bragg grating induced cladding mode coupling caused by ultraviolet light absorption. *Elec. Lett.* 34:1007–1009.

[41] Digonnet, M. 2001. *Rare-Earth–Doped Fiber Lasers and Amplifiers.* Marcel Dekker, New York.

[42] Bilodeau, F. et al. 1990. Ultraviolet-light photosensitivity in Er^{3+}-Ge–doped optical fiber. *Optics Lett.* 15:1138–1140.

[43] Douay, M. et al. 1997. Densification involved in the UV-based photosensitivity of silica glasses and optical fibers. *JLT* 15:1329–1342.

[44] Kanellopoulos, S. E. et al. 1991. Polarization properties of permanent and nonpermanent photorefractive gratings in Hi-Bi fibers. *IEEE PTL* 3:345–347.

[45] Kaino, T. et al. 1984. Preparation of polymer optical fibres. *Rev. Elec. Comm. Lab.* 32:478.

[46] Zubia, J. et al. 2001. Plastic optical fibers: An introduction to their technological processes and applications. *Opt. Fiber Tech.* 7:101–140.

[47] Tomlinson, W. T. et al. 1970. Photoinduced refractive index increase in poly(methyl methacrylate) ands its applications. *Appl. Phys. Lett.* 16:486–488.

[48] Kaminov, I. P. et al. 1971. Poly (methyl metha-crylate) dye laser with internal diffraction grating resonator. *Appl. Phys. Lett.* 18:497.

[49] Welker, D. J. et al. 1998. Fabrication and characterization of single-mode electro-optic polymer optical fiber. *Opt. Lett.* 23:1826–1828.

[50] Peng, G. D. et al. 1999. Photosensitivity and grating in dye-doped polymer optical fibers. *Opt. Fiber Technol.* 5:242–251.

[51] Koike, Y. 1992. High bandwidth and low loss polymer optical fiber. In: *Conference Proceedings of the First International Conference on Plastic Optical Fiber and Applications—POF'92,* pp. 15–19. Paris, France. IGi Group, Boston, MA.

[52] Xiong, Z. et al. 1999. 73 nm wavelength tuning in polymer optical fiber Bragg gratings. In: *Proceedings of the 24th Australian Conference on Optical Fibre Technology ACOFT'99,* pp. 135–138.

[53] Chu, P. L. 2005. Polymer optical fiber Bragg gratings. *Optics and Photonics News* Jul/Aug, pp. 53–56.

[54] Peng, G. D. and P. L. Chu. 2004. Polymer optical fiber gratings. In: *Polymer Optical Fibers* (Nalwa, ed.), pp. 51–71. American Scientific Publishers. Valencia, CA.

[55] Liu, H. B. et al. 2003. Different types of polymer fiber Bragg gratings and their strain/thermal properties. *Optical Memory and Neural Networks* 12:147–155.

[56] Liu, H. B. et al. 2004. Novel growth behaviors of fiber Bragg gratings in polymer optical fiber under UV irradiation with low power. *IEEE Phot. Tech. Lett.* 16:159–161.

[57] Tanio, H. et al. 2000. What is the most transparent polymer. *Polymer J.* 32:43–50.

[58] Liu, H. Y. et al. 2000. *Proceedings of the 25th Australian Conference of Optical Fibre Technology,* pp. 48–50.

[59] Liu, H. B. et al. 2003. Strain and temperature sensor using a combination of polymer and silica fiber Bragg gratings. *Opt. Commun.* 219:139–142.

[60] Yu, JM. et al. Fabrication of UV sensitive single-mode polymeric optical fiber. *Opt. Materials* 28:181–188, 2006.

[61] Poulain, M. et al. 1975. Verres fluorés au tetrafluorure de zirconium, propriétés optiques d'un verre dopé au Nd^{3+}. *Mat. Res. Bull. Col.* 10:243–246.

[62] Le Verre Fluoré, Campus KerLann. F-35170 Bruz, Brittany. Available at http://www.leverrefluore.com/.

[63] IRphotonics, Inc., 627 Rue McCaffrey, Montreal, (Québec), Canada, H4T-1N3. Available at http://www.irphotonics.com/.

[64] Poignant, H. et al. 1994. Efficiency and thermal behavior of cerium-doped fluorozirconate glass fibre Bragg gratings. *Elec. Lett.* 30:1339–1341.

[65] Taunay, T. et al. 1994. Ultraviolet-induced permanent Bragg gratings in cerium-doped ZBLAN glasses or optical fibers. *Opt. Lett.* 19:1269–1271.

[66] Williams, G. M. et al. 1997. Photosensitivity of rare-earth–doped ZBLAN fluoride glasses. *J. Lightwave Technol.* 15:1357–1361.

[67] Zeller, M. et al. 2005. UV-induced index changes in undoped fluoride glass. *J. Lightwave Technol.* 23:624–627.

[68] Xiong, Z. et al. 2003. 10-W Raman fiber lasers at 1248 nm using phosphosilicate fibers. *JLT* 21:2377–2381.

[69] Canning, J. et al. 2001. Low-temperature hypersensitization of phosphosilicate waveguides in hydrogen. *Opt. Lett.* 26:1230–1232.

[70] Larionov, Y. V. et al. 2002. Peculiarities of the photosensitivity of low-loss phosphosilica fibres. *Quantum Elec.* 32:124–128.

[71] Croteau, A. et al. 2002. All-silica double-clad hexagonal Yb:Er co-doped photosensitive fibers. *ECOC'02* 3:P1.4.

[72] Dong, L. et al. 1997. Efficient single-frequency fiber lasers with novel photosensitive Er/Yb optical fibers. *Optics Lett.* 22:694–696.

[73] Sahu, J. K. et al. 2004. Photosensitivity in germanium-free antimony doped aluminosilicate optical fiber prepared by MCVD. *ECOC'04, Th3.3.5.* 908–909.

[74] Liu, H. Y. et al. 2002. Polymer fiber Bragg grating with 28 dB transmission rejection. *IEEE Phot. Technol. Lett.* 14:935–937.

Chapter 10

Hollow-Core Fibers

Steven A. Jacobs,[1] Burak Temelkuran,[1] Ori Weisberg,[1] Mihai Ibanescu,[2] Steven G. Johnson,[2] and Marin Soljačić[2]

[1] OmniGuide Inc., Cambridge, Massachusetts
[2] MIT, Cambridge, Massachusetts

10.1 INTRODUCTION

The history of the development of optical fibers has been largely determined by various constraints stemming from the materials used for the fiber core. For example, in the case of long-haul telecom applications, it is essential to have very low losses at the wavelength of a reliable and commercially available laser. Because silica has very low losses over wavelengths ranging from the visible to the near infrared (IR), which coincide with the operating wavelengths of a number of readily available lasers, it became the material of choice for the fiber core; most of the optical power travels through the core, so the properties of the core material determine the loss properties of such a fiber. Furthermore, the carrier wavelength in telecom applications has migrated from 1.31 to 1.55 μm, to match the wavelength for which the loss of the (mostly) silica core material is minimized. For other important applications, such as surgery, sensing, and industrial welding, it is often desirable to operate at longer wavelengths (>2 μm) where the losses of silica are prohibitively high. Consequently, many other potential core materials that have moderate losses at mid-IR wavelengths have been explored and used, including chalcogenide glasses [1, 2], heavy-metal fluoride glasses [3], and polycrystalline materials [4, 5].

However, instead of looking for a fiber core material with the best properties for a particular application or wavelength, there is another approach: one can make a fiber that guides light within a hollow core. In hollow-core fibers, the cladding is designed to act as a "mirror," reflecting light incident on it back into the core. In contrast to the solid-core fibers, the vast majority of optical power now travels through air, whose optical properties are dramatically different than the optical properties of *any* solid material. Because of the small evanescent tails

315

of the guided light, which penetrate into the cladding, the optical properties of the materials of the cladding still have an influence on the overall fiber transmission characteristics, but these optical properties become much less of a constraint. In particular, the overall losses or nonlinearities of the fiber can now be orders of magnitude smaller than the intrinsic losses and nonlinearities of the solid materials used to manufacture the fiber. It has been predicted, for example, that hollow-core silica fibers could someday lead to even lower transmission losses [6] than solid-core silica fibers, whose losses—0.2 dB/km at $\lambda = 1.55\,\mu$m—are already extremely low. Similarly, the high-peak-power material breakdown can now be shifted to much stronger peak intensities. Moreover, because optical properties of the materials are not as much of a constraint any more, it becomes possible to use many materials with poor optical properties but attractive other properties (e.g., thermal or mechanical properties) to implement optical fibers.

This chapter discusses the properties, applications, and manufacture of hollow-core fibers: Section 10.2 provides a discussion of their transmission properties, Section 10.3 discusses their applications, Section 10.4 discusses various ways hollow-core fibers are being manufactured, and Section 10.5 presents brief concluding remarks. The remainder of the current section provides a short review of the principles of operation and history of solid-core fibers, followed by a description of the principles of operation and history of each of the main kinds of hollow-core fibers.

10.1.1 Wave-Guiding by Total Internal Reflection

Total internal reflection (TIR) fibers are the most commonly used optical fibers. For example, almost all fibers used for telecom applications are of the TIR type. The principle of operation of a TIR fiber is explained in Fig. 10.1.

The fact that TIR can be used to confine and guide light through objects of substantial length but small cross-sectional area was realized and experimented with as early as the nineteenth century; even some applications were proposed and were being explored at that time. In 1956, Larry Curtis made the first successful fiber with a glass core and a glass cladding; similar fibers have been used in endoscopes to look inside the human body ever since. In 1965, C. K. Kao and G. Hockham published an analysis showing that an optical fiber could be a suitable medium for long-distance communication if the losses could be brought down below 20 dB/km. They proceeded to show that a highly purified glass should easily have losses below this limit. Finally, in 1970, four employees of Corning Glass Works—R. Maurer, D. Keck, P. Schultz, and F. Zimar—demonstrated an optical fiber with losses of 17 dB/km, thereby starting the era of optical fiber communications.

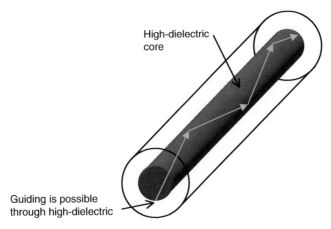

High-dielectric core

Guiding is possible through high-dielectric

Figure 10.1 Optical fiber operating on the principle of total internal reflection (TIR). The core of the fiber has index of refraction $n_{core} > n_{clad}$. Imagine a ray of light propagating in the core that impinges on the interface between the core and the cladding; if the incidence angle on the interface is below the critical angle for TIR, the ray is perfectly reflected back into the core and is, therefore, trapped and guided inside the core.

Interestingly enough, there even exist hollow-core fibers ($n_{core} = 1$) that operate based on TIR; for certain frequency ranges, there are materials (e.g., sapphire at $\lambda = 10.6\,\mu$m) that have an index of refraction $n < 1$ and can, therefore, serve as a suitable cladding [7]. However, the range of applicability of such fibers is limited because these materials turn out to have other highly undesirable properties (e.g., loss, stiffness).

10.1.2 Wave-Guiding by Reflection Off a Conducting Boundary

This class of waveguides has a hollow core surrounded by a conducting boundary that acts as a mirror, reflecting the guided light back toward the core center, thereby providing confinement for light. Conductors can have good, but not perfect, reflection properties; more than 90% reflection for normal incidence is not unusual. Nevertheless, when the angle of incidence of the ray onto the conductor is far from normal and the core is large, the number of reflections per meter of fiber length can be quite small, so the losses can often be acceptably small. Often, the properties of the fiber can be further improved by depositing one or more layers of dielectric onto the conducting surface. Interference behavior associated with the dielectric layers improves the guiding properties of the fiber.

Between the 1930s and the 1970s, Bell Labs had a substantial research program [8] aimed at developing a hollow-core waveguide to transport microwave (GHz) electromagnetic signals for long-distance communications. It developed a 60-mm-diameter hollow-core tube, the inside of which was coated with copper. The range of operational frequencies was aimed to be 40–110 GHz, with losses less than 1 dB/km. Interestingly enough, one of the major obstacles that precluded it from deploying the cable was economic; the price of deployment was substantial, so only deployment for large markets could pay off, yet there were very few markets at the time that could make anywhere close to the full use of the 274-Mbps capacity of such a cable. In the 1970s, Corning developed its optical silica fiber, whose properties were highly superior to the Bell Labs concept, so the interest in microwave transmission through metal waveguides for long-distance communication subsided.

The first optical-frequency hollow-core metallic waveguides were built in 1980 [9]. These rectangular aluminum waveguides could guide approximately 1 kW of $\lambda = 10.6$-μm light (from a CO_2 laser) with losses as low as approximately 1 dB/m. Their rectangular shape prevented them from being uniformly bent, but since then, many circular designs have also been explored [10, 11]. Such fibers can today guide fairly high powers (10–1000 W), anywhere from visible wavelengths all the way to $\lambda > 10.6\,\mu$m, with losses as low as 0.1 dB/m.

10.1.3 Wave-Guiding by Photonic Band-Gaps

In addition to TIR and reflection from a conducting boundary, there is yet another physical mechanism that produces a highly reflecting surface: the photonic band-gap [12–14]. Photonic band-gap crystals are artificially created materials in which the index of refraction varies periodically between high-index values and low-index values. When light of wavelengths comparable to the periodicity tries to propagate inside such a material, it experiences very strong scattering: it cannot propagate and is reflected out of it. In more precise terms, in photonic crystals, a so-called *photonic band-gap* opens for certain frequency regimes; Maxwell's equations do not have propagating solutions for the frequencies inside the photonic band-gap, so light with those frequencies is prohibited from propagation through the material. The material, therefore, acts as a perfect mirror. Note that it is the periodicity length scale of the photonic crystal that determines the frequencies for which the photonic band-gaps occur; by adjusting the periodicity (while using the same constituent materials), one can select the frequency range in which the photonic crystal will act as a reflector.

The main advantage of using a photonic-crystal mirror (compared to a conducting-surface mirror) lies in the fact that photonic crystals can be implemented from dielectric materials, which can often have dramatically lower losses than any conductor (especially in the optical regime). As a result, reflection from a photonic-crystal mirror can incur much less absorption than reflection from any conducting-surface mirror. Hollow-core photonic band-gap fibers have photonic band-gap crystals surrounding their cores, which act as mirrors so that light is confined to propagate within the core. The two main kinds of photonic band-gap fibers are those that use one-dimensional (1D) photonic band-gaps and those that use two-dimensional (2D) photonic band-gaps; both kinds are described in the subsections that follow.

10.1.3.1 Wave-Guiding by 1D Photonic Band-gaps

A 1D photonic band-gap fiber uses a 1D periodic photonic crystal (periodicity is only in a single direction) as its mirror; an example of such a fiber is shown in Fig. 10.2. The first such fiber, called a *Bragg* fiber (not to be confused with a fiber Bragg grating, which is periodic in the axial instead of the radial direction), was proposed by Yeh et al. [15]. Since then, many authors have explored Bragg fibers [16–18]. One particular type of Bragg fiber, called an *OmniGuide fiber,* has an *omnidirectional* reflector surrounding its core. An omnidirectional reflector uses a 1D photonic band-gap to reflect light (within a certain frequency range) perfectly, except for material absorption, for all polarizations and angles of incidence. Needless to say, such a perfect mirror acts as an excellent mechanism to confine light to the core. OmniGuide fibers have been built for many frequency regimes, extending from IR ($\lambda > 10\,\mu$m), all the way down to ultraviolet (UV) ($\lambda < 350$ nm) [19]. In Fig. 10.2, we show an OmniGuide fiber designed to guide CO_2 laser light ($\lambda = 10.6\,\mu$m); it has losses less than 1 dB/m.

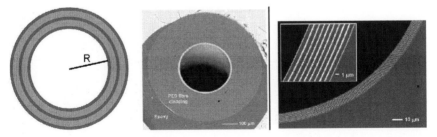

Figure 10.2 A one-dimensional photonic band-gap fiber. (Left) A schematic of a transverse cross-section of such a fiber; the index of refraction is periodic in the radial direction, thereby creating a photonic band-gap that acts as a mirror, confining light to the core. (Middle) An example of a fabricated OmniGuide fiber, and (Right) a magnified detail of the middle panel, clearly showing the one-dimensional photonic crystal.

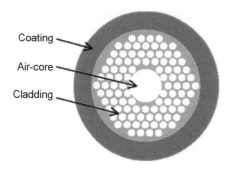

Figure 10.3   A hollow-core two-dimensional periodic photonic band-gap fiber. (Left) A schematic, and (Right) a cross-section of one such fabricated structure.

10.1.3.2 Wave-Guiding by 2D Photonic Band-gaps

The confining mechanism of a 2D photonic band-gap fiber is a mirror implemented by a 2D periodic photonic crystal, like the one shown in Fig. 10.3. Photonic-crystal fibers (PCFs) were pioneered by P. Russell [20] and have since been investigated by many researchers around the globe [21, 22] (such fibers can even be purchased from Crystal Fibre A/S). Such fibers enable many important applications (to be discussed in greater detail in Section 10.3), but it is worth noting here that they are predicted to have lower losses than solid-core silica telecom fibers [6]. The current minimum measured transmission loss in such fibers is 1.7 dB/km at $\lambda = 1.5\,\mu$m [23].

10.2 LIGHT TRANSMISSION IN HOLLOW-CORE FIBER

Hollow-core fibers guide light by means of a reflective cladding. Because the index of refraction of the hollow core is smaller than that of the cladding materials,[1] the guiding mechanism cannot be based on TIR, as is the case for traditional optical fibers. Instead, three major types of reflective cladding are used: (1) a metal tube with optional dielectric coating [11], (2) a multilayer dielectric Bragg mirror (i.e., a 1D photonic crystal) [15, 24, 25], or (3) a 2D photonic crystal [26, 27]. These are schematically depicted in Fig. 10.4.

The following subsections describe transmission in fibers that use these different reflection mechanisms for the cladding. We first, however, introduce a few

---

[1] Hollow-core fibers that have a larger refractive index in the core than in the cladding and can make use of TIR are encountered in rare situations. One example is hollow-core fiber with a sapphire cladding for use at $\lambda = 10.6\,\mu$m [7]; a second example is silica PCFs used for biosensing that have a liquid flowing through the hollow core [28a].

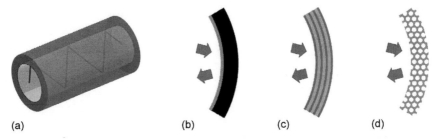

Figure 10.4 (a) A hollow-core fiber guides light by means of a reflective cladding. This cladding can be (b) a metal cladding with an optional dielectric coating, (c) a multilayer dielectric mirror, or (d) a two-dimensional photonic-crystal cladding.

physical quantities that are common to the electromagnetic fields of any axially uniform waveguide, whatever its confinement mechanism. We consider mono-chromatic light of angular frequency ω, with a time dependence of the form $e^{-i\omega t}$. The translational invariance of the waveguide along its axis (which we take to be the z-axis) suggests that the electromagnetic fields of the waveguide will have a z-dependence of the form $e^{i\beta z}$, where β is the as-yet undetermined axial wave vector (also called the *propagation constant*). Thus, at frequency ω, we expect that an electromagnetic field in the waveguide can be written as a plane wave along the z-direction times a field profile in the transverse (x,y) plane:

$$\mathbf{E}(x,y,z,t) = \mathbf{E}(x,y)e^{i(\beta z - \omega t)} \tag{10.1a}$$

$$\mathbf{B}(x,y,z,t) = \mathbf{B}(x,y)e^{i(\beta z - \omega t)}. \tag{10.1b}$$

It is well known that Eqs. (10.1a) and (10.1b) are valid solutions to Maxwell's equations, and represent fields confined to the waveguide core, only for certain discrete values of the axial wave vector. These discrete solutions are known as the *guided modes* of the waveguide and, of course, depend on the particular distri-bution of the dielectric profile $\varepsilon(x,y)/\varepsilon_0 = n^2(x,y)$ in the cross-section. Except in certain simple cases, they must be evaluated numerically. For a given mode, the axial wave vector varies with the angular frequency: $\beta = \beta(\omega)$. This functional dependence is known as the *dispersion relation* for the mode.

Although the evaluation of the dispersion relations for a particular fiber cross-section must, in general, be performed numerically, one can still draw important conclusions about the nature of a waveguide mode simply based on the values of ω, β, and the refractive indices of the materials that make up the core and the cladding. The transverse wave vector in a material with index of refraction n is given by $k_t^2 = (n\omega/c)^2 - \beta^2$. Depending on whether β is smaller or larger than $n\omega/c$, the transverse wave vector k_t will take real or imaginary values, thus determining whether light can or cannot propagate transversely through that material. The line that separates these two regimes is given by $\beta = n\omega/c$ and is usually called the *light line* of a material with index of refraction n.

Based on this simple argument, in Fig. 10.5, we compare the guiding mechanisms in a traditional index-guiding optical fiber, a hollow metal waveguide, and a generic photonic band-gap fiber. The light lines of relevant materials are shown as dashed lines, and a schematic dispersion relation of a typical guided mode confined to the core is shown as a black curve. Regions of the (ω, β) diagram where modes can propagate through the cladding (so-called *radiation* modes) are shaded in dark gray. Light gray regions correspond to (ω, β) pairs for which light can propagate through the core material, but not through the cladding, thus allowing for core-confined guided modes. In all three cases, core guided modes can only exist above the light line corresponding to the index of refraction in the core. The differences between the three cases stem from the specific properties of the claddings. For the traditional index-guiding fiber, core guided modes are situated between the light lines corresponding to the core and cladding materials and can only exist if the core has a refractive index larger than that of the cladding. In the case of the hollow metal waveguide, light is not allowed in the cladding for any pair (ω, β), which means that core guided modes can exist in the entire region above the light line of vacuum. Finally, for a photonic band-gap fiber, light can be guided in the core at (ω, β) values that are above the light line of vacuum and inside a photonic band-gap of the 1D or 2D photonic-crystal cladding.

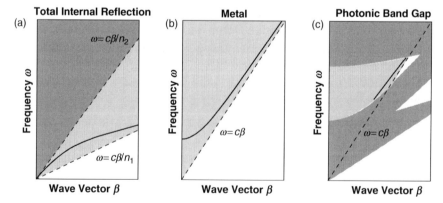

Figure 10.5 Schematic band diagrams for (a) a traditional index-guiding fiber with refractive index n_1 in the core and n_2 in the cladding, (b) a hollow metal waveguide as shown in Fig. 10.4b, and (c) a generic hollow-core photonic band-gap fiber as shown in Fig. 10.4c and d. A typical dispersion relation for a guided mode in each structure is shown as a solid black curve.

10.2.1 Hollow Metal Waveguides

Perhaps the simplest method for guiding light in a hollow core is by enclosing the core with a highly reflective metal. The metal acts as a mirror, so that fields from the core incident on the metal are reflected back into the core, providing the confinement mechanism. For a perfect metal, the evaluation of the transverse dependence of the mode fields from Maxwell's equations is fairly straightforward and is covered in most standard textbooks [28]. When the interior of the waveguide consists of a single homogeneous dielectric material, the mode fields can be separated into two polarizations: TE (transverse electric) and TM (transverse magnetic) with vanishing axial components of the electric and magnetic fields, respectively. The analysis of these modes is particularly simple for the case of a hollow metal waveguide with a circular cross-section. For TM mode fields, the allowed values of the axial wave vector at the frequency ω are

$$\beta_{mn}^2 = \frac{\omega^2}{c^2} - \frac{x_{mn}^2}{R^2}, \tag{10.2a}$$

where m is an index (the "angular momentum") denoting the angular dependence $e^{im\theta}$ of the mode, n is an index denoting the radial dependence of the mode, R is the core radius, and x_{mn} is the nth root of the Bessel function $J_m(x)$. For TE modes, the allowed axial wave vectors are

$$\beta_{mn}^2 = \frac{\omega^2}{c^2} - \frac{y_{mn}^2}{R^2}, \tag{10.2b}$$

where y_{mn} is the nth root of $dJ_m(y)/dy$. Equations (10.2a) and (10.2b) form the dispersion relations for the hollow metal fiber.

Each mode possesses a *cutoff* frequency given by $\omega_c = c\frac{x_{mn}}{R}$ for the TM polarization and $\omega_c = c\frac{y_{mn}}{R}$ for the TE polarization. Below the cutoff frequency, the axial wave vector becomes imaginary and the mode decays exponentially instead of propagating; it is "cutoff." However, it remains a valid solution to Maxwell's equations and is needed, for example, in the modal decomposition of an arbitrary field in the waveguide. At a frequency below its cutoff frequency, a mode is said to be *evanescent*. Note that the sequences of Bessel function roots x_{mn}, y_{mn} are increasing and unbounded with index n. Thus, at a given frequency, only a finite number of propagating modes exist for each angular index m. Indeed, the smallest Bessel function root is $y_{11} \approx 1.84$, associated via Eq. (10.2b) with the TE$_{11}$ mode. For frequencies $\omega < c\frac{y_{11}}{R}$, no propagating modes exist for the hollow metal waveguide. In contrast, TIR fibers always have at least one propagating mode for any frequency. As the core radius R increases, the number of propagating modes at a given frequency will also increase.

For a perfect metal, propagating modes are loss free. Actual metals, of course, are not perfect; they have finite conductivities [29], so a propagating mode will penetrate to a small extent into the metal and thereby become lossy. To reduce this loss, dielectric layers may be placed on top of the metal layer [30–34]. By proper choice of the dielectric layer thicknesses, the reflected waves established at the dielectric layer interfaces will interfere destructively with the transmitted waves, reducing the amplitude of the mode field in the vicinity of the lossy metal and reducing the loss of the mode. This process, which is only operational within a limited frequency range, essentially places a dielectric mirror in front of the metal mirror. Of course, with a sufficient number of properly chosen dielectric layers, a 1D photonic band-gap can be formed, which eliminates the need for the metal layer entirely.

10.2.2 Wave-Guiding in Bragg and OmniGuide Fibers

Bragg and OmniGuide fibers confine light to the hollow core with a mirror-like multilayer cladding, made up of concentric cylindrical rings having alternating low and high indices of refraction [24, 26]. In certain frequency ranges, waves reflected from the multiple dielectric interfaces interfere destructively with the transmitted waves, prohibiting transmission of light through the cladding and providing a confinement mechanism for light in the hollow core. A more detailed understanding of this mechanism can be obtained by analyzing a planar dielectric mirror formed from the same pattern of alternating high- and low-index layers.

Consider a planar Bragg mirror with alternating layers having indices of refraction $n_1 = 2.7$, $n_2 = 1.6$ and thicknesses $d_1 = 0.33a$, $d_2 = 0.67a$, where $a = d_1 + d_2$ is the thickness of one pair of high- and low-index layers, which we will refer to as a *bilayer*. A schematic view of this structure is shown in the inset of Fig. 10.6. We again consider monochromatic light of angular frequency ω. The translational invariance of this system in the z-direction (the direction indicated by the β vector in the inset), like the waveguide situation considered earlier, suggests that the electromagnetic fields of this system will have a z-dependence $e^{i\beta z}$. For some ω and β pairs, light can propagate through the mirror; for others, light decays exponentially in the mirror. The ω–β plane can be divided into regions called *bands,* corresponding to propagation through the mirror, and regions called *band-gaps,* corresponding to exponential decay through the mirror. Light incident from the outside (which will also have a time and z dependence of $e^{i(\beta z - \omega t)}$) on a half-infinite mirror will be partially transmitted for the ω, β pairs outside of the band-gaps of the mirror and will be 100% reflected for those ω, β pairs within the band-gaps of the mirror. We plot the band structure of the mirror in the left panel of Fig. 10.6. Band-gaps (the unshaded regions in the figure) appear in frequency ranges that are

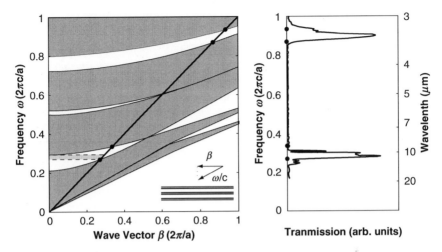

Figure 10.6 (Left) Band structure of the multilayer dielectric mirror depicted in the inset. The dark gray areas represent (ω, β) pairs for which light can propagate through the Bragg mirror. The dashed horizontal lines and the light gray area represent the frequency range of omnidirectional reflection. (Right) Measured transmission spectrum of a hollow-core OmniGuide fiber with a multilayer cladding with parameters very similar to those used for the simulation on the left.

approximately equidistant (e.g., at $\beta = 0$, corresponding to normal incidence on the mirror, they are centered at 0.25, 0.5, and 0.75). The fundamental band-gap also includes a frequency range of omnidirectional reflectivity (shown in light gray between dashed horizontal lines); for frequencies in this range, no propagating modes with $\omega > c\beta$ exist in the cladding, so light incident from air onto the mirror at any angle of incidence must be totally reflected.

An OmniGuide fiber with a diameter much larger than the bilayer thickness a has a transmission spectrum closely related to the band structure of the planar mirror. This can be seen in the right panel of Fig. 10.6, where we plot the measured transmission through a fiber with core radius of 250 μm and cladding layers made of As_2Se_3 (high index, $n_1 = 2.71$ at $\lambda = 10.6\,\mu m$, $d_1 = 1\,\mu m$) and polyether sulfone (low index $n_2 = 1.68$ at $\lambda = 10.6\,\mu m$, $d_2 = 2\,\mu m$). Note that the transmission peaks for a large core fiber will correspond to the portion of the planar band-gap close to the light line of air (delimited by the black dots in the figure). Modes with dispersion relations in this region have mostly grazing angles of incidence with the mirror (i.e., axial wave vectors close to ω/c) and will have the lowest losses. Note that the axial wave vectors in this figure are plotted in units of $(2\pi/a)$, while the angular frequency is plotted in units of $(2\pi c/a)$. The motivation for using dimensionless axes lies in the scale-invariance of Maxwell's equations: ignoring material dispersion, given a wave-guiding structure at one wavelength, a wave-guiding structure with identical properties at another wavelength can be obtained by scaling all dimensions by the change in wavelength [24]. Hence, there is no need (except for discussions of a specific application or where material dispersion is critical) to tie the results plotted to any particular wavelength or bilayer thickness.

A cylindrical OmniGuide fiber is obtained conceptually by wrapping a periodic dielectric mirror around a hollow core. Even though the optical properties of the dielectric mirror are somewhat modified by this change of geometry, the band structure of the planar mirror still largely determines the wave-guiding properties of the OmniGuide fiber. This can be understood by noting that at a large distance from the origin, the curvature of the layers is very small. Thus, an OmniGuide fiber with an infinite number of bilayers will prevent light from escaping the core region for those frequency–axial wave vector pairs that lie within a band-gap of the planar dielectric mirror. This confinement property supports guided modes within the fiber core and, sometimes, surface modes that have the majority of the field at the interface between the core and the cladding.

For an analysis of the mode structure of hollow-core OmniGuide fibers, we focus on a fiber with a small core radius, $R = 3a$, such that the number of guided modes in the fundamental band-gap is not exceedingly large. It is instructive to compare and contrast the modes of the OmniGuide fiber with those of a much simpler structure: a hollow metal waveguide having the same internal diameter (Fig. 10.7). Both waveguides confine light in the core by a reflective cladding, but the simplicity of the hollow metal waveguide allows for analytical solutions. Rearranging Eqs. (10.2a) and (10.2b), the bands have the form $\omega^2 = \omega_c^2 + c^2\beta^2$, where ω_c is the cutoff frequency. Starting from low frequencies, the modes of the metal waveguide are in order TE_{11}, TM_{01}, TE_{21}, TE_{01} and TM_{11} (degenerate modes), TE_{31}, and so on. The first of the two subscript indices is the angular index m of the mode and the second is the radial index n, introduced in Section 10.2.1. As can be seen in Fig. 10.7, the OmniGuide fiber supports guided core modes in the band-gap that have dispersion relations very similar to those of the metal waveguide [35]. The intensity patterns of the modes of the two waveguides are also similar in appearance. For the OmniGuide fiber, however, only modes with angular index $m = 0$ maintain their pure transverse electric or transverse magnetic character, while all other modes become hybrid (e.g., TE_{11} becomes HE_{11}). Also, as a result of the frequency-dependent phase shift upon reflection from the dielectric mirror, the OmniGuide modes are pulled towards the mid-gap frequency when compared to their metallic counterparts. Finally, the degeneracy of the TE_{01} and TM_{11} modes of the metal waveguide is lifted in the all-dielectric fiber.

Two important modes of an OmniGuide fiber, HE_{11} and TE_{01}, are analyzed in more detail in Fig. 10.8. The lowest guided mode, HE_{11}, is a doubly degenerate mode that resembles the fundamental mode of an index-guiding fiber. Its electric field distribution becomes very close to being linearly polarized for large-core

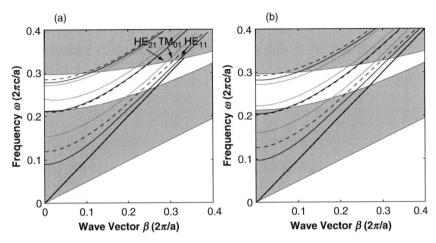

Figure 10.7 (a) Band diagram for an OmniGuide fiber with a core radius $R = 3a$. Modes with angular index $m = 0$ are shown with dashed lines, those with $m = 1$ are shown with solid black lines, and other modes are shown in gray. The thick diagonal line is the light line of air. The dark shaded region corresponds to fields that can propagate through the cladding (the bands of the planar mirror). (b) Band diagram for a hollow metal waveguide with an inner radius $R = 3a$. Starting from low frequencies, the modes are in order TE_{11}, TM_{01}, TE_{21}, TE_{01}, and TM_{11} (degenerate modes), TE_{31}, and so on. For comparison, the lightly shaded regions of this figure correspond to the dark shaded regions of (a).

diameters, so this mode couples very well with a linearly polarized input laser beam. The flux distribution for the HE_{11} mode shown in Fig. 10.8(c) is somewhat elongated in the direction parallel to the axis of polarization of the mode. Note, however, that for fibers with larger cores (radius larger than $30a$ or so), this asymmetry becomes much less pronounced and the flux distribution becomes almost azimuthally symmetric. The TE_{01} mode is a nondegenerate azimuthally symmetric mode for which the electric field is always oriented in the azimuthal direction, as can be seen in Fig. 10.8b. This property, together with the boundary condition at the interface between the core and the dielectric mirror, results in a very small overlap of the field of the TE_{01} mode with the cladding, as can be seen in Fig. 10.8d. As we will see in the following section, this reduction in field overlap with the cladding results in advantageous loss properties for the mode.

10.2.3 Loss Mechanisms in OmniGuide Fibers

After studying the band structure and basic mode properties of the Omni-Guide fiber, we now focus on the various loss mechanisms that arise in this type of hollow-core fiber.

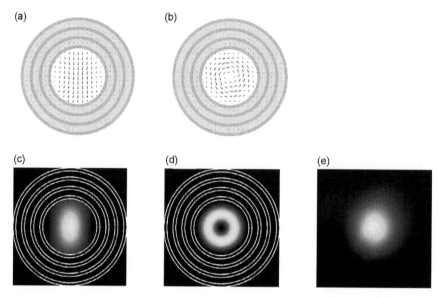

Figure 10.8 (a, b) Quiver plots of the transverse electric field distribution for the (a) HE_{11} (vertical polarization) and (b) TE_{01} modes of the OmniGuide fiber with radius $R = 3a$. The outlines of the dielectric interfaces of the cladding are shown as light gray circles. The field of the TE_{01} mode is azimuthally symmetric and in the azimuthal direction, while that of the HE_{11} mode is closer to a linearly polarized mode. (c, d) The axial component of the Poynting vector for the HE_{11} and TE_{01} modes is plotted as a grayscale surface plot with lighter shades indicating larger intensity. (e) Measured far-field intensity profile at the output of an OmniGuide fiber with $R = 80a$, excited by a linearly polarized laser beam.

10.2.3.1 Effect of a Finite Mirror: Radiation Loss

With an infinite number of layers, the cladding of an OmniGuide fiber forms a perfect reflector. Any realistic OmniGuide fiber, of course, will only contain a finite number of layers in its cladding, which results in an intrinsic radiation leakage loss. Fortunately, this radiation loss, which comes from the tiny exponential tails of the field profiles in the outermost cladding, scales exponentially with the number of cladding layers and can easily be made negligible when dielectric contrast is high.

More quantitatively, Fig. 10.9 shows the radiation loss spectrum of several modes in an OmniGuide fiber with a core radius of $80a$ and with a cladding composed of 25 layers with alternating indices of refraction $n_H/n_L = 2.7/1.6$ and alternating thicknesses $d_H/d_L = 0.33a/0.67a$. The units of dB/m on the vertical axis correspond to the specific choice $a = 3.2\,\mu m$, appropriate for a fiber with a desired operating wavelength of $10.6\,\mu m$. More generally, the units on the vertical axis will scale inversely with wavelength as the bilayer thickness is

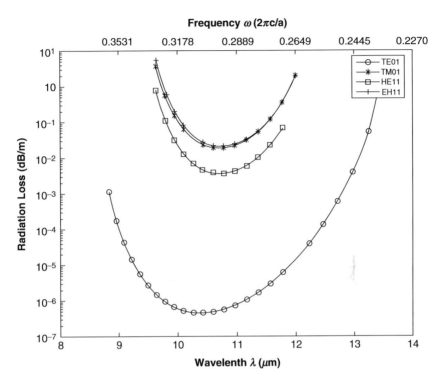

Figure 10.9 Radiation loss spectrum for OmniGuide fiber with mid-gap wavelength $\lambda = 10.6\,\mu\text{m}$, due to leakage through 25 layers, for a core radius of $80a$ $(24\,\lambda)$.

scaled for operation at a different wavelength. Each of the loss curves has a characteristic shape; the losses are at a minimum near the middle of the band-gap, where the exponential confinement of the field is strongest, and increase towards the gap edges. The TE_{01} mode (and any TE_{0n} mode) is different from the others; because this mode is purely TE polarized, it sees the larger TE band-gap of the corresponding planar mirror and hence has larger bandwidth and smaller radiation loss [24]. The other modes (the TM modes and the HE and EH hybrid modes) have a TM component and, hence, are limited by the smaller TM gap of the planar mirror, with consequently smaller bandwidth and larger radiation loss.

Regardless of the mode polarization, however, the radiation loss decreases exponentially as one increases the number of layers, simply because the fields are exponentially attenuated within the multilayer cladding. This is apparent in Fig. 10.10, which shows the loss at the roughly mid-gap frequency as a function of the number of layers. The TE-mode losses decrease by about a factor of four per bilayer at this index contrast ($n_H/n_L = 2.7/1.6$), while the other modes

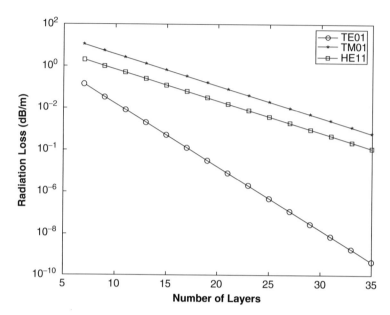

Figure 10.10 Scaling of radiation loss with number of layers in an OmniGuide fiber, at a wavelength $\lambda = 10.6\,\mu m$ for a core radius $80a$ ($24\,\lambda$).

(because of the smaller TM band-gap of the planar mirror) decrease by about a factor of two per bilayer. By 25 layers (12 bilayers), the HE_{11} mode has radiation losses less than 0.01 dB/m for these parameters, and in practice radiative losses are not the limiting factor for high-contrast material systems where sufficiently many layers are easily achievable.

In fact, the index contrast need not be so large to achieve negligible radiation loss. In Fig. 10.11, we show the number of layers required to obtain 0.01 dB/m losses for this core radius ($80a$) and wavelength (10.6 μm), as a function of the index difference Δn between the high-index and low-index material, for a fixed low-index material with $n = 1.6$. (The previous graphs correspond to $\Delta n = 1.1$.) For each index contrast, we look at the glancing-angle mid-gap frequency $a/\lambda = (\tilde{n}_{hi} + \tilde{n}_{lo})/4\tilde{n}_{hi}\tilde{n}_{lo}$, where $\tilde{n} = \sqrt{n^2 - 1}$ (this gives roughly the minimum attenuation). As can be seen from the plot, even an index contrast of $\Delta n = 0.4$ is sufficient to require only 60 layers for HE_{11} and 25 layers for TE_{01}. On the other hand, a 1% contrast as in doped silica fibers would require hundreds or even thousands of layers.

The computation of the radiation losses is straightforward and can exploit any of several standard methods to implement open boundary conditions in electromagnetism. With finite-element [36] or finite-difference methods [37] that require a finite computational cell, one can use a perfectly matched layer (PML)

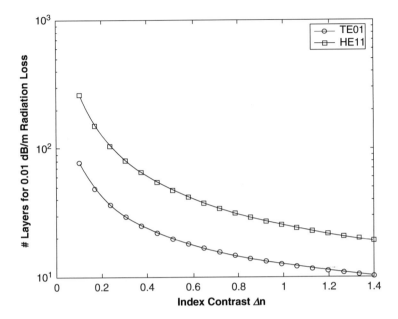

Figure 10.11 Number of layers to obtain 0.01 dB/m mid-gap radiation loss, at a wavelength $\lambda = 10.6\,\mu$m for a core radius 80a (24λ), as a function of the index contrast Δn between high-index and low-index ($n = 1.6$) layers.

to implement a reflectionless absorbing boundary to emulate radiative loss [37]. Alternatively, for multilayer systems, efficient transfer-matrix methods exist that allow one to directly solve for "leaky-mode" solutions [24]; complex-β modes for a real ω that satisfy the boundary condition of zero incoming waves from outside the fiber, where the imaginary part of β gives the radiation loss rate.[2]

10.2.3.2 Material Absorption Losses

Of more common concern is loss due to material absorption, especially at wavelengths like 10.6 μm where transparent materials are unavailable. Hollow-core fibers, by design, greatly suppress such losses because most of the light is

---

[2] There is some mathematical subtlety because such leaky modes are only an approximation for the true behavior of the system, but the accuracy of the leaky-mode approximation is well established in describing the loss rate and the field pattern near the waveguide as long as the loss is sufficiently small (Im $\beta \ll$ Reβ) [42].

confined in the hollow core, but one must still evaluate the losses that result from the small portion of the mode that penetrates the cladding. As is discussed later in this chapter, such losses decrease rapidly with the core radius and quickly become several orders of magnitude smaller than the bulk absorption rates of the cladding materials.

Although a variety of computational methods allow one to evaluate directly the electromagnetic modes in absorbing materials, modeled by a complex refractive index at a given wavelength, more general insights can be obtained by perturbative methods. In particular for all dielectric materials considered here, the imaginary part of the refractive index is much smaller than the real part ($<1\%$ even for the polymer at 10.6 μm), in which case the following formula is essentially exact [24, 38]:

$$\text{absorption loss} = \sum_{\text{Materials}} (\text{material bulk absorption loss}) \cdot (\text{fraction of } \varepsilon|E|^2 \text{ in material})/(v_g/c),$$

where (v_g/c) is the group velocity in units of c (usually $\sim$1.0 for large-core hollow fibers). For example, if the clad materials have a bulk absorption loss of 1000 dB/m and a mode of interest has 99.9% of its electric-field energy in the core and a velocity approximately c, then the waveguide's absorption loss for that mode will be only 1 dB/m. In fact, for the large-core fibers used to transmit 10.6-μm light, many modes have much more than 99.9% of their energy in the core and the bulk absorption losses can be suppressed by four or five orders of magnitude. The effect of material absorption on transmission loss in an Omni-Guide fiber is graphically demonstrated in Fig. 10.12, which compares the measured transmission spectrum of an OmniGuide fiber with the material absorption of the low-index component of the cladding (the polymer polyether sulfone [PES]). Dips in the transmission spectrum clearly correspond with peaks in the material absorption. The material absorption of the high-index component of the cladding is negligible relative to the low-index material.

Because the absorption loss arises from the small exponential tail of the field in the innermost mirror layers, it is not surprising that this loss is a rather sensitive function of the thicknesses of these layers. Small changes to their thicknesses, while having very little effect on the field structure in the core, can introduce a substantial change to the small exponential tail that determines the absorption loss. This is particularly evident in an OmniGuide fiber where the high-index and low-index materials have vastly different bulk absorption losses: a small change to the thickness of the first layer (assumed to be of a high-index, low-loss material), for example, can push substantially more of the exponential tail into the next layer (assumed to be of a low-index, high-loss material), sharply raising the loss. This effect should be mode dependent; a change that pushes more of the tail into the high-loss material for one mode may have the opposite effect on the field

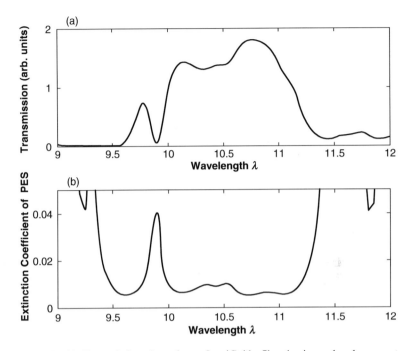

Figure 10.12 (a) Transmission through an OmniGuide fiber having a band-gap centered around 10.6 μm. (b) Extinction coefficient k of polyether sulfone (PES), the low-index material in the cladding. The extinction coefficient of the high-index material is negligible in comparison.

distribution and loss for a different mode. Figure 10.13 shows the dependence of absorption loss on the first layer thickness for several modes of an OmniGuide fiber with indices $n_H = 2.7 + 1.94 \times 10^{-6}i$, $n_L = 1.6 + 7.77 \times 10^{-3}i$, and layer thicknesses $d_H = 1.05\,\mu$m, $d_L = 2.10\,\mu$m. The imaginary parts of the indices of refraction correspond to bulk absorption losses of 10 dB/m and 40,000 dB/m, respectively, at the operating wavelength of 10.6 μm. The core radius of the fiber is 250 μm. The horizontal axis gives the ratio of the first layer thickness to d_H, the thickness of all other high-index layers.

The dependence of mode loss on the thicknesses of the innermost layers in general, and the first layer in particular, strongly suggests that the structure of an OmniGuide fiber be optimized for particular applications. For example, in applications where the HE_{11} mode is the desired operating mode, the Omni-Guide fiber described earlier should be constructed with a first-layer thickness of about one-third the thickness of the other high-index layers (which is also close to optimal for the TM_{01} and EH_{11} modes). Alternatively, if the TE_{01} mode is the desired operating mode, all high-index layers should have the same thickness.

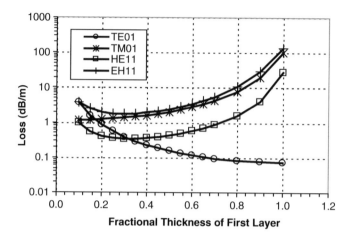

Figure 10.13   The dependence of the losses at $\lambda = 10.6\,\mu$m on the thickness of the first layer for several low-order modes in the OmniGuide fiber described in the text.

Figure 10.14 shows the absorption loss suppression (absorption loss/bulk absorption) as a function of wavelength (and dimensionless frequency) for several modes in an OmniGuide fiber with a core radius of $80a$ ($\sim$24 wavelengths). The thickness of the first layer (high index) is taken to be half the thickness of the other high-index layers. As for the radiative loss, the TE_{01} mode has the lowest absorption loss and the widest bandwidth; this is not only because of the larger TE band-gap in the corresponding planar mirror but also because of boundary-condition considerations discussed later in this chapter. Even the HE_{11} fundamental mode, however, suppresses absorption losses by five orders of magnitude for this core radius.

This absorption-loss suppression is a dimensionless quantity; it is invariant as the whole structure is rescaled with wavelength. It is also almost independent of the number of layers, because the exponential field decay in the cladding means that almost all of the contribution to absorption loss comes from the first few layers. It does depend on two other quantities, however: the index contrast and the core radius. One would expect, for example, that the larger the index contrast, the stronger the confinement of light in the core and the greater the absorption suppression. The dependence on index contrast Δn for a fixed low-index material ($n = 1.6$), however, turns out to be fairly weak, as shown in Fig. 10.15; halving the index contrast increases absorption loss by less than a factor of two. Therefore, the use of more lossy materials to obtain greater index contrast at some point ceases to be advantageous. Index contrast is more important in determining bandwidth and radiation loss, as described in the previous section.

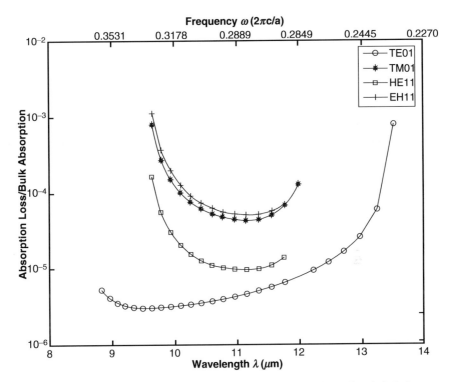

Figure 10.14 Absorption loss spectrum, normalized to the bulk polymer (low-index) absorption loss, for a fiber with mid-gap wavelength of 10.6 μm, a core radius of $80a$ (24λ), and 25 layers.

Another way to reduce absorption loss is to increase the core radius R, and there is a general argument that the absorption losses will decrease as $1/R^3$. This is demonstrated directly in Fig. 10.16, which shows the scaling of the absorption loss versus core radius at the mid-gap frequency.

The origin of this $1/R^3$ behavior, which has been observed in many other systems including hollow metallic waveguides [39], can be seen in an analysis of the field-energy fraction in the cladding. Since the depth of the field penetration into the cladding is determined by the band-gap and is independent of R, a simple surface-area/volume analysis would suggest that the fraction of the field in the cladding and, hence, the absorption, scales as $1/R$. This neglects the boundary conditions on the field, however; similar to a metallic waveguide, the TE$_{01}$ mode's electric field (azimuthally polarized) goes approximately to zero at the boundary of the core. Because of this, its **E** amplitude in the cladding scales

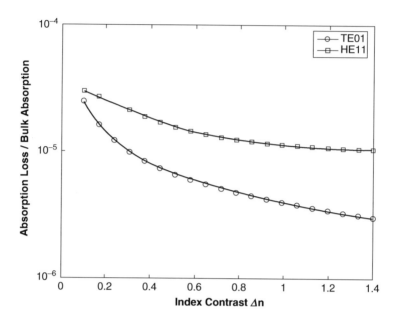

Figure 10.15 Absorption loss, normalized to the bulk polymer (low-index) absorption loss, for an OmniGuide fiber with core radius of $80a$, as a function of the index contrast Δn between high-index and low-index ($n = 1.6$) layers.

as the *slope* dE/dr at R, which goes as $1/R$ for a given mode pattern. This gives an additional $1/R^2$ factor in the $\varepsilon |E|^2$ energy in the cladding and, thus, the $1/R^3$ behavior of the absorption loss [24]. The explanation of the $1/R^3$ behavior for the non-TE modes, for which the analogous metallic boundary condition does *not* force a node in the field at R even for metal waveguides, is more subtle. In the large-R limit, the penetration of the field into the cladding becomes negligible compared to R or to the transverse wave vector $\sqrt{\omega^2/c^2 - \beta^2}$. One can, therefore, employ a *scalar* approximation of the field in the interior (where the index is uniform) and simply impose *zero* boundary conditions at R instead of the finite penetration (this is related to the LP mode categorization that applies for doped-silica fibers with low-index contrast [40] and is mirrored by a similar scalar approximation that applies in the high-frequency limit for holey silica fibers [41]). Given this scalar approximation, the $1/R^3$ behavior follows as for the TE mode. Because this boundary condition only applies asymptotically, however, the TE mode tends to have an advantage in the strength of its field confinement. Also, for smaller R, one can see that the TM and HE mode losses in Fig. 10.15 begin to flatten; the losses are starting to change from $1/R^3$ scaling

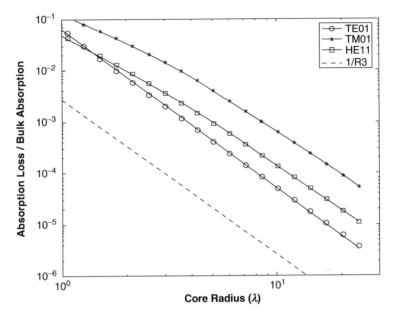

Figure 10.16 Absorption loss, normalized to the bulk polymer (low-index) absorption loss, versus core radius, for 25 layers at mid-gap frequency. The dashed line shows a pure $1/R^3$ dependence, for comparison; all of the curves asymptotically approach this slope.

to the $1/R$ scaling that one expects in the absence of zero-field boundary conditions.

In fact, similar $1/R^3$ scaling occurs for any other loss phenomenon associated with the penetration of the field into the cladding, including radiation loss and surface-roughness scattering, for exactly the same reasons.

10.2.3.3 Loss Discrimination and Single-Mode Behavior

A waveguide with a large hollow core, whether metallic or dielectric, is intrinsically multimode. However, in practice the observed behavior is typically that all of the power is guided in one or a few modes. The reason for this can be seen in the absorption loss spectra of Fig. 10.14: different modes have different penetrations into the cladding and consequently greatly differing losses. This causes the high-loss (high-order) modes to be filtered out, with the effective number of modes being determined by the ratio of the intermodal coupling to the differential loss rates. In early work on microwave communications via metallic waveguides, precisely this loss discrimination was employed to preferentially guide light in the low-loss TE_{01} mode [39].

10.2.3.4 Bending Loss

The wide utility of fiber waveguides derives partly from their ability to bend and guide optical energy to positions that are not in the direct line of sight of the optical source. However, the bending introduces coupling between the low-order and low-loss operating mode (usually the HE_{11} or TE_{01} mode) and higher order, higher loss modes are supported by the waveguide. Thus, the bend causes additional attenuation of the guided field. It is also possible to couple energy into reflected fields, but this generally requires bends with extremely small radii of curvature that cannot be implemented without breaking the fiber. Coupled-mode theory [42] provides the simplest and most efficient framework for evaluating the additional attenuation associated with a bend. Maxwell's equations are expressed in terms of a Serret–Frenet coordinate system based on the curve defined by the axis of the fiber [43, 44]. These equations can be separated into a portion that has the same form as the equations of a straight waveguide and a perturbation, proportional to the inverse of the radius of curvature of the bend. The field in this perturbed waveguide is represented as an infinite series of modes (in general containing both forward and backward propagating components) of the straight waveguide with amplitudes that depend on arc length along the fiber axis. Following standard manipulations, the orthogonality of the modes is employed to derive an infinite set of coupled ordinary differential equations for the arc length–dependent mode amplitudes. In practice, coupling to higher order modes decays rapidly with mode order, so this infinite system can be truncated to obtain a computationally tractable problem. A perfectly matched layer, tailored for the bent waveguide [45], can be used to simulate radiation loss.

A straight circular waveguide with a confinement mechanism that involves dielectric materials supports TE, TM, and hybrid (HE, EH) modes, where each hybrid mode can have two degenerate polarizations. For planar bends (i.e., where the axis of the fiber remains in a plane), the degenerate polarizations of the hybrid modes are not coupled by the bend. One hybrid polarization, corresponding to almost linear polarization perpendicular to the plane of the bend, couples only to other hybrid modes with the same polarization and to the TE modes. The second hybrid polarization, corresponding to almost linear polarization in the plane of the bend, couples only to other hybrid modes with the same polarization and to the TM modes. We refer to these two polarization states as the "low-loss" and "high-loss" polarizations, respectively.

Figure 10.17 shows the theoretical local attenuation, defined as the derivative with respect to arc length of the axial power flow, of the electromagnetic field in an OmniGuide fiber as it propagates through a 90-degree circular bend with radius 10 cm. The bend used in this coupled mode calculation is obtained by winding the fiber, modeled as a stiff beam, under tension around a cylindrical mandrel. The fiber cladding is composed of 40 alternating layers with indices of

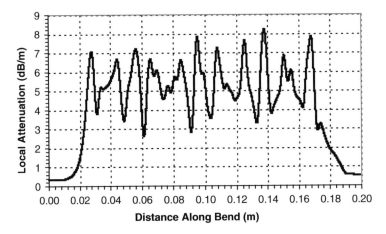

Figure 10.17 The local attenuation (defined as the derivative of the axial power flow) in an OmniGuide fiber (described in the text) as a function of position along a circular bend with 10-cm radius. An HE_{11} mode in the high-loss polarization is incident on the bend.

refraction $n_H = 2.7 + 1.94 \times 10^{-6}i$, $n_L = 1.7 + 7.77 \times 10^{-3}i$, corresponding to a chalcogenide glass and a polymer, respectively. The imaginary parts of the indices correspond to bulk absorption losses of 10 dB/m and 40,000 dB/m, respectively. The high- and low-index layers have thicknesses 0.93 and 1.87 μm, but the first high-index layer has half the thickness of the other high-index layers. The core radius of the fiber is 250 μm, and the operating wavelength is 10.6 μm. An HE_{11} mode with the high-loss polarization (i.e., in the plane of the bend) is incident on the bend.

Initially, only the HE_{11} mode is present, so the local attenuation is constant and small. The fiber then undergoes a rapid transition from straight to bent, and the local attenuation rises rapidly. It then exhibits a number of peaks and valleys, corresponding to destructive and constructive interference of all the modes that get excited by the bend. The fiber then makes a second transition from bent back to straight, and the loss drops quickly. Note that the loss level reached after the end of the bend is higher than the loss level before the bend; some of the higher order and higher loss modes excited by the bend continue to propagate, raising the local attenuation.

Figure 10.18 shows the predicted bend loss, obtained via coupled-mode theory, for an OmniGuide fiber following a 90-degree circular bend as a function of the radius of curvature of the bend. The bend loss is defined to be the difference between the loss (in dB) after propagation through the bend and the loss after propagation through a straight fiber of the same total arc length. The structure of the fiber used in this calculation is the same as Fig. 10.17. The

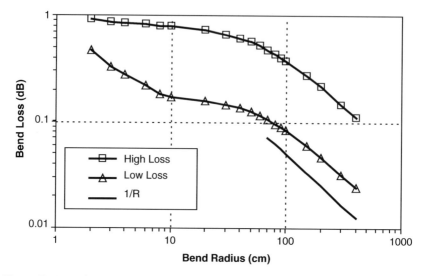

Figure 10.18 Variation of the calculated bending loss with bend radius at a wavelength of 10.6 μm for the OmniGuide fiber described in the text. The marked curves show the bend loss associated with an incident HE_{11} mode in the "high-loss" and "low-loss" polarizations. For comparison, the unmarked curve displays $1/R_{Bend}$ behavior.

bend losses for an incident HE_{11} mode in the low-loss and high-loss polarizations are given by the curves marked with open triangles and open squares, respectively.

The strength of the coupling to higher order modes scales inversely with the bend radius. However, the arc length of the bend scales directly with the bend radius so the total bend loss, which scales as the product of these two effects, should be roughly independent of bend radius, at least for relatively small-radius bends that couple many modes together. At higher bend radii (in this example, more than ∼100 cm), where very few modes are coupled by the bend (i.e., where second-order perturbation theory in $1/R_{Bend}$ becomes applicable), the coupling to higher order modes scales with the inverse of the square of the bend radius [24], so the bend loss becomes proportional to $1/R_{Bend}$. The bend loss for the high-loss polarization generally follows this two-regime dependence on bend radius. The bend loss for the low-loss polarization also follows this same general behavior, although some deviation is observed for small bend radii (in this example, less than ∼10 cm).

The theoretical bend attenuation evaluated with coupled-mode theory for a hollow metal waveguide, with a single-layer dielectric coating, is shown in Fig. 10.19, along with measured results [32]. The fiber core radius is 125 μm and the incident field has the low-loss polarization (i.e., almost perpendicular to

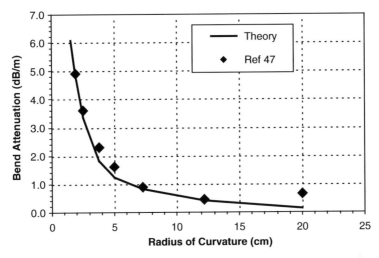

Figure 10.19 Theoretical (solid curve) and measured [32] bend attenuation of a hollow metal waveguide coated with a single dielectric layer of AgI. The hollow-core radius is 125 μm.

the plane of the bend). Note that bend attenuation is obtained by dividing the bend loss by the arc length of the bend; with the removal of the effect of the arc length, the bend attenuation scales with the inverse of the bend radius. The generally good agreement between theory and measurement attests to the validity of the coupled-mode analysis.

Figure 10.20 displays the variation of bend loss with core radius for an OmniGuide fiber with the same cladding as in Figs. 10.17 and 10.18. The bend radius and wavelength used in this calculation are fixed at 10 cm and 10.6 μm, respectively, while the bend angle is kept at 90 degrees. Two competing factors determine the dependence of bend loss with core radius. As the core radius increases, the losses of all modes propagating in the bend decrease (like $1/R_{core}^3$ as discussed in Section 10.2.3.2), which tends to reduce the bend loss. However, the strength of the coupling to higher order modes increases with core radius, which tends to increase the bend loss. The net effect is shown in Fig. 10.20; the bend loss initially increases with core radius but eventually levels off.

10.2.4 Wave-Guiding in 2D Photonic-Crystal Fiber

Another form of periodic cladding that supports band-gaps and can, therefore, serve to guide modes in a hollow core is a 2D periodic arrangement of materials with a significant index contrast. In particular, the most common

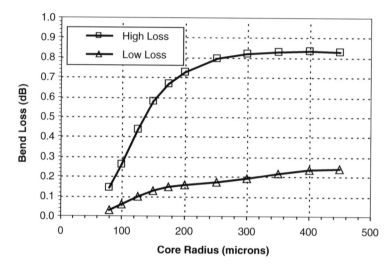

Figure 10.20 Variation of calculated bend loss with core radius for the OmniGuide fiber described in the text. The bend radius and wavelength are fixed at 10 cm and 10.6 μm, respectively.

realization of such a 2D crystal is a triangular lattice of air holes in silica glass, which was first demonstrated to guide light in a hollow core by Knight et al. [26]. Such "holey" fibers have the advantage that they are only fabricated from a single material, conventional silica; the main disadvantage is the lack of cylindrical symmetry, which allows modes to couple to one another more easily and, as is described later in this chapter, makes large-core holey fibers more susceptible to problems with surface states.

Conceptually, the analysis of holey PCFs follows the same outline as in Fig. 10.21. First, we evaluate the band diagram of the infinite cladding, plotting all possible propagating states as a function of frequency ω and axial wave vector β; this band diagram is called the "light cone" of the crystal (although it is only a cone for a homogeneous material). Then, if we find gaps—regions in which there are no solutions that propagate in the infinite cladding—above the air light line, we can use the photonic crystal as a reflective cladding to guide light in a hollow core. This analysis is more complicated than for Bragg fibers because the computations are inherently 2D instead of 1D but can be performed in a few minutes on a workstation, for example, by a plane-wave–expansion method [46].

In particular, we consider the structure shown in the inset of Fig. 10.21, a triangular lattice (with period a) of air holes (radius $0.47a$) in silica ($n = 1.45$), similar to fabricated structures [22]. The light cone and band-gaps of this infinite

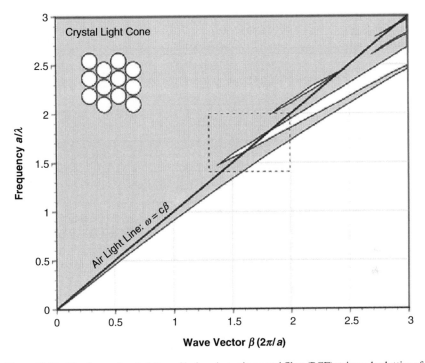

Figure 10.21 Band-gaps for cladding of holey photonic-crystal fiber (PCF): triangular lattice of air holes (period a, radius $0.47a$) in silica ($n = 1.45$). Gray shading indicates regions where there exist propagating solutions in the PCF cladding, whereas open spaces indicate regions where cladding propagation is forbidden (the band-gaps). The presence of band-gaps above the light line of air indicate that the PCF cladding can produce confinement in a hollow core. Dashed box indicates region that is plotted in Fig. 10.22.

holey cladding are shown in Fig. 10.21. In addition to the space below the crystal's light cone (corresponding to conventional TIR guiding), there are finger-like gaps extending to the left, most of which open monotonically as β goes to infinity. These are the band-gaps, and it can be shown that they *always* appear for sufficiently large β, regardless of the hole radius or the index contrast. However, most of these gaps are useless for guiding in a hollow core; only the gaps that lie above the air light line ($\omega = c\beta$) are applicable. Below the air light line, the field would decay exponentially in the core, rather than propagate. Therefore, we focus on the region of the first (fundamental) gap that lies above the air light line, and in particular the region outlined by the dashed lines in Fig. 10.21.

Given the band-gap of the perfect cladding, we expect that it will support confinement of guided modes in a hollow core, which we form by removing from the crystal any dielectric within a radius 1.2*a* or 1.4*a*. The resulting structures are shown in the insets of Fig. 10.22. The former (core "radius" 1.2*a*) is similar to a structure that was fabricated by Smith et al. [22]. The latter (core "radius" 1.4*a*) differs not only by being larger, but also by having a

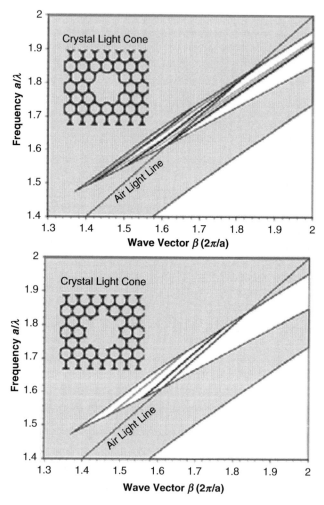

Figure 10.22 Guided modes supported by a hollow core in the holey photonic-crystal fiber (PCF) of Fig. 10.21. (Top) A hollow core created by removing any dielectric up to a radius of 1.2*a*. (Bottom) A hollow core with a slightly larger radius 1.4*a*. Guided modes that have the right symmetry to couple to linearly polarized input light are shown as dark black lines, while other symmetry modes are shown as gray lines.

different *termination* of the surrounding crystal, which has beneficial effects predicted by West et al. [47].

Figure 10.22 plots the dispersion relations of the guided modes in these two structures. It is remarkable that the *larger* core structure has *fewer* modes, in contrast to the behavior of TIR, OmniGuide, and hollow metal fibers, and this is due to the influence of the crystal termination—what part of the crystal lies at the core boundary. For certain crystal terminations, it is known that the crystal supports *surface states* [14]; these are modes that lie within the band-gap but below the air light line and are thus localized around the *boundary* of the crystal rather than within the core. Such states below the air light line are visible in Fig. 10.22 (top). These states continue as they cross the air light line and influence the other guided modes, and the result is that the radius $1.2a$ core exhibits a large number of guided modes that cross one another. In contrast, by changing the termination, in Fig. 10.22 (bottom), we see that *all* of the surface modes disappear, and only a small number of modes remain above the air light line: two linearly polarized modes and four nearly degenerate modes of different symmetry (similar to an LP_{11} mode).

The intensity patterns for some of these modes are shown in Fig. 10.23. At the top right, note the surface-localized pattern of the surface state lying below the light line. Even the other two modes of the $1.2a$ radius core at top left and middle, which lie above the air light line, have a significant fraction of their energy localized near the cladding; only about 50% of their energy is localized within the $1.2a$-radius core, corresponding to a factor of two suppression of, for example, cladding absorption loss, and a strong sensitivity to surface roughness. Moreover, where the guided mode crosses over modes of other symmetry in Fig. 10.22, any symmetry-breaking perturbation can couple the modes and increase losses [47]. This depends sensitively on the surface geometry, however. If we consider the fundamental mode in the $1.4a$ radius core, at bottom left, it is almost entirely localized in the air; 95% of its energy is within the $1.4a$ radius core, corresponding to more than an order of magnitude suppression of cladding absorption loss and much less sensitivity to surface effects. A different symmetry mode in the same $1.4a$ core is shown at the bottom right of Fig. 10.23, resembling the TE_{01} mode of a cylindrical metallic waveguide.

To further reduce losses, for example to match the 10^5 suppression of cladding absorption for the multilayer fibers of the previous sections, and more importantly to reduce scattering from surface roughness, one would have to increase core radius. Similar $1/R - 1/R^3$ scaling laws should apply. However, as core radius is increased, the number of surface states also increases, unless care is taken with the crystal termination. In this way, Mangan et al. [23] were able to reduce the losses from 13 to 1.7 dB/km by increasing the core radius, but because the number of surface states increased, the contiguous bandwidth (determined by

$\beta = 1.6 \; 2\pi/a$ $\beta = 1.7 \; 2\pi/a$

$\beta = 1.6 \; 2\pi/a$

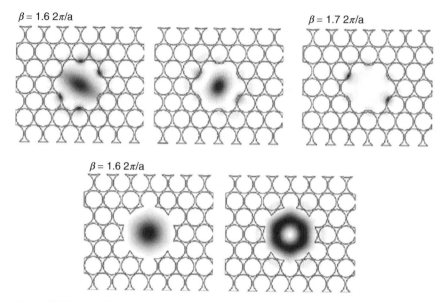

Figure 10.23 Intensity patterns of modes supported by hollow cores in the holey photonic-crystal fiber (PCF) of Fig. 10.22. (Top) Three modes in the $1.2a$-radius core at $\beta = 1.6$ and $\beta = 1.7$, where the latter is a surface state, corresponding to the solid black modes of Fig. 10.22 (top). (Bottom) Two modes in the $1.4a$-radius core at $\beta = 1.6$, corresponding to a linearly polarized mode (left) and an "azimuthally polarized" TE_{01}-like mode (right).

the frequency separation of the surface-state intersections with the fundamental guided mode) was also decreased.

All hybrid modes of a circular dielectric waveguide, including the fundamental HE_{11} mode, have two orthogonal polarizations that travel at the same speed (i.e., are degenerate), but at different speeds when the circular symmetry is broken by imperfections. This phenomenon produces birefringence and polarization mode dispersion (PMD). A very similar phenomenon occurs for PCFs, but the analysis of the symmetry is much more complicated because of the sixfold symmetry of the fiber structure. In particular, it might seem impossible that one would have pairs of degenerate modes, because the structure does not have 90-degree rotational symmetry; one cannot simply rotate a mode by 90 degrees and get an independent mode solution. Nevertheless, pairs of degenerate modes are precisely what one gets with sixfold symmetry (and never trios or sextuples of degenerate modes). This result follows from group representation theory; technically, the structure has the C_{6v} symmetry group, and there are six possible representations of this group that modes can fall into, one of which is the doubly degenerate "two orthogonal polarizations" case [48]. In simple terms, one degenerate mode is the average of 60- and 120-degree rotations of the other mode. We will not go into more detail here but will simply note that the existence of these doubly degenerate modes means that PMD and birefringence occur when the symmetry of a PCF is broken; indeed, these effects occur much more strongly because the index contrast is so much larger than in traditional doped-silica fiber. Most experimental PCF work has relied on such doubly degenerate modes, because they are the fundamental modes of the fiber. Alternatively, it may be possible to employ a higher order nondegenerate operating mode analogous to the TE_{01} mode of Bragg fibers or metal waveguides.

10.3 APPLICATIONS OF HOLLOW-CORE FIBERS

10.3.1 Hollow-Core Fibers for Medical Applications

Optical fibers are used in medicine for a diverse range of applications, from sensing and diagnostics to therapeutics [49]. The main advantage of hollow-core fibers in medicine is their ability to transmit wavelengths for which traditional solid-core fibers are not transparent. For example, traditional silica-based fibers cannot transmit wavelengths above approximately 2.1 μm because the material absorption of silica becomes too large. These longer wavelengths, such as the Er:YAG laser wavelength at 2.94 μm and the CO_2 laser wavelength at 10.6 μm, have significant clinical advantages over the shorter wavelengths accessible to silica-based fiber.

The interaction of light and the different components of biological tissue encountered in medical applications is determined mainly by the absorption of the light by these components. For wavelengths above 2 μm, the absorption of water, the main constituent of biological tissue, rises sharply. This strong absorption of laser light by water can be used for very efficient cutting and ablation of soft tissue. The operating wavelengths of CO_2 lasers (10.6 μm) and Er:YAG lasers (2.94 μm) offer particularly strong water absorption and are, therefore, especially well suited for precise cutting and ablation; the strong absorption prevents the laser energy from penetrating tissue much beyond the point of application. In contrast, wavelengths less than 2 μm penetrate as much as 1 cm beyond the intended target. CO_2 lasers, in particular, are reliable and commercially available and have been used in medicine for more than 30 years [50]. In addition, the CO_2 laser wavelength offers excellent coagulation capability to stop bleeding during laser-based surgery, a capability not shared by the Er:YAG laser wavelength. This unique combination of precise cutting, limited penetration beyond the point of application, and coagulation is found only with the CO_2 laser wavelength, making it the optimal wavelength for many surgical procedures.

In spite of the advantages associated with CO_2 lasers, their use in medicine has been relatively limited because of the lack of a flexible medium to transmit the laser power to a target within the human body. Without a flexible delivery medium, the use of the CO_2 laser was confined to procedures in which a direct line of sight could be established between the laser and the target, primarily in dermatology and ear, nose, and throat (ENT) surgery with a rigid laryngoscope. Thus, since 1976, various academic and commercial groups have attempted to develop alternative waveguides and fibers that would allow the use of CO_2 lasers in a much wider variety of medical applications. For the most part, these attempts have been met with only limited success [51]. Products were developed that used either complicated and bulky articulated arms or restricted line-of-sight micromanipulators. Some commercial hollow metallic waveguides were developed (e.g., by Sharplan, Luxar, and Clinicon), mainly as an articulated-arm substitute. Some solid-core fibers were also developed, based on materials that are relatively transparent in the mid-IR [51], but these fibers proved either too lossy or too brittle.

Hollow-core fibers using an omnidirectionally reflective 1D photonic bandgap to confine light [25] have been developed and used successfully in a number of minimally invasive procedures with CO_2 lasers [52–54]. These OmniGuide fibers, as can be seen in Fig. 10.24, are flexible enough to be introduced into the body through flexible endoscopes, delivering CO_2 laser radiation to regions not accessible by such lasers before. In one particularly interesting example, an OmniGuide fiber was used through the working channel of a flexible endoscope

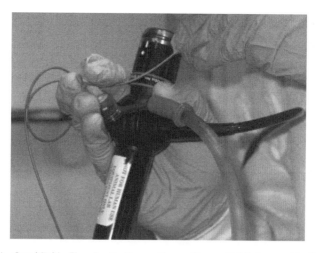

Figure 10.24 OmniGuide fiber (inside the small gray fiber cable) being used inside a flexible endoscope for minimally invasive surgery.

to successfully remove a large malignant tumor from the bronchus intermedius of a patient with lung cancer [53].

10.3.2 Potential Telecom Applications

Various oxide glasses have material absorption losses below 100 dB/km and indices of refraction near 1.6 at the typical telecom wavelengths. When combined with the high-index glass formed from As_2Se_3, a very low loss OmniGuide fiber can be produced. Indeed, with approximately 20 bilayers and the proper choice of the hollow-core radius, the theoretical loss for the TE_{01} mode is below 0.01 dB/km [24]. Losses this low would reduce the number of relatively expensive amplifier modules needed in long-haul systems, changing the fundamental economics of this market. In addition, nonlinear effects in the hollow core would be greatly reduced, allowing for higher launch power and dense wavelength division multiplexing without the usual concerns raised by four-wave mixing.

Johnson et al. [24] proposed that a relatively large hollow-core radius be employed in OmniGuide fiber for telecom applications to achieve the benefits of extremely low loss and nonlinearity. Of course, an OmniGuide fiber with this core size is multimoded but, because of large differential mode attenuation, behaves as essentially single moded, with the lowest loss mode serving as the operating mode. The lowest loss mode for OmniGuide fiber is the TE_{01} mode; the use of a large core radius, therefore, necessitates the use of this mode as the operating mode. The TE_{01} mode, unlike the doubly degenerate HE_{11} mode, has the advantage of being free from PMD but does not couple at all to the TEM_{00} field output from conventional telecom lasers. Thus, its use in telecom systems will require the coincident development of efficient $HE_{11} \rightarrow TE_{01}$ mode converters. In addition, this proposed telecom fiber would need to have very low surface roughness to prevent scattering of power from the operating mode to higher order lossy modes.

Two-dimensional, hollow-core PCF also offers the possibility of extremely low loss and nonlinearity. In addition, because the cladding matrix of this fiber is low-loss silica glass, there is no need for the use of a large core to achieve the desired loss level; true single-mode behavior can be obtained, with the doubly degenerate HE_{11}-like operating mode coupling efficiently to the output from typical telecom lasers. On the other hand, as discussed in Section 10.2.4, this fiber supports lossy surface modes, which couple easily to the operating mode and provide an additional loss mechanism [47]. In addition, the double degeneracy of the operating mode will produce PMD in the presence of perturbations that split the degeneracy. The complexity of the 2D photonic-crystal mirror, with attendant difficulty in maintaining the desired structure during manufacture, enhances the likelihood of the existence of perturbations that both induce coupling between the operating and surface modes and split the degeneracy of the operating mode, giving rise to PMD.

The wealth of parameters that can be modified to fine-tune the confinement mechanism in OmniGuide fiber, combined with the virtual absence of material dispersion, suggest that OmniGuide fiber can be tailored to have certain desirable dispersion properties. Engeness et al. [55], for example, describe a simple modification to the multilayer cladding structure of an OmniGuide fiber to obtain a dispersion-compensating fiber (DCF) with a theoretical dispersion/ loss figure of merit up to five times larger than conventional DCFs. Similar design freedom in 2D PCF has also been exploited to provide theoretically high-performance DCF [56–57]. (Hollow-core 2D PCF for dispersion compensation is also commercially available. See, for example, www.crystal-fibre.com/products/airguide.shtm.)

10.3.3 Hollow-Core Fibers as Gas Cells

Hollow-core fibers also find application as gas cells. Interaction of light with gases leads to a variety of complex and important phenomena of both academic and commercial interest in fields such as nonlinear optics [58], chemical sensing [59], quantum optics [60], and frequency measurement [61]. Because of the low density of gaseous media, maximization of interactions with light is often a challenge. Typically, the signature strength of the phenomena being explored increases with the intensity of the probing beam and with the interaction length. To enhance intensity in a conventional gas cell, one can focus the laser beam more tightly. Unfortunately, that leads to a smaller diffraction length, which in turn implies shorter interaction lengths.

The use of hollow-core fibers as gas cells eliminates these restrictions on intensities and interaction lengths. If the laser beam is confined as a guided mode of a hollow-core fiber, diffraction is prevented, and one can explore very small beam profiles and, therefore, very high intensities. Moreover, simply for practical considerations, lengths of conventional gas cells are typically limited to less than approximately 1 m. In contrast, the lengths of gas cells implemented via hollow-core fibers are much less limited in that respect; the only fundamental limitation is the transmission loss of the fiber. For example, losses of approximately 0.3 dB/m allow for propagation lengths of about 10 m. In addition, as we have seen in the previous sections, hollow-core fibers can often be bent into loops of fairly small radii, thereby enabling very compact gas cells that nevertheless have very long interaction lengths.

In chemical sensing applications [62–64], hollow-core fibers offer significant benefits over conventional gas cells because they require a drastically smaller volume of the sample gas. Moreover, the response time is faster since such smaller volumes fill up faster. For example, Charlton et al. [65] used a photonic band-gap hollow-core fiber to detect ethyl chloride at concentration levels of 30 ppb with a sample volume of 1.5 ml, and a response time of 8 seconds, representing an increase in sensitivity by almost three orders of magnitude over conventional gas cells.

Hollow-core fibers have also enabled many impressive accomplishments in nonlinear optics. For example, stimulated Raman scattering has been observed in hollow-core PCFs filled with hydrogen gas, with up to 92% quantum efficiency and threshold pulse energies six orders of magnitude lower than what was previously reported [66, 67]. In addition, slow-light effects have been observed in acetylene-filled hollow-core PCFs [68]; similar fibers filled with acetylene have also been used to implement frequency locking of diode lasers [69].

10.3.4 Applications of Hollow-Core Fibers for Remote Sensing

Hollow-core optical fibers can alternatively be used as passive elements to deliver radiation from a source being sensed or studied, to the remote sensor, which then analyzes the radiation. Applications of interest include temperature sensing (analysis of black-body radiation) [78] and chemical sensing [70]. This kind of remote sensing can be useful in hazardous environments (e.g., battlefields) or when the point of sensing is difficult to reach (e.g., inside a machine). It is also useful when many points within an environment need to be analyzed often, but not continuously (bottoms of oceans, industrial plants); in that case, the number of sensors that needs to be deployed is significantly reduced compared to the situation in which one sensor is placed at each point that needs monitoring. The motivation to use hollow-core fibers is that they can often transmit broader bandwidths and access wavelength regimes that would be difficult to explore using any other solid-core fiber (e.g., for chemical sensing, the 3- to 20-μm regime is of particular interest). Applications of interest include environmental sensing (pollution, etc.), homeland security, process monitoring, and biomedical sensing (e.g., breath analysis for detection of asthma).

10.3.5 Industrial Applications

CO_2 lasers are used in many industrial applications such as cutting, welding, and marking. It is a challenge, however, to bring the laser beam from the source to the working area, as the path is usually obstructed by other equipment. The use of articulated arms is the only method for beam delivery, even though these bulky systems require a large working space and use mirrors that require constant maintenance and alignment. Early efforts at using solid-core fibers to deliver high-power CO_2 beams have suffered from thermal damage, particularly at the air–fiber interfaces at the input or the output end of the fiber [71–73]. Some of the promising solid-core fibers also suffered from short lifetimes because of degradation of the guiding material when carrying high-power laser beams [71]. To overcome these problems, researchers started to investigate high-power CO_2 laser delivery systems based on hollow-core fibers.

Initial attempts based on rectangular and circular metal-coated hollow-core fibers succeeded in transmitting up to 3 kW of laser power [74, 75]. However, these power levels were achieved with fibers having large core radii, which exhibited poor output mode quality when subjected to movements and bends. Most of the industrial applications require well-defined, low-order output modes so that the cut or mark is clean and sharp. The only way to achieve this type of modal quality with hollow-core fibers is to have a smaller core size that will filter the higher order modes and transmit only the fundamental mode. Of course, as the core size is reduced, the losses of the fiber increase and the power capacity decreases. These problems limited the utilization of hollow-core fibers to lower power industrial applications, like marking or cutting paper or plastic products [76], applications that require powers in the range of 1–100 W, which can be delivered with the small core fibers. The flow of gas through the fiber core and the use of a water-circulating jacket are often employed to improve the power handling capacity of these fibers, but they still find little use in industrial processing. An interesting solution to the problem of fiber-motion–induced mode mixing in a marking application was investigated by Harrington, in cooperation with Domino Laser Corporation [51]. The method involved using eight computer-controlled lasers coupled to eight stationary fibers to form the desired mark.

10.4 HOLLOW-CORE FIBER MANUFACTURING

10.4.1 OmniGuide Fiber Manufacturing

Because of their flexibility, low cost, and relative ease of manufacture, polymer fibers are widely used in applications that do not involve the optical properties of the fiber (e.g., textile fabrics). On the other hand, structures with refined abilities to control and manipulate light, such as Bragg mirrors, are costly to produce and have been mostly limited to planar geometries. OmniGuide fibers, however, combine the best properties of polymer fibers with the unique optical properties of Bragg mirrors. Because OmniGuide fibers consist of approximately 98% polymer, they are extremely flexible, yet they have the required optical properties to guide high-power laser light. The two materials that are used to fabricate the omnidirectional reflective cladding of OmniGuide fibers, the chalcogenide glass As_2Se_3 and the polymer PES, are chosen for their unique properties; they have very different indices of refraction critical for the efficient formation of a 1D omnidirectional photonic band-gap, and they can be thermally co-drawn into layered structures without cracking or delamination. The preform/draw process used to manufacture OmniGuide fibers, similar to the process used for silica optical fiber manufacturing, allows for high-volume production with good geometrical control. For typical applications, a proximal-end connector (for ease of attachment to a laser) and distal tip are added to the OmniGuide fiber.

10.4.1.1 OmniGuide Fiber Manufacture: Preform Construction and Draw

The manufacturing processes of OmniGuide fiber are shown schematically in Fig. 10.25. This manufacturing technique enables easy tuning of the operational band-gap of the OmniGuide fiber, allowing it to transmit light of almost any desired wavelength. For example, the technique described here has been successfully used to manufacture OmniGuide fibers operational at the $CO_2(\lambda = 10.6\,\mu m)$ and Er:YAG ($\lambda = 2.94\,\mu m$) laser wavelengths, using the same raw materials.

The high-index material, the chalcogenide glass As_2Se_3, is synthesized by melting the glass components inside a clean, evacuated, fused silica tube at

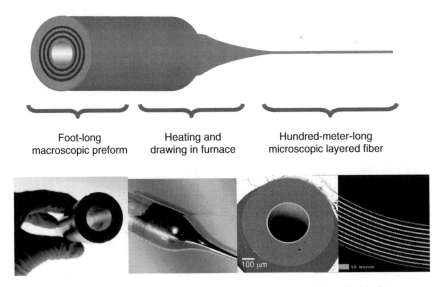

| Foot-long macroscopic preform | Heating and drawing in furnace | Hundred-meter-long microscopic layered fiber |

Figure 10.25 Schematic view of the manufacturing process of OmniGuide fibers.

600 °C. The chemical uniformity of the glass is ensured by rocking the entire furnace assembly while at the melt temperature, thoroughly mixing the glass components.

A multilayer preform is produced by first evaporating a film of As_2Se_3 glass onto both sides of a sheet of the polymer PES. The coated sheet of polymer is then rolled into a cylinder, adding more PES sheets to form an outer cladding for mechanical strength. Heating under vacuum then consolidates the rolled structure. This method allows the production of hollow-core preforms with more than 40 layers. Depending on the desired operating wavelength range, the PES sheet thickness varies from 25 to 50 μm and the deposited As_2Se_3 film thickness is 12–25 μm. These layer thicknesses can be scaled up as desired, although for thicker glass layer deposition, care must be taken to prevent cracking during the rolling process.

After consolidation, the preform is drawn into fiber. At this step, the draw-down ratio can be adjusted to control the layer thicknesses in the resulting fiber and thereby produce fibers that transmit at a desired wavelength. A single preform of approximately 30-cm length can yield more than 100 m of CO_2 fiber ($\lambda = 10.6\,\mu$m) or 300 m of Er:YAG fiber ($\lambda = 2.94\,\mu$m). Continuous monitoring of the outer diameter, as well as band-gap and inner diameter measurements performed on sample fibers during the draw, is used to ensure that the fiber has the desired optical properties.

Another preform/draw technique was successfully used to fabricate Bragg-like hollow-core fibers. These fibers are composed of concentric cylindrical silica rings separated by nanometer-scale support bridges [77]. Because these support bridges are so small, the region between concentric silica rings behaves like a low-index layer (i.e., a layer of air). These alternating air–silica rings form a 1D photonic band-gap. Although these fibers are useful for transmitting wavelengths below 2 μm, the use of silica prevents their application in the mid-IR. The use of a glass transparent in the mid-IR should eliminate this limitation, but the formation and preservation of the thin support bridges with such a glass would remain a challenge.

10.4.1.2 OmniGuide Fiber: Proximal End

For easy connection to a laser source, a standard ST connector is added to the proximal end of the OmniGuide fiber. The ferrule inside diameter (ID) in the connector is chosen to be slightly smaller than the fiber ID to protect the end-face of the fiber (which is not reflective) from the incident laser beam. The connector is attached to the outer surface of the fiber using a high thermal-conductivity epoxy for improved heat transfer. For CO_2 fibers, a short hollow zirconia tube is used as the ferrule to protect the fiber input and reduce the heat generated in the fiber near the coupling plane by filtering out higher order modes. For similar purposes, copper tubes are used to couple into Er:YAG fibers.

10.4.1.3 OmniGuide Fiber: Distal End Termination

For surgical applications, where the fiber will encounter a semi-aquatic environment, precautions are necessary to prevent liquids from entering the hollow core. For CO_2 fibers, continuous helium flow through the hollow core is used to solve this problem. To avoid damage to the fiber tip that results from clogging when in contact with tissue, a short stainless-steel tube is attached to the end of the fiber, with side holes to aid in gas flow venting. However, not all applications allow for gas flow through the core. For example, in Er:YAG laser endoscopic lithotripsy (kidney stone destruction), which takes place in the closed aqueous environment of the urinary tract, gas flow cannot be used. In such cases, the distal end needs to be sealed off so liquid and debris do not penetrate into the fiber core. Distal-end pieces sealed with a low-OH silica window have been used successfully for this purpose. In either case, tips are attached to the outer diameter of the fiber with epoxy.

10.4.2 Techniques Used in the Manufacture of Other Hollow-Core Fibers

10.4.2.1 Hollow-Core Metal-Coated Fibers

There are many approaches used in the manufacture of hollow-core metal fibers [51]. Early attempts to make rectangular metal waveguides by separating the metal strips with a metal or plastic spacer were successful; the resulting waveguides were able to transmit CO_2 laser powers on the order of kilowatts [78]. The use of an additional dielectric layer coating to lower the metal waveguide losses was also used with rectangular waveguides and was combined with new fabrication techniques to reduce the core size for better mode quality [79]. However, because the core was rectangular, these guides suffered from increased losses due to twists and had limited flexibility. Also, the core sizes achieved (1 × 1 mm) were still not small enough for good beam quality. Circular metallic guides, which do not have the problem associated with twisting, were pioneered by Miyagi et al. [10]. The fabrication of this type of fiber typically starts with an aluminum or glass mandrel, which is then coated by a dielectric (Ge, AsSe, or ZnS) and a metal film (Ag, Cu, or Au). Finally, a thick nickel layer is deposited on top of the metal layer. The mandrel is etched away to leave a hollow core. Bhardwaj et al. [80] used a different technique, starting with an extruded Ag tube with hollow core. The dielectric layer inside the tube, which is typically AgBr or AgCl, is deposited inside the core using liquid- or gas-phase reaction with Ag. The fibers fabricated with this technique suffered from surface roughness that increased the losses, despite polishing of the Ag core with an acid solution. Alaluf et al. [81] used a different technique, starting with polyethylene and Teflon tubing, to obtain flexible and low-cost fibers. They coated the inside surface of the tube with Ag and used wet or liquid chemistry to convert some of the Ag to AgI to form the dielectric layer. George and Harrington [82] improved the surface roughness on this type of fibers using polycarbonate or similar tubing that has better surface quality. Harrington [11] also worked with silica glass tubing to improve the surface roughness. In this technique, the starting Ag film needs to be thick to form both the conducting metal boundary and the desired AgI film after reaction with iodine. However, the Ag surface roughness also increases with film thickness. In addition to this roughness/thickness problem, it is also difficult to achieve a uniform film thickness over long lengths of fiber. Miyagi et al. [10] used a different approach to produce a dielectric film over the metal that does not depend on subtraction of the metal layer. They used liquid-phase techniques to deposit polymer films on a thin Ag layer. Because they did not need a thick Ag film to start with, they were able to obtain much less surface roughness, lowering the losses of the fiber. However, the use of a relatively lossy polymer layer (17,000 dB/m at $\lambda = 10.6\,\mu$m) limits the power handling capacity of this fiber.

Two important drawbacks are associated with all these techniques. The first is that the manufacturing process is not readily scalable. The metallic hollow-core fibers have to be processed individually to have a metallic and dielectric layer inside the small bore tubing, in contrast to a preform/draw process, which is easily scaled up simply by increasing the size of the preform. Indeed, the preform/draw process associated with silica fiber manufacture has been scaled up from preforms yielding 12 km to preforms yielding more than 2000 km. The second drawback associated with the manufacture of hollow metal waveguides, with or without dielectric coatings, is the inability to easily reduce inhomogeneities or geometric non-uniformities formed during the macroscopic processing. With a preform/draw process, non-uniformities introduced during the macroscopic preform production are greatly elongated during the subsequent draw process, decreasing their impact on the optical performance of the resulting fiber.

10.4.2.2 Hollow-Core 2D Photonic-Crystal Fibers

Hollow-core 2D PCFs are manufactured by drawing bundles of silica capillary tubes, where the hollow core is formed by removing some tubes from the center [20, 27]. The bundle surrounding the hollow core forms the 2D photonic band-gap, which enables the guiding of light in the core. Fibers were successfully fabricated with this technique, but with the limitations on the operational wavelength imposed by the material absorption of silica (little transmission above 2 μm). There have been efforts to manufacture similar structures using glasses transparent to mid-IR wavelengths, but preserving the uniformity of the structure during draw, which is necessary to keep losses low, is a challenge.

10.5 CONCLUSIONS

Hollow-core fibers offer a number of significant advantages over traditional solid-core fibers. They greatly ease the constraints—absorption, nonlinearity, material dispersion—associated with propagation through the core material and, by proper choice of cladding materials and geometry, are capable of guiding radiation of almost any wavelength. Of particular interest for both commercial and academic applications are hollow-core photonic band-gap fibers, which, in different implementations, may be capable of exceeding the performance characteristics of silica-based solid-core fiber in the near-IR and are capable of the flexible low-loss transmission of mid- to far-IR radiation. This latter capability is already opening up exciting new frontiers for minimally invasive laser surgery. Although the measured performance of these fibers has not yet approached theoretical limits, steady improvements in manufacturing technology and fiber design have reduced this gap and suggest that these fibers will play an increasingly important role in wave-guiding applications.

REFERENCES

[1] Kapany, N. S. and R. J. Simms. Recent developments of infrared fiber optics. 1965. *Infrared Phys.* 5:69.
[2] Nishii, J. et al. 1992. Recent advances and trends in chalcogenide glass fiber technology: A review. *Non-Cryst. Sol.* 140:199–208.
[3] Carter, S. F. et al. 1990. Low loss fluoride fiber by reduced pressure casting. *Electron. Lett.* 26:2115–2117.
[4] Artjushenko, V. G. et al. 1986. Mechanisms of optical losses in polycrystalline KRS-5 fibers. *J. Lightwave Technol.* LT-4:461–465.
[5] Sa'ar, A. et al. 1986. Infrared optical properties of polycrystalline silver halide fibers. *Appl. Phys. Lett.* 49:305–307.
[6] Roberts, P. et al. 2005. Ultimate low loss of hollow-core photonic crystal fibers. *Optics Express* 13:236–244.
[7] Harrington, J. A. and C. C. Gregory. 1990. Hollow sapphire fibers for the delivery of CO_2 laser energy. *Opt. Lett.* 15(10):541.
[8] Alsberg, D. A. et al. 1977. The WT4/WT4A millimeter-wave transmission system. *Bell System Techn. J.* 56:1829.

[9] Garmire, E. et al. 1980. Flexible infrared waveguides for high-power transmission. IEEE J. Quantum Electron. *IEEE J. Quantum Electron.* QE-16:23–32.

[10] Miyagi, M. et al. 1983. Fabrication of germanium-coated nickel hollow waveguides for infrared transmission. *Appl. Phys. Lett.* 43:430–432.

[11] Harrington, J. A. 2000. A review of IR transmitting, hollow waveguides. *Fiber Integrated Optics* 19:211–217.

[12] Yablonovitch, E. 1987. Inhibited spontaneous emission in solid-state physics and electronics. *Phys. Rev. Lett.* 58:2059–2062.

[13] John, S. 1987. Strong localization of photons in certain disordered dielectric superlattices. Phys. *Phys. Rev. Lett.* 58:2486–2489.

[14] Joannopoulos, J. D. et al. 1995. *Photonic Crystals: Molding the Flow of Light.* Princeton University Press, Princeton, NJ.

[15] Yeh, P. et al. 1978. Theory of Bragg fiber. J. *Opt. Soc. Am.* 68:1196–1201.

[16] Doran, N. J. and K. J. Bulow. 1983. Cylindrical Bragg fibers: A design and feasibility study for optical communications. *J. Lightwave Technol.* 1:588–590.

[17] Lazarchik, A. N. 1988. Bragg fiber lightguides. *Radiotekhnika Electron.* 1:36–43.

[18] de Sterke, C. M. and I. M. Bassett. 1994. Differential losses in Bragg fibers. *J. Appl. Phys.* 76:680–688.

[19] Kuriki, K. et al. 2003/04. UV Hollow Photonic Bandgap Transmission Fibers. *Annu. Progress Rep. RLE MIT* 146:31–33.

[20] St.-J. Russell, P. 2003. Photonic crystal fibers. *Science* 299:358–362.

[21] Kerbage, C. and B. J. Eggleton. 2002. Microstructured optical fibers: Enabling integrated tenability for photonic devices. *Optics and Photonics News,* September issue, pp. 38–43.

[22] Smith, C. M. et al. 2003. Low-loss hollow-core silica/air photonic bandgap fiber. *Nature* 424:657–659.

[23] Mangan, B. J. et al. 2004. Low-loss (1.7 dB/km) hollow core photonic bandgap fiber. In: *Proceedings of the Optical Fiber Communications Conference, Los Angeles.* Post-deadline Paper PDP24. OSA – Optical Society of America. Washington, DC.

[24] Johnson, S.G. et al. 2001. Low-loss asymptotically single-mode propagation in large-core OmniGuide fibers. *Opt. Express* 9:748–779.

[25] Temelkuran, B. et al. 2002. Wavelength-scalable hollow optical fibers with large photonic bandgaps for CO_2 laser transmission. *Nature* 420:650–653.

[26] Knight, J. C. et al. 1998. Photonic band gap guidance in optical fibers. *Science* 282:1476–1478.

[27] Cregan, R. F. et al. 1999. Single-mode photonic band gap guidance of light in air. *Science* 285:1537–1539.

[28] Collin, R. 1991. *Field Theory of Guided Waves.* IEEE Press. Princeton, NJ.

[28a] J. M. Fini. 2004. *Measurement Science and Technology* 15:1120.

[29] Miyagi, M. and S. Kawakami. 1984. Design theory of dielectric-coated circular metallic waveguides for infrared transmission. *J. Lightwave Technol.* LT-2(2):116.

[30] Kato, Y. et al. 1993. New fabrication technique of fluorocarbon polymer-coated hollow waveguides by liquid-phase coating for medical applications. In: *Biomedical Fiber Optic Instrumentation* (J. A. Harrington, D. M. Harris, A. Katzir, and F. P. Milanovich, eds.). Proc. Soc. Photo-Opt Instrum. Eng. 2131:66.

[31] Matsuura, Y. and M. Miyagi. 1993. Er:YAG, CO, and CO_2 laser delivery by ZnS-coated Ag hollow waveguides. *Appl. Opt.* 32:6598.

[32] Matsuura, Y. et al. 1995. Polymer-coated hollow fiber for CO_2 laser delivery. *Appl. Opt.* 34(30):6642.

[33] Abe, Y. et al. 1998. Polymer-coated hollow fiber for CO_2 laser delivery. *Opt. Lett.* 23(2):89.

[34] Gopal, V. and J. A. Harrington. 2003. Deposition and characterization of metal sulfide dielectric coatings for hollow glass waveguides. *Opt. Express* 11(24):3182.

[35] Ibanescu, M. et al. 2003. Analysis of mode structure in hollow dielectric waveguide fibers. *Phys. Rev.* E67:046608.

[36] Koshiba, M. 1992. *Optical Waveguide Theory by the Finite Element Method.* KTK Scientific, Tokyo.

[37] Taflove, A. and S. Hagness. 2000. *Computational Electrodynamics: The Finite-Difference Time-Domain Method.* Artech House, Boston, MA.

[38] Weber, W. H. et al. 1974. Design consideration for a 3-D laser Doppler velocimeter for studying gravity waves in shallow water: Comments on (T). *Appl. Opt.* 13:715.

[39] Warters, W. D. 1977. WT4 millimeter waveguide systems: Introduction. *Bell Syst. Techn. J.* 56:1825.

[40] Gloge, D. 1971. Weakly guiding fibers. *Appl. Opt.* 10(10):2252.

[41] Birks, T. A. et al. 1997. Endlessly single-mode photonic crystal fiber. *Opt. Lett.* 22(13):961.

[42] Snyder, A. W. and J. D. Love. 1983. *Optical Waveguide Theory.* Chapman and Hall, London.

[43] Tang, C. H. 1970. An orthogonal coordinate system for curved pipes (correspondence). *IEEE Trans. Microw. Theory Tech.* 18:69.

[44] Kath, W. L. and G. A. Kriegsmann. 1988. Optical tunneling: Radiation losses in bent fiber-optic waveguides. *IMA J. Appl. Math.* 41:85.

[45] Teixeira, F. L. and W. C. Chew. 1998. Analytical derivation of a conformal perfectly matched absorber for electromagnetic waves. *Microwave Opt. Tech. Lett.* 17(4):231.

[46] Johnson, S. G. and J. D. Joannopoulos. 2001. Block-iterative frequency-domain methods for Maxwell's equations in a planewave basis. *Opt. Express* 8(3):173.

[47] West, J. et al. 2004. Surface modes in air-core photonic band-gap fibers. *Opt. Express* 12(8):1485.

[48] Inui, T. et al. 1996. *Group Theory and Its Applications in Physics.* Springer-Verlag, Heidelberg.

[49] Katzir, A. 1993. *Lasers and Optical Fibers in Medicine.* Academic Press, New York.

[50] Vo-Dinh, T., editor. 2003. *Biomedical Photonics Handbook.* CRC Press, New York.

[51] Harrington, J. A. 2004. *Infrared Fibers and Their Applications.* SPIE Press, Bellingham, WA.

[52] Devaiah, A. K. et al. 2005. Surgical utility of a new carbon dioxide laser fiber: Functional and histological study. *Laryngoscope* 115:1463.

[53] Bueno, R. et al. 2005. Flexible delivery of carbon dioxide lasers through the omniguide photonic bandgap fiber for treatment of airway obstruction: Safety and feasibility study. *Chest Meeting Abstracts* 128:497S.

[54] Holsinger, F. C. et al. Use of the photonic band gap fiber assembly CO_2 laser system in head and neck surgical oncology. *Laryngoscope* 116(7):1288.

[55] Engeness, T. et al. 2003. Dispersion tailoring and compensation by modal interactions in OmniGuide fibers. *Opt. Express* 11:1175.

[56] Mogilevtsev, D. et al. 1998. *Optics Lett.* 23(21):1662.

[57] Birks, T. A. et al. 1999. Dispersion compensation using single-material fibers. *IEEE Photonics Technol. Lett.* 11(6):674.

[58] Delone, N. B. and V. P. Krainov. 1988. *Fundamentals of Nonlinear Optics of Atomic Gases, Wiley Series in Pure and Applied Optics.* John Wiley and Sons. Somerset, NJ.

[59] Measures, R. M. 1992. *Laser Remote Sensing: Fundamentals and Applications.* Krieger Publishing Company. Melbourne, FL.

[60] Scully, M. O. and M. S. Zubairy. 1997. *Quantum Optics.* Cambridge University Press.

[61] Udem, T. et al. 2002. Optical frequency metrology. *Nature* 416:233.

[62] Saito, M. and K. Kikuchi. 1997. Infrared optical fiber sensors. *Optical Rev.* 4:527.

[63] Saggese, S. J. et al. 1992. Novel lightpipes for infrared spectroscopy. *Appl. Spectroscopy* 46:1194.

[64] Sato, S. et al. 1993. Infrared hollow waveguides for capillary flow cells. *Appl. Spectroscopy* 47:1665.

[65] Charlton, C. et al. 2005. Mid-infrared sensors meet nanotechnology: Trace gas sensing with quantum cascade lasers inside photonic band-gap hollow waveguides. *Appl. Phys. Lett.* 86:194102.

[66] Benabid, F. et al. 2004. Ultrahigh efficiency laser wavelength conversion in a gas-filled hollow core photonic crystal fiber by pure stimulated rotational raman scattering in molecular hydrogen. *Phys. Rev. Lett.* 93:123903.

[67] Benabid, F. et al. Stimulated raman scattering in hydrogen-filled hollow-core photonic crystal fiber. 2002. *Science* 298:399.

[68] Ghosh, S. et al. 2005. Resonant optical interactions with molecules confined in photonic band-gap fibers. *Phys. Rev. Lett.* 94:093902.

[69] Benabid, F. et al. 2005. Compact, stable, and efficient all-fiber gas cells using hollow-core photonic crystal fibers. *Nature* 434:488.

[70] Worrell, C. A. et al. 1992. Remote gas sensing with mid-infra-red hollow waveguide. *Elec. Lett.* 28:615.

[71] Ikedo, M. et al. 1986. Preparation and characteristics of the TlBr-TlI fiber for a high power CO_2 laser beam. *J. Appl. Phys.* 60:3035.

[72] Sakuragi, S. et al. 1981. KRS-5 optical fibers capable of transmitting high-power CO_2 laser beam. *Opt. Lett.* 6:629.

[73] Takahashi, K. et al. 1987. Silver halide infrared fiber. *Sumitomo Electric Tech. Rev.* 26:371.

[74] Hongo, A. et al. 1992. Transmission of Kilowatt-class CO_2 laser light through dielectric-coated metallic hollow waveguides for material processing. *Appl. Opt.* 31:5114.

[75] Nubling, R. K. and J. A. Harrington. 1996. Hollow-waveguide delivery systems for high-power, industrial CO_2 lasers. *Appl. Opt.* 34(3):372.

[76] Dekel, B. et al. 2000. Hollow glass waveguides and silver halide fibers as scanning elements for CO_2 laser marking systems. *Opt. Eng.* 39:1384.

[77] Vienne, G. et al. 2004. Ultra-large bandwidth hollow-core guiding in all-silica Bragg fibers with nano-supports. *Opt. Exp.* 12(15):3500.

[78] Garmire, E. et al. 1980. Flexible infrared waveguides for high-power transmission. *IEEE J. Quantum Electron.* 16:23.

[79] Machida, H. et al. 1992. Transmission properties of rectangular hollow waveguides for CO_2 laser light. *Appl. Opt.* 31:7617.

[80] Bhardwaj, P. et al. 1993. Performance of a dielectric-coated monolithic hollow metallic waveguide. *Mater. Lett.* 16:150.

[81] Alaluf, M. et al. 1992. Plastic hollow fibers as a selective infrared radiation transmitting medium. J. *Appl. Phys.* 72:3878.

[82] George, R. and J. A. Harrington. 2001. Hollow plastic waveguides for sensor applications. *Proc. SPIE* 4204:230.

[83] Abe, Y. et al. 2000. Flexible small-bore hollow fibers with an inner polymer coating. *Opt. Lett.* 25:150.

Chapter 11

Silica Nanofibers and Subwavelength-Diameter Fibers

Limin Tong[1] and Eric Mazur[2]

[1] *Zheijiang University, Hangzhou, P.R. China*
[2] *Harvard University, Cambridge, Massachusetts*

11.1 NANOFIBER AT A GLANCE

Air-clad silica (SiO_2) nanofibers, named for their submicrometer diameters, have a large core-cladding index contrast for efficient optical confinement. For single-mode operation, these fibers are usually thinner than the wavelength of the light they carry and are, therefore, also called subwavelength-diameter fibers. The small diameter of a nanofiber and the large core-cladding index contrast yield a number of interesting optical properties such as tight optical confinement, large evanescent fields, strong field enhancement, and large waveguide dispersions. The nanofibers are fabricated by taper drawing of standard optical fibers and have extraordinary diameter uniformity and low surface roughness, making them ideal for low-loss optical wave-guiding. They can also be made very long and have high mechanical strength and pliability, facilitating assembly and patterning. Because of their compactness and their optical and mechanical properties, these nanofibers find applications in a variety of fields, including photonic devices, optical sensors, and nonlinear optics.

11.2 INTRODUCTION

In the past 30 years, optical fibers with diameters larger than the wavelength of the guided light have found broad applications in optical communication, optical sensing, and optical power delivery systems [1–3]. Advances in

361

microtechnology and nanotechnology for optoelectronics and photonics [4–6] and the demand for improved performance, wider applications, and higher integration density, however, have spurred efforts for the miniaturization of photonic devices and waveguides. A major step toward the miniaturization of devices is reducing the diameter of the optical fiber or waveguide. Therefore, an important motivation for fabricating optical-quality nanowires or nanofibers is their potential usefulness as building blocks in future micrometer- or nanometer-scale photonic devices and as tools for mesoscopic optics research.

There are several methods for fabricating one-dimensional (1D) optical nanostructures, including bottom-up chemical growth and top-down photo or electron beam lithography [7–9]. The silica nanowires (referred to as "nanofibers" in the following text) introduced in this chapter are fabricated from standard optical fibers by a taper drawing method. Taper drawing of glass fiber is a top-down process that permits the fabrication of nanowires with diameters down to 50 nm [10]. Compared to other techniques, the taper-drawing approach not only provides a simple fabrication method but also yields nanowires with extraordinary diameter uniformity, atomic-level surface smoothness, and ultralow wave-guiding loss that cannot be achieved by subwavelength-width structures obtained by other methods [11–13].

Generally, when the diameter of a nanofiber is smaller than the wavelength of the guided light, the fiber can operate as a single-mode subwavelength-diameter waveguide with air cladding [14, 15]. Because the index difference between the silica core and the surrounding medium (usually air) is large, the small index difference between the high-index center used as core in a standard fiber and the low-index cladding inherited from the starting fiber can be ignored in a nanofiber. The tight optical confinement, large evanescent fields, strong field enhancement, and large waveguide dispersions of the silica nanofibers [16–19] have generated broad interest in their potential for applications in a variety of fields such as microscale and nanoscale photonic devices [20–23], nanofiber optical sensors [24–26], nonlinear interactions and supercontinuum generation [13, 27–34], and atom trapping and guidance [17, 35–37].

This chapter begins with a theoretical modeling of the optical wave-guiding properties of nanofibers, followed by a description of the taper-drawing fabrication technique and electron microscopy of the nanofibers, and then an experimental investigation of the nanofibers with an emphasis on micromanipulation and optical losses. Finally, we briefly review current and potential applications of the nanofibers.

11.3 MODELING OF SINGLE-MODE WAVE-GUIDING PROPERTIES OF SILICA NANOFIBERS

Although the wave-guiding theory and properties of conventional optical fibers have been investigated extensively [3, 14, 38–40], subwavelength-diameter nanofibers have not been modeled until recently. Theoretically, the optical wave-guiding properties of subwavelength-diameter nanofibers can be analyzed using Maxwell's equation using boundary conditions analogous to those used for standard glass optical fibers. However, unlike the weakly guiding optical fibers that have a small refractive index difference between the doped core and undoped cladding, the index contrast between the silica core and air cladding of a nanofiber is much higher, so exact analysis (i.e., without approximations) becomes necessary. This section is devoted to the modeling of single-mode wave-guiding properties of subwavelength-diameter nanofibers based on the exact solutions of Maxwell's equations and numerical calculations.

11.3.1 Basic Model

The mathematic model of an air-clad nanofiber is shown in Fig. 11.1. The fiber is assumed to have a circular cross-section, a uniform diameter, an infinite air-cladding, and a step-index profile as follows:

$$n(r) = \begin{cases} n_1, & 0 < r < a, \\ n_2, & a \le r < \infty \end{cases}, \tag{11.1}$$

where a is the radius of the nanofiber, n_1 and n_2 are refractive indices of the fiber material and the air, respectively.

Within their transparent range (250 nm–2.5 μm), silica nanofibers are non-dissipative and source free, so Maxwell's equations can be reduced to the following Helmholtz equations:

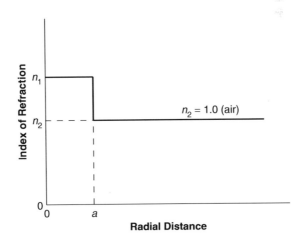

Figure 11.1 Index of refraction profile of an air-clad nanofiber waveguide.

$$(\nabla^2 + n^2k^2 - \beta^2)\,\vec{e} = 0,$$
$$(\nabla^2 + n^2k^2 - \beta^2)\,\vec{h} = 0,$$

(11.2)

where $k = 2\pi/\lambda$ is the wave vector and β is the propagation constant.

Exact solutions for this model have been provided by Snyder and Love [14], yielding the following eigenvalue equations for the HE_{vm} and EH_{vm} modes:

$$\left\{ \frac{J_v'(U)}{UJ_v(U)} + \frac{K_v'(W)}{WK_v(W)} \right\} \left\{ \frac{J_v'(U)}{UJ_v(U)} + \frac{n_2^2 K_v'(W)}{n_1^2 WK_v(W)} \right\} = \left(\frac{v\beta}{kn_1} \right)^2 \left(\frac{V}{UW} \right)^4, \quad (11.3)$$

for the TE_{0m} modes:

$$\frac{J_1(U)}{UJ_0(U)} + \frac{K_1(W)}{WK_0(W)} = 0,$$

(11.4)

and for the TM_{0m} modes:

$$\frac{n_1^2 J_1(U)}{UJ_0(U)} + \frac{n_2^2 K_1(W)}{WK_0(W)} = 0,$$

(11.5)

where J_v is the Bessel function of the first kind, and K_v is a modified Bessel function of the second kind, $U = D(k_0^2 n_1^2 - \beta^2)^{1/2}/2$, $W = D(\beta^2 - k_0^2 n_2^2)^{1/2}/2$, $V = k_0 a(n_1^2 - n_2^2)^{1/2}$, and $D = 2a$ is the diameter of the nanofiber.

Numerically solving these eigenvalue equations after substituting the indices of refraction for air ($n_2 = 1$) and silica ($n_1 = 1.46$ for $\lambda = 633$ nm), we obtain the propagation constants β for an air-clad nanofiber. Figure 11.2 shows the diameter-dependent β at a wavelength of 633 nm, where the fiber diameter D is directly related to the V-number $[V = k_0 D(n_1^2 - n_2^2)^{1/2}/2]$. The figure clearly shows that at a given wavelength, the number of modes that can be supported by the nanofiber is determined by its diameter. When the fiber diameter is reduced to a certain value (denoted by D_{SM} and corresponding to $V = 2.405$), we obtain single-mode operation and only the HE_{11} mode is supported.

The single-mode condition of an air-clad fiber, marked by a dashed line in Fig. 11.2, can be obtained from Eqs. [11.4] and [11.5], yielding

$$V = 2\pi \frac{a}{\lambda_0} (n_1^2 - n_2^2)^{\frac{1}{2}} \approx 2.405.$$

(11.6)

Figure 11.3 shows the single-mode and multimode regimes of the air-clad silica nanofiber obtained from Eq. [11.6] after substituting $n_2 = 1.0$ for the index of refraction of the air and using a Sellmeier-type dispersion formula for the fused silica [41]:

$$n^2 - 1 = \frac{0.6961663\lambda^2}{\lambda^2 - (0.0684043)^2} + \frac{0.4079426\lambda^2}{\lambda^2 - (0.1162414)^2} + \frac{0.8974794\lambda^2}{\lambda^2 - (9.896161)^2}, \quad (11.7)$$

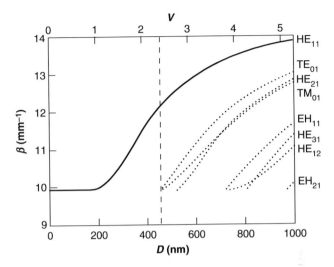

Figure 11.2 Numerical solutions of propagation constant β for an air-clad silica nanofiber at a wavelength of 633 nm. Solid curve: fundamental mode; dotted curves: higher order modes; dashed vertical line: critical diameter D_{SM} for single-mode propagation. (Adapted from reference [15].)

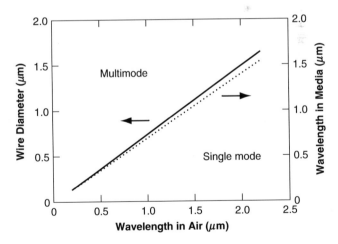

Figure 11.3 Single-mode condition for air-clad silica nanofibers. Solid line: critical diameter for single-mode operation; dotted line: wavelength in silica.

with the wavelength λ in units of μm. The region beneath the solid line in Fig. 11.3 corresponds to single-mode operation. For example, at the He-Ne laser wavelength of 633 nm, a silica nanofiber with a diameter smaller than 457 nm is a single-mode waveguide; in the near infrared at 1.55 μm, the diameter of the silica nanofiber should be less than about 1.1 μm for single-mode operation. The dashed line in Fig. 11.3, representing the wavelength of the propagating light in silica (that is, $\lambda = \lambda_0/n_1$), shows that the nanofiber is always single mode when the fiber diameter is smaller than the wavelength of the light in the silica. Taking into consideration that the UV absorption edge in silica is around 200 nm [41], the minimum critical diameter D_{SM} for silica nanofibers is about 129 nm. Because single-mode operation of silica nanofibers is preferable for most applications, we will concentrate henceforth on the guiding properties of the fundamental modes.

The propagation constant β of the fundamental HE_{11} mode can be obtained by setting $\nu = 1$ in Eq. [11.3] and numerically solving the resulting eigenvalue equation

$$\left\{ \frac{J_1'(U)}{UJ_1(U)} + \frac{K_1'(W)}{WK_1(W)} \right\} \left\{ \frac{J_1'(U)}{UJ_1(U)} + \frac{n_2^2 K_1'(W)}{n_1^2 WK_1(W)} \right\} = \left(\frac{\beta}{kn_1} \right)^2 \left(\frac{V}{UW} \right)^4. \quad (11.8)$$

Writing the electromagnetic fields in the form

$$\begin{cases} \vec{E}(r,\phi,z) = (e_r\hat{r} + e_\phi\hat{\phi} + e_z\hat{z})e^{i\beta z}e^{-i\omega t} \\ \vec{H}(r,\phi,z) = (h_r\hat{r} + h_\phi\hat{\phi} + h_z\hat{z})e^{i\beta z}e^{-i\omega t}, \end{cases} \quad (11.9)$$

we obtain for the electric fields of the fundamental modes inside the core $(0 < r < a)$ [14]:

$$e_r = -\frac{a_1 J_0(UR) + a_2 J_2(UR)}{J_1(U)} \cdot f_1(\phi), \quad (11.10)$$

$$e_\phi = -\frac{a_1 J_0(UR) - a_2 J_2(UR)}{J_1(U)} \cdot g_1(\phi), \quad (11.11)$$

$$e_z = \frac{-iU}{a\beta} \frac{J_1(UR)}{J_1(U)} \cdot f_1(\phi) \quad (11.12)$$

and outside the core $(a \leq r < \infty)$:

$$e_r = -\frac{U}{W} \frac{a_1 K_0(WR) - a_2 K_2(WR)}{K_1(W)} \cdot f_1(\phi), \quad (11.13)$$

$$e_\phi = -\frac{U}{W} \frac{a_1 K_0(WR) + a_2 K_2(WR)}{K_1(W)} \cdot g_1(\phi), \quad (11.14)$$

$$e_z = \frac{-iU}{a\beta} \frac{K_1(WR)}{K_1(W)} \cdot f_1(\phi),$$ (11.15)

where $f_1(\phi) = \sin(\phi)$, $g_1(\phi) = \cos(\phi)$,

$$a_1 = \frac{F_2 - 1}{2}, \ a_3 = \frac{F_1 - 1}{2}, \ a_5 = \frac{F_1 - 1 + 2\Delta}{2}, \ a_2 = \frac{F_2 + 1}{2}, \ a_4 = \frac{F_1 + 1}{2},$$

$$a_6 = \frac{F_1 + 1 - 2\Delta}{2},$$

$$F_1 = \left(\frac{UW}{V}\right)^2 [b_1 + (1 - 2\Delta)b_2], \ F_2 = \left(\frac{V}{UW}\right)^2 \frac{1}{b_1 + b_2},$$

$$b_1 = \frac{1}{2U} \left\{ \frac{J_0(U)}{J_1(U)} - \frac{J_2(U)}{J_1(U)} \right\}, \ b_2 = -\frac{1}{2W} \left\{ \frac{K_0(W)}{K_1(W)} + \frac{K_2(W)}{K_1(W)} \right\}.$$

Because the *h*-components can readily be obtained from *e*-components with some calculations [14], they are not presented here.

Figure 11.4 shows the normalized electric components of the fundamental modes in cylindrical coordinates for silica nanofibers at a wavelength of 633 nm. The dashed line in the radial distribution graph shows the Gaussian profile for reference; the dotted lines represent the electric fields in a silica fiber with critical diameter D_{SM}. As can be seen in the graph, because of the high index contrast between the air and silica, air-clad silica fiber tightly confines the electric field at a diameter of about 400 nm. When the diameter is reduced further, however, a significant portion of the electric field extends far outside the nanofiber, indicating that the field is no longer tightly confined inside or around the fiber. A similar behavior is obtained at other wavelengths.

11.3.2 Power Distribution: Fraction of Power Inside the Core and Effective Diameter

For nanofibers with uniform diameters, there is no net flow of energy in the radial (r) or azimuthal (ϕ) directions, so we need to consider only the energy flow in the *z*-direction. The *z*-component of the Poynting vector inside the core ($0 < r < a$) is [14]

$$S_{z1} = 1/2 \left(\frac{\varepsilon_0}{\mu_0}\right)^{\frac{1}{2}} \frac{kn_1^2}{\beta J_1^2(U)} \left[a_1 a_3 J_0^2(UR) + a_2 a_4 J_2^2(UR) \right.$$

$$\left. + \frac{1 - F_1 F_2}{2} J_0(UR) J_2(UR) \cos 2\phi \right],$$ (11.16)

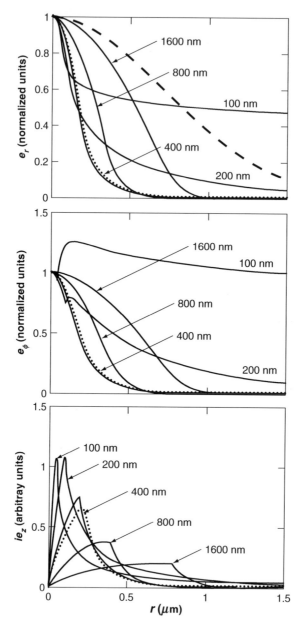

Figure 11.4 Radius dependence of the electric field component of the HE_{11} mode in cylindrical coordinates in silica nanofibers at a wavelength of 633 nm. The fields are normalized as follows: $\varepsilon e_r(r=0) = 1$ and $e_\phi(r=0) = 1$. Results are shown for fiber diameters ranging from 100 to 1600 nm. (Adapted from reference [15].)

and the one outside the core ($a \leq r < \infty$) is

$$S_{z2} = 1/2 \left(\frac{\varepsilon_0}{\mu_0}\right)^{\frac{1}{2}} \frac{kn_1^2}{\beta K_1^2(W)} \frac{U^2}{W^2} \left[a_1 a_5 K_0^2(WR)\right.$$

$$\left. + a_2 a_6 K_2^2(WR) - \frac{1 - 2\Delta - F_1 F_2}{2} K_0(WR) K_2(WR) \cos 2\phi\right]. \tag{11.17}$$

Figure 11.5 shows the Poynting vectors for a 200- and a 400-nm diameter silica nanofiber at a wavelength of 633 nm; the mesh profile represents the fields propagating inside the fiber and the gradient profile stands for the evanescent field. As one can see, the 400-nm fiber confines most of the light inside the fiber, whereas for the 200-nm fiber, a large amount of light is guided outside the wire in the form of an evanescent wave.

To obtain a more intuitive understanding of the power distribution in the radial direction, we calculate two additional parameters. The first is the fractional power inside the core,

$$\eta = \frac{\int_0^a S_{z1} dA}{\int_0^a S_{z1} dA + \int_a^\infty S_{z2} dA}, \tag{11.18}$$

where $dA = a^2 R dR d\phi = r dr d\phi$. The second one is the effective diameter of the light field D_{eff}—the diameter within which $1 - e^2$ (86.5%) of the total power is confined—which can be obtained from

$$\begin{cases} \dfrac{\int_0^{D_{eff}} S_{z1} dA}{\int_0^a S_{z1} dA + \int_a^\infty S_{z2} dA} = 86.5\%, \quad (D_{eff} < a) \\[4mm] \dfrac{\int_0^a S_{z1} dA \int_a^{D_{eff}} S_{z1} dA}{\int_0^a S_{z1} dA + \int_a^\infty S_{z2} dA} = 86.5\%, \quad (D_{eff} > a). \end{cases} \tag{11.19}$$

Figure 11.6 shows the fractional power in the core as a function of the fiber diameter D for silica nanofibers at wavelengths of 633 nm and 1.5 μm. At the critical diameter D_{SM} (dashed lines), η is around 80% at both wavelengths. When the diameter drops below $0.5D_{SM}$, more than 80% of the energy is guided in the evanescent wave outside the silica core. Because η varies so steeply around the range of diameters of interest, it is easy to tailor the amount of confinement to a particular application. Tight confinement, obtained at diameters around D_{SM}, is important for reducing the modal width and increasing the integrated density of the optical circuits with less cross-talk [14, 42], while the weaker confinement obtained at smaller diameters is helpful for exchanging energy between nanofibers within a short interaction length [43] and for improving the sensitivity of evanescent wave–based fiber optic sensors [24, 25, 44].

Figure 11.7 shows the effective diameter D_{eff} of the fundamental mode in a silica nanofiber at 633-nm wavelength. The dotted line shows the real diameter of the nanofibers for comparison. As expected, D_{eff} is large when the fiber diameter is very small and the two curves intersect near the critical diameter (dashed line). The intersection point is the minimum diameter for which it is possible to confine 86.5% of the light energy within the wire at the given wavelength. Note that at

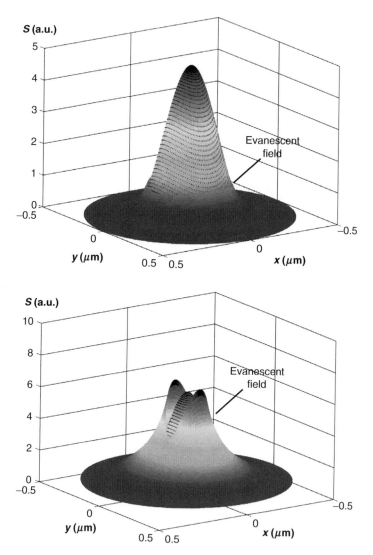

Figure 11.5 z-component of the Poynting vector of 633-nm wavelength light guided by silica nanofibers with a diameter of 400 nm (top) and 200 nm (bottom). Mesh: field inside the nanofiber; gradient: field outside the nanofiber. (Adapted from reference [15].)

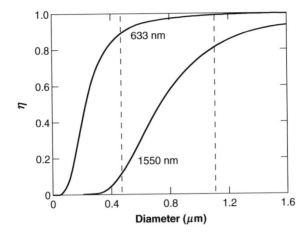

Figure 11.6 Fraction of the power in the fundamental modes carried inside the silica nanofibers at 633-nm and 1550-nm wavelengths. Dashed lines: critical diameters for single-mode operation. (Adapted from reference [45].)

this point the diameter is smaller than the wavelength of the light (450 vs 633 nm). For small fiber diameters, D_{eff} becomes very large. For example, for a nanofiber diameter of 200 nm, D_{eff} is about 2.3 μm, which is more than 10 times the fiber diameter. Maintaining a steady guiding field in such a situation may be difficult; any small deviation (such as surface contamination and/or

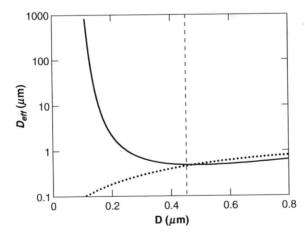

Figure 11.7 Effective diameter of the light field of the fundamental modes in silica nanofibers at a wavelength of 633 nm. Solid line: D_{eff}; dotted line: physical diameter of nanofiber; dashed line: critical diameter for single-mode operation. (Adapted from reference [15].)

microbends) from the ideal condition leads to a change in propagating fields and radiation loss. On the other hand, the high sensitivity of such a guiding fiber to small perturbations may be useful in sensing applications that require high sensitivity.

11.3.3 Group Velocity and Waveguide Dispersion

The diameter-dependent group velocity of the HE_{11} mode for the air-clad silica nanofiber is given by [14]

$$v_g = \frac{c}{n_1^2} \frac{\beta}{k} \frac{1}{1 - 2\Delta(1 - \eta)} \tag{11.20}$$

and shown in Fig. 11.8 for two wavelengths. When the fiber diameter D is very small, v_g approaches the speed of light in vacuum c because most of the light energy propagates in air. As D increases, an increasing fraction of the energy is guided in the silica core and v_g decreases until it reaches a minimum value that is smaller than c/n_1, the group velocity of a plane wave in silica. As D continues to increase, v_g increases again, approaching c/n_1 at large values of D.

Figure 11.8 shows the wavelength dependence of the group velocity for various fiber diameters, also obtained from Eq. [11.20]. For a given fiber diameter D, the group velocity is c when the wavelength λ is very large and approaches c/n_1 when λ is very small, with a minimum value somewhat smaller than c/n_1. Similarly, the wavelength dependence of the group velocity with fiber diameter can be seen in Fig. 11.9.

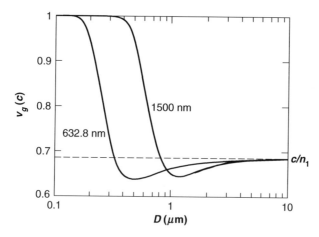

Figure 11.8 Diameter dependence of the group velocity of the fundamental mode in air-clad silica nanofibers at 633-nm and 1.5-μm wavelengths. (Adapted from reference [15].)

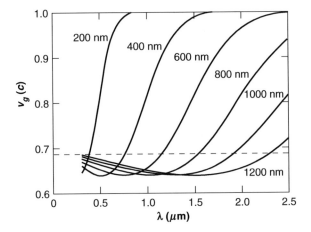

Figure 11.9 Wavelength dependence of the group velocity of the fundamental modes in air-clad silica nanofibers with diameters ranging from 200 to 1200 nm. (Adapted from reference [15].)

From the group velocity in Eq. [11.20], one can obtain the waveguide dispersion [46]:

$$D_w = \frac{d(v_g^{-1})}{d\lambda}. \tag{11.21}$$

Figures 11.10 and 11.11 illustrate the diameter and wavelength dependence of this waveguide dispersion; the dotted line in Fig. 11.11 also shows the material

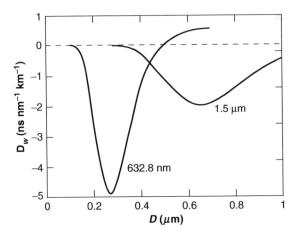

Figure 11.10 Diameter dependence of the waveguide dispersion in air-clad silica nanofibers at 633-nm and 1.5-μm wavelengths. (Adapted from reference [15].)

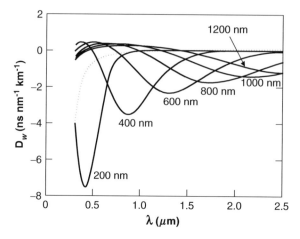

Figure 11.11 Wavelength dependence of the waveguide dispersion in air-clad silica nanofibers with diameters ranging from 200 to 1200 nm. Dotted line: material dispersion. (Adapted from reference [15].)

dispersion of fused silica obtained from Eq. [11.6]. As can be seen, the waveguide dispersion D_w of the nanofibers can be very large compared with those of weakly guiding fibers and bulk material. For example, for an 800-nm diameter silica fiber at a wavelength of 1.5 μm, $D_w = -1400\,\text{ps nm}^{-1}\,\text{km}^{-1}$, which is about 70 times larger than that of the material dispersion. Note also that the total dispersion (the combined material and waveguide dispersions) of a nanofiber can be made positive, zero, or negative within a given spectral range by choosing the appropriate fiber diameter. Controlling light propagation by tailoring the dispersion is widely used in optical communications and nonlinear optics [47–49], so nanofiber waveguides present an opportunity to miniaturize devices in these fields.

11.4 FABRICATION AND MICROSCOPIC CHARACTERIZATION OF SILICA NANOFIBERS

Quite a few techniques can be used to fabricate silica nanowires or nanofibers such as photo or electron beam lithography, chemical growth, and taper drawing of optical fibers [7–11]. Among these techniques, the taper-drawing method exhibits not only simplicity in fabrication, but also the ability to fabricate nanofibers with extraordinary diameter uniformities, atomic-level surface smoothness, and long length that are difficult to achieve by any other means. Diameter uniformity and surface smoothness are particularly critical for low-loss optical wave-guiding in subwavelength–width waveguides [50–53]. This section focuses on taper-drawing fabrication of silica nanofibers.

11.4.1 Two-Step Taper Drawing of Silica Nanofibers

The fabrication of thin silica fibers using a high-temperature taper-drawing technique was first reported in the nineteenth century, when the mechanical properties of the fibers were studied, but their optical properties and applications remained uninvestigated [54, 55]. It was not until a century later, when optical waveguide theory had become well established, that researchers began to investigate the optical applications of very thin silica fibers made by laser- or flame-heated taper drawing of optical fibers [56–60]. Laser heating provides highly stable and repeatable conditions for fiber drawing, but the laser power required for drawing silica fibers with uniform diameters smaller than 1 μm is impractically large [57, 60]. As discussed in Section 11.2, for single-mode operation in the optically transparent range of silica (250–2000 nm), the fiber diameter has to be smaller than 1 μm. Therefore, flame heating is the only practical technique for the taper drawing of single-mode silica nanofibers. Silica nanofibers have been obtained by one-step and two-step taper-drawing methods [10–13, 20]. The one-step approach is simple and convenient. However, when drawing fibers directly from a flame-heated melt, turbulence and convection usually make it difficult to control the temperature gradient in the drawing region and to maintain stable drawing conditions. Consequently silica nanofibers with diameters of less than 200 nm are difficult to obtain with a one-step draw. A two-step technique circumvents these difficulties, making it possible to draw silica nanofibers with diameters as small as 20 nm.

A schematic view of the two-step taper-drawing method is shown in Fig. 11.12a. As in the one-step method, a bare silica fiber is flame-heated and first drawn down to a micrometer-sized diameter taper. A low-carbon fuel such as CH_3OH or hydrogen is recommended to avoid contamination of the fiber with incompletely burned carbon particles. To obtain sufficiently steady to reduce the fiber diameter below 1 μm, we use a tapered sapphire fiber with a tip diameter around 100 μm to absorb the thermal energy from the flame. The sapphire fiber taper (fabricated using a laser-heating growth method [61]) confines the heating to a small volume and helps maintain a steady temperature distribution during the drawing. As long as the working temperature is kept below the melting temperature of sapphire (~2320 K), the sapphire tip can be used repeatedly. One end of the previously drawn fiber with a micrometer-sized diameter is placed horizontally on the sapphire tip, and the flame is adjusted until the temperature of the sapphire tip is just above the drawing temperature (~2000 K). The sapphire tip then is rotated around its axis to wind the silica

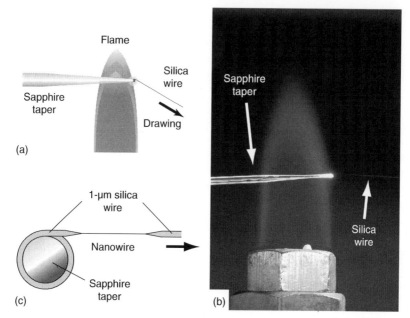

Figure 11.12 Two-step taper drawing of silica nanofibers. After drawing a standard fiber down to a diameter of about 1 μm (step 1), the resulting fiber is wound around a sapphire taper. (a) The coil is heated by thermal conduction of the sapphire taper and then a nanofiber is drawn (step 2). (b) Closeup photograph of the second-step taper drawing of silica nanofibers. (c) Schematic of the nanowire drawing from the silica coil wound around the sapphire taper.

fiber around it and the resulting fiber coil is moved about 0.5 mm out of the flame to prevent melting, as shown in Fig. 11.12b. Finally, a nanofiber is drawn from the coil at a speed of 1–10 mm/sec in the horizontal plane in a direction perpendicular to the axis of the sapphire tip. With this two-step technique, the diameter of a silica fiber can be reduced to about 50 nm, thinner than required for most optical applications.

To obtain even thinner nanofibers for the investigation of the structural, dynamic, and catalytic properties of silica nanowires [62–64], we used a self-modulated drawing force [65] instead of the constant drawing force in the two-step taper-drawing process described earlier. The self-modulated force is obtained by the technique illustrated in Fig. 11.13. The silica fiber is held parallel to the sapphire taper and the elastic bend in the taper area of the fiber generates the tensile force in the microfiber between the silica and sapphire tapers. During the initial stage of the drawing, when the fiber is still thick, the bending center occurs in the thicker part of the taper, causing a relatively large force; as the fiber is elongated and its diameter is reduced, the bending center moves towards the

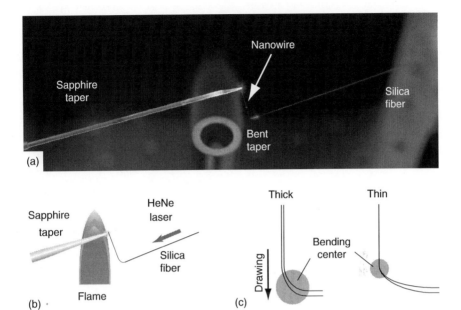

Figure 11.13 Self-modulated taper drawing of silica nanofibers. (a) Closeup photograph of the nanofiber drawing. The red light visible around the nanofiber and tapers is from a He-Ne laser. (b) Schematic diagram of the self-modulated taper drawing setup. A three-dimensional stage is used to mount and adjust the silica fiber taper to form a 90-degree bend, and a He-Ne laser is launched into the silica fiber for illuminating the nanofiber and monitoring the drawing process. (c) The self-modulation of the drawing force is due to the shifting of the bending center as the fiber is drawn. (Adapted from reference [65].)

thin end of the taper, reducing the tensile force that causes the drawing. This self-modulation not only permits the drawing of fibers with diameters as small as 20 nm, but also counteracts the effects of temperature fluctuations by buffering the drawing force and avoids any sudden changes in fiber diameter. To monitor the drawing process, we launch a continuous-wave He-Ne laser (633-nm wavelength) along the silica fiber to illuminate the taper and nanofiber, as illustrated in Fig. 11.13b. When the drawing is completed, the nanofiber is connected to the starting fiber at one end and freestanding on the other end.

11.4.2 Electron Microscope Study of Silica Nanofibers

The nanofibers obtained with the two-step taper-drawing techniques described in the previous section consist of three parts: a millimeters-long taper that is connected to the starting microfiber, a uniform nanofiber with a length up

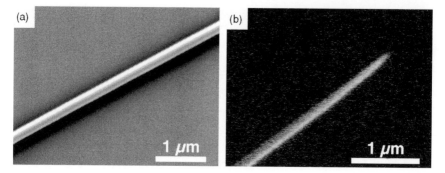

Figure 11.14 SEM images of the (a) uniform part and (b) end of a silica nanofiber with a nominal diameter of 200 nm.

to tens of millimeters, and an abruptly tapered end that is usually several to tens of micrometers in length. Figure 11.14a shows an SEM image of the uniform parts of a nanofiber with a uniform diameter of about 200 nm; Fig. 11.14b shows the abruptly tapered end. In the remainder of this chapter, we focus on the uniform part of the nanofiber, which can be used as a subwavelength-diameter waveguide for low-loss optical wave-guiding. Depending on the experimental conditions such as drawing temperature, force, and speed, the diameter of the taper-drawn silica nanofiber ranges from tens of nanometers to one micrometer. Figure 11.15 shows SEM images of silica nanofibers with diameters ranging from 50 to 400 nm, illustrating the range of dimensions and uniformity of the silica nanofibers.

The surface tension of the molten silica during the drawing process ensures that the cross-sections of the taper-drawn nanofibers are perfectly circular. Figure 11.16 shows an SEM image of the cross-section of a 480-nm diameter

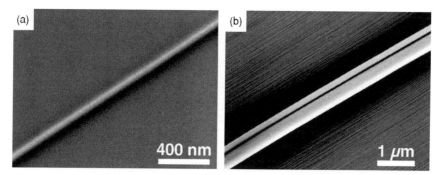

Figure 11.15 SEM images of (a) a silica nanofiber with a diameter of about 50 nm and (b) two parallel 170-nm and 400-nm diameter nanofibers. (Adapted from reference [10].)

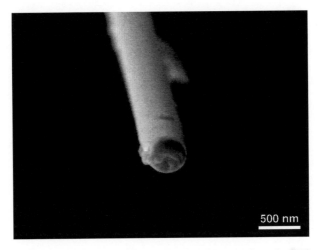

Figure 11.16 SEM image of the perfectly cylindrical cross section of a 480-nm diameter nanofiber.

nanofiber. The resulting cylindrical geometry of the nanofibers makes it possible to obtain exact expressions of the guided modes by solving Maxwell's equations analytically (see Section 11.2) [14, 15].

The length of the nanofibers depends on their diameter. Typically, nanofibers with diameters smaller than 200 nm can have length up to 1 mm; nanofibers with larger diameters can be as long as several hundreds of millimeters. For example, Fig. 11.17 shows an SEM image of a 4-mm long nanofiber with a diameter of 260 nm; the nanofiber is coiled up on the surface of a silicon wafer to show its length.

In addition to being long, taper-drawn nanofibers also provide excellent diameter uniformity and surface roughness (see, e.g., Figs. 11.14b and 11.15). We determined the diameter uniformity of the nanofibers by measuring the diameter variation ΔD along the entire length L with a scanning or transmitting electron microscope. Figure 11.18 shows the measured diameter D and diameter uniformity $U_D = \Delta D/L$ of a thin nanofiber along its length (starting from the thin end). Although the nanofiber exhibits an overall monotonic tapering, the central part of the nanofiber shows a very high uniformity. For example, in the region where $D = 30$ nm, $U_D = 1.2 \times 10^{-5}$, which means that in more than an 80-μm length of nanofiber the maximum diameter difference between the two ends is less than 1 nm. Thicker nanofibers show even better uniformities. For example, the 260-nm diameter nanofiber shown in Fig. 11.17 has a maximum diameter variation ΔD of about 8 nm over its 4-mm length, giving $U_D = 2 \times 10^{-6}$. Brambilla et al. [12] reported U_D as small as 5×10^{-7}.

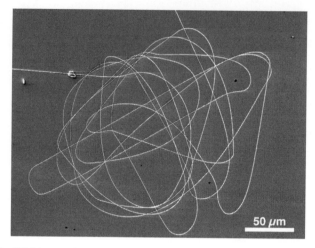

Figure 11.17   SEM image of a 4-mm long nanofiber with a diameter of 260 nm coiled up on the surface of a silicon wafer. (Adapted from reference [10].)

The small diameter of the nanofiber makes it possible to investigate the surface roughness with a TEM. The TEM image in Fig. 11.19 shows that the edge of a 330-nm diameter nanofiber has no irregularities or defects. The typical sidewall root-mean-square roughness of these fibers can be as small as 0.2 nm [65], approaching the intrinsic roughness of melt-formed glass surfaces [66, 67]. Considering that the length of Si-O bond is about 0.16 nm [68], such a roughness

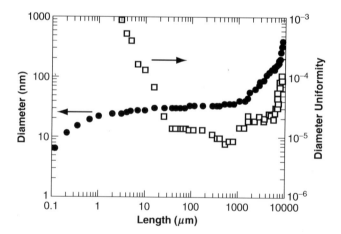

Figure 11.18   Diameter and diameter uniformity of a taper-drawn silica nanofiber measured along its length starting from the distal end. (Adapted from reference [65].)

Figure 11.19 TEM image of the surface of a 330-nm diameter silica nanofiber. The inset shows the electron diffraction pattern demonstrating that the nanofiber is amorphous. (Adapted from reference [10].)

represents an atomic-level smoothness of the nanofiber surface and is much lower than those of silica nanowires, tubes, or strips obtained using other fabrication methods [9, 69–74].

11.5 PROPERTIES OF SILICA NANOFIBERS

For optical applications, the most important properties of silica nanofibers are optical loss, mechanical strength, and pliability. In this section, we review the mechanical properties and optical losses of silica nanofibers and discuss techniques for micromanipulation and assembly.

11.5.1 Micromanipulation and Mechanical Properties

The ability to manipulate nanofibers individually is critical to their characterization and application. Because of their long length, silica nanofibers obtained by the taper-drawing method can be seen under an optical microscope, even when the fiber diameter is less than 100 nm. For example, Fig. 11.20 shows a photograph of a 60-nm diameter silica nanofiber taken under an optical microscope in dark-field reflection mode. The nanofibers can also clearly be

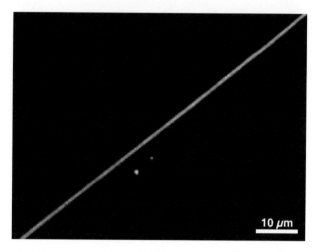

Figure 11.20   Optical microscope image of a 60-nm diameter silica nanofiber supported by a silicon wafer. The image is taken in dark-field reflection mode.

seen when they are supported on a silicon wafer. This optical visibility makes it possible to manipulate single nanofibers, greatly facilitating the handling, tailoring, and assembly of these nanofibers.

A typical experimental setup for the micromanipulation of silica nanofibers is shown in Fig. 11.21. An optical microscope objective is used to image the nanofiber onto a CCD camera for real-time monitoring. To hold and manipulate the nanofibers, probes from a scanning tunneling microscope (STM) are mounted on micromanipulators and placed as shown in Fig. 11.21a; Fig. 21b shows the bending of a nanofiber using the tips of two probes. The nanofibers

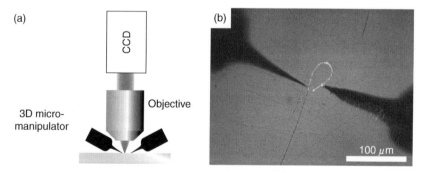

Figure 11.21   (a) Experimental setup for micromanipulating silica nanofibers. (b) Microscope image showing the bending of a silica nanofiber with two STM probes.

can be either freestanding in air or supported by high-index substrates for better visibility (e.g., silicon or sapphire wafers). Using micromanipulation under an optical microscope, the nanofibers can be cut, positioned, bent, and twisted with high precision. To cut a nanofiber to a desired length, a bend-to-fracture method can be applied by holding the fiber with two STM probes on silicon or sapphire substrate and using a third probe to bend the nanofiber to fracture at the desired point. This process leaves flat end faces at the fracture point, as shown in Fig. 11.22.

Because of their excellent uniformity, taper-drawn silica nanofibers show high mechanical strength and pliability. By rubbing the two ends of nanofibers on a finely polished substrate using a tilted probe, they can be twisted together without breaking. Figure 11.23 shows the ropelike twist obtained with a 480-nm diameter silica nanofiber on a silicon wafer. The "nanorope" retains its shape when it is lifted up from the substrate, indicating that the nanofiber can withstand shear deformation.

To position and bend a nanofiber, it is first placed on a finely polished substrate such as a silicon wafer, where it is tightly held in place by the van der Waals or electrostatic attraction between the nanofiber and the substrate, and then pushed by STM probes to a desired bending radius or position. The shape of elastic bends is retained after removing the STM probes because of the attraction between the nanofiber and the substrate. Figure 11.24 shows a 280-nm diameter nanofiber bent to a radius of 2.7 μm. The sharp bend in Fig. 11.24 indicates that the nanofiber has excellent flexibility and mechanical properties.

Figure 11.22 SEM images of the cut end faces of 140-, 420-, and 680-nm diameter silica nanofibers. (Adapted from reference [21].)

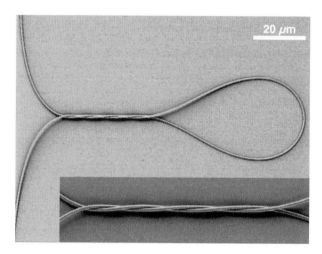

Figure 11.23   SEM image of a twisted 480-nm diameter silica nanofiber.

Using the Young modulus of ordinary silica fibers (73.1 GPa), we find that the tensile strength of the bent nanofiber in Fig. 11.24 is at least 4.5 GPa [75, 76]. Silica nanofibers can also be shaped into more complex forms and tied in knots. Figure 11.25 shows a 15-μm diameter knot assembled with a 520-nm diameter silica nanofiber. Such a ringlike structure can be used as a ring resonator in micro-optical components (see Section 11.5).

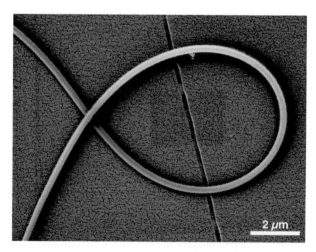

Figure 11.24   SEM image of a 280-nm diameter nanofiber elastically bent to a radius of 2.7 μm. (Adapted from reference [10].)

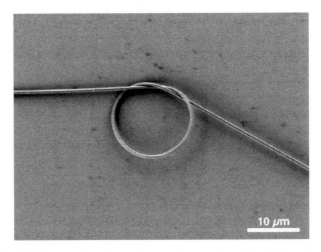

Figure 11.25 SEM image of a 15-μm diameter knot made with a 520-nm diameter silica nanofiber. (Adapted from reference [10].)

To avoid long-term fatigue and fracture due to bending stress [77, 78], the elastically bent nanofibers can be annealed around 1400 K to form permanent plastic deformation, without change in surface smoothness or diameter uniformity. Because the elastically bent fiber is held tightly on the substrate, the geometry of the assembly is not affected by the annealing, making it possible to lay out a final design before annealing. The annealing-after-bending process can also be performed repeatedly to obtain very tight bends or multiple bends as shown (see Figs. 11.26 and 11.27).

11.5.2 Wave-Guiding and Optical Loss

To investigate the optical and wave-guiding properties of silica nanofibers, it is necessary to couple light into and out of them. If a nanofiber still is connected to the starting fiber, one can couple light into the starting fiber in the standard manner to launch a guided wave into the nanofiber through the tapered region. To launch light into nanofibers with freestanding ends, one can use evanescent coupling between a pair of fibers, as shown in Fig. 11.28a. Light is first sent into the core of a single-mode fiber that is tapered down to a nanofiber and the nanotaper is then used to evanescently couple the light into another nanofiber by overlapping the two in parallel. Because of electrostatic and van der Waals forces, nanofibers attract one another, making a parallel contact connection. Fig. 11.28b shows an optical micrograph of the coupling of light between a

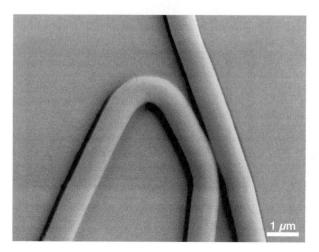

Figure 11.26 SEM image of a sharp plastic bend in an 800-nm diameter silica nanofiber. (Adapted from reference [21].)

390-nm diameter launching taper and a 450-nm diameter nanofiber. The coupling efficiency of this evanescent coupling can be as high as 90% when the fiber diameter and overlap length are properly selected. This method can also be used to couple light out of a nanofiber.

Because of the nanofibers' extraordinary uniformity, their optical losses are low. Figure 11.29 shows a 360-nm diameter nanofiber guiding light of 633-nm

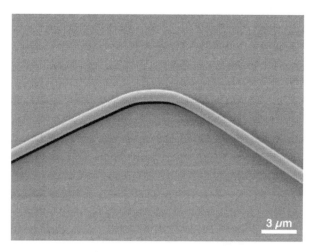

Figure 11.27 SEM image of a double plastic bend in a 940-nm diameter silica nanofiber. (Adapted from reference [21].)

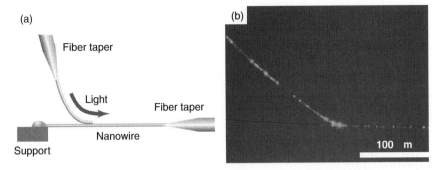

Figure 11.28 Launching light into a silica nanofiber. (a) Schematic diagram for launching light into a silica nanofiber using evanescent coupling. (b) Optical microscope image of a 390-nm diameter nanofiber coupling light into a 450-nm diameter silica nanofiber. (Adapted from reference [10].)

wavelength from the left. The scattering of light along its length is due to nanoparticles that are stuck to the wire and that scatter the evanescent wave. The light guided by the nanowire is intercepted at the right by a supporting 3-μm diameter taper to show qualitatively that the amount of light scattered by the fiber is small compared to that guided by it.

To quantify the optical losses of silica nanofibers, we measure the nanofibers' transmission as a function of the length [10, 12, 13]. The optical loss for free-standing nanofibers in air is shown in Fig. 11.30. In single-mode operation, the

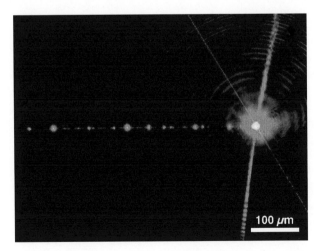

Figure 11.29 Optical micrograph of 633-nm wavelength light guided by a 360-nm diameter silica nanofiber in air. The nanofiber is intercepted by a 3-μm diameter fiber on the right to show the relative intensities of guided and scattered light. (Adapted from reference [10].)

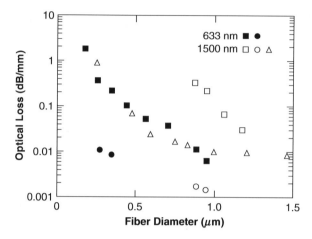

Figure 11.30 Optical loss in freestanding silica nanofibers measured in air at 633-nm and 1550-nm wavelengths. (From references [10, 12, 13].)

optical loss can be as low as 0.0014 dB/mm [13], which is much lower than the optical loss of other subwavelength structures such as metallic plasmon waveguides, nanowires, or nanoribbons [79–83]. The increasing loss with decreasing fiber diameter can be attributed to surface contamination; as the fiber diameter is reduced below the wavelength, more light is guided outside the fiber as an evanescent wave and becomes susceptible to scattering by surface contamination and/or microbends.

Because of their low-loss optical wave-guiding properties, these nanofibers are ideal building blocks for microphotonic applications. For example, low optical loss is essential to obtain a high Q-factor in an optical microcavity resonator [84], to maintain the coherence of the guided light in optical waveguide/fiber sensors using coherent detection [85–87], to reduce the noise or crosstalk in high-density optical integration, and to reduce energy consumption when many devices are connected in series.

11.6 APPLICATIONS AND POTENTIAL USES OF SILICA NANOFIBERS

As shown in the previous section, taper-drawn silica nanofibers can serve as low-loss subwavelength-diameter single-mode optical waveguides. Their high mechanical strength and pliability allows assembly into complex structures. Because silica is one of the fundamental materials for photonics, taper-drawn silica nanofibers hold great promise for nanoscale optical sensors [24–26], for low-energy nonlinear interactions and supercontinuum generation [13, 27–34], and for atom trapping and guiding [17, 35–37]. In this section, we discuss these applications in further detail.

11.6.1 Microscale and Nanoscale Photonic Components

Silica nanofibers have been used as building blocks in the assembly of a variety of microscale or nanoscale photonic components or devices such as linear waveguides, waveguide bends, optical couplers, and ring resonators [20–23]. Because of their small dimensions, low optical losses, evanescent wave-guiding, and mechanical flexibility, nanofiber-assembled photonic devices have a number of advantages over conventional photonic devices.

To assemble microphotonic devices from silica nanofibers, the fibers must be supported by a substrate. The relatively low index of silica (~1.45) necessitates a substrate with an index much lower than 1.45. A microphotonic device consisting of an assembly of silica nanofibers on a silica aerogel substrate has been reported [21]. Silica aerogel is a tenuous porous network of silica nanoparticles with a diameter of about 5 nm, much smaller than the wavelength of the guided light, and has a transparent optical spectral range similar to that of silica [88, 89]. Because the aerogel is mostly composed of air, its refractive index is similar to that of air (1.03–1.08). Figure 11.31 shows a close-up view of a 450-nm diameter

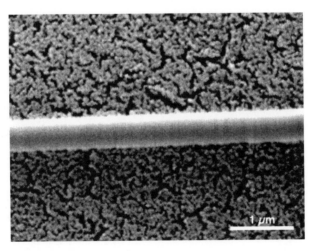

Figure 11.31 SEM image of a 450-nm diameter silica nanofiber supported by silica aerogel. (Adapted from reference [21].)

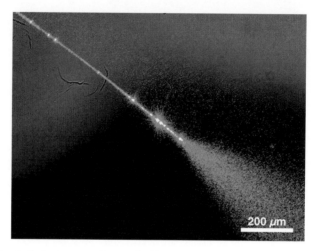

Figure 11.32 Optical microscopy image of a 380-nm diameter silica nanofiber guiding 633-nm wavelength light on the surface of silica aerogel. At the end of the fiber, the light spreads out and scatters on the aerogel surface. (Adapted from reference [21].)

silica fiber supported on a substrate of silica aerogel. Because the index difference between the silica aerogel and air (0.03–0.08) is much lower than the index difference between the silica nanofiber and air (~0.45), the optical guiding properties of aerogel-supported nanofibers are virtually identical to those of air-clad ones. Figure 11.32 shows a 380-nm diameter silica nanofiber guiding 633-nm wavelength light on the surface of a silica aerogel substrate. The uniform and virtually unattenuated scattering along the 0.5-mm length of the fiber and the strong output at the end face show that the scattering is small relative to the guided intensity. Figure 11.33 shows the measured optical loss of silica nanofibers supported by an aerogel substrate. The low loss provides further evidence that the aerogel substrate does not degrade the guiding of light through the nanofibers. For fibers with a diameter near the single-mode cutoff diameter, the loss is less than 0.06 dB/mm, much lower than the optical loss in other subwavelength structures and acceptable for most photonic applications. These data show that silica aerogel supported silica nanofibers can be used as low-loss single-mode linear waveguides, as well as building blocks for assembling micro-photonic devices.

By transferring plastic nanofibers bends (i.e., nanofibers that have been bent and annealed) onto a silica aerogel substrate, one can fabricate microscale waveguide bends with subwavelength diameters. Figure 11.34 shows a plastically bent 530-nm diameter silica nanofiber supported by a silica aerogel substrate. The nanofiber was first bent to a radius of about 8 μm on a sapphire wafer, annealed, and then transferred to silica aerogel. The aerogel-supported plastic

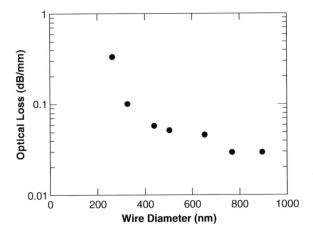

Figure 11.33 Optical loss of aerogel-supported nanofibers measured at a wavelength of 633 nm. (Adapted from reference 21].)

bends show excellent optical wave-guiding with good confinement of the light. Figure 11.35 shows an optical microscope image of 633-nm wavelength light guided through such an aerogel-supported plastic bend. The measured bending losses through a 90-degree bend in a 530-nm diameter nanofiber are shown in Fig. 11.36 as a function of bending radius. For example, the optical loss around a 5-μm radius bend in a 530-nm diameter fiber is less than 1 dB—acceptable

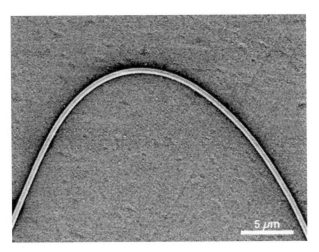

Figure 11.34 SEM image of an aerogel-supported 530-nm diameter nanofiber with a bending radius of 8 μm. (Adapted from reference [21].)

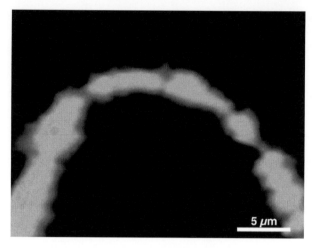

Figure 11.35 Optical microscopy image of an aerogel-supported 530-nm diameter nanofiber guiding light around a bend with a radius of 8 μm. (Adapted from reference [21].)

for use in photonics devices. In contrast, the bending of light by planar photonic crystal structures not only requires much more complex fabrication techniques, but also suffers from inevitable out-of-plane losses [90–92]. Aerogel-supported nanofiber bends thus offer the advantage of compact overall size, low coupling loss, simplicity, and easy fabrication. Furthermore, contrary to

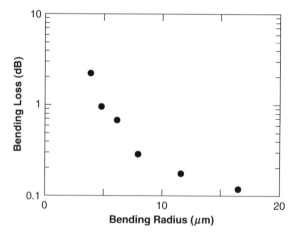

Figure 11.36 Bending loss around a 90-degree bend in an aerogel-supported 530-nm diameter nanofiber measured at a wavelength of 633 nm. (Adapted from reference [21].)

wavelength-specific photonic crystal structures, nanofiber bends can be used over a broad range of wavelengths, from the near-infrared to ultraviolet wavelengths.

Using these waveguide bends as building blocks, one can readily assemble an optical coupler. Figure 11.37 shows an X-coupler assembled from two 420-nm diameter silica fiber bends. When 633-nm wavelength light is launched into the bottom left arm, the coupler splits the flow of light in two. By changing the overlap between the two bends, it is possible to tune the splitting ratio of the coupler. With an overlap of less than 5 μm, the device works as a 3-dB splitter with an excess loss of less than 0.5 dB. In contrast, microscopic couplers such as fused couplers made from fiber tapers using conventional methods require an interaction length on the order of 100 μm [59]. Couplers assembled with silica nanofibers, thus, reduce the device size by more than an order of magnitude.

Another microphotonic device that can readily be fabricated from taper-drawn silica nanofibers is a microring resonator [10, 19, 20, 22, 23, 47, 93]. Figure 11.38 shows a 150-μm diameter microring made by tying a knot in an 880-nm diameter nanofiber. The measured transmittance of this microring for wavelengths near 1.55 μm is shown in Fig. 11.39. The transmittance clearly shows optical resonances with an extinction ratio that corresponds to a Q-factor of more than 1000. Microcoil/loop resonators with Q-factors as high as 95,000 have been realized [93], and a proposed microcoil resonator with self-coupling turns is expected to display a Q-factor as high as 10^{10} [19].

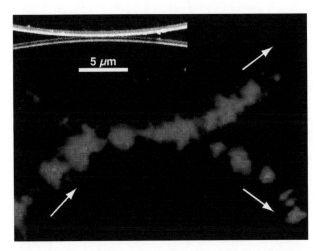

Figure 11.37 Optical microscope image of an X-coupler assembled from two 420-nm diameter silica nanofibers. The two fibers overlap less than 5 μm at the center (see SEM image in inset). The assembly acts as a 3-dB splitter for light launched into the bottom left branch. (Adapted from reference [21].)

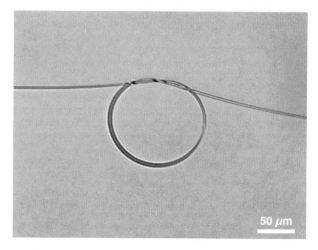

Figure 11.38 Optical microscope image of a 150-μm diameter microring fabricated from an 880-nm diameter silica fiber.

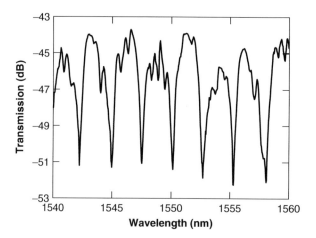

Figure 11.39   Transmission spectrum of the microring in Fig. 11.38.

11.6.2 Nanofiber Optical Sensors

As discussed in Section 11.2, one of the prominent optical properties of a subwavelength-diameter silica nanofiber is its ability to guide light with a large fraction of power propagating outside the solid core. This evanescent wave is highly sensitive to index changes in the environment and to microscale bending of the nanofiber. At the same time, the coherence of the guided light can be maintained over a considerable length because of the low wave-guiding loss of these nanofibers. These properties make silica nanofibers ideal for high-sensitivity nanoscale optical sensing.

A schematic diagram of a silica nanofiber sensing element is illustrated in Fig. 11.40. A single-mode silica nanofiber is exposed to or immersed in a gaseous or liquid environment containing the molecules to be detected. The nanowire can readily be functionalized with the appropriate receptors for the molecules to be detected. If the nanowire is guiding light, any index change around the fiber due to the binding of molecules to the receptors or a temperature change affects the guided light's optical phase and intensity. By detecting the signal at the output, one can thus obtain information about the environment of the nanofiber. Numerical simulations show that if a Mach-Zehnder interferometer is used to detect phase shifts in the guided light, the sensitivity of a nanofiber sensor can be more than one magnitude higher than that of a conventional fiber/waveguide optical sensor [24]. Most importantly, the size of the sensing element is greatly reduced using nanofibers.

Several optical sensors based on subwavelength- or nanometer-diameter silica fibers have been experimentally realized. A nanofiber optical sensor for measuring the refractive index of liquids propagating in microfluidic channels was made from a 700-nm diameter fiber taper that was tapered from a standard single-mode fiber and immersed in a transparent curable soft polymer [25]. A channel for the liquid analyte was created in the immediate vicinity of the taper waist. Light propagating through the nanotaper extends into the channel, making the optical loss in the system sensitive to the refractive-index difference between the

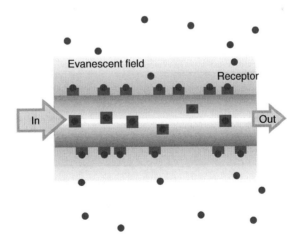

Figure 11.40 Schematic diagram of a silica nanofiber-sensing element. (Adapted from reference [24].)

polymer and the liquid. The estimated sensitivity of this refractive-index sensor is about 5×10^{-4}.

A miniature hydrogen sensor consisting of a subwavelength diameter tapered optical fiber coated with an ultrathin palladium film has also been reported [26]. The hydrogen changes the optical properties of the palladium layer and consequently the absorption of the evanescent waves. Measurements at a wavelength of 1550 nm show that the sensor's response time ($\sim$10 seconds) is several times faster than that of several optical and electrical hydrogen sensors reported so far. Moreover, the sensor is small, reversible, and suitable for detection of hydrogen in the lower explosive limit.

11.6.3 Additional Applications

Besides integration into microphotonic devices and optical sensing, silica nanofibers have been applied in nonlinear optics and supercontinuum generation [13, 27–34] and in atom trapping and guidance [17, 35–37]. Nonlinear optical interactions in nanofibers have been extensively investigated [27, 29–34]. Supercontinuum generation was reported in submicrometer fibers and in microstructured optical fibers with subwavelength core diameter [13, 28]. Because of the tight mode confinement and strong waveguide dispersion, subwavelength-diameter nanotapers or fibers exhibit nonlinear optical properties at relatively low-power and short interaction length. In another promising application, silica nanofibers were used to trap and guide atoms by the optical force of the evanescent field around the fibers [17, 35–37]. It was shown that the gradient force of a red-detuned evanescent-wave field in the fundamental mode of a silica nanofiber can balance the centrifugal force of the atoms [35]. Likewise, using a two-color evanescent light field around a nanofiber a net potential with large depth, coherence time, and trap lifetime can be produced [36].

REFERENCES

[1] Yamane, M. and Y. Asahara. 2000. *Glasses for Photonics*. Cambridge University Press, Cambridge.

[2] Murata, H. 1996. *Handbook of Optical Fibers and Cables*. Marcel Dekker, New York.

[3] Mynbaev, D. K. and L. L. Scheiner. 2001. *Fiber-Optic Communications Technology*. Prentice Hall, New York.

[4] Huang, M. H. et al. 2001. Room-temperature ultraviolet nanowire nanolasers. *Science* **292**, 1897–1899.

[5] Gudiksen, M. S. et al. 2002. Growth of nanowire superlattice structures for nanoscale photonics and electronics. *Nature* 415:617–620.

[6] Prasad, P. N. 2004. *Nanophotonics*. John Wiley & Sons, Hoboken.

[7] Morales, A. M. and C. M. Lieber. 1998. A laser ablation method for the synthesis of crystalline semiconductor nanowires. *Science* 279:208–211.

[8] Romanato, F. et al. 2003. X-ray and electron-beam lithography of three- dimensional array structures for photonics. *J. Vacs. Sci. Technol.* B21:2912–2917.

[9] Xia, Y. et al. 1999. Unconventional methods for fabricating and patterning nanostructures. *Chem. Rev.* 99:1823–1848.

[10] Tong, L. et al. 2003. Subwavelength-diameter silica wires for low-loss optical wave guiding. *Nature* 426:816–819.

[11] Domachuk, P. et al. 2004. Photonics: Shrinking optical fibers. *Nat. Mater.* 3:85–86.

[12] Brambilla, G., et al. 2004. Ultra-low-loss optical fiber nanotapers. *Opt. Express* 12: 2258–2263.

[13] Leon-Saval, S. G. et al. 2004. Mason. Supercontinuum generation in submicron fibre waveguides. *Opt. Express* 12:2864–2869.

[14] Snyder, A. W. and J. D. Love. 1983. *Optical Waveguide Theory*. Chapman and Hall, New York.

[15] Tong, L. et al. 2004. Single-mode guiding properties of subwavelength-diameter silica and silicon wire waveguides. *Opt. Express* 12:1025–1035.

[16] Bures, J. and R. Ghosh. 1999. Power density of the evanescent field in the vicinity of a tapered fiber. *J. Opt. Soc. Am.* A16:1992–1996.

[17] Le Kien, F. et al. 2004. Field intensity distributions and polarization orientations in a vacuum-clad subwavelength-diameter optical fiber. *Opt. Commun.* 242:445–455.

[18] Zheltikov, A. M. 2005. Birefringence of guided modes in photonic wires: Gaussian-mode analysis. *Opt. Commun.* 252:78–83.

[19] Sumetsky, M. 2004. Optical fiber microcoil resonator. *Opt. Express* 12:2303–2316.

[20] Sumetsky, M. et al. 2004. Fabrication and study of bent and coiled free silica nanowires: Self-coupling microloop optical interferometer. *Opt. Express* 12:3521–3531.

[21] Tong, L. et al. 2005. Assembly of silica nanowires on silica aerogels for microphotonic devices. *Nano Lett.* 5:259–262.

[22] Sumetsky, M. et al. 2005. Optical microfiber loop resonator. *Appl. Phys. Lett.* 86:161108.

[23] Sumetsky, M. 2005. Uniform coil optical resonator and waveguide: Transmission spectrum, eigenmodes, and dispersion relation. *Opt. Express* 13:4331–4340.

[24] Lou, J. et al. 2005. Modeling of silica nanowires for optical sensing. *Opt. Express* 13: 2135–2140.

[25] Polynkin, P. et al. 2005. Evanescent field-based optical fiber sensing device for measuring the refractive index of liquids in microfluidic channels. *Opt. Lett.* 30:1273–1275.

[26] Villatoro, J. and D. Monzón-Hernández. 2005. Fast detection of hydrogen with nano fiber tapers coated with ultra thin palladium layers. *Opt. Express* 13:5087–5092.

[27] Foster, M. A. et al. 2004. Optimal waveguide dimensions for nonlinear interactions. *Opt. Express* 12:2880–2887.

[28] Lize, Y. K. et al. 2004. Microstructured optical fiber photonic wires with subwavelength core diameter. *Opt. Express* 12:3209–3217.

[29] Kolesik, M. et al. 2004. Simulation of femtosecond pulse propagation in sub-micron diameter tapered fibers. *Appl. Phys.* B79:293–300.

[30] Magi, E. C. et al. 2005. Air-hole collapse and mode transitions in microstructured fiber photonic wires. *Opt. Express* 13:453–459.

[31] Moison, J. M. et al. 2005. Light transmission in multiple or single subwavelength trefoil channels of microstructured fibers. *Opt. Express* 13:1193–1201.

[32] Brambilla, G. et al. 2005. Compound-glass optical nanowires. *Electron. Lett.* 41:400–402.

[33] Qing, D. K. and G. Chen. 2005. Nanoscale optical waveguides with negative dielectric claddings. *Phys. Rev.* B71:153107.

[34] Zheltikov, A. 2005. Gaussian-mode analysis of waveguide-enhanced Kerr-type nonlinearity of optical fibers and photonic wires. *J. Opt. Soc. Am.* B22:1100–1104.

[35] Balykin, V. I. et al. 2004. Atom trapping and guiding with a subwavelength-diameter optical fiber. *Phys. Rev.* A70:011401.

[36] Le Kien, F. et al. 2004. Atom trap and waveguide using a two-color evanescent light field around a subwavelength-diameter optical fiber. *Phys. Rev.* A70:063403.

[37] Le Kien, F. et al. 2005. State-insensitive trapping and guiding of cesium atoms using a two-color evanescent field around a subwavelength-diameter fiber. *J. Phys. Soc. Jpn.* 74:910–917.

[38] Snitzer, E. 1961. Cylindrical dielectric waveguide modes. *J. Opt. Soc. Am.* **51**, 491–498.

[39] Marcuse, D. 1974. *Theory of Dielectric Optical Waveguides*. Academic Press, New York.

[40] Ghatak, A. and K. Thyagarajan. 1998. *Introduction to Fiber Optics*. Cambridge University Press, Cambridge.

[41] Klocek, P. 1991. *Handbook of Infrared Optical Materials*. Marcel Dekker, New York.

[42] Manolatou, C. et al. 1999. High-density integrated optics. *J. Lightwave Technol.* 17:1682–1692.

[43] Kakarantzas, G. et al. Miniature based on CO_2 laser microstructuring of tapered fibers. *Opt. Lett.* 26:1137–1139.

[44] Qi, Z. M. et al. 2002. A design for improving the sensitivity of a Mach-Zehnder interferometer to chemical and biological measurands. *Sensors Actuat.* B81:254–258.

[45] Tong, L. and E. Mazur. 2005. Subwavelength-diameter silica wires for microscale optical components. *SPIE Proc.* 5723:105–112.

[46] Saleh, B. E. A. and M. C. Teich. 1991. *Fundamentals of Photonics*. John Wiley & Sons, New York.

[47] Bishnu, P. P. 1993. *Fundamentals of Fibre Optics in Telecommunication and Sensor Systems*. John Wiley & Sons, New York.

[48] Birks, T. A. et al. 2000. Supercontinuum generation in tapered fibers. *Opt. Lett.* 25:1415–1417.

[49] Mollenauer, L. F. 2003. Nonlinear optics in fibers. *Science* 302:996–997.

[50] Marcuse, D. 1969. Mode conversion caused by surface imperfections of a dielectric slab waveguide. *Bell Syst. Tech. J.* 48:3187–3215.

[51] Marcuse, D. and R. M. Derosier. 1969. Mode conversion caused by diameter changes of a round dielectric waveguide. *Bell Syst. Tech. J.* 48:3217–3232.

[52] Ladouceur, F. 1997. Roughness, inhomogeneity, and integrated optics. *J. Lightwave Technol.* 15:1020–1025.

[53] Lee, K. K. et al. 2000. Effect of size and roughness on light transmission in a Si/SiO2 waveguide: Experiments and model. *Appl. Phys. Lett.* 77:1617–1619.

[54] Boys, C. V. 1887. On the production, properties, and some suggested uses of the finest threads. *Phil. Mag.* 23:489–499.

[55] Threlfall, R. 1898. *On Laboratory Arts*. Macmillan, London.

[56] Knight, J. C. et al. 1997. Phase-matched excitation of whispering-gallery mode resonances by a fiber taper. *Opt. Lett.* 22:1129–1131.

[57] Dimmick, T. E. et al. 1999. Carbon dioxide laser fabrication of fused-fiber couplers and tapers. *Appl. Opt.* 38:6845–6848.

[58] Cai, M. and K. Vahala. 2001. Highly efficient hybrid fiber taper coupled microsphere laser. *Opt. Lett.* 26:884–886.

[59] Kakarantzas, G. et al. 2001. Miniature all-fiber devices based on CO_2 laser microstructuring of tapered fibers. *Opt. Lett.* 26:1137–1139.

[60] Grellier, A. J. C. et al. 1998. Heat transfer modeling in CO_2 laser processing of optical fibers. *Opt. Commun.* 152:324–328.

[61] Labelle, H. E. and A. I. Mlavsky. 1967. Growth of sapphire filaments from melt. *Nature* 216:574–575.

[62] Wang, Z. L. 2003. *Nanowires and Nanobelts: Materials, Properties and Devices*. Kluwer—Academic, New York.

[63] Zhu, T. et al. 2003. Deformation and fracture of a SiO_2 nanorod. *Mol. Simul.* 29:671–676.

[64] Hu, J. T. et al. 1999. Chemistry and physics in one dimension: Synthesis and properties of nanowires and nanotubes. *Acc. Chem. Res.* 32:435–445.

[65] Tong, L. et al. 2005. Self-modulated taper drawing of silica nanowires. *Nanotechnology* 16:1445–1448.

[66] Jackle, J. and K. Kawasaki. 1995. Intrinsic roughness of glass surfaces. *J. Phys. Condens. Matter* 7:4351–4358.

[67] Radlein, E. and G. H. Frischat. 1997. Atomic force microscopy as a tool to correlate nanostructure to properties of glasses. *J. Non-Cryst. Solids* 222:69–82.

[68] Bansal, N. P. and R. H. Doremus. 1986. *Handbook of Glass Properties*. Academic Press, Orlando.

[69] Wang, Z. L. et al. 2000. Silica nanotubes and nanofiber arrays. *Adv. Mater.* 12:1938–1940.

[70] Wang, Z. L. 2004. Functional oxide nanobelts: Materials, properties and potential applications in nanosystems and biotechnology. *Ann. Rev. Phys. Chem.* 55:159–196.

[71] Pan, Z. W. et al. 2002. Molten gallium as a catalyst for the large-scale growth of highly aligned silica nanowires. *J. Am. Chem. Soc.* 124:1817–1822.

[72] Wang, J. C. et al. 2003. Silica nanowire arrays. *Solid State Commun.* 125:629–631.

[73] Hu, J. Q. et al. 2003. Fabrication of germanium-filled silica nanotubes and aligned silica nanofibers. *Adv. Mater.* 15:70–73.

[74] Sun, S. H. et al. 2003. Preparation and characterization of oriented silica nanowires. *Solid State Commun.* 128:287–290.

[75] Matthewson, M. J. et al. 1986. Strength measurement of optical fibers by bending. *J. Am. Ceram. Soc.* 69:815–821.

[76] Krause, J. T. et al. 1979. Deviations from linearity in the dependence of elongation upon force for fibers of simple glass formers and of glass optical light guides. *Phys. Chem. Glasses* 20:135–139.

[77] Matthewson, M. J. and C. R. Kurkjian. 1987. Static fatigue of optical fibers in bending. *J. Am. Ceram. Soc.* 70:662–668.

[78] Annovazzi-Ledi, V. et al. 1997. Statistical analysis of fiber failures under bending-stress fatigue. *J. Lightwave Technol.* 15:288–293.

[79] Takahara, J. et al. 1997. Guiding of a one-dimensional optical beam with nanometer diameter. *Opt. Lett.* 22:475–477.

[80] Maier, S. A. et al. 2003. Local detection of electromagnetic energy transport below the diffraction limit in metal nanoparticle plasmon waveguides. *Nat. Mater.* 2:229–232.

[81] Maier, S. A. et al. 2002. Observation of coupled plasmon-polarization modes in Au nanoparticle chain waveguides of different lengths: Estimation of waveguide loss. *Appl. Phys. Lett.* 81:1714–1716.

[82] Law, M. et al. 2004. Nanoribbon waveguides for subwavelength photonics integration. *Science* 305:1269–1273.

[83] Barrelet, C. J. et al. 2004. Nanowire photonic circuit elements. *Nano Lett.* 4:1981–1985.

[84] Vahala, K. J. 2003. Optical microcavities. *Nature* 424:839–846.

[85] Abel, A. P. et al. 1996. Fiber-optic evanescent wave biosensor for the detection of oligonucleotides. *Anal. Chem.* 68:2905–2912.

[86] Qi, Z. M. et al. 2002. A design for improving the sensitivity of a Mach-Zehnder interferometer to chemical and biological measurands. *Sens. Actuators* B81:254–258.

[87] Prieto, F. et al. 2003. *Sens. Actuators* B92:151–158.

[88] Pierre, A. C. and G. M. Pajonk. 2002. Chemistry of aerogels and their applications. *Chem. Rev.* 102:4243–4266.

[89] Akimov, Y. K. 2003. Fields of application of aerogels [Review]. *Instrum. Exp. Technol.* 46:287–299.

[90] Joannopoulos, J. D. et al. 1995. *Photonic Crystals: Molding the Flow of Light.* Princeton University Press, Princeton.

[91] Moosburger, J. et al. 2001. Enhanced transmission through photonic-crystal-based bent waveguides by bend engineering. *Appl. Phys. Lett.* 79:3579–3581.

[92] Augustin, M. et al. 2004. High transmission and single-mode operation in low-index-contrast photonic crystal waveguide devices. *Appl. Phys. Lett.* 84:663–665.

[93] Sumetsky, M. et al. 2005. Demonstration of the microfiber loop optical resonator. Optical Fiber Communication Conference, Postdeadline papers. Paper PDP10, Anaheim.

Chapter 12

Chiral Fibers

Victor I. Kopp* and Azriel Z. Genack[†]

*Chiral Photonics, Inc., Clifton, New Jersey
[†]Department of Physics, Queens College of CUNY, Flushing, New York

12.1 INTRODUCTION

Specialized fibers are increasingly being used to manipulate light and to couple light of different wavelengths into and out of fibers in telecommunications and sensing applications. The development of new communication modalities, such as cellular, satellite, and cable communications, has only spurred the growth of optical fiber networks. Wavelength selectivity is achieved by imposing a periodic modulation of the refractive index along the fiber. This is accomplished most often by exposing photosensitive fiber to modulated ultraviolet (UV) light [1]. In gratings with periods greatly exceeding the wavelength, the refractive index may also be modulated by microbending [2], such as may be produced by squeezing the fiber between corrugated plates or by local heating with a CO_2 laser [3] or with an electric arc [4].

Chiral fibers employ an alternative means of implementing periodicity into a glass fiber, which allows for polarization and wavelength selectivity. This extends the functionality of optical fibers and is advantageous in a variety of filter, polarizer, sensor, and laser applications. An example of a double-helix chiral fiber is shown in Fig. 12.1. Glass fibers with cores that are either concentric and birefringent, or nonconcentric, are twisted at a high rate as they are passed through a miniature oven in a drawing tower such as the one shown in Fig. 12.2. The fiber preform, typically from 100 to 200 μm in diameter, is held between a twisting motor affixed to a translation stage on top and a second translation stage on bottom. The speeds of the translation stages, the twisting rate, and the temperature of the short heat zone are varied under computer control to control the diameter and pitch of the fiber along its length. The fibers are heated as they pass through a microfilament or a mini–torch-based oven in which the temperature is monitored by either an infrared (IR) camera or a

401

402 noChiral Fibers

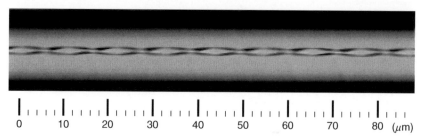

Figure 12.1 Side-view of a double-helix chiral fiber grating.

thermocouple. A right- or left-handed structure is produced depending on the sense of the twisting motor rotation. This flexible fabrication approach produces a stable structure that has double-helix symmetry in the case of twisted concentric birefringent fibers or single-helix symmetry in the case of twisted nonconcentric fibers.

High-contrast chiral gratings can be implemented in a broad range of glass materials free of the constraint that these be photosensitive. In double-helix structures, resonance interactions only occur for co-handed circularly polarized light with the same handedness as the structure, whereas in single-helix structures with nonbirefringent cores, resonance interactions are polarization insensitive. In addition to gratings with constant pitch, gratings with smoothly varying pitch, which naturally arise in twisting fiber preforms to a specific pitch, can be used to produce broadband in-fiber linear polarizers. In addition to a gradual pitch variation, the phase of the helical fiber may be disrupted by an abrupt twist that creates localized states [5, 6] suitable for narrow-band filter and laser applications.

12.2 THREE TYPES OF CHIRAL GRATINGS

Chiral fibers have distinct functionalities in each of three ranges of the ratio of the pitch of the chiral fiber to the optical wavelength in the fiber [7–9], $P/\lambda \equiv Q$. The three types of chiral fiber gratings are (1) resonant chiral short-period gratings (CSPGs) [7] with pitch equal to the optical wavelength of the order of 1 μm which reflect light within the fiber core, (2) nonresonant chiral intermediate-period gratings (CIPGs) [8] with pitch on the order of 10 μm, which scatters light out of the core, and (3) chiral long-period gratings (CLPGs) [8] with pitch on the order of 100 μm, which resonantly couples core modes into co-propagating cladding optical modes. While the single-helix structures are polarization insensitive, the double-helix structures only interact with co-handed circularly or elliptically polarized light, which have the same handedness as the grating and

Figure 12.2 Infrared camera-based twisting tower producing chiral gratings.

freely transmit cross-handed light of the orthogonal polarization in the CSPG and CLPG, as well as at the edge of the nonresonant scattering band in the CIPG.

Double-helix chiral fiber gratings and their optical interactions are illustrated schematically in Fig. 12.3. First, CSPGs reflect co-handed light within the fiber core within a stop band corresponding to a range of wavelength within the

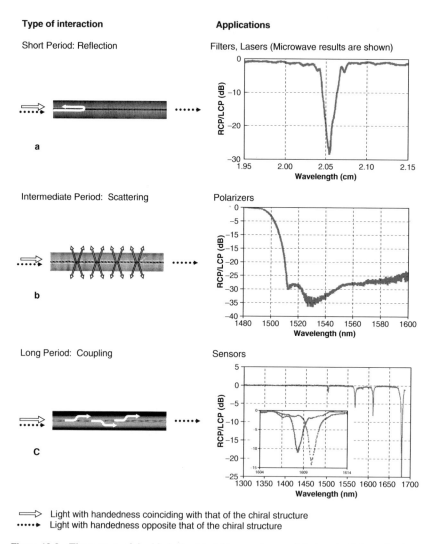

Figure 12.3 Three types of double-helix chiral fiber gratings and their potential applications.

fiber for which $Q = 1$. Cross-handed light of the orthogonal polarization is freely transmitted. A series of long-lived modes occur at the edges of the stop band for co-handed light. Such fibers can serve as polarization-selective spectral filters with a bandwidth relative to the band-center wavelength equal to the fractional fiber birefringence. Because the lowest order mode at the two edges of the stop band are significantly longer lived than other modes [10], these fibers may serve as the basis for lasers when appropriately doped and pumped [11]. It may also be possible to create lasing in spectrally isolated long-lived defect modes in fibers in which an additional twist is introduced [5, 6, 12]. Second, CIPGs with values of Q over a broad range between those of the CSPG and the CLPG scatter light out of the fiber. Near the edges of this band, only co-handed light is scattered. As a result, these may serve as polarization and wavelength-selective filters over this polarization-selective scattering band. Third, CLPGs with ratio $Q \approx 100$ resonantly couple core modes into co-propagating cladding optical modes when the difference between the propagation constants of the core and various cladding modes is compensated by the grating constant. The sensitivity of these resonances to the optical characteristics of the cladding and its surroundings makes the fibers ideal for sensing pressure, temperature, and liquid fluid level [13].

An overview of the performance of chiral fiber gratings with double-helix symmetry is presented in Fig. 12.4. The figure gives the variation of the ratio of right-to-left circularly polarized transmission with Q using data that are shown on the right side of Fig. 12.3, which is discussed in detail later in this chapter. Measurements for small values of Q covering the resonant gap and the long-

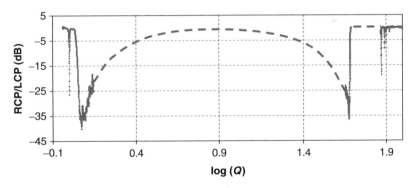

Figure 12.4 Overview of performance of a double-helix chiral fiber grating giving the ratio of right-to-left circularly polarized transmission versus $Q = P/\lambda$. Measurements for small values of Q covering the resonant band-gap and the long-wavelength side of the nonresonant scattering gap were obtained for microwave transmission through a chiral rod, while measurements at the short-wavelength side of the nonresonant band and at larger values of Q were obtained from optical transmission. The dashed line connects the spectral ranges in which experimental results were obtained.

wavelength side of the nonresonant scattering gap were obtained for microwave transmission through a chiral rod, whereas measurements at the short-wavelength side of the intermediate nonresonant band were obtained for IR radiation in the telecommunications range. Because neither measurements nor calculations were carried out in the central portion of the nonresonant scattering gap, this region is indicated by the dashed curve. This curve reflects our belief that scattering will persist, and polarization selectivity will be lost in this region in which light is scattered at a substantial angle from the fiber axis.

A similar study was conducted for single-helix chiral fibers in two ranges of the parameter Q corresponding to the CIPG and CLPG in the optical spectral range and for $Q \sim 1$ corresponding to the CSPG in the microwave range. The results for single-helix chiral fibers show that transmission is polarization insensitive with an overall performance for any polarization, which is similar to that for RCP waves used in Fig. 12.4. Thus, the performance of single-helix gratings is similar to that of isotropic FBGs [1] and long-period gratings (LPGs) [14].

12.3 CHIRAL SHORT-PERIOD GRATING: IN-FIBER ANALOG OF CLC

12.3.1 Fabrication Challenges

CIPG and CLPG structures are produced with sufficient precision that high-quality optical polarizers, filters, and sensors have been manufactured. The precision of short-pitch optical CSPGs that have been produced so far is lower, reflecting the greater fabrications challenge. Although optical CSPGs appear uniform under microscopic inspection, polarization-selective stop bands with spectral shape appropriate for filter applications have not been produced. The properties of CSPGs will, therefore, be illustrated via microwave propagation in scaled up versions of the structure shown in Fig. 12.1.

12.3.2 Analogy to 1D Chiral Planar Structure

The polarization and wavelength-selective properties of chiral fibers rely on the symmetry of the structure. Because specific features of optical interactions arise as the symmetry of structures is lowered, an appreciation for the operation of chiral gratings can be obtained by comparing the optical interactions in chiral fibers to those in periodic structures with higher symmetry. The highest symmetry periodic structure with refractive index modulation in a single direction consists of alternating isotropic layers. Propagation of waves directed along the normal to the layers may be described using a one-dimensional (1D) model. The symmetry is lower in anisotropic planar chiral structures such as cholesteric liquid crystals (CLCs) and in periodic isotropic fibers such as fiber Bragg gratings (FBGs). Chiral short-period fiber gratings are obtained by lowering the symmetry in either of these structures; they lack the transverse translation symmetry of CLCs and the axial rotation symmetry of FBGs.

12.3.3 Comparison of 1D Chiral to 1D Isotropic Layered Structures

In the simplest periodic dielectric structure, composed of alternating layers with different refractive indices, orthogonal modes in different polarization states propagating normal to the structure are degenerate and polarization is maintained throughout the structure. Light may, therefore, be taken to be linearly polarized without loss of generality. As a result of interference in the periodic structure, light cannot propagate normal to the planes for wavelengths within certain bands. Light in these forbidden bands or band gaps penetrates the sample as a standing exponentially decaying evanescent wave. Because there are no propagating optical modes within the band gap, the density of photon states vanishes as the sample size increases. The energy density of the standing wave component of the light in the first mode at the high-frequency band edge has maxima in the low-index layers and nodes in the high-index layers. This leads to a concentration of energy in regions with low refractive index, which is consequently referred to as the *air band* [15]. The opposite situation, in which nodes fall within the low-index layer while maxima coincide with the high-index layers, prevails at the low-frequency edge. This leads to a concentration of energy in the region with high refractive index. This spectral range is, therefore, called the *dielectric band.* The band structure in the vicinity of the stop band and the energy density near the center of the sample for light resonant with the first modes at the high- and low-frequency edges of the stop band are illustrated schematically in Fig. 12.5. The wavelength of the evanescent wave within the band gap and at the first modes at the band edges is twice the period, $\lambda = 2a$, as illustrated in Fig. 12.5b. Higher order band gaps occur at values of $k = 2\pi/\lambda$, which are multiples of π/a.

We next consider optical propagation in samples in which full-rotation symmetry is replaced by double-helix symmetry. This occurs in anisotropic-layered CLCs and in structured thin films (STFs). CLCs occur naturally in certain beetles and can be synthesized at different pitches in mixtures containing chiral moieties between glass plates in which the molecules are oriented and anchored on the inside surfaces of the plates [16]. STFs are produced by oblique deposition of dielectric materials on a rotating substrate [17, 18]. The molecular ordering for CLCs is illustrated in Fig. 12.6b. Rod-shaped molecules, represented by short

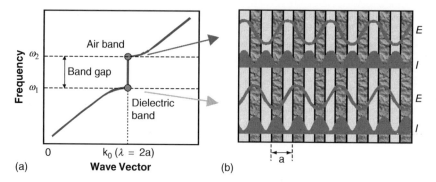

Figure 12.5 (a) Photonic band structure of a layered dielectric system with period *a*. (b) Dark and light layers correspond to high and low refractive indices, respectively. The electric field (E) and intensity (I) near the center of the sample are shown.

bars, are partially ordered in each layer with an average polarization along a direction called the *molecular director*. The director rotates with displacement perpendicular to the birefringent planes to form a periodic helical macrostructure, which can be either right- or left-handed with period *a* and pitch $P = 2a$. The pitch of the helix can be as small as 100 nm.

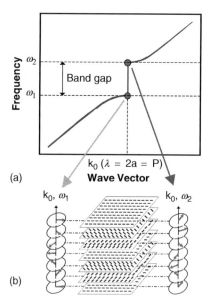

Figure 12.6 (a) Photonic band structure of a CLC with period *a* and pitch $P = 2a$. (b) Arrows indicate the electric field direction being aligned along or perpendicular to the director.

For sufficiently thick films, normally incident co-handed circularly-polarized light is nearly totally reflected within a band centered at vacuum wavelength λ_c such that the wavelength within the medium equals the pitch, $\lambda_c/n = P$, where n is the average of the extraordinary and ordinary refractive indices of the medium, n_e and n_o, respectively. The bandwidth corresponding to the difference in vacuum wavelength between the first modes on the long- and short-wavelength sides of the band is $\Delta\lambda = \lambda_c \Delta n/n$, where $\Delta n = n_e - n$, and $n = (n_e + n_o)/2$ [19]. Over this wavelength range, the standing evanescent wave is composed of counter-propagating waves of the same sense of circular polarization with equal amplitudes at any depth within the sample. In any plane of the structure, the electric field oscillates in a line with a fixed orientation relative to the director [20] (Fig. 12.6b). At the first mode at the long-wavelength edge of the band, the polarization is parallel to the ordinary axis, while at the other extreme of the stop band, it is aligned along the extraordinary axis. As the wavelength is tuned across the stop band, the polarization rotates by 90 degrees in every plane.

The reflection spectrum of co-handed light is qualitatively similar to that in an isotropic-layered sample. In contrast, however, cross-handed radiation is freely transmitted by the chiral structure.

Propagation near the stop band of isotropic binary samples and in anisotropic chiral structures may be compared using a scattering matrix computer simulation. In the simulation, the CLC is treated as a set of equal-thickness anisotropic layers with thickness significantly less than the wavelength of light with the direction of the axes of the optical indicatrix in successive layers rotated by the same small angle within the plane of the layer. The simulation gives the transmittance and reflectance spectra of the CLC and the layered dielectric structures, as well as the intensity distribution inside the sample.

The results of the computer simulation of electromagnetic energy density within the sample for linearly polarized light in a layered dielectric sample and for co-handed circularly polarized light in a CLC structure are compared at the first and second modes at the band edge in Fig. 12.7 and Fig. 12.8. Results are for structures with indices of refraction 1.47 and 1.63, period $a = 0.2\,\mu m$, and total thickness 16 mm. In the layered binary material, the indices correspond to those of two equal-thickness layers, whereas they correspond to ordinary and extraordinary indices in the CLCs. Interference in layered structures leads to strong modulation of light in layered structures on the scale of a wavelength. In contrast, wavelength-scale oscillations of the electromagnetic energy within the CLC sample are nearly absent within the stop band.

This difference between the intensity distribution within the sample in layered and CLC structures is displayed for the first and second modes at the stop-band edge. The low-frequency modulation of intensity in Fig. 12.7 and Fig. 12.8 reflects the Bloch wave vector of these modes. Resonances at a particular

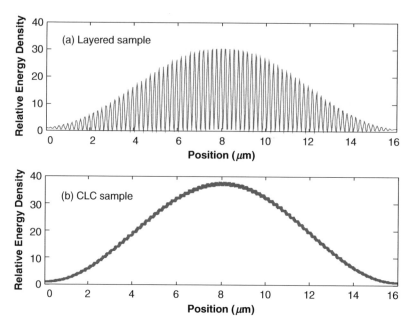

Figure 12.7 Distribution of the energy density of the electromagnetic field inside a one-dimensional periodic sample at the wavelength of the $n = 1$ mode for layered (a) and CLC samples (b). The refractive indices are 1.47 and 1.63 for the layers of the layered dielectric structure and for the ordinary and extraordinary indices of the CLC. The sample thickness is 16 μm and the period is 0.2 μm. The energy density of the incident wave is unity.

thickness occur at multiples of half of the Bloch wavelength. The number of peaks inside the medium gives the mode number from the band edge.

The differences in the nature of light propagation through layered and chiral structures lead to quantitative changes in transmission and reflection spectra. Figure 12.9 shows transmittance spectra calculated for layered and CLC structures with the same parameters as in Fig. 12.7 and Fig. 12.8. The stop band is wider for co-handed light in the CLC than in the binary layered structure because the field for the modes at the two band edges in CLC is aligned along or perpendicular to the director throughout the structure, as seen in Fig. 12.6, so the full index difference is experienced. In contrast, the field experiences an index of refraction in binary samples at the band edges, which lies between values in each of the layers because the intensity is not strictly confined to one or another of the dielectric layers. At the same time, the integrated intensity of the wave on resonance with band-edge modes is seen in Fig. 12.7 and Fig. 12.8 to be larger in the CLC than in the binary structure. First, the intensity is not modulated in the chiral structure, which contributes a factor of 2 in integrated intensity relative to that in the binary sample, and second, the peak intensity is somewhat greater in the chiral sample. The enhanced

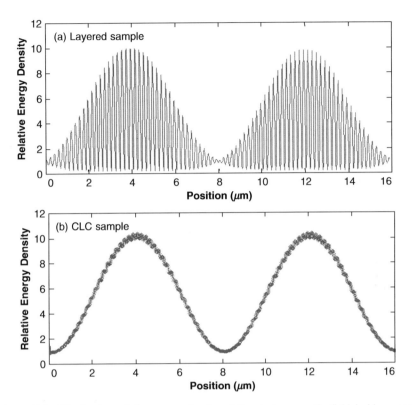

Figure 12.8 Distribution of the energy density of the electromagnetic field inside a one-dimensional periodic sample at the wavelength of the $n = 2$ mode for layered (a) and CLC samples (b). The parameters of the samples are the same as in Fig. 12.6.

integrated intensity reflects the lengthened dwell time for light within the CLC structure. The lengthened photon dwell time at the band edge corresponds to a narrower line width for corresponding modes. Because the photon dwell time in the structure varies as $1/n^2$, for the nth mode from the band edge, the lifetimes of the first band-edge mode are substantially longer than those for other modes, lasing is initiated in the first band-edge mode in presence of gain. Low-threshold lasing has been observed for the first mode at the band edge of CLCs [10].

12.3.4 Microwave Experiments

The principles of operation of a CSPG are most readily demonstrated using a scaled-up model that functions in microwave frequency range in which perfectly periodic structures are readily produced [7]. Plastic rods with dimensions com-

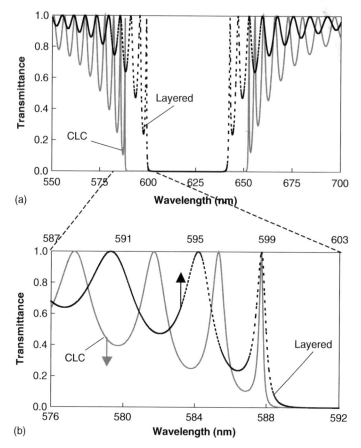

Figure 12.9 Comparison of transmittance for binary layered and CLC structures in a large spectral range (a) and at the band edge (b). The spectra are shifted in (b) so that the peaks of the first mode at the band edge coincide.

parable to the wavelength of the microwave radiation were either milled or twisted as the rod was passed through a heat zone. A band gap was not seen for $\lambda \sim P$ when a single thin plastic rod was wound around the thicker rod, as shown in Fig. 12.10a. But when a second helix displaced by one-half the pitch was wound around the thicker rod so that the chiral structure had the symmetry of a double helix, as shown schematically in Fig. 12.10b, a stop band was observed.

In analogy with planar chiral structures, a reflection band for co-handed radiation is formed over the range of vacuum wavelengths over which the wavelength in the fiber is equal to the pitch, $\lambda_{vac}/n_{eff} = P$, where n_{eff} is the effective index that reflects the distribution of propagating energy between

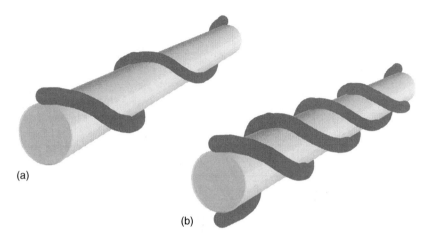

(a)

(b)

Figure 12.10 Examples of fibers possessing (a) single- and (b) double-helix symmetry.

the core and cladding of the fiber. The width of the gap in the vacuum wavelength is then given by $\Delta\lambda_{vac} = P\Delta n/< n >$ where Δn is the birefringence, which is the difference in effective index of refraction for linearly polarized light aligned along the slow and fast axes of the fiber, and $<n>$ is the average of these indices. Because the birefringence in these structures is high, the bandwidth is wide and the attenuation coefficient for co-handed light is large.

Measurements of microwave complex microwave field transmission coefficient, $E \exp(i\varphi)$ were carried out using a vector network analyzer. Here E is the amplitude and φ the phase, so that the real and imaginary parts of $E \exp(i\varphi)$ are the in- and out-of-phase components of the field transmission coefficient. Measurements of transmission of co- and counter-handed radiation within the 1.5- to 2.3-cm wavelength range through a rod of a length equal to 78P placed between a series of Teflon rods, which couple the grating shown in Fig. 12.11 to a source and detector. Horns with polarization states set to produce either right or left circularly polarized (RCP or LCP) radiation were used to launch and detect the microwave radiation. A stop band is observed in transmission with $\Delta\lambda \sim 0.01\lambda_c$ with a reflection peak over the same range centered at $\lambda_c = 2.09\ cm$. At shorter wavelengths, the beginning of a broad dip is observed for co-handed radiation. The drop in transmission is not associated with a corresponding increase in reflection. The broad peak in LCP radiation reflects the frequency dependence of coupling of the horns to the Teflon rod and does not change appreciably as the length of the Teflon rods in the system increases.

The transit time of the wave passing through the Teflon rods sandwiching the central chiral rod is given by the spectral derivative of the phase, $d\varphi/d\omega$, shown in Fig. 12.12b. The high value of the delay time at the band edge is associated with a resonance with a spectrally narrow long-lived state at the band edge. In

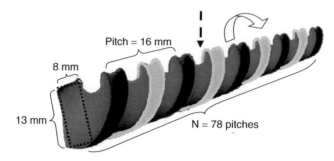

Figure 12.11 Double-helix polymeric rod with a rectangular cross-section. Arrow in the middle indicates a place at which the fiber was cut and twisted around its axis to produce a chiral twist, which gives rise to a defect state in the center of the stop band.

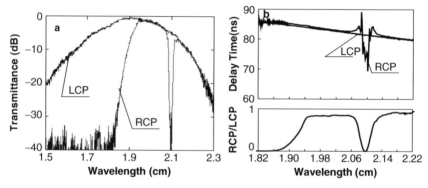

Figure 12.12 Broadband polarized spectra of microwave radiation transmitted through the polymeric rod, shown in Fig. 12.8. (a) Transmission spectrum on semilog scale. (b) Delay time obtained from the phase derivative of the transmitted RCP and LCP fields and the ratio of the transmission shown on linear scale.

this case, φ increases by π as the wavelength is tuned thorough a single narrow mode. The broad gap in transmission below 1.9 cm is due to scattering out of the fiber and is discussed later in the section on the CIPG.

12.3.5 Optical Measurements

Despite that CSPG with uniformity sufficient for filter applications has not been made yet, CSPGs with substantial reflectivity were fabricated. The presently achieved level of the pitch variation is near 1%. This level of uniformity is apparently already sufficient to obtain lasing from highly birefringent-doped fibers but is not high enough for filter applications. The quality of optical CSPGs has been improving rapidly and we anticipate it will shortly be the basis of useful devices.

12.4 CHIRAL INTERMEDIATE-PERIOD GRATING

12.4.1 Symmetry of CIPG Structures

We next consider the differences between propagation in a planar and in a fiber geometry. Because the sample is no longer uniform in the transverse direction, core modes may be scattered into modes with propagation vector pointing away from the fiber axis when the difference between the propagation constant of the core mode and the scattered wave is compensated by the grating constant associated with the periodic structure. In FBGs, the forward and backwards propagating core modes are coupled via the grating constant, which is $2\pi/a$, where a is the grating period. In chiral fibers, core modes are similarly coupled via the grating condition above. In single-helix chiral gratings $a = P$, whereas in double-helix gratings, $a = P/2$, giving grating constants of $2\pi/P$ and $4\pi/P$, respectively. We focus here on double-helix gratings because these are polarization selective.

In short-period gratings, the gratings constant is so large that the forward propagating core mode can only couple into a single other mode, the backwards-scattered core mode. As a result, reflection is analogous to that of chiral planar materials such as cholesteric liquid crystals or STFs. Propagation near the stop band of CSPGs is, therefore, essentially 1D and closely mimics the propagation in planar chiral structures. Similarly, propagation in isotropic FBGs and in single-helix chiral fibers is similar to that in binary-layered structures.

In weakly twisted fibers, the grating constant is too small to couple the core mode to the backward propagating core modes because the propagation constant of these modes differs by more than $2\pi/a$. For many years, twisted fibers with pitch of tens of centimeters have been studied because of their potential for reducing polarization mode dispersion, which limits the bandwidth of transmission [21]. Optical fibers with similar pitch were also used either to create circularly birefringent fibers for maintaining circularly polarized light or for converting the state of polarization [22, 23]. In all of these applications, the goal is to have light of different polarization interact with the fiber in a similar manner to reduce the polarization sensitivity of the fiber. It was shown that reducing the pitch leads to further reduction of the polarization sensitivity. In contrast, fibers that are twisted to a much shorter pitch are highly polarization selective.

12.4.2 Microwave Experiments

A broad gap in the transmission spectrum can be clearly seen at the short-wavelength side of Fig. 12.12. This gap exists only for co-handed polarization and corresponds to the double-helix CIPG. The transit time shown in Fig. 12.12b is found to be the same for left- and right-handed circular polarization in a right-handed structure. This is in contrast to a drop in the transit time within the stop band and an increase at band-edge modes for co-handed relative to cross-handed radiation. The transit time in the intermediate band is consistent with the linear dispersion at lower frequencies. These results confirm that the dip in the CIPG is not the result of coherent resonant reflection within the chiral fiber but is the result of scattering out of the fiber.

12.4.3 Optical Measurements

High-quality CIPGs have been fabricated for applications from 980 to 2000 nm. This range may be expanded to shorter or longer wavelength by scaling the structure. These structures are of particular interest because of the multiplicity of applications for which they would be well suited.

In CIPGs with pitches of order 10, or 100 microns co-handed circularly polarized light can be selectively scattered from the fiber, whereas the orthogonal polarization is freely transmitted over a wide range of wavelengths near the edges of this band.

For a specific wavelength, the resonant pitch for short- or long-period chiral gratings is defined by the corresponding phase-matching condition. In contrast, light is scattered out of the fiber by CIPGs for a broad range of pitch. Over this range, the direction of the scattered light changes from the near-backward to the near-forward direction. The polarization selectivity of the optical CIPG is most readily demonstrated near the short-wavelength edge of the scattering band. Polarization selectivity was described earlier for microwave radiation near the long-wavelength edge of the scattering band

Figure 12.13 shows the spectrum of the extinction ratio between co- and cross-handed circularly polarized components at the short-wavelength edge of the nonresonant scattering band. The structure has a pitch of 45 μm and a length of 11 mm. Appreciable scattering of cross-handed radiation occurred in the middle of the band, leaving regions near the long- and short-wavelength band edges in which scattering of cross-handed radiation was negligible. Polarizers may, therefore, be produced at the band edges of the CIPG.

Chiral optical fibers at the long-wavelength edge of the scattering band may also be produced. A portion of the fiber with a pitch of 0.95 μm and a length of 500 μm has an extinction ratio close to 9 dB at 1460 nm.

12.4.4 Synchronization of Optical Polarization Conversion and Scattering

In the process of producing stretches of fiber with uniform pitch from an untwisted preform, which can be spliced to standard fibers, the preform twist must be accelerated to reach the desired pitch and finally decelerated. This process introduces new functionality to chiral fibers. We find that it is possible to create an optical fiber with variable pitch in which linearly polarized light along the slow axis is transmitted without attenuation while orthogonally polarized incident light is strongly attenuated as it is scattered out of the fiber. This indicates that the two processes of polarization conversion and scattering are synchronized [24]. Thus, an elliptically polarized mode deriving from the slow linearly polarized mode of the untwisted fiber is not scattered at any point along the fiber but is freely transmitted as its state of polarization is continuously transformed. At the same time, light with the orthogonal polarization is strongly scattered from the sample. As a result, when the final pitch of the right-handed sample falls in the range of the CIPG, linearly polarized light incident along the slow axis is efficiently converted to left circularly polarized light while light polarized along the fast axis is scattered out of the fiber. Because the reverse process converts one component of elliptically polarized light to linear polarized light, decelerating the twist will restore the light to its initial state of linear polarization along the slow axis. This structure serves as an efficient linear polarizer with low insertion loss.

We first analyze the polarization evolution along the adiabatically twisted optical fiber with birefringent core using a 4×4–transfer matrix approach

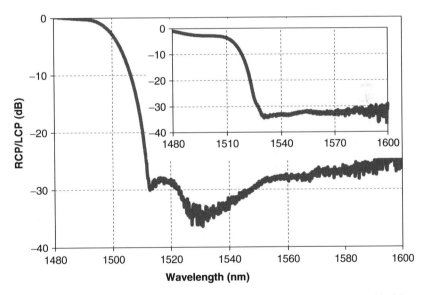

Figure 12.13 Ratio of right-to-left circularly polarized transmission through a chiral intermediate period grating. Inset shows broadband in-fiber polarizer based on an intermediate chiral grating with nonuniform pitch.

[25–27], designed for planar chiral media. This calculation can account for the longitudinal propagation, but not for scattering out of the core mode, which occurs in samples with transverse structural variation. In this approach, the grating period is not constrained to be constant. We confirmed that the transfer matrix method converges as the thickness of individual anisotropic layers with ordinary and extraordinary refractive indices, n_f and n_s, along the fast and slow axes, respectively, and the angle between consecutive layers are proportionately reduced and the number of layers correspondingly increases. Measurements of polarized transmission demonstrate the synchronization of conversion and scattering in twisted fibers, such as the fiber shown in Fig. 12.14.

We calculate the conversion of linearly polarized 1.5-μm-wavelength radiation for different values of the birefringence in a 16-mm long stack of 1-μm thick anisotropic layers structure twisted to a final right-handed pitch of 110 μm. The linear twist acceleration is close to that of the fibers in which measurements are presented later in this chapter. The incoming linearly polarized wave is oriented along the fast fiber axis. The conversion into a right circularly polarized wave is displayed in the plot in Fig. 12.15 by the residual transmission of left-circularly polarized light. The wave is linearly polarized as it enters the sample at $z = 0$, so that half of the energy is left-circularly polarized, corresponding to the -3 dB level on the graph. For small values of birefringence $\Delta n = n_s - n_f$ conversion is negligible. For higher values of $\Delta n = 0.004$, the residual left-circularly polarized light falls as P decreases, with a conversion into the right-circularly polarized wave of more than 99% conversion at $z = 16$ mm.

For larger values of Δn, the conversion is again not optimal. At $\Delta n = 0.013$, the substantial intensity of left-circularly polarized light indicates that the resulting polarization is far from circular. This can also be seen from the variation of the effective refractive indices in Fig. 12.16. In the case of a complete conversion to the circularly polarized wave, both curves would converge to $n_{av} = (n_s + n_f)/2$. The measured birefringence in the fiber used was about 0.013. In accord with results in Fig. 12.15, the polarization state of the wave at $z = 16$ mm at a final pitch of 110 μm is not circular.

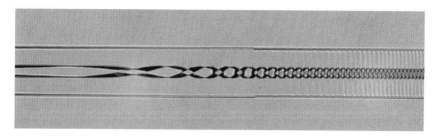

Figure 12.14   Side-view of an adiabatically twisted optical fiber with rectangular core.

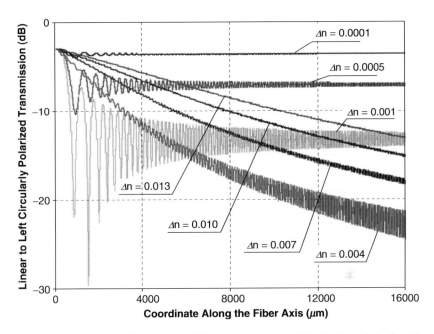

Figure 12.15 Calculation of conversion of linear polarization to right circular polarization. The residual left circularly polarized light is shown in the graph for different values of the birefringence of the untwisted fiber.

We studied the polarization evolution for this value of Δn for a fiber in which the pitch is accelerated in the first half and decelerated in the second half of the structure. Figure 12.17 illustrates the polarization evolution in a right-handed 32-mm structure with $\Delta n = 0.013$, with accelerating and then decelerating twist. Figure 12.17a shows the evolution of the wave initially polarized along the fast fiber axis on the Poincaré sphere in the laboratory frame. The surface of the Poincaré sphere represents all possible states of polarization. Points on the equator correspond to linearly polarized waves, whereas north and south poles represent right- and left-circularly polarized waves, respectively. It can be seen that an initially linearly polarized wave moves towards the north pole, as illustrated by the arrow originating from the point F. The polarization trajectory rotates around the north pole without actually reaching it, in agreement with Fig. 12.15. Then, after passing the fiber midpoint, the polarization state moves back to the equator, reaching the point indicated by the second arrow. This point on the sphere depends on the precise angle of rotation of the fiber modulo 2π and represents a linearly polarized wave. The evolution of the linearly polarized wave initially oriented along the slow fiber axis is a symmetrical curve originating from the point S and moving towards the south pole.

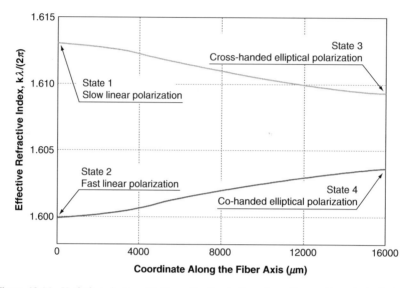

Figure 12.16 Variations in the effective refractive indices for orthogonally polarized eigenmodes in adiabatically twisted birefringent optical fiber. Linear polarizations along the fast and slow axes and right and left circularly polarized radiation are intrinsic modes of the untwisted fiber and of the fiber twisted with short constant pitch. While the cross-handed circular polarization is not scattered, the co-handed circular polarization is scattered out of the fiber.

To get a better physical picture of the polarization evolution, we recalculated the path on the Poincaré sphere in a coordinate system, which rotates with the fiber, in which the X and Y-axes coincide with the slow and fast fiber axes at any cross-section. The results of the calculations are shown in Fig. 12.17b.

The initially linearly polarized wave represented by the point F evolves towards the north pole along a sphere meridian, stops at the same distance from the pole as in Fig. 12.17a, and then returns to the initial point. Again, scattering is not taken into account and the evolution of the linearly polarized wave oriented along the slow axis is symmetrical with respect to the center of the sphere.

Measurements were made on a twisted fiber prepared from a custom preform with refractive indices in the visible spectral range for the core and cladding of 1.69 and 1.52, respectively. The fiber preform was drawn down to a 125-μm diameter with a 10- $\times$ 5-μm rectangular core. This preform was further drawn and twisted as it passed through a miniature oven. The control of the twisting tower was integrated into a single computer program in which the precise translation, drawing and twisting of the fiber, and its temperature were coordinated. A side image of the twisted fiber is shown in Fig. 12.14. We compare the performance of three samples prepared with linear twist acceleration and deceleration, with shortest pitch of 110 μm. These samples are (1) final cladding

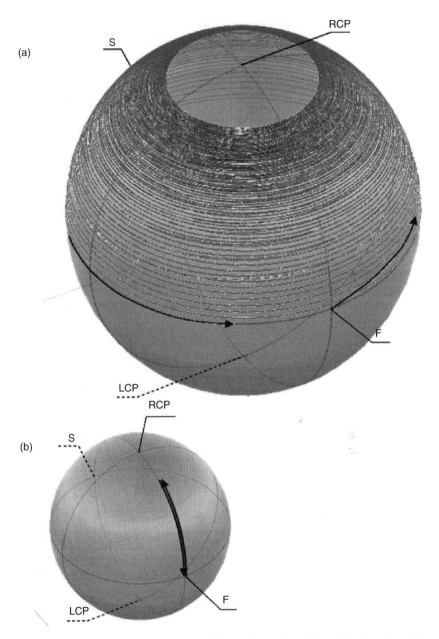

Figure 12.17 Calculation of conversion of linear (along fast axis) polarization in a right-handed optical fiber with $\Delta n = 0.013$. (a) Evolution of the polarization in the laboratory's coordinate system shown on the Poincaré sphere. (b) Evolution of the polarization in the structure's coordinate system.

diameter of 14.5 μm and total length of 32 mm; (2) final cladding diameter of 14 μm and total length of 44 mm, and (3) final cladding diameter of 14 μm and total length of 51 mm. At a wavelength of 1.5 μm, the untwisted core supports two modes with orthogonal polarizations with a difference in effective refractive index difference of approximately 0.013.

An Agilent 83437A broadband unpolarized light source was used for spectral measurements between 1450 and 1600 nm. The incident light was polarized along either the slow or the fast fiber axis using a fiber-connected walk-off linear polarizer. The output light was guided to a fiber connected to an Agilent 86145B optical spectrum analyzer without having its polarization analyzed.

Measurements of optical transmission through the three twisted fiber samples are shown in Fig. 12.18. Curves of transmission for linearly polarized light initially oriented along the slow fiber axis, for which scattering is minimal, are shown for samples 1, 2, and 3. The corresponding extinction ratios for these samples, which are the ratios of transmission for waves polarized along the fast and slow fiber axes, are also shown.

The splice-induced insertion losses were near 1.5 dB and may be seen in transmission at a wavelength of 1450 nm for all three samples in Fig. 12.18. At longer wavelengths, there are relatively small twist-induced insertion losses, which are approximately 0.2 dB for sample 1 and 0.5 dB for samples 2 and 3 at $\lambda = 1550$ nm. This indicates that the wave polarized along the slow axis very closely follows the

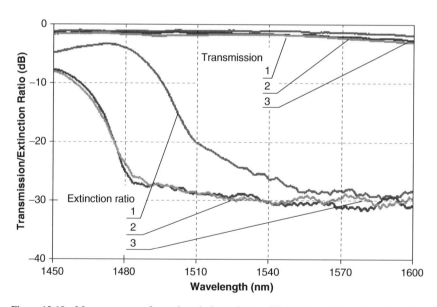

Figure 12.18 Measurements of wavelength dependence of light transmission through optical fibers twisted with different pitch profile.

evolution path that originates at the point S and is symmetrical to the trajectory shown in Fig. 12.17. In contrast, the wave initially polarized along the fast axis is strongly suppressed at a wavelength of 1550 nm in all three samples. This indicates that the wave at 1550 nm initially polarized along the fast axis does not follow the trajectory shown in Fig. 12.17 but approaches the center of the Poincaré sphere as it is scattered. This behavior is possible only if scattering from the fiber is fully synchronized with polarization conversion. This indicates that at any point along the trajectory shown in Fig. 12.17, one elliptically polarized wave is scattered and an orthogonal wave is not. This conclusion is bolstered by the comparison of these measurements and transfer matrix calculations.

The high extinction ratio over a 100-nm range, seen from Fig. 12.18, indicates that synchronization takes place over a broad spectral range. This range is independent of the twist acceleration or deceleration, as can be seen from the comparison of results for samples 2 and 3, but the range shifts with fiber diameter, as can be seen from the measurements in sample 1. This shift is expected because the product of the shortest pitch and the effective refractive index of the core mode determines the spectral position of the scattering band in CIPGs [7]. Increasing the core diameter results in a larger effective refractive index. This, in turn, shifts the position of the intermediate scattering gap to longer wavelengths. The synchronization of polarization conversion and wave scattering along a chiral optical fiber with varying pitch makes possible a broad-band, low-loss, and high extinction ratio linear and circular in-fiber polarizers.

12.5 CHIRAL LONG-PERIOD GRATING

In both isotropic LPGs and CLPGs, the core mode is resonantly coupled to co-propagating cladding modes. This is the basis of their use as sensors. The double-helix CLPG may have a particular advantage because the cross-handed wave, which is not scattered by the CLPG, is available for monitoring the integrity of the fiber system. In addition, the sinusoidal modulation of the dielectric constant gives rise to a single set of dips without harmonics for each cladding mode.

12.5.1 Optical Measurements

12.5.1.1 Double-Helix CLPG

A double-helix CLPG was made with pitch of 78 μm and was 55 mm long. Sharp dips in transmission were observed at wavelengths associated with coupling of the core mode to distinct co-propagating cladding modes. In conventional fiber LPGs, the wavelength of the resonant dips is determined by the condition, $k_{core}-k_{clad} = 2\pi/a$, where k_{core} is the propagation constant of the core mode, k_{clad} is the propagation constant of a particular cladding mode, and a is the period of the grating [14]. A similar phase-matching condition applies to CLPGs. The wavelength of strongest extinction of the CLPG can be readily adjusted by changing the chiral pitch. The sensitivity of these resonances to the optical characteristics of the cladding and its proximate environment makes the fibers ideal for ambient sensing. The ratio of right- to left-circularly polarized transmission through the sample is shown in Fig. 12.19. The results show that while LCP light is not affected by the right-handed structure, RCP light displays dips in transmission. The inset shows the shift of the transmission dip when the CLPG is surrounded by gasoline.

12.5.1.2 Single-Helix CLPG

Single-helix CLPGs are also capable of coupling core modes with the co-propagating cladding modes. The single-helix structure does not match the geometry of lightwave of any particular polarization. This leads to polarization insensitive interaction of the lightwave with the structure. The preform for

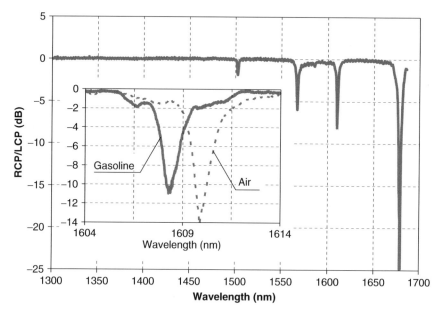

Figure 12.19   Ratio of right-to-left circularly polarized transmission through a double-helix chiral long-period grating.

single-helix CLPG could be made of low numerical aperture (NA) fiber with a nonconcentric core. The performance of these structures is similar to that of regular fiber LPGs produced by microbending. As a result, the index contrast achievable cannot be larger than the difference in index of the core and cladding, which is smaller than the grating strength available in double-helix CLPGs. The performance of the single-helix CLPG made of a custom silica fiber with non-concentric core is shown in Fig. 12.20. Both single- and double-helix CLPGs have their advantages for sensor applications. The main similarities and differences between single- and double-helix CLPGs are summarized in Table 12.1.

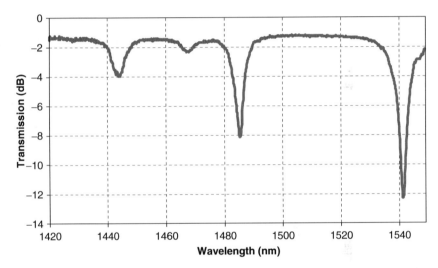

Figure 12.20 Polarization insensitive transmission through a single-helix chiral long-period grating.

Table 12.1

Key parameters of single- and double-helix CLPGs

	Polarization insensitivity	Independent use of orthogonal polarizations for multiplexing	Fabricated in low NA fibers	Easily coupled to standard fiber	Ultra-narrow transmission dips	Temperature, strain, and twist sensitivity
Single helix	Yes	No	Yes	Yes	No	Yes
Double helix	No	Yes	No	No	Yes	Yes

12.6 CONCLUSION

In conclusion, we have demonstrated the distinctive polarization and wave-length-selective properties of optical CLPGs and intermediate-period gratings that make them useful as sensors, filters, and polarizers. We have also discussed the desirability of chiral short-period gratings as lasing sources and discussed progress towards this goal. Because of the flexible method of fabrication, which does not require photosensitive glass, chiral fiber gratings can be produced to perform a wide array of functions over a broad wavelength range including IR and THz.

ACKNOWLEDGMENTS

We thank V. Churikov, G. Zhang, C. Draper, N. Chao, J. Singer, and D. Neugroschl for help in preparing twisted fibers. This work was supported by the U.S. Department of Commerce, National Institute of Standards and Technology, Advanced Technology Program, Cooperative Agreement Number 70NANB3H3038, and the National Science Foundation under Grants No. DMI-0340149 and DMI-0450551.

REFERENCES

[1] Othonos, A. and K. Kalli. 1999. *Fiber Bragg gratings: Fundamentals and applications in telecommunications and sensing.* Artech House, Norwood, MA.
[2] Probst, C. B. et al. 1989. Experimental verification of microbending theory using mode coupling to discrete cladding modes. *J. Lightwave Technol.* 7:55–61.
[3] Davis, D. D. et al. 1998. Long-period fibre grating fabrication with focused CO_2 laser pulses. *Electron. Lett.* 34:302–303.
[4] Rego, G. et al. 2001. High-temperature stability of long-period fiber gratings produced using an electric arc. *J. Lightwave Technol.* 19:1574–1579.
[5] Kopp, V. I. and A. Z. Genack. 2002. Twist defect in chiral photonic structures. *Phys. Rev. Lett.* 89:033901.
[6] Kopp, V. I. et al. 2003. Transmission through chiral twist defects in anisotropic periodic structures. *Opt. Lett.* 28:349.
[7] Kopp, V. I. and A. Z. Genack. 2003. Double-helix chiral fibers. *Optics Lett.* 28:1876.
[8] Kopp, V. I. et al. 2004. Chiral fiber gratings. *Science* 305:74.
[9] Kopp, V. I. et al. 2004. Chiral fiber gratings polarize light. *Photonics Spectra* 38:9, 78.
[10] Kopp, V. I. et al. 1998. Low-threshold lasing at the edge of a photonic stop band in cholesteric liquid crystals. *Opt. Lett.* 23:1707.
[11] Digonnet, M. J. E., Ed. 2001. *Rare-earth–Doped Fiber Lasers and Amplifiers.* Marcel Dekker, New York.
[12] Hodgkinson, I. J. et al. 2003. Supermodes of chiral photonic filters with combined twist and layer defects. *Phys. Rev. Lett.* 91:223903.

[13] Krohn, D. A. 2000. *Fiber Optic Sensors: Fundamentals and Applications.* Instrument Society of America, Research Triangle Park.

[14] Bhatia, V. and A. M. Vengsarkar. 1996. Optical fiber long-period grating sensors. *Opt. Lett.* 21:692.

[15] Joannopoulos, J. D. et al. 1995. *Photonic Crystals: Molding the Flow of Light.* Princeton University Press, Princeton, NJ.

[16] Chandrasekhar, S. 1977, 1994. *Liquid Crystals.* Cambridge University Press, Cambridge.

[17] Robbie, K. et al. 1999. Chiral nematic order in liquid crystals imposed by an engineered inorganic nanostructure. *Nature* 399:764.

[18] Lakhtakia, A. and R. Messier. 2005. *Sculptured Thin Films: Nanoengineered Morphology and Optics.* SPIE Press.

[19] de Vries, H. 1951. Rotatory power and other optical properties of certain liquid crystals. *Acta Cryst.* 4:219.

[20] Kopp, V. I. et al. 2003. Lasing in chiral photonic structures. *Progress Quant. Electr.* 27:369.

[21] Rashleigh, S. C. 1983. Origins and control of polarization effects in single-mode fibers. *IEEE J. Lightwave Technol.* LT-1:312.

[22] Nolan, D. A. and M. N. Islam. 2001. U.S. Patent 6,229,937.

[23] Ulrich, R. and A. Simon. 1979. Polarization optics of twisted single-mode fibers. *Appl. Opt.* 18:2241.

[24] Kopp, V. I. et al. 2006. Synchronization of optical polarization conversion and scattering in chiral fibers. *Optics Lett.* 31.

[25] Teitler, S. and B. Henvis. 1970. Refraction in stratified anisotropic media. *J. Opt. Soc. Am.* 60:830.

[26] Berreman, D. W. 1972. Optics in stratified and anisotropic media: 4X4-Matrix Formulation. *J. Opt. Soc. Am.* 62:502.

[27] Wohler, H. et al. 1988. Faster 4×4 matrix mathod for uniaxial inhomogeneous media. *Opt. Soc. Am.* A5:1554.

Chapter 13

Mid-IR and Infrared Fibers

James A. Harrington

Departments of Ceramics and Materials Engineering,
Rutgers University, Piscataway, New Jersey

13.1 INTRODUCTION

Infrared (IR) optical fibers are fibers that transmit radiation from 2 to approximately 20 μm. The first IR fibers were fabricated in the mid-1960s from a rather special class of IR transparent glasses called *chalcogenide glasses*. It was well known that mixing chalcogen elements, for example, arsenic and sulfur, can form a dark red glass that is transparent well beyond 2 μm. In 1965, this arsenic trisulfide (As_2S_3) glass was first drawn into crude optical fiber by Kapany et al. [1], but the losses were more than 10 dB/m from 2 to 8 μm. A loss of 10 dB/m means that a 1-m length of As_2S_3 fiber would transmit only 10% of the incident light. Energy would also be lost because of reflection from the fiber end-faces, which for this high refractive index glass ($n = 2.3$) amounts to an additional 31%. Furthermore, their As_2S_3 fiber was quite brittle. During the mid-1970s, there was interest in developing an efficient and reliable IR fiber to link broadband long wavelength radiation to remote photodetectors in military sensor applications. For example, researchers at Hughes Research Laboratories in Malibu, California, studied various IR fibers that could be used in a surveillance satellite application to transmit IR radiation in the 3- to 5-μm, 8- to 12-μm, and longer wavelength bands to an interior IR detector array [2]. In addition, there was an ever-increasing need for a flexible fiber delivery system for transmitting CO_2 laser radiation in surgical applications. Based on these needs, various IR materials and fibers were developed. These fibers include the heavy metal fluoride glass (HMFG) and polycrystalline fibers, as well as hollow rectangular waveguides. Although none of these fibers had physical properties even approaching those of conventional silica fibers, they were, nevertheless, useful in lengths less than 2–3 m for a variety of IR sensor and power delivery applications.

IR fiber optics may logically be divided into three broad categories: glass, crystalline, and hollow waveguides. These categories may be further subdivided

429

based on the fiber material, structure, or both, as shown in Fig. 13.1. Over the past 30 years, many novel IR fibers have been made in an effort to fabricate a fiber optic with properties as close to silica as possible, but only relatively few have survived. An excellent review of the types of IR fibers may be found in a book by Harrington [3]. Other sources of general information on IR fiber types may be found in the literature [4–8]. In this chapter, only the best, most viable, and in most cases, commercially available IR fibers are discussed. In general, both the optical and the mechanical properties of IR fibers remain inferior to silica fibers, so the use of IR fibers is still limited primarily to non-telecommunication short-haul applications requiring only tens of meters of fiber rather than kilometer lengths common to telecommunication applications. The short-haul nature of IR fibers results from the fact that most IR fibers have losses in the few dB/m range. An exception is the fluoride glass fibers, which can have losses as low as a few dB/km. In addition, IR fibers are much weaker than silica fiber and, therefore, are more fragile. These deleterious features have slowed the acceptance of IR fibers and restricted their current use to applications in chemical sensing, thermometry, and laser power delivery.

The obvious key property of IR fibers is their ability to transmit wavelengths longer than most oxide glass fibers. While some IR fibers can transmit well beyond 20 μm, most applications do not require the transmission of IR wavelengths longer than about 12 μm. A summary of the spectral loss for five of the six subcategories of fibers listed in Fig. 13.1 is shown in the composite data [3] in Fig. 13.2. Some of the important optical and mechanical properties of IR fibers are listed in Table 13.1. For comparison, the properties of silica fibers are also listed. The data in Table 13.1 and in Fig. 13.2 reveal that, compared to silica, IR fibers usually have higher loss; larger refractive indices and dn/dT; lower melting or softening points; and greater thermal expansion. From these data, it may also be seen that there is a wide variation in the range of transmission for the different IR fibers and that the loss of most of the IR fibers is quite high compared to silica fibers. Remembering that 1 dB/m is a bulk loss of about 20% per meter, it is again evident that this high loss will restrict applications to meter-long lengths. All of the solid-core fibers shown in Fig. 13.2 have a much lower theoretical or intrinsic loss. The reason that the losses are so high is that

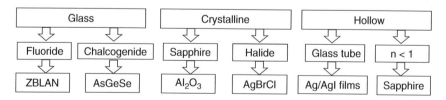

Figure 13.1 Major categories of infrared fiber optics and an example of each fiber type.

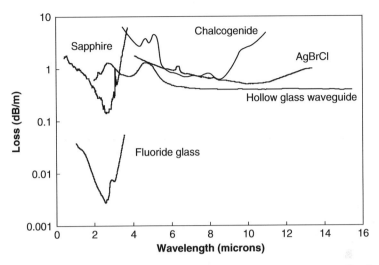

Figure 13.2 Composite loss spectra for some common infrared fiber optics: ZBLAN fluoride glass, SC sapphire, chalcogenide glass, PC AgBrCl, and hollow glass waveguide.

the fibers contain impurities and imperfections that give rise to a large extrinsic absorption and scattering. Some of these extrinsic absorption bands are evident in Fig. 13.2. The absorptions shown in Fig. 13.2 for the hollow waveguide are not due to impurities, rather they are due to interference effects resulting from the thin-film coatings used to make the guides. Finally, many IR fibers do not have a proper cladding analogous to conventionally clad oxide glass fibers. Nevertheless, core-only IR fibers such as sapphire and chalcogenide fibers can still be useful. This is because their refractive indices are sufficiently high that there is less evanescent wave energy outside the core. As long as the unclad fiber does not come in contact with an absorbing medium, the fiber can operate reasonably well, as there will be very little leakage of light from the core to the surrounding air [9].

The motivation to develop a viable IR fiber stems from many proposed applications [3]. A summary of the most important current and future applications and the associated candidate IR fiber that will best meet the need is given in Table 13.2. There are several noteworthy trends seen in this table. The first is that hollow waveguides are an ideal candidate for laser power delivery at all IR laser wavelengths. The air core of these waveguides gives an inherent advantage over solid-core fibers because IR materials used in solid-core fibers have laser damage thresholds that are frequently very low. The air-core waveguides are capable of delivering close to 3000 W of cw CO_2 laser power, far in excess of any IR solid-core fiber. However, solid-core IR fibers are ideal evanescent-wave sensors for

Table 13.1

Selected physical properties of key infrared fibers compared to conventional silica fiber

Property	Glass			Crystal		Hollow
	Silica	HMFG ZBLAN	Chalcogenide AsGeSeTe	PC AgBrCl	SC sapphire	Hollow silica waveguide
Glass transition or melting point, °C	1175	265	245	412	2030	150 (useable T)
	1.38	0.628	0.2	1.1	36	1.38
Thermal	0.55	17.2	15	30	5	0.55
conductivity,	70.0	58.3	21.5	0.14	430	70.0
W/m °C	2.20	4.33	4.88	6.39	3.97	2.20
Thermal expansion coefficient, 10^{-6} °C^{-1}	1.455 (0.70)	1.499 (0.589)	2.9 (10.6)	2.2 (10.6)	1.71 (3.0)	NA
	+1.2 (1.06)	−1.5 (1.06)	+10 (10.6)	−1.5 (10.6)	+1.4 (10.6)	NA
Young's modulus, GPa	0.24–2.0 ~800	0.25–4.0 0.08	4–11 5	3–16 3	0.5–3.1 0.4	0.9–25 0.5
Density, g/cm^3	NA	NA	2	0.5	NA	0.4
Refractive index (λ, μm)						
dn/dT, 10^{-5} °C−1 (λ, μm)						
Fiber transmission, range, μm						
Loss* at 2.94 μm, dB/m						
Loss* at 10.6 μm, dB/m						

* Typical measured loss. NA, not applicable.

monitoring chemical processes in the sensitive fingerprint region of the infrared spectrum. In these applications, the chemical or biological agent surrounds the fiber core and some portion of the light is coupled out of the core into the surrounding medium. This type of chemical sensor is potentially very sensitive and selective. Chalcogenide and silver halide fibers are particularly good for this application, as they are quite inert and their high refractive index means that only a small portion of the light is out-coupled from the core into the absorbing medium. A good fiber for gas sensing is the hollow waveguide, as the core of this fiber can be filled with gas so that light propagating through the waveguide is partially absorbed by the gas. Temperature measurements using long wavelength transmissive fibers like the silver halides or hollow waveguides are possible over a

Table 13.2

Examples of infrared fiber candidates for various sensor and power delivery applications

Application	Comments	Suitable IR fibers
1. Fiber optic chemical sensors	Evanescent wave principle: liquids	AgBrCl, sapphire, chalcogenide, HMFG
2. Fiber optic chemical sensors	Hollow-core waveguides: gases	Hollow glass waveguides
3. Radiometry	Blackbody radiation, temperature measurements	Hollow glass waveguides, AgBrCl, chalcogenide, sapphire
4. Er/YAG laser power delivery	3-μm transmitting fibers with high damage threshold	Hollow glass waveguides, sapphire, germanate glass
5. CO_2 laser power delivery	10-μm transmitting fibers with high damage threshold	Hollow glass waveguides
6. Thermal imaging	Coherent bundles	HMFG, chalcogenide
7. Fiber amplifiers and lasers	Doped IR glass fibers	HMFG, chalcogenide

large temperature range. Normally blackbody radiation from a source is transmitted through the fiber and the temperature determined by calibration to a blackbody of known temperature. Because blackbody radiation from room-temperature objects is peaked near 10 μm, IR fibers are excellent candidates for use in measuring temperatures less than 50 °C.

13.2 HALIDE AND HEAVY METAL OXIDE GLASS FIBER OPTICS

There are two IR transmitting glass fiber systems that resemble conventional silica-glass fibers in that they are drawn from glass preforms and they have a core region surrounded by a cladding layer. One is the heavy metal fluoride glasses (HMFGs), or fluoride glass for short, and the other is heavy metal germanate glass fibers based on GeO_2. The germanate glass fibers generally do not contain fluoride compounds. They also do not contain silica (SiO_2), rather they contain heavy metal oxides to shift the IR absorption edge to longer wavelengths. The advantage of germanate fibers over HMFG fibers is that germanate glass has a higher glass transition temperature and, therefore, a higher laser-damage threshold. But the loss for the HMFG fibers is lower.

13.2.1 Fluoride Glass Fibers

Poulain et al. [10, 11] discovered the HMFGs based on zirconium fluoride, also called *fluorozirconate glasses,* accidentally in 1975 at the University of Rennes. In general, the typical fluoride glass has a glass-transition temperature, T_g, four times less than silica, is considerably less stable, and has practical failure strains of only a few percent compared to silica's greater than 5%. While an enormous number of multicomponent fluoride glass compositions have been fabricated, comparably few have been drawn into fiber. This is because the temperature range for fiber drawing is normally too small in most HMFGs to permit fiberization of the glass. The most popular HMFGs for fabrication into fibers are the fluorozirconate and fluoroaluminate glasses, of which the most common are ZBLAN (ZrF_4-BaF_2-LaF_3-AlF_3-NaF) and AlF_3-ZrF_4-BaF_2-CaF_2-YF_3, respectively. The key physical properties that contrast these two glasses are summarized in Table 13.3. An important feature of the fluoroaluminate glass is its higher T_g, which largely accounts for the higher laser damage threshold for the fluoroaluminate glasses compared to ZBLAN at the Er/YAG laser wavelength of 2.94 μm.

The fabrication of HMFG fiber is similar to any glass fiber–drawing technology except that the preforms are made using some type of melt-forming method rather than by a vapor-deposition process common with silica fibers. Specifically, a casting method based on first forming a clad glass tube and then adding the molten core glass is used to form either multimode or single-mode fluorozirconate-fiber preforms. The cladding tube is made either by a rotational casting technique in which the clad tube is spun in a metal mold or by merely inverting and pouring out most of the molten clad glass contained in a metal mold to form a tube [12]. The clad tubing is then filled with a higher index core glass. Other preform fabrication techniques include rod-in-tube and crucible techniques. The fluoroaluminate fiber preforms have been made using an unusual extrusion

Table 13.3

Comparison between fluorozirconate and fluoroaluminate glasses of some key properties that relate to laser power transmission and durability of the two HMFG fibers; other physical properties are relatively similar

Property	Fluorozirconate ZBLAN	Fluoroaluminate AlF_3-ZrF_4-BaF_2-CaF_2-YF_3
Glass transition temperature, °C	265	400
Durability	Medium	Excellent
Loss at 2.94 μm, dB/m	0.01	0.1
Er/YAG laser peak output energy, mJ	300	850
	300-μm core	500-μm core

technique in which core and clad glass plates are extruded into a core/clad preform [13]. All methods, however, involve fabrication from the melted glass rather than from the more pristine technique of vapor deposition used to form SiO_2-based fibers. This process creates inherent problems such as the formation of bubbles, core–clad interface irregularities, and small preform sizes. Most HMFG fiber drawing is done using preforms rather than the crucible method. A ZBLAN preform is drawn at about 310 °C in a controlled atmosphere (to minimize contamination by moisture or oxygen impurities which significantly weaken the fiber) using a narrow heat zone compared to silica. Either UV acrylate or Teflon coatings are applied to the fiber. In the case of Teflon, heat shrink FEP fluoride is generally applied to the glass preform before the draw.

The theoretical or intrinsic attenuation in HMFG fibers is predicted to be about 10 times less than that for silica fibers [14]. Based on extrapolations of the intrinsic losses resulting from Rayleigh scattering and IR multiphonon absorption, the minimum loss is projected to be about 0.01 dB/km at 2.55 μm [15]. Refinements of the scattering loss have modified this value slightly to be 0.024 dB/km or about eight times less than that for silica fiber [15]. In practice, however, extrinsic loss mechanisms still dominate fiber loss. The lowest measured loss for a 60-m long ZBLAN fiber is 0.45 dB/km at 2.3 μm [16]. This loss is dominated by extrinsic loss mechanisms due to scattering (crystallites, oxides, and bubbles) and impurities such as Ho^{3+}, Nd^{3+}, Cu^2, and OH^-.

In Fig. 13.3, losses for two ZBLAN fibers are shown. The data from British Telecom (BTRL) represents state-of-the-art fiber 110 m in length [15]. The other curve is more typical of commercially available (Infrared Fiber Systems, Silver Spring, MD) ZBLAN fiber. The lowest measured loss for a BTRL, 60-m long fiber is 0.45 dB/km at 2.3 μm. Some of the extrinsic absorption bands that contribute to the total loss shown in Fig. 13.3 for the BTRL fiber are Ho^{3+} (0.64 and 1.95 μm), Nd^{3+} (0.74 and 0.81 μm), Cu^{2+} (0.97 μm), and OH^- (2.87 μm). Scattering centers such as crystals, oxides, and bubbles have also been found in the HMFG fibers. In their analysis of the data in Fig. 13.3, the BTRL group separated the total minimum attenuation coefficient (0.65 dB/km at 2.59 μm) into an absorptive loss component equal to 0.3 dB/km and a scattering loss component equal to 0.35 dB/km. The losses for the fluoroaluminate glass fibers are also shown for comparison in Fig. 13.3 [13]. Clearly, the losses are not as low as those for the BTRL-ZBLAN fiber, but the AlF_3-based fluoride fibers do have the advantage of higher glass-transition temperatures and, therefore, are better candidates for laser power delivery.

The reliability of HMFG fibers depends on protecting the fiber from attack by moisture and on pretreatment of the preform to reduce surface crystallization. In general, the HMFGs are much less durable than oxide glasses. The leach rates for ZBLAN glass ranges between 10^{-3} and 10^{-2} g/cm^2/day. This is about five orders of magnitude higher than the leach rate for Pyrex glass. The

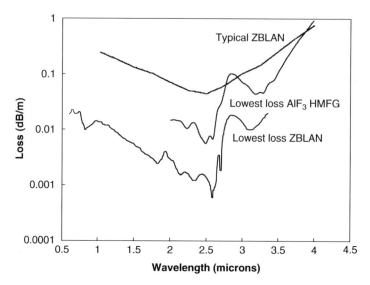

Figure 13.3 Losses in the best British Telecom (BTRL) [14] and typical (Infrared Fiber Systems, Silver Spring, MD) ZBLAN fluoride glass fibers compared to fluoroaluminate glass fibers.

fluoroaluminate glasses are more durable, with leach rates that are more than three times lower than those for the fluorozirconate glasses. The strength of HMFG fibers is less than that for silica fibers. From Table 13.1, we see that Young's modulus E for fluoride glass is 51 GPa compared to 73 GPa for silica glass. Taking the theoretical strength to be about one-fifth that of Young's modulus gives a theoretical value of strength of 11 GPa ($R = 1.198r\frac{E}{\sigma_{max}}$) for fluoride glass. The largest bending strength measured has been about 1.4 GPa, well below the theoretical value. To estimate the bending radius R, we may use the approximate expression, where σ_{max} is the maximum fracture stress and r is the fiber radius [17].

13.2.2 Germanate Glass Fibers

Heavy metal oxide glass fibers based on GeO_2 have shown great promise as an alternative to HMFG fibers for 3 μm laser power delivery [18]. Today, GeO_2-based glass fibers are composed of GeO_2 (30–76%)–RO (15–43%)–XO (3–20%), where R represents an alkaline-earth metal and X represents an element of Group IIIA [19]. In addition, small amounts of heavy metal fluorides may be added to the oxide mixture. The oxide-only germanate glasses have glass-transition temperatures as high as 680 °C, excellent durability, and a relatively high refractive index of 1.84. In Fig. 13.4, loss data are given for a typical germanate glass fiber. Although the losses are not as low as they are for the fluoride glasses shown in Fig. 13.3, these fibers have an exceptionally high damage threshold at 3 μm. Specifically, more than 20 W (2 J at 10 Hz) of Er/YAG laser power has been launched into these fibers.

13.2.3 Chalcogenide Glass Fibers

Chalcogenide glasses are composed of two or more chalcogen elements normally selected from the small group including As, Ge, Sb, P, Te, Se, and S. When these elemental materials are heated and mixed in an oxygen-free environment, some very stable and simple glasses can result. One of the oldest chalcogenide glasses studied is the binary glass arsenic trisulfide, As_2S_3. This glass is deep red in color, and it is very stable. In the mid 1960s, this glass was drawn into the first IR fiber by Kapany [1]. It was not until some 10–15 years later, however, that these materials were studied seriously as viable IR fiber candidates. The reticence to pursue these materials in the early days came in part from the toxic nature of some of the elements used in the glasses. Today they are a popular IR fiber material, as they are readily drawn into fiber with a broadband IR transmission but are much more delicate in nature than the oxide glass fibers. They are finding many applications in chemical and temperature sensor systems and as IR image bundles [20].

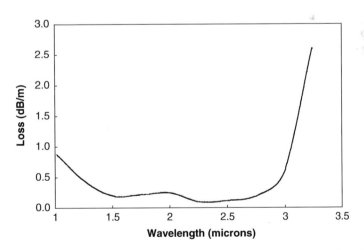

Figure 13.4 Germanate glass fiber manufactured by Infrared Fiber Systems, Silver Spring, MD.

   Chalcogenide fibers fall into three categories: sulfide, selenide, and telluride. Within these categories, one usually finds that the binary and ternary glasses are excellent choices for fiberization. That is, unlike the fluoride glasses where it is commonplace to have five or more components, most chalcogenide glasses have only two or three elemental components. In general, these glasses have softening temperatures comparable to fluoride glass. They are very stable, durable, and largely insensitive to moisture. A distinctive difference between these glasses and the other IR fiber glasses is that they do not transmit well in the visible region and their refractive indices are quite high. Additionally, most of the chalcogenide glasses, except for As_2S_3, have a rather large value of dn/dT [21]. This fact limits the laser power handling capability of the fibers.

   Chalcogenide glass is made by combining highly purified (>6 nines purity) raw elements in a sealed ampoule that is heated and mixed in a rocking furnace for about 10 hours. After melting and mixing, the glass is quenched and a glass preform fabricated using rod-in-tube or rotational casting methods. Fiber can be drawn using a preform or from the melt using the double-crucible method. As in the fluoride glass fibers, a buffer polymer coating is applied over the cladding using a UV acrylate or by first applying a Teflon heat-shrink tube over the preform and then drawing into fiber.

   The transmission range for chalcogenide fibers depends heavily on the mass of the constituent elements. The lighter element glasses such as arsenic trisulfide have a transmission range from 0.7 to about 6 μm [22]. This glass and some phosphorous-containing and Ge-S–based glasses are the only ones transmitting visible radiation [21]. Longer wavelength transmission is possible through the addition of heavier elements like Te and Se. When these elements are present, the glasses take on a silvery metallic appearance, and they become essentially opaque in the visible region. This trend is evident from the loss spectra shown for the most important chalcogenide fibers in Fig. 13.5. A key feature of essentially all chalcogenide glasses is the strong extrinsic absorption resulting from contaminants such as hydrogen, H_2O, and OH^-. For example, there are invariably strong absorption peaks at 4.0 and 4.6 μm due to S-H or Se-H bonds, respectively, and at 2.78 μm and 6.3 μm due to OH^- (2.78 μm) and/or molecular water. As a result, typical chalcogenide loss spectra are normally replete with extrinsic absorption bands, as is clearly seen from the data in Fig. 13.5. This would seem at first glance to be sufficiently deleterious that the applications for these fibers would be limited. However, many applications for these fibers are possible simply by working outside these extrinsic bands. Another important feature of most of the chalcogenide fibers is that their losses are usually much higher than the fluoride glasses. In fact at the important CO_2 laser wavelength of 10.6 μm, the lowest loss is still slightly above 1 dB/m for the Se-based fibers [23].

   A more recent chalcogen-type glass is based on a combination of chalcogen elements mixed with halides such as iodine. These so-called chalcohalide glasses

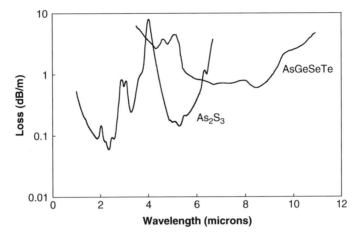

Figure 13.5 Two common chalcogenide glass fibers: As$_2$S$_3$ and an AsGeSeTe fiber [23]. Note the many impurity bands pervasive in these fiber systems.

afford the advantage of longer wavelength transmission than pure chalcogenide glass fibers. The most popular compositions studied today are the quaternary systems based on tellurium. These are the TeX glass systems and one of the most popular is Te-Se-As-I. For this TeX glass, the halide X = I$^-$. In thin window-type samples, TeX glasses transmit from 1 to 20 μm. Thus, their transparency range extends further than the standard telluride glasses fabricated without any halogen. Most of these glasses have rather low values of T_g usually about 150 °C or less.

The preparation of the TeX glasses is very similar to that for the chalcogenides. The starting materials are first purified and then a rod-in-tube preform is made by a rotational casting method. Blanchetiere et al. [24] at the University of Rennes has done much of the work on these glasses and on fiber drawing. For the core glass composition, they have chosen Te$_2$Se$_{3.9}$As$_{3.1}$I, and for the clad glass, Te$_2$Se$_4$As$_3$I. The refractive indices for the two glasses are 2.8271 and 2.8205 at 10.6 μm for the core and clad, respectively, giving a fiber numerical aperture (NA) of about 0.2. Fiber drawing from the preform was done at about 200 °C, with a drawing speed of 0.5–3.5 m/min. An online, UV acrylate coating was applied to the fiber for protection against moisture and, of course, to improve the strength of the fiber.

The losses for a core-only and a core/clad TeX fiber, made by Lucas' group at the University of Rennes, are presented in Fig. 13.6. From the data in Fig. 13.6, it can be seen that the minimum attenuation occurs between 7 and 9 μm. The minimum loss for the core/clad fiber is about 1 dB/m at 9 μm. A core-only fiber with an acrylate coating has a slightly lower loss of 0.5 dB/m at 9 μm, but the

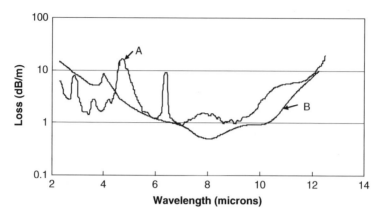

Figure 13.6   Chalcohalide or TeX glass fibers. (A) is a core/clad Te-Se-As-I fiber and (B) is an unclad Te-Se-I fiber.

acrylate coating on this fiber exhibits absorption especially in the 8- to 10-μm region. The core/clad TeX fiber shows some strong extrinsic absorption bands. From Fig. 13.6, there is strong absorption due to OH$^-$ (3 μm), Se-H (4.6 μm), and H_2O (6.3 μm). The data also show that the loss at 10.6 μm is rather large, but it is significantly better at 9.3 μm where a CO_2 laser can operate. Using a 9.3 μm CO_2 laser, they were able to transmit 2.6 W of laser power through a 600-μm diameter, 1-m long fiber.

13.3 CRYSTALLINE FIBERS

Crystalline IR fibers are an attractive alternative to glass IR fibers because most nonoxide crystalline materials can transmit longer wavelength radiation than IR glasses and, in the case of sapphire, exhibit some superior physical properties [3]. One disadvantage over glass fibers is that crystalline fibers are somewhat difficult to fabricate because crystalline materials do not have a glassy region, so they cannot be drawn into fiber as is done with glasses. Crystalline fibers must be fabricated either using modified crystal-growth techniques in which a fiber is pulled from the melt or by heating the crystal to temperatures below the melting point and then applying significant pressure to extrude the material through a die.

There are two types of crystalline fiber: polycrystalline (PC) and single-crystal (SC) fiber. Historically, the first crystalline fiber made was the PC fiber, KRS-5 (TlBrI). This fiber was fabricated by a hot-extrusion technique at Hughes Research Labs in 1976 [2]. KRS-5 was chosen because it is very ductile and because it can transmit beyond the 20 μm range required for the intended military surveillance satellite application. Today the best PC fibers, made from silver halide crystals, have losses in the 0.3 dB/m range at 10.6 μm. Nevertheless, the Ag-halide PC fibers continue to be popular today for short-length applications in sensor systems and for limited use in low-power laser delivery.

There has been comparatively less work on SC fiber optics. One reason for this is that they are much harder to fabricate than PC fibers. Only a few crystalline host materials have been studied, with the most important being the refractory oxides. Of these, sapphire fiber (Al_2O_3) is the most studied SC fiber, and it has the lowest loss [8, 25–27].

13.4 POLYCRYSTALLINE (PC) FIBERS

There are many halide crystals that have excellent IR transmission, but only a few have been fabricated into fiber optics. The technique used to make PC fibers is hot extrusion. As a result, only the silver and thallium halides have the requisite physical properties such as ductility, low melting point, and independent slip systems to be successfully extruded into fiber. In the hot extrusion process, a single-crystal billet or preform is placed in a heated chamber and the fiber extruded to net shape through a diamond or tungsten carbide die at a temperature about one-half the melting point. The final PC fibers are usually from 500 to 900 μm in diameter with no buffer jacket. The polycrystalline structure of the fiber consists of grains on the order of 10 microns or larger in size. The billet may be clad using the rod-in-tube method. In this method, a mixed silver halide such as AgBrCl is used as the core and then a lower index tube is formed using a Cl$^-$ rich AgBrCl crystal. The extrusion of a high-quality core–clad fiber is not easy because the extrusion process distorts the core–clad interface, often leaving a highly irregular core region. Artjushenko et al. [28] at the General Physics Institute in Moscow have achieved excellent clad Ag-halide fibers with losses nearly as low as the core-only Ag-halide fiber. At Tel Aviv University, high-quality multimode [29] and single-mode [30] Ag-halide fibers have been produced. Today, the PC Ag-halide fibers represent the best PC fibers. KRS-5 is no longer a viable candidate largely because of the toxicity of Tl and the greater flexibility of the Ag-halide fibers.

The optical losses in PC fibers are well above the intrinsic loss of the bulk material. In general, the best PC fibers made have losses between 0.3 and 0.5 dB/m around 10 μm. This is typically the lowest loss region, which is fortuitous because many applications for these fibers involve the transmission of CO_2 laser radiation. The losses for both core-only and core–clad silver halide fibers are shown in Fig. 13.7. The core-only fiber (A) is the AgBrCl fiber extruded by Moser et al. [31] at Tel Aviv University, while fiber (B) is a core–clad AgBrCl

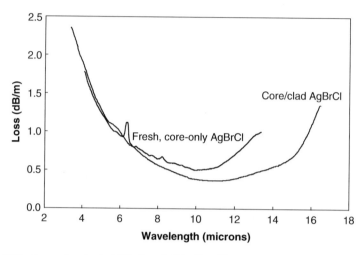

Figure 13.7 Losses in a typical PC silver halide fiber [3] compared to core-clad silver halide fiber.

fiber fabricated by Artjushenko [32] in Moscow and ART Photonics in Berlin. Both loss curves represent the current technology of silver halide fibers and, in the case of fiber (B), the quality of commercially available fiber. The core–clad fiber is available with an NA of 0.15 or 0.3. The core diameters of the fibers range from 500 to 900 μm. The lengths of the fibers typically do not exceed 3 or 4 m but may extrude fibers as long as 20 m.

There are interesting features of the loss data in Fig. 13.7. First, there are several impurity absorption bands due to water at 3 and 6.3 μm and sometimes a SO_4^{--} absorption near 9.6 μm. These bands are seen in the core-only fiber but are less evident in the core–clad fiber. Presumably this is due in part to the presence of the clad layer, which protects the core from contamination by water and other ions during extrusion. Furthermore, we note the increasing attenuation as the wavelength decreases. This is a result of λ^{-2} scattering from strain-induced defects in the extruded fiber. An important feature of the data is that the minimum loss is near 10.6 μm. These fibers have been used to transmit about 100 W of CO_2 laser power, but the safe limit seems to be 20 to 25 W [33]. The higher powers can more easily damage the fiber as a result of the low melting point of the fibers.

There are several difficulties in handling and working with PC fibers. One is an unfortunate aging effect in which the fiber transmission is observed to decrease in time [34]. Normally the aging loss, which increases uniformly over the entire IR region, is a result of strain relaxation and possible grain growth as the fiber is stored. Another problem is that Ag halides are photosensitive and exposure to visible or UV radiation creates colloidal Ag, which in turn leads to increased losses in the IR. Finally, the AgBrCl is corrosive to many metals. Therefore, the fibers should be packaged in dark jackets and connectorized with materials such as Ti, Au, or ceramic materials.

The mechanical properties of these ductile fibers are quite different from those of glass fibers. The fibers are weak, with ultimate tensile strengths of about 80 MPa for a 50/50 mixture of AgBrCl. The main difference, however, between the PC and glass fibers is that the PC fibers plastically deform well before fracture. This plastic deformation leads to increased loss as a result of increased scattering from separated grain boundaries. Therefore, in use, the fibers should not be bent beyond their yield point; too much bending can lead to permanent damage and a high loss region in the fiber.

13.5 SINGLE-CRYSTAL (SC) FIBERS

The most common and viable SC fiber developed is sapphire [8]. Sapphire is an insoluble uniaxial crystal with a melting point of 2053 °C. It is an extremely hard and chemically inert material that may be conveniently melted and grown in air. The usable fiber transmission is from about 0.5 to 3.2 μm. Other important properties include a refractive index of 1.75 at 3 μm, a thermal expansion about 10 times higher than silica, and a Young's modulus approximately 6 times greater than silica. These properties make sapphire an almost ideal IR fiber candidate for applications less than about 3.2 μm. In particular, this fiber has been used to deliver more than 10 W of average power from an Er/YAG laser operating at 2.94 μm [25]. Oxide materials like Al_2O_3 have the advantage of high melting points and chemical inertness, and they may be conveniently melted and grown in air [35–37].

Sapphire fibers are fabricated using either the edge-defined, film-fed growth (EFG) [38] or the laser-heated pedestal growth (LHPG) [35] techniques. In either method, some or all of the starting sapphire material is melted and an SC fiber is pulled from the melt. In the EFG method, a capillary tube is used to conduct the molten sapphire to a seed fiber, which is drawn slowly into a long fiber. Multiple capillary tubes, which also serve to define the shape and diameter of the fiber, may be placed in one crucible of molten sapphire so that many fibers can be drawn at one time. The LHPG process is a crucibleless technique in which a small molten zone at the tip of a SC sapphire source rod (<2 mm diameter) is created using a CO_2 laser. A seed fiber slowly pulls the SC fiber as the source rod continuously moves into the molten zone to replenish the molten material. Both SC fiber growth methods are very slow (several mm/min) compared to glass fiber drawing. The EFG method, however, has an advantage over LHPG methods because more than one fiber can be continuously pulled at a time. LHPG methods, however, have produced the cleanest and lowest loss fibers because no crucible is used, which can contaminate the fiber. The sapphire fibers grown by these techniques are unclad, pure Al_2O_3 with the C axis usually aligned along the fiber axis. Fiber diameters range from 100 to 300 μm and lengths are generally less than 2 m. In general, the fibers are all unclad, but it is possible to add a polymer coating such as Teflon using heat-shrink tubing.

The optical properties of the as-grown sapphire fibers are normally inferior to those of the bulk starting material. This is particularly evident in the visible region and is a result of color-center type defect formation during the fiber draw. These defects and the resulting absorption can be greatly reduced if the fibers are postannealed in air or oxygen at about 1000 °C. In Fig. 13.8, the losses for LHPG fiber grown at Rutgers University [37] and EFG fiber grown by Saphikon, Inc. (Milford, NH), now St. Gobain Crystals, are shown. Both fibers have been annealed at 1000 °C to reduce short-wavelength losses. We see that the LHPG fiber has the lowest overall loss. In particular, LHPG fiber loss at the important Er/YAG laser wavelength of 2.94 μm is less than 0.3 dB/m compared to the intrinsic value of 0.15 dB/m. There are also several impurity absorptions beyond 3 μm that are believed to be due to transition metals like Ti or Fe. Sapphire fibers have also been used at temperatures up to 1400 °C without any change in their transmission.

Young's modulus for sapphire is very high. In fact, the modulus for sapphire is about six times greater than that for silica. In practice, this means that SC sapphire fibers are rather stiff, a feature readily observed when one bends

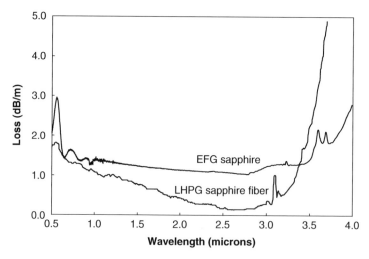

Figure 13.8 SC sapphire fibers grown by the edge-defined, film-fed growth (EFG) (Saphikon, Inc., Milford, NH) and laser heated pedestal growth (LHPG) methods [37].

equal-diameter sapphire and silica fibers. There has been only limited strength data taken on the optical SC sapphire fibers. Wu et al. [39] have measured 110-μm diameter LHPG fiber under tension. Their measurements yielded failure strains between 1.20 and 1.85%. These are measurements on only two samples, so their statistics are quite limited. Photran, Inc. (formerly Saphikon) claims that its 325 μm diameter fiber can be bent into a 60-mm loop. Jundt et al. [35] indicate that their 150-μm diameter LHPG fiber can be bent to a 4-mm radius.

13.6 HOLLOW-CORE WAVEGUIDES

The first optical frequency hollow waveguides were similar in design to microwave guides. Garmire et al. [40] made a simple rectangular waveguide using aluminum strips spaced 0.5 mm apart by bronze shim stock. Even when the aluminum was not well polished, these guides worked surprisingly well. Losses at 10.6 μm were well below 1 dB/m and Garmire et al. demonstrated early on the high-power handling capability of an air-core guide by delivering more than 1 kW of CO_2 laser power through this simple structure. These rectangular waveguides, however, never gained much popularity primarily because their overall dimensions ($\sim$0.5 $\times$ 10 mm) were quite large in comparison to circular cross-section guides and because the rectangular guides cannot be bent uniformly in any direction. As a result, hollow circular waveguides with diameters of 1 mm or less fabricated using metal, glass, or plastic tubing are the most common guide today. In general, hollow waveguides are an attractive alternative to conventional solid-core IR fibers for laser power delivery because of the inherent advantage of their air core. Hollow waveguides not only enjoy the advantage of high laser power thresholds but also low insertion loss, no end reflection, ruggedness, and small beam divergence. A disadvantage, however, is a loss on bending that varies as $1/R$, where R is the bending radius. In addition, the losses for these guides vary as $1/a^3$, where a is the radius of the bore. Unfortunately, this means that the flexibility of the very small bore (approximately $<$250 μm) guides is somewhat negated by their higher loss. However, the loss can be made arbitrarily small for a sufficiently large core. The bore size and bending radius dependence of all hollow waveguides is a characteristic of these guides not shared by solid-core fibers. Initially these waveguides were developed for medical and industrial applications involving the delivery of CO_2 laser radiation, but they have been used to transmit incoherent light for broadband spectroscopic and radiometric applications [41]. They are today one of the best alternatives for power delivery in IR laser surgery and industrial laser delivery systems with losses as low as 0.1 dB/m and transmitted cw laser powers as high as 2.7 kW [42].

Hollow-core waveguides may be grouped into two categories: (1) those whose inner core materials have refractive indices greater than one (leaky guides) and

(2) those whose inner wall material has a refractive index less than one (attenuated total reflectance, i.e. ATR, guides). Leaky or $n > 1$ guides have metallic and dielectric films deposited on the inside of metallic [43], plastic [44], or glass tubing [45]. ATR guides are made from dielectric materials with refractive indices less than one in the wavelength region of interest [46]. Therefore, $n < 1$ guides are fiber-like in that the core index ($n \approx 1$) is greater than the clad index. Hollow sapphire fibers operating at 10.6 μm ($n = 0.67$) are an example of this class of hollow guide [47].

13.6.1 Hollow Metal and Plastic Waveguides

The earliest circular cross-section hollow guides were formed using metallic and plastic tubing as the structural members. Tubing made from stainless steel or nickel was used in many of the early guides. One of the most successful approaches was used by Miyagi and his group in Japan. In their method they first sputter deposited Ge [48], ZnSe, and ZnS [49] coatings on aluminum mandrels. Then a layer of Ni was electroplated over these coatings before the aluminum mandrel was removed by chemical leaching. The final structure was then a flexible Ni tube with optically thick dielectric layers on the inner wall to enhance the reflectivity in the infrared. These guides had losses near 0.2 dB/m at 10.6 μm, but the hardness and springy character of the nickel tubing can be a disadvantage because it is less flexible than glass or plastic tubing. Plastic tubing seems almost ideal in that it is very flexible and inexpensive. At Tel Aviv University, Dahan et al. [50] applied Ag followed by AgI coatings on the inside of polyethylene and Teflon tubing to make a very flexible waveguide. The problem with these plastic materials is that they tend to be too soft and the inside surfaces are somewhat rough. The softness leads to deformation of the circular cross-section on bending and the roughness increases the scattering losses. Better results are obtained when harder polymers like polycarbonate tubing is used. George and Harrington [51] made very low loss (0.05 dB/m at 10.6 μm) waveguides using polycarbonate tubing.

13.6.2 Hollow Glass Waveguides

One of the most popular hollow waveguides today is the hollow glass waveguide (HGW) developed by Harrington [41] at Rutgers University. The advantage of the hollow glass structure over other hollow structures is that it is simple in design, flexible, and most important, it has a very smooth inner surface. HGWs have a metallic layer of Ag on the inside of silica glass tubing and then a dielectric layer of either AgI or a polymer-like cyclic olefin polymer (COP) is deposited over the metal film. Figure 13.9 shows a cross-section of an HGW with Ag/AgI

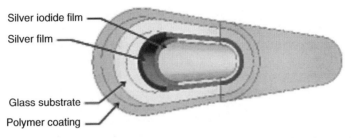

Silver iodide film

Silver film

Glass substrate

Polymer coating

Figure 13.9 Structure of the hollow glass waveguides (HGWs) showing the metallic and dielectric films deposited inside silica glass tubing using liquid-phase chemistry techniques.

coatings. The fabrication of the Ag/AgI HGW begins with silica tubing, which has a polymer (UV acrylate or polyimide) coating on the outside surface. A liquid-phase chemistry technique is used to deposit the Ag and AgI films inside the glass tubing [52]. This technique is similar to that used by Croitoru et al. [53] to deposit metal and dielectric layers on the inside of plastic tubing. The first step involves depositing a silver film using standard Ag plating technology. Generally, the Ag film is between 0.2 and 1 μm thick, and it is deposited slowly over about 1 hour. Immediately after the silver is deposited, an iodine solution is pumped through the tubing, and through a subtraction process, a layer of AgI is formed. By controlling the concentration of the iodine solution and the reaction time, an AgI film of the correct optical thickness can be deposited. Using these methods, HGWs with bore sizes ranging from 100 to 1200 μm and lengths as long as 13 m have been made.

The measured spectral response for two different 700-μm bore, 1-m long HGWs is given in Fig. 13.10. One guide has a 0.3-μm thick AgI film and the other a 0.8-μm film [41, 45]. The spectral response of the HGW with the thinner AgI layer is appropriate for the shorter IR wavelengths as, for example, for transmission of Er/YAG laser energy. The thicker AgI film gives the lowest loss at 10.6 μm. Clearly, for broadband IR applications, it is desirable to have a thin AgI layer. It would seem, therefore, that merely making even a thinner dielectric layer would yield a waveguide that would transmit well into the visible region of the spectrum. It is possible in principle to tailor the optical response to achieve short wavelength transmission, but this is very difficult using AgI films. The reason is that AgI (purple color) does not transmit well in the visible region. A better approach for visible wavelengths is to use a transparent polymer dielectric film over the Ag layer. Miyagi and his group have successfully used COP for visible and IR transmission [48]. An advantage of this approach is that the hollow waveguide will transmit visible radiation from, for example, a visible laser and this will provide an aiming beam along with the IR energy.

The straight-guide loss data for several bore sizes using CO_2 and Er/YAG lasers are shown in Fig. 13.11. The solid curves are theoretical calculations of the

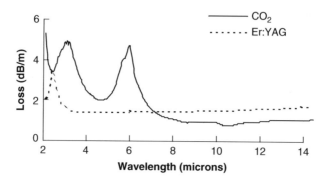

Figure 13.10 Typical spectra of hollow glass waveguides (HGWs) with Ag/AgI coatings. The guide with the thinner AgI layer (dotted line) is suitable for Er/YAG laser transmission at 3 μm, whereas the thicker AgI layer (solid curve) has the lowest loss at CO_2 laser wavelengths.

losses for the lowest order HE_{11} mode showing the $1/a^3$ dependence predicted by Marcatili and Schmeltzer [54]. At the CO_2 laser wavelengths, the calculated losses agree quite well with those measured. However, at 3 μm the calculated losses are much lower than the measured values. This is a result of increased scattering losses at the shorter wavelengths and the multimode character of the Er/YAG laser used in the measurements.

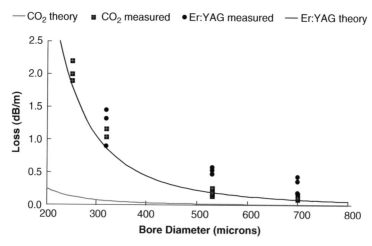

Figure 13.11 Measured and calculated losses for straight hollow glass waveguides (HGWs) with Ag/AgI coatings. The solid line near the data points is the calculated loss for 10.6 μm and it can be seen that there is good agreement with the measured values. The lower solid curve is the calculated loss for 2.94 μm where there is poor agreement with theory.

The increased loss on bending HGWs is shown for two 530-μm bore guides in Fig. 13.12 [45]. These data show that the loss increases as the curvature or $1/R$ increases. All data were taken with a constant length of waveguide under bending. A curvature of $20\,\text{m}^{-1}$ represents a bend diameter of only 10 cm! This is sufficiently small for most applications. It is important to note that while there is an additional loss on bending for these hollow guides, it does not necessarily mean that this restricts their use in power delivery or sensor applications. Normally most fiber delivery systems have rather large bend radii and, therefore, a minimal amount of the guide is under tight bending conditions and the bending loss is low. From the data in Fig. 13.12, one can calculate the bending loss contribution for an HGW link by assuming some modest bends over a small section of guide length. An additional important feature of hollow waveguides is that they are nearly single mode. This is a result of the strong dependence of loss on the fiber mode parameter. That is, the loss of high order modes increases as the square of the mode parameter so even though the guides are very multimode, in practice only the lowest order modes propagate. For example, a less than 300-μm bore guide will operate virtually single mode at 10 μm. HGWs have been used quite successfully in IR laser power delivery and, more recently, in some sensor applications. Modest CO_2 and Er/YAG laser powers below about 80 W can be delivered without difficulty. At higher powers, water-cooling jackets have been placed around the guides to prevent laser damage. The highest CO_2 laser power delivered through a water-cooled, hollow metallic waveguide with a bore of 1800 μm was 2700 W and the highest power through a water-cooled 700-μm bore HGW was 1040 W [55]. Sensor applications include gas and temperature measurements. A coiled HGW filled with gas can be used in place of a more

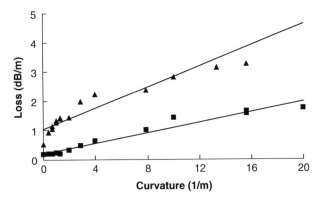

Figure 13.12 Bending losses for two Ag/AgI, 530-μm bore hollow glass waveguides (HGWs). (▲) Data taken at the Er:YAG laser wavelength of 2.94 μm and (■) data taken at CO_2 laser wavelength of 10.6 μm.

complex and costly White cell to provide an effective means for gas analysis. Unlike evanescent wave spectroscopy in which light is coupled out of a solid-core–only fiber into media in contact with the core, all of the light is passing through the gas in the hollow guide cell, making this a sensitive, quick response fiber sensor. Temperature measurements may be aided by using an HGW to transmit blackbody radiation from a remote site to an IR detector. Such an arrangement has been used to measure jet engine temperatures.

13.7 SUMMARY

During the past 30 years of the development of IR fibers, there has been a great deal of fundamental research designed to produce a fiber with optical and mechanical properties close to that of silica. IR fibers are still far from that Holy Grail, but some viable IR fibers have emerged that can be used to address some of the needs for a fiber that can transmit greater than 2 μm. Yet the current IR fiber technology is still limited by high loss and low strength. Nevertheless, more applications are being found for IR fibers as users become aware of their limitations and, more importantly, how to design around their properties.

There are two near-term or short-length applications of IR fibers: laser power delivery and sensors. An important future application for these fibers, however, is in active fiber systems like the Er- and Pr-doped fluoride fibers and doped chalcogenide fibers. As power-delivery fibers, the best choice seems to be hollow waveguides for CO_2 lasers and either SC sapphire, germanate glass, or HGWs for Er/YAG laser delivery. Chemical, temperature, and imaging bundles make use of mostly solid-core fibers. Evanescent wave spectroscopy (EWS) using chalcogenide and fluoride fibers is quite successful. A distinct advantage of an IR fiber EWS sensor is that the signature of the analyte is often very strong in the infrared or fingerprint region of the spectrum. Temperature sensing generally involves the transmission of blackbody radiation. IR fibers can be very advantageous at low temperatures, especially near room temperature where the peak in the blackbody radiation is near 10 μm. Finally, there is an emerging interest in IR imaging using coherent bundles of IR fibers. Several thousand chalcogenide fibers have been bundled by Amorphous Materials (Garland, TX) to make an image bundle for the 3- to 10-μm region.

REFERENCES

[1] Kapany, N. S. and R. J. Simms. 1965. Recent developments of infrared fiber optics. *Infrared Phys.* 5:69–75.
[2] Pinnow, D. A. et al. 1978. Polycrystalline fiber optical waveguides for infrared transmission. *Appl. Phys. Lett.* 33:28–29.

[3] Harrington, J. A. 2004. *Infrared Fiber Optics and Their Applications.* SPIE Press, Bellingham.

[4] Harrington, J. A. 2001. Infrared fiber optics. In: *Handbook of Optics; Fiber and Integrated Optics* (M. Bass et al., eds.). McGraw-Hill, New York.

[5] Katsuyama, T. and H. Matsumura. 1989. *Infrared Optical Fibers.* Adam Hilger, Bristol.

[6] Aggarwal, I. and G. Lu. 1991. *Fluoride Glass Optical Fiber.* Academic Press, New York.

[7] France, P. et al. 1990. *Fluoride Glass Optical Fibres.* Blackie, London.

[8] Sanghera, J. and I. Aggarwal. 1998. *Infrared Fiber Optics.* CRC Press, Boca Raton, FL.

[9] Kaiser, P. et al. 1975. Low loss FEP-clad silica fibers. *Appl. Opt.* 14:156.

[10] Poulain, M. et al. 1975. Verres fluores au tetrafluorure de zirconium proprietes optiques d'un verre dope au Nd^{3+}. *Mat. Res. Bull.* 10:243–246.

[11] Poulain, M. et al. 1977. New fluoride glasses. *Mat. Res. Bull.* 12:151–156.

[12] Tran, D. et al. 1984. Heavy metal fluoride glasses and fibers: A review. *J. Lightwave Technol.* LT-2:566–586.

[13] Itoh, K. et al. 1994. Low-loss fluorozirco-aluminate glass fiber. *J. Non-Cryst. Solids* 167:112–116.

[14] France, P. W. et al. 1987. Progress in fluoride fibres for optical communications. *Br. Telecom. Tech. J.* 5:28–44.

[15] Carter, S. F. et al. 1990. Low loss fluoride fibre by reduced pressure casting. *Elect. Lett.* 26:2115–2117.

[16] Szebesta, D. et al. 1993. OH absorption in the low loss window of ZBLAN(P) glass fibre. *J. Non-Cryst. Solids* 161:18–22.

[17] Matthewson, M. J. et al. 1986. Strength measurement of optical fibers by bending. *J. Am. Cer. Soc.* 69:815–821.

[18] Kobayashi, S. et al. 1978. Characteristics of optical fibers in infrared wavelength region. *Rev. Electrical Comm. Lab.* 26:453–467.

[19] Tran, D. 1993. Heavy metal-oxide glass optical fibers for use in laser medical surgery. Infrared Fiber Systems, Inc., US Patent no. 5,274,728; issued 12/28/1993.

[20] Nishii, J. et al. 1992. As_2S_3 fibre for infrared image bundle. *Int. J. Optoelectronics* 7:209–216.

[21] Nishii, J. et al. 1992. Recent advances and trends in chalcogenide glass fiber technology: A review. *J. Non-Cryst. Solids* 140:199–208.

[22] Kanamori, Y. et al. 1984. Preparation of chalcogenide optical fiber. *Rev. Electrical Comm. Lab* 32:469–477.

[23] Nishii, J. et al. 1987. Transmission loss of Ge-Se-Te and Ge-Se-Te-Tl glass fibers. *J. Non-Cryst. Solids* 95-96:641–646.

[24] Blanchetiere, C. et al. 1995. Tellurium halide glass fibers: Preparation and applications. *J. Non-Cryst. Solids* 184:200–203.

[25] Nubling, R. and J. A. Harrington. 1998. Single-crystal LHPG sapphire fibers for Er:YAG laser power delivery. *Appl. Opt.* 37:4777–4781.

[26] Chang, R. S. F. et al. 1991. Fabrication of laser materials by laser-heated pedestal growth. *Proc. SPIE.* 1410:125–132.

[27] Harrington, J. A. 1990. *Selected Papers on Infrared Fiber Optics, Milestone Series,* Vol. MS-9. SPIE, Bellingham, WA.

[28] Artjushenko, V. et al. 1995. Infrared fibers: Power delivery and medical applications. *Proc. SPIE* 2396:25–36.

[29] Shalem, S and Katzir, A. 2005. Core-clad silver halide fibers with few modes and a broad transmission in the mid-infrared. *Opt. Lett.* 30:1929–1931.

[30] Shalem, S. et al. 2005. Silver halide single-mode fibers for the middle infrared. *App. Phys. Lett.* 87:91103-1–91103-3.

[31] Moser, F. et al. 1996. Medical applications of infrared transmitting silver halide fibers. *IEEE J. Selected Topics Q. Electron.* 2:872–879.

[32] Artjushenko, V. 2003. From ART Photonics, Berlin, Germany (personal communication).

[33] Takahashi, K. et al. 1984. Optical fibers for transmitting high-power CO_2 laser beam. *Sumitomo Electric Tech. Rev.* 23:203–210.

[34] Wysocki, J. A. et al. 1988. Aging effects in bulk and fiber TlBr-TlI. *J. Appl. Phys.* 63:4365–4371.

[35] Jundt, D. H. et al. 1989. Characterization of single-crystal sapphire fibers for optical power delivery systems. *Appl. Phys. Lett.* 55:2170–2172.

[36] Chang, R. S. F. et al. 1995. Recent advances in sapphire fibers. *Proc. SPIE* 2396:48–53.

[37] Nubling, R. and J. A. Harrington. 1997. Optical properties of single-crystal sapphire fibers. *Appl. Opt.* 36:5934–5940.

[38] LaBelle, H. E. 1980. EFG, the invention and application to sapphire growth. *J. Cryst. Growth* 50:8–17.

[39] Wu, H. F. et al. 1991. Mechanical characterization of the single crystal alpha-Al_2O_3 fibers grown by laser-heated pedestal technique. *Light Metal Age* 49:97–98.

[40] Garmire, E. et al. 1980. Flexible infrared waveguides for high-power transmission. *IEEE J. Quantum Electron.* QE-16:23–32.

[41] Harrington, J. 2000. A review of IR transmitting, hollow waveguides. *Fiber Integr. Optics* 19:211–227.

[42] Hongo, A. et al. 1992. Transmission of kilowatt-class CO_2 laser light through dielectric-coated metallic hollow waveguides for material processing. *Appl. Opt.* 31:5114–5120.

[43] Matsuura, Y. and M. Miyagi. 1993. Er:YAG, CO, and CO_2 laser delivery by ZnS-coated Ag hollow waveguides. *Appl. Opt.* 32:6598–6601.

[44] Alaluf, M. et al. 1992. Plastic hollow fibers as a selective infrared radiation transmitting medium. *J. Appl. Phys.* 72:3878–3883.

[45] Matsuura, Y. et al. 1995. Optical properties of small-bore hollow glass waveguides. *Appl. Opt.* 34:6842–6847.

[46] Gregory, C. C. and J. A. Harrington. 1993. Attenuation, modal, polarization properties of n < 1, hollow dielectric waveguides. *Appl. Opt.* 32:5302–5309.

[47] Harrington, J. A. and C. C. Gregory. 1990. Hollow sapphire fibers for the delivery of CO_2 laser energy. *Opt. Lett.* 15:541–543.

[48] Miyagi, M. et al. 1986. Fabrication and transmission properties of electrically deposited germanium-coated waveguides for infrared radiation. *J. Appl. Phys.* 60:454–456.

[49] Matsuura, Y. et al. 1990. Fabrication of low-loss zinc-selenide coated silver hollow waveguides for CO_2 laser light. *J. Appl. Phys.* 68:5463–5466.

[50] Dahan, R. et al. 1992. Characterization of chemically formed silver iodide layers for hollow infrared guides. *Mater. Res. Bull.* 27:761–766.

[51] George, R. and J. A. Harrington. 2002. New coatings for metal-dielectric hollow waveguides. *Proc. SPIE* 4616:129–134.

[52] Matsuura, K. et al. 1996. Evaluation of gold, silver, and dielectric-coated hollow glass waveguides. *Opt. Eng.* 35:3418–3421.

[53] Croitoru, N. et al. 1990. Characterization of hollow fibers for the transmission of infrared radiation. *Appl. Opt.* 29:1805–1809.

[54] Marcatili, E. A. J. and R. A. Schmeltzer. 1964. Hollow metallic and dielectric waveguides for long distance optical transmission and lasers. *Bell Syst. Tech. J.* 4:1783–1809.

[55] Nubling, R. K. and J. A. Harrington. 1996. Hollow-waveguide delivery systems for high-power, industrial CO_2 lasers. *Appl. Opt.* 34:372–380.

Chapter 14

Hermetic Optical Fibers: Carbon-Coated Fibers

Paul J. Lemaire[1] and Eric A. Lindholm[2]

[1]*General Dynamics Advanced Information Systems, Florham Park, New Jersey*
[2]*OFS Specialty Fiber Division, Avon, Connecticut*

14.1 INTRODUCTION

Hermetically coated optical fibers have a thin layer of an impervious material applied over the surface of the glass fiber. Hermetic coatings are used to improve the reliability of fibers by preventing strength degradation caused by moisture attack on the fiber surface, and by preventing the diffusion of hydrogen into the core of the fiber. The typical polymer coatings that are used on optical fibers are able to prevent liquid water from contacting the glass surface but are not able to stop the diffusion of H_2O molecules or even-smaller H_2 molecules. Water molecules accelerate crack growth on the glass–fiber surface in a strength degradation process known as *fatigue*. Hydrogen molecules can quickly diffuse into the core of a nonhermetic optical fiber where they can cause significant increases in optical loss. Although the effects of water and hydrogen permeation are minor for typical telecommunication environments, they can cause significant problems when optical fibers are used in environments in which the mechanical stresses and/or hydrogen levels are higher than normal. For instance, fibers used in oil-well data logging are exposed to hydrogen and high stress levels at high temperatures, conditions that can quickly cause failure of standard fibers [1]. For fibers exposed to demanding conditions, it is often necessary to "seal" the fiber with a hermetic coating. Although many materials have been considered for use as hermetic coatings on optical fibers, carbon coatings have been the most successful solution and are the focus of this chapter.

Figure 14.1 shows a schematic representation of a single-mode optical fiber with a hermetic carbon coating and a single-layer polymer coating. When the hermetic layer is a nonductile material such as carbon, a polymer coating, similar

453

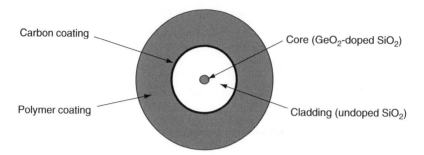

Figure 14.1 End-view schematic of a single-mode fiber with a hermetic carbon coating and a single-layer polymer coating. The carbon layer thickness is exaggerated in this figure.

or identical to that used on nonhermetic fibers, is used to protect the thin hermetic layer from scratches and other mechanical damage.

Many materials have been evaluated for use as hermetic coatings on optical fibers. Because an effective hermetic coating has to block the diffusion of small molecules—H_2O and H_2—the material used for the coating should have a close-packed structure. Early efforts to develop hermetic coatings for glass fibers focused on metallic coatings [2]. Although fibers drawn through a molten pool of metal have been shown to be resistant to static fatigue and hydrogen ingression, it can be difficult to achieve long lengths of low-loss fiber with thin pinhole-free metal coatings. Thicker metal layers can be used to avoid pinholes, but the increased thickness usually causes microbending losses to increase to unacceptable levels. Fibers with short metallized sections, on the other hand, are successfully used for sealing fibers into hermetic packages because the loss increases are acceptably low when the metallized fiber sections are short.

Carbon-coated optical fibers are made by depositing thin carbon layers onto the surface of the silica during the fiber-draw process [3, 4]. Layer thicknesses of only 20–50 nm of carbon have been shown to be sufficient to make the fiber hermetic against both water and hydrogen diffusion, without causing microbend losses. Because of these advantages, carbon-coated optical fibers provide a good engineering solution for specialty fibers used in demanding applications. Carbon-coated optical fibers have been used to solve hydrogen loss problems in underwater cables [5], to protect sensitive erbium-doped amplifier fibers [6], to prevent fracture and darkening in fibers used in sensor systems for oil drilling, to protect fibers in automotive applications [7], and to prevent fatigue in tightly routed fibers in avionics. While optically transparent diamond-like carbon coatings have been used to provide hard abrasion-resistant coatings for fiber devices [8], this type of carbon coating is not able to block water or hydrogen diffusion and, thus, is not useful for providing hermetic protection in most applications.

The target audience for this chapter is the end-user (i.e., the engineer who needs to specify a hermetic fiber to solve a reliability problem). The different materials that have been evaluated as candidate hermetic coatings are described to give historical perspective. However, the primary emphasis is on carbon-coated hermetic fibers, because this was the primary type of hermetic fiber as this chapter was being written. The chapter describes the primary reliability risks—fatigue and hydrogen losses—and how these problems are solved using hermetic fibers. The processes used to deposit carbon coatings onto fibers are discussed, as are some of the material studies done to characterize the nature of the carbon structure in carbon-coated fibers. Measurement techniques used to evaluate mechanical and optical properties of carbon hermetic fibers are described because these methods are used to evaluate the specifications for a hermetic fiber, and to determine its suitability for use. Hydrogen testing is discussed in detail because this places the greatest demands on a carbon coating. Techniques for handling, cleaving, and splicing hermetic fibers are also described. Finally, we discuss some of the applications of hermetic fibers to solve critical reliability problems, as well as the methods that can be used to specify the properties of hermetic fibers.

14.2 HISTORY

Hermetic coatings are used to protect fibers against fatigue and hydrogen-induced loss increases (i.e., against degradation by water and hydrogen, respectively). The development history of hermetic coatings, therefore, has been tied to the need to address these reliability issues.

Hermetic coatings for modern optical fibers have been investigated since the 1970s. It was recognized early on that the mechanical reliability of fibers was controlled by moisture-assisted bond breaking at the tips of microscopic cracks on the glass surface. Since it was recognized that this was an inherent reliability issue for silica fibers, a variety of hermetic coatings were considered potential ways of reducing or eliminating the water-related fatigue of fibers. Several early patents described, in general terms, the advantages of protecting the glass surface using metals, ceramics, or carbon [9, 10]. A paper gives a good historical and scientific overview of strength and fatigue issues relating to hermetic carbon-coated fibers [11]. Fatigue of fibers is discussed later in this chapter, and in Chapter 24.

Metal coatings were one of the first approaches to be considered. The general approach was to apply a metal coating during fiber drawing. While metal-deposition methods such as vacuum evaporation, sputtering, and deposition from glow discharges were mentioned in early work [9], most of the practical methods were based on drawing fibers through molten metals. Pinnow et al. [2]

reported 15- to 20-μm aluminum coatings that were applied during fiber draw. Although the average fiber strength was low (~3.4 GPa [500 kpsi]), the fracture stress was found to be independent of the strain rate, indicating that the metal coating was effectively blocking water from reaching the fiber surface. Some individual samples had strengths as high as 6.3 GPa (900 kpsi). Some success was achieved by others in stabilizing the coating deposition, thus eliminating coating discontinuities ("pinholes") and improving the mechanical properties [12].

One important problem with these metal-coated fibers was increased microbending loss. Problems associated with microbending losses were widespread in metal-coated fibers for several fundamental reasons: typical metal coatings were relatively thick. Therefore, bends imparted to the fiber during normal handling tended to cause semipermanent deformations of the metal coatings and of the fiber, resulting in localized microbends and associated losses. In addition, the thermal contraction of a metal coating as it cooled from the melting point resulted in compressive forces on the fiber, which in turn caused small lateral displacements leading to microbend losses. Winding and coiling the metal-coated fibers also tended to change the fiber loss. These losses were particularly problematic for long-length telecommunication applications with tight loss budgets.

An example of work on metal-coated fibers is that on indium-coated fibers. Indium coatings, with their lower melting points (156 °C), were investigated as a way to lower the microbend losses [13]. They showed fatigue parameters of $n = 32$, significantly higher than the $n = 20$ value usually seen for nonhermetic fibers.[1] The excess optical losses were 0.1–0.2 dB/km, tolerable for some short-distance applications. The fibers showed as-drawn fiber strengths of about 3.5 GPa (500 kpsi).

Although metal coatings were successfully made and did provide a way to prevent water diffusion, the optical losses were too high for many applications. Nonmetallic coatings based on materials such as Si_3N_4 [9], SiC [14], SiON [15], TiC [16], and carbon [10] were all investigated as potential low-loss hermetic coatings. These high-temperature coatings were typically applied by using different types of vapor-phase deposition during fiber drawing. The development of these and other hermetic coatings was accelerated in the 1980s by the realization that hydrogen-induced loss increases could occur in deployed optical fibers [17, 18]. Because H_2 molecules are smaller than H_2O molecules, the requirements on coatings became more stringent. Metals such as aluminum could stop H_2 diffusion, but they suffered from the previously mentioned problem of microbending

[1] Fiber fatigue is discussed in more detail later in this chapter, and in Chapter 24 of this book. Higher values of the fatigue parameter, n, are desirable in that they indicate an improved resistance to delayed fiber fracture.

losses. Appropriately deposited films of inorganic coatings such as SiC, Si_3N_4, SiO_xN_y, and several forms of carbon, were shown to be capable of blocking H_2 (and H_2O) without causing high losses.

Work at Hewlett Packard Laboratories used mixtures of various gases to deposit coatings of the general formula $Si_xC_yN_zO_w$ [14]. One of the goals was the protection of fibers from fatigue in oil-well data-logging applications, where fibers can see 2% strains and temperatures in excess of 200 °C in a high-humidity corrosive environments [1]. Mixtures of SiH_4, CO_2, NH_3, hydrocarbon gases, N_2, and He were used to deposit thin films of various compositions. Although not all of the coating compositions were successful, several compositions showed good fatigue resistance with little or no adverse effect on fiber loss. The patent discussed the importance of the Si-C bond in achieving fatigue resistant hermetic fibers. The inclusion of O and N in the coating composition was thought to help match the physical properties of the film to the underlying silica fiber, but the coatings were basically SiC according to the claims in their patent. The reported values for the fatigue parameter, *n*, were between 8 and 256, with film thicknesses ranging from about 10 to 67 nm. As was seen in much of the later work on ceramic and carbon coatings, there was a noticeable reduction in the as-drawn strength of some of the fibers. Results in one paper described 3.2 GPa (464 kpsi) fiber strengths for hermetic fibers [19]. Although this was acknowledged to be lower than the typical 5 GPa (725 kpsi) as-drawn strength for nonhermetic fibers, it was pointed out that the hermetic fiber strength was maintained at this 3.2 GPa level and that of the nonhermetic fiber steadily degraded due to moisture attack. Analysis showed that after about 2.5 hours, the strength of the hermetic fiber would exceed that of the nonhermetic fiber. The coatings were also shown to slow the diffusion of hydrogen into the fiber core [19]. No hydrogen was detected after 1000 hours at room temperature, but some H_2 diffusion was noted at 75 °C, albeit at a much lower level than in nonhermetic fibers.

British Telecomm Labs developed practical silicon oxynitride hermetic coatings in the early 1980s [15]. These coatings were tested at both low (0.74 atm) and high (65 atm) H_2 partial pressures. In neither case was it possible to detect H_2 absorption peaks in the fibers. This allowed the maximum room temperature diffusion coefficient for H_2 to be estimated as being about 5×10^{-19} cm^2/sec. This H_2 diffusivity is more than seven orders of magnitude lower than that for SiO_2. Because the experiments were done only at 21 °C, it was not possible to predict the H_2 diffusivity at higher temperatures. It is likely that the same coatings would have been effective barriers to H_2O molecules and, thus, would have improved fatigue and prevent H_2 losses.

A study in 1986 described fatigue and strength results for SiON, C, TiC, SiC, and for C-SiON and C-TiC composite coatings [16]. The SiON fibers had low strengths of 1.4–1.7 GPa (200–250 kpsi) but had high *n* values of about 90.

The carbon-coated fibers in that study had strengths of 4.1–4.5 GPa (600–650 kpsi), but n values in the range of 23–25 indicated that these carbon coatings were only partially effective in blocking H_2O diffusion. The results of that study were typical in demonstrating the need to tradeoff fracture strength against hermetic properties.

Starting in about 1988, there was significant progress in carbon-coated fibers. The carbon-coated fibers developed by a number of companies showed good engineering properties with acceptable fracture strengths and good resistance to fatigue and hydrogen loss.

Corning reported results for thin (<50-nm) amorphous carbon coatings, showing good fatigue and hydrogen properties and usable fiber strengths [4]. Dynamic fatigue values of $n \sim 110$ were achieved with fiber strength of about 3.5 GPa (500 kpsi). Analysis showed that the improved fatigue characteristics of hermetic fibers allowed the hermetic fibers to be used at 80% of their proof-test level, versus only about 30% for a nonhermetic fiber. Hydrogen diffusion in the same fibers resulted in loss changes of only about 0.25 dB/km at 1240 nm after 200 days at 21 °C, indicating that the coatings were effective in retarding H_2 diffusion at room temperature. As expected, H_2 diffusion was faster at higher temperatures. Carbon coatings are still used on Corning erbium-doped fibers, both to permit tight coiling of the Er fibers in compact amplifier modules and to stabilize the fibers against long-term loss changes [6].

In 1985 AT&T reported results for hermetic fibers with approximately 50- to 100-nm carbon coatings having a pyrolytic graphite structure [3]. The coatings were deposited by thermally decomposing carbon-containing gases onto the fiber surface during fiber draw. These fibers had relatively high as-drawn strengths of 4.1 GPa (600 kpsi). Fiber fatigue was minimal as evidenced by dynamic fatigue values from 350 to 500 and static fatigue values of about 200. Hydrogen permeation through the coatings was not measurable at room temperature but could be detected by exposing fibers to high pressure H_2 at 100–150 °C [20]. By characterizing the diffusion constant as a function of temperature, it was possible to quantitatively predict fiber loss increases in H_2 atmospheres as a function of time and temperature. Similar to the work on SiON, the room temperature diffusion coefficient for H_2 in the film was extremely low, about $1 \times 10^{-20} \, cm^2/sec$ at 21 °C for the carbon films [20]. These fibers incorporated reactive gettering sites in the optically inactive silica cladding layers. These sites aided in protection against hydrogen loss by reacting with and immobilizing the trace levels of H_2 that penetrated the carbon coating at high temperatures.

Several Japanese groups investigated the relationship between the properties of carbon-coated fibers and the carbon morphology. Workers at Sumitomo predicted long-term loss increases using accelerated H_2 testing and correlated the predicted behavior with the surface morphology of the carbon [21]. They concluded that a smooth surface structure leads to improved performance of the fiber in hydrogen. Work at NTT Laboratories came to a similar conclusion regarding fiber strength [22]. Fiber strengths of more than 4.8 GPa (700 kpsi) were obtained, with dynamic fatigue n values as large as 670. These high values were attributed to an ultrasmooth carbon surface (as measured by a Scanning Tunneling Microscope). Hydrogen measurements at 75 °C showed that the coatings did have the ability to retard H_2 diffusion, although long-term predictions were not made. Field-test experiments evaluating cabled carbon-coated fibers were conducted at Furukawa and showed that no hydrogen losses occurred over a 1.5-year period even for fibers in a water-filled cable [23]. Various mechanical tests on the cabled fibers show that loss increases did not occur even under simulation of harsh handling conditions. A similar study by NTT concluded that the performance of carbon-coated fibers was excellent in field tests, and that the fibers offered improved reliability against fatigue [24].

A study by BTRL pointed out, in a 1991 paper, the differences between carbon-coated fiber obtained from different manufacturers, and discussed the materials properties of the carbon films as well as the testing techniques used to characterize strength, hydrogen permeation, and electrical resistivity of the coatings [25]. This paper discussed the use of electrical resistivity measurements to characterize carbon coating properties, and the correlations of resistivity with coating properties affecting H_2O and H_2 diffusion. The authors discussed the sensitivity of the coating properties to the details of the carbon bonds—diamond-like versus graphitic—and the methods used to deposit the coating. They concluded that of the fibers studied, only a subset would be suitable for use in environments where H_2 losses might be a problem. They did find a good correlation between electrical resistivity and hydrogen performance for some of the fibers in the study. They also concluded that thinner coatings tended to be associated with higher fiber strengths. Overall, the paper highlighted the importance of having clearly defined test procedures to quantitatively determine the suitability for use of a carbon-coated "hermetic" fiber.

By about 1990, the engineering of carbon-coated fibers had resulted in fibers that met stringent optical performance requirements and had significantly improved reliability. There were, however, several engineering and perception tradeoffs associated with these carbon-coated fibers. The black color of the carbon coatings altered the visual appearance of colored inks applied over the polymer coating for fiber identification purposes. Fusion splicers required frequent cleaning when used to splice hermetic fibers. Finally, there was a perception that all carbon-coated fibers had inadequate fracture strengths, probably stemming from the fact that early hermetic fibers had been much weaker than both nonhermetic fibers and the improved carbon-coated fibers that came later.

Since 2005, hermetic fibers have been used predominantly in specialty applications. One reason that the coatings are not used more widely is that problems of fatigue and hydrogen aging have been solved for many applications without the use of hermetic fiber coatings. Design rules limit the stresses seen by fibers, greatly decreasing fatigue-induced weakening of nonhermetic fibers. Similarly, optical fiber cable designs and materials have been improved so that H_2 evolution in a cable is a relatively uncommon event, and the extra protection offered by a carbon coating is not usually required. However, there are a growing number of specialty applications that do require that the fiber be protected from H_2 aging and/or fatigue. Fiber sensors used in oil-well data logging need the protection offered by carbon coatings, to protect the fibers from both fatigue and hydrogen aging. Tightly routed fibers, for instance, in airframes or ultra-compact fiber modules, can avoid fatigue and maintain their reliability even at very small bend radii when they have hermetic carbon coatings. Finally, in some applications (e.g., space or undersea), the stringent system reliability requirements are more readily met when the fiber reliability is enhanced with a protective carbon coating.

14.3 DEPOSITION OF CARBON COATINGS ON FIBERS

The deposition of the carbon layer on an optical fiber occurs during the fiber-draw process. Immediately after the glass fiber reaches its final size (e.g., 125 μm), it is introduced into a hydrocarbon gas such as dilute acetylene, where the retained heat of the fiber causes the hydrocarbon to "crack" (thermally decompose) in a pyrolytic reaction. This reaction on the surface of the glass leads to the chemical vapor deposition of carbon onto the fiber. Although the deposition process is relatively straightforward, it has many possible permutations based on the deposition reactor design, precursor hydrocarbon gas, and fiber-draw speed. All of these factors affect the pyrolysis of the reactant gas, the deposition rate, and the hermeticity of the resulting carbon layer.

Because the carbon layer is a thin brittle coating, it does not protect the fiber from mechanical damage such as scratches. For this reason, carbon-coated fibers are always protected by polymer coatings, which are applied using standard coating applicators located below the hermetic coating reactor on the draw tower. Standard dual-layer acrylate coatings can be used, as can specialty polymer coatings such as polyimide coatings for high-temperature applications. The draw speed needs to be compatible with both the carbon and the polymer coating process. For instance, a carbon coating process that requires a high draw speed will not be compatible with a polymer coating process that needs a low draw speed.

A carbon deposition reactor is designed to strip the boundary layer of air from around the glass fiber and to deliver a hydrocarbon gas to the glass surface at a reaction temperature around 700–900 °C. For fibers drawn at high speeds (>5 m/sec), the fiber exits the draw furnace at a high temperature and the reaction zone is very long (tens of centimeters). Carbon reactors used at this speed are typically elongated chambers located below the draw furnace, where an organic gas is introduced at one end and unreacted gases are exhausted from the other. As the draw speed decreases, the reaction zone shrinks and the carbon reactor may need to be attached directly to the draw furnace to ensure the fiber has enough retained heat for carbon deposition.

Reactant gases include a variety of hydrocarbons, and indeed, many different carbon-containing gases have been investigated. Reactants have included methane, acetylene, ethylene, propane, butadiene, trichloroethylene, and benzene [26, 27]. In addition, some researchers have reported that the addition of chlorine gas to the reactor improves the carbon fiber's strength and hermetic characteristics, possibly by acting as a hydrogen scavenger during the pyrolytic reaction. Hydrogen, which is a byproduct of the pyrolysis reaction, can, in some cases, become trapped under the carbon coating, resulting in increased fiber loss. The choice of precursor gas depends on the reactor design and the temperature of the fiber in the reaction zone, which in turn is determined by the fiber draw speed.

The carbon layer applied to the fiber is black and is usually shiny in appearance. It is electrically conductive, which allows measurement of the coating's electrical resistance and inference of the coating thickness. The carbon layer thickness can be estimated from the electrical resistance with the following equation:

$$R = \frac{\rho}{2\pi r \delta}, \tag{14.1}$$

where R is the linear electrical resistivity (Ω/cm), ρ is the resistivity of the applied carbon layer (Ω-cm), r is the fiber radius, and δ is the carbon layer thickness.

By correlating dynamic fatigue measurements with measurement of electrical resistance, it is sometimes possible to use the linear resistivity as a measure of the coating's ability to block water diffusion. For instance, one manufacturer has shown that a linear resistivity of $R < 25\,k\Omega/cm$ is sufficient to guarantee that the carbon layer will be hermetic from the standpoint of fatigue [28]. The measured resistance in this case corresponded to a thickness of about 20 nm. The correlation between thickness and resistance depends on the resistivity of the applied carbon, which in turn is a function of graphitic crystallite size, orientation, and perfection [29]. Because the material properties are sensitive to processing conditions, the resistance values for a given coating thickness may be different for fibers made by different manufacturers.

Coating resistivity can be measured directly using an ohmmeter and liquid metal contacts. The polymer coating needs to be stripped from the fiber for these off-line measurements. Online measurements, conducted during the drawing of a hermetic fiber, are advantageous because they give immediate information about the coating properties and do not require destructive testing (i.e., polymer coating removal) after the fiber is drawn. Noncontact measurements of electrical resistance and laser scattering techniques have been used to measure resistivity and coating thickness, respectively [28, 30].

One analysis of the applied carbon material indicated that the layer consists of disordered graphitic platelets that are randomly oriented and bonded on the surface of the glass fiber in a continuous structure [5]. Thus, to close down pathways between the graphite platelets for water or hydrogen ingression, the carbon layer needs to have a thickness more than some minimum value. One study concluded that the carbon thickness should be more than 20 nm to prevent fatigue and 25 nm to delay hydrogen ingression [28]. It is important to realize that that thickness per se is not always a useful parameter for coating characterization, especially when comparing fibers made by different processes. The direct measurement of coating thickness requires careful sample preparation and examination of the fiber in an electron microscope. In addition, the material properties of a carbon coating depend on the processing conditions. For instance, a thick diamond-like carbon coating will offer much less protection from water and H_2 than a thinner coating with a graphitic structure.

14.4 FATIGUE PROPERTIES OF CARBON-COATED FIBERS

By excluding water from the surface of the glass, the carbon layer prevents the onset of moisture-assisted crack growth, known as *fatigue*. Resistance to fatigue is measured by the stress corrosion resistance parameter n. A high n value indicates a greater resistance to fatigue and, thus, a greater capability for a fiber to maintain its original strength. The static fatigue parameter, n_s, can be determined by measuring time to failure for fibers subjected to different static loads. Alternatively the dynamic fatigue parameter, n_d, can be determined by measuring failure stress as a function of strain rate. The values for n_s and n_d typically have similar values for a given fiber. Because dynamic fatigue tests are quicker to perform, they are more often used in characterizing fibers. Conventional fibers tend to have n values around 20, whereas hermetic fibers demonstrate n values of 100 or more. The area of fiber strength and fatigue in nonhermetic optical fibers has been studied extensively. For more information on this subject, see Chapter 24 in this book, as well as review articles on the subject [31–33].

The strength retention of carbon-coated fibers makes them ideal for use in adverse environments, in which water or other corrosive chemicals would normally lead to premature failure in unprotected fibers. Carbon-coated fibers immersed in water have demonstrated no loss in strength, even after many months at elevated temperatures [34]. In laboratory experiments, carbon coatings have been shown to be impervious to concentrated hydrofluoric acid and to hot sodium hydroxide solutions [26, 45].

In addition to providing a long lifetime at standard application stresses, hermetic fiber will permit operation at elevated mechanical stresses. A rule of thumb for nonhermetic fibers is that the application stress should not exceed about 20–30% of the proof-test level [4, 35]. However, because the carbon layer prevents the crack growth that leads to failure, hermetic fibers can operate at up to 80% of the proof-test level [4, 6]. This is a very attractive feature for fibers used in high stress applications such as compact fiber-based components that require tight fiber bend radii.

Fatigue is defined as crack growth on a stressed fiber in the presence of water. The stress applied to a fiber during use may be tensile stress, such as in aerial cables, or bending stress, such as those found in tight enclosures. On a molecular level, the mechanics of fatigue are straightforward: an H_2O molecule at a crack tip in the SiO_2 glass matrix will rupture the silicon–oxygen bond, leading to crack growth. Fatigue occurs preferentially at the crack tip because this is the point of maximum stress concentration (i.e., where the silicon–oxygen bonds are most strained). Thus, to mitigate the effect of fatigue, it is best to minimize (1) the size of cracks on the surface of the glass fiber, (2) the stress applied to the fiber, and (3) the presence of water. The role of the hermetic coating is to affect factor (3) by preventing water molecules from reaching the fiber surface and the crack tip.

The fatigue resistance of hermetic fibers can be assessed using dynamic tensile testing [36]. In this procedure, fibers are strained to failure at different strain rates. Nonhermetic fibers show decreased strengths at lower strain rates because the lower strain rates give more time for water reaction at the crack tip. Figure 14.2 shows data for hermetic carbon-coated fibers and a typical nonhermetic fiber. The fracture stresses were determined for both fiber types using a range of strain rates: 25%/min, 2.5%/min, 0.25%/min, and 0.025%/min [29]. The fracture stresses were plotted at each strain rate, as shown in Fig. 14.2. The slope (m) of each line is calculated by linear regression and the dynamic fatigue resistance factor n_d for each fiber was derived from Eq. (14.2):

$$n_d = \frac{1}{m} - 1, \tag{14.2}$$

where n_d is the dynamic fatigue factor and m is the slope of stress–strain curve.

A flatter slope (i.e., a slope approaching zero) will produce a higher dynamic fatigue value, n_d. The graph demonstrates the advantages of a high n value. For the nonhermetic fiber, the breaking strength declines as the strain rate is decreased and the fiber spends more time under strain. The longer time under strain increases the time for water molecules to react at the crack tip. In actual use, a nonhermetic fiber that is held under stress will gradually weaken. For

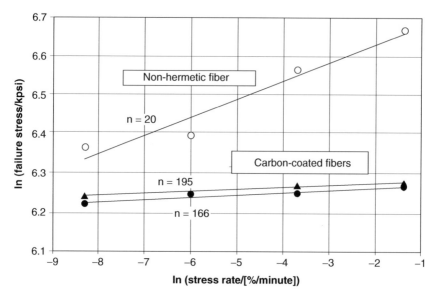

Figure 14.2 Dynamic fatigue characteristics of hermetic and nonhermetic fibers.

hermetic fibers, the absence of water at the crack tip prevents fatigue, so there is no appreciable decrease in fracture strength at slower strain rates. Similarly, the hermetic fiber will maintain its strength over long times even when exposed to water or humidity in a stressed state.

Although there are several models to estimate a fiber's lifetime under stress, they all generally derive from the Weiderhorn model [37]:

$$t_f = BS_{int}^{n-2}\sigma^{-n}, \tag{14.3}$$

where t_f is the time to failure, B is a crack growth parameter, S_{int} is the intrinsic strength of the fiber, and σ is the applied stress.

The difficulty in forming a standard equation for fiber lifetime lies in the fact that, for any given fiber, B and S_{int} are unknown. However, if a fiber is proof-tested, then the term S_{int} may be reasonably replaced in Eq. (14.3) by σ_p, the proof-test stress. Long-term stress levels for nonhermetic fibers, with n values of about 20, have to be maintained at much lower levels than those for high n-value hermetic fibers.

The ability of hermetic fibers to retain their strength over long periods is demonstrated in Figs. 14.3 and 14.4. In this study, standard nonhermetic acrylate-coated fibers and carbon-coated fibers (also with acrylate coatings) were exposed to various aqueous environments over a 9-month period [38]. The nonhermetic fibers exhibited strength degradation that was accelerated with water

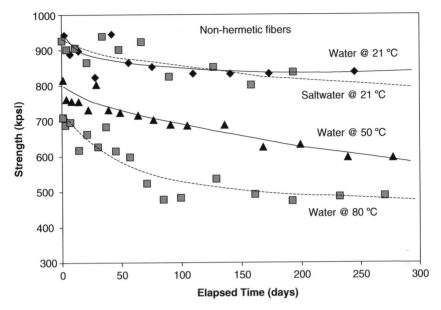

Figure 14.3 Aging of acrylate-coated fibers in aqueous environments.

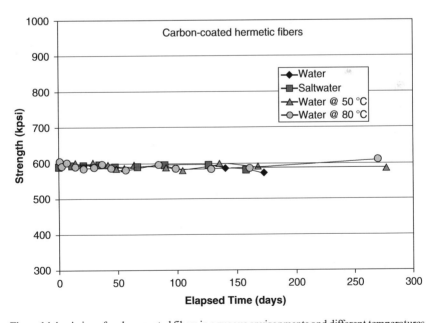

Figure 14.4 Aging of carbon-coated fibers in aqueous environments and different temperatures.

temperature. For fibers hermetically sealed with a thin layer of carbon, no change from initial strength was observed for any of the aged samples (Fig. 14.4).

Similar results were observed for fibers soaked in conventional chemicals, indicating that carbon coatings can protect fibers from harsh environments and elevated temperatures more effectively than standard polymer coatings.

14.5 HYDROGEN LOSSES IN OPTICAL FIBERS

14.5.1 Hydrogen-Induced Losses in Nonhermetic Fibers

In the early 1980s, it was realized that hydrogen-induced losses could impair optical fiber systems. Fibers exposed to hydrogen, even at low levels, showed increased losses that, in some cases, caused fiber losses to exceed the system-aging margin, resulting in system failure. Hydrogen was found to originate from certain silicone-based polymers [17] and from the generation of H_2 by corrosion of metal parts inside the cables [18]. H_2 molecules can diffuse through polymer materials and through a fiber's silica cladding in a matter of days. At room temperature, the H_2 is detectable at the core in about a day and losses reach their equilibrium levels in less than 2 weeks [39]. As hydrogen gas diffuses into the light-carrying portion of the optical fiber, it can lead to both reversible and permanent optical loss increases. Figure 14.5 shows hydrogen-induced loss changes in an accelerated experiment with a nonhermetic fiber exposed to pure H_2 for 3 days at 150 °C [39].

The reversible loss increases are due to unreacted H_2 molecules dissolved in the silica matrix. H_2 molecules are not infrared active in the gas phase but become absorbing when dissolved in SiO_2. There are several localized loss peaks in the 1000- to 1700-nm spectral region, including a prominent first overtone absorption at about 1240 nm. Figure 14.6 shows the molecular H_2 spectrum for a fiber that was equilibrated in a high-pressure H_2 gas environment [40]. There is a loss edge that increases at wavelengths longer than about 1500 nm, affecting losses in the 1550-nm transmission window. The absorbing strength of H_2 in the 1550-nm region is about 0.6 dB/km/atm at room temperature. The effects in the 1310-nm window are about three times lower than at 1550 nm. The molecular H_2 losses are proportional to the local H_2 partial pressure. The losses are reversible in that the H_2 losses will recover if the hydrogen is removed from the fiber environment. The time scale for loss recovery is the same as for growth; 95% of the H_2 will leave the fiber core in 2 weeks at room temperature once the fiber is removed from the H_2 atmosphere. Hydrogen diffusion rates increase with temperature, but the strength of the absorption actually falls off with increases in temperature because the solubility of H_2

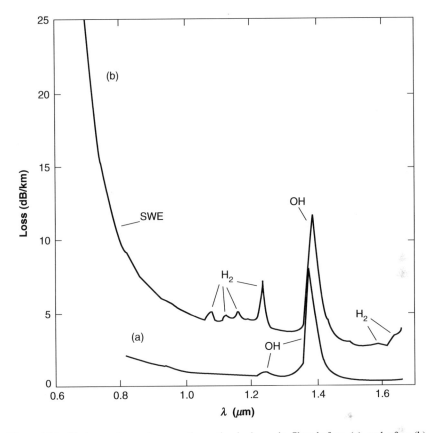

Figure 14.5 Hydrogen losses in a nonhermetic single-mode fiber before (a) and after (b) exposure to H_2 at 150 °C for 3 days. Loss features labeled "H_2" are reversible, whereas loss features labeled "SWE" and "OH" are permanent.

decreases as the temperature is raised. Molecular hydrogen losses are approximately the same for all silica-based fibers. These losses are avoided only by carefully controlling hydrogen in the fiber environment (e.g., the cable) or by using a hermetically coated fiber that is designed to block H_2 diffusion.

At elevated temperatures, H_2 can react with point defects in the fiber core, giving rise to OH absorption peaks at 1390 nm, 1240 nm, and other wavelengths, and to a short wavelength edge (SWE) that causes spectrally broad loss increases in Ge-doped fibers [41] (Fig. 14.5). Unlike molecular H_2 losses, the reaction rates for hydrogen in different fiber types can differ greatly and tend to increase rapidly as the temperature is increased. These reactions are usually not reversible. The absorbing strength of some of the lossy species can be quite

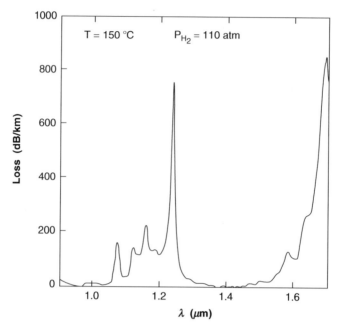

Figure 14.6   Molecular H_2 losses in a nonhermetic single-mode fiber exposed to H_2 at high pressure.

high. For instance, 1 ppm[2] of OH will cause about 15 dB/km of loss increase at 1390 nm. By lowering the core dopant levels (e.g., Ge and P), it is sometimes possible to lower the fiber's reactivity, but such dopant changes are not always practical. For instance, fibers with undoped silica cores are less prone to hydrogen reactions, but the variety of fiber designs that can be achieved without Ge and/or P doping is limited. As with molecular H_2 losses, it is possible to eliminate or at least decrease the losses due to hydrogen reactions by using a hydrogen-blocking carbon coating.

14.5.2 Hydrogen Losses in Carbon-Coated Hermetic Fibers

At room temperature, most carbon coatings can block H_2 to such an extent that no loss changes will be detectable over the time scale of days. However, at elevated temperatures (e.g., 100 °C), some H_2 diffusion may be detectable in accelerated

---

[2]   1 ppm is defined as 10^{-6} mol of OH per mole of SiO_2.

aging experiments or under aggressive use conditions. Whether a carbon-coated fiber is suitable for use in a hydrogen environment depends on the tolerable loss changes, the operating temperature, the concentration of hydrogen around the fiber, and the material properties of both the fiber and the carbon coating.

Determining whether a coating is adequately hermetic requires a quantitative analysis. No carbon fiber coating made to date has been shown to be perfectly hermetic. Increases in ambient hydrogen levels, pressure, temperature, and fiber length all lead to increased demands on the fiber coating. Similarly, if the fiber's core glass material is highly reactive with hydrogen (e.g., some erbium-doped fibers and Ge-P co-doped multimode fibers), the demands on the hermetic coating are increased.

14.5.3 Testing of Hermetic Fibers in Hydrogen

To characterize a hermetic coating's ability to retard hydrogen diffusion, it is necessary to use accelerated aging experiments. In these experiments, a length of hermetic fiber is exposed to hydrogen, usually at an elevated temperature and sometimes using high-pressure hydrogen. Loss changes are detected either by measuring the changes in fiber loss in real time (*in situ* measurements) or by measuring the fiber loss before and after the exposure of the fiber to hydrogen. Measurement of loss changes can be done via OTDR loss measurements (1310, 1550, and/or 1625 nm), by using a single-wavelength laser source and an optical detector, or by using a broadband light source and an optical spectrum analyzer (OSA) to obtain a full loss spectrum. Diffusion of hydrogen through carbon depends strongly on the temperature. Therefore, unless the temperature dependence is already known, experiments need to be done at two or more temperatures to characterize the temperature dependence of H_2 diffusion through the coating.

Figure 14.7 shows an example of an experimental setup that can be used for *in situ* characterization of a hermetic fiber. In this case, the measurement equipment allows a choice between OTDR and optical loss measurements, although typically only one method would be used. Loss-change data are obtained periodically throughout the accelerated aging experiment, generally by computer-controlled test equipment. Before testing, one should wind the fiber onto a spool that will not be damaged at the test temperature and that will not impose stresses on the fiber due to thermal expansion of the spool material.

Because of the flammable and explosive nature of hydrogen gas, it is essential to recognize that hydrogen gas must be handled with appropriate safety precautions, especially if it is used at high pressures. Because hydrogen can cause brittleness in some types of steel, it is important to verify that the materials and fittings used to handle the hydrogen are compatible with the temperatures and pressures used in the experiments.

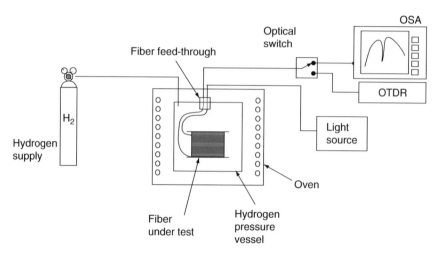

Figure 14.7 Experimental setup for measuring *in situ* loss changes in a hermetic fiber exposed to hydrogen.

A fiber with a "good" hermetic coating may exhibit only small loss changes when tested in hydrogen, especially if the test temperature is under about 100 °C. Improved measurement sensitivity can be obtained by increasing the fiber length and/or by measuring the loss change at a wavelength where the hydrogen absorption is strong. Fiber lengths of about 1 km are usually sufficient for experiments run in the 100–200 °C range. The 1240-nm H_2 peak is distinct and readily measured. It is generally the easiest feature to monitor if spectral measurements are being made. (There is small contribution at the same wavelength from an OH absorption band. The OH contribution at 1240 is about 1/20 that of the OH overtone at 1390 nm.) When an OTDR is used to measure loss changes, it is best to use 1550- or 1625-nm wavelengths because the H_2 absorption is significantly stronger at these wavelengths than at 1310 nm. For weakly guiding fibers, microbend loss increases can sometimes complicate the data analysis because these losses may also contribute to 1550- and 1625-nm losses. For this reason, it is usually preferable to characterize fibers with spectral loss measurements since this allows the tracking of hydrogen-specific features such as the 1240-nm H_2 overtone.

Figure 14.8 shows before-and-after loss measurements for a carbon-coated fiber tested at 152 °C for 15 hours in 144 atm of H_2 [42]. The increase in the 1240 H_2 peak and the rising loss beyond 1600 nm are clear evidence of H_2 in the fiber core. The overall upward shift in loss values is due to spectrally broad SWE loss increases, associated with hydrogen reaction at Ge sites. Figure 14.9 shows the growth of the 1240-nm H_2 peak in three sections of fiber tested at different

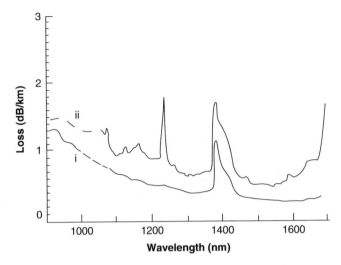

Figure 14.8 Loss spectra for a carbon-coated fiber before (i) and after (ii) exposure to 144 atm H_2 at 152 °C for 15 hours.

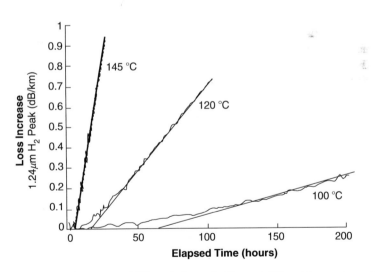

Figure 14.9 H_2 loss increases at 1240 nm in hermetic fibers at different temperatures, all at about 144 atm H_2 pressure.

temperatures, with 140 atm of H_2 pressure. As is typical for carbon-coated fibers, there is an initial "lag" period followed by a period where the losses increase linearly with time. The explanation for this behavior is discussed in the following section.

14.5.4 Diffusion of Hydrogen in Hermetic Fibers

The time and temperature dependencies for hydrogen diffusion through carbon coatings can be accurately predicted using classic diffusion equations once the coating properties have been determined. When a hermetic fiber is exposed to a hydrogen-containing atmosphere, the hydrogen will initially diffuse into the carbon and later into the silica part of the fiber. Characteristic time constants, τ_i and τ_f, can be used to describe the durations of the initial and final stages of diffusion. Loss increases occur only when the H_2 reaches the silica core of the fiber. Figure 14.10 shows hydrogen concentration profiles at different times for a hermetic fiber exposed to hydrogen. In Fig. 14.10a, the fiber is first exposed to H_2 and no diffusion has yet occurred. In Fig. 14.10b, the hydrogen has partially diffused through the carbon but has not reached the silica part of the fiber. In Fig. 14.10c, the H_2 concentration has started to increase in the silica, and H_2 losses will start to become measurable in the fiber. In Fig. 14.10d–f, the concentration of H_2 in the fiber gradually reaches equilibrium (Fig. 14.10f).

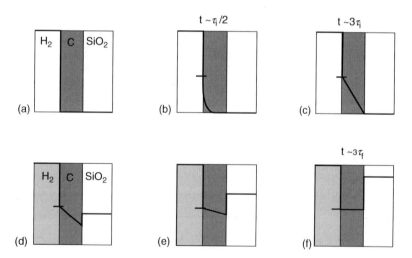

Figure 14.10   Development of hydrogen concentration profile in a carbon-coated fiber. The left side of each figure corresponds to the outside of the fiber, while the right side corresponds to the centerline of the fiber. The thickness of the carbon layer is greatly exaggerated.

Because of the slow diffusion of H_2 in carbon coatings near room temperature, the values of τ_i and τ_f can be quite large, from months to years for τ_i and more than 10^4 years for τ_f.

The H_2 concentration gradients are flat in the gas phase and in the silica because H_2 diffusion in these materials is much faster than in the carbon layer. Because of differing solubilities of H_2 in SiO_2 and carbon, there are offsets in the concentrations at the interfaces, as seen, for instance, in Fig. 14.10d–f.

The rate of hydrogen diffusion into the fiber can be mathematically formulated in a manner similar to that described by Crank [43]. For the early stages (Fig. 14.10a–c), the loss changes associated with molecular H_2 losses are proportional to the concentration of H_2 in the fiber core and are given by [20]

$$\Delta \alpha_{H_2}(t) = L_{H_2} \, P_{H_2} \, K_s \left[\frac{t}{\tau_f} - \frac{\tau_i}{\tau_f} - \frac{12\tau_i}{\pi^2 \tau_f} \sum_{n=1}^{\infty} \frac{(-1)^n}{n^2} \exp\left(\frac{-n^2 \pi^2 t}{6\tau_i} \right) \right], \quad (14.4)$$

where L_{H_2} is the optical loss due to a given concentration of H_2, P_{H_2} is the hydrogen partial pressure, and K_s is the solubility of H_2 in SiO_2 per unit partial pressure. The elapsed time is given by t, and the two characteristic time constants, τ_i and τ_f, depend on material properties and the carbon thickness as follows:

$$\tau_i = \frac{\delta^2}{6D_c}, \quad (14.5)$$

$$\tau_f = \frac{r\delta K_s}{2D_c K_c}, \quad (14.6)$$

where δ is the coating thickness, D_c is the diffusivity of H_2 in the carbon-coating material, r the fiber radius, and K_c is the solubility of H_2 in carbon per unit partial pressure.

The value of τ_i indicates how long it takes H_2 to penetrate the coating, whereas τ_f is a measure of how fast the hydrogen losses will increase once the H_2 is present in the fiber. The molecular H_2 losses follow a simple exponential time dependence after an initial diffusion lag time, τ_i:

$$\frac{\Delta\alpha_{H_2}(t)}{\Delta\alpha_{H_2}(\infty)} = 1 - \exp\left[\frac{-(t - \tau_i)}{\tau_f} \right], \quad (14.7)$$

where $\Delta\alpha_{H_2}(\infty)$ is the equilibrium hydrogen loss and is equal to $L_{H_2} P_{H_2} K_s$. To predict the hydrogen loss for a carbon-coated fiber, the values for τ_i and τ_f need to be known. Because the coating properties (δ, D_c, and K_c) that determine τ_i and τ_f depend on manufacturing methods and will typically not be known, it is necessary to have a practical method for experimentally determining τ_i and τ_f. By doing *in situ* loss measurements on a fiber exposed to hydrogen, one can determine the values for τ_i and τ_f based using short-duration accelerated experiments [20]. A loss that is associated with molecular H_2, typically the 1240-nm peak, is measured at different times. The results will show an initial lag period followed by a period where the H_2 losses increase linearly with time. A schematic representation is shown in Fig. 14.11. Extrapolation of the straight portion of the curve back to the time axis gives τ_i. The slope of the curve is inversely proportional to τ_f and directly proportional to the value of the hydrogen loss, $\Delta\alpha_{H_2}(\infty)$, which would be seen in a nonhermetic fiber (or for a hermetic fiber at infinite time). The value for τ_f is therefore $\tau_f = \Delta\alpha_{H_2}(\infty)/m$, where m is the linear slope of the loss change versus time curve.

The value for $\Delta\alpha_{H_2}(\infty)$ depends strongly on the wavelength, as shown in Fig. 14.6. It decreases weakly with temperature and is proportional to the H_2 partial pressure used in the experiment. Molecular H_2 losses are similar for different fiber types and values can generally be obtained from published literature [39, 44]. A useful expression for estimating equilibrium hydrogen losses is

$$\Delta\alpha_{H_2}(\infty) = P_{H_2}A_{H_2} = P_{H_2}\left[a_{H_2}\exp\left(\frac{8.67\,kJ/mole}{RT}\right)\right], \qquad (14.8)$$

where a_{H_2} is a wavelength-dependent hydrogen absorption with units of dB/(km-atm). Values for a_{H_2} at 1240 nm (the first H_2 overtone peak) and at 1550 nm are 0.355 and 0.017 dB/(km-atm), respectively [39]. Exact loss values depend on the resolution bandwidth setting of the optical spectrum analyzer. Values at other wavelengths can be obtained by scaling the a_{H_2} value using the spectral shape of the H_2 spectrum (e.g., Fig. 14.6).

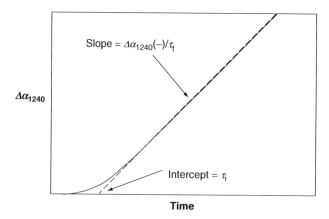

Figure 14.11 Schematic of accelerated test data showing the determination of values for τ_i and τ_f.

Typical results from *in situ* testing of a carbon-coated fiber are shown in Fig. 14.12. Sections of a typical carbon-coated fiber were tested at high hydrogen pressures (137–144 atm) at different temperatures. Results like those in Fig. 14.9 were obtained by monitoring the growth of the 1240-nm peak at temperatures from 50 to 150 °C and analyzing the data to determine τ_i and τ_f. As shown in Fig. 14.12, the temperature dependencies for τ_i and τ_f were consistent with Arrhenius relations and had similar activation energies [20]. The values for τ_i and τ_f as functions of temperature were

$$\tau_i = 6.91 \times 10^{-7} \exp\left[\frac{82.07\,kJ/mole}{RT}\right] \text{sec}, \qquad (14.9)$$

$$\tau_f = 1.7 \times 10^{-5} \exp\left[\frac{99.21\,kJ/mole}{RT}\right] \text{sec}, \qquad (14.10)$$

where R is the gas constant (8.314 J/mol-K).

By using Eq. (14.7) with the experimental values for τ_i and τ_f, the effective hydrogen level in the fiber core can be calculated as a fraction of the external hydrogen pressure. Figure 14.13 shows predictions based on the data in Fig. 14.12, compared to a nonhermetic fiber. The H_2 loss increase is the ratio from the vertical axis in Fig. 14.13 multiplied by the outside hydrogen pressure

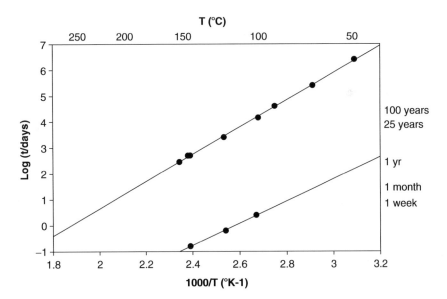

Figure 14.12 Data for a typical carbon-coated fiber obtained from high-pressure *in situ* experiments.

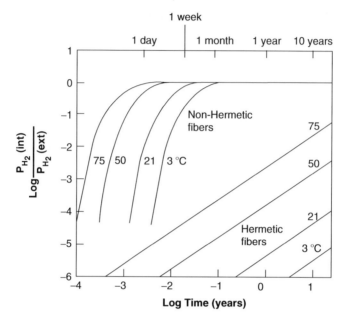

Figure 14.13 Predicted internal hydrogen levels for hermetic and nonhermetic fibers at differ-ent temperatures. Vertical axis shows internal hydrogen pressure as a ratio to the external level. Predictions are based on the data for the fiber in Fig. 14.12.

and the hydrogen absorption (A_{H_2}) at the wavelength and temperature of interest (Eq. [14.8]).

The hydrogen-blocking properties of a specific carbon coating depend on how the coating is deposited. The reactants and the processes used are specific to individual fiber manufacturers. The range of τ_i and τ_f values measured for different carbon coatings can be quite large. A survey was made of some commercially available carbon-coated fibers and of fibers made by varying the processing conditions on a single draw tower [45]. The results showed that most of the fibers had very good hydrogen-blocking properties, with $\tau_f s$ (at 150 °C) ranging from 100 days to more than 10,000 days. The range of activation energies within this grouping of fibers was fairly tight: 84–105 kJ/mol (20–25 kcal/mol). Values of τ_f at room temperature for this group of fibers were from 8×10^6 to 2×10^9 days. Some of the experimental fibers, though having black carbon coatings of normal appearance with high n values, had significantly lower τ_f values. The τ_f values for these fibers ranged from 1 to 8 days at 150 °C and 2000 to 30,000 days at 21 °C. The activation energies for these low τ_f fibers were 63–67 kJ/mol (15–16 kcal/mol). The lower τ_f and activation energy values suggested that the carbon layers for these fibers were different in structure or below a critical thickness. The initial lag times, τ_i, increased as τ_f increased. This was the expected behavior because both τ_i and τ_f increase as the coating thickness (δ) increases and as the diffusivity of H_2 in carbon coating (D_c) decreases. (See Eqs. [14.5] and [14.6].)

14.5.5 Effects of Glass Composition on Hermetic Fiber Behavior

The diffusion analysis in the previous section assumes that H_2 molecules diffuse inertly through the fiber's silica cladding. It also assumes that there are no sources of hydrogen in the fiber's cladding material. Most of the glass in a single-mode fiber is made up of the undoped SiO_2 that is outside the fiber's core. The properties of this optically inactive glass can vary and depend on the materials and the manufacturing processes used in making the fiber.

In some cases, silica glasses derived from the fusion of natural quartz crystals can contain metastable OH impurities [46], which can later convert into mobile forms of hydrogen that can diffuse into the core and cause loss. When fibers made using such glasses are hermetically coated, the hydrogen impurities are trapped under the hermetic coating. At high temperatures, the hydrogen can be liberated from the cladding glass. It can then diffuse into the fiber core and cause hydrogen-related loss increases such as OH growth or SWE increases. The magnitude of the trapped-hydrogen losses will depend on time, temperature, wavelength, and the composition of both the core and the cladding glasses. Although the loss increases due to trapped hydrogen may be acceptably small for some applications, it is best to avoid fibers made with these natural fused-quartz glasses when a hermetic coating will be applied to the fiber.

Other types of silica have the opposite effect and can scavenge trace levels of hydrogen. When the cladding of a hermetic fiber is made of silica that contains reactive "gettering" sites, the fibers can show improved resistance to hydrogen losses. The small amounts of H_2 that diffuse through the coating quickly react with the gettering sites and are immobilized in the cladding. Drawing-induced silica defects are known to exist at concentrations of about 100 ppb in some types of undoped silica [42]. These defects react quickly with H_2, even at room temperature [42, 47], and thus provide a practical secondary hydrogen barrier to the hermetic coating. The duration of this gettering-induced lag time in a hermetic fiber, t_{gh} is

$$t_{gh} = \frac{\tau_f C_g f_g}{P_{H_2}(ext)K_S},$$ (14.11)

where C_g is the concentration of gettering sites and f_g is the fraction of the silica material that contains the gettering sites. When coupled with a good hermetic coating, the reactive sites can greatly increase the duration of the lag period (i.e., the period where loss increases are zero). For instance, assuming coating properties like those in Fig. 14.12, 100 ppb of reactive sites will result in a lag time of about 7.5 years at 100 °C and $P_{H_2} = 0.01$ atm. The normal lag time without reactive gettering sites, τ_i, would be about 2 days at this temperature.

The protective effect provided by the reactive silica material will remain until the reactive sites are depleted. The duration of this extended lag period, there-fore, depends on the concentration of reactive sites in the glass, the coating properties (i.e., the value of τ_f), and the external hydrogen pressure. For a hermetic fiber that does *not* contain reactive sites, the length of the lag period will be τ_i, which does not depend on hydrogen pressure. For a hermetic fiber that *does* have reactive sites in the cladding, the length of the lag period will be inversely proportional to the hydrogen pressure used in the accelerated test experiment (or to the ambient hydrogen pressure seen by the hermetic fiber in actual use).

Whether a hermetic fiber contains a useful concentration of reactive sites is most easily determined by monitoring the *in situ* loss changes at two different hydrogen pressures. For example, if a fiber is tested under 1.0 atm and 0.1 atm H_2 pressure, the lag period should be approximately 10 times longer for the fiber tested at 0.1 atm. Similar lag periods at both pressures indicate that reactive sites are either absent or low in concentration. The figure of merit for hermetic fibers that contain reactive sites is the product $\tau_f C_g$, where C_g is the concentration of reactive sites in the cladding. The length of the reactive lag time is proportional to the quantity $\tau_f C_g / P_{H_2}$. Once the lag period, $t_{gh,expt}$, is measured in an accelerated experiment, the results can be used to estimate the duration of the lag period at the anticipated use condition, $t_{gh,use}$:

$$t_{gh,\,use} = t_{gh,\,expt} \frac{P_{expt}}{P_{use}} \exp\left[\frac{(E_f - E_s)}{R}\left(\frac{1}{T_{use}} - \frac{1}{T_{expt}}\right)\right], \qquad (14.12)$$

where P_{expt} and P_{use} are the hydrogen pressures for the experiment and for the use condition, E_f is the activation energy in the exponential term that charac-terizes τ_f (e.g., 99.21 kJ/mol for the fiber of Fig. 14.12), and E_s (8.67 kJ/mol) is the term accounting for the decreasing H_2 solubility with temperature. For example, if an experimental lag time of 21 hours was observed at test conditions of 150 °C and 1.0 atm of H_2, the expected reactive lag period would be 6 years for assumed use conditions of 75 °C and a hydrogen pressure of 0.1 atm. While there can be significant differences in E_f values for different carbon-coating types, the value is often in the range of 84–105 kJ/mol for high-quality carbon coatings.

Ideally the value for E_f should be determined experimentally for a new coatings or for coatings with unknown properties.

14.6 USE AND HANDLING OF CARBON-COATED HERMETIC FIBERS

Carbon-coated fibers can be worked with in much the same way as standard nonhermetic fibers. However, there are a few differences, as discussed in the following section.

14.6.1 Fiber Strength

As mentioned earlier, the fracture strengths of carbon-coated fibers are lower than those of standard nonhermetic fibers. The root cause of the strength reduction is at least partially understood [11, 22], and it is reasonably clear that some tradeoff in fiber strength is necessary to achieve improved hermetic properties, especially with respect to hydrogen diffusion. Nonetheless, fibers with high strength and good hydrogen properties can be made. For instance, the carbon-coated fibers whose hydrogen data are shown in Fig. 14.12, had reasonably high strengths of about 4.1 GPa (600 ksi) and were capable of blocking hydrogen diffusion over a wide range of conditions. Higher strength carbon-coated fibers can be made, but care should be taken to verify that their hermetic properties are adequate for the application.

14.6.2 Fiber Handling

Bend radii guidelines for nonhermetic fibers are determined by fiber fatigue considerations. Typical fiber handling practice for nonhermetic fibers requires that the long-term stresses imposed on a fiber be limited about 20% of the proof-test level [35]. Because a carbon-coated fiber is virtually immune to fatigue, it is permissible to use tighter bend radii when storing fibers on spools or in splice trays. One application note on hermetic fibers allows the fiber to be used at 80% of its proof-test value [6]. In this case, the specification of a value less than 100% was done to allow for less-than-perfect control of the stress values seen by a deployed fiber. For a carbon-coated fiber proof-tested at 0.7 GPa (100 kpsi), the allowable stress would, therefore, be 0.55 GPa (80 kpsi), corresponding to a minimum bend radius of about 8 mm. The corresponding value for a non-hermetic fiber would be about 33 mm. Different manufacturers may provide different handling guidelines for their hermetic fibers.

14.6.3 Fiber Stripping, Cleaving, and Connectorization

Polymer coatings can be stripped using normal fiber stripping tools. The carbon coating itself will usually adhere well to the glass fiber and, therefore, will not be removed by the stripping tool. Once the polymer coating is removed, the fiber can be cleaved using a conventional fiber cleaver. Because the carbon coating is very thin, the overall fiber diameter of a stripped fiber will be very close to the standard 125-μm size, allowing the fiber end to be inserted and bonded into standard fiber ferrules. In general, the only reason to remove the carbon coating is for fusion splicing, described in the next section.

14.6.4 Fusion Splicing

Carbon-coated fibers can be fusion spliced using the same commercial splicers that are used for nonhermetic fibers. Both the polymer coating and the black carbon coating should be removed from each fiber end in preparation for splicing. The thin carbon layer is best removed using a pre-fusion arc. This "burn-off" of the carbon coating allows the optics in an automatic fusion splicer to "see" the fiber core and align the two sides. Once the carbon is removed from each end, standard fusion splicer programs can be used to splice the fiber. A proof-test should be done on the splice region, identical to the way that a nonhermetic fusion splice would be tested for strength.

In principle, it is desirable to have some way to reapply the carbon coating in the splice region. Although some work was done to develop methods for local deposition of carbon onto bare fibers [48], the equipment for carbon recoating is not readily available. Further, it is rarely necessary to reapply the carbon on the short, approximately 10-mm, section of spliced fiber. Hydrogen losses on such a short section of fiber are negligible. H_2 molecules that diffuse into this section of fiber will be confined to the splice region because lengthwise diffusion away from the unprotected splice region will be slow. Mechanical fracture of the unprotected splice region can be avoided by use of a standard splice protector or by simply avoiding the application of high stresses to this region. It is important to note that the minimum bend radius for the spliced section of hermetic fiber will be the same as for a nonhermetic fiber unless the carbon coating is reapplied.

Fusion splicers that are used to splice carbon-coated fibers may require more frequent cleaning, particularly of the electrodes. Some users have found it advantageous to have a separate piece of equipment that is used to remove the carbon from the fiber ends before fusion splicing, helping to reduce extra maintenance on the fusion splicer itself. An example of such equipment is a small high-temperature oven that heats the fiber end to a temperature at which the carbon reacts with oxygen from the air, vaporizing the carbon as CO_2.

14.6.5 Fiber Color

Semi-opaque inks are commonly used to color-code and identify optical fibers. These fiber colors are applied as thin coatings (several micrometers) on top of the acrylate polymer coatings. The black color of a carbon coating can affect the appearance of a color-coded fiber, causing the fiber to appear to be darker. This may require that special color charts be used for hermetic fibers or can require that certain colors be avoided to prevent misidentification.

14.7 SPECIFYING CARBON-COATED FIBERS

When specifying the properties of a carbon-coated fiber, one must remember that different carbon-coated fibers have different properties and that no coating will be 100% impervious to all species, especially small molecules like H_2O or H_2. For instance, it is possible to have a fiber with a layer of carbon of normal appearance that is nonetheless as susceptible to hydrogen loss as a nonhermetic fiber. Therefore, some level of testing needs to be conducted to show that a fiber is adequately hermetic. The properties of carbon coatings are known to be sensitive to the processing conditions used in drawing and coating the fiber. This sensitivity may require that individual lots of fibers be tested to verify that they meet minimum requirements for resistance to fatigue and/or hydrogen sensitivity. The extent of the testing depends on how sensitive the coating properties are to normal variations in manufacturing parameters.

A fiber manufacturer will typically have qualified a hermetic fiber during its development. Qualification experiments will have been used to show that the fiber and its coating are capable of limiting fatigue and/or hydrogen-induced loss increases. It is desirable that the qualification experiments be done under accelerated test conditions. For characterizing a fiber's resistance to fatigue, accelerated test conditions will be elevated temperatures and humidity levels. For characterizing a fiber's resistance to hydrogen aging, the accelerating factors will be elevated temperature and hydrogen partial pressure. Qualification tests tend to be time consuming, often requiring weeks to months to complete, making them expensive to conduct and not repeated for each lot of manufactured fiber. However, because the properties of hermetic coatings depend on the details of the processing steps (e.g., draw conditions), it is common to conduct certification tests as part of the manufacturing process. In a certification test, a subset of tests is conducted to verify that the fiber meets some minimum standards. Examples of certification tests for carbon-coated hermetic fibers are room-temperature measurements of the dynamic fatigue parameter n_d and measurement of fiber loss before and after exposure to hydrogen at an elevated temperature for a designated time.

A primary question is whether the fiber needs to be protected from fatigue or from hydrogen-induced optical loss. In some cases, both types of protection are needed. A coating that is capable of blocking H_2 diffusion will generally be impervious to H_2O diffusion and fatigue. The converse, however, is not true. There are carbon coatings that can effectively eliminate fatigue but that are not cable of blocking H_2 diffusion. Improving a coating's H_2 blocking ability may have an adverse effect on the fiber's fracture strength, so overspecifying H_2 resistance may not be advisable.

The ability of a carbon coating to prevent fatigue is usually ensured by showing that the fiber has a high n value. A high n value means that the fiber's fracture stress is insensitive of strain rate or the time that the fiber is under load. Typical n values for nonhermetic fibers are on the order of 18–22. The value of n for a hermetic fiber will typically be determined using dynamic fatigue measurements. Good hermetic carbon coatings usually have n values that are 80 or higher. Values much in excess of 100 become sensitive to measurement error and can be misleading [49]. For instance, a carbon-coated fiber with a nominal n value of 250 should not automatically be assumed to be better than one with an n value of 150 unless significant care has been taken in the data acquisition and analysis. Dynamic fatigue measurements should be obtained using at least three strain rates. In most cases, dynamic measurements are done at room temperature. Obtaining fatigue data at elevated temperature and humidity is most readily done via static fatigue measurements. However, because static fatigue measurements are time consuming, they are not commonly used as certification tests.

A primary engineering tradeoff that needs to be made in improving the hermetic properties of a carbon-coated fiber is that of initial fiber strength. The general trend is that a carbon-coated fiber's strength will decrease as the hermetic properties improve. It is reasonable to expect fiber strengths of about 4.1 GPa (600 kpsi) for carbon-coated fibers that have excellent fatigue resistance ($n_d > 100$) and very low hydrogen permeability ($t_f \sim 10^6$ days at 50 °C). Significant improvements in hydrogen blocking ability can be achieved if fiber strengths of 2.8–3.4 GPa (400–500 kpsi) are acceptable. In almost all cases, a reasonable compromise between fiber strength and hermetic properties can be reached for a given application.

Showing that a hermetic coating can limit hydrogen losses requires that the fiber be tested in a controlled hydrogen environment and that loss changes be below some critical value. While a full characterization of the coating (i.e., τ_i and τ_f as functions of temperature) is advantageous, it is not always feasible for a fiber manufacturer to provide such detailed data. A more common procedure is to test a hermetic fiber in hydrogen for a given time and temperature and show that loss increases are below some specified value at a designated wavelength. Ideally, the basis of the test (i.e., the rationale for the pass–fail criterion) will be provided by the fiber's manufacturer. Alternatively, the passing criterion for such a test can be set using the approximation shown in Eq. (14.13). Equation (14.13) is based on Eq. (14.7) and is derived using the approximation that $\exp[-(t-\tau_i)/\tau_f] \sim t/\tau_f$ for small values of t, and assuming that τ_i is short in comparison to t. This expression is appropriate when losses due to molecular H_2 are expected to be the dominant loss change. The predicted change in loss (dB/km) at system conditions, $\Delta\alpha_{sys}$, is calculated based on measured results in an accelerated experiment.

$$\Delta\alpha_{sys} = \Delta\alpha_{\exp t}\frac{t_{sys}}{t_{\exp t}}\frac{a_{sys}}{a_{\exp t}}\frac{P_{sys}}{P_{\exp t}}\exp\left[\frac{(E_f - E_s)}{R}\left(\frac{1}{T_{\exp t}} - \frac{1}{T_{sys}}\right)\right], \qquad (14.13)$$

The *sys* and *exp t* subscripts refer to the deployed system and to the accelerated experiment, respectively. $\Delta\alpha_{\exp t}$ is the loss change (dB/km) measured in the experiment. The parameters a_{sys} and $a_{\exp t}$ are the pre-exponential terms from Eq. (14.8) and express the H_2 absorption strength at the wavelengths of interest. P_{sys} and $P_{\exp t}$ are the hydrogen partial pressures for the system conditions and for the experiment. In most cases, the molecular H_2 peak will be monitored during the experiment and $a_{\exp t}$ will be the value for 1240-nm peak (0.355 dB/km-atm). Similarly, a typical value for a_{sys} is 0.017 dB/km-atm, corresponding to a system wavelength of 1550 nm. The exponential term accounts for the temperature dependence of τ_f and of H_2's solubility in silica.

The acceptable loss change in an experiment, $\Delta\alpha_{\exp t}$, is determined by calculating the predicted system loss change, $\Delta\alpha_{sys}$ and verifying that it is less the value allocated for system aging. If it is not, the acceptable value for $\Delta\alpha_{\exp t}$ needs be lowered, which in turn will usually require an improved fiber coating. If the pass–fail criterion is "no" loss change, then $\Delta\alpha_{\exp t}$ should be the minimum detectable loss change, which will depend on the sensitivity and stability of the loss test set and on the length of the fiber under test. The only term in Eq. (14.13) that might not be known is E_f, the activation energy characterizing t_f's dependence on temperature. This value needs to be established during the qualification process of a new coating because it has a major influence on the predictions. For the carbon coatings described earlier, E_f ranged from 84 to 105 kJ/mol. Note that Eq. (14.13) is not appropriate for predicting loss increases if the system losses are dominated by hydrogen reactions, for instance, by OH formation or reaction of H_2 at Ge sites in the core. These reactions become increasingly important factors at high temperatures.

When a fiber is used at an elevated temperature, or when it has a core composition that is known to be reactive with hydrogen, there are several approaches that can be used to qualify the hermetic fiber. The most thorough approach is to fully model the hydrogen diffusion through the coating and its reaction in the fiber's core. The loss change is integrated over time, accounting for the gradually increasing level of hydrogen inside the fiber. This requires detailed knowledge of the reaction kinetics for the particular fiber core composition, values for τ_i and τ_f, and information about the concentration of gettering sites in the cladding. When done correctly, this approach results in an accurate prediction of loss change as a function of time, temperature, P_{H_2} and wavelength. However, the amount of work entailed can be significant, and for this reason such studies are carried out only when necessary.

A more practical approach is to separately determine a tolerable partial pressure of H_2, P_{tol}, such that loss changes in a nonhermetic version of the fiber would be suitably small at the operating conditions. The hermetic coating then needs to ensure that the internal hydrogen partial pressure, P_{int}, stays below this critical P_{H_2} value. A modification of Eq. (14.13) permits the internal H_2 pressure to be predicted on the basis of an experiment that monitors changes, $\Delta\alpha_{exp\,t}$, in an H_2-related loss feature (e.g., the 1240-nm peak):

$$P_{int} = P_{sys}\frac{\Delta\alpha_{sys}(t)}{\Delta\alpha_{sys}(\infty)} = \frac{\Delta\alpha_{exp\,t}}{a_{exp\,t}}\frac{t_{sys}}{t_{exp\,t}}\frac{P_{sys}}{P_{exp\,t}}exp\left[\frac{(E_f - E_s)}{RT_{exp\,t}} - \frac{E_f}{RT_{sys}}\right]. \qquad (14.14)$$

The value for P_{int} is calculated for t_{sys} equal to the system lifetime and using the observed loss change, $\Delta\alpha_{exp\,t}$, from an accelerated test. As long as the value calculated for P_{int} is less than the tolerable P_{H_2} value, P_{tol}, the carbon coating will provide a useful barrier.

Yet another alternative is to show that the net lag time, due to reactive sites in the cladding, (t_{gh}) and due to the diffusion lag (τ_i), is greater than the system lifetime. For instance, Eq. (14.12) can be used to calculate the expected lag time, t_{gh}, caused by reactive sites in the cladding. If this lag time is longer than the system lifetime, the combination of the coating and the reactive sites is sufficient to protect even a highly reactive core composition. For the example given immediately after Eq. (14.11), the lag time attributable to reactive cladding sites was predicted to be 7.5 years at 100 °C and $P_{H_2} = 0.01$ atm, sufficiently long for many applications. If the reliability of a carbon-coated fiber depends on the presence of reactive gettering sites in the fiber, periodic hydrogen tests should be conducted by the fiber manufacturer to verify the $\tau_f C_g$ figure of merit for the carbon-coated fibers. The concentration of the reactive sites in the glass is not readily measured by other means.

As previously discussed, the electrical resistance of a carbon coating can be used to measure carbon-coating properties and as an indirect measure of coating thickness and quality. If qualification experiments show a good correlation between a fiber's electrical resistance and its fatigue and/or hydrogen properties, then the measurement of electrical resistance can be used as a valid quality-control parameter. There is limited value in comparing electrical resistance measurements when comparing fibers made by different methods or by different manufacturers. Some types of carbon can have low electrical resistance (suggesting a thick carbon layer) but can still have poor hermetic properties. The material properties of carbon layers deposited on fibers depend on the details of the gas reactants used and the temperature of deposition. Because these factors will not be the same for different manufacturers but will influence the coating's electrical resistance, the comparison of resistance values across fiber types is of limited value. On the other hand, used as a quality-control metric for a stable coating deposition process, the electrical resistance can be a useful parameter.

14.8 APPLICATIONS FOR CARBON-COATED HERMETIC FIBERS

14.8.1 Fibers in Underwater Cables

Fibers in underwater cables have sometimes exhibited optical loss increases due to the evolution of hydrogen gas. Failures have been due to H_2 gas evolved from the galvanic corrosion of metallic components in the submerged cables [50] or from outgassing of H_2 from certain types of silicone-based polymers. In most cases, these problems were solved by redesigning the cables to avoid dissimilar metals and by avoiding the use of hydrogen-generating polymers. However, hermetic carbon coatings were successfully used in some cables and were shown to be a viable solution to the hydrogen-aging problem. One advantage of using carbon-coated fibers to protect against hydrogen is that it avoids the expense and delay associated with the redesign and qualification of a cable. The low fatigue properties of carbon-coated fibers are an additional factor influencing the decision to use carbon-coated fibers in underwater cables.

Because the typical temperature of an underwater cable is low (3–20 °C), hydrogen reactions in the fiber core will be negligible for most types of single-mode fibers. The only losses that are likely to affect the fibers are those due to molecular H_2. The low temperatures also result in slow hydrogen diffusion through the carbon coatings. For instance, the results for the carbon-coated fiber shown in Fig. 14.13 at 10 °C are $\tau_i = 30$ years and $\tau_f = 8 \times 10^5$ years. This fiber would be immune to H_2 losses over times much longer than a typical 25-year lifetime.

Equation (14.13) can be used to estimate long-term loss changes in underwater cables. The exact hydrogen level in the cable will usually not be precisely known. Nonetheless, because the H_2 permeation through most carbon coatings is very slow, it is usually possible to assume highly pessimistic values for the hydrogen pressure inside the cable and still justify very low hydrogen aging losses over the system lifetime. Likewise, even though carbon coatings with low E_f values may not be adequate for high-temperature applications, they may be perfectly adequate for the low temperatures encountered in underwater cables.

14.8.2 Amplifier Fibers

Carbon coatings have been used to protect erbium-doped amplifier fibers. These fibers play a critical role in commercial telecommunications and in specialty applications. In their nonhermetic version, these fibers have the same bending limitations as other fibers. This limits the physical size of amplifier modules because the bend radii used in winding and routing the fiber cannot be less than a critical value, typically about 33 mm. By using hermetic carbon coating, the problem of fatigue is eliminated, allowing tighter fiber coils and more compact amplifier modules. The carbon coating also protects the Er-doped fiber from hydrogen aging. Many Er-doped fibers use core compositions that are highly reactive with hydrogen [51]. A hermetic carbon coating is one way to protect the fiber from loss increases that could otherwise occur due to unexpectedly high H_2 levels.

14.8.3 Avionics

The enhanced mechanical reliability of carbon-coated fibers also makes them attractive for avionics applications, where restricted spaces can require that fibers be routed with tight bend radii. In addition, the fibers see significant environmental stresses due to cycling between temperature and humidity extremes. Although these harsh conditions can cause fatigue-induced failures in nonhermetic fibers, carbon-coated fibers will retain their strength even in the presence of high humidity and stress levels. Carbon-coated fibers are preferable to metal-coated fibers in these applications both because of their low optical loss and because they are significantly lighter than metal-coated fibers.

14.8.4 Geophysical Sensors

Arguably the most challenging environment for specialty optical fibers is the environment encountered in oil-well down-hole data logging [1, 52, 53]. Fibers used in sensor systems in oil wells can be exposed to temperatures up to 300 °C, high pressures, water, corrosive chemicals such as H_2S, and high levels of H_2 gas. The sensor fibers are often installed into stainless-steel conduits using high-pressure water or other liquids. For the purpose of strength retention and chemical resistance, the carbon coating plays a vital role. When H_2 gas is present at the high temperatures encountered in the down-hole environment, the optical aging of the fibers can be rapid. The temperatures encountered in the application of these fibers can be comparable to the highest temperatures used in accelerated aging experiments. At these temperatures, H_2 diffusion becomes rapid and the dopants in the fiber core become increasing reactive with hydrogen. The losses due to reacted hydrogen can greatly exceed those due to unreacted molecular H_2. The extent to which H_2 molecules react in the fiber core to form lossy OH and Ge defects is highly dependent on the core glass composition. The use of a carbon coating with low H_2 permeability can greatly extend the useable lifetime of fibers used for data logging.

Expressions such as Eq. (14.13) are generally *not* useful for predicting the loss changes in a fiber used in a high-temperature environment because the only losses that are accounted for are those due to unreacted molecular H_2. An alternative approach is to identify a carbon-coated fiber and show that its properties are sufficient to give a reliability advantage in the actual field environment. Once the desirable fiber parameters are identified, they need to be tightly controlled in the manufacturing process. By establishing an appropriate certification test, where the loss changes are shown to be below a critical limit, it is possible to detect changes in the manufacturing process that could adversely affect the fiber reliability.

As discussed earlier, it is possible to make carbon-coated fibers that have relatively slow H_2 permeation even at high temperatures. When a fiber must survive H_2 exposure in a high-temperature down-hole environment at 150 °C, a carbon coating that has a τ_f of 10,000 days will be a better choice than a coating with τ_f in the 1- to 100-day range, The increased hydrogen protection might require a tradeoff in fiber fracture strength of up to 0.7 GPa (100 kpsi), but this is likely to be acceptable in view of the improved optical lifetime. Using low-dopant low-reactivity fiber core compositions and increasing the concentration of reactive sites in the cladding can also improve the reliability of a fiber exposed to hydrogen at high temperatures.

14.9 CONCLUSION

Carbon-coated optical fibers play an important role among specialty optical fibers. The thin carbon layer provides a hermetic barrier to water and hydrogen without inducing microbending losses. A well-designed carbon coating stops water from diffusing to the silica surface of the fiber and, thus, prevents latent fiber fractures associated with fatigue. Because of this, carbon-coated fibers can be configured with tight bend radii, without the risk of fatigue-induced fracture. Carbon-coated fibers can also be designed to prevent hydrogen-induced loss increases, an effect that can be problematic for fibers used in aggressive applications such as sensor systems in oil-well data logging. The properties of carbon-coated fibers depend on the properties of both the carbon and the underlying glass, which in turn depend on the processes used in making the hermetic fiber. By establishing suitable qualification and certification tests, we can ensure that carbon-coated fibers will meet a user's requirements for providing enhanced long-term fiber reliability.

REFERENCES

[1] Normann, R. et al. 2001. Development of fiber optic cables for permanent geothermal wellbore deployment. In: *Proceedings, Twenty-Sixth Workshop on Geothermal Reservoir Engineering, Stanford University, Jan. 29–31, 2001*. Stanford University, Palo Alto, CA.

[2] Pinnow, D. A. et al. 1979. Reductions in static fatigue of silica fibers by hermetic jacketing. *Appl. Phys. Lett.* 34(1):17–19.

[3] Huff, R. G. et al. 1988. Amorphous carbon hermetically coated optical fibers. In: *Conference Proceedings of Conference on Optical Fiber Communication*, Paper TUG2.

[4] Lu, K. E., et al. 1988. Recent developments in hermetically coated optical fiber. *J. Lightwave Technol.* 6(2):240–244.

[5] Moore, K. et al. 1995. Review of characteristics and applications of commercially available carbon-coated hermetic fiber. In: *Conference Proceedings of International Wire and Cable Symposium*, pp. 305–308.

[6] Kohli, J. T. and G. S. Glaesemann. 2005. Corning's hermetically coated erbium-doped specialty fibers. Corning white paper, available at www.corning.com/photonicmaterials/pdf/Hermetic_WP.pdf; accessed 11/25/2005.

[7] Lindholm, E. A. et al. 2005. Strength and reliability of silica optical fibers for automotive communication networks. In: *Photonics in the Automobile* (T. P. Pearsall, ed.), *SPIE Proc.* 5663:129–134.

[8] Brennan, J. F., III, et al. 2000. Diamond-like film encapsulated fibers for long-length fiber grating production. In: *Proceedings of Optical Fiber Communication Conference*, Post Deadline paper PD1.

[9] DeVita, S. and J. R. Vig. 1977. Method of treating optical waveguide fibers. U.S. Patent 4,028,080.

[10] Kao, C. K. and M. S. Maklad. 1980. U.S. Patent 4,183,621.

[11] Kurkjian, C. R. and H. Leidecker. 2000. Strength of carbon-coated fibers. In: *Optical Fiber and Fiber Component Mechanical Reliability and Testing* (M. J. Matthewson, ed.), *Proceedings of SPIE*. SPIE Int. Soc. Opt. Eng. 4215:134–143.

[12] Wysocki, J. A. and A. Lee. 1981. Mechanical properties of high-strength metal-coated fibers. *Digest of IOOC*, Paper MG4.

[13] Sato, M. et al. 1981. Mechanical and transmission properties of high strength indium coated optical fibers. In: *Conference Proceedings of International Wire and Cable Symposium*, pp. 45–49.

[14] Hanson, E. G. et al. 1985. Optical fiber with hermetic seal and method for making same. U.S. Patent 4.512,629.

[15] Beales, K. J. et al. 1984. Practical barrier to hydrogen diffusion into optical fibres. *Elect. Lett.* 20(4):159–161.

[16] Chaudhuri, R. and P. C. Schultz. 1986. Hermetic coating on optical fibers. In: *Reliability Considerations in Fiber Optic Applications, SPIE Int. Soc. Ont. Eng. Proceedings of SPIE Int. Soc. Opt. Eng.*

[17] Uchida, N. et al. 1983. Infrared loss increase in silica optical fiber due to chemical reaction of hydrogen. In: *Postdeadline Conference Proceedings Ninth European Conference on Optical Communication.* (ECOC), OSA Opt. Soc. Am.

[18] Uesugi, N. et al. 1983. Infra-red optical loss increase for silica fiber in cable filled with water. *Electron. Lett.* 19(19):762–764.

[19] Hiskes, R. et al. 1984. High performance hermetic optical fibers. In: *Conference Proceedings of Conference on Optical Fiber Communication,* Paper WI6. OSA—Opt. Soc. of Am.

[20] Lemaire, P. J. et al. 1990. Diffusion of hydrogen through hermetic carbon films on silica fibers. *Mat. Res. Soc. Symp. Proc.* 172:85–96.

[21] Aikawa, H. et al. 1993. Characteristics of carbon-coated optical fibers and structural analysis of the carbon film. In: *Conference Proceedings of International Wire and Cable Symposium,* pp. 374–380.

[22] Yoshizawa, N. and Y. Katsuyama. 1989. High strength carbon-coated optical fibre. *Electr. Lett.* 25(21):1429–1431.

[23] Akiyama, T. et al. 1991. Long-term reliability of a carbon-coated optical fiber cable. In: *Conference Proceedings of International Wire and Cable Symposium,* pp. 151–159.

[24] Katsuyama, Y. et al. 1991. Field evaluation result on hermetically coated optical fiber cables for practical application. *J. Lightwave Technol.* 9(9):1041–1046.

[25] Sikora, E. S. R. et al. 1991. Examination of the strength characteristics, hydrogen permeation, and electrical resistivity of the carbon coatings of a number of "hermetic" optical fibres. In: *Conference Proceedings of International Wire and Cable Symposium,* Paper 14–3.

[26] DiMarcello, F. V. et al. 1990. High speed manufacturing process for hermetic carbon coated fibers. In: *Conference Proceedings on Optical Fiber Communication,* Paper THH5. OSA—Opt. Soc. Am.

[27] DiMarcello F. V. et al. 1991. Hermetically sealed optical fibers. U.S. Patent 5,000,541.

[28] Tuzzolo, M. R. et al. 1993. Hermetic product performance: Ensuring the uniformity of the carbon layer. In: *Conference Proceedings of International Wire and Cable Symposium,* pp. 381–384.

[29] Lindholm, E. A. et al. 1999. Low-speed carbon deposition process for hermetic optical fibers. In: *Conference Proceedings of International Wire and Cable Symposium.*

[30] Atkins, R. M. et al. 1990. Measuring and controlling the thickness of a conductive coating on an optical fiber. U.S. Patent 5,057,781.

[31] Kurkjian, C. R. et al. 1989. Strength and fatigue of silica optical fibers. *J. Lightwave Technnol.* 7(9):1360–1370.

[32] Kapron, F. P. and H. H. Yuce. 1991. Theory and measurement for predicting stressed fiber lifetime. *Optical Eng.* 30(6):700–708.

[33] Matthewson, M. J. 1994. Optical fiber reliability models. In: *Fiber Optics Reliability and Testing* (D. K. Paul, ed.), pp. 3–31, Vol. CR50, *SPIE Critical Reviews.* SPIE Int. Soc. Opt. Eng.

[34] Krause, J. T. et al. 1988. Mechanical reliability of hermetic carbon coated optical fibers. In: *Proceedings of Sixth EFOC/LAN, Amsterdam.* OSA—Opt. Soc. of Am.

[35] Castilone, R. J. 2005. Mechanical reliability: Applied stress design guidelines. Corning white paper WP5053. Available at www.corning.com/docs/opticalfiber/wp5053_07-01.pdf; accessed 10/21/2005.

[36] TIA/EIA Fiber Optic Test Procedure: Method for measuring dynamic tensile strength and fatigue parameters of optical fibers by tension. TIA/EIA-455-28C, 1999.

[37] Wiederhorn, S. M. 1972. A chemical interpretation of static fatigue. *J. Am. Ceramic Soc.* 55:81–85.

[38] Lindholm, E. A. et al. 2004. Aging behavior of optical fibers in aqueous environments. In: *Reliability of Optical Fiber Components, Devices, Systems and Networks* (H. G. Limberger and M. J. Matthewson, eds.), pp. 25–32, SPIE Proc. 5465. SPIE Int. Soc. Opt. Eng.

[39] Lemaire, P. J. 1991. Reliability of optical fibers exposed to hydrogen: Prediction of long-term loss increases. *Optical Eng.* 30(6):780–789.

[40] Lemaire, P. J. 1994. Hydrogen induced losses and their effects on optical fiber reliability. In: *Fiber Optics Reliability and Testing* (D. K. Paul, ed.), pp. 80–104, Vol. CR50, *SPIE Critical Reviews*. SPIE Int. Soc. Opt. Eng.

[41] Tomita, A. and P. J. Lemaire. 1984. Hydrogen-induced loss increases in germanium doped single-mode fibers. *Electron. Lett.* 20(12):512–514.

[42] Lemaire, P. J. and M. D. deCouteau. 1987. Optical spectra of silica core optical fibers exposed to hydrogen. In: *Optical Fiber Materials and Properties* (S. R. Nagel et al., eds.), pp. 225–232, Vol. 88, *MRS Symposia Proceedings*. MRS Mat. Res. Soc.

[43] Crank, J. 1975. *The Mathematics of Diffusion*. Oxford University Press, London, 1975.

[44] Beales, K. J. et al. 1983. Increased attenuation in optical fibres caused by diffusion of molecular hydrogen at room temperature. *Electron. Lett.* 19(22):917–919.

[45] Huff, R. G. et al. 1991. AT&T Bell Laboratories, unpublished results.

[46] Bell, T. et al. 1962. Water in vitreous silica: Part 2: Some aspects of hydrogen-water-silica equilibria. *Physics Chem. Glasses* 3(5):141–146.

[47] Blankenship, M. G. et al. 1987. Short-term transient attenuations in single-mode optical fibers due to hydrogen. In: *Proceedings of Conference on Optical Fiber Communication*, Paper WA3. OSA—Opt. Soc. of Am.

[48] Inniss, D. and J. T. Krause. 1991. Hermetic splice overcoating. *Optical Eng.* **30**(6):776–779.

[49] Bubnov, M. M. et al. 1992. Maximum value of fatigue parameter n for hermetically coated silica glass fibers. In: *Proceedings of Conference on Optical Fiber Communication*, Paper ThF2. OSA—Opt. Soc. of Am.

[50] Anderson, W. T. et al. 1988. Hydrogen gas effects on installed submarine single-mode fiber cables. In: *Conference Proceedings of International Wire and Cable Symposium*, pp. 188–199.

[51] Lemaire, P. J. et al. 1994. Hydrogen-induced-loss increases in erbium-doped amplifier fibers: Revised predictions. In: *Proceedings of Conference on Optical Fiber Communication*, Paper FF1. OSA—Opt. Soc. of Am.

[52] Mendez, A. et al. 1999. Applications of optical fiber sensors in subsea and downhole oil well environments. *Proc. SPIE* 3852:16–28.

[53] Skinner, N. and J. Maida. 2004. Downhole fiber-optic sensing: The oilfield service provider's perspective. In: *Fiber Optic Sensor Technology & Applications* (M. A. Marcus et al., eds.), Vol. III, *Proc. SPIE,* 206–220.

Chapter 15

Metal-Coated Fibers

Vladimir A. Bogatyrev and Sergei Semjonov

Fiber Optics Research Center, Geneva Physics Institute of the Russian Academy of Sciences, Moscow, Russia

15.1 INTRODUCTION

An optical fiber has to be defended by some protective coating from mechanical damage during handling and from environmental factors during its use. In many cases, a polymer coating is appropriate for enough protection. However, there are a number of special applications of optical fibers in which ordinary polymer-coated fibers cannot be used. These applications can be divided into several groups:

1. Increased reliability (hermeticity of the coating is important)
2. High vacuum (when outgassing from the coating is undesirable)
3. Possibility of soldering (embedded fibers, pigtails, inlets to high vacuum)
4. Delivery of high-power laser radiation (polymer can inflame by scattered light)
5. Medical applications (metal-coated fibers can be sterilized using ETO, steam, e-beam, or γ-radiation)
6. Harsh environments
   - High-temperature environments ($>350\ ^{\circ}$C)
   - Nuclear radiation (polymer coating decays under radiation)
   - Chemicals (if they do not cause corrosion of the metal)

Real-life applications (such as sensors, aerospace, chemical industry, deep-well oil-field industry) can belong to several groups simultaneously.

In contrast to carbon (another type of hermetic coating), metal coatings do not need an additional protective polymer coating. Thus, metal coatings have no contender in applications for which the presence of a polymer coating is undesirable.

The known specific applications for metal-coated fibers are as follows:

- Radiation-resistant fiber optic systems intended for use in the nuclear industry (e.g., plasma diagnostic systems in thermonuclear reactors, image guides for visual inspection of nuclear installations). To increase radiation resistance, the fiber can be heated up to approximately 400 °C [1]. Alternatively, its glass can be loaded with molecular hydrogen [2].
- High-temperature alarm systems remaining functional in accidental conditions (e.g., in case of fire).
- Fiber optic sensors of temperature, vibration, and so on integrated into complicated devices (e.g., jet engines, turbines).
- High-temperature fiber optic systems resistant to hydrogen penetration meant for applications in the chemical and oil-field industries.
- Enhanced-reliability fiber optic devices in which fibers are soldered to connectors (e.g., devices for the space industry) [3].
- Coolable, incombustible fibers for laser-power delivery.

The fiber can be coated by a metal film after drawing in a separate process (off-line) or during drawing (in-line).

Off-line metal-deposition processes, for example, sputtering of trimetal coatings (Ti/Pt/Au), were reported [4]. Another example is an electrolytic plating process for application of Ni/Au coatings [5].The aforementioned methods are feasible if only a short length of a polymer-coated fiber (several inches) is to be coated by metal. Such a length is enough if the fiber is to be soldered at the seal location during pig-tailing or packaging. The off-line slow deposition process can guarantee precise thickness of the metal layer (typical thickness is a few microns), which is important to adjust and to fix the fiber to a high accuracy by soldering.

An attractive possibility is to apply a metal coating in-line (during the drawing process). Attempts to apply a metal (Ni, Mo, Cu, or Ag) coating in-line using sputtering in a vacuum [6, 7], magnetron sputtering (Cu) [8], ion-plasma deposition (Sn, or In) [9] were not quite successful; the coatings obtained were not hermetic and the fiber strength was low. In addition, the aforementioned methods required rather expensive equipment and subsequent electrolytic plating was necessary to solder the fiber ends. The continuous plating process of Ni and Cu application on a carbon-coated fiber has also been realized [10]. A high fiber strength is hardly to be achieved with this method. At present, only the "freezing" method allows application of a metal coating in-line during the drawing process of the fiber. In this case, the fiber passes through a layer (approximately a few millimeters) of molten metal. If the temperature of the melt is close to the melting point of the metal and the temperature of the fiber is lower, then some layer of the metal can "freeze" on the surface of the fiber. For this technique, a usual drawing tower can be used with just one modification; a specially constructed metal applicator should replace the polymer die.

It was demonstrated that metal-coated fibers fabricated by the "freezing" technique were indeed hermetically sealed. It means that because of the absence of water vapor under the metal, the fiber strength can be twice as high as that of polymer-coated fibers (5.5 GPa) [11]. This value for metal-coated fibers approaches the glass strength in liquid nitrogen ($\sim$14 GPa). Fatigue parameter n in this case will be also high ($>$100), as compared to $n \sim 20$ for polymer-coated fibers. Carbon-coated fibers also demonstrate such a high fatigue parameter but cannot reach such a high strength because of cracking of the brittle carbon coating during the tests at high elongation (approximately $>$5%) [12].

Unfortunately, the freezing method of the metal-coating application has some restrictions:

1. Only metals with a comparative low ($\leq$1400 °C) melting point (In, Sn, Pb, Zn, Al, Ag, Cu, Au, Ni) can be applied without technological problems resulting in strength reduction. It appears that it is impossible to achieve satisfactory application of such metals as Ti, Co, and Pd by the "freezing" method.
2. There are only a very limited number of alloys that have been optimized for application by this method. Compositions of such optimized alloys are usually far from the conventional alloys, such as corrosion-resistant alloys.
3. Stable metal application is possible in a limited range of the coating thickness (e.g., 15–25 μm for a fiber diameter of 125 μm, or approximately 50 μm for the fiber diameter of 250 μm).

Our 20 years of experience shows that for many special applications Al-coated fibers are quite suitable. In some cases, related to soldering or extremely high temperatures, Cu-coated or Au-coated fibers are better candidates. Ni-coated fibers could be used as sensors of magnetic field.

Use of long-length metal-coated fibers is associated with the problem of microbending optical losses because of a high expansion modulus of the metal and a very high difference in the thermal expansion coefficients of silica and metals. Preliminary thermocycling is usually used to stabilize optical losses in the reduced temperature range. Fibers of a thick diameter and/or a high aperture are less sensitive to the microbending effect.

In the following section, we provide more detailed information on the "freezing" technology and the properties of metal-coated fibers obtained by this technique.

15.2 FREEZING TECHNIQUE

In the 1960s, Arridge et al. [13] and Arridge and Heywood [14] demonstrated, for the first time, aluminum application on silica fibers using the freezing process. These nonoptical fibers were meant for use in fiber-reinforced aluminum constructions. Later, this method was used for aluminum coating deposition on optical fibers [15–18]. At present, only the freezing method allows one to apply a metal coating in-line during drawing of a fiber of any length. For this technique, usual drawing towers can be used with just one modification: A specially constructed metal applicator should replace the polymer die (Fig. 15.1).

In this technique, the fiber passes through a layer (approximately few millimeters) of molten metal. If the temperature of the melt is close to the melting point of the metal and the temperature of the fiber is somewhat lower, a layer of the metal can freeze on the surface of the fiber. To obtain a stable uniform metal film, the duration of the contact of the fiber with the molten metal in the metallizer should be shorter than the time of fiber heating to the metal melting point. Otherwise, the frozen layer will melt again and the fiber will pass through the metallizer without any coating.

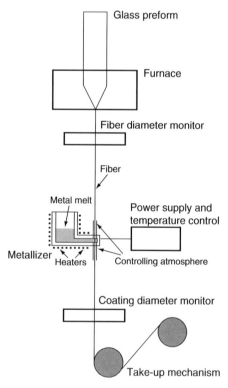

Figure 15.1 Schematic diagram of a typical metal-coating setup: a standard drawing tower equipped with a metallizer.

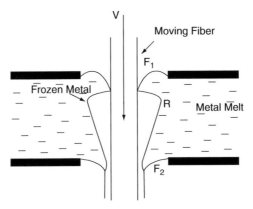

Figure 15.2 Schematic drawing of coating tip of a metallizer for explanation of the freezing technique. F_1, inlet meniscus; F_2, outlet meniscus; R, axisymmetrical body corresponding to the metal being frozen on the fiber.

A detailed schematic of the process is given in Fig. 15.2. At the moment of the first contact of the fiber with the melt, the layer of the frozen metal arises on the fiber surface. Its thickness rapidly grows to the value depending on the energy the fiber can take from the melt. In other words, the coating thickness depends on the fiber thickness and its temperature, as well as on the temperature of the melt. After the appearance of a coating, during the further passage through the melt, the thickness of the metal layer gradually decreases as a result of melting of the frozen metal. Because all metals are well wetted with their melts, some amount of molten metal is carried out from the metallizer as a liquid film on the solid surface of the frozen metal.

A mathematical description of the process is quite complicated. Nevertheless, there are a few publications on this issue, in which the calculated results are close to the experimental ones [14, 19–21]. The main parameters defining the thickness of the metal coating are the diameter and the temperature of the fiber, the temperature of the melt (i.e., how far it is from the melting point), the longitudinal length of the bath of the melt, and the speed of the fiber drawing. Typical results are shown in Fig. 15.3.

In contrast to the polymer coating process, the diameters of the inlet and outlet do not significantly influence the diameter of the coating. These diameters may be much greater than that of the resultant metal-coated fiber. Molten metal does not stream up or down from the metallizer through the inlet or outlet owing to surface tension. Surface tension keeps metal from streaming, only if the melt does not wet the material of the metallizer and does not react with it. An oxidizing film can also be a problem for surface tension, so an oxygen-free atmosphere is very desirable, at least, at the outlet.

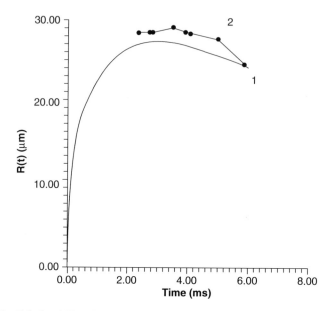

Figure 15.3 Calculated (1) and experimental (2) dependences of thickness of aluminum coating as a function of the time of contact fiber with the melt metal (fiber diameter is 125 μm, T = 661, 2°C) (used, with permission, from reference [19]).

The maximum thickness of the coating strongly depends on the fiber diameter and its temperature, that is, on the energy the fiber can take from the melt when heated to the melting point. For a 125-μm fiber, we obtained a maximum thickness of approximately 25 μm for most of the studied metals; for a 250-μm fiber, it was about 60 μm. Changing the process parameters (e.g., the temperature of the melt), we could change the thickness of the coating in the range between the maximum value and half of that. It is possible to maintain the chosen coating thickness along the fiber length with a typical accuracy of ± 2 μm.

In accordance with the theory, we could reduce the coating thickness down to a few microns. Nevertheless, if we tried to obtain the thickness less than half of the maximum value, the application process became rather unstable and sensitive to small perturbations of the drawing parameters. Under abnormal coating application conditions, uncoated fiber spans (several millimeters in length) became possible. In addition, with a coating thickness smaller than a certain critical value, there arose holes in the coating (Fig. 15.4). Figure 15.5 presents a view of a metal-coated fiber after stripping the coating from the front side. More or less regular cavities in the metal are seen through the transparent fiber glass. Figure 15.6 is an SEM picture of a copper coating taken off of a fiber. Cavities can be seen as well forming orthagonal bands along the inner surface of the copper metal coating.

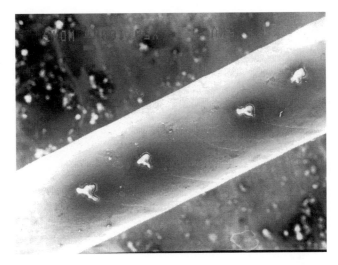

Figure 15.4 Holes in metal coating with minimal thickness. First holes arise in the place of cavities on the boundary glass–metal.

Cavities in the glass–metal interface are inherent in the freezing technique. Usually they appear regularly with a period comparable to the fiber diameter. This phenomenon is due to hydrodynamic instability of the melt's flow near the inlet meniscus [22, 23]. Oscillations of the meniscus can be described as standing

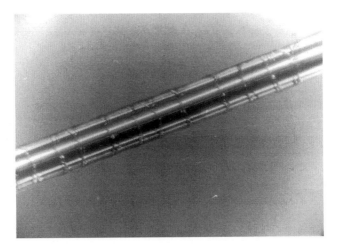

Figure 15.5 Photo of a section of a typical metal-coated fiber taken after stripping the coating from the front side. Through the fiber glass, one can see the inner surface of the coating, where cavities show up as narrow bright bands.

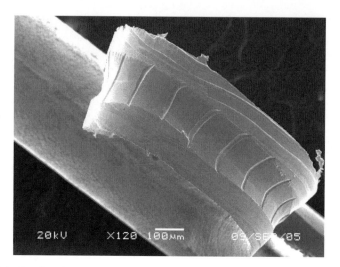

Figure 15.6 Interior of a metal coating contacting with a fiber. (Part of the metal coating was sliced by a knife.)

capillary waves (Fig. 15.7). These standing waves are excited by fluctuation of the parameters of the process, such as the fiber-drawing speed, diameter, position, and so on. The depth of the cavities can be about 5 μm. Thus, this effect limits the minimum possible thickness of a continuously applied metal layer onto a fiber.

The presence of cavities makes the strength and fatigue of a metal-coated fiber sensitive to the atmosphere over the metallizer, because the cavities are filled with

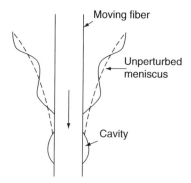

Figure 15.7 A schematic explaining the appearance of cavities at the fiber–metal interface due to oscillations of the inlet meniscus of the metal melt.

the corresponding gases (see the next section). Moreover, microbending optical losses in metal-coated fibers can be caused by the cavities or, at least, increased by their presence.

Only metals with a comparatively low melting point (In, Sn, Pb, Zn, Al, Ag, Cu, Au) can be applied by the freezing method without a noticeable problem with the fiber strength due to the reaction of the melt with silica. Very high stresses can arise in the metal film and in the silica fiber during cooling after metal application because of a big difference in thermal expansion coefficients of silica and the metal. Fortunately, it is not a problem for pure metals, thanks to quick stress relaxation at a high temperature due to mobile dislocations. However, for nonoptimal alloys, the difference in thermal expansion coefficients may be a problem, because mobility of dislocations may be significantly lower. As a result, the fiber or the metal film can crack during cooling (Fig. 15.8).

A metal coating usually has a glossy smooth surface, but its structure can be various. If this surface is etched in a special way, the structure of the metal can be brought out. We found that in some cases (at some regimens), the metal film consisted of small polycrystalline grains (Fig. 15.9), whereas at other regimens, it looked like a surface of a monocrystal (Fig. 15.10). X-ray analysis confirmed this visual observation (Fig. 15.11). Our experience shows that both mechanical and optical properties of metal-coated fibers are significantly better in the case of a "monocrystalline" structure of the coating [11, 24].

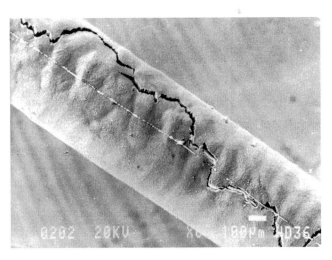

Figure 15.8 Crack in a metal coating, which arose immediately after the coating application.

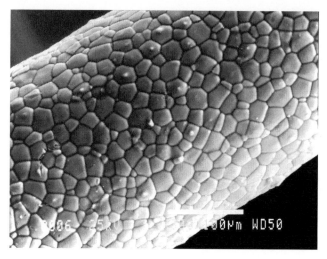

Figure 15.9 SEM photo of the surface of an Al coating with shallow "polycrystalline" grain structure.

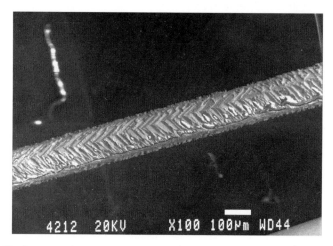

Figure 15.10 Structure of a tin coating surface after selective crystallographic etching of a high-strength hermetically coated optical fiber (used, with permission, from reference [24]).

15.3 STRENGTH AND RELIABILITY

The overall strength and reliability of silica-based optical fibers depend on the fatigue effect. It means that flaws in the glass subjected to a tensile stress in the presence of moisture grow subcritically before failure [25]. Because of this, the strength of standard polymer-coated fiber (a typical value is ~5.5 GPa at 50% RH

Figure 15.11 (Top) Transmission white x-ray beam Laue photo for an Al coating applied under optimum conditions "monocrystalline" structure (subgrains with small misorientation). (Bottom) Laue photo for an Al coating with "polycrystalline" structure.

and room temperature) is significantly less than that at the liquid nitrogen temperature ($\sim$14 GPa), when the influence of water vapor is minimized [26]. Fatigue is the reason of limited reliability of polymer-coated optical fiber under static stress. A power law with fatigue parameter $n \sim 20$ is usually used to describe the fatigue effects and to predict the time to failure of the fiber in service.

In the case of strength testing at different loading rates (dynamic fatigue), the power law gives

$$\frac{\sigma'_1}{\sigma'_2} = \left(\frac{\sigma_{d1}}{\sigma_{d2}}\right)^{n+1}, \tag{15.1}$$

where σ_{d1} and σ_{d2} are the tensile strength of similar samples at loading rates σ_1' and σ_2', respectively. That is, for $n = 20$, the tensile strength increases by a factor of approximately 1.12 if the loading rate increases by a factor of 10.

In the case of tests at a constant stress (static fatigue), time to failure t will increase by a factor of 10 if applied stress σ_s is decreased by a factor of about 1.12 (for $n = 20$), according to the following relation:

$$\frac{t_2}{t_1} = \left(\frac{\sigma_{s1}}{\sigma_{s2}}\right)^n. \tag{15.2}$$

It was predicted that in the absence of moisture on the fiber surface (e.g., under a hermetic coating), slow crack growth still could take place under stress, because of thermofluctuations. Time to failure t under static stress σ_s in that case can be evaluated by the following expression:

$$t = t_0 \exp\left[\frac{U_0}{kT}\left(1 - \frac{\sigma_s}{S_i}\right)\right], \tag{15.3}$$

where t_0 is the value close to the period of atomic thermal fluctuations ($\sim 10^{-13}$s); U_0 is the energy of Si-O bonds in silica glass (~ 110 kcal/mol); S_i is the fiber initial strength (in the absence of thermofluctuations at $T = 0°K$).

The thermofluctuation model predicts that the strength of a hermetically coated fiber must be about 0.84 S_i at room temperature and about 0.95 S_i in liquid nitrogen. Because these estimations are based on the exponential description, parameter n depends on time and applied stress. We can estimate the n value using (15.3)

$$n = -\frac{d(\ln[t_s])}{d(\ln[\sigma_s])} = \frac{U_o}{kT}\frac{\sigma_s}{S_i}. \tag{15.4}$$

It can be calculated from Eqs. (15.3) and (15.4) that the n value decreases under laboratory conditions from 155 to 135, when time to failure changes from 1 second to 30 years [27].

Thus, the main features of an ideal hermetically coated fiber are a high strength (approaching that in liquid nitrogen) and a high n value.

In real cases, there exists a problem of correctly measuring the strength of metal-coated fibers. It is quite difficult to measure such a high strength by a usual tensile testing machine because of the difficulty of fixing the fiber ends during the test without damaging the fiber surface. Because all the fiber length between the holders usually molders away after the failure, it is impossible to detect the position of the initial place of break, whether it is in the holder or between the holders. The two-point bending technique is free of this problem, but in this case, only a very short fiber length (~ 1 mm) is subjected to stress [28]. In addition, correct calculation of the bending radius and the respective strength in the case of the two-point bending technique is hampered by a nonlinear behavior of

Young's modulus of silica at elongations up to 8% and by the absence of experimental data at higher elongations (failure in liquid nitrogen occurs at ~15% elongation).

Nevertheless, it was demonstrated that metal-coated fibers fabricated by the "freezing" technique could be really hermetically sealed. It means that the strength at room temperature is close to that in liquid nitrogen and n value is higher than 100. Bending tests show that to obtain the best result, the metal film has to be of "monocrystalline" structure [24] (Fig. 15.12) and water vapor has to be excluded from the atmosphere over the metallizer, where the fiber is still uncoated [29] (Fig. 15.13). Tensile tests of fibers coated by various metals do not give so high strength levels (Fig. 15.14). Nevertheless, all those fibers were stronger than the conventional polymer-coated fibers.

Carbon-coated fibers also demonstrate fatigue parameter $n > 100$, but their strength is less than that of standard fibers [30, 31]. The reason is brittleness of the carbon film. It cannot survive a high elongation (approximately >5%). Carbon-coating cracks during strength tests at a higher elongation were followed by a failure of the fiber itself [12]. Metal-coated fibers made by other methods (sputtering, plating, etc.) are usually nonhermetic or may contain water under the coating. Thus, only metal-coated fibers fabricated by the "freezing" technique demonstrated a strength higher than that of standard fibers.

Unfortunately, it is difficult to take advantage of this unique strength of metal-coated fibers in the case of a long fiber length. The problem to obtain a

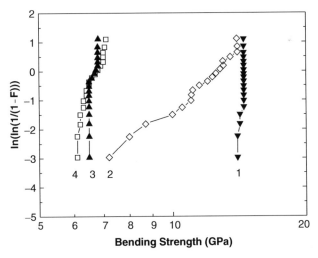

Figure 15.12 Weibull plot of bending strength of tin-coated fibers with "monocrystalline" structure of coating (1) and "polycrystalline" structure (2) and after coating removed for both types of fiber (3, 4) (used, with permission, from reference [24]).

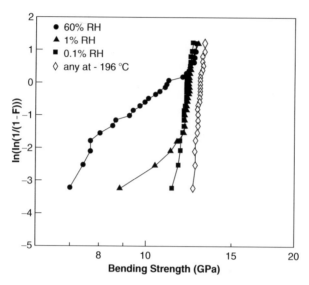

Figure 15.13 Weibull plot of bending strength of tin-coated fibers drawn under different relative humidity of atmosphere over inlet meniscus (used, with permission from reference [29]).

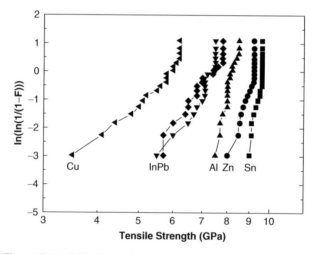

Figure 15.14 Weibull plot of tensile strength of metal-coated fibers.

long-length metal-coated fiber of a uniform high strength is significantly more complicated than that solved for polymer-coated fibers [32]. In addition to usual sources of possible defects, a number of new types of defects, such as inclusions of hard particles in the metal, pores or holes in the metal film, and so on, should be identified, systematized, and eliminated.

In addition, to ascertain a high strength of the whole length, the fiber is to be proof-tested (rewound from one reel to another) at some load (strain). However, the process of proof-testing itself at loads over approximately 5–10 N (for a 125-μm fiber) can damage the metal coating and the fiber. Moreover, tin- or indium-coated fibers survive proof-testing only in the case of an additional polymer coating applied over the metal. This problem is less acute, if only bending at low tension on a certain set of wheels is used as the proof-test procedure instead of tensile load.

15.4 DEGRADATION AT HIGH TEMPERATURE

Ideally, in the absence of corrosion or oxidation effects, a metal coating remains at work at temperatures close to the melting temperature of the metal. However, in reality, reaction of metal with silica can drastically reduce the fiber strength at an elevated temperature. It should be noted that we did not observe this effect for metals with a low melting point (In, Sn, Pb, Zn). Some reduction of strength of tin-coated fibers at high temperatures was explained by thermofluctuation effects [27]. However, for an Al coating (one of the best candidates for most applications), a reaction between aluminum and silica significantly reduces the working temperature.

Initially an interaction between silica and aluminum at high temperatures was observed in bulk aluminum reinforced by silica fibers [33]. Molten aluminum easily reacts with silica in the following way:

$$4\,Al + 3\,SiO_2 \rightarrow 2\,Al_2O_3 + 3\,Si \qquad (15.5)$$

This reaction also takes place at the interface at temperatures lower than the melting point ($\sim$660 °C) with a rate decreasing with decreasing the temperature. Because of this, the strength of Al-coated fibers degrades rapidly at temperatures of about 500 °C. Even a carbon film at the interface cannot significantly slow this process [34, 35] (Fig. 15.15).

Activation energy of 250–290 kJ/mol was obtained from experiments on strength degradation at different temperatures. A conclusion can be drawn from extrapolation to lower temperatures that Al-coated fibers can be used in long-term applications only at temperatures less than 400 °C.

Unfortunately, the process at the metal–silica interface was not so well studied for metals with a higher melting point (Au, Ag, Cu, Ni). These metals do not react chemically with silica at a high temperature, in contrast to aluminum. However, for fibers coated with copper with protection from oxidizing, we observed strength degradation after aging at temperatures of 600–1000 °C during a few hours. The reason for this degradation was crystallization of the surface of silica glass [36, 37]. The effect of degradation was higher if oxygen

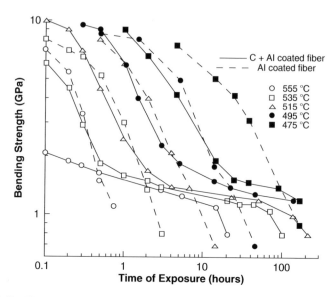

Figure 15.15 Time dependence of room temperature bending strength of fibers coated with C + Al and Al after heat treatment at various temperatures (used, with permission, from reference [35]).

penetrated into the cavities at the silica–copper interface to form a copper oxide film at the silica surface. Strength degradation of Ni-coated fibers at 600 °C was even stronger than that of copper-coated fibers [38].

This phenomenon is in qualitative agreement with the data on crystallization of bulk silica, which is significantly accelerated by contaminations on the silica surface [39]. In the case of metal coatings, all the surface of the fiber is "contaminated" by a metal. Thus, crystallization of the silica surface at elevated temperatures (>600 °C) is a fundamental factor resulting in strength degradation. Significant reduction of strength at 600 °C usually occurs during approximately 1 week or faster depending on the specific metal.

15.5 OPTICAL PROPERTIES OF METAL-COATED FIBERS

Because the drawing technology of metal-coated fibers differs from that of polymer-coated fibers only in the fact that the fiber is heated for a short time to the melting temperature of the metal, metal-coated fibers feature approximately the same optical loss as polymer-coated fibers. In some cases, the loss in metal-coated fibers is even lower. For example, in Al-coated low-OH pure silica core fibers, the 0.63-μm band inherent in polymer-coated fiber disappears because of thermal annealing in the metallizer [40, 41]. However, a relatively thick metal coating causes additional microbending loss. The physical mechanisms of this effect include

1. A high Young modulus of metals, close to that of silica
2. A large difference in the thermal expansion coefficients of silica and metals
3. A low threshold of plastic deformation of pure metals (<10 MPa)

Losses in as-drawn normal numerical aperture (NA) Al-coated fibers can be as high as 20–100 dB/km at room temperature. Certain treatment or temperature cycling can change the level of microbending loss. For example, rewinding the fiber to another reel can either increase or decrease the excess loss depending on a number of parameters (rewinding tension, diameter of the guiding rollers, the reel diameter, etc.). Although the excess loss can be minimized, it poses a problem for some applications of metal-coated fibers.

It is known that high-NA metal-coated fibers demonstrate virtually no excess loss due to microbending.

A properly selected regimen of temperature cycling in the range from -20 to $+60$ °C led to a reduction of microbending loss to approximately 0.1 dB/km in a multimode Al-coated fiber (cladding diameter 125 μm, NA ~ 0.2) [42]. Further thermal treatment of this fiber in the range 5–40 °C gave an excess loss of no more than 0.2 dB/km. The increase of loss in such a fiber due to rewinding can be suppressed by a subsequent temperature cycling.

A typical example of loss variation during temperature cycling in the range from -120 to $+300$ °C is given in Fig. 15.16.

It is seen that the added loss strongly depends on the thermal prehistory. The temperature range for each type of metal-coated fiber can be roughly divided into three regions (Figs. 15.16 and 15.17).

Region 1. Loss in this region does not depend on the thermal prehistory and does not exceed several decibels per kilometer.

Region 2. Loss depends on temperature and thermal prehistory in a complicated way. Loss relaxes to an acceptable level of several decibels per kilometer during a period of 1 hour to 1 month.

Region 3. The excess loss level is high (>20–40 dB/km) and does not change in the course of thermal annealing.

Region 2 corresponds to temperatures of -70 to -20 °C for Sn coating, to -50 to -10 °C for Pb coating, to $+20$ to $+200$ °C for Al coating, to $+20$ to $+250$ °C for Cu coating, and to $+20$ to $+150$ °C for Au coating.

At higher temperatures (*Region 1*), the microbending loss problem does not exist. At lower temperatures (*Region 3*), loss relaxation takes too long

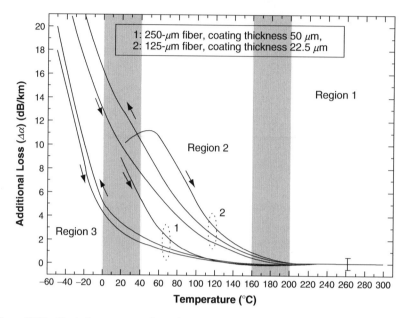

Figure 15.16 Typical temperature dependences of additional loss of an Al-coated graded-index multimode fibers (core diameter, 62.5 μm).

a time. Although the boundaries of the regions are not determined exactly, they do not strongly vary for different fiber types and are governed by the type of coating.

Zeroth excess loss occurs in a fiber when no mechanical stress is applied to the fiber by the coating—in other words, when no microbending takes place.

In this case, the internal stresses in the metal must be close to zero.

Metals are subject to plastic deformation in a wide temperature range, from zero to the melting point. The free energy of a crystalline material rises during deformation because of the presence of dislocations and interfaces. A material containing such defects is thermodynamically unstable. If the material is subsequently heated to a high temperature (annealed), thermally activated processes, such as solid-state diffusion, promote elimination of defect. Many papers have been devoted to this problem (e.g., see reference [43]). In the course of recrystallization, because of the motion of dislocations, their annihilation, and migration of the grains' boundaries, the structure of grains is restored or a new structure arises with a very low density of dislocations.

This means that the residual stresses in the coating are minimal after recrystallization. The relationship between the recrystallization rate and the temperature is given by the Arrhenius equation: The temperature of recrystallization

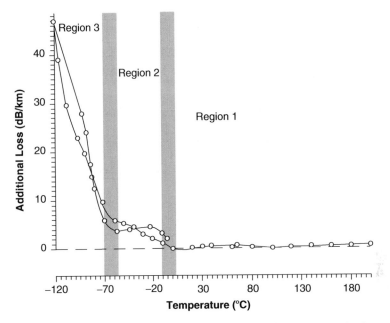

Figure 15.17 Typical temperature dependence of the additional optical loss in tin-coated single-mode fibers at $\lambda = 1.3\,\mu$m (used, with permission, from reference [19]).

decreases with increasing the annealing time. Data on the recrystallization temperature for some metals is given [43, 44]:

Al, 130–200 °C; Cu, 180–240 °C; Au, 160–200 °C; Sn, −70 °C; Pb, −50 °C

The recrystallization temperature ranges coincide well with boundary of *Region 2,* in which a noticeable reduction of losses occurs in metal-coated fibers. However, the recrystallization temperature and rate may strongly vary depending on the specific crystalline structure of the metal, composition of impurities, degree of deformation, and the sample shape. It was found that fibers with a polycrystalline structure of the metal coating have a high excess loss and relaxation is not so efficient.

For this reason, it is desirable to apply a metal coating with a monocrystalline structure. Another way to overcome the microbending loss problem is to increase the fiber diameter. In this case, the coating cannot bend the fiber so strongly, despite the increased thickness of the coating. For Al-coated fibers with NA ~ 0.2 and cladding diameter more than 250 μm, losses larger than several decibels per kilometer are observed only at $T < 20$ °C. For fibers with a cladding diameter of 300–1000 μm, the microbending loss problem does not virtually exist.

Apart from the reversible microbending loss, irreversible loss growth occurs in metal-coated fibers in the temperature range 300–1000 °C. Such effects have been detected but have not been investigated thoroughly [41, 45]. It is known that the irreversible loss strongly depends on the core and cladding chemical composition and increase as a result of hydrogen penetration from outside [31].

15.6 SUMMARY

During the last 20 years of research on metal-coated fiber technology, many fundamental works have been performed aimed at obtaining coatings with genuine hermeticity. With such coatings, the static fatigue phenomena can be minimized and the fiber strength can be close to the theoretical limit for silica glass.

Although some problems remain unsolved, hermetically metal-coated fibers have gained recognition as an important fiber type for applications in harsh environments, where fibers with ordinary coatings cannot be used.

A variety of metals with the melting points of 1400 °C or less have been mastered as coating materials to endow the fibers with unique properties.

A variety of applications have been determined for which hermetically metal-coated fibers is the best-suited fiber type. Such applications include

- Radiation-resistant fiber optic systems intended for use in the nuclear industry (e.g., plasma diagnostic systems in thermonuclear reactors, image guides for visual inspection of nuclear installations). To increase radiation resistance, the fiber can be heated up to approximately 400 °C. Alternatively, its glass can be loaded with molecular hydrogen.
- High-temperature alarm systems remaining functional in accidental conditions (e.g., in case of fire).
- Fiber optic sensors of temperature, vibration, and so on integrated into complicated devices (jet engines, turbines).
- High-temperature fiber optic systems resistant to hydrogen penetration meant for applications in the chemical and oil-field industries.
- Enhanced-reliability fiber optic devices in which fibers are soldered to connectors (e.g., devices for the space industry).
- Coolable, incombustible fibers for laser-power delivery.

REFERENCES

[1] Tangohan, G. L. et al. 1984. Optical, mechanical and radiation performance of metal-coated fibers at high temperature. *Tech. Digest*. OFC-84, paper WF6, pp. 102–104.
[2] Tomashuk, A. L. et al. 1999. Radiation-induced absorption and luminescence in specially hardened large-core silica optical fibers. In: *Proceedings of 5th European Conference on*

Radiation and Its Effects on Components and Systems, Fontevraud, France, 13–17 September, 1999.

[3] Simpkins, P. et al. 1995. Aluminium-coated silica fibers strength and solderability. *Electron. Lett.* 31(9):747–749.

[4] Bubel, G. M. et al. 1989. Mechanical reliability of metallized optical fiber for hermetic terminations. *J. Lightwave Technol.* 7(10):1488–1493.

[5] Filas, R. W. 1998. Metallization of silica optical fibers. In: *Materials Research Society Symposium Proceedings,* Vol. 531, pp. 263–272. MRS spring meeting 1998.

[6] Hale, P. G. et al. 1981. *Physics of Fiber Optics, Advances in Ceramics* (B. Bendow and S. S. Mitra, eds.), Vol. 2, pp. 115–123. American Ceramics Society, Columbus. Ohio.

[7] Almaida, J. B. et al. 1979. On line metal coating of optical fibers. *Optik* 53(3):231–234.

[8] Rogers, H. N. 1991. High temperature coating for optical fiber. *SPIE* 1580:64–67.

[9] Stein, M. L. et al. 1981. *Physics of Fiber Optics, Advances in Ceramics* (B. Bendow and S. S. Mitra, eds.), Vol. 2, pp. 124–133. American Ceramics Society, Columbus, Ohio.

[10] Nozawa, T. et al. 1992. Novel metal coated solderable optical fiber. In: *Proceedings of OFC 92.* Paper ThF3, p. 217. *OSA—Opt. Soc. of Am.*

[11] Bogatyrjov, V. A. et al. 1991. Super-high-strength hermetically metal-coated optical fibers. *Sov. Lightwave Commun.* 1:227–234.

[12] Semjonov, S. L. et al. 1994. Mechanical behavior of low- and high-strength carbon-coated fibers. *Proc. SPIE* 2290:74–78.

[13] Arridge, R. G. C. et al. 1964. Metal coated fibers and fiber reinforced metals. *J. Sci. Inst.* 41:259–261.

[14] Arridge, R. G. C. and D. Heywood. 1967. The freeze-coating of filaments. *Br. J. Appl. Phys.* 18:447–457.

[15] Pinnow, D. A. et al. 1979. Reductions in static fatigue of silica fibers by hermetic jacketing. *Appl. Phys. Lett.* 34:17–19.

[16] Pinnow, D. A. et al. 1979. Advances in high-strength metal-coated fiber-optical wave-guides. In: *Proc. OFC* pp. 16–18 *OSA—Opt. Soc. of Am.* Washington, DC.

[17] Wysocki, J. A. and A. Lee. 1981. Mechanical properties of high-strength metal-coated fibers. In: *Proc. IOOC* p. 24. *IEEE—Inst. Elect. Elec. Eng.* San Francisco, CA.

[18] Inada, K. and T. Shiota. 1986. Metal coated fibers. *Int. Soc. Opt. Eng.* 584:99–106.

[19] Birukov, A. S. et al. 1993. Calculation of the thickness of a metal coating for fibre produced by the freezing technique. *Sov. Lightwave Community* 3:235–246.

[20] Simpkins, P. G. 1994. Thermal response of optical fibers to metallization processing. *Materials Sci. Eng.* B23:L5–L7.

[21] Biriukov, A. S. et al. 1998. Theoretical investigation of metal coating deposition on optical fibers by freezing technique. The model of the process. *Materials Research Society, Symposium Proceedings,* Vol. 531, pp. 273–284, MRS Spring Meeting 1998, San Francisco.

[22] Biriukov, A. S. et al. 1995. On the origin of periodic macrostructures in metals frozen on a moving substrate [in Russian]. Preprint GPI 4.

[23] Biriukov, A. S. et al. 1997. On the origin of periodic inclusions in metals frozen on a moving substrate. *J. Applied Phys.* 81(10):7018–7023.

[24] Bohatyrjov, V. A. et al. 1991. High-strength hermetically tin-coated optical fiber. OFC'91. *San Diego Tech. Digest,* Paper WL9, p. 115.

[25] Kurkjian, C. R. et al. 1989. Strength and fatigue of silica optical fibers. *J. Lightwave Technnol.* 7:1360–1370.

[26] France, P. W. et al. 1980. Liquid nitrogen strengths of coated optical glass fibers. *J. Mat. Sci.* 15:825–830.

[27] Bohatyrjov, V. A. et al. 1991. Mechanical reliability of polymer-coated and hermetically coated optical fiber based on proof testing. *Optical Eng.* 30(6):690–698.

[28] Matthewson, M. J. et al. 1986. Strength measurement of optical fibers by bending. *J. Amer. Ceram. Soc.* 69:815–825.

[29] Bubnov, M. M. et al. 1992. Influence of residual water on the strength of metal coated optical fibers. *Mat. Res. Sym. Proc.* 244:97–101.

[30] Huff, R. G. and F. V. DiMarcello. 1987. Hermetically coated optical fibers for adverse environments. *Proc. SPIE* 867:40–45.

[31] Lu, K. E. et al. 1990. Mechanical and hydrogen characteristics of hermetically coated optical fibre. *Opt. Quant. Electron.* 22:227–231.

[32] DiMarcello, F. V. et al. 1985. Multikilometer lengths of 3,5 GPa (500ksi) prooftested fiber. *J. Lightwave Technol.* LT3(5).

[33] Standage, A. E. and M. S. Gani. 1967. Reaction between vitreous silica and molten aluminium. *J. Am. Ceram. Soc.* 50:101–105.

[34] Bohatyrjov, V. A. et al. 1992. Heat resistant optical fibers hermetically sealed in aluminum cladding. *Sov. Tech. Phys. Lett.* 18(11):698–699.

[35] Bubnov, M. M. et al. 1993. Strength of dual-hermetic coated fibers at high temperatures (>400). OFC/IOOC'93 *Technical Digest* paper WA3, p. 77.

[36] Bogatyrjov, V. A. et al. 1997. Performance of high-strength Cu-coated fibers at high temperatures. In: *Optical Fiber Communication Conference,* Vol. 6, *OSA Technical Digest Series,* pp. 182–183. Optical Society of America, Washington, DC.

[37] Biriukov, A. S. et al. 1998. Reliability and optical losses of metal-coated fibers at high temperatures. *Materials Research Society, Symposium Proceedings,* Vol. 531, pp. 297–300, MRS Spring Meeting 1998, San Francisco. MRS Mat. Res. Soc.

[38] Biriukov, A. S. et al. 1998. Magnetosensitive Ni-coated optical fibers. In: *Materials Research Society, Symposium Proceedings,* Vol. 531, pp. 291–295, MRS Spring Meeting 1998, San Francisco. MRS Mat. Res. Soc.

[39] Leko, V. K. and O. B. Mazyrin. 1985. *Properties of Silica Glass* [in Russian]. Nauka, Leningrad.

[40] Bohatyrjov, V. A. et al. 1995. Super high strength metal coated low hydroxyl low chlorine all silica optical fibres. In: *Conference Proceedings of RADECS'95, Arcachon, France,* p. 503.

[41] Shiota, T. et al. 1986. High temperature effects of aluminum coated fiber. *J. Lightwave Technol.* LT-4(8):1151–1156.

[42] Abramov, A. A. et al. 1993. Optical performance of low loss aluminium coated fibers exposed to hydrogen and temperature cycling. *OFC/IOOC 93 Technical Digest,* Paper WA3, p. 76. *OSA—Opt. Soc. of Am.* Washington, DC.

[43] Humphreys, F. J. and M. Hatherly. 1995. Recrystallization and related annealing phenomena. Pergamon Press, New York, NY.

[44] Martienssen, W. and H. Warlimont (eds.). 2005. Springer Handbook of Condensed Matter and Materials Data. Springer. Heidelberg, Germany.

[45] Tanaka, S. et al. 1984. Loss increase characteristics of various kinds of hermetic coated fiber at high temperature. In: *Conference Proceedings ECOC '84.* pp. 312–313. IEEE—Inst. Elect. Elec. Eng.

Chapter 16

Elliptical Core and D-Shape Fibers

Thomas D. Monte, Liming Wang, and Richard Dyott[*]

KVH Industries, Inc., Tinley Park, Illinois

16.1 OVERVIEW

16.1.1 Elliptical Core Optical Fiber

Most polarization-maintaining (PM) fibers work on the basic principle of creating two decoupled paths by introducing anisotropic stress generated by the different thermal expansions of the glass materials across the optical core region. In comparison, the elliptical core single-mode fiber employs the distinct property of geometry or form birefringence, rather than stress, to achieve its polarization-preserving characteristics. By separating the propagation constants of the two fundamental modes, the elliptical core fiber is able to reduce intermodal coupling. As a result, the polarization is maintained over a significant distance to be used in interferometric sensors.

In 1961 the first solution of an elliptical dielectric rod was published by Lyubimov et al. [1]. Later the next year, Yeh [2] outlined the extensive analysis of the elliptical core waveguide. When fiber optics emerged as the likely candidate for long-distance transmission lines, Dyott and Stern [3] analyzed the slightly elliptical core optical fiber as an annoying limitation in telecommunication links based on the group delay difference between the two orthogonal fundamental modes, which was followed by a study by Schlosser [4] of delay distortion due to elliptical deformation. Ramaswamy et al. [5] and Dyott et al. [6] soon realized that when the ellipticity and the core-to-cladding index difference are made sufficiently large, intermodal coupling is reduced, polarization is maintained, and elliptical core fiber applications eventually emerged generally concentrated in interferometric sensors. This larger core-to-cladding index

[*]*Deceased*

513

contrast is useful not only to maximize the birefringence, but also to prevent radiation when the fiber is bent around a small radius. Another requirement particularly useful in interferometers is that extraneous light launched into the cladding is rapidly attenuated, even in short fiber lengths, minimizing spurious interference [7]. This is satisfied by a nonguiding cladding, a cladding with an index lower than the surrounding medium, often referred to as a *depressed index cladding*. The elliptical fiber's azimuthally stable modes are exploited in a new set of higher order mode sensors not available with the standard circular core fiber [8].

16.1.2 D-Shape Elliptical Core Fiber and Variations with Assessable Regions

The D-shape elliptical core fiber satisfies two additional and important requirements for polarization-preserving fiber: the accurate angular location of the birefringent axes and access to the optical fields [9, 10]. Because the flat is close to the cladding, the evanescent optical fields are assessable after removing only a small amount of silica material. The D shape with the flat of the D parallel to the major axis of the elliptical core is generally preferred because of the tighter mode field of the oHE_{11} mode, compared to the eHE_{11} mode [2]. The axes of birefringence are positioned by locating the flat of the D typically against a flat surface or by the glint from the reflection off of the flat. This inherent feature of the D-shape fiber locates the birefringence axes no matter how far down the fiber from a reference point and circumvents the lack of angular reference exhibited by circular clad commercial PM fibers that exhibit some degree of internal twist. These unique features of the D-shape fiber facilitate PM component fabrication described in subsequent sections.

Several key shape and orientation variations of the standard D-shape fiber find interesting applications. Figure 16.1a and b show the standard orientation and 90-degree rotated elliptical core versions of the D-shape fiber. The vertical core orientation D-shape fiber is used in partially etched replaced core devices [11]. The rotated core orientation near 45 degree (Fig. 16.2c) was investigated with surface relief Bragg gratings [12]. The quadrant fiber (Fig. 16.1d) was proposed in 1985 to make co- and cross-polarized couplers [13]. Because much of the work in integrated optics is based on Si substrate platforms, modified circular clad (Fig. 16.1e), and D-shape fiber with the flats at the appropriate 70.5-degree (Fig. 16.1f) mate perfectly with the standard anisotropic wet etched Si V groove. The main advantage is coupling to integrated optics without the need for active polarization alignment. A notched D-shape fiber (Fig. 16.1g) enables higher electric fields across the core region for all fiber electro-optic devices [14].

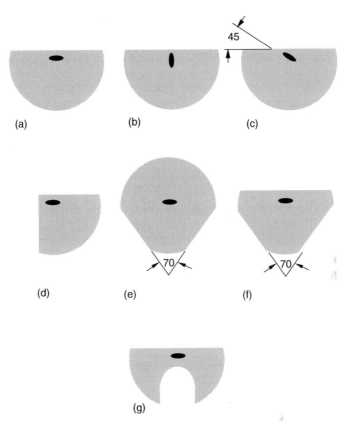

Figure 16.1 Some variations of D-shape elliptical core fiber. (a) Standard orientation with the elliptical core's major axis parallel to the flat, (b) 90-degree rotated core orientation, (c) 45-degree rotated core orientation, (d) quadrant fiber, (e) elliptical core fiber with V-flats, (f) D-shape fiber with additional V-flats, and (g) the notched D-shape fiber.

16.2 MANUFACTURING OF ELLIPTICAL CORE AND D-SHAPE FIBERS

One method to manufacture an elliptical core preform is to fabricate a tube with non-uniform thickness and collapse the tube by heating it to the softening point. The surface tension in the shaped walls during the collapse and subsequent draw cause the fiber core to be nearly elliptical in cross-section [7, 15]. In this way, a familiar method to make circular core fiber using modified chemical vapor deposition is tailored to create a noncircular core. Referring to Fig. 16.2, the initial circular silica tube is ground with two opposing slight flats on the outer

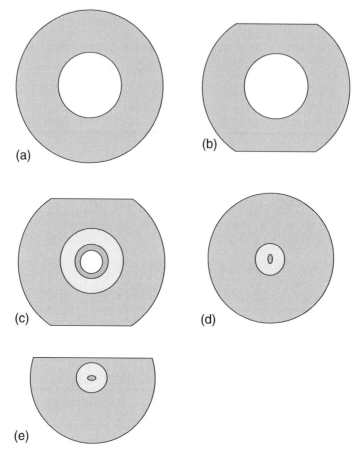

Figure 16.2   Preform stages. (a) Silica tube, (b) grinding slight opposing flats, (c) deposition of depressed index cladding and core glasses, (d) collapse under vacuum, and (e) grinding the flat.

periphery. Then, after cleaning the inner surface, a series of fluorine-doped silica glass layers are deposited inside to create the depressed index cladding. The cladding index is approximately 0.003 less than the index of pure silica, which is enough to prevent interaction between the guided light and the undesirable cladding modes. A final deposition of germania-silica is deposited for the core before the preform collapse. The nominal core-to-cladding index difference is 0.035. During the outer collapse of the preform, surface tension pulls the outside surface circular and the resulting inner index profile exhibits a noncircular cross-section. The depth of the opposing ground flats and the magnitude of the vacuum control the geometry of the core during the preform collapse. The target is an approximate ellipse with the core's major to minor axis ratio of 0.5. A small vapor spot or dip in the index profile centered on the fiber axis occurs during the preform vacuum collapse. Subsequently, some diffusion of the core–cladding boundary occurs during the fiber pulling.

To produce D-shape fibers and other custom-shaped fibers, the circular preform is mounted, with the proper angular orientation of the elliptical axis, inside a grooved surface plate fixture and ground. The initial coarse grind removes the bulk of the material. Several fine grinds allow for a smooth finish on the flat of the D-shape preform, thereby reducing the potential for scattering and generally improving the polarization holding. Usually the flat does not intersect the outer edge of the depressed-index cladding boundary. With the noncircular outer surface of the preform and the core in a predetermined geometric relationship to the core, the drawing rate and temperature are controlled to produce a fiber with a cross-section similar to that of the preform.

16.3 ELLIPTICAL CORE FIBERS: CHARACTERISTICS AND PROPERTIES

The core boundary is defined by an elliptical cross-section with major and minor radii, a and b, and with ellipticity b/a. Using n_1 and n_2 for the core index and cladding index, respectively, the normalized frequency is defined by

$$V_b = k_o b \sqrt{(n_1^2 - n_2^2)}, \qquad (16.1)$$

where $k_o = 2\pi/\lambda_o$. Exact closed-form solutions to calculate propagation constants of the elliptical core step-index fiber modes are not available because of the azimuthally asymmetry of the elliptical geometry. Instead, the propagation constant of each mode is solved by setting up a determinant of a truncated set of infinite Mathieu functions [2]. Other approaches using equivalent rectangular waveguide or equivalent circular guide approximations are established as well [16–18]. Although the results from these approaches agree fairly with the measurements, it is commonplace to employ modern numerical techniques to solve for the propagation constants and mode fields [19, 20].

The fundamental circular core HE_{11} mode splits into two modes: odd $HE_{11}(oHE_{11})$ and even $HE_{11}(eHE_{11})$ derived from the odd and even Mathieu functions, respectively, as the ellipticity is introduced. Both oHE_{11} and eHE_{11} modes have no cutoff. The oHE_{11} mode has transverse electric fields along the major axis of the guide. Likewise, the eHE_{11} mode has transverse electric fields along the minor axis, as shown in Fig. 16.3. The propagation constant of the oHE_{11} modes is greater than that of the eHE_{11} mode because the oHE_{11} is bound tighter inside the core compared to the eHE_{11} mode. A greater percentage of

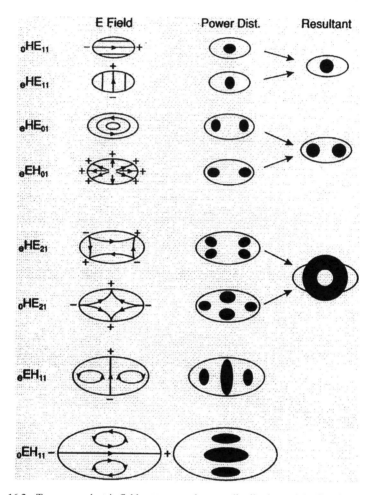

Figure 16.3 Transverse electric field patterns and power distributions of the first few modes of an elliptical core fiber.

power is in the core for the oHE_{11} mode [2]. Consequently, the eHE_{11} evanescent tails extend further into the cladding than those of the oHE_{11} mode. These features are exploited in many useful devices. The oHE_{11} mode is usually preferred because its smaller velocity is more resistant to bend-induced radiation. Frequently, these two fundamental modes are referred to as the *slow* and the *fast* mode, for the oHE_{11} and the eHE_{11} mode, respectively. In the elliptical core waveguide, all of the modes are hybrid, HE_{nm} or EH_{nm}, in comparison to the circular core waveguide that allows circular symmetrical nonhybrid modes $TE_{0n}(H_{0n})$ and $TM_{0n}(E_{0n})$.

16.3.1 Birefringence

The difference between the normalized propagation constant of the two fundamental modes, $_o\bar{\beta} = _o\beta/k_o$ and $_e\bar{\beta} = _e\beta/k_o$ is the fiber birefringence,

$$\Delta\bar{\beta} = _o\bar{\beta} - _e\bar{\beta}. \tag{16.2}$$

The fiber birefringence is of great value serving to de-couple the propagation constants and maintain the polarization. For most practical fibers, $\Delta\bar{\beta}$ is proportional to the square of the index difference between the core and the cladding. Therefore, elliptical core fibers with large core–cladding index differences will exhibit higher birefringence. Usually the beat length characterizes this birefringence, $L_B = \lambda_o/\Delta\bar{\beta}$. Elliptically core fiber has a birefringence strongly dependent on wavelength [6, 21–23]. The maximum achievable normalized birefringence is based on the ellipticity and V_b, as shown in Fig. 16.4 [6]. KVH Industry's elliptical core fiber typically exhibits a normalized birefringence of at least 1.5×10^{-4}.

Zhang and Lit [24] investigated the temperature and strain sensitivities of the birefringence between three types of PM fibers. The two stress-induced birefringence fibers exhibited temperature sensitivities that were seven times larger than

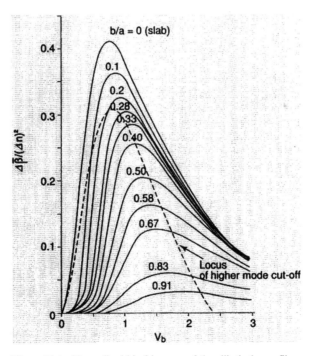

Figure 16.4 Normalized birefringence of the elliptical core fiber.

that of the elliptical core fiber. This residual sensitivity of the elliptical core fiber is due to the small component of stress-induced birefringence caused by the expansion mismatch between the core and the cladding during manufacturing as the fiber cools. The variation of the locked-in stress with temperature changes the birefringence. In regards to the strain sensitivity, the two stress-induced birefringence fibers were 27 times more sensitive than the elliptical core fiber. Because the birefringence is based on the shape of the core, the elliptical core fiber exhibits stable characteristics under adverse conditions of bending, twisting, and wide temperature variations, which can seriously degrade the performance of stress-induced birefringence fibers. Similar measurements of modal birefringence and its sensitivity to temperature and hydrostatic pressure have been reported [25].

The difference in the group index $n_g = \frac{c}{v_g}$ for each fundamental mode is termed the *group birefringence*, $\Delta n_g =_e n_g - _o n_g$ and is related to the birefringence, as shown in Eq. (16.3):

$$\Delta n_g = \left[\Delta\bar{\beta} + V_b \frac{\partial(\Delta\bar{\beta})}{\partial V_b} \right] \tag{16.3}$$

Elliptical core fibers exhibit a group birefringence that is highly dependent on the operating wavelength [26]. The group birefringence is used to calculate the differential time delay between the two fundamental modes for a given length of fiber. The point of vanishing group birefringence, where $\Delta\bar{\beta} = -V_b \frac{\partial(\Delta\bar{\beta})}{\partial V_b}$, is always in the overmoded region for the elliptical core fiber regardless of the ellipticity [7].

16.3.2 Polarization Holding

PM fibers are important to fiber sensors, integrated optic interconnections, fiber jumpers, source pigtails, and many other devices. An important characteristic of birefringent fibers is the ratio of the power leakage into the unexcited mode to that of the total input power. The statistical determination of the polarization transfer as a function of length is characterized by the h parameter [27, 28]. The inverse $1/h$ is the characteristic distance for the polarization transfer. Measurements of the h parameter of KVH's elliptical core fiber are routinely 50 dB/m. The elliptical core fiber generally has a reduced intrinsic sensitivity because the birefringence axes are controlled during the drawing process.

16.3.3 Ellipticity and Higher Order Modes

Elliptical core fibers have nearly twice as many modes as the circular core counterpart because of the broken circular symmetry. The first set of higher order modes, the TE_{01} (H_{01}), and TM_{01} (E_{01}) modes of the circular core fiber, transform into eHE_{01} and oEH_{01} hybrid modes without circular symmetry [29–31]. The first higher order mode of the elliptical core fiber, eHE_{01}, depends on the ellipticity and the normalized cutoff wavelength is defined by an empirical expression from Dyott [32]:

$$V_{bcutoff} = 2.405(b/a)^{0.6275} \qquad (16.4)$$

The second higher order mode depends on the ellipticity and Δn. For practical values of Δn and ellipticity $0.38 < b/a < 1$, the oEH_{01} mode is the second higher order mode. However, when $b/a < 0.38$, a mode is borrowed from the third set of higher order modes. The third set of higher order modes is based on the splitting of the circular HE_{21} mode into the eHE_{21} and oHE_{21} modes. Consequently, when $b/a < 0.38$, the eHE_{21} mode is the second higher order mode. The power distributions and transverse electric fields of the pertinent modes are shown in Fig. 16.3. The power distribution splits from a central region for the eHE_{11} and oHE_{11} modes into double regions for the eHE_{01} and oEH_{01} modes. These double regions are aligned with the major axis of the ellipse. The first two sets of common modes are grouped together and often called the LP_{01} and even LP_{11} modes [33–38]. The sequence of modes of an elliptical core fiber for two values of ellipticity, $b/a = 0.8$ and $b/a = 0.375$, were experimentally confirmed [32] (Fig. 16.5).

When operating a two-mode device, usually the dominant modes and the first set of higher order modes are used. The next higher order mode is avoided by proper selection of the ellipticity and the operating wavelength with the largest bandwidth noted near $b/a = 0.5$. The main advantage of the elliptical core fiber is that the higher order mode-field pattern orientations are stable. In a circular core fiber, the modes are degenerate and the small external perturbations or waveguide perfections introduce phase shifts and uncontrollable coupling between the modes. The differences in the propagation constants between the dominant modes and the first higher order modes with common polarizations, are proportional to the core-to-cladding index, Δn.

16.4 D-SHAPE FIBERS: CHARACTERISTICS AND PROPERTIES

The extraordinary advantage that the D-shape fiber supplies is the opportunity for manufacturing all fiber optical components, which in general possess intrinsic benefits such as reduced back-reflection, low insertion loss, and yet polarization preservation. These components and devices are so-called *evanescent field devices* that can be constructed on a D-shape optical fiber waveguide

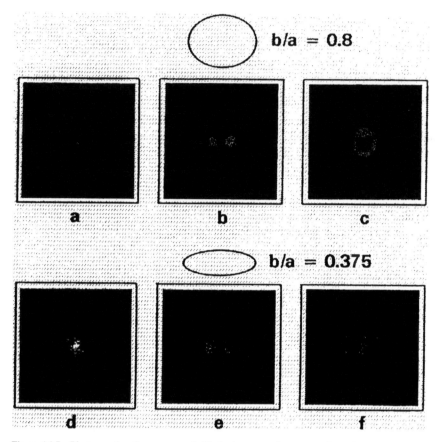

Figure 16.5 Photographs of sequences of elliptical core mode patterns for values of ellipticity, b/a = 0.8 and b/a = 0.375.

substrate. Wet-etching away the silica-based cladding materials enables access to the guiding waveguide field [39, 41].

16.4.1 Accessing the Optical Fields: Fiber Etching

The field of the propagating wave distributed in the elliptical core extends beyond the core–cladding interface decaying exponentially from the interface into the cladding. Generally, a waveguide with a small V value has larger magnitude of the field in the cladding. For instance, the diameter of KVH's standard D-shape fiber used in FOG is 85 μm and the distance of the flat from the core center is 10 μm. Wet etch removes the cladding silica and fluorine-doped silica isotropically. By fine control of the etching process, the tail of the evanescent wave can be accessed from the direction of the fiber flat. By modifying the cladding material close to the core–cladding interface, the waveguide characteristics will be changed and consequently the propagation characteristics and that of the guiding light will be also changed.

Besides the wet-etching fiber method, there are several other techniques to access the evanescent wave in D-shape fibers, which include: tapering the fiber in heat [40], heating the fiber to defuse the dopants in the core (e.g., germanium) into cladding [42], etching to reduce the whole cladding diameter [43, 44], and side-polishing a region of the one side of D-shape or circular fiber [45, 46]. Among all these techniques, wet-etching the D-shape fiber is convenient and accurate. The degree of the evanescent wave access is among the best, as explained in the following subsections.

16.4.2 Wet Etching of Silicon Dioxide–Based Cladding and Germanosilicate Core

Hydrofluoric (HF) acid and buffered oxide etch (BOE) are widely using in the fabrication of micro-electromechanical systems, integrated circuits, fiber probes for near-field optics, and all-fiber devices. The exact chemical mechanisms are complex. The main reaction between silicon dioxide and HF acid are [47, 48]

$$SiO_2 + 6HF \rightarrow H_2SiF_6 + 2H_2O. \tag{16.5}$$

The etch rate depends on etchant concentration, agitation, time, and temperature. Addition of ammonium fluoride (NH_4F) creates a buffered HF solution (BFH), or BOE, that maintains a more stable etching rate, because addition of ammonium NH_4F to HF controls the pH value and replenishes the depletion of fluoride ions.

$$NH_4F \leftrightarrow NH_3 + HF \tag{16.6}$$

In a BOE solution, the H_2SiF_6 further react with NH_3 as the following:

$$H_2SiF_6 + 2NH_3 \rightarrow (NH_4)_2SiF_6 \tag{16.7}$$

Similarly, GeO_2 undergoes analogous reactions. Nevertheless, the rates of reactions and the dissolution of germanium-related products—H_2GeF_6, $(NH_4)_2$ GeF_6, and so on—are different with those of silica ones, which results in the different etching rate in the GeO_2-doped region depending on the proportion of the constituents of the etchant. BOE consists of a mixture of 40% ammonium fluoride and 49% HF solution in a ratio of X:1, etching characteristics of different regions of the fiber depends on this ratio [49, 50]. For lower values of X ($X < 5$), the GeO_2-doped core etches at faster rate compared to fluorine-doped silica cladding and pure silica cladding. As the concentration of HF in the mixture decreases ($X > 5$), GeO_2-doped core etches at a slower rate. The fluorine-doped silica region etched at a faster rate than GeO_2-doped core and pure silica cladding and the rate is independent of the proportion of the constituents of the etchant. The relatively different etch rates enable a selective etch to the cladding and core according to the needs.

16.4.3 Standard Etching (Etch to Reach Evanescent Field)

Because the etching rate is strongly dependent on the etching conditions, accurate control of the degree of the etching to the fiber is crucial for a predictable manufacturing process. The etching conditions, such as acid concentration and temperature, have to be maintained accurately. To relieve the requirements for the strict condition control, a real-time monitor of the etching process was proposed [7, 51, 52]. The PM properties of the D-shape fiber enables this type of monitor by the use of an *in situ* optical polarimetric measurement.

The etching setup with an optical monitor is shown diagrammatically in Fig. 16.6. Light from a coherent source is launched at 45 degrees to the birefringent axes of the fiber, exciting both fundamental modes equally. The source wavelength is chosen to produce a strong evanescent field in the fiber, that is, the V value should be at the lower end of its operating range. The coherence length

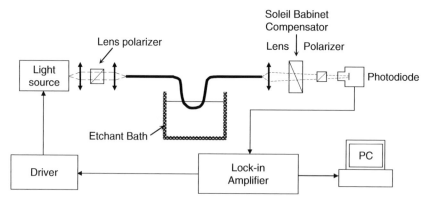

Figure 16.6 An experimental setup for an *in situ* monitor of the etching in a polarization-maintaining D-shape fiber.

L_s of the source should be long enough so that the de-coherence length L_D is much longer than the fiber length L, where

$$L_D = \frac{L_s}{\Delta\bar{\beta} + V \frac{\partial(\Delta\bar{\beta})}{\partial V}}. \qquad (16.8)$$

Light emerging from the other end of the fiber passes through a Soleil–Babinet compensator and another polarizer onto a detector and amplified using a lock-in amplifier. The data are then stored and displayed in a PC.

During the etch, first the silica boundary layer and then the cladding of the fiber-guiding region (having a refractive index of ~1.5) are replaced by the acid/water solution with an index of about 1.33. When the acid reaches the evanescent field region, the phase velocity of both fundamental modes is increased. However, the even $e_{HE}11$ mode has an evanescent field extending further than that of the odd $o_{HE}11$ mode so that a phase shift is introduced between the modes causing a rotation of the polarization of the light emerging from the fiber, which is converted by the output polarizer into a change in amplitude of the detected signal.

Figure 16.7 shows a typical plot of detector output versus time. The periodic variations become more and more closely spaced until eventually the acid reaches the core and the amplitude of the signal becomes zero.

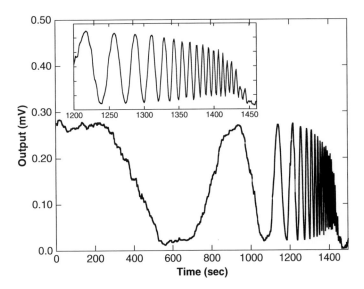

Figure 16.7 A typical output plot of the polarimetric etch monitor for the removal of fiber core. The fiber is etched until the core is thoroughly removed. The insert is the same plot with an enlarged time scale.

The method can control the etching depth very accurately with excellent repeatability. The accuracy of etching also depends on the length of the etched section. Estimated accuracy for etching D-shape fiber of 3-cm long is several tens of nanometers.

The polarimetric monitor provides a sensitive measurement on the relative etching degree so long as the cladding-water interface reaches the evanescent field tail. As described later, the degree of etching required to manufacture fiber polarizers and couplers is slight, as only the tail of the evanescent field is barely exposed. On the other hand, to achieve a better manipulation of the guided light, a deeper and selective etching on the fiber cladding enables the exposure of the fiber core.

16.4.4 Exposing the Core

Using a BOE of $NH_4F:HF$ ratio of 20:1, the etch rate of the cladding is more than 10 times larger than that of the core. The upper half of the cladding can be removed and the core can be exposed from the upper side of the fiber [53]. A typical cross-section of a core-exposed D-shape fiber is shown in Fig. 16.8 [54]. The etching was done in two steps. The first step is to pre-etch the D-shape fiber

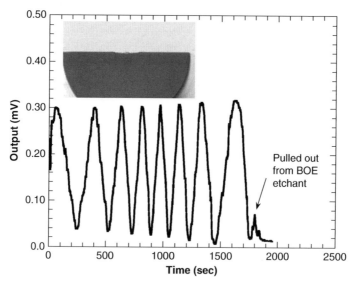

Figure 16.8 A typical output plot of the polarimetric etch monitor for the exposure of fiber core. The insert is an optical microscope photograph of the cross-section of the etched fiber.

using 25% HF acid to remove the relatively bulky cladding material quickly because the etch rate is high for HF acid to the fluorine-doped and pure silica. The second step is a fine-etch of the cladding material to expose the core using BOE, which has a relatively slower etch rate to cladding than the HF acid. Figure 16.9 shows the polarimetric etch monitor output as a function of etch time during the second step. The periodic variation continues to become more closely separated at the first five recorded periods, and then the spacing of the peaks becomes eventually larger. At this time, the fiber flat position is etched close to the position of the major axis of the core ellipse, and the group velocity of odd $o_{HE}11$ mode is more influenced by the removal of the cladding than that of the even $e_{HE}11$ mode.

Because the air in the core-exposed section replaces approximately half of the upper side of the cladding material, the V value of the section is significantly raised. The insertion loss increase is mainly due to the mode-field mismatch at the two transition areas, although surface scattering loss from the core–air interface also contributes. If a material with a refractive index close to the original $F:SiO_2$ cladding is used to replace the cladding, the waveguide characteristics can be recovered. This creates an exciting opportunity for novel on-fiber devices.

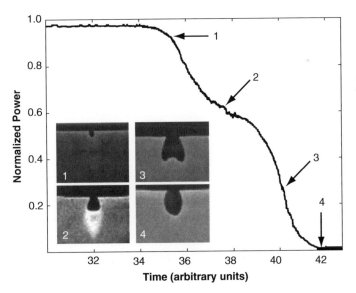

Figure 16.9 transmitted light power as a function of etch time during the core removal process and the corresponding SEM cross-sectional images.

16.4.5 Partial and Full Core Removal

In principle, the wave-guiding field can be further manipulated if the fiber core is fully, or partially, removed, and the empty core is refilled with certain functional material that replaces the removed core and rebuilds the waveguide characteristics. The method and technique for partially and fully removing the core of the D-shape fiber have been proposed [39]. This technique enables new and exciting opportunities in the field of in-fiber electro-optic devices by refilling the removed core with an electro-optic active polymer to fabricate, for example, all-fiber modulators. The selective removal of the core is based on the differential etching of HF acid to the fiber materials. The pure HF acid etches the germanosilicate core with a rate 11 times faster than the rate for etching the pure thermal silica and approximately 8 times faster than the fluorine-doped silicon dioxide cladding.

When the cladding is etched away and the interface is tangent to the core ellipse, the HF etches the germanosilicate core with a much higher rate than the cladding, and the core is selectively removed. During the core removal, the relatively higher refractive index core ($n_{core} = 1.4756$) is gradually replaced by the lower index etchant ($n_{HF} = 1.33$). The decrease of effective index of refraction of the waveguide also decreases the V value of the waveguide and the power transmission decreases correspondingly, because of transition and scattering loss. Finally, the transmission is thoroughly terminated. Although the polarimetric etching monitor technique explained in Section 16.4.3 is applicable in this situation as well (Fig. 16.7), an easier method to control the *in situ* etching process is to monitor the power change directly because of the dramatic power change during the core removal [11].

Figure 16.9 shows a recorded transmission of power as a function of etching time. The power starts to rapidly decrease when the core etching begins. The etching rate is then hindered by the so-called *vapor spot,* which is relatively poor in germanium, and resumes to the high rate until the whole core is removed. The cross-sections at different intermediates are shown as inserts in Fig. 16.9.

16.5 D-SHAPE FIBER COMPONENTS

The elliptical core D-shape fibers are used to create rudimentary PM components. These components are used extensively in all types of interferometers; however, a long list of other applications includes fiber amplifier pump couplers, reflectors, filters, polarization monitors, and various sensors. The use of elliptical core D-shape fiber opens the possibilities of constructing spliceless interferometers and fiber optic assemblies that eliminate additional losses, spurious reflections, and any distinct polarization coupling points at the splices.

16.5.1 Couplers

A true evanescent field coupler can be made by two D-shape fibers positioned flat-to-flat in comparison to the tapered optical fiber couplers with a small-diameter pulled-down region acting as the core with a free space cladding. The main advantages of the D-shape fiber couplers are that the light remains guided completely in the cores and the guiding regions are completely enclosed. As a result, there is no coupling dependence or loss due to the surrounding materials [55, 56].

PM couplers are characterized by insertion loss, power split ratio, and extinction ratio. Practical PM couplers exhibit stable properties over temperature. Performances of D-shape elliptical core couplers include less than 0.3-dB loss and better than 25-dB extinction ratio. With tuning, the coupling is typically held to within 1–2% of the desired splitting ratio. One method of tuning the coupler is by heating and pulling to slightly reduce the diameter of the fibers and extend the evanescent fields [57]. Another method, diffusion tuning, is preferred. With diffusion tuning, heat is applied to the coupler to diffuse the germania-doped cores into the common cladding region [58]. The power ratio between the two output coupler legs is monitored at the desired wavelength to determine when to terminate the diffusion process.

The thin outer layer of silica between the depressed index cladding and the flat of the D-shape fiber and a small portion of the depressed index cladding are removed typically by wet etching throughout the coupler region before coupling. In this way, the amount of required diffusion to produce a 50/50 coupler is reduced. Excessive diffusion of the cores enlarges the effective optical mode fields, which will eventually overlap with the silica index cladding outer boundary resulting in high insertion loss. Index-matched–clad D-shaped fibers can be used to make both types of couplers, diffusion, and pulled tapered types, without the preliminary wet etching step [59, 60].

With the D-shaped flats aligned and fused together using a heat source, the temperature stability and thermal mismatch concerns are addressed. Surrounding the coupler in a Vycor glass tube or channel mitigates variable coupling and polarization performance due to a change in the refractive index of the coupling region from mechanical strain. Such mechanical strains are inherent with bending or twisting of the fiber leads and when using packing materials with mismatched coefficients of thermal expansion. Because the Vycor glass softens approximately 100 °C lower than silica and exhibits a thermal coefficient of expansion matching that of silica, the surrounding Vycor fixes and protects the fiber coupler leads [7]. Figure 16.10 shows the cross section of a D-shaped elliptical core diffusion tuned fiber coupler in a vacuum-collapsed Vycor tube.

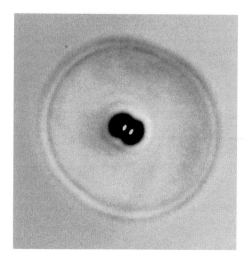

Figure 16.10   Cross-sectional view of a D-shape elliptical core fiber coupler.

16.5.2 Loop Mirrors

A D-shaped elliptical core fiber is particularly suitable to fabricate a polarization preserving all fiber loop mirror or reflector by looping back the flats of the fiber to make a 50/50 coupler [56]. If the 50/50 coupler is lossless then all of the incident power is reflected along the input fiber. The loop acts as a wavelength dependent reflector that is sensitive to the variation in coupling with wavelength [61]. In one application, the loop mirror at the end of a fiber laser will let the pump power in and block the light at the lasing wavelength [62]. The D-shape elliptical core fiber is the perfect candidate to be used in the construction of birefringent all fiber cascaded higher order filters [63] and all fiber reflection Mach–Zehnder interferometers [64].

16.5.3 Polarizers

A fiber optical polarizer is one of the fundamental components in all-fiber optical interferometers and many other fiber systems that are highly dependent on the polarization of the light guided by the optical fiber. Various methods to make relatively short fiber polarizers often involve grinding or polishing circular clad fiber laterally and exposing the core to a birefringent crystal [65, 66], a birefringent film [67], or using a thin metal film or a thick metal film to induce differential attenuation in the orthogonal modes of a birefringent fiber

[45, 68–72]. Another type of all-fiber polarizer is based on longer lengths of single-polarization fibers often wrapped in small-diameter coils to induce differential polarization bending loss [73].

When a proper birefringent crystal is used to replace the portion of removed cladding, the refractive index of the crystal is less than that of the effective index of the waveguide for the transmitted mode and greater than that of the orthogonal mode, causing the unwanted light to escape into the bulk crystal. Birefringent properties of liquid crystals have been used to make an all-fiber polarizer in this manner [74]. A similar approach, but without the birefringent material, uses a thin multimode planar dielectric overlay of zinc sulfide, vacuum deposited directly onto the fiber, and relies on differential coupling between the TE and TM like modes of the asymmetrical structure [75]. In this arrangement, the TE and TM resonances are sufficiently shifted to act as a polarizer across a limited wavelength range.

Thick metal film polarizers with a low index dielectric buffer layer between the optical fiber core and the metal interface rely on bulk absorption caused by the imaginary part of the refractive index of the metal at optical frequencies above the plasma resonance frequency [45]. Other types of fiber polarizers use very thin layers of single metal or bimetal film configurations (5–30 nm) in close proximity to the core (1–2 μm) to create surface-plasmon waves, which selectively couple the waveguide's evanescent waves. Fiber polarizers of varied configurations using aluminum, chromium, gold, silver, nickel, indium, and platinum have been constructed with high polarization extinction ratios [76–80].

Polishing even short lengths of circular clad fiber consistently to the required submicron accuracy and simultaneously aligned with the birefringent axes to maintain low loss and high polarization extinction is difficult. One of the most efficient and easiest ways to manufacture fiber optical polarizers is based on the D-shape elliptical core fiber [71, 72]. An etched D-shape fiber, typically 30–70 mm long, covered with a thin, yet optically thick, metal of indium film will selectively suppress modes with the electric field orthogonal to the fiber metal surface and results in high extinction ratio (>50 dB) and low-loss (0.5 dB) fiber optic polarizers. The polarizer can be fabricated at any point along the fiber and cascaded with other PM fiber components.

16.5.4 Butt Coupling to Active Devices

The first step for all the fiber optic applications is to couple a maximum amount of light from a single-mode laser diode to the fiber. For a PM fiber, the coupling efficiency is dependent not only on the regular alignment parameters such as lateral, axial, and angular offsets, but also on the rotational alignment between the major axes of the mode fields of fiber and laser. D-shape elliptical core fiber is beneficial for this type of alignment.

Commonly used techniques to coupling laser diodes to standard circular fibers use either a focusing lens or a microlens fiber tip [81–84]. The latter is preferable because it produces highly efficient coupling and compact packaging. Chemical etching and subsequent heat melting generally form the fiber end microlenses. However, these techniques have practical problems for the fibers with stress-induced birefringence, such as bowtie and PANDA-type fibers. The microscale stresses at the core region do not allow the uniform etching and melting required for making quality microlenses. Several techniques have been proposed and demonstrated to overcome the problems, but processes were cumbersome or offered limited coupling efficiency. For the KVH D-shape elliptical fibers, the birefringence is significantly attributed to the geometric factor (i.e., the elliptical shape of the core), and the contribution from the local stress is minor. Therefore, the non-uniform etching and melting problem induced by stress is not as critical as the other types of PM fibers. One practical issue to overcome is that the axis of microlens formed through the wet-etching and flame, or laser, melting does not coincide with core axis of the asymmetrical D-shape fiber. Although there is a possible solution for this problem, making the fiber symmetrical by wedge polishing the opposite side of the flat of a D-shape fiber tip before etching, more procedures are required and reproducibility is a concern. Instead of the commonly used pig-tailing techniques, direct butt-coupling to the facet of a laser diode has proven to be a simple and convenient technique to make a highly efficient and stable laser coupling to a D-shape fiber.

The pigtail efficiency can be estimated as coupling two elliptical mode fields that are represented by two mismatched Gaussian beams. If the major axes of the two elliptical mode fields are overlapped as x- and y-axes and the coupling efficiency can be treated as composed of two coupling components in the x and y directions as [85–87]

$$\eta = \eta_x \eta_y, \tag{16.9}$$

where

$$\eta_\alpha = \frac{\int_0^\infty \psi_{1\alpha}^*(\alpha)\psi_{2\alpha}(\alpha)d\alpha}{\sqrt{\int_0^\infty |\psi_{1\alpha}(\alpha)|^2 |\psi_{2\alpha}(\alpha)|^2 d\alpha}}, \alpha = x, y, \tag{16.10}$$

where, $\psi_{1,2}(\alpha) = \exp(-\alpha^2/\omega_{1\alpha,2\alpha}^2)$ and $\omega_{1\alpha,2\alpha}$ are the $1/e$ amplitude radii of the field of the two beams in the x, or y, direction. The power coupling efficiency is expressed as

$$|\eta|^2 = \frac{2\omega_{1x}\omega_{2x}}{\omega_{1x}^2 + \omega_{2x}^2} \frac{2\omega_{1y}\omega_{2y}}{\omega_{1y}^2 + \omega_{2y}^2}. \tag{16.11}$$

If we directly butt-couple a fiber that has a mode-field size of $\omega_{1x} = 4.2\,\mu m$ and $\omega_{1y} = 2.8\,\mu m$ to a laser that has a mode-field size of $\omega_{2x} = 3.8\,\mu m$ and

$\omega_{2y} = 1.0\,\mu$m, the theoretical power coupling efficiency is $|\eta|^2 = 63\%$, according to Eq. (16.11), in which the limitation from the modal asymmetry is outstanding because $|\eta_x|^2 = 99\%$ and $|\eta_y|^2 = 63\%$. The influences of the astigmatism of the diode lasers on the coupling efficiency are also studied theoretically [85]. It is found that the astigmatism is important only at high coupling efficiencies. The influence is negligible for coupling efficiencies less than 50%.

A UV-curable epoxy is used as an adhesive to attach a D-shape fiber to the laser diode facet. The epoxy has a refractive index close to that of silica, $n \sim 1.45$ at 820 nm. With the outer medium changed from air to epoxy, the outer index of refraction changes from 1.00 to 1.46. Therefore, the reflection of the output cavity of the laser diode changes to some extent. This reflection change results in furthering the lasing threshold change to a certain degree.

It has been demonstrated at KVH that a direct butt-couple of D-shape fiber to the facet of a laser diode results not only in a reasonable high efficiency but also good thermal stability. This method has resulted in a 65% coupling efficiency [7, 82]. The pigtail is also reasonably stable over temperature. Figure 16.11 shows a typical pigtail stability evaluation chart measured in KVH production. The horizontal axis is temperature and the vertical axis is a normalized output

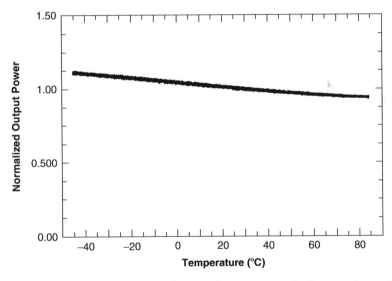

Figure 16.11 Forward monitor power to back-monitor power normalized at room temperature in the configuration using the constant forward detector power configuration. The pigtail output power was set to approximately 68 μW. The slope of the curve is related to the back-monitor resposivity change with temperature. The forward monitor diode was located outside the temperature chamber.

power with respect to the backward power measured from the photodiode attached with the laser diode. The normalized power changes 0.5 dB when the temperature change spans −46 to 62 °C due to the small change on the reflection when operating below threshold. With the proper choice of adhesives, the lifetime of the laser does not degrade as tested over several years.

16.5.5 Coupling to Integrated Optics

Fiber-to-waveguide connection is one of the key technologies in the realization of guided-wave optical devices. Optical telecommunication systems are the leading force for these technologies. Most of the applications in optical telecommunication are based on the use of random polarized light, because the available PM fibers are impractical and expensive for long transmission. Inside the network nodes, devices of well-defined polarization are one option to handle the polarization problems such as polarization-dependent loss (PDL) and the polarization mode dispersion (PMD) [88]. The field of optical fiber sensors is another area where PM coupling to integrated optics is also needed. In such components using polarization modes, the rotational alignment between the devices and the inputs—and to a lesser extent the outputs—PM fibers must be very accurate. Furthermore, when using different PM components in the systems, all the interconnections must be rotationally aligned with high precision to minimize polarization cross-talk between different polarization modes. The circular outer shape of the regular PM fiber does not give enough alignment accuracy through visual examination. Several methods for rotational alignment of a PM fiber have been proposed [89–93]. They are typically based on changing the phase difference between the polarization modes continuously and simultaneously rotating a polarizer in from the fiber. The phase difference can be varied to generate strain on the fiber such as heating, stretching, and pressing or by tuning the wavelength. Accuracies between 0.2 and 1.0 degree are typically achievable with these methods. Slightly improved accuracy—less than 0.1 degree—was demonstrated with an interferometric technique using a broadband source [94]. All of these methods are somewhat complicated and time consuming.

The D-shape fiber guarantees a good alignment between the elliptical core axes and the flat. The accuracy can be controlled within 1 degree during preform grounding. This makes the rotational alignment simple and accurate according to the alignment of the fiber flat visually.

For a sensing system using interferometric architectures, for instance, in a fiber optic interference gyroscope, the spurious sub-interferometers are particularly harmful. These spurious signals are created by Fresnel back-reflections due to the index mismatch between the integrated optics circuit and the fiber coil. In practice, a slant angle of 10 degrees on the $LiNbO_3$ modulator circuit reduces the back-reflected power to below −60 dB. Because the axis of the butt-coupled fiber must be oriented according to refraction laws to preserve a low loss connection, the slant-polished angle of 15 degrees of SiO_2 fiber is required [95, 96].

During the preparation of the input and output fiber subassemblies, a small tube or block is attached to the end of each fiber. This tube is angle-cut and polished to minimize back-reflections from the interfaces. The tube increases the surface area and bond strength of the pigtail and liability. UV or thermal curable adhesives are used to attach the fiber subassemblies to the chips. To de-couple the strain induced by the differential thermal expansion among the integrated optic chip, the fiber, and the packaging material, subassemblies with a strain-absorbing bow and a compliant adhesive are used [88, 97].

16.6 SPLICING

16.6.1 D-Shape to D-Shape Fiber Splicing

The D-shaped elliptical core fiber can be readily fusion spliced to itself with low loss, high strength, and excellent polarization properties. The flat is used for polarization alignment of the PM fiber core without the usual cumbersome fiber side-viewing techniques to align the birefringent axes. The D-shape of the PM fiber aids in orientation alignment when the components are spliced into a fiber optic gyroscope and other interferometers that require high polarization extinction levels. Using an industry standard fusion splicer—the Ericsson Model 975—the V-groove chucks and the mechanical flat top clamps perfectly aligning both fibers with the flats upright. As a result, the splice insertion losses are, at most, 0.5 dB with the best splice losses being too low to measure. Typically, the splice extinction ratios are better than 25 dB. The D-shape fiber's self-aligning feature enables efficient and economical fusion-spliced PM fiber assemblies. The orientation of the major axis of the fiber core is readily recognized and aligned for the splice because the fiber flat has been ground parallel with the major axis of the elliptical core during the preform fabrication.

16.6.2 D-Shape to Circular Clad Fiber Splicing

Connection to other types of optical fibers is sometimes necessary. Splicing D-shaped fiber to circular clad fiber is feasible as well. Dissimilar optical fiber cross-sections are difficult to splice together with simultaneous high-strength and low-loss properties. Large transverse movements or offsets with respect to the optical mode-field size between the two fibers result because of the asymmetrical forces from surface tension during the fusion molten glass stage. Furthermore, if a tolerable insertion loss is achieved, the fusion splice is weak because the dissimilar cross-sections are subject to high stresses at the junction.

One straightforward method to remedy these difficulties is to edge-polish the cleaved end of the circular clad fiber before the fusion process. In this way, the circular clad fiber takes on the *D* shape just at the cleaved surface. Figure 16.12 shows the edge of the circular clad fiber undergoing polishing. The cleaved end of the fiber extends a predetermined distance from the holder and is lowered gradually to meet the polisher. After the initial setup, the edge-polishing process is very reproducible. The resultant edge-polished fiber is shown in Fig. 16.13. In Fig. 16.14, the side-view of the edge-polished circular clad fiber is shown mated up to the D-shape fiber. The splice loss, which usually depends on the mismatch between the fiber mode fields (Eq. [16.11]), is lessened because of thermal diffusion of the core to clad boundary of the higher index difference optical fiber. Routine splice losses less than 0.8 dB are achievable when splicing KVH's elliptical core D-shaped fiber to Corning's SMF-28 fiber. To maintain the polarization through a splice between a D-shape and a circular clad PM fiber, the angular orientation of the circular PM fiber must be prealigned with the fiber holder before edge-polishing.

16.7 IN-FIBER DEVICES

Three unique characteristics of the elliptical core D-shape fiber facilitate the developments of in-fiber devices: The D-shape fiber clad simplifies a long-distance access to the evanescence field through selective wet etching, the

Figure 16.12   Edge polishing the cleaved end of a circular clad fiber.

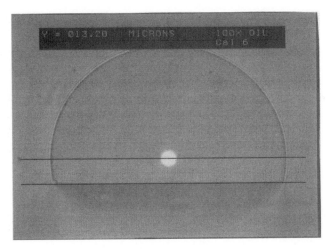

Figure 16.13 Cross-section at the cleaved end of an edge polished circular clad fiber.

elliptical core maintains light polarization, which is crucial for interferometric sensors and devices, and the high-level germania constituent enhances the photo-sensitivity to the grating on the fiber core. Utilizing one, or a combination, of these characteristics, different types of the devices and sensors have been proposed and demonstrated.

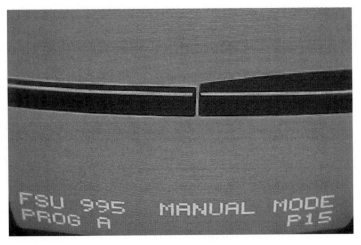

Figure 16.14 Side view in fusion splicer of D-shape fiber (left) and edge-polished circular clad fiber (right) before fusion.

16.7.1 Electro-Optic Overlay Intensity Modulators

Extensive theoretical and experimental studies have been carried out on the fiber optic components based on the principle of asymmetrical directional coupling. In these devices, one of the waveguides can be a D-shape fiber with its flat etched, or an optical fiber with its side cladding polished, to close proximity of the core, and the other is a planar waveguide, which regularly has higher refractive index. Efficient evanescent coupling from the fiber to the overlay waveguide occurs when one of the film waveguide modes, usually highest order mode, is matched to that of the fiber. Light coupled into the planar waveguide might not return back to the fiber waveguide and is absorbed, scattered, or irradiated away. The theoretical analysis on the coupling between a curved fiber and a planar waveguide is conducted either analytically [98–100] using the coupled-mode theory [101, 102] or numerically using the beam propagation method [103].

To understand the modulator characteristics governed by the physical parameters, let us omit the structure details such as the fiber curvature and derive analytical fields in the two compound waveguides using coupled-mode equations [101]

$$\frac{\partial}{\partial z} A_f(z) = -i\kappa A_{wg}(z) \exp{(i\Delta\beta z)}$$
$$\frac{\partial}{\partial z} A_{wg}(z) = -i\kappa A_f(z) \exp{(-i\Delta\beta z)} - \alpha_{wg} A_{wg}(z),$$

(16.12)

where, A_f is the amplitude of the field in the fiber, A_{wg} is the amplitude of the field in the planar waveguide, $\Delta\beta = \beta_f - \beta_{wg}$ is the difference between the propagation constant in the two waveguides, κ is the coupling coefficient between the waveguides, and α_{wg} is loss coefficient induced by the planar waveguide.

Assuming that all the light is in the fiber initially, that is, $A_f(0) = 1$ and $A_{wg}(0) = 0$, the solution to Eq. (16.12) is

$$A_f(z) = \frac{1}{K_r} \exp{[-(z/2)(\alpha_{wg} - i\Delta\beta)]} \cdot [(\alpha_{wg} - i\Delta\beta)\sinh{(zK_r/2)} + K_r\cosh{(zK_r/2)}]$$
$$A_{wg}(z) = \frac{-2i}{K_r} \exp{[-(z/2)(\alpha_{wg} + i\Delta\beta)]}\kappa^* \sinh{(zK_r/2)},$$

(16.13)

where, $K_r = [(\alpha_{wg} - i\Delta\beta)^2 - 4|\kappa|^2]^{1/2}$. The dependence of the modulator transmission, $A_f(L)$ on $\Delta\beta$ at different coupling efficiencies, κ, calculated according to Eq. (16.13) is plotted as Fig. 16.15. The transmission dip happens when the two waveguide modes match, $\Delta\beta$. The planar waveguide mode can be changed by varying one of the waveguide parameters such as thickness, refractive index, or wavelength. For EO modulation, by altering the refractive index of the EO

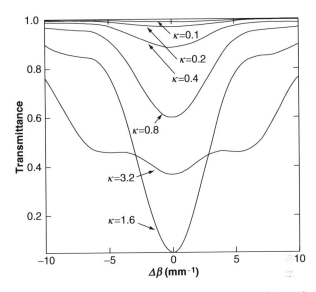

Figure 16.15 The calculated modulator transmittance as a function of $\Delta\beta$, using Eq. (16.12) with different k value.

waveguide, the phase-matching condition is changed, resulting in the modulation of the electric field in the fiber. Applied voltage V alters the refractive index n as $\Delta n = (1/2)\pi n^3 \lambda r(V/d)$, where r is the EO coefficient and d is the electrode's gap.

The overplayed EO waveguides were reportedly using polymer [103–105], lithium niobate crystal [106], liquid crystal [107, 108], and a GaAs/AlAs multiple quantum-well waveguides [103, 109]. Figure 16.16 shows a wavelength and switching response of this type of modulator [104]. A polished fiber half-coupler was made from a standard single-mode 1.3-μm optical fiber and a 1.5-mm thick corona poled EO polymer film, sandwiched with ITO electrodes, was fabricated on the flat. The indices for TM and TE polarization are 1.626 and 1.6339, respectively. The intensity dips correspond to the resonant phase matching of the fiber mode with the highest order mode of the film. Applying 400 V across the 1.5-μm thick film results in a 12-nm red shift on the resonant dip of the TM mode and gives a modulation depth of 15.5 dB (Fig. 16.16).

16.7.2 Replaced Cladding Phase Modulators

The idea for a replace-clad modulator on a core-exposed D-shape fiber is that the phase of the guiding light can be electrically manipulated if the material that replaced the original cladding material is EO active.

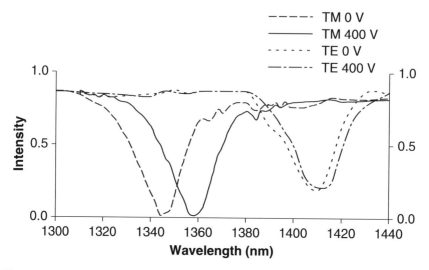

Figure 16.16 Wavelength and switching response for 1.5 μm polymer film, with bottom electrode thickness of 30 nm and top electrode thickness of 70 nm.

The insert of Fig. 16.17 shows the structure of the phase modulator. The upper half of the fiber clad of length L is replaced by an EO material, and two parallel metal electrode strips are fabricated along each side of the core with a separation gap d. When a modulating voltage, V, is applied between the electrodes, a phase delay $\Delta\phi$ is induced as

$$\Delta\phi(V) = \pi n^3 \lambda r \Gamma L(V/d). \tag{16.14}$$

To achieve a low insertion loss and a desirable phase modulation, the refractive index of the new cladding material should be lower than the index of the core but still reasonably large to maintain a low V value. The Γ value (<1) is determined by overlaps among the evanescent fields, the EO cladding material, and the applied electric field. If the value of V_b is close to 1, up to 30% of the modal power can be distributed in the upper EO material replaced cladding [54]. Either reducing the core size through wet-etching or replacing the cladding with a relatively higher index material can realize a lower V_b value waveguide.

Using a fluorine-containing polymer that incorporates EO chromophores (DR1) as a cladding material, low insertion loss phase modulators are fabricated [54]. The phase modulation produced by this modulator was measured using an open-loop interference fiber optic gyroscope (FOG). A π-phase modulation on a 25-mm long modulator was achieved. In this type of FOG configuration, a phase modulator is incorporated within the Sagnac interference loop asymmetrically with respect to the directional coupler. A sinusoidal modulating voltage

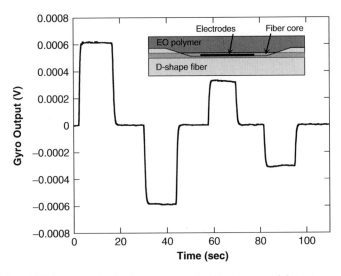

Figure 16.17 A FOG output signal using a replaced cladding phase modulator to supply a non-reciprocal dynamic phase bias for the measurement of Sagnac phase shift. The signal is measured when the rate table turns at +40, −40, +20, and −20 degrees/sec, sequentially. The insert schematically shows the side view of the modulator.

$V = V_0 \sin(\omega t)$ is applied to the modulator and results in a time-varying phase modulation according to Eq. (16.14) and can be expressed as $\phi(t) = \phi_0 \sin(\omega t)$. A differential phase delay between the counter-propagating waves is generated:

$$\Delta\phi = 2\phi_0 \sin(\omega\tau/2) \sin(\omega t), \qquad (16.15)$$

where τ is the light transition time through the loop. The interference signal induced by Sagnac phase shift ϕ_s, at the modulating frequency, ω, is

$$I(\omega)/I_{01} = 2J_1(\Delta\phi) \cos(\omega t) \sin\phi_s, \qquad (16.16)$$

where I_{01} is the light intensity propagating in one direction of the fiber and J_1 is the first order of Bessel function of first kind. The output of the FOG is a function of the phase modulation $\Delta\phi$ and the Sagnac phase shift. The gyro output is shown in Fig. 16.17. The intensity modulation in the phase modulation is better than 100 ppm.

16.7.3 Partial and Full Core-Replaced Devices

The D-shape of the elliptical core fiber also implies opportunities to make in-fiber devices by replacing the core by a functional material [39]. The techniques for partial and full core removal by HF acid wet-etch have been demonstrated

[41, 110]. To fabricate this type of device, the core of a section of the D-shape fiber is removed by dipping into the HF acid etchant. To deposit the new functional material, usually a polymer, into the core trough, a spin-coating technique is used to generate a smooth finishing surface (Fig. 16.18). Because regularly the refractive index of the material is higher than that of the fiber core, the process should be optimized for a low insertion loss device. It was discovered, through waveguide simulations and experiments, that a partially etched and partially filled core with high index polymer reduces the insertion loss. The etch depth of the core is chosen to allow for the formation of a single-mode polymer waveguide near the center of the fiber. The thickness of the polymer in the core should not be too thick; otherwise, a slab mode might be launched in the

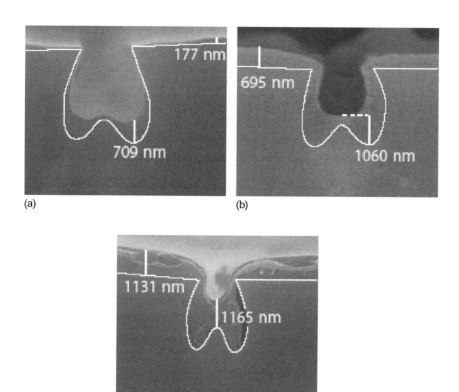

Figure 16.18 Cross-sectional SEM images of the partially removed core fibers filled with the EO polymer with its viscosity increasing from (a) to (c). The white lines were added to show the interfaces between the glass and polymer.

polymer film coated on the flat. The experiments also demonstrated that long transition distances between the unetched fiber and the polymer waveguide section reduce the transition loss, because the graduated transition allows an adiabatic energy transfer between two waveguides with different V values.

A 2-cm long partially removed core fiber was filled with a dispersed red 1 (DR1) chromophore blended polymethylmethacrylate (PMMA). The best loss achieved is approximately 1.6 dB. The polymer blend used is of DR1:PMMA ratio of 0.075:1 by weight and has a refractive index of 1.54 at a wavelength of 1550 nm. Figure 16.18 shows the SEM images of the cross-sections of the waveguides filled with different viscosities when the polymers are spin-coated. The insertion loss for (a), (b), and (c) are 1.6, 36, and ∞ dB, respectively [110]. Once the low-loss replaced core fiber has been achieved, the device can be used for phase, birefringence, or intensity modulation.

16.7.4 Fiber Bragg Grating Devices

The benefits of using D-shape fibers for fiber Bragg grating (FBG) fabrication and applications are twofold: The relative high germanium doping level in the fiber core is sensitive to the UV irradiation and etching the cladding from the flat direction enables direct access of the evanescent light. Both characteristics of the D-shape fiber are used to fabricate FBG devices [111, 112].

Bend sensors with direction recognition based on long-period gratings (LPGs) written in D-shape fiber is demonstrated [113, 114]. The fiber was photo-sensitized by H_2 loading before the inscription of a LPG structure with a period of 381 mm by use of a 244-nm UV laser and the point-by-point method [113]. The inscription results in two series of transmission peaks, corresponding to those of the two polarization modes. The retained transmission peak for the light polarization in the fast axis is at 1629.66 nm and has an extinction ratio of 8.48 dB. The transmission peak for the orthogonal slow axis is at 1645.43 nm and has an extinction ratio of 8.34 dB. The spectral positions of the peaks shifted when the grating fiber is bended, and the shifts linearly depend on the curvature of the bending (Fig. 16.19). The slope of the shift-curvature lines was found to be strongly dependent on the fiber orientation, because of the asymmetrical location of the core relative to the geometrical center of the cladding cross-section.

Short-period FBGs were etched on the flat of D-shape fiber for high-temperature sensing [115, 116]. To fabricate these gratings, the flat side cladding was first removed by wet etching to reach to the evanescent wave using BOE, to an extent that the distance from the flat to the top of the core is approximately 0.4 μm. A grating pattern of 534-nm period was written on a previously spun photoresist layer on the flat, by means of a two-beam interference using a

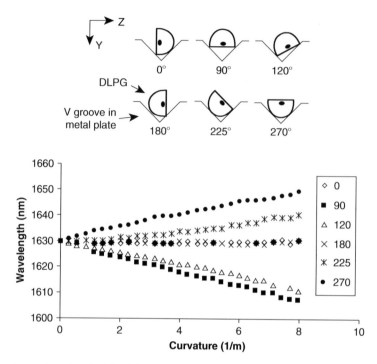

Figure 16.19 Wavelength shift of the transmission peaks of the long-period gratings inscribed in D-shape fiber versus curvature at different fiber orientations.

363.8-nm laser. The developed grating pattern was dry-etched using DRIE, and the sinusoidal pattern was then transferred onto the flat of the fiber (Fig. 16.20). The short-period gratings degenerate the two retained transmission peaks corresponding to the two light polarizations into one peak, which is spectrally located at 1552.6 nm with a extinction ratio of approximately 8 dB. In the temperature range 200–1100 °C, the temperature sensitivity of the spectrum shift is 16.2 pm/°C.

High-sensitivity optical chemsensors based on the LPG on D-shape fiber have also been demonstrated [117, 118]. The guided light was made accessible through wet-etching the D-shape fiber with LPG written to the core. As the flat is etched to a distance of 7.8 μm from the center of the core, the retained transmission peak shifts when the surrounding-medium refractive index changes. This characteristic enables concentration measurements of known chemicals when the index-percentage relation is calibrated. The peak location changes from 1545.4 to 1546.5 nm when the concentration of aqueous solutions of sugar changes from 0 to 60%, which corresponds to the refractive index from 1.33 to 1.44.

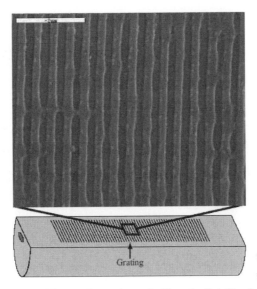

Figure 16.20 SEM image of a grating etched into the flat side of a D-fiber.

16.7.5 Variable Attenuators

An in-fiber variable attenuator can be built on a D-shape fiber as a type of the evanescent field device. If a bulky external material, whose refractive index is greater than the mode effective index, replaces a part of the evanescent field reachable cladding, the mode can become leaky and some of the optical power can be radiated [119]. If the index of the external material can be changed with a controllable mean, through the effects such as thermo-optic, electro-optic, or acoustic-optic, a device with controllable attenuation is achievable.

The combination of the fiber and the external bulk overlay has been modeled as an equivalent planar four-layer waveguide structure, so that the circular fiber can be replaced with an equivalent planar waveguide if the modal fields and propagation constant of the planar waveguide match those of circular fiber. Analytical expressions for the real and the imaginary parts of the propagation constant of the complex leaky mode have been derived [120].

Chandani and Jaeger [121] have studied the power-loss dependency on the refractive index of the external medium on a D-shape elliptic core fiber [121]. The fiber has indices of 1.441 and 1.475 at the operation wavelength of 1550 nm for cladding and core, respectively. Figure 16.21a shows the calculated power loss for this fiber according to Sharma et al. [120] and Thyagarajan et al. [122]. A 1-cm long section of a D-shape fiber is wet-etched in a 10% HF solution until the flat is

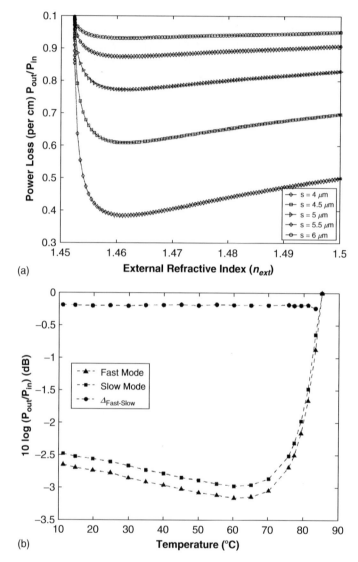

(a)

(b)

Figure 16.21 (a) Calculated insertion loss for TE mode versus the refractive index of the overlay materials, *n*, for different core-flat distance, *s*. (b) Measured fiber insertion loss for both modes for a fiber etched for 180 min as a function of temperature of immersion oil.

4–6 μm from the center of the elliptical core. The etched fiber was then immersed into oil with its index and its thermo-optic coefficient known. By controlling the oil temperature, the index of the oil is fine-tuned. The loss-temperature, therefore, loss-index, curve is well fitted using Sharma's theory in a temperature range of 10 ~ 90 °C, which corresponds to the index change of 1.4808 ~ 1.4499, respectively. The distance of the flat from the core center (4.705 μm) has been determined as a least-fit parameter. The power loss for the fast mode is approximately 0.2 dB larger than that for the slow mode in the whole index range, as shown in Fig. 16.21b. The purpose of the study is for a temperature sensor, although in-fiber optic attenuators have been proposed under the same principle.

16.7.6 Optical Absorption Monitoring

The D-shape fiber, with its evanescent field accessible by exposing the fiber core, is a reasonable approach to monitor the absorption spectra of gas and contaminations of an aqueous solution. Fiber optic evanescent wave absorption spectroscopy has been of research interest because of the many potential advantages over the conventional chemical-monitoring methods such as speed of response, selectivity, *in-situ* testing, and safety [123–125].

In the design of the sensors, several basic aspects have to be considered. First, most of the gases of interest for the environmental and safety measurements possess much stronger absorption lines in the mid-infrared (MIR) than in the near infrared. Although the low attenuation transmission window of silica-based optical fibers is in the range 0.6–2 μm—with particularly good transparency between 1 and 1.8 μm—special fibers, such as unclad silver halide fibers [126], chalcogenides [127], and fluoride glass fibers, were fabricated to extend an MIR transparent window to the gas absorption region, such as hexane, trichloro-trifluoro-ethane, methane, acetone, and others [128–132]. Chapter 13 of this handbook describes these types of fibers. However, the technological benefits then point towards the near-infrared system as offering many advantages when compared to the MIR equivalent despite the radically stronger line strengths in the longer wavelength region. Second, most systems operate with fairly small values of absorption and the detected strength is linearly proportional to the value of the line strength. The short exposure length of the side-clad polished fibers does not supply enough sensitivity. D-shape fibers enable a much longer length and higher degree of core exposure, which enhances the sensitivity in the sense of gas-field overlap and interaction length. It is also possible to draw D-shape fibers that make the evanescent field intrinsically exposed by grinding the flat of the preform to close proximity to the core. Bending a D-shape fiber further increases the sensitivity. A methane-sensing sensitivity at 1.66 μm is calculated to be increased as much as 30% by bending a D-shape fiber to a 12-cm radii curvature with its flat surface to the outside [130, 131].

An alternative method to enhance the sensitivity is by using air-guiding photonic band-gap (PBG) fibers [132, 133]. The gaseous species filled into the air holes of the fiber absorb the guided light through a long interaction length. A drawback of the PBG fiber sensors, comparing to the D-shape fiber sensors, is that the response time for the former should be much longer because of the small diameter of the air holes for the gas to be filled in. Third, the sensitivity also depends on the resolution or the spectral detection scheme. A narrow line width and wavelength-tunable light source of a high spectral resolution spectrometer is crucial. When the source line width exceeds the absorption line width, the maximum detected signal expressed as a fraction of the detected optical power decays linearly with the increase in source line width signal. Diode lasers with distributed feedback (DFB) architectures are the most convenient. However, their tuning ranges are relatively modest. The extended cavity air path tunable laser facilitates source tuning throughout the entire usable gain of the optically active elements. In the tunable fiber laser, the long cavity ensures narrow line widths and low noise operation.

Culshaw et al. [134] used a D-shape fiber evanescent wave optical fiber gas sensor to realize distributed methane detection using the single absorption line of 1.66 μm with resolutions of the order of 100 ppm methane. The D-shape fiber used was 5 m in length and specially designed with near-zero core/flat distance. The dependency of the detection sensitivity on the core size and the core/clad index difference was investigated.

16.7.7 Intrinsic Fiber Sensors

A unique characteristic of the highly elliptical core fibers is their supporting to two spatial modes in a range of wavelength in the two polarization direction, respectively [8, 135–137]. The two modes experience different phase shifts when the fiber is strained. The interference between the output lights in two modes supplies a sensitive way to measure the strain. The two-mode optical fiber is useful for building new types of fiber devices such as sensors that selectively sensitized for perturbations of interest and were immune to others. These perturbations may include strains, twists, temperature change, and so on.

16.7.7.1 Strain Sensor

In a range of wavelength elliptical core fibers exclusively support the fundamental LP_{01} and even LP_{11} spatial mode with Eigen-polarizations of both modes parallel to the major and minor core axes. When the LP_{01} and LP_{11} modes are launched equally in an elliptical core fiber, the far field output pattern is a superposition of the contributions from the two modes as a double loped modal interference pattern, and the its intensity distribution is depended on the phase difference, $\Delta\phi$, between the two modes (Fig. 16.22). Applied strain alters the differential phase and gives rise to a corresponding change in pattern. Because the orientation of the double lopes is determined by the orientation of the core axes, and both modes in the identical fiber without separate reference arms as needed in Mach–Zehnder or Michelson interferometers, the interference pattern is stable and less susceptible to the environmental noises. These make this type of sensor applicable for the remote sensing in the harsh conditions [135–144].

For a strain measurement, light with a known polarization state is launched into an elliptical core two-mode fiber with approximately equal intensities in each of the two spatial modes. After passing through the stretched section of the fiber, the light is projected onto a screen for the far-field interference pattern. The local variation of the intensity, as a function of differential phase, of the interference pattern can be written as

$$I \sim 1 + v\cos(\Delta\phi), \tag{16.17}$$

where v is the fringe visibility, depending on the launching conditions, detection area, and the location. The differential spatial modal phase $\Delta\phi$ is modified upon

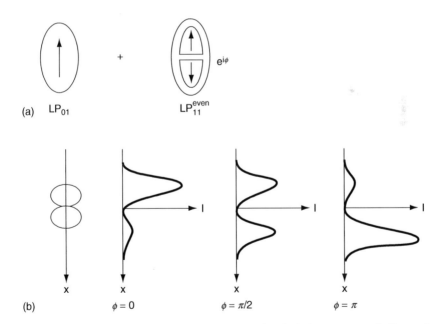

Figure 16.22 Evolution of the two-lope output pattern in elliptical-core, two-mode fiber-optic sensor. ϕ describes the phase difference between the LP_{01} and the LP_{11} modes.

the applied stretch and related to the modal birefringence $\Delta\beta = \beta_{01} - \beta_{11}$ and the length of the sensing fiber, l, by

$$\delta(\Delta\phi) = \Delta\beta\delta l + l\delta(\Delta\beta), \tag{16.18}$$

where the subscript *01* and *11* stand for the LP_{01} and the LP_{11} mode for a two-mode fiber, respectively. For a change in $\Delta\phi$ of 2π, there will be one complete oscillation of the intensity pattern. Measurement of the elongation $\delta l_{2\pi}$ required in a sensing fiber for a 2π change in $\Delta\phi$ calibrates the measurement sensitivity, the coefficient $\delta(\Delta\phi)/\delta l$ is then calculated from the measurement $2\pi/\delta l_{2\pi}$. Because $\delta\beta$ is a function of deformation δl, under a stable ambient temperature, the light intensity is solely a function of dl

$$I \sim 1 + V\cos\left(2\pi\frac{\delta l}{\delta l_{2\pi}} + \Delta\phi_0\right), \tag{16.19}$$

where $\Delta\phi_0$ is the unperturbed phase difference between the two spatial modes at the fiber end. Huang et al. [140] have measured β and $\Delta\beta$ as functions of the fiber elongation, dl, and the wavelength on an elliptical core fiber. By launching the two modes into both Eigen polarizations of the elliptical fiber, two independent spatial mode interferometers were effectively built, one for each polarization state. Because the strains coefficients induced by stretch and temperature change on the two spatial mode interferometers are different, simultaneous measurement for both fiber elongation and temperature change are possible.

Devices based on the two-mode optical fiber sensing are reported using the selective measurement of one type of strain perturbations, or simultaneous measurement of two types of the perturbations. Bohnert et al. [142] used two dual-mode fibers in tandem acting as unbalanced sensor and recovery interferometers. Using homodyne phase tracking, small optical-phase modulations induced by periodic strain were recovered. A stability of the detected AC signal is achieved to within $\pm 10^{-3}$ in the presence of additional quasi-static phase drift of large amplitude. The minimum detectable differential modal-phase modulation is $5\,\mu\text{rad rms}/\sqrt{\text{Hz}}$ at 70 Hz. A piezoelectric quartz high-voltage transducer has been demonstrated. A simultaneous strain and temperature measurement was realized by incorporation of both polarimetric and two-mode differential interferometric schemes in an elliptical core fiber. Using elliptical core fibers, strains due to stretch and temperature change can be measured simultaneously with resolutions of 10 μm/m and 5 °C, respectively [138]. A technique based on the evaluation of the condition number of a matrix is shown to be useful in evaluating comparative merits of multi-parameter sensing schemes.

Individually measuring the polarization changes in the LP_{01} and LP_{11} modes of an elliptical core fiber, a simultaneous recovery of temperature and the strain was realized [139]. To couple only the LP_{11} mode out of the fiber, an in-line mode splitter was developed by side-polishing the fiber and overlaying a prism with film interlay. When the index of the interlay is chosen correctly, 35-dB modal separation was enabled while preserving the polarization properties of both modes. The polarization changes of both modes were analyzed by inserting a polarizer with 45 degrees with respect to the core axes, respectively. The scheme has been demonstrated to be sensitive to changes of 1 °C and 5 $\mu\varepsilon$.

16.7.7.2 Twist Sensor

Twisting fiber induces circular birefringence, which can be used to measure angular displacements [145–150],

$$\Delta\alpha = g\tau \ \text{(rad/m)} \tag{16.20}$$

where g is a photoelastic coefficient, $g = -0.5n_0^2(p_{11} - p_{12}) \approx 0.146$ for silica fibers [147], p_{11} and p_{12} are components of the strain-optical sensor of the fiber, and n_0 is the average refractive index of the fiber.

The polarization evolution of the fundamental mode in a twisted single-mode fiber was obtained by using either the Poincaré sphere [146] or the coupled-mode equations [145]. If x′,y′ is the local coordinate that follows the principle axes of the twisted fiber, and x,y is that for the fixed principle axes of the fiber. Under assumptions that there is no coupling between the LP_{01} and LP_{11} modes and that fiber is uniformly twisted with a rate ϕ. The electric field components E'_x and E'_y for each spatial mode, at the output end of the fiber, are represented as [140, 145]

$$E^i_{x'} = \exp\left(\frac{-j}{2(\beta_{ix} - \beta_{iy})l}\right)$$
$$\times \left[E^i_{x0} \cos\frac{K_i l}{2} - j\frac{\Delta\beta_i}{K_i} E^i_{x0} \sin\frac{K_i l}{2} + \frac{\phi(2 - g_i)}{K_i} E^i_{y0} \sin\frac{K_i l}{2} \right] \tag{16.21}$$

and

$$E^i_{y'} = \exp\left(\frac{-j}{2(\beta_{ix} - \beta_{iy})l}\right)$$
$$\times \left[E^i_{y0} \cos\frac{K_i l}{2} + j\frac{\Delta\beta_i}{K_i} E^i_{y0} \sin\frac{K_i l}{2} - \frac{\phi(2 - g_i)}{K_i} E^i_{x0} \sin\frac{K_i l}{2} \right] \tag{16.22}$$

with $K_i = \sqrt{(\Delta\beta_i)^2 + [\tau(2 - g_i)]^{-2}}$, $I = 1,2$, where l is the fiber length and g_i is a photoelastic coefficient for ith mode. The E^i_{x0} and E^i_{y0} are electric fields in the two polarizations for the ith mode at the input end of the fiber. The local intensity at the output fiber end is

$$I = I_{x'} + I_{y'} = \left|E^1_{x'} + E^2_{x'}\right|^2 + \left|E^1_{y'} + E^2_{y'}\right|^2, \tag{16.23}$$

where I is a function of the twist rate, τ. Huang et al. [140] calculated and measured $I_{y'}$ of one lope of the pattern as a function of twist rate ϕ, on a 19.3-cm long elliptical core two-mode fiber, when only LP_{01}^x mode was excited [140]. With the twist rate ϕ increase, LP_{01}^y and LP_{11}^y modes were coupled and form interference fringes (Fig. 16.23). Mancier et al. [149] developed an optical fiber twist sensor for measuring angular displacements at low temperature. The sensing part is composed of a fiber coil rotated between two points, which induces a twist of two sections of the fiber. It is demonstrated that the sensor was able to take angular measurement over a 100-degree range with an accuracy of 0.2 degree. The thermal sensitivity is studied and it was concluded that the sensor has to work in stabilized temperature environments so $\Delta T < 5°C$ to keep the 0.2-degree accuracy.

16.7.8 D-Shape Fiber Opto-Electronic Devices

Novel D-shape fiber–based opto-electronic devices have been proposed and patented. For example, two-dimensional (2D) photonic crystal structures can be microfabricated on the D-shape fiber. If the silica vacant locations are filled with functional materials, exciting new opportunities for the D-shape fibers are expected [151]. Fiber optic filters and modulators based on the D-shaped fiber

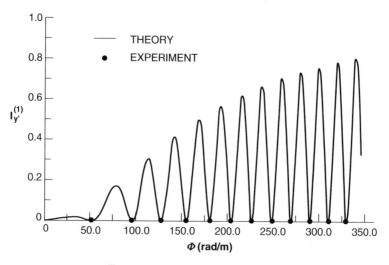

Figure 16.23   Normalized $I_{y'}^{(1)}$ as a function of twist rate ϕ for a 19.3-cm long Polaroid fiber at $\lambda = 514.5$ nm. The solid line represents the theoretical curve. The dots represent experimental nulling points.

are patented. A photonic crystal structure as an overlay on the side-polished D-shape fiber is proposed [152]. The devices can be tunable if electro-optic active materials are used. A method to fabricate electrodes on D-shape fibers has also been proposed and patented [153].

16.8 RARE EARTH-DOPED ELLIPTICAL CORE FIBER

Among the several methods reported to introduce rare earth dopant into the fiber, the solution doping technique, developed by Townsend et al. [154] in 1987, is the simplest and, therefore, the most popular. The MCVD process for the germania core deposition is carried out at a reduced temperature so that a partially sintered germania layer inside the quartz tube is formed. A solution of salt of the rare earth (usually aqueous or alcoholic solutions of nitrate of chloride salts) soaks into porous soot and then is glassed at a higher temperature. The tube is then collapsed to form preform. Passing chlorine through the heated tube before the glassification reduces the OH content considerably. The concentration is limited by the clustering of the rare earth ions, which quench the photo-excited ions. Co-doping with aluminum isolates the rare earth ions by forming a salvation shell at each neodymium ion and allows for a high doping level of the rare earth ions without clustering [155]. A 33-dB small signal gain was reported on a 23-m long Al–Nd co-doped fiber under 50-mW pump power, whereas the small signal gain was only 7 dB for the Nd-only–doped fiber under the same pumping conditions [156].

Several milliwatts of superfluorescent light at approximately 1080 nm were measured when a 9-m long fiber was forwardly pumped by an 820-nm LD that delivers approximately 20-mW pump power into the fiber. A fiber optic gyroscope for finding true north was demonstrated at KVH using this superfluorescent light source. Gain anisotropy was observed in a 300-ppm Nd-doped fiber with core size of 2.5 by 1.25 μm and core-cladding index difference of 0.032. Under the measurement conditions of 21-mW pump power at 810 nm and 450 μW seeding power at 1088 nm, the small signal gains were 3.4 and 3.1 dB, respectively, when the polarization directions for the pump and seeding light are in both the major and the minor axes [157–159]. The anisotropic behaviors of the gains are consistent with a model based on stronger confinement of the odd HE_{11} mode. Microsecond optical-optical switching in this type of fiber was demonstrated [160]. In this device, the effective indices of a two-mode fiber interferometer operated at 633 nm were resonantly modulated using an 807-nm pump laser, which results in switching between the two interferential lopes.

Erbium- and ytterbium-doped PM fibers were fabricated and studied, mainly motivated as a potential, but expensive, candidate to solve signal distortion problem caused by the polarization dispersion [161]. A 14-dB net gain was reported in a 27-m long erbium-doped PANDA-type PM fiber. The fiber transmitted only one polarization mode in the emission wavelength of 1.535 μm while supporting two polarization modes at the pump wavelength of 1.485 μm [162]. A high-power PM amplifier was demonstrated using an Yb-doped bowtie fiber. Backward-pumped by a high-power 974-nm diode bar, a 9.5-m long fiber had an amplified seed signal at a wavelength of 1050 nm with approximately 40 dB small-signal gain. The saturated output power of 2.3 W and extinction ratio of 17 dB was demonstrated at pump power of 5.7 W [163].

One of the important applications of the rare earth-doped fiber is as a broadband superfluorescent light source, particularly for navigation-grade FOG. The advantage of the rare earth-doped, especially erbium-doped, fiber over the other existing broadband light sources is its excellent mean wavelength stability. Together with its low pump power requirement, the broadband light produced by rare earth-doped fiber meets the stringent requirements for inertial navigation-grade FOG [164–167].

REFERENCES

[1] Lyubimov, L. A. et al. 1961. Dielectric waveguide with elliptical cross-section. In: *Radio Engineering and Electronics*. 6:1668–1677.

[2] Yeh, C. 1962. Elliptical dielectric waveguides. *J. Applied Physics* 33:3343–3352.

[3] Dyott, R. B. and J. R. Stern. 1971. Group delay in glass fiber waveguides. *Electronics Lett.* 7:82–84.

[4] Schlosser, W. O. 1972. Delay distortion in weakly guiding optical fibers due to elliptical deformation of the boundary. *Bell System Technical J.* 51:487–492.

[5] Ramaswamy, V. et al. 1978. Polarization characteristics of non-circular-core single mode fibers. *Applied Optics* 17:3014–3017.

[6] Dyott, R. B. et al. 1979. Preservation of polarization in optical fiber waveguides with elliptical cores. *Electronics Lett.* 15:380–382.

[7] Dyott, R. B. 1995. *Elliptical Fiber Waveguides.* Artech House, Massachusetts.

[8] Kim, B. Y. et al. 1987. Use of highly elliptical core fibers for two-mode fiber devices. *Opt. Lett.* 12:729.

[9] Dyott, R. B. and P. F. Shrank. 1982. Self-locating elliptical cored fiber with an accessible guiding region. *Electronics Lett.* 18:980–981.

[10] Dyott, R. B. 1987. Method for making self-aligning optical fiber with accessible guiding region. U.S. Patent 4,668,264, granted May 26, 1987.

[11] Markos, D. J. et al. 2003. Controlled core removal from a D-shaped optical fiber. *Applied Optics* 42:7121–7125.

[12] Lowder, T. L. et al. Surface relief Bragg gratings using a rotated core D-fiber. Submitted.

[13] Dyott, R. B. 1990. Method of joining self-aligning optical fibers. U.S. Patent 4,950,318, granted August 21, 1990.

[14] Monte, T. D. 1998. Grooved optical fiber for use with an electrode and a method for making same. U.S. Patent 5,768,462, granted June 16, 1998.

[15] Pleibel, W. and R. H. Stolen. 1981. Polarization-preserving optical fiber. U.S Patent 4,274,854, granted June 23, 1981.

[16] Kumar, A. and R. K. Varshney. 1984. Propagation characteristics of highly elliptical core optical waveguides: A perturbation approach. *Optical Quantum Electronics* 16: 349–354.

[17] Eyes, L. et al. 1979. Modes of dielelectric waveguides of arbitrary cross sectional shape. *J. Optical Soc. Am.* 69:1226–1235.

[18] Skinner, I. M. 1986. Simple approximation formulae for elliptical dielectric waveguides. *Optical Quantum Electronics* 18:345–353.

[19] Eguchi, M. and M. Koshiba. 1994. Accurate finite-element analysis of dual-mode highly elliptical core fibers. *J. Lightwave Technol.* 12:607–613.

[20] Koshiba, M. 1990. *Optical Waveguide Analysis*. McGraw-Hill, New York.

[21] Adams, M. J. et al. 1979. Birefringence in optical fibers with elliptical cross section. *Electronics Lett.* 15:62–63.

[22] Varshney, R. K. et al. 1988. Characterization of highly elliptical submicron core polarization preserving fibers: Theory and experiment. *Applied Optics* 27:3114–3120.

[23] Rashleigh, S. C. 1982. Wavelength dependence of birefringence in highly birefringent fibers. *Optic Lett.* 7:294–296.

[24] Zhang F. and J. W. Lit. 1993. Temperature and strain sensitivity measurements of high-bi-refringent polarization-maintaining fibers. *Applied Optics* 32:2213–2218.

[25] Urbanczyk, W. et al. 2001. Dispersion effects in elliptical-core highly birefringent fibers. *Applied Optics* 40:1911–1920.

[26] Shibata, N. et al. 1982. Polarization mode dispersion measurement in elliptical core single mode fibers by a spatial technique. *IEEE J. Quantum Electronics* 18:53–58.

[27] Kaminow, I. P. 1981. Polarization in optical fibers. *IEEE J. Quantum Electronics* 17:15–22.

[28] Rashleigh, S. C. et al. 1982. Polarization holding in birefringent single-mode fibers. *Optic Letters* 7:40–42.

[29] Yamashita, E. et al. 1979. Modal analysis of homogeneous optical fibers with deformed boundaries. *IEEE Trans. Microwave Theory Techn.* 27:352–356.

[30] Saad, S. M. 1985. On the higher order modes of elliptical optical fibers. *IEEE Trans. Microwave Theory Techn.* 33:1110–1112.

[31] Rengarajan, S. R. 1989. On higher order mode cut-off frequencies in elliptical step index fibers. *IEEE Trans. Microwave Theory Techn.* 37:1244–1248.

[32] Dyott, R. B. 1990. Cut-off of the first higher order modes in elliptical dielectric waveguide: An experimental approach. *Electronics Lett.* 26:1721–1722.

[33] Shaw, J. K. et al. 1991. Direct numerical analysis of dual-mode elliptical-core optical fibers. *Opt. Lett.* 16:135–137.

[34] Klein, K. F. and W. E. Heinlein. 1982. Orientation- and polarisation-dependent cutoff wavelengths in elliptical-core single-mode fibres. *Electronics Lett.* 18:640–642.

[35] Blake, J. N. et al. 1987. Elliptical core two-mode fiber strain gauge. *SPIE Fiber Optic Laser Sensors* 838:332–339.

[36] Dyott, R. B. 1994. Composition of LP_{11} modes in elliptically cored fibre. *Electronics Lett.* 30:993–994.

[37] Blake, J. et al. 1992. Splitting of the second order mode cut-off wavelengths in elliptical core fibers. *IEEE Conf. Proc. 8th Optical Fiber Sensors*, pp. 125–128. OSA—Opt. Soc. of Am.

[38] Eguchi, M. and M. Koshiba. 1995. Behavior of the first higher-order modes of a circular core optical fiber whose core cross-section changes into an ellipse. *J. Lightwave Technol.* 13:127–136.

[39] Dyott, R. B. 2001. Method of incorporating optical materials into an optical fiber. U.S. Patent 6,718,097.

[40] Moar, P. N. et al. 1999. Fabrication, modeling, and direct evanescent field measurement of tapered optical fiber sensors. *J. Appl. Phys.* 85:3395–3398.

[41] Smith, M. S. 2001. The fabrication of an integrated active waveguide in a D-shaped optical fiber using the centrifugal casting method. Master degree dissertation of Brigham Young University.

[42] Chan, K. B. et al. 1970. Propagation characteristics of an optical waveguide with a defused core boundary. *Electronics Lett.* 6:748–750.

[43] Lin, C. Y. and L. A. Wang. 1999. Loss-tunable long period fibre grating made from etched corrugation structure. *Electronics Lett.* 35:1872–1873.

[44] Dyott, R. B. et al. 1987. Indium-coated D-shaped-fiber polarizer. *Optic Lett.* 12:287–289.

[45] Eickoff, W. 1980. In-line fibre-optic polariser. *Electronics Lett.* 16:762–764.

[46] Moodie, D. G. and W. Johnstone. 1993. Wavelength tunability of components based on the evanescent coupling from a side-polished fiber to a high-index-overlay waveguide. *Opt. Lett.* 18:1025–1027.

[47] Pangaribuan, T. et al. 1992. Reproducible fabrication technique of nanometric tip diameter fiber probe for photon scanning tunneling microscope. *Jpn. J. Applied Phys.* 31:L1302–L1304.

[48] Stockle, R. et al. 1999. Wild, high quality near-field probes by tube etching. *Appl. Phys. Lett.* 75:160.

[49] Ohtsu, M. 1998. *Near-Field Nano/Atom Optics and Technology.* Springer-Verlag, Tokyo.

[50] Mononobe, S. and M. Ohtsu. 1996. Fabrication of a pencil-shaped fiber probe for near-field optics by selective chemical etching. *J. Lightwave Technol.* 14:2231–2235.

[51] Mermelstein, M. D. 1986. All-fiber polarimetric sensor. *Appl. Opt.* 25(8):1256–1258.

[52] Jensen, M. A. and R. H. Selfridge. 1992. Analysis of etching induced birefringence changes in elliptic core fibers. *Appl. Opt.* 31:211–216.

[53] Ashkin, A. and R. H. Stolen. 1986. Exposed core optical fibers, and method of making same. U.S. Patent no. 4,630,890.

[54] Wang L. and T. D. Monte. "All Fiber Replaced Optical Cladding Phase Modulator for FOG Applications and Beyond. In: *Proceedings of in 18th International Conference on Optical Fiber Sensors.* OSA—Opt. Soc. of Am. Caneun, Mexico.

[55] Dyott, R. B. and J. Bello. 1983. Polarization holding directional coupler made from elliptically cored fiber having a D section. *Electronics Lett.* 19:607–608.

[56] Dyott, R. B. et al. 1984. Polarization holding directional coupler using D fiber. *Proc. SPIE* 479:23–27.

[57] Nayar, B. K. and D. R. Smith. 1983. Monomode-polarization-maintaining fiber directional couplers. *Optic Lett.* 8(10):543–545.

[58] Handerek, V. H. and R. B. Dyott. 1985. Fused D-fibre couplers. *SPIE Proc.* 574.

[59] Georgiu, G. and A. C. Boucouvalas. 1985. Low-loss single-mode optical couplers. *IEEE Proc.* 132(5).

[60] Cryan, C. V. et al. 1993. Low loss fused D fibre couplers. *Electron. Lett.* 29:1432–1433.

[61] Mortimore, D. B. 1988. Fiber loop reflectors. *J. Lightwave Technol.* 6(7):1217–1224.

[62] Miller, I. D. et al. 1987. A Nd+ doped c.w. fiber laser using all-fiber reflectors. *Electronic Lett.* 23:329–332.

[63] Dickinson, G. et al. 1992. Properties of the fiber reflection Mach-Zehnder interferometer with identical couplers. *Optic Lett.* 17(17):1192–1194.

[64] Han, Y. et al. 1999. Architecture of high-order all-fiber birefringent filters by the use of the Sagnac interferometer. *IEEE Photon. Technol. Lett.* 11(1):90–92.

[65] Bergh, R. A. et al. 1980. Single-mode fiber-optic polarizer. *Opt. Lett.* 5:479–481.

[66] Hosaka, T. et al. 1982. Single-mode fiber-type polarizer. *IEEE J. Quantum Electron.* 18:1569–1572.

[67] Dyott, R. B. and R. Ulrich. 1986. U.S. Patent no. 4,589,728. *Optical Fiber Polarizer.*

[68] Polky, J. N. and G. L. Mitchell. 1974. Metal-clad planar dielectric waveguide for integrated optics. *J. OSA* 64(3):274–279.

[69] Johnstone, W. et al. 1990. Surface plasmon polarizations in thin metal films and their role in fiber optic polarizing devices. *J. Lightwave Technol.* 8(4):538–544.

[70] Feth, J. R. and C. L. Chang. 1986. Metal-clad fiber-optic cutoff polarizer. *Opt. Lett.* 11:386.

[71] Dyott, R. B. et al. 1987. Indium-coated D-shaped-fiber polarizer. *Optic Lett.* 12(4): 287–289.

[72] Dyott, R. B. 1987. Indium-clad fiber optic polarizer. U.S. Patent no. 4,712,866; granted December 15, 1987.

[73] Varnham, M. P. et al. 1984. Coiled-birefringent-fiber polarizers. *Optic Lett.* 9:306–308.

[74] Liu, K. W. et al. 1986. Single-mode-fiber evanescent polarizer/amplitude modulator using liquid crystals. *Optic Lett.* 11(3):180–182.

[75] Creaney, W. et al. 1994. Low loss fibre optic polarisers using differential coupling to dielectric waveguide overlays. *Electronics Lett.* 30(4):349–351.

[76] Aervas, M. N. and I. P. Giles. Optical-fibre surface-plasmon-wave polarizers with enhanced performance. *Electronics Lett.* 25(5):321–323.

[77] Kumar, A. et al. 1989. Relative transmission loss of TE- and TM-like mode is metal coated coupler halves. *Electronics Lett.* 25(5):301–323.

[78] Zervas, M. N. and I. P. Giles. 1990. Performance of surface-plasma-wave fiber-optic polarizers. *Optic Lett.* 15(9):513–515.

[79] Zervas, M. N. 1990. Surface plasmon-polarization fiber-optic polarizers using thin chromium films. *IEEE Photonics Technol. Lett.* 2(8):597–599.

[80] Todd, D. A. et al. 1993. Polarization-splitting polished fiber optic couplers. *Optical Engineering* 32(9):2077–2082.

[81] Presby, H. M. and C. A. Edwards. 1992. Efficient coupling of polarization-maintaining fiber to laser diodes. *IEEE Photon. Technol. Lett.* 4:897–899.

[82] Hunziker, W. et al. 1992. Elliptically lensed polarization maintaining fibres. *Electronic Lett.* 28:1654–1656.

[83] Jopson, R. M. et al. 1985. Microlens for stressed cladding polarization preserving fiber. *Electronic Lett.* 21:758–759.

[84] Eisenhauser, W. and H. Richter. 1987. Light coupling from a semiconductor laser into a polarization-maintaining single-mode fiber. *Electron. Lett.* 23:201–202.

[85] Helleso, G. et al. 1992. Diode laser butt coupling to K+ diffused waveguides in glass. *IEEE Photon. Technol. Lett.* 4:900–902.

[86] Donhowe, M. N. and R. G. Hunsperger. 1988. Coupling non-circular core optical fibers to laser diode. *Proc. SPIE* 988:201–208.

[87] Kogelnik, H. and T. Li. 1966. Laser beams and resonators. *Appl. Opt.* 5:1550.

[88] Wooten, E. L. et al. 2000. A review of lithium niobate modulators for fiber-optic communications systems. *IEEE J. Selected Topics Quan. Electron.* 6:69–82.

[89] Caponio, N. and C. Svelto. 1994. A simple angular alignment technique for a polarization-maintaining-fibre. *IEEE Photonics Technol. Lett.* 6:728–729.

[90] Aalto, T. T. et al. 2003. Method for the rotational alignment of polarization-maintaining optical fibers and waveguides. *Opt. Eng.* 42:2861–2867.

[91] Walker, G. R. and N. G. Walker. 1987. Alignment of polarization maintaining fibres by temperature modulation. *Electron. Lett.* 23:689–691.

[92] Ebberg, A. and R. Noe. 1990. Novel high precision alignment technique for polarization maintaining fibres using a frequency modulated tunable laser. *Electron. Lett.* 26:2009–2010.

[93] Ida, Y. et al. 1985. New method for polarization alignment of birefringent fibre with laser diode. *Electron. Lett.* 21:18–21.

[94] Takada, K. et al. 1987. Precise method for angular alignment of birefringent fibers based on an interferometric technique with a broadband source". *Appl. Opt.* 26:2079–2987.

[95] Lefevre, H. C. and H. J. Arditty. 1993. Integrated optical components and digital-processing techniques. In: *Optical Fiber Rotation Sensing* (W. K. Burns, ed.). Academic Press, Boston, San Diego, New York, London, Sydney, Tokyo, Toronto.

[96] Kincaid, B. E. 1988. Coupling of polarization-maintaining optical fibers to Ti:LiNbO3 waveguides with angled interfaces. *Opt. Lett.* 13:425–427.

[97] Moyer, R. S. et al. 1998. Design and qualification of hermetically packaged lithium niobate optical modulator. *IEEE Trans. Components Packaging Manufact. Technol. B* 21:130–135.

[98] Marcuse, D. 1989. Investigation of coupling between a fiber and an infinite slab. *J. Lightwave Technol.* 7:1222–1301.

[99] Panajotov, K. P. et al. 1994. Distributed coupling between a single-mode fiber and a planar waveguide. *J. Opt. Soc. Am. B* 11:826–834.

[100] Panajotov, K. P. 1994. Polarization properties of a fiber-to-asymmetric planar waveguide coupler. *J. Lightwave Technol.* 12:983.

[101] Hamilton, S. A. et al. 1998. Polymer in-line fiber modulators for broadband radio-frequency optical links. *J. Opt. Soc. Am. B* 15:740–750.

[102] Yariv, A. 1973. Coupled-mode theory for guided-wave optics. *IEEE J. Quant. Electron.* QE-9:919.

[103] Arft, C. et al. 2000. In-line fiber evanescent field electrooptic modulators. *J. Nonlinear Opt. Phys. Mater.* 9:79–94.

[104] Fawcett, G. et al. 1992. In-line fiber-optic intensity modulator using electro-optic polymer. *Electron. Lett.* 28:985–986.

[105] Pan, F. et al. 1999. Waveguide fabrication and high-speed in-line intensity modulation in 4-N,N-4′-dimethylamino-4′-N′-methyl-stilbazolium tosylate. *Appl. Phys. Lett.* 74: 492–494.

[106] Creaney, S. et al. 1996. Continuous-fiber modulator with high-bandwidth coplanar strip electrodes. *Photon. Technol. Lett.* 8:355–357.

[107] Nayar, B. K. et al. 1986. Electro-optic monomode fibre devices with liquid crystal overlays. *Tech. Digest 12th European Conf. On Opt. Comm.* 1:178.

[108] Ioannidis, Z. K. et al. 1991. All-fiber optic intensity modulators using liquid crystals. *Appl. Opt.* 30:328–333.

[109] Mao, E. et al. 1999. GaAs/AlGaAs multiple-quantum-well in-line fiber intensity modulator. *Appl. Phys. Lett.* 75:310–312.

[110] Smith, K. H. et al. 2004. Fabrication and analysis of a low-loss in-fiber active polymer waveguide. *Appl Opt.* 43:933–939.

[111] Jensen, M. A. and R. H. Selfridge. 1992. Analysis of diffraction gratings based on D-shape fibers. *J. Opt. Soc. Am. A* 9:1086–1090.

[112] Psaila, D. C. et al. 1997. Comb filters based on superstructure rocking filters in photo-sensitive optical fibers. *Opt. Commun.* 141:75–82.

[113] Zhao, D. et al. 2004. Bend sensors with direction recognition based on long-period gratings written in D-shaped fiber. *Appl. Opt.* 43:5425–5428.

[114] Allsop, T. et al. 2006. The spectral sensitivity of long period gratings fabricated in elliptical core D-shaped optical fibre. *Opt. Comm.* 259:537–544.

[115] Lowder, T. L. et al. 2005. High-temperature sensing using surface relief fiber Bragg gratings. *IEEE Photon. Technol. Lett.* 17:1926–1931.

[116] Smith, K. H. et al. 2006. Surface relief fiber Bragg gratings for sensor applications. *Appl. Opt.* 45:1669–1775.

[117] Zhou, K. et al. 2004. High-sensitivity optical chemsensor based on etched D-fibre Bragg gratings. *Electron. Lett.* 40:232–234.

[118] Chen, X. et al. 2004. Optical chemsensors utilizing long-period fiber Gratings UV-inscribed in D-fiber with enhanced sensitivity through cladding etching. *IEEE Photon. Technol. Lett.* 16:1352–1354.

[119] Wagoner, G. A. et al. 1999. Fiber optic attenuators and attenuation systems. U.S. Patent no. 5,966,493.

[120] Sharma, A. et al. 1990. Analysis of fiber directional couplers and coupler half-blocks using a new simple model for single-mode fibers. *J. Lightwave Technol.* 8:143–151.

[121] Chandani, S. M. and N. A. F. Jaeger. 2005. Fiber-optic temperature sensor using evanescent fields in D fibers. *IEEE Photon. Technol. Lett.* 17:2706–2708.

[122] Thyagarajan, K. et al. 1987. Analytical investigations of leaky and absorbing planar structures. *Optical Quantum Electron.* 19:131–137.

[123] Jin, W. et al. 1997. Prospects for fibre-optic evanescent-field gas sensors using absorption in the near-infrared. *Sensors Actuators B* 38–39:42–47.

[124] Jin, W. et al. 1995. A liquid contamination detector for D-fibre sensors using white light interferometry. *Meas. Sci. Technol.* 6:1471–1475.

[125] Jin, W. et al. 1993. Absorption measurement of methane gas with a broadband light source and interferometric signal processing. *Opt. Lett.* 18:1364.

[126] Messica, A. et al. 1994. Fiber-optic evanescent wave sensor for gas detection. *Opt. Lett.* 19:1167.

[127] Aggarwal, I. D. et al. 2002. Development and applications of chalcogenide glass optical fibers at NRL. *J. Optoelectronics Adv. Mat.* 4:665:678.

[128] Tai, H. et al. 1987. Fiber-optic evanescent-wave methane-gas sensor using optical absorption for the 3.392-Mu m line of a He-Ne laser. *Opt. Lett.* 12, 437.

[129] Taga, K. et al. 1994. Fiber optic evanescent field sensors for gaseous species using MIR transparent fibers. *Analyt. Bioanalyt. Chem.* 348:556–559.

[130] Muhammad, F. A. and H. S. Al-Raweshidy. 1995. Analysis of curved D-fiber for methane gas sensing. *IEEE Photon. Technol. Lett.* 7:538–539.

[131] Muhammad, F. A. et al. 1998. Polarimetric optical D-fiber sensor for chemical applications. *Microwave Opt. Technol. Lett.* 19:318–321.

[132] Ritari, T. et al. 2004. Gas sensing using air-guiding photonic bandgap fibers. *Optics Express* 12:4080–4087.

[133] Jensen, J. B. et al. 2004. Photonic crystal fiber based evanescent-wave sensor for detection of biomolecules in aqueous solutions. *Opt. Lett.* 29:1974–1976.

[134] Culshaw, B. et al. 1992. Evanescent wave methane detection using optical fiber. *Electron. Lett.* 28:2232–2234.

[135] Uttam, D. et al. 1985. Interferometric optical fibre strain measurement. *J. Phys. E: Sci. Instrum.* 18:290–293.

[136] Eftimov, T. A. and W. J. Bock. 1998. Analysis of the polarization behavior of hybrid modes in highly birefringent fibers. *J. Lightwave Technol.* 16:998–1005.

[137] Blake, J. N. et al. 1987. Strain effects on highly elliptical core two-mode fibers. *Opt. Lett.* 12:732.

[138] Vengsarkar, A. M. et al. 1994. Fiber-optic dual-technique sensor for simultaneous measurement of strain and temperature. *J. Lightwave Technol.* 12:170–177.

[139] Thursby, G. et al. 1995. Simultaneous recovery of strain and temperature fields by the use of two-moded polarimetry with an in-line mode splitter/analyzer. *Opt. Lett.* 20:1919.

[140] Huang, S.-Y. et al. 1990. Perturbation effects on mode propagation in highly elliptical core two-mode fibers. *J. Lightwave Technol.* 8:23–33.

[141] Murphy, K. A. et al. 1990. Elliptical-core two mode optical-fiber sensor implementation methods. *J. Lightwave Technol.* 8:1688–1696.

[142] Bohnert, K. et al. 1995. Coherence-tuned interrogation of a remote elliptical-core, dual-mode fiber strain sensor. *J. Lightwave Technol.* 13:94–103.

[143] Wang, A. et al. 1995. Two-mode elliptical core optical fiber sensors for strain and temperature measurement. *Smart Mater. Struct.* 4:42–49.

[144] Bock, W. J. and T. A. Eftimov. 1993. Simultaneous hydrostatic pressure and temperature measurement employing a LP01-LP11 fiber-optic polarization-sensitive intermodal interferometer. *Proc. IEEE Instrum. Meas. Technol. Conference, Irvine, CA,* pp. 426–429.

[145] Okoshi, T. et al. 1981. Measurement of polarization parameters of a single-mode optical fiber. *J. Opt. Commun.* 2:134.

[146] Ulrich, R. and A. Simon. 1979. Polarization optics of twisted single-moded fibers. *Appl. Opt.* 18:2241.

[147] Rashleigh, S. C. 1983. Origins and control of polarization effects in single-mode fibers. *J. Lightwave Technol.* LT-1:312–331.

[148] West, S. T. and C.-L. Chen. 1989. Optical fiber rotary displacement sensor. *Appl. Opt.* 28:4206.

[149] Mancier, N. et al. 1995. Angular displacement fiber-optic sensor: Theoretical and experimental study. *Appl. Opt.* 34:6489.

[150] Wolinski, T. R. 1994. Stress effects in twisted highly birefringent fibers. *Proc. SPIE* 2070:392–403.

[151] Koops, H. W. P. and G. Meltz. 2000. In-fiber photonic crystals and systems. U. S. Patent no. 6,075,915.

[152] Chen, H. K. et al. 2006. Fiber-optic tunable filter and intensity modulator. U.S. Patent no. 7,024,072.

[153] Park, S. et al. 2003. Method for forming two thin conductive films isolated electrically from each other on a fiber. U.S. Patent no. 6,625,361.

[154] Townsend, J. E. et al. 1987. Solution-doping technique for fabrication of rare-earth–doped optical fibres. *Electron. Lett.* 23:329–331.

[155] Kiiveri, P. and S. Tammela. 2000. Design and fabrication of erbium-doped fibers for optical amplifiers. *Opt. Eng.* 39:1943–1950.

[156] Miyazaki, T. et al. 1994. Neodymium-doped fibre amplifier at 1.064 μm. *Electron. Lett.* 30:2142–2143.

[157] Koplow, P. et al. 2000. Polarization-maintaining, double-clad fiber amplifier employing externally applied stress-induced birefringence. *Opt. Lett.* 25:387–389.

[158] Srinivasan, B. et al. 1994. *Polarization Properties of Fiber Lasers Based on Rare-Earth-Doped Polarization Preserving Fibers,* CLEO 94, *Technical Digest,* Paper CTuK79, 8:131.

[159] Srinivasan, B. et al. 1995. Direct measurement of gain anisotropy in elliptical core fiber amplifiers, CLEO '95, *Technical Digest,* Paper CMB8 15, p. 9.

[160] Sadowski, R. W. et al. 1993. Microsecond optical–optical switching in a neodymium-doped two-mode fiber. *Opt. Lett.* 18:927.

[161] Saito, S. et al. 1990. An over 2,200 km coherent transmission experiment at 2.5 Gbit/s using erbium-doped-fiber amplifier. Technical Digest, OFC'90, San Francisco, Paper PD-2.

[162] Tajima, K. 1990. Er^{3+}-doped single-polarisation optical fibers. *Electron. Lett.* 26:1498–1499.

[163] Kliner, D. A. V. et al. 2001. Polarization-maintaining amplifier employing double-clad bow-tie fiber. *Opt. Lett.* 26:184–186.

[164] Shi, L. et al. 2006. Broadband Er3+–Yb3+ co-doped superfluorescent fiber source. *Opt. Comm.* 257:270–276.

[165] Flaquler, D. G. et al. 2002. Polarization and wavelength stable superfluorescent sources. U.S. Patent Application Publication no. 2002/0154384.

[166] Digonnet, M. J. F. and D. G. Falquler. 2002. Polarization and wavelength stable superfluorescent sources using Farady rotator mirrors. U.S. Patent no. 6,483,628.

[167] Wysocki, P. F. et al. 1994. Characteristics of erbium-doped superfluorescent fiber sources for interferometric sensor applications. *J. Lightwave Technol.* 12:550–567.

Chapter 17

Multimode, Large-Core, and Plastic Clad (PCS) Fibers

Bolesh J. Skutnik and Cheryl A. Smith

Ceram Optec Industries, Inc., East Longmeadow, Massachusetts

17.1 INTRODUCTION

After years of playing ensemble roles, large-core multimode fiber has stepped into the spotlight of fiber optic technology and innovation. From the smallest of veins in the human body, to the vastness of the universe, when the need for every photon matters, the advantages of large-core (>200 micron) multimode specialty fibers are taking the lead. As the name implies, multimode fibers are those types of fibers designed to carry multiple rays of light or modes. There are two types of multimode fibers: step index and graded index. For purposes of this chapter, we discuss the types and applications of large-core step-index multimode optical fibers.

Many industrial and medical applications require a range of geometries, clad–core ratios, and numerical apertures (NAs) for step-index multimode fibers depending on whether the end-use is for laser surgery, illumination, or sensing. Fiber core geometries can range from 100 μm to more than 1000 μm, and the clad–core ratios can range from 1.05 to more than 1.20. In general, the larger the NA available, the smaller the clad–core ratio or the smaller the fiber core can be. Smaller cores and core–clad ratios lead to lesser expense for materials and more flexible fibers.

Smaller dimensioned optical fibers also permit the use of smaller catheters, enabling associated surgery procedures to be less invasive. Small systems also can require broader illumination from optical fibers that may be minimized in number or in size. For ultraviolet (UV) applications, pure silica core all-silica optical fibers are the more reliable and have the best transmission. Generally high-power transmission also requires the excellent chemical stability of all-silica optical fibers. In the past, all-silica fibers were restricted to NAs of 0.22 or less. Early on, pure silica core and doped silica clad fibers of this NA were not very thermally stable for large diameter sizes, for example, much above 800-μm cores. The thermal problems were related to the interface between the doped and undoped silicas and, over time, were solved so that today 0.22 NA fibers with cores much greater than 1 mm are available with suitable thermal stability. An NA of 0.22 has an acceptance angle of about 25 degrees.

Medical applications for lasers and optical fibers continue to grow and evolve over the years. Much of this growth is spurred by the development of more minimally invasive procedures, which can benefit from small-diameter fibers to deliver high radiation energy from laser sources in a variety of emission patterns. These applications also benefit from using the low intrinsic loss character of silica-based core material, as well as the power capability of a silica/silica construction. Pure silica core all-silica optical fibers are now available with an NA of 0.30 + 0.02. Variations include fibers with nonsolarizing UV transmission, as well as fibers with transmission through the near-infrared (NIR) region of the electromagnetic spectrum. Additionally ultrahigh NA fibers with silica cores and silica/silica structures are now available for use in the visible and NIR regions with effective NAs higher than 0.6. Properties of these fibers are presented and the advantages over other fibers and potential medical applications are discussed in the following sections.

To produce a step-index multimode fiber, a core material of silica (either pure or doped) is clad with a lower index material (doped silica, hard plastic, plastic) to form a waveguide, as illustrated in Fig. 17.1.

These fibers will have a protective jacket beyond the cladding that does not effect the transmission of light through the fiber, although there are additional

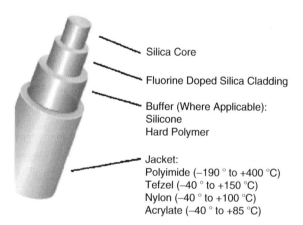

Figure 17.1   Schematic representation of a step-index fiber.

coatings and buffer layers that can be added to change the NA of a fiber. The NA of the fiber is calculated as

$$NA = [n_{core}^2 - n_{clad}^2]^{1/2}. \qquad (17.1)$$

Step-index fibers will only propagate light that enters the fiber within its acceptance angle. For certain applications, there are great benefits to either increasing or reducing the NA and thereby changing the acceptance angle.

The fiber-pure synthetic fused silica core can be of a high OH content for applications in the deep UV to visible (VIS) wavelengths or a low OH content for use in the VIS to NIR wavelengths. The low OH silica core can be doped to produce fibers with very high NAs. Silica is a good material based on its optical and thermal properties. It can be produced synthetically with ultrahigh purity and can operate from less than 200 to more than 2400 nm with little absorption (Fig. 17.2).

The cladding materials can be doped silica, hard plastic, or plastic. The combination of the silica core and various cladding options offers a multitude of fiber products for a wide range of applications. The upper limit number of modes that can be carried in a step-index fiber is known as the normalized frequency parameter, or V number. It is calculated as follows:

$$V = (2\pi a/\lambda)NA \qquad (17.2)$$

17.2 LARGE-CORE SILICA/SILICA (ALL-SILICA) FIBER

A pure fused silica core with doped silica clad produces an industry-standard fiber with an NA of 0.22. These fibers are available with core diameters from 50 to 2000 microns. Silica/silica fibers can offer excellent transmission from the deep UV to the NIR, along with a good focal ratio degradation, which is important in

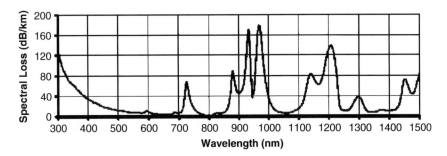

Figure 17.2 Ultraviolet–visible–near-infrared (UV-VIS-NIR) transmission spectrum for a high-OH silica/silica fiber.

spectroscopy applications, especially in astronomy. Individual silica/silica fibers with polyimide coatings can handle temperatures of 400 °C and in fused bundle configurations can reach close to 1500 °C at the fused ends. Applications in medical/pharmaceutical, forensic, sensors, remote detection, or monitoring of hazardous environments all benefit from the use of optical fibers, and all-silica fibers are best suited to provide this mechanism, especially in the UV region. The use of excimer lasers and strong UV light sources has grown in medical and industrial fields, as has the number of spectroscopic techniques that use UV absorbance and luminescence measurement to characterize material. Although some commercially available fibers can handle transmission of low intensities of laser radiation, there still exist difficulties for high-power radiation transmission. These standard fibers offer low attenuation and high transmission in the 215- to 254-nm spectral range, but on exposure to unfiltered deuterium lamps sources, the fibers drop to less than 50% transmission within 24 hours of continuous irradiation. Standard UV fibers develop color centers when subjected to pulsed excimer laser radiation (193 nm). This solarization issue has been virtually eliminated by the development of Optran UVNS UV Non-Solarizing fibers. The UVNS fibers exhibit only minor changes in transmission when exposed to unfiltered deuterium lamp sources, and while the standard synthetic silica fibers developed color centers within 10,000 pulses of excimer laser radiation, the UVNS fibers remained virtually unchanged [1], as can be seen in Fig. 17.3. Although these fibers are drawn using standard techniques, the preforms are produced using a proprietary procedure for the modified plasma chemical vapor deposition of silica. Although the first Optran UVNS fibers had an NA of 0.22, fibers have been developed with NAs of 0.26–0.30, allowing sampling of larger areas and greater collection of transmitted or reflected beams from material under test. The acceptance circle at a fixed distance from the fiber end increases dramatically from the 0.22 to the 0.30 NA, with the 0.30-NA fiber having an acceptance circle that is 86% larger. The long-term stability in UV applications

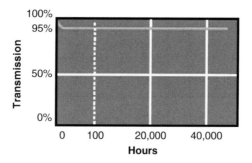

Figure 17.3   Transmission versus time for UVNS nonsolarizing silica/silica fiber.

and larger NAs extend the range of usefulness for these fibers in medical and spectroscopic applications. The larger NA allows the use of a smaller fiber core, which reduces the cost and increases the flexibility of the fibers, especially for remote detection, sensing, and medical applications requiring high-power densities such as excimer laser angioplasty and the perforation of the heart muscle [2].

Nonsolarizing silica/silica optical fibers can be fused to form a pattern of fibers slightly deformed into a hexagonal shape that produces a tight-packed structure with minimal dead space. These bundles are excellent replacements for epoxy bundles or liquid-light guides, as they provide higher transmission over the wavelength range and can withstand temperatures up to 1500 °C; the maximum for an epoxy bundle is 400 °C, and the liquid-light guides withstand less than 50 °C. The bundles can be used with high-power lasers, pump diode lasers, and high-intensity UV light sources. Applications include high-temperature sensing, illumination, spot curing, and wafer fabrication. When these bundles are produced with nonsolarizing fibers, the lifetime for the bundles is greatly improved over that of liquid-light guides and the transmission can be as much as 50% higher (Figs. 17.4 and 17.5). Because the bundles can be produced in lengths of up to 20 m, they offer an efficient cost-effective solution for remote spectroscopy. The active area for these bundles can be as small as 0.8 mm and as large as 20 mm.

Fiber optic bundles with epoxy ends have almost limitless room for design. From a common end of bundled fibers (virtually no limit to the size of the active area), almost any number of legs can be broken out for either the distribution of light to or the collection of light from a source. These bundles can be produced with fibers with core diameters from 50 μm to 1.5 mm in randomized or mapped distribution. The design of the bundle depends on its intended use and can be configured with the fibers in a spot, rectangular, linear, circular, or almost any other geometry. The available options allow for applications from spectroscopy to instrumentation to industrial monitoring and sensing. Rugged jacketing materials are available for field applications including mining and downhole

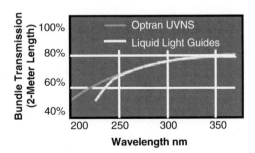

Figure 17.4 Optical transmission of UVNS bundle versus a liquid-core light guide.

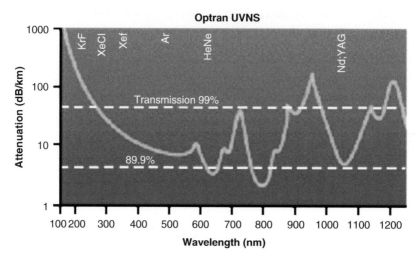

Figure 17.5   Transmission spectrum of a UVNS silica/silica fiber.

sensing. Bundles of small fibers are very flexible, with the bend radius being based on the diameter of a single fiber, not the entire bundle. This allows instrumentation designers the flexibility of locating equipment out of harms way and routing only the fibers. Industrial, spectroscopic, aircraft, military, space, tactical, and hazardous sensing applications all benefit from the limitless design potential available.

17.3 HIGH NA AND LOW NA SILICA/SILICA FIBERS

Many medical and sensing applications have need of "broad" irradiation patterns but benefit from small-diameter fibers to provide minimal invasive surgery. These applications also benefit from using the low intrinsic loss character of silica-based core material and the power capability of a silica/silica construction. Pure silica core all-silica optical fibers are now available up to an NA of 0.30 ± 0.02. Variations include fibers with nonsolarizing UV transmission and fibers with transmission through the NIR region of the electromagnetic spectrum. Additionally ultrahigh NA fibers with silica cores and silica/silica structures are now available for use in the VIS and NIR regions, with effective NAs higher than 0.6. Medical applications require a range of geometries, clad–core ratios, and NAs for step-index multimode fiber depending on whether the end-use is for laser surgery, illumination, or sensing. Fiber core geometries can range from 100 μm to more than 1000 μm, and the clad–core ratios can range from 1.05 to more than 1.20. In general, the larger the NA available, the smaller the clad–core ratio or the smaller the fiber core can be. Smaller cores and core–clad ratios lead to lesser incurred materials expense and more flexible fibers. Smaller dimensioned optical fibers also permit the use of smaller catheters, enabling associated surgery procedures to be less invasive. Small systems also can require broader illumination from optical fibers, which may be minimized in number or in size. For UV applications, pure silica core all-silica optical fibers are the most reliable and have the best transmission. Generally high-power transmission also requires the excellent chemical stability of all-silica optical fibers. In the past, all-silica fibers were restricted to NAs of 0.22 or below. Early on, pure silica core and doped silica clad fibers of this NA were not very thermally stable for large-diameter sizes (e.g., much above 800-μm cores). The thermal problems were related to the interface between the doped and undoped silicas and over time were solved, so today 0.22-NA fibers with cores much greater than 1 mm are available with suitable thermal stability. An NA of 0.22 has an acceptance angle of about 25 degrees. Medical applications for lasers and for optical fibers continue to grow through time. Much of this growth is spurred by the development of more minimally invasive procedures, which puts greater demands on using the smallest feasible optical fibers and systems. At the other end of the spectrum of uses are the new medical applications/procedures that use short pulsed radiation at very high power levels and power densities. Large-diameter fibers are often used because of the power densities. Even here, the ability to lower core sizes is welcomed because of their improved handling characteristics [2].

All-silica fibers for use in the VIS to NIR wavelengths can now be produced with NAs as high as 0.53. For NAs up to 0.30, the fiber construction employs a pure silica core with doped silica clad. For NAs of 0.37 and higher (up to 0.53), the silica core is doped. High-power laser diodes typically operate in the VIS to NIR ranges of the spectrum, and as a result, fibers with doped cores can be used for most applications. The increased NAs of these fibers correspond to an acceptance area that is up to 550% larger (comparing 0.22–0.56; see Fig. 17.6). The ability to provide a smaller fiber that is able to capture all of the lasers' output power without the use of lenses or additional optical components allows for a more reliable system (fewer components) within a smaller package. The higher NA of these fibers offers a benefit in photodynamic therapy (PDT) and diagnostic applications. The diffusers used to distribute light to the diseased tissue take light from the higher order modes traveling in a fiber, and the higher the NA, the greater the potential for harvesting these modes near the cladding–core interface. On the diagnostic side, the broader acceptance angle of the fibers allows for the most efficient collection of the luminescence.

The ultrahigh NA fibers described can be used as delivery fibers, especially for high-power diode laser systems. The benefits arise because of requirements of

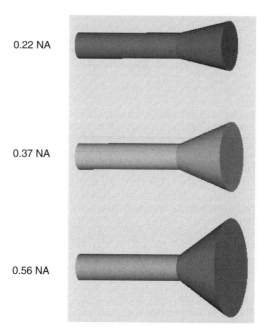

0.22 NA

0.37 NA

0.56 NA

Figure 17.6   Acceptance cones for 0.22, 0.37, and 0.56 numerical aperture fibers.

phase space to allow reduction in size from the dimension of a bundle of coupling fibers to a delivery fiber size. In other words, the product of NA and fiber bundle size is equal to the product of the delivery fiber size and its NA [3].

Other uses of the large NA fibers are in illumination applications, especially in cases in which the fiber end is not in air. For example, fewer illumination fibers might be used in an ophthalmology application if the NA of the fiber can be more than 0.50. Hands-free helmet-type illuminators are another area that benefits from being able to use these ultrahigh NA optical fibers.

Medical applications that generally need to treat larger areas and whose light sources are not in the UV can be more efficiently performed with larger NA fibers, as more area is covered by the fiber output. Examples might be PDT, wound healing, and general interstitial radiation therapy. Whether the medical action is by photons directly or indirectly as converted to thermal phonons, these applications benefit greatly from large NA optical fibers. Some aspects of tissue welding that are shared with wound healing such as a need to treat areas much larger than the output of a standard optical fiber can easily be seen to benefit from fibers with larger NA values.

A special point with reference to the ultralow OH grades of these fibers is that they can be used in medical applications with lasers or other sources operating at wavelengths above 2 μm. Figure 17.7 depicts the mid-IR transmission spectrum for one such fiber. Some variations have also been used to transmit radiation at wavelengths as high as 2.4 μm. They provide good transmission in a very desirable wavelength and with the ability to maintain high-power densities.

Photons are available from sources other than lasers. Coupling photonic energy in many cases using lamps, high-brilliance LEDs, or other high, power LEDs can be a challenge, because the sources often have broad beams and are projected in highly divergent beams from the source. Rather obviously, optical fibers with large to ultralarge NAs would be a benefit in capturing the photons and transmitting them to some remote application area, such as inside a patient or to several patients in adjacent stations, rooms, or beds.

In summary, optical fibers are now available for use or in the design of photonic treatment systems that have the following properties: NA values up to 0.30 for pure silica core, fluorosilica-doped cladding, high or low OH, in nonsolarizing UV grades; NA values up to 0.56 for germanium-doped core, fluorosilica-doped cladding, low to ultralow OH grades. These open up more efficient uses of fiber optics and photonics in a wide range of medical applications and treatments [4].

There are all-silica fibers available with very low NAs (down to 0.11), which allows for the coupling of narrow active area devices. These have applications in delivery of power from laser diodes and narrow band devices. The advantages of silica coupled with the options of very low to very high NA allow for the development of an ever-expanding list of applications never considered.

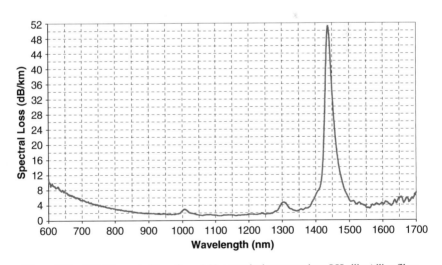

Figure 17.7   Mid-IR spectral loss for a 0.28 numerical aperture low-OH silica/silica fiber.

17.4 PLASTIC AND HARD POLYMER CLAD SILICA FIBERS

17.4.1 Plastic Clad Silica Fibers

Plastic clad silica (PCS) fibers have structures with a silica glass core surrounded by a thin plastic (silicone) cladding material (Fig. 17.8). Oftentimes, a protective jacket made from polymeric materials such as Tefzel is also applied. As the cladding material is not UV cured in these fibers, they will have better transmission in the UV-wavelength range while offering the advantage of being less expensive than all-silica fiber designs. PCS fibers can be more difficult to terminate, as the fiber core can piston from within the cladding. The NA of this fiber is 0.40 in short lengths.

17.4.2 Hard Polymer Clad Silica

Hard polymer clad silica (HPCS) fibers have emerged over the last few decades as an option for many applications in the medical, industrial, scientific, and military markets. The fiber structure is generally a pure fused silica core with a cladding of a thin hard polymer material and an outer jacket.

The HPCS fibers are less expensive than all-silica fiber constructions and offer the benefits of high strength, lower static fatigue, less strain at the core–clad interface, high core-to-clad ratios, and lighter weights. The hard polymeric cladding remains on the fiber during terminations, thereby maintaining the high strength of the fiber.

The HPCS optical fibers function well over a wide range of temperatures. Samples of fiber exposed to liquid nitrogen temperatures (−196 °C) and below were used to carry spectroscopic information from materials held at these

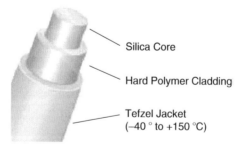

Figure 17.8 Schematic representation of a plastic clad silica (PCS)–hard polymer clad silica (HPCS) fiber.

temperatures. On the other end, the fibers are usable up to 125 °C almost continuously. Of more interest in medical applications is that the static fatigue behavior of these fibers remains predictable and unchanged even in moist (steam) environments.

The fibers are available in both high and low OH for operation in the UV, VIS, and NIR regions. The HPCS fibers will transmit UV wavelengths but use at less than 400 nm is hampered by the absorption of the hard cladding, as depicted in Figs. 17.9 and 17.10. However, developments have led to HPCS fibers with attenuations less than 1 dB/m even at 300 nm and less than 1.5 dB/m at 275 nm [5].

The properties of this fiber design offer advantages in a variety of applications. The ease of termination and high strength make the fiber suitable for short data links, especially in harsh environments including those with exposure to radiation. Because the fibers have a high core-to-clad ratio and the cladding remains on during termination, it is possible to do terminations on the fiber with very little loss due to fiber core mismatches. Fibers can be cut and connectorized in field environments, allowing for great flexibility in military, mining, and oil-field sensing applications. In the medical market, the mechanical, optical, and structural properties of the fiber are especially useful in the design of laser delivery, endoscopic, and biosensing systems. High core-to-clad ratio provides better coupling, ability to accept higher energy density, and reduced losses due to bending or flexing. In the competitive world of medical disposables, HPCS fibers offer an advantage in cost as well. Industrial applications for the fibers include sensors for indicating distance, temperature, proximity, liquid levels, and

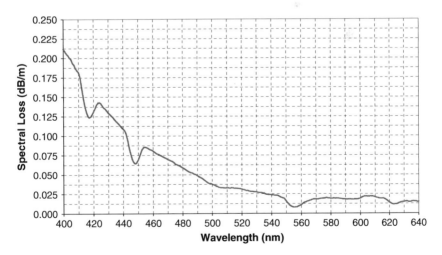

Figure 17.9 Ultraviolet (UV) optical transmission for a hard polymer clad silica (HPCS) fiber.

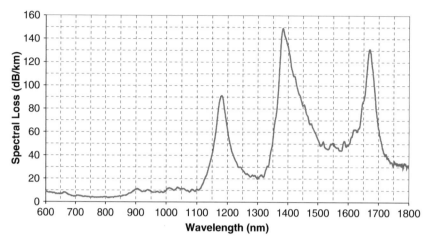

Figure 17.10   Near-infrared (NIR) optical transmission for a hard polymer clad silica (HPCS) fiber.

short-haul data links, often to areas with hazardous or extremely harsh environments. Use of HPCS fiber in the military includes the initiation of explosives, tactical short-link communications, and vehicular systems. The initiation of explosives with fiber optics was one of the earliest applications for HPCS fibers and remains in use in both mining and military applications. The automotive industry has investigated the use of HPCS fibers for applications that require higher temperature and strength than is available in all-plastic fiber designs.

The HPCS fibers are available with NAs of 0.37–0.48, and there is ongoing research for increasing these numbers. Constant investigation of new cladding materials offers the opportunity for even higher reliability fibers in terms of strength and fatigue. The unique properties of HPCS fibers will continue to offer benefits to an ever-increasing number of innovative new applications.

17.5 SILICA FIBERS WITH NANO-POROUS CLADDING/COATING

Unlike all-silica fibers, which require a jacket or buffer material to protect the outer layer from potential damage due to environmental exposure to moisture, fibers are being produced with a nano-porous cladding that requires no additional jacket. Using modified sol-gel technology, the cladding is produced on line from an oligimeric organo-silicate. "Sol-gel" technologies transition a liquid "sol" into solid "gel" form and allows for great flexibility in design. The objectives of this research are the development of radiation-resistant multimode fibers for space applications, fiber optic lines that could have signal taken off without creating a break in the main line, and biomedical/technological applications in which the fibers could be used as a tip for PDT. Placement of sensing materials, even high toxic or active ones, could be achieved by including them in the nano-porous cladding. They could be placed in an inactive form and then remotely activated by a photonic signal. Some specific applications could include direct treatment of body tissues and fluids, sensing the delivery dosages of radiation procedures, and locating specific tissues, such as cancerous tissue with minimally invasive techniques. Tests have indicated that complexes such as rhodamine can be activated by signals traveling in the fiber core after incorporation of the complex in a section of the modified sol-gel clad fiber. The guided waves within the fiber core extend into the cladding for some distance—where the complex is incorporated into the cladding—thus, interacting with it and activating it. Further studies on these effects will allow for the design of sensors for various chemical and biochemical moieties. Development of high-strength fibers with the potential of having modified sections interact with the surrounding environment is an exciting advancement for sensing and biomedical applications [6].

17.6 UNLIMITED APPLICATION POTENTIAL

In this chapter, we have indicated the construction of specialty, large-core, step-index multimode fibers and the variations of manufacture that make possible fibers with choices for wavelength performance and lifetime, NA, strength and fatigue, and high coupling efficiency in everything from a single strand to a bundle of thousands of fibers. Much like an artist choosing a palette, designers now have options to use fibers for applications never considered. On the most basic level, a fiber optic moves light, for a reason as simple as illumination to as complex as the composition of a star.

In the same way, every human has a fingerprint unique to him or her, every element has a signature that can be defined spectrally as belonging only to that element. A spectrometer can identify the elements present in an object by studying the makeup of the light emitted. Fiber can be the bridge between the mystery of an object's composition and the key to unlocking that mystery, the spectrometer. By deciphering the spectral response, whether it be the emission of light from a star, a sample of groundwater, or human cells, we can begin to identify the physical properties of the material. The study of these spectra (spectroscopy) is used in physical and analytical chemistry to identify a substance by the spectra emitted. In astronomy, a telescope can use a spectrograph to measure the chemical composition and physical properties or to measure the speed of astronomical objects. The various disciplines of spectroscopy (UV, VIS, IR, Raman, etc.) all benefit from the availability of specialty optical fibers and assemblies. The advancements in specialty optical fibers make it possible to engineer probes capable of delivering and collecting light in completely new and innovative ways.

Regardless of the end-use, delivery of light from a source is the first application of optical fibers. Specialty step-index multimode fibers offer the means of coupling very wide or very narrow beams of light over a wide spectrum for delivery to the other side. Individual fibers or bundles of fibers can be coupled to light sources for the illumination of materials or samples or the delivery of a laser beam. The use of fiber optics in the medical world is explosive and the most widely known use of the specialty fiber to the general public.

Specialty fiber optics coupled with laser technology now allow for minimally invasive surgical procedures that would have required open surgeries. Urology (soft tissue and lithotripsy), dentistry, ear, nose, and throat, ophthalmic, orthopedic, gynecology, and vascular applications all benefit from the use of specialty fibers. Shorter recovery times with less pain, better wound healing and blood coagulation, and less chance of infection are all byproducts of the development of these fibers.

PDT is generally used to treat hyperproliferative tissue diseases, including dysplasia. Such diseases commonly affect extended volumes of tissue but from a patient standpoint are relatively localized. In the normal application of PDT to these diseases, the light, with appropriate wavelength for the photosensitizer being used, is transmitted to the treatment site through optical fibers that are terminated at their distal end with diffusers. Diffusers may be lenses, elongated sections used to scatter light sideways or special tips that deflect energy primarily around the fiber tip rather than forward. In normal operation, the diffusers used to distribute light from the optical fiber transmission medium to the diseased tissue take light from the higher order modes traveling in the fiber and disperse them into the surrounding tissue. The higher the NA of the optical fiber, the greater the potential for having most of the light in higher order modes near the cladding–core interface, where they can be more easily harvested for treating the diseased tissue. Overlaunching treatment light into the fibers helps to populate the higher order modes. The number of modes in the fibers grows much faster than linearly as the NA increases for a fixed core size and operating wavelength(s). Small gains NA, thus, can have great benefits.

PDTs will offer the world a better standard of care for many types of cancer. Photosensitive drugs delivered to a site in the body and activated by a laser delivered through a fiber will revolutionize the way we view the treatment of disease [3].

Specialty fiber optic products are a means to many ends. From the scientific investigation of everything from groundwater to the makeup of the universe to the opportunity to initiate explosives without the loss of human life to the health concerns of our world, specialty fiber plays a leading role, a role that will continue to expand to meet the ever-changing requirements of the future.

REFERENCES

[1] Skutnik, B. J. and B. Foley. 1999. All silica, non-solarizing optical fibers for UV medical application. *SPIE Proc.* 3595:133–139.
[2] Skutnik, B. J. et al. 2004. Reliability of high NA, UV non-solarizing optical fibers. reliability of optical fiber components, devices, systems, and networks II. *SPIE Proc.* 5465:250–256.
[3] Skutnik, B. J et al. 2004. Optical fibers for improved light delivery in photodynamic therapy and diagnosis. Optical methods for tumor treatment and detection: Mechanisms and techniques in photodynamic therapy XIII. *SPIE Proc.* 5315:107–112.
[4] Skutnik, B. J. et al. 2004. High numerical aperture silica core fibers. Optical fibers and sensors for medical applications IV. *SPIE Proc.* 5317:39–45.
[5] Skutnik, B. J. et al. 2005. Hard plastic clad silica for near UV applications. Optical fibers and sensors for medical applications V. *SPIE Proc.* 5691:23–29.
[6] Skutnik, B. J. 1999. *Micro Pourous Silica: The All New Silica Optical Fiber.* American Chemical Society, Fan National Meeting, Conference Proceedings, Paper 694, New Orleans, LA.

Chapter 18

Tapered Fibers and Specialty Fiber Microcomponents

James P. Clarkin

Polymicro Technologies, LLC, Phoenix, Arizona

18.1 INTRODUCTION

In applications that utilize specialty optical fibers, very often there are requirements placed on the optical fiber design that are not conducive to the launch or delivery requirements placed on the proximal and distal ends of the fiber.

For example, in high-power medical applications, such as laser angioplasty or laser lithotripsy, a small-fiber diameter may be required so the fiber probe assembly will be able to bend easily inside the small blood vessels of the body. However, this may cause difficulty on the laser input launch (proximal) end of the fiber because the smaller core diameter offers a smaller target for the laser. Adjusting the optical path via lenses can make compensation, but this often increases the cost of the system and makes the laser-to-fiber alignment less robust for day-to-day use.

A solution to this problem can be modification of the fiber proximal and/or distal (output) ends into a shaped optical microcomponent such that the desired optical path is achieved without addition of expensive, bulky, or otherwise troublesome components to the system. The microcomponents discussed here are not an addition of discrete optical components, but actual changes made to the optical fiber material shape. Although the microcomponent modification may increase the cost of the fiber optic probe assembly, it may decrease the overall system cost because of fewer optical components or smaller fiber size, may increase in system reliability/performance, or may simply be an enabling technology where larger optical components are not feasible in the application's environment.

579

Shaped fiber microcomponents are useful devices for medical and industrial applications that require high-power laser delivery (material or tissue cutting), even light distribution over a broad area (tissue ablation or photodynamic therapy), modified beam divergence or spot size (materials processing and communications links), or optical power redirection from the axis of the fiber in an area with small space restrictions (tissue ablation or perforations inside the human body).

Fiber microcomponents have been successfully used for many years. Various fiber microcomponent shapes and sizes can be used to reshape the beam pattern of the light entering or exiting an optical fiber. The fiber core diameters used range from 200 to more than 1000 μm and are typically fabricated from fibers with glass core–glass clad designs, although glass core–plastic clad silica (PCS) designs can also be used. Most commonly the fiber end tips themselves are machined or sculpted using the glass material of the fiber itself. No additional glass material is needed in the process. The process can be either mechanical or thermal, with the latter being primarily but not limited to laser machining. Because the fiber end shapes are fabricated directly from the glass material of the fiber, the interface between the shape and the fiber itself is eliminated. Thus, there are no coupling losses between the microcomponent and the fiber. They are materially continuous. Furthermore, having no interface, there is no potential for contamination that might exist if the shape were bonded by fusion splicing or epoxy to the fiber. This serves to reduce optical losses and dramatically increase the mechanical strength and durability of the device.

A wide variety of fiber microcomponents have been used. The basic fiber microcomponent design categories are

- Tapers
- Lenses
- Diffusers
- Side-fire and angled ends

There is, of course, a great deal of design variations within these categories (Fig. 18.1).

Common factors in the target application that may dictate the use of any of these microcomponents include

- Minimum bend radius the fiber must be capable of achieving
- Space restrictions in the laser work area
- Launch numerical aperture (NA) of the source
- Size, shape, and optical power density of the input spot
- Wavelength of operation
- Output pattern required
- Direction of output beam

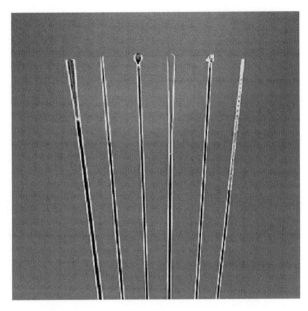

Figure 18.1 Examples of fiber microcomponents.

The operational parameters first depend on whether the fiber microcomponent will be used as an input or an output device. Obviously, the launch conditions will be most critical when the tip is used on the proximal (input) end and output conditions more critical when used on the distal (output) end. The key operational parameters of concern are summarized in Table 18.1.

A detailed explanation of each microcomponent design and function follows. Emphasis is made on tapers, because they are most commonly used with specialty optical fibers.

Table 18.1

Key operational parameters for proximal and distal fiber microcomponents

Launch conditions (proximal)	Output conditions (distal)
Input NA	Output NA
Spot size	Spot size
Power density	Power density
Modal distribution	Modal distribution
Alignment	Fiber NA
Back-reflection	Space restrictions

A discussion of specialty fiber microcomponents would not be complete without commenting on their roll in microfluidic detection technologies. This is discussed in Section 19.6, later in this chapter. Of key interest here is the application of fiber microcomponents for detection, identification, and quantification of biomolecules. Inclusive to this is the microfabrication of the detection window itself, which results in a flow cell.

18.2 TAPERS

A taper is either an enlargement or reduction in the fiber core diameter over a length of the fiber and can be used on either the proximal or the distal end. The purpose of the taper can be to passively alter the input or output divergence (i.e., NA) with regard to the optical fiber or to alter the optical power density at the fiber's proximal surface or output target area.

Despite the funnel-like appearance of the taper, the well-known optical concept of conservation of brightness prevents the taper from behaving as a magical light "funnel" that forces light from a large fiber into a small fiber. There is a price to be paid in transmission through the taper, and this price is paid in NA. The tapers actually change the NA of the light as it travels down the taper, losing light that exceeds the critical angle for total internal reflection in the optical fiber [2, 3]. When light travels down a taper from larger to smaller diameter ("down" taper), the angle the light makes with the fiber axis will increase (Fig. 18.2).

In most cases, the tapers are fabricated from a fiber with a glass core and glass cladding (glass/glass). The clad diameter–to–core diameter ratio generally remains constant through the taper and the fiber, so the taper actually has a glass cladding layer. Tapers can also be made on PCS fibers. In this case, the plastic cladding is removed in the taper area. Because the cladding around the taper is, therefore, air or epoxy, these tapers tend to be more limited in power capability, transmission, and spectral range as compared to a comparably sized glass/glass fiber. However, in many cases the performance of the PCS fiber tapers is good enough and the cost of the PCS fiber can be 2–10 times less expensive as compared to the glass/glass.

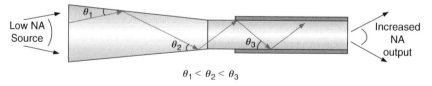

$$\theta_1 < \theta_2 < \theta_3$$

Figure 18.2   Down taper: numerical aperture increases as light travels through the taper.

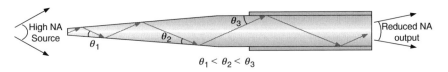

$$\theta_1 < \theta_2 < \theta_3$$

Figure 18.3 Up taper: numerical aperture decreases as light travels through the taper.

Conversely, as light travels up a taper from smaller to larger diameter ("up" taper), the angle will decrease (Fig. 18.3). This is important when considering the NA of the other components of the system and the desired output NA.

Laser systems, which produce high peak powers, can be difficult to couple into fiber because of the high-power densities involved. Typically, the bulk fiber material can withstand the high-power density, but the fiber end surface is the weak link and may become damaged due to surface contamination, end-face defects, or dielectric breakdown of the surrounding medium, initiating breakdown of the fiber surface. The degradation can occur very rapidly with nearly instantaneous catastrophic failure of the fiber end.

Therefore, the most common fiber microcomponent used in medical and industrial applications is the enlarged core taper on the proximal fiber end (Fig. 18.2). Proximal end tapers are useful in medical and industrial applications where high-power density yet small flexible fiber sizes are desirable. This type of taper allows a reduction in power density at the fiber end-face and increases the size of the fiber core "target" for the incoming laser power.

Using a proximal end down-taper, the spot size of the incoming optical power can be enlarged proportionally to the taper. Adjusting the focal point to be inside the taper can perform this. This results consequently in a reduction of the optical power density impinging on the front surface of the taper without a reduction in the total power. The power density at the larger input surface can be reduced to levels well below the damage threshold, thereby avoiding catastrophic failure of the fiber end. It should be noted that power may be lost as light travels through the taper because of high angle input modes further increasing in angle, eventually exceeding the fiber NA and refracted out of the fiber. The launch optics can be designed to minimize this loss. The loss, however, is typically less than the power gained by being able to increase or maintain the total power input into a taper versus the smaller fiber core without a taper.

18.2.1 Design of a Fiber Taper

The well-known concept of conservation of brightness states that if light losses are negligible, the spatial and angular content of the light anywhere within or at either end of a taper are described by

$$A_i n_i^2 \sin^2 \Theta_i = A_o n_o^2 \sin^2 \Theta_o, \qquad (18.1)$$

where subscript i refers to the input and o refers to the output of the taper and

A = cross-sectional area of the light distribution normal to the taper or fiber axis
Θ = maximum angular extent of the light distribution
n = refractive index of the medium, where Θ is measured (Fig. 18.4).

Because $n \sin \Theta = \mathrm{NA}$ and $A_i/A_o = d_i^2/d_o^2$, where

NA = numerical aperture
d_i = input taper diameter
d_o = output taper diameter,

it follows that as light transmits through a sufficiently long taper, the following equation applies:

$$\frac{NA_o}{NA_i} = \frac{d_i}{d_o}. \qquad (18.2)$$

However, in the case discussed here where the taper is integral with the optical fiber, if the product of the input NA and the ratio of the diameters exceeds the greatest NA that the taper can support (which can occur in the case of the proximal end down-taper when the input NA is too high), light will escape into the cladding and be lost. In this situation, this relation will no longer be valid.

Therefore, if one applies an input NA to the end of the taper that would be equal to the NA for the base fiber, the light throughput would be inversely proportional to the square of the taper ratio. For a 2:1 diameter ratio taper, the throughput would be 25%, exactly the same as butting a large fiber directly to the small fiber with no taper at all!

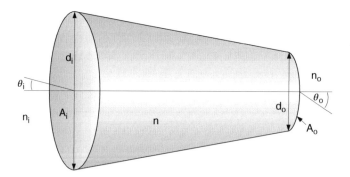

Figure 18.4   Basic fiber taper.

Therefore, the recommended maximum input NA is

$$NA_i \leq NA_o(d_o/d_i). \tag{18.3}$$

If this relationship is adhered to, the NA increase down the length of the taper will not exceed the NA of the fiber, thereby minimizing loss in the taper. For example, if the source applied to the large end of the 2:1 taper has an NA half that of the base fiber, then perfect transmission can be expected through a perfect taper. Of course, the NA of the light in the fiber will now be twice that of the original source. Note that there still may be optical losses from imperfections in the taper geometry or taper surface plus any Fresnel losses from the front surface of the taper.

As another example, with a down-taper with a 3:1 diameter ratio attached to a fiber with a 0.22 NA, the maximum input NA can be calculated as

$$NA_i = 0.22^*(1/3)$$
$$NA_i = 0.073.$$

Therefore, to obtain the best possible coupling efficiency into the fiber, the launch NA must be 0.073 or less. This relationship can also be used to calculate the required taper ratio if the launch and fiber NA are already established.

In summary, the key concerns regarding the use of fiber tapers are optical loss versus taper and fiber geometry, input NA, and the transformation in NA over the length of the taper.

Note that fiber microcomponent tapers are typically manufactured in taper diameter ratios of 1.1–5.0 on fibers that are 200-μm core diameter or larger. It is possible but difficult to fabricate tapers with good geometrical tolerancing on smaller fibers, because of the fiber easily overheating in the tapering process, causing the glass material of the fiber to quickly flow and making it difficult to control the geometry of the taper.

As expected from the previous equations, the performance of the taper is relatively independent of taper length. This was confirmed in actual measurements [1] in which the taper loss was found to depend strongly on input NA but to be relatively independent of taper length and fiber diameter. This work was supported by an optical modeling ray trace model [5], which agreed with the general trends of loss being strongly dependent on input NA and relatively independent of fiber diameter and taper length, even for very long length (>1-m) tapers. Figure 18.5 displays these ray trace model results for a 2:1 taper (core diameters of 400–200μm) with the actual measured results overlaid.

However, for launch conditions that use significantly non-uniform beam profiles, the model does predict that a longer taper will help to smooth (homogenize) the beam profile of the output power.

Figure 18.6 is a diagram showing some important taper parameters. The most significant of the physical parameters is the ratio of the diameter of the taper end to the diameter of the base fiber.

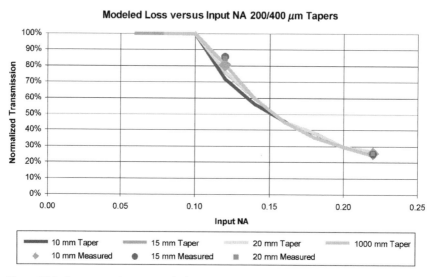

Figure 18.5 Loss versus input numerical aperture (NA) for different taper lengths in an optical ray trace model (with measured data overlaid).

The discussion, thus far, has concentrated on the use of proximal end down-tapers for the purpose of reducing the optical power density impinging on the optical fiber surface. In some cases, the same type of taper is used not for reducing the optical power density, but for cases in which the minimum beam waist from the laser is larger than the fiber diameter or the laser system is difficult to maintain in focus. The use of a taper can make the coupling tolerances much more forgiving with minimal compromise to coupling performance and allow a loosening of tolerances in the optical system.

An optical taper can also be used on the output end of an optical fiber using its angle-changing property to alter the angular distribution of the output

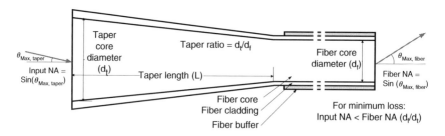

Figure 18.6 Diagram of important fiber taper parameters.

intensity. For example, if a lower output divergence (smaller spot size) than the fiber normally exhibits is desired, an up-taper (Fig. 18.7) can be used.

Alternately, if a larger divergence is required, a down-taper (Fig. 18.8) can be used.

In situations in which it is desirable for the optical power to spill out of the fiber abruptly at a spot size larger than the fiber NA alone can attain, but there is only a short distance to the target (such as a laser scalpel), a very short taper length can be used. This decreases the output spot size versus the design shown in Fig. 18.9, and thereby increasing the power density of the spot. In addition, the shorter tip is mechanically more robust and the sharp point can be used for perforating membranes.

18.3 LENSES

Various lenses such as concave, convex, and spherical (ball) can be fabricated as fiber microcomponents. These lenses are useful for modifying beam divergence and spot size. The shaped lenses are used for improved coupling from laser diodes to fibers, reduction in overall Fresnel losses, reducing or increasing the depth of focus, increasing or decreasing output spot size, and collimating or decollimating light. Applications are very broad, from low-power communications links to

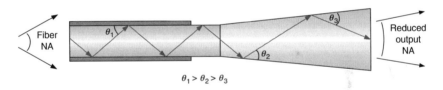

Figure 18.7 Distal up taper: output light converges.

Figure 18.8 Distal down taper: output light diverges.

Figure 18.9 Short-length distal down taper.

high-power industrial lasers. The lenses can be combined with other shaped tips as well, such as a convex lens on the end of a taper.

Convex and ball lenses can be fabricated on the end of an optical fiber by simply heating the fiber end until the glass softens and surface tension rounds the fiber end. Precise control of the heating conditions will result in a good piece-to-piece repeatability. Concave lenses can also be fabricated but often entail more complex machining processes.

The design of the simple fiber lens is similar to the normal design of a Plano convex or Plano concave lens, with the flat side obviously being the optical fiber itself. Because there is no interface between the lens and the fiber, there will be no Fresnel losses to account for in the system optical budget.

Standard lens design equations also apply to the design of fiber microcomponents. In the case of a spherical lens on the fiber end, the radius, R, of the lens surface must be determined. Starting with the thin lens equation [6]:

$$1/d_o + 1/d_i = (n_2/n_1 - 1)(1/R_1 - 1/R_2), \qquad (18.4)$$

where Fig. 18.10 defines the equation's parameters.

In the case of collecting collimated light from a source and coupling it into the optical fiber (Fig. 18.11), the lens must be designed so that the coupled light does not exceed the fiber NA and thereby become lost.

In this situation,

d_o = infinity
R_2 = infinity,

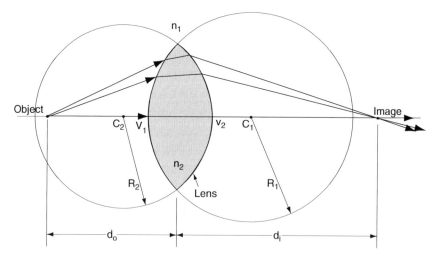

Figure 18.10 Defining the thin lens equation.

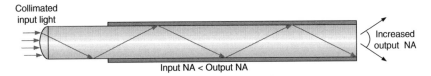

Figure 18.11 Integral positive (convex) lens fiber microcomponent.

and assuming $n_1 = 1$ (air), the thin lens equation reduces to

$$1/d_i = (n_2 - 1)/R_1. \tag{18.5}$$

The distance d_i can then be calculated through the definition of NA:

$$NA_2 = n_2 \sin \Theta_2 \tag{18.6}$$

Because the NA_2 of the fiber and the glass fiber core index of refraction, n_2, are both known, $\sin \Theta_2$ can be calculated. Assuming a (thin) lens where the lens thickness is insignificant, per Fig. 18.10,

$$d_i = h/\sin \Theta_2, \tag{18.7}$$

where h is the input beam radius. The radius of curvature of the lens, R_1, can then be calculated by inserting d_i and n_2 into the reduced thin lens equation, above. Similar calculations can be performed for light either entering or exiting at various angles.

In some cases in which there is an input NA greater than the fiber NA, coupling efficiency can be increased by the use of a concave lens (Fig. 18.12).

In cases in which the collection efficiency of the lens is limited by the lens (fiber) diameter, a ball lens can be utilized (Fig. 18.13). Again, the thin lens

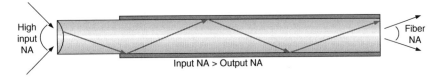

Figure 18.12 Integral negative (concave) lens fiber microcomponent.

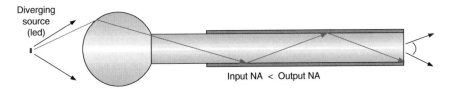

Figure 18.13 Ball lens fiber microcomponent.

equation can be used for calculating the ball lens radius. A common application of such a ball lens is in coupling to laser diodes. The ball lens transforms the light emitted by the laser at a high acceptance angle (high NA) to a smaller angle that will be accepted by the fiber NA. Coupling improvement of three to five times are typically obtained over straight fibers (0.16 NA) butt-coupled to the laser diode. For comparison, a separate ball lens and fiber can achieve an 18–20 times coupling improvement over straight fibers, but this can be offset by higher material and assembly costs [7].

Lenses described here are not typically good for absolute beam collimation. The spherical aberration of the lenses creates a diffuse focal point, thereby making it extremely difficult to absolutely collimate the beam (Fig. 18.14).

Note that more complicated designs are possible, where tapers and lenses can be combined, although such structures are rarely cost effective because of the complexity of fabrication.

18.4 DIFFUSERS

Diffusers are generally used on the distal end as a means of redirecting and scattering the optical power in an even 360-degree cylindrical output along the length of the tip (Fig. 18.15). This is typically performed by machining grooves or threads into the glass of the fiber deep enough to extract and scatter light traveling through the fiber core. The scattered light bathes an area with the optical power, making it useful for applications such as photodynamic therapy or tissue ablation (e.g., prostate reduction and urology procedures) (Fig. 18.16).

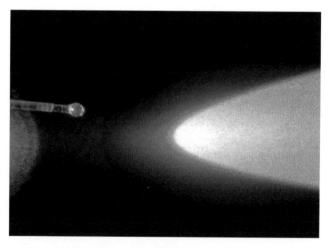

Figure 18.14   Output from a ball lens fiber microcomponent.

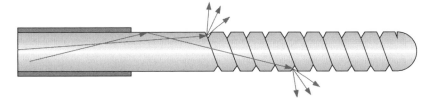

Figure 18.15 Diffuser microcomponent.

Because coupling into a diffuser would be very high loss, they are generally limited to distal end use only. The diffuser tip, having the rather deep grooves, often has a silica cap placed over it for additional mechanical durability and protection from contamination.

The design of diffusers varies depending on the output length and uniformity required. There are no straightforward equations (as in the case of lenses or tapers) that can calculate the design of the diffuser. A detailed ray diagram can be used as a first approximation however. It is important to "budget" the amount of optical power being withdrawn versus diffuser length. As the light traveling down the fiber approaches the diffuser, it can be assumed that the light occupies the full NA of the fiber. The light at the higher order modes exits the diffuser tip preferentially over the modes traveling at low angle or straight down the fiber core. If the diffuser design is uniform down its length, then the output intensity will be higher at the beginning of the diffuser than at the end. If this is not acceptable in the application, the output from the diffuser can be made more uniform down the length of the diffuser by adjusting the diffuser design so that

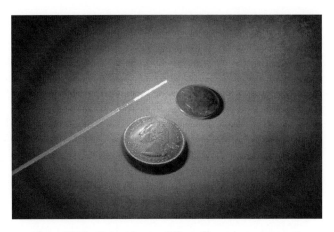

Figure 18.16 Output from a diffuser fiber microcomponent.

the light traveling at low angles is progressively stripped more severely down the diffuser length. This can be performed by progressively increasing the groove depth, width, and pattern so light is more aggressively stripped (scattered) from the fiber as it travels down the diffuser.

18.5 SIDE-FIRE AND ANGLED ENDS

The side-fire consists of an angle machined into the distal end of the fiber. In most side-fire designs, this angle is 40–43 degrees from the fiber axis. Optical power impinging on the angle is redirected approximately 90 degrees from the fiber axis. Note that an angle slightly less than 45 degrees is most optimal. This is because at 45 degrees or more, a significant portion of the light impinging on the angled end exceeds the critical angle for total internal reflection and then exits the fiber in an undesirable direction.

The side-fire microcomponents are made by first machining in the desired angle into the fiber, then attaching a glass cap over the angled end. The side-fire requires a medium of lower refractive index around the backside of the angled end to operate. Commonly, a glass cap is placed over the diffuser. The cap provides mechanical protection and an area of low index (i.e., air) on the backside of the angled end, thereby creating the conditions necessary for reflection out the side of the fiber.

The side-fire is particularly useful in invasive surgical procedures in which the optical power needs to be redirected in a very confined space, such as tissue ablation, cutting, and perforations (e.g., transmyocardial revascularization [TMR]). The side-fire is primarily used for such *in vivo* medical applications, so the protective glass cap also serves to protect the fiber end from damage and contamination of the angled fiber end (Fig. 18.17).

Angled fiber ends are also very useful in reducing back-reflection down the fiber. Putting a 7- to 10-degree angle (depending on fiber NA) into the fiber will cause the Fresnel back-reflection off the fiber end-face to reflect at an angle that will not be accepted by the fiber NA and, therefore, not be propagated back down the fiber. In high-power laser cutting and welding applications, this

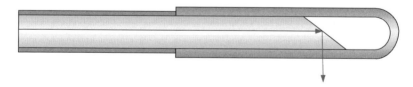

Figure 18.17   Side-fire fiber microcomponent.

back-reflection can cause the fiber distal end termination or a sharp bend point in the fiber itself to overheat and self-destruct. Similarly, when used on the input end, the angle can dramatically reduce back-reflection into the source laser. Such back-reflection can damage optical components, generate signal noise, and create instability in the laser source. In either case, the reflected power is dumped into some type of absorbing heat sink, which effectively dissipates the energy without destroying the optical fiber assembly or creating a safety hazard (Fig. 18.18).

18.6 OPTICAL DETECTION WINDOWS FOR MICROFLUIDIC FLOW CELLS

Optical detection windows are used in microfluidic flow cell spectroscopy. Although not necessarily optical fibers in nature, optical detection windows are produced in a similar fashion to many of the fiber microcomponents and are often partnered with specialty optical fibers to complete the flow cell device. In this application, a fluid (gas or liquid) to be analyzed is transferred down a small glass capillary similar in size to an optical fiber. While traveling down the capillary, the fluid is transformed or reacted or otherwise undergoes a separation process. The fluid then passes through the optical detection window through which the fluid is scanned for fluorescence or spectral absorbance. In some cases, specialty optical fibers discussed earlier in this chapter are added to serve as

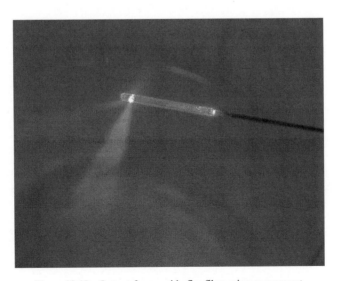

Figure 18.18 Output from a side-fire fiber microcomponent.

either the conduit for the exciting radiation wavelength(s) of light from the source to the window or the conduit for the output radiation from the window to the analyzing spectrometer. Figure 18.19 shows a schematic of a typical design that comprises an optical detection window being used as a microfluidic flow cell.

The detection windows themselves are typically composed of a 2- to 10-mm region of the glass capillary where the exterior protective plastic coating has been carefully removed (Fig. 18.20), or it can be an enlarged length of the capillary (Fig. 18.21) where the fluid speed decreases and the illumination volume increases, thereby increasing the sensitivity of the flow cell.

The advantage of these optical detection windows and flow cells include the following:

• Small sample sizes
• High sensitivity and throughput speed
• Low dead volume, as no flow cell connectors are typically needed
• Analyte processing in a confined, controlled, and safe area (within capillary)
• Continuous sample processing versus batch
• Compact flow cell footprint for multiplexed sample processing

An application that greatly benefited from such detection window technology is DNA sequencing. Many of the DNA sequencing instruments are capillary based and use optical detection windows in 1- to 384-channel arrays for very

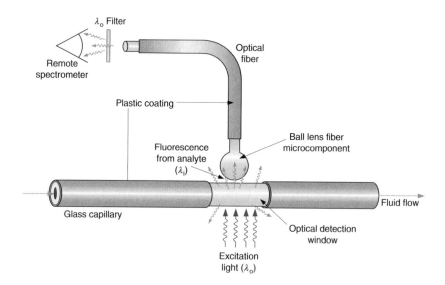

Figure 18.19   Schematic of optical detection window for microfluidic flow cells.

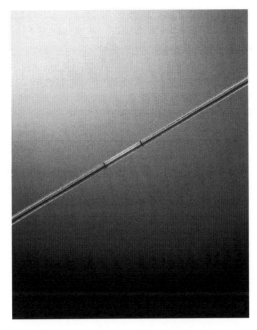

Figure 18.20   Glass capillary optical detection window cut through plastic coating.

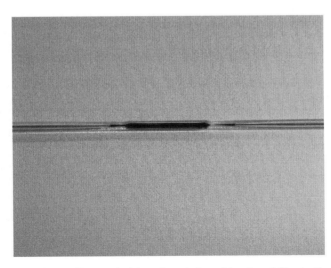

Figure 18.21   Glass capillary optical detection window with enlarged illumination volume.

high throughput processing. This is an obvious requirement for efficient sequencing of a genome and was of particular significance during the Human
Genome Sequencing Project, as the human genome contains more than 3 billion
base pairs [8].

An interesting new technology is using a fiber capillary (a specialty optical
fiber with a small hole down its center) as a high sensitivity fluidic sensor cell.
The fiber capillary (light-guiding capillary) consists of a core, which is an
annulus around the ID of the capillary. Designing an appropriate (low) refractive index layer of glass around the OD of the annular glass core creates an outer
cladding. The ID of the annular core interfaces with the (lower index) fluid in the
capillary. Because there is material of low index on either side of the annular
core, light launched into the fiber end will travel via total internal reflection
down the annular core. The evanescent field of the light impinging on the surface
of the ID interacts with the fluidic material, creating a high efficient evanescent
field sensor microcomponent. The output of the annular core is monitored for
spectral absorbance through the fiber end or through an optical detection
window cut into the fiber plastic coating (as discussed earlier). Figure 18.22
shows a cross-section of such a microcomponent.

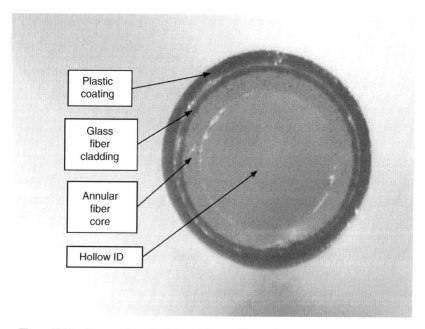

Figure 18.22   Cross-section of a light-guiding capillary microcomponent (200-μm ID).

ACKNOWLEDGMENTS

The author thanks Richard Timmerman, John Shannon, Joseph Macomber, and Gary Nelson for their valuable technical input and assistance with the ray diagrams.

REFERENCES

[1] Clarkin, J. P. et al. 2004. Shaped fiber tips for medical and industrial applications. Optical fibers and sensors for medical applications IV. *SPIE* 5317.

[2] *The Book of Polymicro Technologies*, 2001, pp. 5-2–5-5. Polymicro Technologies, LLC, Phoenix, AZ.

[3] Hecht, J. 1999. *Understanding Fiber Optics*, pp. 573–574. Prentice Hall, Upper Saddle River, NJ.

[4] Allard, F. 1989. *Fiber Optics Handbook*, pp. 3.56–3.57. McGraw Hill, New York, NY.

[5] Lambda Research Corporation. 2001. Tracepro-Software for Opto-Mechanical Modeling, Version 2.3.

[6] Hecht, E. 1987. *Optics*, 2nd ed., pp. 137–140. Addison-Wesley, Boston, MA.

[7] Howes, M. J. and D. V. Morgan. 1980. *Optical Fiber Communications: Devices, Circuits, and Systems*, pp. 40–42. Wiley & Sons Somerset, NJ.

[8] Cantor, C. R. and C. L. Smith. 1999. *Genomics: The Science and Technology Behind the Human Genome Project*. Wiley & Sons. Somerset, NJ.

Chapter 19

Liquid-Core Optical Fibers

Juan Hernández-Cordero

Instituto de Investigaciones en Materiales, Universidad Nacional Autónoma de Mexico, Mexico City, Mexico

19.1 INTRODUCTION

It is widely recognized that the need for larger bandwidths for communication systems was the main driving force for the development of optical fibers. The invention of the laser in 1960 triggered great expectations regarding the possibility of increasing the amount of information carried by a modulated wave using an optical signal. Simultaneously, it was acknowledged that a suitable transmission medium was needed so that optical signals could propagate over long distances with minimum losses. Today several technological barriers have been broken, leading to the development of optical communication systems relying on optical fibers as a transmission medium. Such systems not only have fulfilled the needs foreseen in the early 1960s but also have enabled the development of more sophisticated technologies. Optical communication systems can now manage simultaneously the transmission of video, data, and voice, thereby conveniently exploiting the large bandwidths offered by optical fibers.

Throughout the years, fabrication methods for low-loss silica fibers have improved considerably, and it is now possible to tailor the spectral properties of glass fibers when using the appropriate materials within the core. However, in the early stages, liquids were among the first materials tested as core media for optical fibers. While solid materials fully compatible with silica had not been found, several liquids offered two main features that were attractive enough for fiber fabrication: a low-absorption coefficient and a refractive index higher than that of glass. Thus, liquids provided two essential requirements for optical transmission, namely, low losses and wave-guiding by means of total internal reflection. Although several limitations for long-distance transmission were found, these were regarded as perfect step-index multimode fibers, with a constant refractive index across the core and a sharp transition at the core–cladding boundary [1].

599

The development of liquid-core fibers for long-distance communications was not further pursued once the fabrication methods for low-loss glass fibers were available. Nonetheless, several fundamental concepts and fiber characterization techniques currently in use were developed using liquid-core fibers. A number of liquids were tested in the early days as core materials and these studies led to a better understanding of scattering effects on these waveguides [2, 3]. Besides having a direct influence on fiber losses, scattering in liquid-core fibers proved later on to be a useful mechanism to enhance nonlinear effects in liquids, mainly because of the long interaction lengths offered by the fiber. As we will see in the following sections, this and other interesting features are crucial for all the wide variety of applications that benefit from the use of liquid-core fibers.

19.2 PROPAGATION OF LIGHT IN LIQUID-CORE FIBERS: MODAL FEATURES, DISPERSION, AND POLARIZATION EFFECTS

Generally speaking, liquid-core fibers are multimode waveguides and mode theory provides suitable theoretical background to understand the propagation of light within these fibers. Operation of these waveguides is based on multimode propagation of light within the liquid core, which is contained by a hollow glass or capillary tube used as fiber cladding. Because the liquids used as core materials have a higher refractive index than that of the glass tube, the guided beams propagate by multiple total internal reflections at the core–cladding interface (Fig. 19.1). As the first practical realization of a suitable transmission medium for light over relatively long distances, liquid-core fibers offered a convenient testbed for several experimental studies. Furthermore, they were used in the first demonstration of television broadcasting through optical fibers [1].

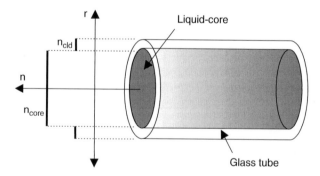

Figure 19.1 Typical waveguide structure for a liquid-core fiber.

Among the fundamental concepts studied with liquid-core fibers, modal effects and dispersion were perhaps the most interesting. Studies regarding mode launching conditions yielded very important information that was extremely useful to understand how light propagates in a cylindrical waveguide. Even though the glass tubes used as cladding in the early days had very high losses, it was demonstrated that total transmission attenuation could be lower than 20 dB/km [4]. Further research later showed that this was due to the numerical aperture (NA) and the multimode nature of the fibers. Because these fibers can have very high NA, high-order modes can be effectively filtered or coupled to lower order modes, thereby showing low transmission losses and high bandwidths. Early studies also demonstrated the effects of cladding losses upon varying the angle of incidence of the probe beam. This method allowed for the identification of mode filtering effects at the cladding, that is, an increase in losses after an angle of incidence smaller than the critical angle indicated that the modes propagating at such angles would experience higher attenuation [5].

Because communications applications were not further pursued with these fibers, information regarding dispersion is limited to the early papers. Among other key features, pulse rates of $200 \, \text{Mbits/km}^{-1}/\text{sec}^{-1}$ were shown to be feasible with liquid-core fibers and it was demonstrated that an increase in the fiber core diameter and a reduction in fiber bend radius both increase the dispersion [5]. As confirmed in these early experiments, mode conversion during propagation is responsible for the dispersion effects observed in these fibers. Further fundamental research on fiber dispersion focused on the effects of bending radius and launching conditions. Because core homogeneity and very low scattering for the wavelength of interest, no bandwidth limit was apparent with these fibers [6]. Remarkably, attenuation did not seem to increase dramatically with bend radius, which is interesting because glass core fibers are susceptible to bend losses due to their weakly guiding structure. Dispersion effects are, thus, related to the multimode nature of the fiber and to the NA.

Other experimental techniques were used to evaluate modal effects in liquid-core fibers and proved very useful to validate mode-coupling theory. As an example, experiments based on far-field observation verified a theoretical model describing coupling between modes supported by optical fibers [7]. Results showed that lateral stress applied to the fiber can generate mode coupling and mode scrambling. Coupling between modes was expressed in terms of a normalized mode conversion coefficient, which turned out to be two orders of magnitude smaller for liquid-core fibers compared to glass core fibers. Liquid-core fibers are, thus, less susceptible to stress effects and the excited modes for a given launching condition will propagate with minimum coupling. In fact, most of the early reports on liquid-core fibers show that single-mode propagation is maintained in these fibers once the fundamental mode is excited.

Further understanding of the multimode propagation of light in fibers was also obtained upon using the high-temperature dependence of the liquid core. A detailed experimental study was carried out in which the mode cutoff frequencies were observed in transmitted and scattered light when varying the temperature of a short segment of fiber [8]. Selective excitation of the modes supported by the fiber was achieved through adjustments on the temperature and the launching conditions. Thus, mode cutoffs were readily observed as a function of temperature. These results were shown to compare well with theoretical predictions obtained with weakly guiding mode theory. Another interesting feature of the experimental results was an oscillating behavior attributed to modal interference. This modal effect can be related to core size and ellipticity, thereby suggesting a characterization method for fiber geometry.

Polarization properties of liquid-core fibers have also been reported [9]. In agreement with the aforementioned studies on stress effects, liquid-core fibers do not exhibit stress birefringence, which is commonly observed in glass core fibers. This yields polarization maintaining behavior for straight fibers as long as single-mode operation is sustained. Nonetheless, bends in the fiber can create axes of birefringence, which are orthogonal to each other, so polarization of the guided beam can be adjusted mechanically. The phase shift induced by these axes can be compensated by rotating the plane of curvature, as is commonly done with fiber polarization controllers with rotatable paddles. Besides bending and stress, no other birefringence sources were reported in the early papers. For most liquid-core fibers, uniformity and purity of the liquid core is most likely to minimize the effects of other sources of birefringence. However, as we will see, liquid crystals can be used in the core to develop polarization-sensitive fiber devices.

19.3 FABRICATION AND CHARACTERIZATION METHODS

The simplest description of a liquid-core fiber is a capillary tube filled with a liquid that has suitable optical properties for a given application. There are, however, some variations on the geometry and even on the cladding material that widens the types of waveguides using a liquid core. Regarding geometry, several studies describe the use of liquid-core planar waveguides. Devices and applications based on this geometry are not within the scope of this chapter. Instead, we focus on fibers with glass cladding. Other designs and materials are further described in the next section (Section 19.4).

Liquid-core fibers are fabricated using a hollow glass tube that serves as a cladding. Because of the high NA and the modal effects observed in these fibers, there are no stringent requirements regarding losses in the glass. In fact, some of the early papers note the low losses obtained with these fibers in spite of the high bulk losses of the glass tubes. Several types of glasses have been reported to be useful for fabrication of these waveguides and even capillaries have been reported to yield adequate features in the transmission properties of the fiber. The glass tube is pulled following the same procedure as that used for pulling glass fibers. As can be seen in early reports on fabrication of liquid-core fibers, the predecessors of the modern draw towers were first developed for pulling hollow glass tubes [10]. As we will see in the following section (Section 19.4), reports have demonstrated that "holey" fibers can also be filled with liquids and, thus, serve as liquid-core fibers. In this particular case, the fabrication process is more complicated than simply pulling a single glass tube, but the pulling method is similar in both cases. Evidently, the goal is to avoid collapsing the glass tube so a properly sized core can be created. Once the tube is pulled to the required dimensions, the cladding is ready to host the liquid core, although some applications may require an extra coating on the inner wall.

The filling of the hollow fiber is generally carried out through hydrostatic pressure and the dimensions of the fiber will determine the time required for this process. A wide range of applications involve the use of small lengths of fibers, so capillary forces are enough to fill short lengths of a hollow fiber. However, for long lengths, the filling process requires an increase in pressure to achieve convenient filling times. A cell can be specially designed for controlling the hydrostatic pressure and to host the hollow fiber, the liquid, and a window so that light can be launched into the fiber core. As an example, early papers on liquid-core fibers report the use of a Monel cell (Fig. 19.2) with a Teflon plunger that allowed filling a 50-m long fiber in half an hour [2]. Evidently, the core size is important in evaluating the time required to fill the fiber. This can be readily seen from formal analysis of laminar flow in small-bore pipes. As an example, the time (T) taken to fill a fiber of length L is given by [10]

$$T = \frac{16\mu}{P}\left(\frac{L}{d}\right)^2, \tag{19.1}$$

where P is the applied pressure, μ is the coefficient of viscosity of the liquid, and d is the core diameter. It is clear from this expression that for a fixed pressure, long lengths of fiber will require longer filling times, and conversely, fibers with larger core diameters require shorter filling times.

As shown in reference [10], the maximum pressure that can be applied to a glass fiber before rupture is related to the tensile strength of the glass (S), the outer diameter of the tube (D), and the core diameter. Explicitly, this can be evaluated approximately as

$$P_{max} = \frac{S(D-d)}{D}. \tag{19.2}$$

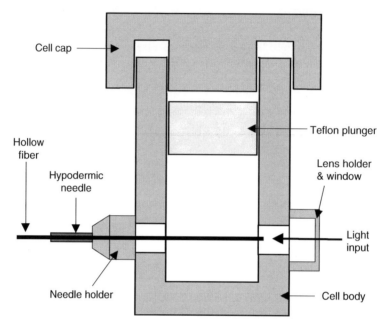

Figure 19.2   Monel cell for hydrostatic pressure filling of liquid-core fibers (used, with permission, from reference [2]).

Both of these equations can be used to determine the optimum core size (d_{opt}) required to fill a fiber of length L in a minimum time. An increase in pressure to reduce filling times is, thus, limited to the mechanical properties of the glass tube. Other methods that may be used focus on reducing the capillary forces upon reducing the viscosity of the liquid, provided that the optical properties remain unaltered.

Several liquids have been used as fiber cores ranging from bromobenzene and o-dichlorobenzene (the first reported in the literature [2]) to water [11] and ethanol [12]. The selection of the liquid for the fiber depends on the required optical features for a given application. Some of the desired characteristics in a liquid are evident; for instance, if a simple capillary or hollow fiber is used as cladding, the liquid has to have an index of refraction higher than that of the glass. However, this condition is not necessary if the liquid is to be enclosed inside a microstructured fiber. For transmission over long lengths of fiber, the liquid must have low loss at the wavelength of interest and, in general, the liquid has to be stable and nonvolatile and must have low viscosity. Also, scattering effects are important in applications involving nonlinear effects; thus, a suitable liquid with proper scattering coefficient must be chosen. The NA of liquid-core fibers is generally high compared to that of glass fibers. Evidently, the value for this parameter depends on the refractive indices of the liquid core and the cladding. Values ranging from 0.2 to 0.6 have been reported, and once again, the optimum value is determined by the requirements of a given application. Finally, the optical properties of liquid-core fibers are determined using what are today the standard characterization methods for optical fibers.

19.4 APPLICATIONS

As mentioned earlier, liquid-core fibers were first envisioned as a suitable transmission medium for optical communications systems. Practical limitations for the fabrication of these waveguides and, more importantly, the development of highly transparent glass core fibers proved that other materials offered better features for communications applications. However, the possibility of selecting a liquid core with very specific optical properties has remained highly attractive for a wide range of applications. Among other interesting features, liquid-core fibers offer the possibility for transmission of ultraviolet (UV) light, enhancement of nonlinear effects in liquids, and even the use of liquid crystals for developing polarization-sensitive fiber devices. The development of microstructured fibers has also opened new possibilities for the fabrication of waveguides with unique features that can be tailored upon selecting a suitable liquid. Development of light sources based on nonlinear effects and sensors using liquids are two broad categories in which applications of liquid-core fibers can be classified. Some specific examples that fall into these categories are reviewed in the following sections.

19.4.1 Waveguides for Special Spectral Regions and Optical Chemical Analysis

Early studies on the spectral features of liquid-core fibers focused on the visible and infrared (IR) regions, mainly because of the availability of laser sources at these wavelengths. Naturally, upon selecting the appropriate liquid for the core, the fiber also offered the possibility of guiding light at other nonstandard wavelengths. As an example, the first reports on transmission losses at the IR region (3.39 μm) used a fiber with a tetrachloroethylene (C_2Cl_4) liquid core [13]. Losses at this particular wavelength region are high for this fiber (10^4–10^5 dB/km) [14], but these reports suggested that other liquids with suitable optical properties were likely to offer better results.

Transmission of UV light has also proven to be feasible with specially designed liquid-core fibers. Absorption spectrometry, in which fibers are regularly used as a means to increase the interaction length of the light with the analyte, has increased its spectral range of operation from the visible to the UV upon the use of aqueous ethanol solutions as fiber core [15]. Essentially, the fibers are used as cuvettes and the transmission spectrum is registered to detect absorption bands for specific analytes. Because light can travel several meters along the fiber, an enhancement in sensitivity is effectively achieved with these waveguides. Aqueous solutions require adding solvents to increase the refractive index of the liquid, although approaches such as adding an inner coating to the glass tube have allowed for the use of simple liquids such as water (Fig. 19.3). The use of Teflon coatings in glass and plastic tubes has yielded good results in highly sensitive detection of pesticides and water pollutants using liquid-core fibers together with chromatographic and spectroscopic techniques [16, 17]. Theoretical analysis has shown that Teflon layers as small as 5 μm are sufficient to confine the light within the liquid core and, thus, avoid environmental effects and scattering by the capillary material [18]. Applications such as pH monitoring have also reported enhancement in detection sensitivity using a simple transmission monitoring scheme [19]. In this case, transmission through the liquid-core fiber is simply monitored as a function of pH, so the absorption spectra can be used to determine the pH value.

Chemical analysis has benefited from the use of liquid-core waveguides in planar and cylindrical geometries. Besides liquid chromatography and absorption spectroscopy, other analytical techniques such as fluorescence and Raman measurements have enhanced their detection limits due to Teflon-coated capillaries. The Teflon AF family of fluoropolymers is perhaps the most widely used for inner coatings of waveguides, because the range of refractive indices available (1.29–1.31) is adequate to generate total internal reflection when aqueous solutions are used as a fiber core. Generally speaking, the liquid-core fiber acts as a flow cell and as a reaction chamber in which light can be generated either by a chemical reaction or by the Raman effect [20, 21]. Light is then guided by the

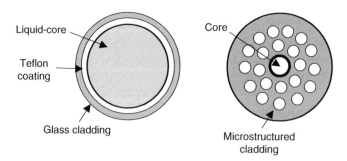

Figure 19.3 Liquid-core fibers with Teflon coating and microstructured cladding.

fiber to a detection system to monitor the spectral and/or the intensity features of the guided beam (Fig. 19.4). AF-polymer tubing is now commercially available and liquid-core fibers are nowadays almost a standard tool in analytical chemistry. Furthermore, availability of miniaturized LEDs and detector arrays has also opened the possibility of developing compact and portable chemical analysis systems using liquid-core fibers [22–24].

19.4.2 Fiber Sensors

The first report on a liquid-core fiber sensor was for voltage monitoring through the Kerr effect [25]. However, detailed analysis of the performance of these fibers as sensors was first carried out in distributed temperature sensing using OTDR [26]. The main advantage of using liquid-core fibers in this configuration is that Rayleigh scattering and the NA are highly dependent on temperature for these types of waveguides. Furthermore, as mentioned earlier, the modal features vary with temperature as well. The fibers used in these first experiments were pulled from silica tubes; other special preparation included deposition of a layer of high-purity silica inside the tubes, and the addition of an outer polyimide coating. The core was filled with hexachlorobuta-1,3-diene using a high-pressure syringe system. With this filling system, the authors could fill 150 m of fibers with 150-μm core in 30 minutes. The fibers had an NA of 0.2 and 0.54 at 900 and 589 nm, respectively, and losses were measured to be 13 dB/km at 900 nm.

Results from the first experiments on distributed temperature sensing were very useful to determine the temperature sensitivity of the fibers. Two major

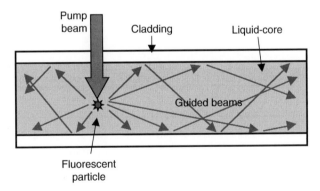

Figure 19.4 Schematic representation of a fluorescence sensor using liquid-core fibers. The same principle is used for nuclear detectors in which the core is a liquid scintillator.

temperature effects are understood to play an important role: changes in NA and in scattering loss. Although other means such as transmission losses monitoring can be used to register temperature changes, the backscattering loss allows for the use of OTDR. As the temperature increases, the scattering loss increases due to the thermal agitation, which is strongly dependent on temperature. Analysis of the setup used by Hartog [26] showed that a resolution of 1 m over fiber lengths of more than 100 m was attainable, with a temperature accuracy of 1 °C. The sensitivity obtained experimentally was 23.3×10^{-3} dB/°C (0.54%/°C) over a temperature range of 5–110 °C. A simpler approach that does not require an OTDR system was later demonstrated based on transmission measurements as a function of temperature [27]. As shown in this report, liquid-core fibers also offer the possibility of creating multiplexed arrays of sensors using different liquids to extend the temperature range of operation.

Nuclear radiation detectors can also be developed using liquid-core fibers. The idea behind these sensors is to fill capillaries with liquid scintillators so that light generated by luminescence is guided by the fiber. Studies on radiation resistance of liquid scintillators and capillaries filled with these liquids have been carried out demonstrating that high scintillating and trapping efficiencies can be achieved [28]. Such arrangements are capable of yielding a track hit density higher than that of detectors based on plastic fibers or semiconductors. Because of the large attenuation lengths (which, in turn, depend on the ID of the capillaries), it is possible to construct detectors with lengths more than 2 m and a spatial resolution of less than 20 microns/hit. Several glass tubes of different grades were used as capillaries for the fibers. The influence of radiation in quartz was investigated upon comparing the attenuation for scintillating liquid-core fibers with quartz tubes before and after exposing it to radiation. Although glass darkening was observed in low-grade quartz, it was found that radiation resistance of the arrangements was limited by the liquid rather than the capillary glass. A comparison between plastic scintillating fibers and scintillating liquid-core fibers showed that radiation resistance is much better for the latter arrangements. From this study, it was suggested that plastic fiber will work well for doses less than 5 Mrad, but that liquid-core fibers can be used for doses higher than 60 Mrad without changing the liquid scintillator.

The use of liquid scintillators as fiber cores has also allowed for the development of novel sensing techniques capable of yielding high-quality imaging of ionizing particle tracks with very high spatial and time resolution [29]. Using a CCD as readout, the fibers act simultaneously as target, detector, and light guides. As in other applications, the liquid composition can be optimized to maximize light output and attenuation length. A passing charged particle creates scintillating light in the liquid core and a fraction of the light is guided by the fiber due to total internal reflection. The amount of light trapped within the core depends on the NA and the ID of the capillaries. The use of CCDs as readout has also allowed for the development of a sensor array that can function as a vertex detector [30]. These high-resolution systems include a set of image-intensifier tubes followed by a CCD or an electron bombardment CCD (EBCCD) camera. Detectors with as much as 10^6 capillaries have been reported to yield resolutions of 20–40 μm and are capable of withstanding radiation levels at least an order of magnitude higher than those of other tracking devices of comparable performance [31, 32]. Other relevant achievements obtained with these vertex liquid-core sensors include the recording of high-quality images of neutrino interactions [33].

19.4.3 Nonlinear Optical Effects

Scattering properties of liquids have always been attractive for observing nonlinear optical effects. However, most of them require long interaction lengths and high power to yield useful features. Because liquid-core hollow fiber systems allow for higher local intensity and longer gain length, they have been used in several applications involving nonlinear processes. Stokes-shifted, super-broadening, stimulated scattering was among the first effects demonstrated with a liquid-core fiber [34]. A laser beam with intensities from 10^6 to 5×10^8 W/cm^2 was fed into a 250-cm long fiber using CS$_2$ as a liquid core. Among other features, the spectral range of the stimulated scattering radiation was very large (> 700 cm^{-1}). Further research on this topic demonstrated that a nonlinear material used as core in a hollow waveguide was useful for generating ultrafast broadband radiation [35]. In both cases, stable and spectrally broadened radiation was effectively generated because of the extended interaction length and the high power excitation attainable with the liquid-core fiber.

The flexibility offered by liquid-core fibers for nonlinear interactions is not only limited to the capability of selecting a suitable liquid, but it is also possible to choose a proper core size to manage high powers. This is particularly important for avoiding problems related to laser-induced breakdown of the liquid. In this sense, the use of these waveguides allowed for detailed experimental analysis of effects such as stimulated Raman scattering (SRS), stimulated Rayleigh wing scattering (SWS), stimulated Raman–Kerr scattering (SRKS), and parametric generation of radiation. Several liquid samples have been used, yielding superbroadened radiation even with tuning capabilities [36, 37]. The results for the liquids used in these examples can be summarized as follows: (1) CS$_2$ and toluene show superbroadening on both the pump line and the SRS lines, (2) benzene exhibits superbroadening predominantly on the SRS lines, and (3) carbon tetrachloride shows no appreciable broadening on any of the lines. These experiments, together with a theoretical model, yielded enough information to determine that the threshold and the specific spectral distribution of the superbroadening effect depend on the molecular structure of the liquid.

The Raman-Kerr scattering process is perhaps the most widely studied optical effect with liquid-core fibers. Upon theoretical modeling, it has been possible to identify the requirements to observe this process [38]: (1) The molecules of the liquid Kerr medium must be anisotropic, (2) liquid samples must be transparent or have a small loss factor, (3) pump intensity must be sufficiently high, and (4) the gain length must be long enough. All these requirements can be fulfilled with a liquid-filled hollow fiber. Kerr liquid-filled hollow-fiber system can, thus, be used as a broadband, multiwavelength coherent light source with spectral and temporal feature that can be tailored for specific requirements. A clear example of this is shown in He et al. [38], in which the liquid and the hollow fiber are selected to enhance the multiorder SRS and SKS that yields the superbroadening effect. In this particular case, CS_2 was recognized as the most efficient liquid for generating SKS and SRS, whereas the optimized fiber parameters were ID 0.1–0.25 mm, length 3–4 m, and focusing length for the input coupling of 10–15 cm. The total output spectral range was measured to be up to $4000 \, cm^{-1}$ (i.e., more than six orders larger that the typical SRS). The location of the output spectrum depends on the frequency of the pump source and can be in the near UV (300–400 nm) to the near IR (0.7–2.0 microns). Spectroscopy is clearly one of the main applications that can benefit from further development of light sources based on these optical effects. Raman spectra of liquids dissolved in CS_2 can also be obtained with liquid-core fibers [39], although a wider variety of liquids can also be analyzed upon using a Teflon-coated hollow fiber [40]. It has also been demonstrated that when compared to conventional measurements using cuvettes, the use of a fiber geometry effectively enhances the Raman bands obtained from liquid samples [41].

Liquid-core fibers have been further used for other applications involving nonlinear optical effects. These include the amplification of amplified spontaneous emission (ASE) signals [42], studies of spectral narrowing of stimulated scattering [43], and optical limiting of laser pulses [44]. Whereas the first two applications are based on the Raman–Kerr scattering effect, limiting of laser pulses is based on the nonlinear absorption processes that occur in the high-index liquid used as fiber core. The choice of liquid depends on the pulse duration and an opaque cladding and organic liquid are generally used for this kind of application. Optical limiting action has been attributed to thermal density effects such as self-defocusing, wide-angle nonlinear scattering, and input coupling and propagation mode losses.

19.4.4 Medical Applications

Optical fibers have been used extensively in medical application for diagnostics and laser delivery. Among other features, medical diagnostics require the development of reliable sensors with *in situ* monitoring capabilities, which in turn implies that noninvasive measurements are also required. Several fiber optic sensors have been proved to be useful for this application, and although a wide variety of materials have been used, most of the reports involve the use of glass fibers [45]. As seen in the previous sections, liquid-core waveguides have been proved to be useful for analytical chemistry, so the realization of liquid-core fiber sensors for *in situ* measurements and medical diagnostics should be considered feasible.

The realization of an instrument based on capillary optrodes has been already reported [46]. An inexpensive setup was shown to be useful for analyzing small liquid samples and the instrument was sought to be useful for emergency medicine. The capillary optrodes were constructed using the commonly used glass capillaries for blood sampling from a pierced fingertip. The sensors were sensitized with polymers and fluorophores and each of them could, therefore, be targeted to detect a different analyte (Fig. 19.5). The principle of operation of these waveguide optrodes is the fluorescence generated in the polymer coating trapped and guided by the wall of the capillary; thus, it can be seen as the equivalent to an evanescent waveguide sensor. Up to three different analytes were detected per sensor because of the capability of coating the inner surface with three different polymers. The realization of compact systems based on other analytical techniques should also be possible with compact light sources and detector arrays.

Laser delivery waveguides have been extensively studied for laser tissue engineering and therapy applications. The wide range of wavelengths required for the broad variety of medical applications have proven to be one of the main challenges for the design and fabrication of suitable optical fibers. As a general rule, wavelengths comprised within the visible and the near IR use silica fibers as waveguides, whereas the UV and mid-IR portions of the spectrum require specialty fibers [45]. There is, however, a report on the use of a liquid-core waveguide for laser tissue ablation using a visible laser, showing that high-peak

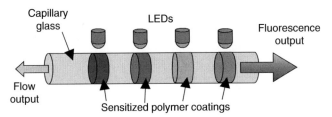

Figure 19.5   Sensor array with sensitized polymers. Each polymer can detect a different analyte, yielding different spectral features.

power transmission is feasible with this arrangement [47]. Although this "optical catheter" was demonstrated in *in vivo* experiments for laser angioplasty, no further improvements on these types of devices have been reported and silica fibers for this spectral region seem to remain the preferred choice. Regarding the UV wavelengths, the use of liquid-core fibers for laser delivery in medical applications has not been reported. Nonetheless, as seen before, these waveguides have proven to operate in this spectral region.

Wavelengths in the IR spectrum can be guided by several types of optical fibers that have proven to be useful for laser delivery. These include hollow waveguides, IR-transmitting glasses, and crystalline fibers [45]. However, complicated fabrication processes, high sensitivity to bending losses, and a low damage threshold are, until now, the principal factors limiting these fibers. As an alternative, several studies have focused on the use of liquid-core fibers for laser delivery systems. Besides being an inexpensive option, liquid-core fibers offer attractive features such as variability in diameter, high flexibility, and mechanical stability. IR absorption effects of water in the core and permeation of atmospheric water and of the solvent through the cladding have been reported, leading to fiber designs suitable to operate at $2.94\,\mu m$ [48]. For this wavelength, carbontetrachloride (CCl_4) has been used as core with plastic tube and quartz capillary as cladding. Bending radii below 10 mm are possible and a minimum transmission loss of 2 dB/m can be reportedly achieved [49]. Further studies have also shown that because of an overlap of the refractive indices of CCl_4 and fused silica between 500 nm and $1\,\mu m$, laser wavelengths in this spectral range (such as those obtained with Nd/YAG and HeNe lasers) cannot be transmitted in this fiber. However, upon using a mixture of CCl_4 and tetrachloroethylene (C_2Cl_4), the fiber becomes transparent from the near-UV (380 nm) up to the near IR ($3\,\mu m$), and consequently, it is suitable also for the Nd/YAG laser [50]. Distal energy densities up to $30\,J/cm^2$ have been achieved, thereby exceeding the ablation threshold of soft tissue. Thus, minimally invasive surgery can potentially be carried out with the aid of these fibers.

19.4.5 Special Waveguide Structures and Devices with Liquid Cores

Thus far, we have reviewed applications involving fibers with simple or diluted liquids and capillary tubes. However, there are other materials and waveguide structures that have increased the applications and driven further research involving liquid-core fibers. As an example, the use of liquid crystals has proven to be useful for fabrication of polarization-sensitive fiber devices. Experimental studies have shown that liquid-crystal core fibers with elliptical geometry yield adequate polarization properties for sensing and communications applications [51]. Fabrication of fiber devices such as long-period gratings based on liquid-crystal core fibers has also been demonstrated, showing that band rejection filters can be fabricated using this type of waveguide [52]. Adequate control of the optical properties of liquids is, thus, extremely important for dynamically adjustable fiber devices. Advances in physics of fluids, and in particular in the development of electrical and magnetically controlled liquids, will continue to spur future developments of devices based on liquid-core fibers.

As we have seen in the previous sections, the waveguide structure also plays an important role in the guiding properties of liquid-core fibers. In this sense, the development of complex waveguide structures has renewed the interest of using liquids as core materials. Theoretical studies have shown that a microstructured silica–air cladding provides excellent confinement for light guided in a liquid core, provided that the average cladding index is sufficiently below the index of water [53]. Realization of such fiber is, thus, limited by the fabrication of the microstructured cladding, an engineering problem that most likely will soon be addressed successfully. Similar structures have already been successfully demonstrated with water [54] and ethanol [55], so the development of compact biosensor, pollutant monitors, and chemical sensors based on liquid-core micro-structured fibers is successfully underway. Moreover, dynamic control of other arrangements based on liquid-core and liquid-cladding waveguide structures has been successfully demonstrated [56]. Reconfigurable optical switches, modulators, and optical couplers should, therefore, be possible with all-liquid optical waveguides, thereby increasing the usefulness of light guides for sensing and communications applications.

19.5 CONCLUSIONS

Liquid-core optical fibers have been extensively studied since the early years of optical fibers. Several useful properties of these fibers have been used in a wide variety of applications such as laser delivery systems, observation of nonlinear phenomena, analytical chemistry and optical biosensing. Some of these applications require rugged and compact measuring systems for field tests, and a number of prototypes have been developed based on liquid-core fibers. Flexibility and versatility are perhaps the most practical features offered by these waveguides. Requirements of specific applications can be fulfilled by selecting an appropriate liquid for the core and a suitable waveguide structure to host the liquid. Advances in physics of fluids and in waveguide design will, therefore, increase the usefulness of this fiber in developing devices for communications and optical measuring systems.

REFERENCES

[1] Gambling, W. A. 2000. The rise and rise of optical fibers. *IEEE J. Selected Topics Quant. Electr.* 6(6):1084–1093.

[2] Stone, J. 1972. Optical transmission loss in liquid-core hollow fibers. *IEEE J. Selected Topics Quant. Electr.* March:386–388.

[3] Stone, J. 1972. Optical transmission in liquid-core quartz fibers. *Appl. Phys. Lett.* 20(7):239–240.

[4] Payne, D. N. and W. A. Gambling. 1972. New low-loss liquid-core fibre waveguide. *Electr. Lett.* 8(15):374–376.

[5] Gambling, W. A. et al. 1972. Dispersion in low-loss liquid-core optical fibres. *Electr. Lett.* 8(23).

[6] Gambling, W. A. et al. 1972. Gigahertz bandwidths in multimode, liquid-core, optical fibre waveguide. *Opt. Commun.* 6(4):317–322.

[7] Gambling, W. A. et al. 1975. Mode conversion coefficients in optical fibers. *Appl. Opt.* 14(7):1538–1542.

[8] Planas, S. A. et al. 1982. Geometrical characterization of liquid core fibers by measurement of thermally induced mode cutoffs and interference. *Appl. Opt.* 21(15):2708–2715.

[9] Papp, A. and H. Harms. 1977. Polarization optics of liquid-core optical fibers. *Appl. Opt.* 16(5):1315–1319.

[10] Payne, D. N. and W. A. Gambling. 1973. The preparation of multimode glass- and liquid-core optical fibres. *Opto-electronics* 5:297–307.

[11] Martelli, C. et al. 2005. Water-core Fresnel fiber. *Opt. Expr.* 13(10):3890–3895.

[12] Yiou, S. et al. 2005. Stimulated Raman scattering in an ethanol core microstructured optical fiber. *Opt. Expr.* 13(12):4786–4791.

[13] Majumdar, A. K. et al. 1979. Infrared transmission at the 3.39 um Helium-Neon laser wavelength in liquid-core quartz fibers. *IEEE J. Selected Topics Quant. Electr.* QE-15(6):408–410.

[14]. Takahashi, H. et al. 1985. Optical transmission loss of liquid-core silica fibers in the infrared region. *Opt. Commun.* 53(3):164–168.

[15] Wang, W. et al. 1998. Spectrophotometry with liquid-core optical fibre in aqueous solution phase in the ultraviolet region. *Analyt. Chim. Acta* 375:261–267.

[16] Gooijer, C. et al. 1998. Detector cells based on plastic liquid-core waveguides suitable for aqueous solutions: One-to-two decades improved detection limits in conventional-size column liquid chromatography with absorption detection. *J. Chromatogr. A* 824:1–5.

[17] Dress, P. et al. 1998. Water-core waveguide for pollution measurements in the deep ultraviolet. *Appl. Opt.* 37(21):4991–4997.

[18] Dress, P. et al. 1998. Physical analysis of teflon coated capillary waveguides. *Sensors Actuators B* 51:278–284.

[19] Dress, P. and H. Franke. 1997. Increasing the accuracy of liquid analysis and *p*H-value control using a liquid-core waveguide. *Rev. Sci. Instr.* 68(5):2167–2171.

[20] Li, J. and P. K. Dasgupta. 1999. Chemiluminescence detection with a liquid core waveguide: Determination of ammonium with electrogenerated hypochlorite based on the luminal-hypochlorite reaction. *Analyt. Chim. Acta* 398:33–39.

[21] Li, J. and P. K. Dasgupta. 1999. Chemiluminescence detection with a liquid core waveguide. Determination of ammonium with electrogenerated hypochlorite based on the luminal-hypochlorite reaction. *Analyt. Chim. Acta* 398:33–39.

[22] Wang, S. et al. 2001. A miniaturized liquid core waveguide-capillary electrophoresis system with flow injection sample introduction and fluorometric detection using light-emitting diodes. *Analyt. Chem.* 73(18):4545–4549.

[23] Wang, S. and Z. Fang. 2005. Integrating functional components into capillary electrophoresis systems using liquid-core waveguides. *Analyt. Bioanalyt. Chem.* 382:1747–1750.

[24] Kostal, V. et al. 2005. Fluorescence detection system for capillary separations utilizing a liquid core waveguide with an optical fibre-coupled compact spectrometer. *J. Chromatogr. A* 1081:36–41.

[25] Kuribara, M. and Y. Takeda. 1983. Liquid core optical fibre for voltage measurement using Kerr effect. *Electr. Lett.* 19:133–135.

[26] Hartog, A. H. 1983. A distributed temperature sensor based on liquid-core optical fibers. *J. Lightwave Technol.* LT-1(3):498–509.

[27] de Vries, M. et al. 1991. Liquid core optical fiber temperature sensors. *IEEE Proc. Southeastcon '91* 2:1135–1138.

[28] Golovkin, S. V. et al. 1995. Radiation damage studies on new liquid scintillators and liquid-core scintillating fibers. *Nucl. Instr. Methods Phys. Res. A* 362:283–291.

[29] Annis, P. et al. 1997. A new vertex detector made of glass capillaries. *Nucl. Instr. Methods Phys. Res. A* 386:72–80.

[30] Konijn, J. et al. 2000. Capillary detectors. *Nucl. Instr. Methods Phys. Res. A* 449:60–80.

[31] Annis, P. et al. 1998. High-resolution tracking using large capillary bundles filled with liquid scintillator. *Nucl. Instr. Methods Phys. Res. A* 386:186–195.

[32] Hoepfner, K. and W. Schmidt-Parzefall. 2000. Application of liquid-core fibres for a radiation-hard vertex detector. *Nucl. Instr. Methods Phys. Res. A* 440:45–46.

[33] Hoepfner, K. et al. 1998. Reconstruction of neutrino interactions observed in a liquid-core fibre detector. *Nucl. Instr. Methods Phys. Res. A* 406:195–212.

[34] He, G. S. and P. N. Prasad. 1990. Stimulated Kerr scattering and reorientation work of molecules in liquid CS_2. *Phys. Rev. A* 41(5):2687–2697.

[35] Zhou, J. Y. et al. 1990. Efficient generation of ultrafast broadband radiation in a submillimeter liquid-core waveguide. *Appl. Phys. Lett.* 57(7):643–644.

[36] He, G. S. et al. 1990. A novel nonlinear optical effect: Stimulated Raman-Kerr scattering in a benzene liquid-core fiber. *J. Chem. Phys.* 93(11):7647–7655.

[37] Zhou, J. Y. et al. 1991. Generation of frequency-tunable ultrashort optical pulses with liquid-core fibers. *Opt. Lett.* 16(23):1865–1867.

[38] He, G. S. et al. 1995. Broadband, multiwavelength stimulated-emission source based on stimulated Kerr and Raman scattering in a liquid-core fiber system. *Appl. Opt.* 34(3):444–454.

[39] Shuquin, G. et al. 2004. Application of liquid-core optical fiber in the measurements of Fourier transform Raman spectra. *Chem. Phys. Lett.* 392:123–126.

[40] Altkorn, R. et al. 1997. Low-loss liquid-core optical fiber for low-refractive-index liquids: Fabrication, characterization, and application in Raman spectroscopy. *Appl. Opt.* 36(34):8992–8998.

[41] Qi, D. and A. J. Berger. 2004. Quantitative analysis of Raman signal enhancement from aqueous samples in liquid core optical fibers. *Appl. Spectrosc.* 58(10):1165–1171.

[42] He, G. S. and G. C. Xu. 1992. Efficient amplification of a broad-band optical signal through stimulated Kerr scattering in a CS_2 liquid-core fiber system. *IEEE J. Quant. Electr.* 28(1):323–329.

[43] Correia, R. R. B. et al. 1999. Dye-induced spectral narrowing of stimulated scattering in CS_2. *Chem. Phys. Lett.* 313:553–558.

[44] Khoo, I. et al. 2001. Passive optical limiting of picosecond-nanosecond laser pulses using highly nonlinear organic liquid core fiber array. *IEEE J. Sel. Topics Quant. Electr.* 7(5):760–768.

[45] Katzir, A. 1993. *Lasers and Optical Fibers in Medicine.* Academic Press, New York.

[46] Kieslinger, D. et al. 1997. Lifetime-based capillary waveguide sensor instrumentation. *Sensors Actuators B* 38-39:300–304.

[47] Gregory, K. W. and R. R. Anderson. 1990. Liquid core light guide for laser angioplasty. IEEE Journal on Quantum Electronics, 26(12):2289–2296.

[48] Diemer, S. et al. 1995. Liquid light guides for 2.94 µm. *Proc. SPIE* 2396:88–94.

[49] Meister, J. et al. Advances in the development of liquid-core waveguides for IR applications. *Proc. SPIE* 2677:120–127.

[50] Diemer, S. et al. 1997. Liquid-core light guides for near-infrared applications. *Appl. Opt.* 36(34):9075–9082.

[51] Wolinski, T. R. and A. Szymanska. 2001. Polarimetric optical fibres with elliptical liquid-crystal core. *Measure. Sci. Technol.* 12:948–951.

[52] Jeong, Y. and B. Lee. 2001. Theory of electrically controllable long-period gratings built in liquid-crystal fibers. *Opt. Eng.* 40(7):1227–1233.

[53] Fini, J. M. 2004. Microstructure fibres for optical sensing in gases and liquids. *Measure. Sci. Technol.* 15:1120–1128.

[54] Martelli, C. et al. 2005. Water-core fresnel fiber. *Opt. Exp.* 13(10):3890–3895.

[55] Yiou, S. et al. 2005. Stimulated Raman scattering in an ethanol core microstructured optical fiber. *Opt. Exp.* 13(12):4786–4791.

[56] Wolfe, D. B. et al. 2004. Dynamic control of liquid-core/liquid-cladding optical waveguides. *Proc. Natl. Acad. Sci. USA* 101(34):12434–12438.

Chapter 20

Polymer Optical Fibers

Olaf Ziemann

Polymer Optical Fiber Application Center, Nürnberg, Germany

20.1 INTRODUCTION

This chapter describes the polymer optical fiber (POF), probably one of the fiber types with the highest loss and the smallest bandwidth. Nevertheless, it is the only optical fiber that can be installed by everyone without any special tool. That is why the potential of POF systems is very high.

20.2 POF BASICS

The first POFs were manufactured by DuPont as early as the late 1960s. Because of the incomplete purification of the source materials used, optical attenuation values remained in the vicinity of 1000 dB/km. During the 1970s, it became possible to reduce losses nearly to the theoretical limit of approximately 125 dB/km at a wavelength of 650 nm. At that point, glass fibers with losses significantly below 1 dB/km at 1300 nm/1550 nm were already available in large quantities and at low prices. Digital transmission systems with a high bit rate were then almost exclusively used in telecommunications for long-range transmissions. The field of local computer networks was dominated by copper cables (either twisted-pair or coaxial) that were completely satisfactory for the typical data rates of up to 10 megabits per second (Mbps) commonly used then. There was hardly any demand for an optical medium for high data rates and small distances, so the development of the POF was slowed down for many years. A significant indicator for this is that at the beginning of the 1990s, the company Hoechst stopped manufacturing polymer fibers.

During the 1990s—after data communication for long-haul transmission had become completely digital—the development of digital systems for private users was started on a massive scale. In many areas of life, we are being increasingly confronted with digital end-user equipment. The CD player has largely replaced analog sound carriers (vinyl records and cassettes). The MP3 format is leading to a revolution in music recording and distribution. The DVD (Digital Versatile Disc) replaces the analog video recorder. Even today more digital television programs are available than analog programs. Decoder boxes have become standardized (MPEG2 format) and will be integrated into television sets. More and more households are using powerful PC and digital telephone connections (ISDN) or triple-play services. With offers such as T-DSL (ADSL and VDSL technology provided by Deutsche Telekom AG), as well as fast internet access via satellite or broadband digital services on the broadband cable network, private users were offered access to additional digital applications even before the start of the new millennium. Likewise, in the automotive field the step towards digitalization has long been made. CD changers, navigation systems, distance-keeping radar, and complex control functions are increasingly part of the standard equipment being provided in all classes of vehicles. The development of electronic outside mirrors, fast network connections—even from within an automobile—and automatic traffic guidance systems will ensure a further increase in the range of digital applications for the motor vehicle. All these examples demonstrate that completely new markets for digital transmission systems are being developed for short-range applications. POFs can meet many of these requirements to an optimum degree and are, therefore, increasingly of interest.

20.2.1 Materials for POF

The majority of all used POFs are made from polymethylmethacrylate (PMMA) as the core material. Due to the Rayleigh scattering and the strong absorption of the C-H-bonds, the smallest attenuation is approximately 100 dB/km (Fig. 20.1).

The absorption peak at 620 nm with a typical loss of 440 dB/km is related to the sixth overtone of the C-H-bond vibration as an example. Loss minima of PMMA are at 520 nm, 570 nm, and 650 nm. The only way to reduce the loss of the material is the substitution of hydrogen by heavier atoms, like fluorine. Figure 20.2 shows the molecule structure of PMMA and CYTOP (a completely fluorinated polymer by Asahi Glass Co.).

The lowest loss ever reported for PF-GI POF is less than 10 dB/km (see reference [1] and Fig. 20.3 for the attenuation spectrum).

Other options for polymer fibers are polycarbonate (PC) or elastomers. The main reason for the search for other materials is the limitation of the operating temperature of PMMA-based POF. Most of the POFs available are

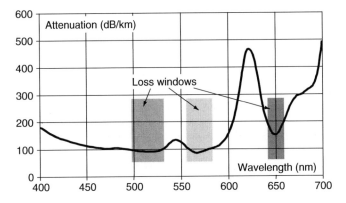

Figure 20.1 Attenuation spectrum of PMMA-POF.

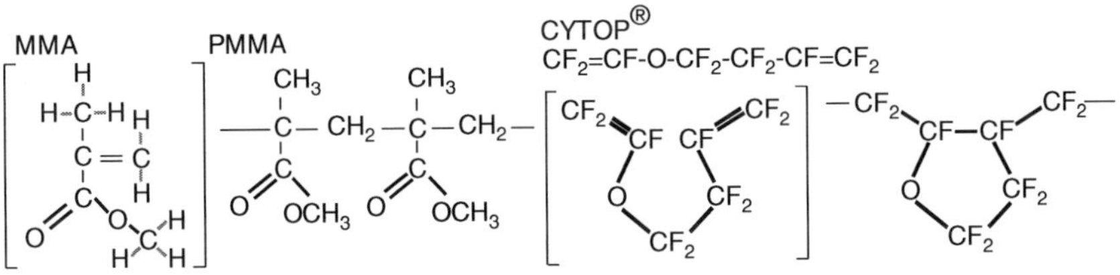

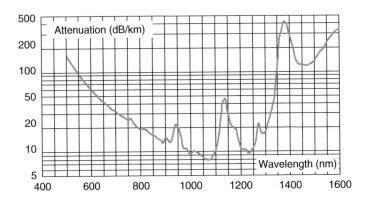

Figure 20.2 Molecule structure of PMMA and PF polymer.

Figure 20.3 Loss of PF-GI-POF (used, with permission, from reference [1]).

specified for a maximum operation temperature of +70 to +85 °C. The temperature range can be increased by cross-linking of the PMMA up to 130 °C (see reference [2] for examples). Much higher temperatures (up to +170 °C) can be realized by using elastomers, but there are no commercial products available.

20.2.2 Light Propagation Effects in POF

Because of the large fiber diameter, the high numerical aperture (NA), and the short operation wavelengths, the POF can guide more modes than every other kind of fiber. The standard 1-mm POF with an NA of 0.50 owns about $2\frac{1}{2}$ million modes. On the other hand, the mode-dependent effects in POF are more important than in other fibers. The reasons are the high loss in the cladding material (several 10,000 dB/km) and the strong mode mixing in the polymer material and mainly at the core–cladding interface.

The effect of the cladding material absorption can be seen clearly by measuring the mode-dependent loss. One result for a 1-mm PMMA POF is shown in Fig. 20.4, measured at 650-nm wavelength [3].

The consequence of the strong mode-dependent loss is that the far-field width under equilibrium mode distribution (EMD) conditions is much smaller than calculated from the NA. The measured bandwidth of most POFs is higher than determined by the NA (under Uniform Mode Distribution [UMD] assumption) and the bend loss is smaller. The influence of mode-dependent loss is shown in Fig. 20.5 [4].

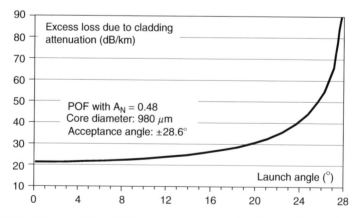

Figure 20.4 Influence of the cladding absorption on the loss (used, with permission, from reference [3]).

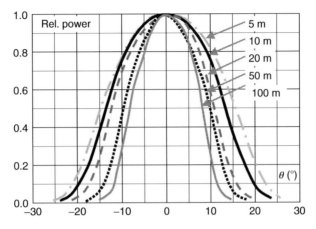

Figure 20.5 Effect of mode-dependent loss on the far-field distribution; 1 mm PMMA-POF (used, with permission, from reference [4]).

The second effect is the strong mode mixing (strongly dependent on the fiber type and the cable manufacturing). Figure 20.6 shows the length-dependent measured attenuation for a 1-mm POF [5]. The equilibrium state can be seen after a length of about 100 m.

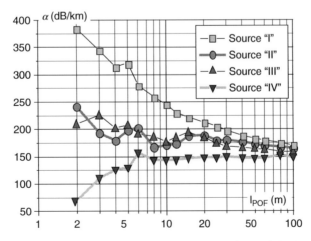

Figure 20.6 Launch and length-dependent loss of PMMA-POF.

20.2.3 Bandwidth of POF

The parameter with the biggest influence of the mode-dependent effects is the bandwidth. Results of length and launch-dependent bandwidth for a 1-mm PMMA POF, measured at 650 nm wavelength, are presented in Fig. 20.7 [4].

The differences of the measured bandwidth for short lengths are more than one order of magnitude and still a factor of 2 at 100 m. The same behavior can be seen for other multimode fibers as 200 μm plastic clad silica (PCS). An example is shown in Fig. 20.8.

20.3 TYPES OF POF

POFs are available with different index profiles. The aim of the modified profiles is an increased bandwidth and a reduced bending sensitivity. As was the case with silica glass fibers, the first POFs were pure step-index profile fibers (SI POF). This means that a simple optical cladding surrounds a homogenous core. For this reason, a protective material is always included in the cable. Figure 20.9 schematically represents the refractive index curve.

Glass multimode fibers usually have an NA of approximately 0.20. Glass fibers with polymer cladding have an NA in the range of 0.30–0.50. The large refractive index difference between the materials that are used for the core and the cladding of polymer fibers allows significantly higher NA values. Most of the

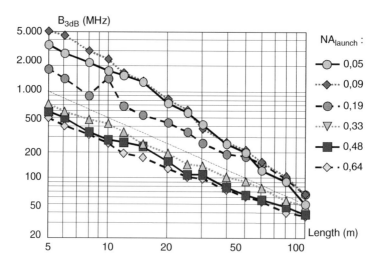

Figure 20.7   Length and launch-dependent bandwidth of 1 mm PMMA-POF.

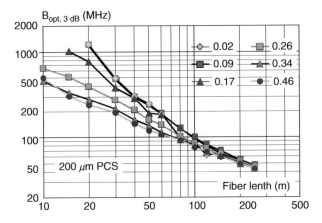

Figure 20.8   Length and launch-dependent bandwidth of 200-μm SI-PCS.

initially produced SI POFs had an NA of 0.50 (e.g., see references [6–8]). SI POF with an NA around this value is nowadays generally called *standard NA POF,* or *standard POF* for short. The bandwidth of such fibers is approximately 40 MHz for a 100-m long link (quoted as the bandwidth-length product 40 MHz · 100 m). For many years, this was a completely satisfactory solution for most applications. However, when it became necessary to replace copper cables with POF to accomplish the transmission of ATM (i.e., asynchronous transfer mode) data rates of 155 Mbps over a distance of 50 m, a higher bandwidth was required for the POF. In the mid-1990s, all three important manufacturers developed the so-called *low-NA POF.*

POFs with a reduced NA (low-NA POFs) feature a bandwidth increased to approximately 100 MHz/100 m because the NA has been reduced to approximately 0.30. The first low-NA POF was presented in 1995 by Mitsubishi Rayon [9]. Usually the same core material as for SI POF is used, but the cladding material has an altered composition.

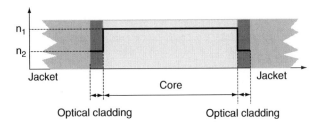

Figure 20.9   Structure of SI-POF.

Unfortunately, practical testing showed that although this fiber met the requirements of the ATM forum [10] with respect to bandwidth, it did not meet the requirements with respect to bending sensitivity. These requirements specify that for a 50-m long POF link, the losses resulting from a maximum of ten 90-degree bends having a minimum bending radius of 25 mm should not exceed 0.5 dB. To meet both these requirements simultaneously, it became necessary to find a new structure.

The double SI (DSI) POF features two claddings around the core, each with a decreasing refractive index (Fig. 20.10). In the case of straight installed links, light conduction is achieved essentially through the total reflection at the boundary surface between the core and the inner cladding. This index difference results in an NA of around 0.30, similar to the value of the original low-NA POF.

When fibers are bent, part of the light will no longer be conducted by this inner boundary surface. However, it is possible to reflect back part of the decoupled light in the direction of the core at the second boundary surface between the inner and the outer cladding. At further bends, this light can again be redirected so that it enters the area of acceptance of the inner cladding. The inner cladding has a significantly higher attenuation than the core. Light propagating over long distances within the inner cladding will be attenuated so strongly that it will no longer contribute to pulse propagation. Over shorter links, the light can propagate through the inner cladding without resulting in too large a dispersion. A schematic illustration is shown in Fig. 20.11. All low-NA POFs offered today are DSI POFs in reality.

As described earlier, the requirements of high bandwidth and low sensitivity to bending are difficult to accomplish together within one fiber having a diameter of 1 mm. Fibers with a smaller core diameter can solve this problem because the ratio to the fiber radius is larger for the same absolute bending radius. However, this contradicts the requirements for easy handling and light launching. As a compromise, Asahi developed a multicore fiber (MC POF) (see references [11–13]). In this fiber, many cores (19 to >200) are put together in production in such a way that they together fill a round cross-section of 1-mm diameter. Figure 20.12 shows the parameters for the percentage of covered area.

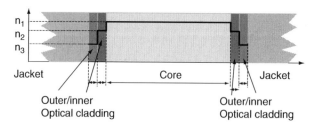

Figure 20.10   Structure of a double step-index profile fiber.

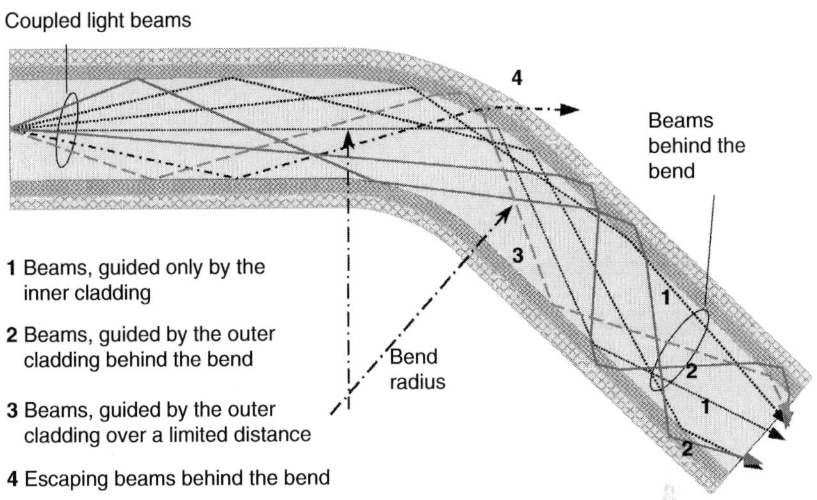

Coupled light beams

4

Beams
behind the
bend

1 Beams, guided only by the
inner cladding

2 Beams, guided by the outer
cladding behind the bend

Bend
radius

3 Beams, guided by the outer
cladding over a limited distance

4 Escaping beams behind the bend

3

1

2

1

2

Figure 20.11 Operation of a bent double step-index profile fiber.

For 37 cores with $d_m = 5\,\mu$m, the partition of core area is only 65.3% and the value is 51.7% for 217 cores. Practical experience shows that a better utilization of the area can be achieved.

During the manufacturing process, the fibers are placed together at a higher temperature, which means that they change their shape and, thus, reduce the gaps between the fibers. Apparently, the resulting deviations from the ideal round shape do not play a significant role in light propagation (the causes for this are not yet completely understood.

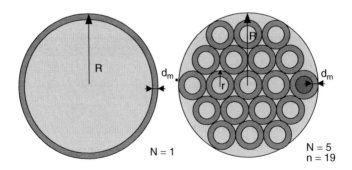

R

d_m

R

r

d_m

N = 1

N = 5
n = 19

Figure 20.12 Schematic arrangement of cores in an MC-POF.

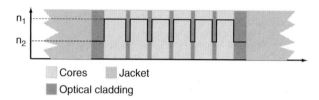

Figure 20.13   Structure of a step-index multicore fiber.

Figure 20.13 shows the refractive index curve of a MC POF, shown as a cross-section through the diameter of the fiber. The index steps correspond to those of a standard POF.

Because the bandwidth only depends on the NA for SI fibers, it should be possible to measure values comparable to the standard POF. However, the fact is that the measured values are actually significantly higher, which has been explained with the aforementioned mode-selective attenuation mechanisms. In the MC POF, too, an increase in bandwidth was achieved by reducing the index difference. Because of the smaller core diameters, it was still possible to avoid an increase in bending sensitivity. Even better values were achieved with individual cores having a two-step optical cladding such as illustrated in Fig. 20.14. The principle is the same as in the DSI POF with an individual core. In this case, a bundle with single cladding is completely surrounded by a second cladding material ("sea/islands" structure).

The MC POF features a noticeably reduced sensitivity to bending and only insignificantly increased attenuation, as well as a significantly increased bandwidth compared to single-core fibers, possibility because of smaller NAs. Whether these fibers can be produced at the same price is still an open question. Should this be possible, data rates of 1000 Mbps over 50 m can easily be achieved. Figure 20.15 shows a cross-section of different MC POFs.

When using graded-index (GI) profiles, an even greater bandwidth becomes possible. In these profiles, the refractive index continually diminishes (as a gradient), starting from the fiber axis and moving outwards to the cladding. Of particular interest are profiles that follow a power law.

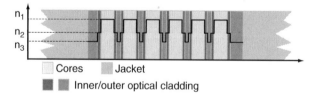

Figure 20.14   Structure of a double step-index multicore fiber.

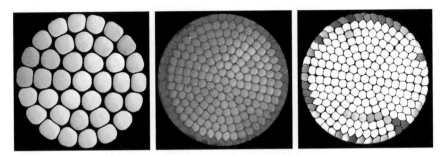

Figure 20.15 Cross-view of different multicore fibers (Asahi Chem.).

$$\text{refractive index } n^2 = n^2_{\text{fiber axis}} \cdot \left[1 - \left(\frac{\text{distance to fiber axis}}{\text{core radius}}\right)^g \cdot \Delta\right] \quad (20.1)$$

Parameter g is referred to as the index coefficient. When $g = 2$, we speak of a parabolic profile. The limiting case of step-index profile fibers is described by $g = \infty$. Parameter Δ signifies the complete index difference between the fiber axis and the edge of the core. Figure 20.16 shows a parabolic index profile.

Because of the continuously changing refractive index, the light beams in a GI fiber do not propagate in a straight line but are constantly refracted towards the fiber axis. Light beams that are launched at the center of the fiber and do not exceed a certain angle are completely prevented from leaving the core area without any reflections occurring at the boundary surface. This behavior is illustrated schematically in Fig. 20.17. The geometric path of the beams running on a parallel axis is still significantly smaller than the path of beams that are introduced at a greater angle.

However, as can be seen, the index is smaller in the regions distant from the core. This means a greater propagation speed. In an ideal combination of parameters, the different path lengths and different propagation speeds may cancel each other out completely so that mode dispersion disappears. In reality, this is only possible in approximation. However, it is possible to increase bandwidths by two to three orders of magnitude compared with the SI fiber.

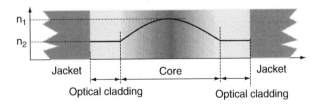

Figure 20.16 Structure of a graded-index profile fiber.

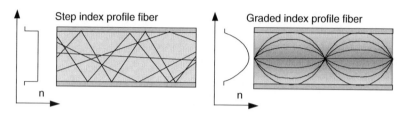

Figure 20.17   Comparison of step- and graded-index profile.

When considering not only the pure mode dispersion but also chromatic dispersion (i.e., the dependence of the refractive index on the wavelength and spectral width of the source), an optimum index coefficient g deviating from 2 is achieved. This has been the subject of comprehensive investigations by the research group around Y. Koike [9, 13–18]. In references [16] and [19], the significance of this effect is particularly pronounced. Because of the smaller chromatic dispersion of fluorinated polymer compared with silica, the bandwidth of GI POF theoretically achievable is significantly higher than that of multimode GI glass fibers. In particular, this bandwidth can be realized over a significantly greater range of wavelengths. This makes the PF-GI POF interesting for wavelength multiplex systems. However, in this case, the index profile must be maintained very accurately, a requirement for which no technical solution has yet been provided.

Another factor involved in the bandwidth of GI POF is the high level of mode-dependent attenuation [20] compared to glass fibers. In this case, modes with a large propagation angle are suppressed resulting in a greater bandwidth. An example is the simulation that was carried out by Yabre [20]: The bandwidth of a 200-m long PMMA-GI POF increases from 1 GHz to more than 4 GHz, taking into account the attenuation of higher modes. This was also shown in practical trials. Mode coupling is less significant for GI fibers than it is for SI fibers because the reflections at the core–cladding boundary do not occur.

Following the many technological problems experienced in the production of GI fibers having an optimum index profile that remains stable for the duration of its service life, an attempt was made to approach the desired characteristics with the multistep-index (MSI) profile fiber. In this case, the core consists of many layers (e.g., four to seven) that approach the required parabolic curve in a series of steps. Here, a "merging" of these steps during the manufacturing process may even be desirable. A diagram of the structure is shown in Fig. 20.18.

In this case, light beams do not propagate along continually curved paths as in the GI POF, but on multiply diffracted paths as demonstrated in Fig. 20.19. However, given a sufficient number of steps, the difference to the ideal GI profile is relatively small so that large bandwidths can nevertheless be achieved. MSI

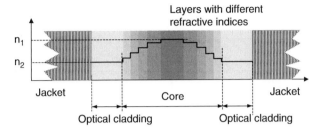

Figure 20.18 Structure of a multistep-index profile fiber.

POFs were presented in 1999 by a Russian institute (Tver near Moscow [21]) and by Mitsubishi (ESKA-MIU, see reference [22]).

The manufacturing of PMMA-based POF was very difficult over a number of years. A number of groups tried to generate the GI profile by dopants. The disadvantage is that the doping reduces the glass-transition temperature of the core. This leads to diffusion of the dopant molecules at higher temperature, ending with a destroyed index profile. A number of products have been announced over the last years but had never become available fibers. Table 20.1 gives an overview of published GI and MSI POFs.

The most interesting development in the field of PMMA-GI POF was made by Optimedia in South Korea in the last years. The manufacturing process is shown in Fig. 20.20. A rotating PMMA tube is filled with liquid monomers. The composition of the reactants can be changed continuously or stepwise. The polymerization process is induced by heat and/or UV radiation. The result is a preform with an MSI or parabolic profile. Because of the use of co-polymers instead of dopants, temperature stability is comparable to PMMA fibers.

The measured refractive index profile is shown in Fig. 20.21. The profile is very close to the ideal parabolic shape. At present, two fibers with $900\,\mu m/1000\,\mu m$ or $500\,mm/750\,\mu m$ core/outer diameter are available.

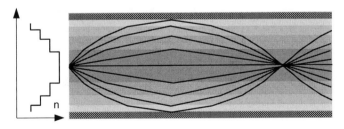

Figure 20.19 Light propagation in the MSI-POF.

Table 20.1

Parameters of MSI- and GI-POF (see references [23] and [24] for details)

Reference	Year	Institution	Material	$Oslash_{core}$, μm	Loss (dB/km)	at λ (nm)	NA	Comments
[25]	1998	Mitsubishi	k.A.	1000	110	650	0.47	80 MHz/100 m
[26]	1982	Keio Univ.	MMA co VPAc	n.a.	1070	670	n.a.	First GI-POF
[13]	1990	Keio Univ.	PMMA	n.a.	n.a.	k.A.	n.a.	670 nm: 300 MHz/km
[26]	1990	Keio Univ.	MMA co VB	n.a.	130	650	n.a.	
[27]	1990	Keio Univ.	MMA-VB	n.a.	134	652	n.a.	260 MHz/1 km
[27]	1990	Keio Univ.	MMA-VPAc	n.a.	143	652	n.a.	125 MHz/1 km
[28]	1992	Keio Univ.	PMMA	200–1500	113	650	n.a.	1.000 MHz/km
[28]	1992	Keio Univ.	PMMA	200–1500	90	570	n.a.	
[29]	1994	Sumitomo	PMMA	400	160	650	0.26	$\Delta n = 0.014, 8\,GHz/50\,m$
[30]	1995	BOF	PMMA	600	300	650	0.19	3 GHz/100 m
[31]	1995	Keio Univ.	PMMA-DPS	500–1000	150	650	n.a.	585 MHz/km
[32]	1997	Keio Univ.	PMMA	n.a.	n.a.	n.a.	n.a.	2 GHz/100 m
[33]	1997	Mitsubishi	PMMA	700	210	650	0.30	500 MHz/50 m, MSI
[34]	1998	Kurabe	PMMA	500	132	650	n.a.	2 GHz/100 m
[34]	1998	Kurabe	PMMA	500	145	650	n.a.	2 GHz/90 m
[34]	1998	Kurabe	PMMA	500	159	650	n.a.	680 MHz/50 m
[34]	1998	Kurabe	PMMA	500	329	650	n.a.	
[21]	1999	RPC Tver	PMMA/4FFA	800	400	650		MSI, 310 MHz/100 m
[35]	2002	Dig. Optr.	Polymer	180	350	685	0.20	No samples
[22]	2002	KIST Korea	PMMA	1000	120	650	0.26	$g = 2.4$; 3.45 GHz/100m
[36]	2002	Huiyuan	PMMA	n.a.	n.a.	n.a.	n.a.	Announced for 2001
[37]	2003	Luvantix	PMMA	n.a.	160	650	0.33	3.5-GHz bandwidth
[38]	2004	Lumistar	PMMA	500	n.a.	n.a.	n.a.	3 Gbps/50 m
[39]	2005	Nuvitech	PMMA	500	180	650	0.25	3 Gbps/50 m
[39]	2005	Nuvitech	PMMA	900	180	650	0.30	3 Gbps/50 m
[40]	2005	Optimedia	PMMA, co-polymer	900	n.a.	650	n.a.	3 GHz/50 m
[41]	2005	Optimedia	PMMA, co-polymer	900	200	650	0.30	3 GHz/50 m
[42]	2004	Lumistar-X	New low loss	120	100	850	n.a.	10 GHz/50 m

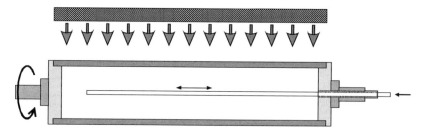

Figure 20.20 Manufacturing of PMMA-GI-POF by Optimedia (used, with permission, from reference [41]).

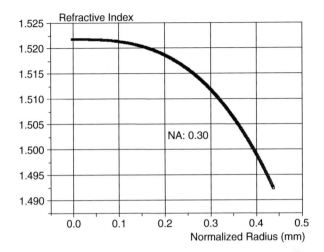

Figure 20.21 Index profile of PMMA-GI-POF by Optimedia (used, with permission, from reference [41]).

We have measured the long-term stability of the 1-mm GI POF. After 5000 hours of aging at 80 °C (dry atmosphere), no degradation of the bandwidth could be observed, indicating a stable index profile. The maximum transmitted bit rate was 2 Gbps over 100 m using a 655-nm edge emitting laser diode and a 800-mm Si-pin photo diode. The loss spectrum of the OM-Giga is shown in Fig. 20.22 (www.fiberfin.com).

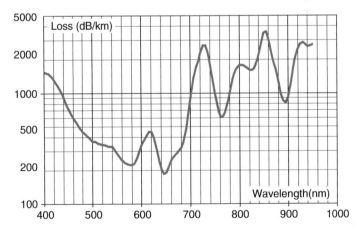

Figure 20.22 Loss spectrum of PMMA-GI-POF by Optimedia.

20.4 POF STANDARDS

The different kinds of POF are specified in the IEC 60793 as the fiber classes A4. Table 20.2 lists the main parameters of the different categories ("IEC 60793-2-40 Ed. 2.0: Optical Fibres; Part 2-40: Product specifications—Sectional specification for category A4 multimode fibres").

- The classes A4a–A4c describe SI-PMMA POFs as used is mobile networks, home applications, and automation.
- The fiber class A4d is a DSI fiber for Fast Ethernet and IEEE1394 applications.
- GI and MSI fibers mainly for high speed home networks are categorized in class A4e.
- The final classes A4f–A4h are GI POFs made of perfluorinated materials for building backbones and local area networks with data rates up to 10 Gbps.

Table 20.2

Parameters of different POFs

Parameter	Unit	Class A4a	Class A4b	Class A4c	Class A4d
Ø core	μm	n.d.	n.d.	n.d.	n.d.
Ø cladding	μm	1000 ± 60	750 ± 45	500 ± 30	1000 ± 60
Ø jacket	mm	2.2 ± 0.1	2.2 ± 0.1	1.5 ± 0.1	2.2 ± 0.1
Core concentricity	%	≤6	≤6	≤6	≤6
Loss at 650 nm	dB/km	≤400	≤400	≤400	≤400
With EMD launch	dB/km	≤300	≤300	≤300	≤180
Bandwidth	MHz/100 m	≥10	≥10	≥10	≥100
Bend loss	dB/10 bend	≤0.5	≤0.5	≤0.5	≤0.5
Numerical aperture	—	0.50 ± 0.15	0.50 ± 0.15	0.50 ± 0.15	0.30 ± 0.05

Parameter	Unit	Class A4e	Class A4f	Class A4g	Class A4h
Ø core		≥500	200 ± 10	120 ± 10	62.5 ± 5
Ø cladding	μm	750 ± 20	490 ± 10	490 ± 10	245 ± 5
Ø jacket	mm	2.2 ± 0.1	n.d.	n.d.	n.d.
Core concentricity	%	≤6	≤4	≤4	≤2
Loss at 650 nm	dB/km	≤180	≤100	≤100	n.d.
Loss at 850 nm	dB/km	n.d.	≤40	≤33	≤33
Loss at 1300 nm	dB/km	n.d.	≤40	≤33	≤33
Bandwidth 650 nm	MHz/100 m	≥200	≥800	≥800	n.d.
Bandwidth 850 nm	MHz/100 m	n.d.	1500–4000	1880–5000	1880–5000
Bandwidth 1300 nm	MHz/100 m	n.d.	1500–4000	1880–5000	1880–5000
Bend loss	dB/10 bend	≤0.50	≤1.25	≤0.60	≤0.25
Numerical aperture	—	0.25 ± 0.07	0.19 ± 0.015	0.19 ± 0.015	0.19 ± 0.015

20.5 POF TRANSMISSION SYSTEMS

The following sections present some selected POF transmission systems with different fibers. A more detailed description can be found elsewhere [23] (second edition planned for end of 2006). The following examples are the best values for bit rate and/or transmission distance for several fibers.

20.5.1 SI-PMMA POF

The typical bandwidth of a PMMA POF with a standard NA of 0.50 is about 40 MHz/100 m. Therefore, the maximum bit rate of a 100-m link should be around 100 Mbps, but there are a number of options for higher capacity. The mode dispersion can be dramatically reduced by low NA launch and detection. Postcompensation and precompensation by high-pass filtering can give further improvements. Finally, if there is sufficient signal-to-noise ratio (SNR), some penalty can be accepted.

High data rate transmission experiments on standard SI POFs were introduced in a series of publications [43–47] spanning the years 1992–1994. With data rates of 265 and 531 Mbps (1994), 100-m POF was covered. Figure 20.23 illustrates the principle of the test setup.

The Mitsubishi ESKA EXTRA EH4001 was used as the fiber medium. It has 139 dB/km of attenuation at 652 nm. A Philips laser diode CQL82 with a wavelength of 652 nm served as the light source. The laser was operated at 290 K (17 °C) with 36-mA bias current. To increase the bit rate, a first order high-pass filter was preconnected as the peaking filter. With the help of input optics, $2.7 \, \text{mW}_{\text{p–p}}$ of power was achieved at launch of NA = 0.11. During

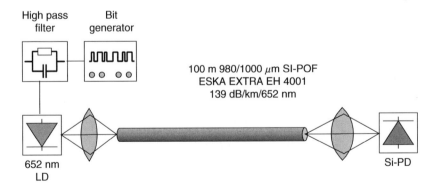

Figure 20.23   High-bit-rate data transmission over SI-POF.

modulation, the average power was −1.7 dBm (0.68 mW); with the peaking filter, the average power fell to −6.7 dBm (0.21 mW). An AEG-Telefunken BPW89 photodiode with 4.9-pF capacity at 20 V of reverse voltage was used as a receiver. The responsivity is 0.4 A/W at 650 nm (76% external efficiency). The coupling to the POF is done with a ball lens. A second high-pass filter was connected behind the receiver as a compensation filter for the mode dispersion. The receiver achieved −22.1 dBm sensitivity at BER = 10^{-9}. As a result, a data rate of 265 Mbps was achieved.

The newest result is the transmission of 580 Mbps over 100 m of standard POF (Mitsubishi MH4001) at the POF-AC in 2006. We used a 650-nm laser with +6 dBm optical power, an 800-μm Si-pin-PD, directly coupling at the receiver and transmitter side, and a passive compensation filter behind the receiver.

20.5.2 PMMA-GI POF

The highest bit rate ever transmitted over PMMA-GI POF is described elsewhere [48]. The transmission wavelength was 645 nm (5-mW optical power), the fiber diameter was about 500 μm, and the receiver based on an Si-APD with −29 dBm sensitivity (Fig. 20.24).

The POF-AC has demonstrated the transmission of 2 Gbps over 100 m of a 1-mm PMMA-GI POF (OM-Giga of Optimedia) in 2005, using a 650-nm laser diode and an Si-pin-PD receiver. The transmission of the complete coaxial cable TV signal (862 MHz) over 50 m of OM-Giga PMMA-GI POF has been demonstrated by the Fraunhofer Institute [49].

20.5.3 PF-GI POF

The perfluorinated material CYTOP (made by Asahi Glass) offers the lowest attenuation of all POFs. Because of the parabolic index profile and the low chromatic dispersion, the bandwidth of PF-GI POF can be very high. In contrast to the SiO$_2$ GI-GOF, the bandwidth is high over a wide wavelength range from

645 nm
LD

200 m PMMA-GI-POF,
Mitsubishi, 164 dB/km
Ø 0.5 mm

Si-APD

Figure 20.24 2.5 Gbps over 200-m PMMA-GI-POF.

600 to 1300 nm. Some of the best results ever reported with the PF-GI POF include the following:

- 1.25 Gbps over 1006 m at 1300 nm [50]
- 2.5 Gbps over 550 m at 1310 nm [51]
- Three-channel WDM 2.5 Gbps over 200 m [48]
- 12.5 Gbps over 100 m at 850 nm (Nexans)

The bit rate time's length world record was realized by the group around D. Khoe in 1999 (see references [48] and [51]). The transmission length was 550 m and the bit rate was 2.5 Gbps. This was made possible by providing a 550-m GI POF piece with a core diameter of 170 μm without any connector (Fig. 20.25).

Experiments with various sources were carried out. The measured attenuation for the wavelengths was as follows:

- 110 dB/km at 650 nm (LD as source)
- 43.6 dB/km at 840 nm (VCSEL as source)
- 31 dB/km at 1310 nm (LD as source)

The VCSEL supplies 1.3 dBm of power at a spectral width of 1 nm. It was possible to couple it directly to the POF (<1-dB loss). A passive filter for the VCSEL frequency response compensation was used.

An Si-APD with 230-μm diameter was used for the receiver at 840 nm. It reached -28.6 dBm sensitivity with a BER $= 10^{-9}$, whereby a budget of 29.9 dB was available. In addition, a 1310-nm DFB laser with a modulation bandwidth of 5 GHz, a spectral width of 0.1 nm, and maximum 0.4 dBm of optical output power (1.1 mW) was used. The laser is a standard transmitter element for single-mode fiber systems and is equipped with a corresponding fiber pigtail for single-mode fiber systems. The single-mode fiber was also used for direct coupling to the GI POF (<0.1-dB loss). With this method, only a small part of the mode field is excited, which increases the bandwidth considerably. The highest transmission length with PF-GI POF, ever reported, was 1006 m [50] (Fig. 20.26).

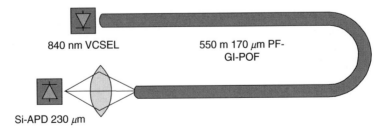

840 nm VCSEL   550 m 170 μm PF-GI-POF

Si-APD 230 μm

Figure 20.25   POF system with record transmission distance according to reference [51].

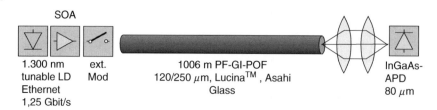

Figure 20.26   POF system with record transmission distance according to reference [50].

20.6 APPLICATIONS OF POF

There is a wide area of POF applications, including the following examples:

- POFs are used in car networks for entertainment systems (Digital Domestic Bus [D2B], Media Oriented System Transport [MOST], IDB 1394 systems) and for control networks (Byteflight). The use of PCS will be specified in the next MOST generation.
- POFs and PCS are widely used in automation (field busses like Profinet, Sercos, and Fast Ethernet).
- A wide area of application exists for illumination and design (side emitting fibers, textiles).
- POFs can be used as sensors (e.g., pedestrian protection, demonstrated by Siemens VDO 2004, see also references [52–54]).

20.6.1 POF in Automobile Networks

DaimlerChrysler introduced the use of POF in automotive networks in 1998. The D2B was designed for the transmission of entertainment data with up to 5.6 Mbps. The next generation of optical car busses was the MOST system, developed under the participation of manufacturers like DaimlerChrysler, Audi, and BMW. The bit rate of the MOST bus is about 25 Mbps, sufficient for DVD data as well. Some of the technical data of the MOST system are as follows [55–57]:

- Transmitter: 650 nm LED or RC-LED
- Wavelength range (over temperature): 630–685 nm
- Spectral width: <30 nm
- Fiber coupled power: −1.5 to −10 dBm
- Optical power budget: 16.5 dB
- Temperature range: −40 to +85 °C

- Receiver dynamic range: 25 dB
- Fiber type: 980 μm PMMA-SI POF, NA 0.50
- Fiber bandwidth; >30 MHz/100 m
- Minimum fiber bend radius: 25 mm

The architecture of the MOST system is a unidirectional active ring. Every device works as a repeater. The maximum link length between two devices is about 8 m. Power consumption is reduced by a sleep mode. The disadvantage of the architecture is the complete loss of operation if one device fails. The first MOST car was the BMW 7 series. In September 2005, the number of cars with MOST equipment was increased to 34, representing 10 Mio devices per year. Cars in alphabetical order include the following:

- Audi A6, A8, Q7
- BWM 1, 3, 5, 6, 7 Series
- Citroen C8
- DaimlerChrysler A, B, C, CLS, E, M, S, SLK classes
- Dodge RAM
- Lancia Phedra
- Landrover Discovery, Range Rover, Freelander
- Maybach
- Mitsubishi Colt
- Peugot 807
- Porsche Boxter, 911, Cayenne
- Rolls-Royce Phantom
- Saab 9-3
- Smart Forfour
- Volvo S40, V50, XC90

Active components for MOST and Byteflight (Infineon Technologies Regensburg, Germany) and a typical hybrid optical/electrical MOST connector (Komax, Switzerland) is shown in Fig. 20.27.

Figure 20.27 Active components and hybrid connector for car networks.

20.6.2 POF Sensors

Fiber optic sensors using POF have been described in a general overview [52, 53]. POF sensors are mostly restricted to multimode fibers, are of intrinsic and extrinsic type, and the change of intensity will be measured in most cases.

In these application areas, additional benefits of POFs are exploited such as ease of handling, robustness, flexibility, and low weight. These new systems benefit from the positive experience gained with more than 4 million cars equipped with optical data bus systems using MOST, D2B, or Byteflight. One possible sensor is the Pedestrian Protection System (PPS), developed and published by Siemens VDO [58–60]. Due to European Union directive 2003/102/EC, pedestrian protection has to be provided for every new car from 2007 on. This can be fulfilled by either of the following:

- Passive means: structural measures such as "soft" front ends and sufficient deformation room between the hood and the engine
- Active means: sensors that identify a pedestrian impact and then trigger protective means such as lifting the hood by means of actuators as shown in Fig. 20.28.

Figure 20.28 Active pedestrian protection system with lifting hood (used, with permission, from reference [61]).

The basic principle behind this approach is a cladding surface treatment of the fiber at discrete zones along the fiber. As shown in Fig. 20.29, bending the fiber in one direction leads to a better transmission, whereas bending in the other direction leads to a lower transmission, compared to the straight position.

The spatial resolution is achieved by using several fiber strands in parallel and numerical signal analysis of the treated zones. Because of its principle, the sensor is able to distinguish between positive and negative bends, and because of its high bandwidth, the sensor can determine the impact with a high temporal resolution. This is necessary to distinguish not only between a human leg and a lamp-pole but also for identifying a collision with an animal.

A different approach for the same and other applications has been published [62] called *Kinotex cavity sensor*. The principle behind it is light scattering dependent on the compression of the scattering medium (e.g., rubber foam). The transducer operates by detecting a change in energy intensity in and around an illuminated integrating cavity. Deformation of the integrating cavity by an external influence such as pressure results in a localized change to the illumination energy intensity. This change is measured. The information can be used to measure the state of deformation.

The *Tactile Sensor* (described in reference [63]) uses the so-called "evanescent field." Although it is a known fact from theory, it used to be often forgotten that

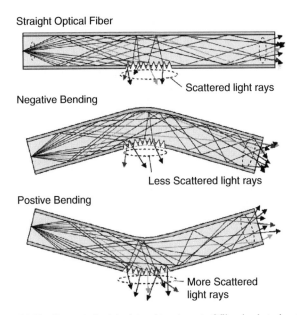

Figure 20.29 Sensor principle: lateral treatment of fiber leads to bending.

the optical rays do not reflect exactly on the boundary between two optical media. In fact, the rays (electromagnetic field) break into the adjacent medium and do not suddenly drop to zero at the core–cladding boundary. They decay instead exponentially within the cladding. The penetration depth d_p depends on the difference between the respective refractive indices of core and cladding material (the higher the difference, the lower d_p), the angle ε of incidence (the steeper ε the higher d_p) and the wavelength λ of the incident light (the higher λ, the higher d_p). In ray optics, this phenomenon is known as "Goos-Hänchen-Shift," indicating that the reentry of the ray into the core region is shifted along the geometrical core–cladding interface.

This type of sensor gives a very high sensitivity without the necessity of actual physical contact and the first coming applications are envisaged as safety protection sensor for car windows or garage doors (to prevent shutting) or as tactile sensors for robot gripper tools [64], but also applications as described in the previous sections are conceivable.

20.6.3 POF in Home Networks

The biggest application area is the use of POF in home networks as an extension of broadband access networks, first of all for FTTB systems. One possible example for a future network is shown in Fig. 20.30. The building is connected by a broadband access (at the beginning ADSL2+, later WiMax, or glass fiber for higher capacity). The building network is based on 470-nm duplex

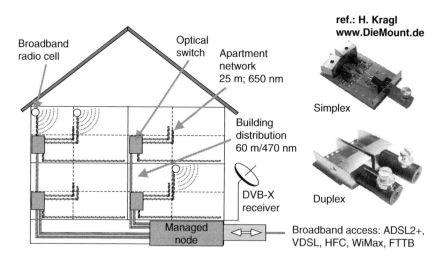

Figure 20.30   Example for POF building network.

POF transceivers. All apartments are equipped with 650-nm simplex transceivers with reduced reach but very simple installation.

Wireless systems are installed in the rooms. The active node realizes the handover. Because of the small required radio power in the rooms, the interferences between the radio cells are very small and the full WLAN capacity can be used. One of the major reasons for the simplicity of installation is the use of connector-less systems (Fig. 20.31). The customer can install all POFs without special tools in seconds. Connector-less solutions are developed by DieMount and Ratioplast/Infineon, for example.

At this time (2006), only very few companies offer end-user devices and components with POF interfaces. That is why the customer requires media converters for the installation of POF links. Two examples of commercially available products are shown in Fig. 20.32 (PC Card by DieMount and Fast Ethernet media converter by Ratioplast). With increasing number of installations, devices with direct POF interfaces will become available, first of all offered by broadband access suppliers.

20.7 POF FABRICATION METHODS

Initially, POF fabrication methods focused on techniques for making SI fibers such as the preform and extrusion techniques. Later on, as the proliferation and

Figure 20.31 Connector-less POF transceivers (examples DieMount, Ratioplast, and Infineon Technologies).

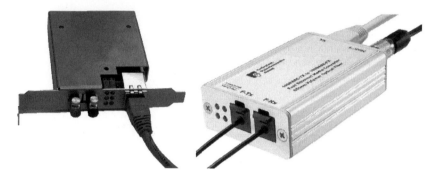

Figure 20.32   Market available POF transceivers.

use of POF for LAN applications increased and there was an interest to further increase their transmission capacity, the focus centered on developing fabrication techniques for GI POFs. Among the techniques available for GI POF manufacturing are gel polymerization, extrusion, centrifuging, photochemical, and others.

20.7.1 SI POF: Preform and Extrusion Method

In the preform method, a cylinder is produced that already has the index profile of core and cladding while having a much larger diameter. During the drawing process, the diameter is reduced until the desired size has been reached (Fig. 20.33).

Ideally, the index profile should be maintained during this process even though proportionally reduced. The length of the fiber per preform is determined as follows:

Length of fiber = preform length · (preform diameter/fiber diameter)^2

This method is applied generally for glass fibers. Automatic processes are then applied to make several 100 km of fiber out of each preform, as the following example shows:

Length of glass fiber = 2 m preform (5 cm preform diameter/125μm)^2 = 320 km

It is easy to see that the large core diameter of common POFs is not favorable for this process because only a few kilometers of fiber can be produced from each preform, for example:

Length of POF = 1 m preform · (5 cm preform diameter/1 mm)^2 = 2.5 km

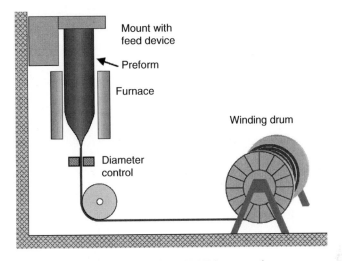

Figure 20.33 Production of POF from a preform.

When extrusion techniques are applied, the POF is produced in a continuous process directly from monomers. For SI POF, this process is very simple. Figure 20.34 shows such an arrangement (see, for example, references [65, 66]).

In addition, two further processes are mentioned in reference [66]. In the thrust extrusion technique, polymerization is carried out in a closed heated container from which the fiber is subsequently expelled through a nozzle at high pressure. The cladding is applied directly within the nozzle. This is a

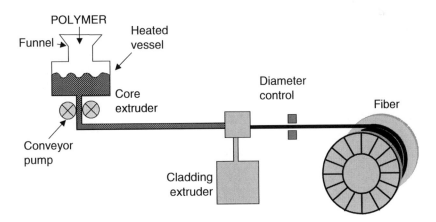

Figure 20.34 Production of SI-POF through extrusion.

noncontinuous process just like the preform technique. In the spin-melt process, a volume of ready-to-use polymer pellets is melted and pressed through a spin head that incorporates many holes. The holes serve to form the core and apply the cladding. This process is very efficient but also very expensive.

20.7.2 Production of Graded-Index Profiles

A number of processes for the manufacture of GI profiles are described in the technical literature:

- Surface gel polymerization technique
- Centrifuging
- Photochemical reactions
- Extrusion of many layers

In most of these techniques, the principle is to initially create a preform of up to 50-mm diameter and then to subsequently draw this preform down to the desired fiber size.

20.7.3 Interfacial Gel Polymerization Technique

The interfacial gel polymerization technique was developed by Y. Koike of Keio University (for an example, see reference [28]). In this process, a tube is initially manufactured with PMMA. This tube is then filled with a mixture of two monomers M_1 (high refractive index and large molecules) and M_2 (smaller refractive index and smaller molecules). Initially, the inner wall of the PMMA tube is slightly liquefied in a stove that has been typically heated to 80 °C. This results in a layer of gel and accelerates polymerization. The smaller molecule M_2 can more easily diffuse into this layer of gel so that the concentration of M_2 increases more and more towards the middle. The index profile is, thus, formed in accordance with the resulting concentration gradient.

For manufacturing a PMMA-GI POF, Koike [28] proposes that MMA (M_1) be supplemented with monomers VB, VPAc, BzA, PhMA, and BzMA. The material that was finally used is BzA because its reactivity is comparable with that of MMA. The 15- to 22-mm thick preform is then drawn at temperatures between 190 and 280 °C to produce fibers ranging from 0.2 to 1.5 mm in diameter. Figure 20.35 illustrates the principle (see also reference [31]).

Koike et al. [26] describes this method in more detail. The PMMA tube is produced by rotating a glass reactor at 3000 min^{-1} at 70 °C that is partially filled with MMA. The polymerization process for the core takes place at a speed of 50 min^{-1} and at a temperature of 95 °C and requires approximately 24 hours to

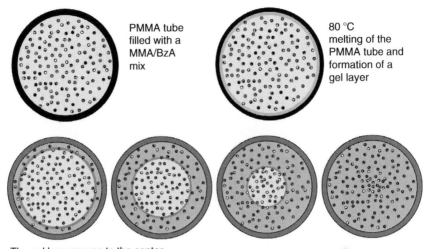

PMMA tube
filled with a
MMA/BzA
mix

80 °C
melting of the
PMMA tube and
formation of a
gel layer

The gel layer moves to the center
Concentration of M_2 increases from outer to the center

Figure 20.35   GI profile formation by gel polymerization technique.

be completed. Shi et al. [30] describes the production of a PMMA-GI POF with DPS as dopants. For traditional materials such as BB or BBP, one obtains fibers with an NA of 0.17–0.21, whereas with DPS, an NA of 0.29 is possible. The greater NA improves the bending characteristics and makes the introduction of light easier.

In principle, GI POFs do not require an optical cladding. On the other hand, it is necessary to find a way to continually increase the refractive index towards the axis. Essentially this can be achieved through doping and co-polymerization. In the case of silica glass, the index variation can be easily achieved by replacing the silicon atoms with germanium because these two substances behave identically within the glass structure. However, the components used for optical fibers do not allow such a simple replacement of individual atoms.

The process of doping involves inserting small molecules between the long chains of the actual core material, which increases the refractive index. What is important is that the dopants do not diffuse out of the polymer material too easily and do not show too strong absorption in the desired wavelength range. The doping process always lowers the glass-transition temperature. It is, therefore, desirable to insert a molecule that accomplishes the required change in the refractive index even at small concentrations (a few percent).

In co-polymerization, one uses chains composed of different monomers. The ratio of monomers determines the refractive index. In this case, although the sequence may be irregular, it is important that no long chains of one

Figure 20.36   Index variation by dopants.

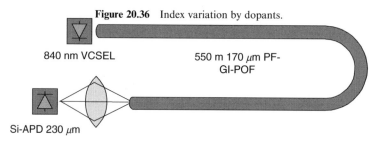

840 nm VCSEL   550 m 170 μm PF-
                                        GI-POF

Si-APD 230 μm

Figure 20.37   Index variation by copolymerization.

monomer are formed because otherwise the losses due to scattering increase considerably. This means that the bonding force of monomers among one another must not be greater than the bonding force to the respective other monomer. Of course, both monomers must have sufficient transparency. Figures 20.36 and 20.37 show schematic illustrations of the principles.

For perfluorinated materials, the dopant process is used mainly. A lot of experimental fibers have been developed with doped PMMA as well. In theory, the dopant method allows lower loss compared with the co-polymerization (due to increased Rayleigh scattering). Unfortunately, the reduction of the glass-transition temperature makes the fibers not usable for higher temperatures. The fiber with the best parameters on the market (by Optimedia, see earlier discussion) is made by co-polymerization. The attenuation is increased by about 50 dB/km, but the temperature stability is quite good.

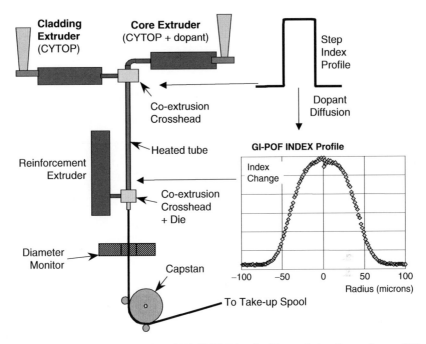

Figure 20.38   Continuous extrusion of PF-GI-POF (used, with permission, from reference [67]).

20.7.4 GI POF Extrusion

A very new method for continuous production of PF-GI POF was developed by Chromis-Fiberoptics (formerly, OFS). The process in shown in Fig. 20.38 [67]. An SI-type POF is extruded conventionally. The GI profile is made by diffusion inside a heated tube (wound around a big cylinder). The best results (loss of 20 dB/km at 1300 nm and about 10 GHz/100 m bandwidth) are close to the values of the PF-GI POF made from a preform.

REFERENCES

[1] Watanabe, Y. et al. *FTTH Utilizing PF-GIPOF in Apartment Complexes, Conference Proceedings, POF'2003, Seattle, 14–17, 09, 2003,* pp. 256–258.

[2] Ziemann, O. and H. Poisel. 2003. *Entwicklung der High-T-POF, 16. ITG-Fachgruppentreffen 5.4.1, München, June, 25, 2003.*

[3] Paar, U. et al. 1992. Excitation-dependent losses in plastic optical fibers. *SPIE* 1799:48–56.

[4] Bunge, C.-A. et al. 2002. *Theoretical and Experimental Investigation of FF and Bandwidth for Different POF Conference Proceedings. POF'2002, Tokyo, Sep. 18–20, 2002,* pp. 217–220.

[5] Luber, M. et al. 2002. *Characterisation of EMD Quality for Different Light Sources. Conference Proceedings POF'2002, Tokyo, Sep. 18–20, 2002,* pp. 161–164.

[6] Asahi Chemical Industry Co., Ltd. *High-Efficiency Plastic Optical Fiber, Luminous.* Datenblatt Nichimen.

[7] Mitsubishi, Rayon Co., Ltd. 1997. "Plastic Optical Fiber for Data Communication, ESKA PREMIER, ESKA MEGA" [data sheet].

[8] "Laser Components. Optische Fasern und passive Komponenten für die LWL-Technik." Katalog, 1995.

[9] Koike, Y. 1998. *POF—From the Past to the Future, Conference Proceedings POF'1998, Berlin, Oct. 5–8, 1998,* pp. 1–8.

[10] Information Gatekeepers, Inc. 1996. ATMF SWG: Physical Layer, RBB. Proposal of 155 Mbps Plastic Optical Fiber PMD Sublayer for Very Low Cost Private UNI. In: *Graded Index POF,* pp. 210–234. Information Gatekeepers, Inc. Boston, MA.

[11] Munkuni, H. et al. 1994. Plastic optical fiber high-speed transmission. In: *Conference Proceedings POF'1994, Yokohama, Oct. 26–28, 1994,* pp. 148–151.

[12] Munekuni, H. et al. 1994. Plastic optical fiber for high speed transmission. In: *Graded Index POF,* pp. 164–167. Information Gatekeepers, Inc. Boston, MA.

[13] Koike, Y. 1996. Status of POF in Japan. *Conference Proceedings POF'1996, Paris Oct. 22–24, 1996,* pp. 1–8.

[14] Koike, Y. and E. Nihei. 1996. Polymer optical fibers. In: *OFC'1996,* pp. 59–60. OSA—Opt. Soc. of Am.

[15] Koike, Y. 1996. POF for high-speed telecommunication [lecture manuscript], pp. 21–26.

[16] Ishigure, T. and Y. Koike. 2000. Potential bit rate of GI-POF. In: *POF'2000, Boston, Sep. 5–8, 2000,* pp. 14–18.

[17] Koike, Y. 1997. Progress in plastic fiber technology. In: *OFC'1997, Dallas,* pp. 103–138. OSA—Opt. Soc. of Am.

[18] Ishigure, T. et al. 1998. High-bandwidth GI-POF and mode analysis. In: *Conference Proceedings POF'1998, Berlin, Oct. 5–8, 1998,* pp. 33–38.

[19] Koike, Y. 2000. Progress in GI-POF—status of high speed plastic optical fiber and its future prospect. In: *POF'2000, Boston, Sep. 5–8, 2000,* pp. 1–5.

[20] Yabre, G. 2000. Theoretical investigation on the dispersion of graded-index polymer optical fibers. *J. Lightwave Technol.* 18(6):869–877.

[21] Levin, V. et al. Production of multilayer optical fibers. In: *Conference Proceedings POF'1999, Chiba, July 14–16, 1999,* pp. 98–101.

[22] Shin, B.-G. et al. 2002. Graded-index plastic optical fiber fabrication by the centrifugal deposition method. In: *Conference Proceedings POF'2002, Tokyo, Sep. 18–20, 2002,* pp. 57–59.

[23] Daum, W. et al. 2002. *POF—Polymer Optical Fibers for Data Communication.* Springer, Berlin.

[24] Daum, W. et al. 2006. *POF—Polymer Optical Fibers for Data Communication—Handbook for Thick Optical Fibers,* 2nd edition. Springer, Berlin.

[25] Koeppen, C. et al. 1998. Properties of plastic optical fibers. *J. Opt. Soc. Am. B* 15(2): 727–739.

[26] Koike, Y. et al. 1995. High-bandwidth graded-index polymer optical fiber. *J. Lightwave Technol.* 13:1475–1489.

[27] Koike, Y. et al. 1990. Low-loss, high bandwidth graded index plastic optical fiber. *Fiber Optics Magazine* 12(6):9–11.

[28] Koike, Y. 1992. High bandwidth and low-loss polymer optical fibre. In: *Conference Proceedings POF'1992, Paris, July 22–23, 1992,* pp. 15–19.

[29] Nonaka, T. et al. 1994. Characteristics of POFs and POF cables. In: *Conference Proceedings POF'1994, Yokohama, Oct. 26–28, 1994*, p. 122.

[30] Shi, R. F. et al. 1995. Measurements of graded-index plastic optical fibers. In: *Conference Proceedings POF'1995, Boston, Oct. 17–19, 1995*, pp. 59–62.

[31] Ishigure, T. et al. 1995. High-bandwidth, high-numerical aperture graded-index polymer optical fiber. *J. Lightwave Technol.* 13(8):1686–1691.

[32] Koike, Y. 1997. GI-POF in high-speed telecommunication. In: *Conference Proceedings POF'1997, Kauai, Sep. 22–25, 1997*, pp. 40–41.

[33] Shimada, K. et al. 1999. Digital home network with POF. *Conference Proceedings POF'1999, Chiba, July 14–16, 1999*, pp. 129–132.

[34] Takahashi, H. et al. 1998. Fabrication techniques of GI-POF towards mass production. In: *Conference Proceedings POF'1998, Berlin, Oct. 5–08, 1998*, pp. 50–54.

[35] Myers, G. W. 2002. Reducing the costs of optical interconnects in the telecommunication market. Digital Optronics. Available at: www.digitaloptronic.com.

[36] Liu, Z. and X. Wu. 2002. The development of POF in China. In: *Conference Proceedings POF'2002, Tokyo, Sep. 18–20, 2002*, pp. 193–195.

[37] Luvantix. 2003. Product News—Preform for Plastic Optical Fiber. Available at: www.luvantix.com.

[38] Komori, S. 2004. LUMISTAR. Press release, pof-info@fujifilm.co.jp, Aug. 2004.

[39] Nuvitech. 2005. Nuvigiga Product Specifications (GI-POF). Nuvitech Co. Ltd.

[40] Park, M. et al. 2005. High-performance GI-POF produced by extrusion. In: *Conference Proceedings OECC'2005, Seoul July 4–8, 2005*, pp. 416–417.

[41] Park, C. W. 2006. PMMA-based GI-POF (OM-Giga), 21. ITG-sub committee 5.4.1 meeting, Oldenburg, 12.05.2006.

[42] Komori, S. 2006. LUMISTAR. Press release, Jan. 2006. Available at: http://home.fujifilm.com /news/n060118.html.

[43] Kuchta, D. M. et al. 1994. High speed data communication using 670 nm vertical cavity surface lasers and plastic optical fiber. In: *Conference Proceedings POF'1994, Yokohama, Oct. 26–28, 1994*, p. 135.

[44] Walker, S. and R. J. S. Bates. 1993. Towards gigabit plastic optical fibre data links: Present progress and future prospects. In: *Conference Proceedings POF'1993, Den Haag, July 28–29, 1993*, pp. 8–13.

[45] Yaseen, M. et al. 531 Mbit/s, 100-m all-plastic optical-fiber data link for customer-premises network application. In: *OFC/IOOC'1993*, pp. 171–172. OSA—Opt. Soc. of Am.

[46] Bates, R. J. S. et al. A 265 Mbit/s, 100 m plastic optical fibre data link using a 652 nm laser transmitter for customer premises network applications. In: *ECOC'1992*, Vol. 1, pp. 297–300. OSA—Opt. Soc. of Am.

[47] Bates, R. J. S. et al. 1996. The limits of plastic optical fiber for short distance high speed computer data links. In: *Graded Index POF*, pp. 173–185. Information Gatekeepers, Inc. Boston, MA.

[48] Khoe, G. D. 1999. Exploring the use of GIPOF systems in the 640 nm to 1300 nm wavelength area design. In: *Conference Proceedings POF'1999, Chiba, July 14–16, 1999*, pp. 36–43.

[49] Junger, S. et al. 2004. Cable TV transmission over POF. In: *Conference Proceedings POF'2004, Nürnberg, Sep. 27–30, 2004*, pp. 35–39.

[50] Khoe, G. D. et al. 2002. High capacity polymer optical fibre systems. In: *Conference Proceedings POF'2002, Tokyo, Sep. 18–20, 2002*, pp. 3–8.

[51] Li, W. et al. Record 2.5 Gbit/s 550 m GI POF transmission experiments at 840 and 1310 nm wavelength. In: *Conference Proceedings POF'1999, Chiba, July 14–16, 1999*, pp. 60–63.

[52] Kalymnios, D. et al. 2004. POF sensor overview. In: *Conference Proceedings POF2004, Nürnberg, Sep. 27–30, 2004,* pp. 237–244.

[53] Poisel, H. et al. 2005. POF sensors for automotive and industrial use come of age [Invited Paper]. In: *Conference Proceedings POF2005, Hong Kong, Sep. 19–22, 2005,* pp. 285–289.

[54] Poisel, H. et al. 2005. Krümmungssensor mit Optischen Polymerfasern. 17. Internationalen Wissenschaftlichen Konferenz der Hochschule Mittweida, Nov. 3, 2005.

[55] Muyshondt, H. 2005. Automotive LAN seminar, Tokyo.

[56] Schönfeld, O. 2000. MOST transceiver components. Presented at the MOST All Members Meeting, June 2000.

[57] Kibler, T. et al. 2004. Optical data buses for automotive applications. *J. Lightwave Technol.* 22(9):2184–2199.

[58] Miedreich, M. and H. Schober. 2005. Pedestrian protection system, featuring fiber optic sensor. *ATZ Worldwide* 107:15.

[59] Miedreich, M. and B. L'Hénoret. 2004. Fiber optical sensor for pedestrian protection. In: *Conference Proceedings POF'2004, Nürnberg, Sep. 27–30, 2004,* pp. 386–392.

[60] Djordjevich, A. 2003. Alternative to strain measurement. *Opt. Eng.* 42(7):1888–1892.

[61] www.siemensvdo.de/de/ pressarticle2003.asp?ArticleID=200309_006.

[62] www.canpolar.com/principles.shtm.

[63] Kodl, G. 2003. Large area optical pressure detecting sensor based on evanescent field. In: *Conference Proceedings POF'2003, Seattle, Sep. 14–17, 2003,* pp. 64–67.

[64] Weiss, C. 2005. Kraftabhängiges Greifen mit neuartigem POF—Sensor auf Basis des Evaneszenten Feldes. In: *Proceedings International Forum Mechatronics, June 15/16, 2005, Augsburg, Germany.*

[65] Raman, R. 1999. "Plastic Optical Fibers—A Primer." Tutorial IWCS'1999, Atlantic City, Sep. 14, 1999.

[66] Weinert, A. 1999. *Plastic Optical Fibers.* Publicis MCD Verlag, Erlangen, München.

[67] Blyler, L. L. et al. 2002. Materials technology for perfluorinated graded-index polymer optical fibers. In: *Conference Proceedings POF'2002, Tokyo, Sep. 18–20, 2002,* pp. 43–44.

Chapter 21

Sapphire Optical Fibers

J. Renee Pedrazzani

Institute of Optics, University of Rochester, Rochester, New York

The potential of single-crystal fibers was recognized 50 years ago, when the inherent strength and perfection of the crystalline matrix of the fibers were determined. Despite this, sustained research into the production of such fibers was delayed until the 1970s. Until then, control over the growth characteristics of the fibers was poor, and there were no well-defined areas of utilization [1]. The first application that drove the production of single-crystal sapphire fibers was use as a reinforcing member for metal-matrix composites [2]. Sapphire, whose chemical formula is Al_2O_3, was thought to be an attractive candidate to employ for this purpose because it possesses a high melting temperature of 2053 °C, has a low solubility in water, is resistant to chemically hostile environments, and has many attractive physical and mechanical properties [3]. However, fiber-reinforcing applications require fibers to possess good mechanical characteristics, but make no restriction on the optical qualities of the fibers.

Interest in developing optical-quality sapphire fibers emerged after the commercial effort to manufacture structural-grade sapphire fibers began. Sapphire optical fibers are not reasonable candidates for long-distance telecommunications lines because of the difficulties involved in manufacturing low-cost fibers having perfect crystalline structure and surfaces as smooth as conventional silica-based glass fibers. Researchers at Stanford University, realizing that sapphire fibers were not likely to replace glass-based fibers for telecommunications applications, embarked on a program to investigate the potential applications and manufacture of optical-quality sapphire optical fibers. Their investigation indicated that fibers grown using the laser-heated pedestal growth (LHPG) method were optically superior to those grown using the edge-defined film-fed growth (EFG) method, which was commonly used in the manufacture of structural-grade sapphire fibers for reinforcing applications [1]. Other researchers concurred, and with the possible exception of Saint-Gobain Saphikon, the best optical-quality sapphire fibers have been grown by researchers using the LHPG method [4, 5]. Optical-quality sapphire fibers have increased in popularity over the years and, now, are used in various fields and applications such as laser-power delivery probes and tips for surgical and treatment lasers; rugged high-temperature thermometers for industrial applications; visible and infrared (IR) transmitting waveguides for spectroscopy applications [6]; and high-temperature–resistant sensing elements in optical fiber Fabry–Perot extrinsic sensors and devices.

This chapter begins by describing the two most popular methods used to grow single-crystal sapphire fiber: LHPG and the EFG. The positive and negative aspects of each method are reviewed. Next, the optical and mechanical properties of sapphire fibers are described. The chapter continues with a discussion of the possibilities that exist for cladding and overcoating the bare sapphire fiber, and the difficulties associated with this procedure are addressed. The fourth section of the chapter provides a brief overview of sapphire fiber applications. Finally, the chapter concludes with a listing of relevant physical properties for Al_2O_3.

21.1 THE GROWTH OF SAPPHIRE FIBER

Single crystals of synthetic corundum, also known as α-Al_2O_3 or "white sapphire," may be grown according to a variety of methods [7]. The most common of these methods are: crystallization from a melt, solution growth, and condensation from the vapor phase. The second of these methods has attracted considerable interest because it permits the continuous and controlled growth of single-crystal fibers of uniform diameter. This method also permits the fibers to be grown along a specific crystallographic axis [1]. However, sapphire fiber, which is a birefringent waveguide, is customarily grown so that the c plane [0001] coincides with the optical axis of the fiber [8]. The two most popular implementations of this technique are the EFG and the LHPG methods.

The LHPG process, favored by university-based researchers, uses a float-zone technique to grow optical-quality sapphire fibers. Researchers at Stanford University [1, 3, 9], Rutgers University [10–12], and the University of South Florida [13, 14] are the most notable to use this method. This method has not been commercially embraced because of cost considerations: the LHPG process allows only a single fiber to be grown at a time. This method has been widely acknowledged to produce better optical-quality sapphire fibers than those made via the EFG process [11, 12], although commercial product data claim that this may no longer be the case [5].

Use of the LHPG method, schematically represented in Fig. 21.1 and likened to the float-zone method of crystal growth, results in a single-crystal fiber being

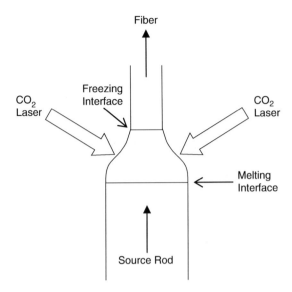

Figure 21.1 The laser-heated pedestal growth process.

grown from the molten tip of a source crystal [12]. The tip of an Al_2O_3 source rod is liquefied when the beam of a CO_2 laser is split and focused onto it. Inserting a seed crystal into the liquefied bead creates a molten zone extending from the source rod to the seed crystal, with the shape of the molten zone determined by surface tension.

The fiber is grown by carefully controlling the speed at which the seed crystal is moved away from the source rod and the speed at which the source rod is advanced into the molten zone. The longest length of fiber that can be grown using this method is determined by the dimensions of the source rod. The purity of the fiber should be determined by the purity of the source rod; the heating method is clean, and the molten zone never comes in contact with a crucible or a die [9]. Under optimal growth conditions, the fiber has a diameter of approximately one-third of the source rod. Because mass is conserved during the growth process, the ratio

$$\frac{d_s}{d_f} = \sqrt{\frac{v_f}{v_s}} \tag{21.1}$$

where d_s is the diameter of the source rod, d_f is the diameter of the fiber, v_s is the velocity of the source rod, and v_f is the velocity of the fiber, is valid [12].

Although this description of LHPG is straightforward, it omits mention of the efforts exerted to produce a fiber of uniform diameter and good optical quality. Surface irregularities in the fiber, including diameter fluctuations, contribute substantially to scattering losses, and the fiber diameter is directly dependent on the shape of the molten zone. The shape of the molten zone is affected by fluctuations in the diameter of the source rod, variations in the power of the CO_2 laser, and the velocities of the source rod and fiber [3]. In addition, the melt possesses a very low viscosity, and the molten zone, stabilized only through surface tension, is easily perturbed by air currents and vibrations [4]. The detrimental effects of air currents are minimized by sealing the source rod and growing fiber end in an atmosphere-controlled chamber [1]. This precaution also permits the growth of the sapphire fibers in an inert atmosphere; it is believed that losses arise when OH^- and other gaseous species are incorporated into the sapphire fiber while it is growing. Researchers at Rutgers University used a computer-controlled feedback system to stabilize the laser output and to monitor and control the fiber diameter. With this system, the authors of the paper were successful in producing sapphire fibers possessing a $\pm 0.5\%$ diameter variation. Without the use of feedback, the diameter variations were greater by a factor of 10. With this stable growth process and tight diameter control, the researchers produced fibers with losses of 0.3 dB/m at a wavelength of 2.94 μm. This is very close to the theoretical minimum of 0.13 dB/m at the same wavelength [4].

The EFG process—patented by Saint-Gobain Saphikon—unlike LHPG, permits the growth of many sapphire fibers simultaneously, making the commercial production of sapphire fiber more economically viable. This method of sapphire fiber growth was developed to meet the demand for sapphire fiber structural reinforcements for metal-matrix composites. The mechanical strength, as opposed to the optical quality, of the fiber was the overriding consideration; the high loss in the early sapphire fibers produced by this method attenuates nearly all injected light over a centimeter of propagation length. When the demand for structural filaments lagged, Saint-Gobain Saphikon investigated the possibility of producing optical-grade sapphire fiber for use in high-temperature–resistant fiber optics [15]. Saint-Gobain Saphikon supplies sapphire fiber for this purpose and for medical and spectroscopy applications [6].

The EFG method uses a reservoir containing molten source material to continuously supply the growth of a single-crystal sapphire structure (Fig. 21.2). Continuous growth of lengths more than 300 m has been demonstrated. Unlike the LHPG method, the melt is contained in a molybdenum crucible, and the cross-section of the sapphire crystal is determined by the surface configuration of the molybdenum die anchored in the melt. During growth, the melt wicks into the die and the fiber growth, initiated by a seed crystal, occurs at the top of the die. The use of an anchored die in the melt presents a stable platform from which to grow the crystal. Because the top of the die extends above the level of the melt in the crucible, thermal variations present in the melt are damped as the melt travels the length of the tube. This helps

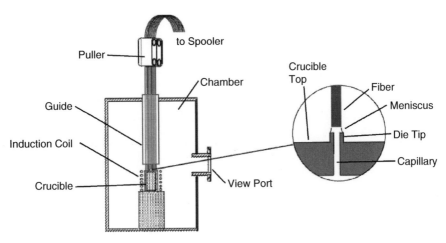

Figure 21.2   The edge-defined film-fed growth (EFG) process.

ensure a stable melt temperature at the solidification interface. Molybdenum is considered to have good chemical compatibility with molten alumina; sapphire crystals grown using molybdenum crucibles and dies contain only 5 ppm of molybdenum, and molten alumina is able to wet the molybdenum. The liquid phase of one material wets the solid phase of another when the contact angle is less than 90 degrees [16]. As in the LHPG method, production of optical-quality sapphire fibers requires isolating the growing crystal from vibrations and atmospheric currents; ensuring a stable temperature at the growth interface; surrounding the growing crystal with an inert atmosphere; and carefully controlling the growth rate of the crystal [15].

The edges of the die are used to control the shape of the cross section of the sapphire crystal. When the die is shaped as a hollow capillary tube and the top of the die makes an angle of 75 degrees or greater with the inner edge of the die, the melt wicks up through the inside of the die and spreads over the top surface of the die. A fiber grown under these conditions will have a diameter equal to the outer diameter of the cylinder. Similarly, a die consisting of three capillary tubes fastened together can be used to grow three-bore tubing. In this case, the melt wicks up, through the approximately triangular interstitial region created by the mating of the three tubes, and spreads to cover the ringed surface formed by the ends of the three capillaries. This results in three-bore tubing as the melt never extends over the inner diameter of any of the three conjoined tubes. The flexibility imparted to the growth process results in the ability to grow a sapphire crystal of nearly any cross-section, including ribbons, single filaments, single-bore tubing, and multiple-bore tubing [16].

21.2 OPTICAL AND MECHANICAL CHARACTERISTICS OF SINGLE-CRYSTAL SAPPHIRE FIBER

The appeal of using sapphire fiber as an optical waveguide is driven by its capability to transmit light radiation at longer wavelengths into the IR than glass-based fibers (up to $\sim$3.5 μm) coupled with some of its superior physical and mechanical properties (melting point in excess of 2000 °C, chemical inertness, high laser-power thresholds, etc.). Sapphire fibers are crystalline in nature and not amorphous glasses. There are two types of crystalline fibers: single-crystal (SC) and polycrystalline (PC). Meter-long lengths of SC fibers have been made only from a handful of materials out of a possible pool of more than 80 IR transmissive crystalline materials. In this context, sapphire is the most popular SC fiber.

Exploration of the optical uses of sapphire fiber began and is sustained because it is capable of operation in chemically harsh and high-temperature environments, it has a large transparency window, and it is a strong and hard material. Sapphire fiber has not been adopted for a wide variety of uses because growth processes, which are difficult to perfect, result in sapphire fibers being lossier than their glass counterparts, and because sapphire fibers will always be more expensive. Additional complications arise from present-day sapphire fibers being unclad and highly multimoded. Sapphire has also been observed to decrease in strength and flexibility at high temperatures.

Sapphire (Al_2O_3) is an insoluble, uniaxial crystal (trigonal structure) material with a melting point more than 2000 °C. It is hard and noncorrosive to organic solvents and acids [17]. It has a refractive index of 1.75 at 3.2 μm and, compared to silica glass, it has a thermal expansion 10 times greater and a Young modulus approximately seven times greater. However, because of this last property, sapphire fibers are very stiff and fragile compared to silica fibers. Sapphire fiber is most transparent over the 240- to 4000-nm range of the spectrum [3] and has an unrealized theoretical loss minimum of 0.13 dB/m at a wavelength of 2.94 μm [4]. The measured transmission spectrum for bulk sapphire fiber (Fig. 21.3) as published by Innocenzi et al. [18], and plotted as discrete squares, differs from the theoretical loss spectrum. The intrinsic loss spectrum is composed of contributions from Urbach, Brillouin, and multiphonon absorption processes [12]. The Urbach process causes an increase in the value of the absorption coefficient as the frequency increases toward the band-gap energy of sapphire [19]. Brillouin scattering occurs when incident light, consisting of photons, is scattered by a mode of vibration of the crystal lattice, composed of acoustic phonons [20]. "Multiphonon" refers to multiphonon absorption processes that occur in the sapphire crystal.

Unlike experimental data taken for fused silica, which typically coincide well with theoretical loss curves, loss measurements of crystalline materials are

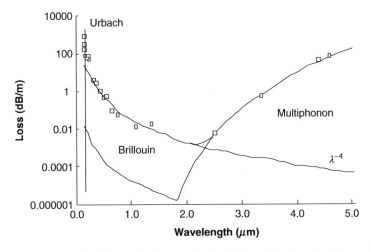

Figure 21.3   Measured and theoretical attenuation in bulk sapphire (used, with permission, from reference [12]).

typically higher than those predicted by theory. These measurements show a minimum measured loss at 1.78 μm that is orders of magnitude larger than the 3.4×10^{-6} dB/m predicted by theory. The greater losses arise from the influence of crystal defects in the material, which result from imperfect growth conditions. Experimental data for wavelengths less than 1.5 μm can be fitted by a power law that varies as λ^{-4}, which implies the presence of Rayleigh scattering caused by inhomogeneities in the crystal [12]. Rayleigh scattering occurs from the presence of particles or material inhomogeneities that are small compared to the wavelength of light [21]. Hence, the optical properties of as-grown sapphire fibers are typically inferior to those of the bulk material. This degradation is particularly evident in the visible region of the spectrum and is the result of color-center defects formed during the drawing process. However, said defects can be greatly reduced by post-annealing of the drawn fibers at 1000 °C in an air or oxygen atmosphere [12].

The loss spectrum of low-loss sapphire fibers grown at Rutgers using the LHPG method, compared with those made by EFG, is shown in Fig. 21.4. In general, the quality of optical-grade sapphire fibers has improved considerably over the past few years. Fibers with losses less than 0.5 dB/m at 2.94 μm and approximately 3.0 dB/m at 0.633 μm are grown routinely with maximum lengths less than 3 m and with diameters of 150, 250, 325, and 425 μm.

Extrinsic loss in sapphire fibers, which should be minimized for optical applications, is largely due to scattering and absorption mechanisms. Impurities and color centers exacerbate extrinsic absorption, and voids, inhomogeneities, and surface perturbations contribute to scattering losses. Impurities in sapphire

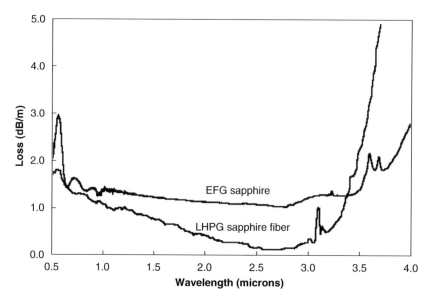

Figure 21.4 Loss transmission for LHPG versus EFG sapphire fibers.

fibers may derive from the source material or gaseous species such as H_2O [3]. In fibers grown using the EFG method, the crucible and die may also be contributors [16]. Measurements suggest that fibers possess OH^- impurities and color centers, specifically V-type (hole) centers that contribute to absorption losses around the wavelengths of 410 and 3000 nm. These defects are likely incorporated during growth of the crystal. V-centers occur when a hole is trapped in the vacancy of an ionic crystal. The absorption at 410 nm is attributed to an O^{2-} ion trapping a hole-pair in the vicinity of an Al^{3+} vacancy. There is also the indication of OH^- impurities partially compensating V-centers. The presence of OH impurities is supported because the absorption corresponding to the stretching band for the OH^- impurity, occurring around 3000 nm, is observed. Also occurring around 3000 nm are three peaks that indicate the presence of transition metal impurities. The magnitude of the absorption in the bands around both 410 and 3000 nm—as mentioned before—can be reduced by annealing.

The majority of extrinsic scattering losses in sapphire fibers are attributed to voids internal to the fiber, inhomogeneities, and surface variations of the fiber. The number and type of scattering sites determine the attenuation of each fiber, and these are dependent on the quality of the growth process. It is well understood that any variations in the growth conditions of the fiber, including the temperature at the growth interface, changes in the speed of the growth, and perturbations of the shape and size of the meniscus, which feeds the growth of the crystal, cause variations in the diameter of the fiber [3]. Unstable growth conditions are also responsible for the incorporation of inclusions into the fiber. It is noted in particular that some sapphire fibers grown using the EFG technique possess bubbles that resulted from a growth rate that was too fast. Annealing does not appear to reduce these losses [2], but it is believed that refinements in the mechanics of the growth process can greatly reduce, if not eliminate, these sources of loss [12]. Sapphire fibers generally do not suffer from twinning and dislocation defects, unlike bulk sapphire. This is because the diameter of sapphire fibers is small; unless a defect propagates parallel to the axis of the fiber, it will eventually grow out of the fiber. Dislocations in sapphire fibers will occur, but they will be of a lesser density than in bulk sapphire [1].

Sapphire fibers are not clad, which results in a number of undesirable characteristics. Single-crystal sapphire fibers cannot be grown as a core–clad structure from rod-in-tube source rods, as convective currents in the molten zone quickly act to destroy the geometrical structure of the source rod [11]. Other techniques, discussed in Section 21.3, have been proposed to clad sapphire fibers, but none represents a universally acceptable solution. In general, the cladding of a fiber exists to mechanically support the core, act as a barrier separating the core from environmental contaminants, and reduce scattering losses by lessening the difference in dielectric constants at the surface of the core [22]. Scattering can result in power from the scattered mode being coupled into other modes or ejected from the fiber as radiation losses [21]. Contaminants on the surface of the sapphire fiber attenuate the intensity of the guided light through a few mechanisms. Any substance having an index of refraction higher than sapphire will, by defeating total internal reflection, cause loss at all points of contact with the sapphire fiber. A material with an index of refraction that is both lower than that of sapphire and different than that of the surrounding atmosphere will also cause loss if it is in sporadic contact with the surface of the sapphire fiber; scattering occurs at the points of discontinuity between dielectric constants [15]. Sapphire fibers have also incorporated impurities into their structure, a behavior observed particularly at elevated temperatures, which causes losses through scattering [23].

Sapphire fibers are not single moded for operation in the 250- to 4000-nm window of transparency and are not likely to be in the future. Without a physical cladding, the entire volume of the sapphire fiber may be considered to be the core, and the surrounding atmosphere the cladding. Sapphire fibers typically have diameters between 100 and 300 μm. For single-mode operation, the diameter of such a fiber is given by

$$d = \frac{\lambda(2.405)}{\pi n_s}, \tag{21.2}$$

where d is the diameter of the fiber, n_s is the index of refraction of sapphire, and λ is the wavelength of operation [22]. This specifies fiber diameters ranging from approximately 1 to 0.1 microns for single-mode operation over the transparency window for sapphire. It is claimed that 40-μm diameter fibers can be grown without modification to existing LHPG machinery. With modified apparatus, 5 μm diameter fibers could potentially be grown, but the cost is projected to be prohibitive [24]. Successfully handling such a fiber without breaking it would also be extremely difficult.

Sapphire is a brittle material, and the flexibility and strength of sapphire fiber decreases at high temperatures. Sapphire is among the hardest of natural minerals and is rated as a 9 on the Mohs scale [7]. Despite this property, because of the brittle nature of sapphire, fiber handling results in surface damage and consequent reduction in strength [25]. Fibers tumbled against one another have suffered a tensile strength reduction of 30% [26]. Saint-Gobain Saphikon has combated this problem by presenting the option of coating the surface of each fiber with an organic polymer to protect the fiber before shipping. While this protection is recognized as necessary for strength preservation, the process used to remove the sizing can both cause damage to and fail to remove all of the impurities on the fiber's surface [25].

Sapphire also suffers a severe degradation in strength as the operating temperature is increased, as is shown in Fig. 21.5. This property affects the minimum diameter bend that the fibers can survive at any given temperature. The bend survival strength σ_{bs}, also shown in Fig. 21.5, is determined by

$$\sigma_{bs} = \frac{d_f E}{2R_s}, \qquad (21.3)$$

where d_f is the fiber diameter, E is the elastic modulus, and R_s is the bend survival radius. Fibers fail when plastic deformation permits the growth of fractures in the fibers. Bend tests conducted at temperatures higher than 1650 °C result in visible plastic deformation of the fibers. This occurs because of a dislocation (pyramidal plane) slip that is activated at temperatures exceeding 1600 °C. The bend behavior is reportedly dependent on surface flaws and the presence of inclusions in the fiber [8].

21.3 CLADDING AND COATING OF SAPPHIRE FIBERS

Identifying suitable materials to act as coating and cladding materials for optical-grade sapphire fibers is challenging because of the number of requirements they must meet. Ideal claddings should enhance the optical characteristics, without limiting the useful operation, of the sapphire fibers. Ideal coatings should act as a barrier between chemically harsh environments and sapphire, and the coatings should be capable of operation over at least the same temperature range as sapphire alone. Obstacles to this goal include the need to match the

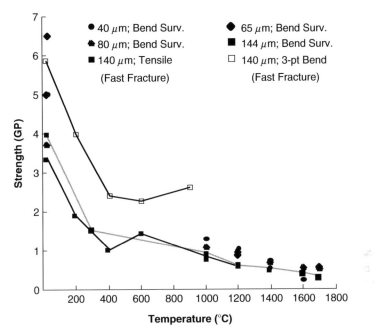

Figure 21.5 One-hour bend survival strengths, fast-fracture tensile strengths, and bend strengths versus temperature (used, with permission, from reference [8]).

coefficient of thermal expansion of the materials and sapphire, the need for a good bond between the fiber and the material, the (so far) insurmountable problem of growing a core–clad sapphire fiber in one process, and the inability to identify materials suitable for a wide range of applications. The benefits of cladding and coating sapphire optical fibers were addressed in Section 21.2.

Claddings will potentially permit single-mode operation, improve the wave-guiding properties of the sapphire fiber core, prevent radiation losses, and provide mechanical support to the core of the fiber. The cladding must have a lower index of refraction than sapphire, possess a coefficient of thermal expansion compatible with that of sapphire, form a layer with low intrinsic stress, and be able to survive large fluctuations in temperature. Because sapphire fibers cannot be clad as they are being grown, claddings must be applied after the growth of the fiber. Sapphire fibers can be clad using either ion exchange techniques or the direct deposition of materials. Few reports exist in the literature that describe the effects dopants have on the index of refraction of sapphire fiber; it is not a method that is being actively investigated. Direct deposition is a more viable method [27].

The direct deposition of a cladding on the surface of sapphire fiber can be effected by metal-organic chemical vapor deposition (MOCVD), physical vapor deposition, or solution-based methods. During physical deposition methods, such as evaporation and sputtering, the fiber must be rotated to ensure a uniform coating. Solution-based methods, such as sol-gel, frequently require an additional process to densify the porous coating. This densification can result in considerable stress between the cladding and the sapphire fiber. MOCVD is presented as a superior technique because the deposited layer is typically dense, homogenous, and pure. A wide variety of materials may be deposited using this method, and the composition and thickness of the deposited layer may be controlled to within tight tolerances. Suggested cladding materials to investigate for cladding deposition include SiO_xN_y, Mg_xSiO_y, and Ti_xSi_yO, whose x/y ratio may be varied to give a range of refractive indices [27]. In general, adequate cladding materials that are both transparent and capable of surviving temperatures higher than 1000 °C are considered so rare as to not exist. There has been some success with polycrystalline alumina (Al_2O_3), but these coatings are considered prohibitively expensive [24, 28].

Coatings isolate the fiber from environmental contaminants and protect the surface from abrasion. This protection will enable wider use of sapphire fiber by extending the useful lifetime of the fiber and increasing the number of potential operating environments. There is no one candidate material that can serve as a sapphire fiber coating in any environment; the coating must be chosen, in part, according to the application [27]. It is also desirable that elements do not interdiffuse between the coating and the fiber materials. Commonly considered materials include polycrystalline alumina (Al_2O_3), metal niobium, silicon carbide (SiC), and zirconia (ZrO_2). Coating techniques include chemical coating, vapor deposition, plasma spraying, and electrochemical plating [29].

These proposed sapphire fiber coatings have different strengths and weaknesses. Alumina and sapphire share the same chemical composition, and the only stable phase of alumina above 1199 °C is α-Al_2O_3; single crystal α-Al_2O_3 is sapphire. α-Alumina has a coefficient of thermal expansion (CTE) of $8.1 \times 10^{-6}/°C$ at 1000 °C, and a melting temperature of 2053 °C. Other properties of polycrystalline alumina, such as the tensile strength and hardness, are dependent on the grain size, porosity, and purity of the alumina. Alumina possesses high chemical stability, but in a sodium atmosphere above 1100 °C, sodium diffusion will occur. This leads directly to chemical reactions and microcracking [29].

Silicon carbide has a CTE of $5.8 \times 10^{-6}/°C$ at 1000 °C and a melting temperature of 2830 °C. It is strong, dense, and thermally shock resistant. While it readily oxidizes to form an outer layer of silica (SiO_2), this layer will protect the material from further oxidation. Silicon carbide is used to best effect in a reducing atmosphere [29].

Zirconia is a less attractive option, because pure, or unstabilized, zirconia undergoes a change in structure, from monoclinic to tetragonal, at 950 °C. This results in approximately a 9% increase in volume [29]. Sapphire fibers coated with unstabilized zirconia experience large strains when the zirconia changes structure leading to coating microcracks [30]. It is recommended that zirconia be stabilized with yttria (Y_2O_3) to force a stable cubic structure [29]. The monoclinic form of zirconia has a CTE of $6.5 \times 10^{-6}/°C$ at 1000 °C and the stabilized cubic form $10.5 \times 10^{-6}/°C$ at 1000 °C [31]. When zirconia is stabilized, the strain is reduced, but residual strain results from the difference in CTEs between the sapphire and stabilized zirconia; this will likely result in strength degradation [30]. Zirconia will react with hydrochloric and nitric acids, but it is stable in oxidizing and moderately reducing atmospheres. At 2000 °C, zirconia will react with refractory carbides and stabilized zirconia will begin to vaporize [29].

Niobium has a CTE of $7.1 \times 10^{-6}/°C$, which makes it an attractive option for a sapphire coating. It resists reactions with sodium, mercury, and other substances at elevated temperatures, but a high sodium vapor atmosphere at temperatures exceeding 1000 °C results in reactions between the niobium and alumina [29].

A research group at Drexel University has clad sapphire fiber with alumina and overcoated the clad fiber with silicon carbonate. The fibers are intended for use as sensors embedded in ceramic matrix composites. The authors of the paper use a novel chemical deposition technique to ensure good adhesion between the sapphire and the polycrystalline alumina. The technique involves combining alumina particles and a polymerizable monomer carrier. The monomer is allowed to polymerize on the sapphire surface, the binder is removed, and the layer is sintered. This process is repeated several times to achieve a cladding layer 20-μm thick [28]. A 0.25-μm layer of silicon carbide is applied, via plasma enhanced CVD, as an outer coating to protect the embedded fiber from the surrounding environment [32]. After thermal cycling tests, performed at 900 and 1300 °C, evidence of microcracks, debonding, or damage to the fiber was observed [33].

21.4 APPLICATIONS OF SAPPHIRE FIBERS

21.4.1 Optical Fiber Sensors

Silica-based optical fiber sensors are generally restricted for use in environments below 700 °C, because the integrity of the glass fiber is adversely affected at higher temperatures. Specifically, at 1000 °C the migration of the dopants from the fiber core becomes significant [34]. In addition, at temperatures exceeding 900 °C, the combination of strain and elevated temperatures will also induce creep and plastic deformation in the silica optical fiber. Most silica-based sensor systems are specified for use at less than 300 °C, which is intended to provide an adequate margin of safety given the upper operating temperature limit of available fiber coatings. For instance, acrylate coatings are rated for 85 °C and degrade at temperatures higher than 150 °C, while polyimide coatings deteriorate between 400 and 500 °C. Some metal coatings are capable of surviving temperatures at which the silica softens.

In general, there is interest in the development of sensors that can operate in high temperature and chemically harsh environments for which most conventional sensors are not suited. Such high-temperature sensors would find use in the control of high-temperature combustion and industrial processes and in the development of advanced high-temperature materials. Platinum gauges are used to perform some measurements, but their applicability is limited. Optical sensors made of sapphire are able to withstand environments in which few other sensors can function. Sapphire-based sensors used for the spectroscopic analysis of liquids and gases have enjoyed the greatest commercial success. Researchers have also constructed interferometric and polarimetric sensors from sapphire fiber, although these have been less of a research and development focus.

21.4.1.1 High-Temperature Sensors

The first proposed design for a sapphire-based temperature sensor relies on the principles of blackbody radiation, and it is the basis of both commercial products and the majority of next-generation sapphire-based temperature sensors under development [35]. In the first sapphire-based sensor design, developed by R. R. Dils and shown in Fig. 21.6, a blackbody cavity is sputtered onto the end of a sapphire rod 0.25–1.25 mm in diameter and 0.05–0.30 m in length. The nonmetalized end of the sapphire fiber is butt-coupled to a standard glass optical fiber, and the output of the glass fiber is collected by a detector. The radiance emitted by the blackbody cavity is used to determine the temperature of the environment of the sensor; as the temperature of the environment increases, the spectrum emitted by the blackbody predictably shifts to shorter wavelengths according to the Planck radiation law. Although the sapphire rod employed in this sensor is flame-polished to smooth the surface, most of the scattering losses arise from the remaining surface imperfections. At temperatures exceeding 1100 °C, scattering, absorption, and reemission at internal defects and surface imperfections also occur. Dils observes that because these sources of loss will likely

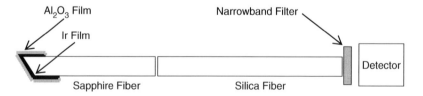

Figure 21.6   Schematic of a blackbody cavity high-temperature sensor.

challenge the accuracy of the sensor, only high-quality crystals should be used. Commercial temperature sensors based on this design claim operating temperatures ranging between 300 and 2000 °C, an accuracy of 0.10% at 1000 °C [36].

This type of sensor has been used to monitor the temperature of internal combustion engines [37], aircraft turbines, and high-velocity combustion flows [38]. While platinum and iridium are commonly sputtered on the end of the sapphire fiber sensor to create a blackbody, these films can deteriorate at temperatures exceeding 1600 °C. Success in doping the sapphire fiber end with Cr_2O_3, as an alternative to coating the end of the fiber in a comparatively fragile metallic film, has been reported [39]. The low-temperature measurement range of this sensor is limited to more than 300 °C because the low-absorption window of sapphire occurs between 0.25 and 6 μm [40]. Use of a hollow sapphire tube, which has a low-absorption window extending from 9.6 to 17.2 μm, has been shown to enable temperature measurements between 45 and 900 °C [41]. Other adaptations of the sensor design aim to improve the accuracy of the measurement; system errors coupled with changing transmission and emission losses can result in inaccurate temperature measurements. One solution is to base the temperature measurement on the ratio of the optical powers detected in two wavelength bands [42].

21.4.1.2 Spectroscopy and Chemical Sensing

Another popular use of sapphire fiber is in fiber optic attenuated total reflectance (FO/ATR) spectroscopy. The analysis of the transmitted IR spectra of a substance can be used to determine its chemical makeup. When a material absorbs infrared radiation too strongly to enable this measurement, FO/ATR can be an alternative.

In FO/ATR, as illustrated in Fig. 21.7, a section of unclad and uncoated fiber is submerged inside a sample of the substance under test. For total internal reflection to confine the source light to the fiber, the fiber must be of a higher index then the sample. When this is the case, an evanescent field extends into the sample and is partially absorbed by it. Spectral analysis of the light exiting the sensor determines the chemical composition of the sample.

Several types of optical fibers have been employed in FO/ATR spectroscopy, including chalcogenide, silver halide, and heavy metal glass fibers, but only

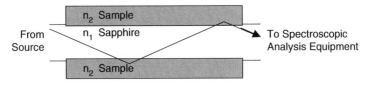

Figure 21.7 Illustration of the principle of a FO/ATR spectroscopic sensor.

sapphire fiber possesses the combination of mechanical strength, chemical resistance, and high-temperature survivability required for a number of applications [43]. Sapphire-based FO/ATR spectroscopy systems have been used to determine the C_2 content of ethylene/propylene co-polymers [17]; to monitor the thermal stability of jet fuel [43]; to monitor coal liquefaction processes [44]; for the on-line analysis of chlorinated hydrocarbons [39]; to measure gaseous hydrocarbons at elevated temperatures [45]; and as a cure-state monitoring device [46].

21.4.1.3 Physical Parameter Sensors

Work towards developing sapphire-based strain, displacement, acoustic wave detection, and other physical measurand sensors has also been performed. The first such reported design describes an extrinsic intensity-based strain sensor, depicted in Fig. 21.8 [47]. The intensity of light captured and guided by the second fiber decreases with the length of the gap; the gap length can be determined by the optical power incident on the detector. The resulting curve of detected power versus gap length is highly nonlinear and is, thus, not easily interpreted.

Although intensity sensors are simple to implement and use, the use of more complex interferometers permits greater measurement accuracy. The report of the first sapphire-based extrinsic interferometric sensor was made by Murphy et al. [34]. The sensor, illustrated in Fig. 21.9, consists of two sapphire fiber rods inserted end-to-end into a sapphire tube. The second rod has a reflective coating (metallic or dielectric). The tube protects the air gap and end-faces from undesired particulate accumulation. This sensing head configuration forms a low-finesse Fabry–Perot cavity, where two light beams produce optical interference. One beam is reflected back from the mirror-coated end-face of one of the rods, while the second beam is the back-reflected light arising from the Fresnel reflection from the uncoated rod's end-face. The interference of these two beams at the detector causes a sinusoidal variation of the detected intensity that is a function of the gap length. Measurements of temperature, pressure, strain, elongation, and other physical parameters can be performed by correlating the changes in the gap separation of this Fabry–Perot cavity, to the changes and shift in the interference fringes.

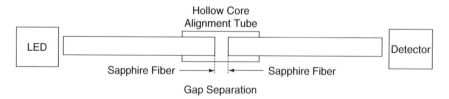

Figure 21.8   Intensity-based sensor schematic.

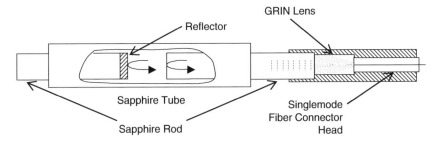

GRIN Lens

Reflector

Sapphire Tube

Sapphire Rod

Singlemode
Fiber Connector
Head

Figure 21.9 Sapphire extrinsic Fabry–Perot interferometer.

21.4.2 Medical Applications

Sapphire fibers have proven to be effective delivery waveguides for high-energy Er/YAG and Er/YSGG lasers. One of the most popular applications has been for hard-tissue dental procedures. Measurements conducted by Nubling and Harrington [12] established the laser damage threshold on sapphire fibers at approximately $1245\,\mathrm{J/cm^2}$. Laboratory experiments have also proven the possibility to deliver up to 150,000 laser pulses at 275 mJ/pulse.

A variety of commercial sapphire fiber laser delivery probes and tips (Fig. 21.10) have been developed that help define a specific output radiation shape for various medical applications such as in ophthalmology, dentistry, endoscopic surgery, and orthopedics.

Figure 21.10 Sapphire fiber laser delivery tips (photo courtesy Photran LLC).

21.5 APPENDIX: MATERIAL PROPERTIES OF Al_2O_3

The data in Table 21.1 are taken from the *Handbook of Optical Materials* by Marvin J. Weber [48], except where indicated.

Aluminum Oxide, Al_2O_3, is an insulator known variously as sapphire, corundum, and alumina. This uniaxial crystal belongs to the trigonal crystal system and space group R-3c. The extraordinary ray, e, is polarized parallel to the c-axis, the axis of anisotropy, and the ordinary ray, o, is polarized perpendicular to the c-axis.

Table 21.1

Material properties of Al_2O_3

Material property	Value(s)
Transmission Window[a] (μm)	0.19–5.2
Band gap (eV)	9.9
Refractive index at 632.8 nm, extraordinary[b] (n_e)	1.7579
Refractive index at 632.8 nm, ordinary[c] (n_o)	1.7659
Melting point (K)	2319
Heat capacity at 300 K (J g^{-1} K^{-1})	0.777
Coefficient of thermal expansion ‖a at 300 K (10^{-6} K^{-1})	6.65
Coefficient of thermal expansion ‖c at 300 K (10^{-6} K^{-1})	7.15
Coefficient of thermal expansion at 1000 K[d] (10^{-6} K^{-1})	8.1
Thermal Conductivity at 250, 300, and 500 K (W m^{-1} K^{-1})	58, 46, 24.2
Molar heat capacity Cp (J mol^{-1} K^{-1}) at 200, 250, 300, 400, 500, and 600 K	51.12, 67.05, 79.45, 88.91, 106.17, 112.55
Density (g cm^{-3})	3.98
Vickers hardness[e] (kg mm^{-2})	1370VH1000
Mohs hardness[f]	9
Cleavage plane	None
Solubility (g/100 g H_2O)	9.8E-5
Poisson's ratio, μ	0.23
Young's modulus, E (GPa)	400
Modulus of rigidity, G (GPa)	162
Bulk modulus, B (GPa)	250

[a] The transmission window is defined to include the range of wavelengths for which the transmission through a 1-mm thick sample at 300 K is greater than 10%.

[b, c] Dispersion formulas for the extraordinary, n_e, and ordinary, n_o, refractive indices allow the computation of n_e and n_o at any wavelength, λ, within the range 0.22–5.0 microns. In these formulas, the value of λ is in micrometer:

$$n_o^2 = 1 + \frac{1.43134936\lambda^2}{[\lambda^2 - (0.0726631)^2]} + \frac{0.65054713\lambda^2}{[\lambda^2 - (0.1193242)^2]} + \frac{5.3414021\lambda^2}{[\lambda^2 - (18.028251)^2]}. \tag{21.4}$$

$$n_e^2 = 1 + \frac{1.5039759\lambda^2}{[\lambda^2 - (0.0740288)^2]} + \frac{0.55069141\lambda^2}{[\lambda^2 - (0.1216529)^2]} + \frac{6.59273791\lambda^2}{[\lambda^2 - (20.072248)^2]} \tag{21.5}$$

[d] From reference [29].

[e] The Vickers hardness has been evaluated to be 1370 kg/mm^2 using an applied load of 1000 kg.

[f] From reference [7].

REFERENCES

[1] Feigelson, R. S. et al. 1984. Single crystal fibers by the laser-heated pedestal growth method. *SPIE Int. Soc. Opt. Eng. Proc. SPIE Int. Soc. Opt. Eng.* 484:133–141.

[2] Merberg, G. N. and J. Harrington. 1993. Optical and mechanical properties of single-crystal sapphire optical fibers. *Appl. Opt.* 32:3201–3209.

[3] Jundt, D. H. et al. 1989. Characterization of single-crystal sapphire fibers for optical power delivery systems. *Appl. Phys. Lett.* 55:2170–2172.

[4] Nubling, R. K. and J. A. Harrington. 1997. Optical properties of single-crystal sapphire fibers. *Appl. Opt.* 36:5934–5940.

[5] Saphikon, Inc. Product Information. Available at *www.saphikon.com*.

[6] Fitzgibbon, J. and G. Hayes. 1999. Sapphire optical fibers: Rugged fiber expands surgical and sensor applications. *Photonics Spectra* 33:118–119.

[7] Steijn, R. P. 1961. On the wear of sapphire. *J. Appl. Phys.* 32:1951–1958.

[8] Morscher, G. N. and H. Sayir. 1995. Bend properties of sapphire fibers at elevated temperatures I: Bend survivability. *Mat. Sci. Engr. A* A190:267–274.

[9] Jundt, D. H. et al. 1989. Growth and optical properties of single-crystal sapphire fibers. *SPIE Int. Soc. Opt. Eng. Proc. SPIE Int. Soc. Opt. Eng.* 1048:39–43.

[10] Merberg, G. and J. Harrington. 1991. Single-crystal fibers for laser power delivery. *SPIE Int. Soc. Opt. Eng. Proc. SPIE Int. Soc. Opt. Eng.* 1591:100–108.

[11] Merberg, G. N. and J. Harrington. 1993. Optical and mechanical properties of single-crystal sapphire optical fibers. *Appl. Opt.* 32:3201–3209.

[12] Nubling, R. K. and J. A. Harrington. 1997. Optical properties of single-crystal sapphire fibers. *Appl. Opt.* 36:5934–5940.

[13] Phomsakha, V. et al. 1994. Novel implementation of laser heated pedestal growth for the rapid drawing of sapphire fibers. *Rev. Sci. Instrum.* 65:3860–3861.

[14] Chang, R. S. F. et al. 1995. Recent advances in sapphire fibers. *SPIE Int. Soc. Opt. Eng. Proc. SPIE Int. Soc. Opt. Eng.* 2396:48–53.

[15] LaBelle, H. E. et al. 1986. Recent developments in growth of shaped sapphire crystals. *SPIE Int. Soc. Opt. Eng. Proc. SPIE Int. Soc. Opt. Eng.* 683:36–40.

[16] LaBelle, H. E., Jr. 1980. EFG, the invention and application to sapphire growth. *J. Crystal Growth* 50:8–17.

[17] Götz, R. et al. 1998. Application of sapphire fibers to and IR fiber-optic sensor for the investigation of polymers at elevated temperature. *Appl. Spectrosc.* 52:1248–1252.

[18] Innocenzi, M. E. and R. T. Swimm. 1990. Room-temperature optical absorption in undoped α-Al_2O_3. *J. Appl. Phys.* 67:7542–7546.

[19] Thomas, M. E. et al. 1993. Vacuum-ultraviolet characterization of sapphire ALON, and spinel near the band gap. *Opt. Engr.* 32:1340–1343.

[20] Omar, M. A. 1975. *Elementary Solid State Physics*. Addison-Wesley Publishing Company, Reading, MA.

[21] Sa'ar, A. and A. Katzir. 1988. Scattering effects in crystalline infrared fibers. *J. Opt. Soc. Am. A* 5:823–833.

[22] Keiser, G. 1991. *Optical Fiber Communications*. McGraw-Hill, New York.

[23] Tong, L. et al. 1999. Performance improvement of radiation-based high-temperature fiber-optic sensor by means of curved sapphire fiber. *Sensors Actuators A Phys.* A75:35–40.

[24] Bates, S. C. 1996. High temperature fiber optic imaging. *SPIE Int. Soc. Opt. Eng. Proc. SPIE Int. Soc. Opt. Eng.* 2839:227–238.

[25] Trumbauer, E. R. et al. 1994. Effect of cleaning and abrasion-induced damage on the Weibull strength distribution of sapphire fiber. *J. Am. Ceram. Soc.* 77:2017–2024.

[26] Yin, H. et al. 1996. Abrasion resistance and strength characteristics of [0001] sapphire fibers modified by 175 KEV Mg$^+$ ion irradiation. *Scripta Materialia* 35:749–754.

[27] Desu, S. B. et al. 1990. High temperature sapphire optical sensor fiber coatings. *SPIE Int. Soc. Opt. Eng. Proc. SPIE Int. Soc. Opt. Eng.* 1307:2–9.

[28] El-Sherif, M. A. et al. 1993. A novel sapphire fiber-optic sensor for testing advanced ceramics. *Ceram. Eng. Sci. Proc.* 14:437–444.

[29] Raheem-Kizchery, A. R. et al. 1989. High temperature refractory coating materials for sapphire waveguides. *SPIE Int. Soc. Opt. Eng. Proc. SPIE Int. Soc. Opt. Eng.* 1170: 513–517.

[30] Davis, J. B. et al. 1993. Fiber coating concepts for brittle-Matrix composites. *J. Am. Ceram. Soc.* 76:1249–1257.

[31] Stanford Materials Company. Available at *www.stanfordmaterials.com*.

[32] El-Sherif, M. A. et al. 1994. Optical response of sapphire multimode optical sensor for ceramic composite applications. *SPIE Int. Soc. Opt. Eng. Proc. SPIE Int. Soc. Opt. Eng.* 2072:244–251.

[33] El-Sherif, M. A. et al. 1994. Development of a fiber optic sensor for ceramic materials characterization. *Ceram. Engr. Sci. Proc.* 15:373–381.

[34] Murphy, K. A. et al. 1991. Sapphire fiber interferometer for microdisplacement measurements at high temperatures. *SPIE Int. Soc. Opt. Eng. Proc. SPIE Int. Soc. Opt. Eng.* 1588:117–124.

[35] Dils, R. R. 1983. High-temperature optical fiber thermometer. *J. Appl. Phys.* 54:1198–1201.

[36] Cooper, J. R. 1985. Optical fiber thermometry: Quest for precision. *Photon. Spectra* 19:71–76.

[37] Zheng, Q. and K. Torii. 1994. Response of optical fiber thermometer with blackbody cavity sensor (aiming to measure the gas temperature inside internal combustion engine). *JSME Int. J. B Fluids Therm. Eng.* 37:588–595.

[38] Tregay, G. W. et al. 1994. Durable, fiber-optic sensor for gas temperature measurement in the hot section of turbine engines. *SPIE Int. Soc. Opt. Eng. Proc. SPIE Int. Soc. Opt. Eng.* 2295:156–163.

[39] Shen, Y. et al. 1994. Study on the doped crystal fiber high temperature sensor. *SPIE Int. Soc. Opt. Eng. Proc. SPIE Int. Soc. Opt. Eng.* 2292:421–428.

[40] Kellner, R. et al. 1995. Recent progress on mid-IR sensing with optical fibers. *SPIE Int. Soc. Opt. Eng. Proc. SPIE Int. Soc. Opt. Eng.* 2508:212–223.

[41] Saggese, S. J. et al. 1990. Hollow waveguides for sensor applications. *SPIE Int. Soc. Opt. Eng. Proc. SPIE Int. Soc. Opt. Eng.* 1368:2–14.

[42] Ye, L. and Y. Shen. 1996. Development of a sapphire fiber thermometer using two wavelength bands. *SPIE Int. Soc. Opt. Eng. Proc. SPIE Int. Soc. Opt. Eng.* 2296:393–400.

[43] Serio, M. A. et al. 1993. *In-situ* fiber optic FT-IR spectroscopy for coal liquefaction processes. *SPIE Int. Soc. Opt. Eng. Proc. SPIE Int. Soc. Opt. Eng* 2069:121–131.

[44] Serio, M. A. et al. 1993. A novel test for fuel thermal stability. *SPIE Int. Soc. Opt. Eng. Proc. SPIE Int. Soc. Opt. Eng.* 2069:20–31.

[45] Edl-Mizaikoff, B. et al. 1995. IR fiber optic evanescent field sensors for gas monitoring. *SPIE Int. Soc. Opt. Eng. Proc. SPIE Int. Soc. Opt. Eng.* 1866:253–264.

[46] Druy, M. A. et al. 1989. Fourier transform infrared (FTIR) fiber optic monitoring of composites during cure in an autoclave. *SPIE Int. Soc. Opt. Eng. Proc. SPIE Int. Soc. Opt. Eng.* 1170:150–159.

[47] Murphy, K. A. et al. 1990. High temperature sensing applications of silica and sapphire optical fibers. *SPIE Int. Soc. Opt. Eng. Proc. SPIE Int. Soc. Opt. Eng.* 1370:169–178.

[48] Weber, M. J. 2003. *Handbook of Optical Materials*. CRC Press, New York.

Chapter 22

Optical Fibers for Industrial Laser Applications

Adrian Carter, Kanishka Tankala, and Bryce Samson

Nufern, East Granby, Connecticut

22.1 FIBER LASERS AND AMPLIFIERS: AN INTRODUCTION

Both optical amplifiers and lasers are based on the process of stimulated emission; a concept first proposed by Albert Einstein in 1916 but not demonstrated until 1954 when Charles Townes used stimulated emission to produce microwave oscillation in the "maser" (microwave amplification by stimulation emission of radiation). It was another 6 years later before Theodore Maiman demonstrated the first laser (light amplification by stimulated emission of radiation). In these devices, an "optically active" species has one of its electrons "excited" into a higher energy level, a passing photon of energy equal to the difference between the electron's energy and a lower energy state "stimulates" the electron to fall into the lower energy state and thereby emit a second photon. This photon will necessarily have the same energy, optical frequency, and phase as the original photon. The waves associated with these two photons constructively interfere and the result is a more intense "amplified" optical beam. If the electrons are reexcited into the higher energy level and feedback of the amplified signal is provided via a resonance cavity, then the "amplifier" may become a "laser."

The lanthanide-doped glass fiber laser was invented in the mid 1960s [1–3], making it almost as old as the laser itself. However, the complications inherent to their early design have until now restricted their real-world applications. Moreover, fiber lasers remained for many years significantly inferior to their Nd/YAG and gas laser alternative technologies, rendering them little more than a scientific curiosity with but a few minor niche applications.

671

22.2 CLADDING PUMPED FIBERS

Early fiber lasers were side-pumped with a flashlamp, but in 1974, Julian Stone and Charles Burrus [4] took the technology a significant step forward when they demonstrated a neodymium-doped multimode fiber laser that was end-pumped with a laser diode. However, at that time the only available technique for achieving an acceptable optical quality of the laser output was to employ a fiber with a geometrically small core (on the order of a few microns). The need to couple excitation energy directly into this small core meant that the total achievable output power of these devices was limited to the milliWatt range. With the advent of cladding pump fiber designs in 1988 [5], the limitation to power scaling fiber devices became the availability of high brightness pump radiation rather than the fiber itself. This trend continued using the available fiber technology, culminating in 1999 with the demonstration of the world's first single-mode fiber laser exhibiting a continuous-wave (CW) output power in excess of 100 W [6].

Traditional optical fiber has a core refractive index raised respective to the surrounding cladding material. The coating has a significantly higher refractive index than either the core or the clad and from an optical perspective is designed to strip out higher order cladding modes that might otherwise re-couple with the core modes. Double-clad fibers (DCFs) differ from traditional "single-clad" optical fiber in the fundamental design of a secondary external wave-guiding structure surrounding the inner core waveguide. For some applications, it is potentially advantageous to be able to splice the fiber directly with the all-glass pump delivery fiber, and in such cases, a triple-clad fiber incorporating a third all-glass 0.23-numerical aperture (NA) cladding between the inner cladding and the polymer coating may be used. The differences in design are shown in Fig. 22.1.

By negating the requirement for excitation energy to be coupled directly into the relatively small single-mode core, DCF makes it possible to employ low-cost, large-area (multimode), high-power semiconductor pump sources. The fundamental concept of the DCF structure is that low-brightness, high-power multimode diodes yielding tens and even hundreds of Watts can be used to provide pump power for lanthanide-doped fibers that will convert that energy into high-brightness, high-power, potentially single-mode output. Fibers for high-power laser and amplifier applications require large claddings with high NAs for efficiently coupling pump energy. Such fibers are typically available with cladding diameters up to around 1 mm but more commonly around 400 μm. A fluorinated polymer optical cladding typically provides an NA of around 0.46 and that is in turn often surrounded by a more standard telecommunications type of jacket (for abrasion resistance). The choice of cladding diameter is

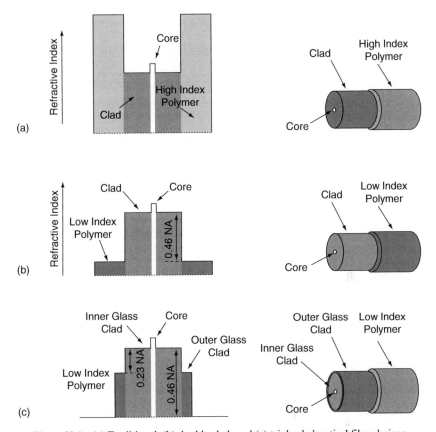

Figure 22.1 (a) Traditional, (b) double-clad, and (c) triple-clad optical fiber designs.

dictated primarily by the brightness of available pump diodes and the total power being coupled. The geometry of the inner cladding is typically shaped to prevent the propagation of skew rays that might otherwise pass down the fiber length without traveling through and being absorbed by the doped core region.

22.3 LARGE-MODE-AREA YTTERBIUM-DOPED FIBERS: THE POWER REVOLUTION

For certain applications, such as ranging and free-space communications, operating in the "eye-safe" 1.5–2.0 μm range is preferred. Furthermore, there are a number of sensing and medical applications that require other specific wavelengths. For such applications, it becomes necessary to employ various

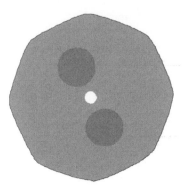

Figure 22.2   Cross-section of a 20-μm core, 400-μm inner-clad PANDA-type ytterbium-doped fiber.

optically active lanthanide ions, such as neodymium, thulium, or co-doped erbium/ytterbium. However, for non–wavelength-specific applications requiring only extremely high-output powers, a number of unique advantages have made ytterbium the dopant of choice. More specifically, ytterbium-doped fibers offer high output powers tunable over a broad range of wavelengths, from around 975 to 1120 nm (typically ~1060 nm) [7].

Ytterbium also has a relatively small quantum defect—that is to say, because the pump wavelength (typically 915–975 nm) is close to the lasing wavelength, very little energy is lost to heating. Furthermore, unlike other lanthanide ions, ytterbium has only a single excited state and thereby is not subject to complications arising from excited state absorption (ESA) and is relatively immune to self-quenching processes. Consequently, high concentrations of ytterbium ions

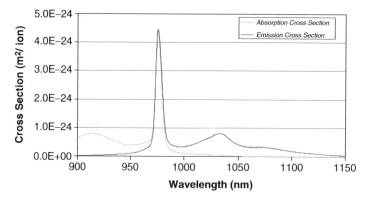

Figure 22.3   Absorption and emission cross-sections for an ytterbium-doped double-clad fiber.

can be incorporated while maintaining excellent conversion efficiencies (typically >75%). For this reason, the industry has focused on the development of ytterbium-doped fibers and the following discussion deals primarily with these fiber designs. It should be noted, however, that neodymium/ytterbium–co-doped fibers have demonstrated power scaling advantages by virtue of the fact that they increase the options for wavelength multiplexing the pump diodes (the neodymium has a peak absorption ~810 nm and gain peak ~1060 nm) [8].

One of the most significant keys to ensuring broad marketability of the fiber laser is to develop a technique for producing ever increasing output powers without sacrificing beam quality. Naturally it is possible to ensure diffraction-limited beam quality from a single-mode core in a DCF geometry. Unfortunately, such a design also limits the total achievable output power and in pulsed laser devices the average power, peak power, and pulse energy. These limitations are the result of low energy storage (for pulsed applications, the energy storage capacity is determined by a combination of the number of active species present and the maximum achievable population inversion, which is in turn determined by the likelihood of amplified spontaneous emission [ASE] [9]) and the effects of parasitic nonlinear processes. More specifically conventional small-core, high NA fiber designs limit the maximum achievable output power because of their fundamental susceptibility to optical nonlinearities, including stimulated Raman scattering (SRS), stimulated Brillouin scattering (SBS), and self-phase modulation. To overcome the limitations imposed by these parasitic nonlinear processes, it has been necessary to develop fibers with high lanthanide concentrations in relatively large-core, low-NA fibers. By increasing the core diameter of a fiber and reducing the core NA, it is possible to maintain single-mode operation while both reducing the fraction of spontaneous emission captured by the core and decreasing the power density in the fiber, thereby increasing the threshold power for the nonlinear processes. Furthermore, the total number of active ions present and so the energy storage capacity increases as the square of the core diameter (for a given glass dopant concentration and cladding diameter). Consequently, it is possible to reduce the length of the fiber device thereby further increasing the threshold for the nonlinear processes.

Of course, there is an upper limit to the core diameter beyond which single-mode operation is not guaranteed. More specifically for a step-index fiber, it is known that single-mode operation requires that the V-value remains below 2.405, where V is proportional to the core diameter (d_{core}) and NA (NA_{core}) and inversely proportional to the wavelength of operation (λ) [10]:

$$V = \frac{\pi d_{core} NA_{core}}{\lambda} \qquad (22.1)$$

At very low NAs (approximately <0.06), fibers begin to exhibit extremely high bend sensitivity. This imposes a practical lower limit on NA and, hence, an upper limit on core diameter. Fortunately, however, there are a number of techniques for the suppression of higher order lasing modes that allow us to use even larger core diameters, wherein essentially multimoded fibers can be made to operate with a diffraction-limited beam quality. These techniques include suitably manipulating the fiber index and dopant profiles [11, 12]; using special cavity configurations [13]; tapering the fiber ends [14]; adjusting the seed launch conditions [15]; and coiling the fiber to induce substantial bend loss for all transverse modes other than the fundamental [16]. Perhaps the simplest and least expensive of these is the coiling technique, which does not require careful matching of the seed mode and does not rely on complex fiber designs. It is only necessary to choose the radius of curvature (based on core diameter and NA) that will discriminate against high-order modes. This technique exploits the fact that the fundamental mode is the least sensitive to bend loss and that the attenuation due to bend loss is exponentially dependant on the bend radius. For example, Fig. 22.4 shows the bend loss as a function of bend radius for a 0.06 NA, 30-μm core diameter fiber. Such a fiber in a linear configuration can support around five modes, but with the appropriate choice of bend radius (say, ~50 mm), the LP11 experiences around 50 dB/m of attenuation (and higher order modes are even more severely attenuated) while the LP01 mode experiences only around 0.01 dB/m. It is important to note that this technique does not involve the stripping of power from higher order modes, but the suppression of those modes along the entire fiber length. As such, power is not attenuated and the efficiency of the laser device is not markedly reduced.

In Fig. 22.5, we show the measured near-field spatial profile of an ytterbium-doped fiber amplifier with a core diameter of 25 μm and an NA of 0.1 when

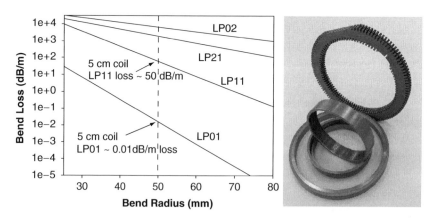

Figure 22.4   Bend loss as a function of bend radius for a 0.06 numerical aperture, 30-μm core diameter fiber and coil forms commonly used to induce the chosen bend diameter in large-mode area fiber.

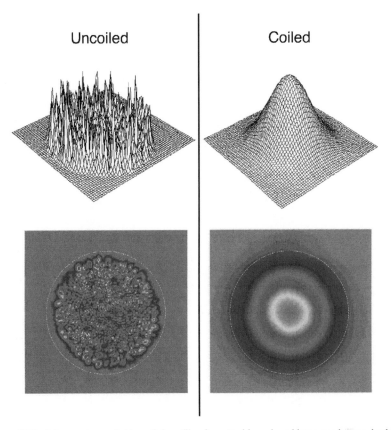

Figure 22.5 Measured near-field spatial profile of an ytterbium-doped large-mode area double-clad fiber amplifier in an uncoiled (left) and coiled configuration (right) (used, with permission, from reference [17]).

seeded with a CW laser at 1064 nm. The profile on the left shows the multimoded ($\sim$27 guided modes) output of the uncoiled fiber and on the right the diffraction-limited (measured M^2 value of 1.08 ± 0.03) output of the coiled fiber [17].

These so-called *large-mode-area (LMA) fibers* are directly responsible for the explosion in demonstrated diffraction-limited beam-quality output powers, now exceeding the kiloWatt level from a single fiber [18–20] (Fig. 22.6). With the advent of this new class of fibers, the power limitations were once again placed on the pump source rather than the fiber.

The advantages of lanthanide-doped LMA fibers are realized by understanding the limiting mechanisms of output power for a typical laser or amplifier. One such mechanism is ASE, which extracts energy from the fiber in an incoherent

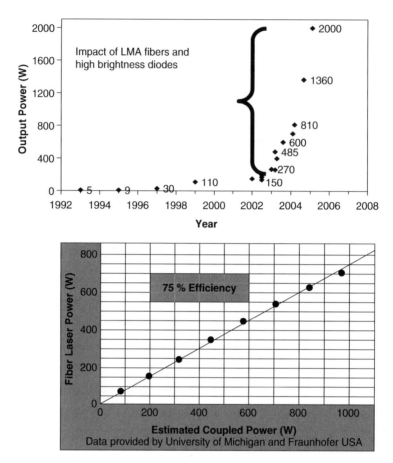

Figure 22.6 Since they were introduced, large-mode area (LMA) fibers have enabled a power scaling revolution and now produce near diffraction-limited beam quality at powers exceeding 1 kW and slope efficiencies of around 75%.

manner. As described earlier, LMA fibers have cores with low NAs, typically smaller than single-mode telecom fibers. This reduction in core NA reduces the amount of fluorescence captured by the core and, thus, the reduction of amplification of that fluorescence. A second mechanism, nonlinear in nature, is SBS, which results from an acoustic wave formed from the superposition of the propagating light wave and the counter-propagating stokes wave generated from the index modulation in the glass created by the propagating wave. The threshold power at which SBS occurs in single-mode fibers is given by Eq. (22.2):

$$\mathbf{P}_{Th} = 21 \cdot \frac{A_{eff}}{L_{eff} \cdot g_B}, \tag{22.2}$$

where A_{eff} is the effective area of the fiber, L_{eff} is the effective length of the fiber given by $L_{eff} = [1\text{-}\exp(-\alpha L)]/\alpha$, and g_B is the Brillouin gain coefficient. The larger core sizes (i.e., effective areas) of LMA fiber raises the SBS threshold, compared to single-mode fibers (by a few orders of magnitude) and enhances the power handling capability of the laser. For example, when pumped at 915 nm a single-mode ytterbium-doped fiber with a 0.15 NA, a core diameter of 5 μm, and an ytterbium ion concentration of around 1 wt% has an SBS threshold of around 40 W at 3-kHz line width, while a 20-μm core 0.06-NA fiber has a threshold of around 340 W and a 30-μm core 0.06-NA fiber has a threshold of around 680 W [21]. In practice, the Brillouin pump line-width can range from a few kiloHertz to several megaHertz, so the actual thresholds may be significantly higher depending on the system configuration. By pumping at 976 nm or using more highly doped fiber, even higher SBS thresholds may be achieved.

22.4 POLARIZATION-MAINTAINING LMA DCF

It is not feasible to indefinitely increase the output power capability of an LMA DCF through scaling of the core diameter. Ultimately there will be some upper limit, above which output beam quality will begin to degrade. To help overcome this hurdle, research is also underway to further refine the design of

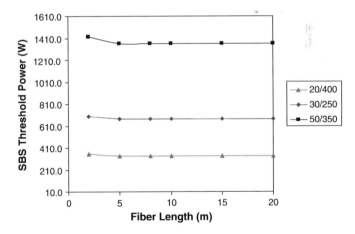

Figure 22.7 SBS thresholds versus fiber length for LMA fibers with 20-, 30-, and 50-μm core diameters and 1 wt% of ytterbium (3-kHz pump spectral line width and 36.5 MHz Brillouin line width) (used, with permission, from reference [21]).

LMA DCFs, through optimization of the glass composition and wave-guiding structure. These include techniques for reducing the peak power density of light propagating in the core, via careful manipulation of the core refractive index profile [22, 23]. Large-flattened-mode (LFM) optical fibers have been designed and fabricated to further increase the threshold for nonlinear processes in LMA fiber by homogenizing the power-intensity profile across the core region (Fig. 22.8). At very high peak powers, optical damage of the fiber and self-focusing are limitations to power scaling. Beam-expanding endcaps reduce surface damage effects and the increased mode-field area (MFA) achieved through LFM fiber designs have enabled greater than 1.5-MW peak powers (at ~1-ns pulse duration) to be achieved. Despite the very large MFA from these fibers, good beam quality (Table 22.1) has been demonstrated in practical systems [24]. Nevertheless, the effectiveness of such techniques is somewhat limited and alternative techniques are required for significant power-scaling requirements.

CW output powers exceeding 1 kW have already been demonstrated in multiplexed fiber devices with poor beam quality [25] and near diffraction-limited output powers exceeding 1 kW have also been demonstrated [19, 20]. However, with the growing need for output powers of several kiloWatts for industrial cutting and welding applications and greater than 100 kW (CW) for military and aerospace applications, the current goal of a number of research groups is to achieve diffraction-limited kiloWatt powers from a single fiber and then to combine the outputs of several such devices. A number of such power-scaling techniques have been demonstrated including coherent beam combining, spectral beam combining, and polarization beam combining. For these extremely high-power applications, operation under stable linear polarization is becoming a requirement [26, 27]. Furthermore there are a number of other applications requiring polarization-maintaining (PM) output including coherent

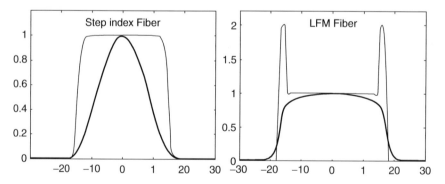

Figure 22.8   Normalized signal power as a function of radial dimension for a conventional step-index and large flattened mode fiber.

Table 22.1

Power amplifier performance of a large flattened mode (LFM) fiber with around 1.5-MW peak power [24]

Parameter	Achieved
Average power	>10 W
Pulse energy	0.75 mJ @ 10 W
Repetition rate	12.16 Kpps @ 10 W
Pulse duration	<0.5 ns @ 10 W
M^2	<1.15 @ 10 W
Spectral line width	25 GHz @ 10 W
SNR ratio in 0.1 nm	−27 dB @ 10 W

optical communications, nonlinear frequency conversion, pumping optical parametric devices, and all manner of mode-locked, Q-switched, and narrow linewidth fiber lasers. Consequently, there has been an increasing demand for PM DCFs.

In the past, different approaches have been suggested to obtain PM operation using non-PM fibers [27, 28]. Such approaches, however, have their limitations and the preferred technology is to use a truly PM DCF. Although passive PM fibers have been commercially available for many years, actively doped PM fibers have not been available until recently [29, 30]. In fact an amplifier employing Yb-doped PM DCF was first reported by Kliner et al. [30] in 2001. This fiber was of bowtie geometry, and though acceptable for proof of concept and research and development, it has substantial limitations in terms of preform manufacturability, uniformity, and scalability. Furthermore, the nonideal refractive index profile inherent to such doped bowtie fibers (Fig. 22.9) makes diffraction-limited operation difficult to achieve.

Figure 22.10a schematically demonstrates the steps involved in making a bowtie type of PM fiber. A high-quality synthetic quartz tube is used as a substrate and several layers of borosilicate glass are first deposited on the inner wall of the rotating substrate. Next the substrate rotation is stopped, and using a specialized ribbon burner, the boron in the glass is volatilized from a selected sector of the deposited layer. The substrate tube is then rotated by 180 degrees and a similar sector is volatilized. Special care has to be taken to ensure that the sectors of glass from which the boron has been volatilized are diametrically opposite to each other and dimensionally equal along the substrate length. Several layers of glass are further deposited before the doped core is deposited. These layers act as a buffer between the borosilicate stress members and the core and ensure that the evanescent field does not propagate in the stress elements to any significant extent. The actively doped core is typically deposited using a solution doping technology. The substrate tube with the various layers of deposited glass is then carefully

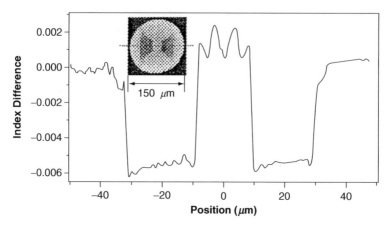

Figure 22.9   Refractive index profile and optical image of the nonideal bowtie PM double-clad fiber [30].

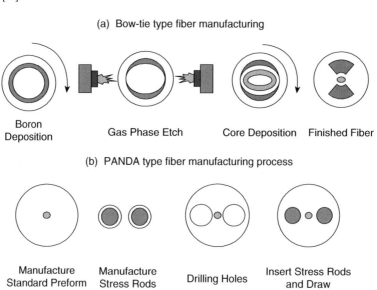

Figure 22.10   Schematic diagrams illustrating the steps involved in the fabrication of (a) bowtie and (b) PANDA-type PM fibers.

collapsed into a rod. The collapsed preform is further processed to obtain the desired inner cladding geometry and drawn into a fiber.

PANDA-type PM DCFs are manufactured in two separate stages, as schematically illustrated in Fig. 22.10b. The actively doped preform is fabricated in a separate process and may employ a manufacturing technology more suitable for yielding highly uniform lanthanide and co-dopant distributions. A high-quality synthetic quartz tube is used to deposit the lanthanide-doped glass. The tube is then collapsed into a rod and further processed so when drawn, the fiber will have the desired core and inner cladding dimensions. In a separate step, two circular stress elements of desired composition are fabricated. Two holes of the desired dimension are drilled, either side of the core, in the lanthanide-doped preform. The circular stress members are inserted into the holes and incorporated into the preform. The preform with the stress members is then drawn into a fiber of desired size and geometry.

The bowtie technology offers the advantage of fabricating the stress members and the lanthanide-doped core in a one-step process. In addition, the distance of the stress members from the core can be precisely controlled by the number of buffer layers deposited between the stress layers and the core. The stress elements can be brought very close to the core, and hence, for a given size and composition of the stress element, a relatively high birefringence may be achieved. However, this technology has several significant disadvantages. The need to deposit stress elements and a lanthanide-doped core within the same substrate tube limits the ability to independently control the polarization and lasing properties of the fiber. Furthermore, although the stress elements can be brought close to the core, the size of the stress elements that may be deposited is restricted and thereby limits the size of the preform that can be made with a desired birefringence. In other words, the technology does not lend itself to volume production. Finally, most DCFs require a noncircular geometry of the inner cladding, which calls for some processing step such as grinding or thermal processing to obtain a desired geometry. In the case of a bowtie type of preform, the grinding (or thermal processing) operation has to be conducted with the stress members in place. PM preforms are relatively fragile because of the large amount of stress incorporated in the preform and are, therefore, prone to fracture on exposure to mechanical (or thermal) shock during a grinding (or thermal processing) operation. The bowtie preform technology is, therefore, not preferred for making volume production of PM DCF.

The technology used to make PANDA-type PM DCF not only offers several advantages but addresses the limitations of the bowtie technology. In this process, both the lanthanide-doped preform and stress member fabrication steps are effectively decoupled, providing independent and highly effective control of the polarization properties and composition of the lanthanide-doped glass. Furthermore, relatively large stress-inducing members may be fabricated, which substantially increases the limit of preform size and makes the process more suitable for preform scale up. Finally, all processing required to achieve a noncircular geometry may be accomplished before incorporating the stress members and, hence, improving production yields. The PANDA-type PM technology is, therefore, amenable to fabricating PM DCF and is the technology of choice for reproducible and uniform volume production.

The PM ability of all PM fibers relies on residual stress anisotropy across the core, which in turn arises from differences in thermal expansion coefficient between the stress members, core and cladding. The composition, location, and geometry of the stress members determine the birefringence in the fiber. In PM DCFs, the core and cladding geometries are very different to standard telecommunications type of PM fibers; more specifically in LMA DCFs, the large diameter of the core negatively affects the achievable birefringence. Before the feasibility of PANDA-type PM-LMA DCFs could first be demonstrated, considerable research had to be performed to optimize the compositional and the geometrical design of the stress members, and in 2003, the results of such detailed experimental and theoretical analyses were reported [31, 32]. Figure 22.11 shows the key dimensional parameters that determine the birefringence that can be obtained in a PM DCF.

These include the size of the stress member (d_s) and the position of the stress member (d_p) relative to the inner cladding diameter (d_f) and the core diameter (d_c). In addition to the geometric factors, the composition of the stress rod determines the birefringence that is achieved in the fiber. Figure 22.12 shows the effect of stress rod size and location on the birefringence (and beat length) of the fiber. As can be seen, the birefringence may be increased (or the beat length reduced) by increasing the size of the stress members (d_s) and keeping all other parameters constant. Similarly, the birefringence may be increased by moving the stress rods closer to the core.

Although it is theoretically possible to use these two geometric parameters to achieve very large values of birefringence, a limiting criterion imposed on d_s and d_p is the distance of the stress members from the core. This limiting distance is

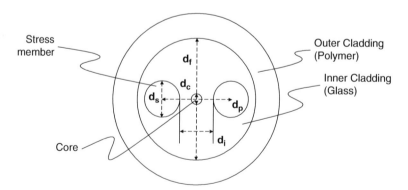

Figure 22.11 Geometric considerations in a PM double-clad fiber (used, with permission, from reference [31]).

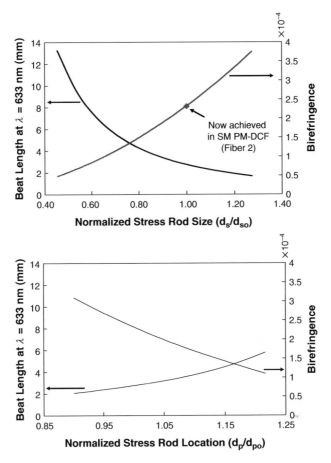

Figure 22.12 Birefringence and beat-length of PM double-clad fibers as a function of stress rod size (left) and location (right) (used, with permission, from reference [31]).

indicated by the distance between the inside edges of the stress members (d_i). If d_i becomes very small, the probability of overlap between the mode field and the stress members increases, resulting in increased attenuation and bend loss of the laser or amplifier signal wavelength. To provide a safety margin for avoiding any overlap between the modal power profile in the fiber and the stress members, it is necessary to determine a critical ratio d_i/MFD so that losses are minimized. For small-core single-mode fibers used in low to medium power applications, it is possible to achieve sufficient birefringence using standard stress member compositions and operate well within the limiting ratio. However, for large-core fibers as needed for high-power applications, achieving sufficient birefringence while operating within the limiting ratio is more challenging. In such cases, a higher coefficient of thermal expansion difference and, hence, higher birefringence can be achieved by adjusting the composition of stress members so they are similar to those used for gyroscope fibers. Indeed a broad range of ytterbium-doped LMA DCFs, whose characteristics are optimized for various output powers, are now commercially available [33]. An optical image showing the cross-section of such a fiber, with a 20-μm core and 400-μm inner-cladding diameter and a birefringence exceeding 3.5×10^{-4}, is presented in Fig. 22.2.

22.5 FIBER LASERS: STATE OF THE ART

LMA fibers with core diameters of 20–30 μm and NAs of around 0.06 have become the industry standard for high-power laser and amplifier devices because of their ability to deliver good beam quality through preferential modal excitation [15] or coiling induced higher order mode losses [16]. The addition of PANDA-type stress elements to make PM-LMA fibers has added to the application space for the fiber technology and enabled high-power linearly polarized fiber amplifiers both in the CW [34] and pulsed regimens [35]. The availability of fibers with large claddings (400 μm) and high cladding NAs (0.46) in conjunction with high brightness pump sources has featured in many of the high-power results. More particularly, they have facilitated the amplification of single-frequency sources into the high-power regimen (hundreds of Watts [36]) and as such are potential building blocks for coherent beam-combining [37] and fiber array phase-locking [38] experiments.

An indicator of maturity in the LMA fiber technology is the availability of standard support components with LMA-compatible fiber pigtails, including the multimode pump combiners, which also serve as signal multiplexers. These components are available with input fibers compatible with industry-standard pigtail fibers on commercial high-power diodes. For example, the (6 + 1)-to-1 design consisting of LMA-compatible 20/400 DCF on the output of the combiner with six 200/220 0.22-NA pump delivery fibers on the input side are commercially available. Furthermore high-brightness, fiber-coupled pump diodes compatible with pump combiners are now commercially available with industrial-grade reliability. These advancements in high-brightness pumps and high-power pump combiners have enabled the high-power, monolithic design shown in Fig. 22.13.

Experimental results for the system are presented in Fig. 22.14 and demonstrate the applicability of these high-power LMA monolithic amplifiers to output powers greater than 200 W CW. Although the power level is well below that demonstrated with broad line-width fiber lasers and amplifiers [18–20], these LMA devices are applicable to amplifying single-frequency input signals with

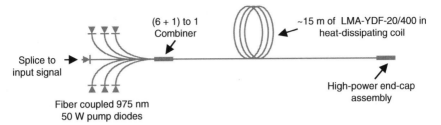

Figure 22.13 Schematic of the monolithic 200-W large-mode area amplifier.

coherence lengths suitable for further beam combining into the multi-kiloWatt regimen [36].

PM versions of these LMA fibers have also been demonstrated to exhibit excellent slope efficiencies and operate at high powers, greater than 400 W pump power limited [34]. Indeed, Nufern scientists have combined the PM-LMA fiber concept with an optimized coil form to deliver a unique linearly polarized fiber laser, as shown schematically in Fig. 22.15. Alternative methods for delivering polarized fiber lasers inevitably include external or free space components, or at the very least extra polarizing components within the cavity. By simply optimizing the fiber and coil combination, it is possible to make a high-power polarized fiber laser by taking advantage of the difference in bend-induced attenuation for each of the two polarization states. The excellent polarization/extinction ratio, greater than 95%, was obtained with diffraction-limited beam quality, and a grating stabilized line width of around 0.1 nm. Importantly, the output power

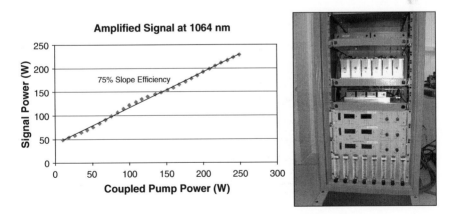

Figure 22.14 Amplified signal power as a function of coupled pump power (left) from the 19-in. rack mounted large mode area monolithic amplifier (right).

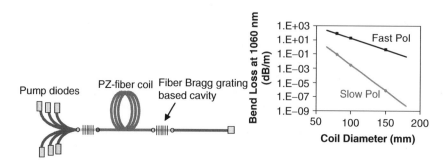

Figure 22.15 Nufern's proprietary technique for making a linearly polarized fiber laser using the coiling induced polarizing effect removes the need for external polarizing elements or extra elements in the cavity.

was limited only by the available pump power, and in fact it is anticipated that the maximum CW output power for this fiber design will be around 2 kW [34].

LMA fibers have also had a dramatic impact on pulsed fiber laser technology where pulse energies are now approaching 100 mJ with multimode beam quality and around 4 mJ single mode [39]. High peak powers, in combination with the short pulse durations (nanoseconds) and diffraction-limited beam quality, are desired in many applications including marking, micromachining, and drilling anything from fuel injectors through to turbine blades. Military applications involve target designation and laser radar, which is also a growing commercial market with applications such as vehicle guidance, robotics/vision systems, metrological, and surveying.

The combination of high average power (over hundreds of Watts), excellent beam quality, and a polarized output has caught the attention of the DPSSL community. Such performance specifications are extremely difficult to achieve in either CW or pulsed YAG/Vanadate lasers where fundamental problems such as thermal lensing push resonator design to the limits. Furthermore, the flexibility of monolithic components make concatenating amplifier stages relatively easy in fiber devices, opening the potential for highly flexible devices generating pulse durations from sub-nanosecond to CW and pulse energies in the milliJoule range.

22.6 LARGE-MODE-AREA EYE-SAFE FIBERS

As discussed earlier, new developments in LMA fibers have led to demonstrations of kiloWatt-level outputs in CW lasers and megaWatt-level peak powers in pulsed amplifiers (with sub-nanosecond pulses). However, development of LMA fibers has largely been restricted to ytterbium-based fibers for use at around 1.0 μm, because of the relative ease of manufacturing LMA ytterbium-doped fibers and high optical efficiencies (~80%), associated with this laser system. The ability to achieve relatively high concentrations of ytterbium with relatively low levels of matrix-modifying co-dopants lends itself to fabricating low-NA, large-core fibers. The low-NA core supports only a few modes and the higher order modes can be easily discriminated against by preferential seeding [15] or bending [16] to achieve diffraction-limited operation. In spite of the numerous advantages, a significant drawback of the ytterbium-based system is the relatively high sensitivity of the human eye to wavelengths in the 1.0-μm region.

The retinal absorption of radiation in the 1.5- and 2-μm wavelength regions is substantially lower than that at 1.0 μm. High-power lasers and amplifiers operating in these wavelength bands are, therefore, of interest in both military and commercial applications such as free-space and satellite optical communications and LIDAR. In military applications, these wavelengths are also of interest because they minimize collateral damage. Similarly, eye-safe lasers are welcome in commercial application because they greatly reduce the challenges associated with laser safety. In addition to eye safety, the 2.0-μm lasers may also find applications as pump source for holmium-doped lasers or nonlinear conversion to longer wavelengths. Although there has been significant interest in eye-safe lasers, progress in developing high-power CW and pulsed lasers at these wavelengths has been limited by the availability of LMA erbium/ytterbium– and thulium-doped fibers. This limitation is itself due to the difficulties in manufacturing optically efficient low-NA optical fibers containing these lanthanide dopants.

Despite the lack of LMA fibers, important work has been conducted in developing high-power erbium/ytterbium lasers using multimode fibers. Koroshetz et al. [40] demonstrated a 40 W, 10-Gbps amplifier for free-space communication and Shen et al. [41] reported 188 W of CW output with an M^2 of 1.9 using a 30-μm, 0.22-NA erbium/ytterbium–co-doped fiber. Furthermore, Yusim et al. [42] have been able to take a 20-μm core erbium/ytterbium fiber, which supports 30 modes, and achieve 100 W output with a near diffraction-limited beam by employing single-mode fiber-based components in the cavity and making careful splices between the single-mode fibers and the multimode active fiber. Although a high-power diffraction-limited output was achieved, such methods are cumbersome and emphasize the need for LMA fibers. Similarly, noteworthy work has been conducted in developing highly efficient thulium-doped fibers [43, 44] by exploiting the two-for-one cross-relaxation process between thulium ions and output powers as high as 85 W have been demonstrated albeit with a multimode output [44].

Unlike ytterbium-doped fibers, the compositional requirements for erbium/ytterbium–co-doped and thulium-doped fibers make fabrication of LMA variants particularly challenging. More specifically, it is well known that sensitizing erbium-doped fibers with ytterbium enhances pump absorption and hence increases the efficiency at the erbium lasing wavelength. Erbium has a very narrow absorption peak and, like any lanthanide ion, cannot be incorporated into silica glass at extremely high concentrations without clustering. Sensitization is accomplished by taking advantage of the broad absorption band and the high cross-section of ytterbium as compared with erbium [45], with the net result that erbium/ytterbium–co-doped fibers have a very broad absorption band with a peak absorption that is more than two orders of magnitude greater than conventional erbium-doped fibers [46]. For efficient energy transfer from the ytterbium to erbium ions, the Raman shift of the base glass is increased by doping it with phosphorus. The presence of $P = O$ bonds increases the phonon energy of the glass host, and the Raman spectrum has a peak at $1330\,cm^{-1}$, as compared to $1190\,cm^{-1}$ for pure silica [47]. This helps in rapid depopulation of the erbium $^4I_{11/2}$ energy level, limiting the back-transfer of energy from erbium to ytterbium ions (Fig. 22.16). Thus, efficient erbium/ytterbium fibers contain phosphorus in the core, which also aids in minimizing the clustering of the lanthanide ions [48]. A substantially high level of phosphorus is required to achieve both of the

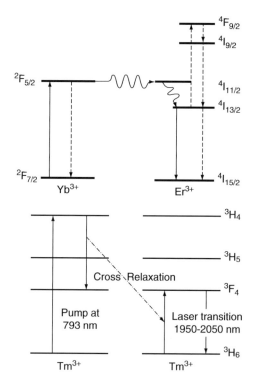

Figure 22.16 Energy transfer processes in erbium/ytterbium- (left) and thulium-doped fibers (right).

aforementioned goals. However, phosphorus also increases the refractive index of the base glass, resulting in relatively high core NAs of around 0.17–0.20 or higher. Thus, the difficulty in producing an LMA fiber, which requires a low NA, becomes apparent.

Thulium-doped fibers have a number of potential pump bands (Fig. 22.17). However, pumping at 793 nm is preferred because of the availability of high-power semiconductor diodes in the 800-nm spectral region. In conventional thulium-doped fibers, the maximum conversion efficiency of a 793-nm pumped laser is about 40%. However, compositionally engineered thulium-doped silica fibers can achieve substantially higher efficiencies (approaching 80%). Important energy transfer processes relevant to the performance of thulium-doped silica fibers have been identified [44, 49, 50]. Figure 22.16 shows the relevant cross-relaxation mechanism observed in thulium-doped silica fibers. Two cross-relaxation processes, namely $^3H_4, {}^3H_6 \rightarrow^3 F_4, {}^3F_4$ and $^3H_4, {}^3H_6 \rightarrow^3 H_5, {}^3F_4$, have been identified with the $^3H_4, {}^3H_6 \rightarrow^3 F_4, {}^3F_4$ being particularly efficient because of the large degree of spectral overlap between the $^3H_4, \rightarrow^3 F_4$ emission and the $^3H_6 \rightarrow^3 F_4$ absorption. This cross-relaxation process results in the generation of two signal photons for every pump (793-nm) photon and may be promoted by using high thulium concentrations. However, energy transfer up-conversion processes, namely $^3F_4, {}^3F_4 \rightarrow^3 H_5, {}^3H_6$ and $^3F_4, {}^3F_4 \rightarrow^3 H_4, {}^3H_6$, have to be kept in check to prevent quenching of the $^3F_4$ energy multiple. This can be minimized by using very high alumina:thulium concentrations and thereby

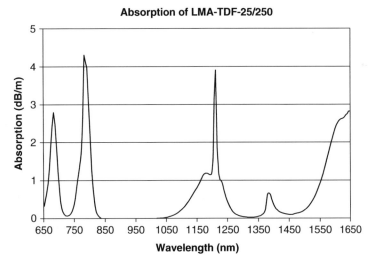

Figure 22.17 Cladding absorption for a large mode area thulium-doped double-clad fiber with a 25-μm core and a 250-μm cladding diameter.

preventing clustering of the thulium ions [48]. Therefore, both high thulium and high alumina concentrations are required to achieve greater than 100% quantum efficiencies in thulium-doped fibers. In fact, careful compositional engineering can yield power conversion efficiencies similar to that of ytterbium-doped fibers, making such fibers extremely desirable for eye-safe laser system. However, these high dopant concentrations substantially increase the refractive index of the core as compared with the pure silica substrate, and typical NAs are in the range 0.18–0.24. Hence, as in the case of erbium/ytterbium–co-doped fibers, the compositional requirements for an efficient thulium-doped fiber also limits the ability to manufacture low NA LMA fibers.

In a typical step-index fiber, the NA of the core is defined by the index of the core and that of the cladding surrounding it. However, if an appropriate pedestal is designed, the core may have an effective index as defined by the index of the core and that of the pedestal, as shown in Fig. 22.18. The pedestal index may be chosen so as to define an effective core NA of less than 0.1. Similarly the pedestal size may be chosen so that further increasing the pedestal diameter has no significant effect on the core modes, particularly the fundamental mode. At this diameter, the pedestal behaves as a "true" cladding to the core, rather than an extended core feature. If the pedestal is made any smaller, the mode-field diameter of the core fundamental mode starts decreasing, and core light may be coupled to the pedestal modes. At the same time, the pedestal should not be made much larger than required, because of increased manufacturing cost, as well as the possibility of trapping pump light in the helical modes in the pedestal.

The first demonstration of highly efficient large core (25 μm) LMA ytterbium/erbium–co-doped and thulium-doped DCFs was recently reported [51]. In both cases, the core NAs were chosen to be around 0.1 and the fibers were drawn to 300 and 250 μm cladding diameters, respectively, with a cladding NA of 0.46. Figure 22.19 shows the absorption and about 30% slope efficiency for the erbium/ytterbium 25/300 pedestal fiber.

The key benefit of the pedestal fiber design is clearly the lower number of modes supported by the doped core of the fiber. Consequently, it should be easier to excite and maintain the fundamental mode in this fiber than in a fiber with a similar diameter high NA ($\sim$0.17) core. The V-number of a 25-μm core fiber at an NA of 0.17 is 8.616, while at an NA of 0.10, it is 5.067. Figure 22.20 shows that incorporation of the pedestal is expected to reduce the core modes from 11 to 4, making it possible to achieve near diffraction-limited beam quality. Importantly, both exciting the fundamental mode (at the expense of higher order modes) and maintaining the power within this mode through the appropriate fiber length should be significantly easier in the few-moded fiber as compared with the highly multimode fiber, because of reduced mode coupling.

Experimentally, good beam quality from the pedestal fiber has been verified in a seeded amplifier configuration, in which care was taken to excite the

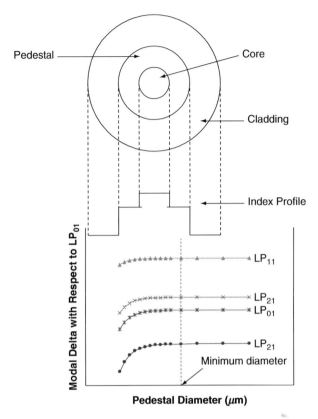

Figure 22.18 Schematic diagram of a large mode area fiber using a pedestal design (top) and modeling data for the determination of minimum pedestal diameter (bottom).

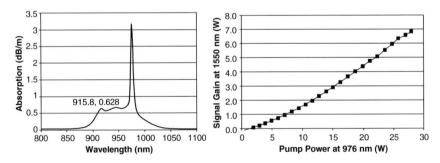

Figure 22.19 Absorption spectra (left) and slope efficiency for the large mode area EYDF-25/300 (right).

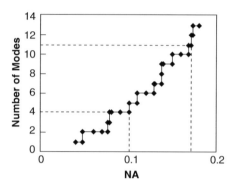

Figure 22.20 The influence of numerical aperture (NA) on the number of transverse modes supported by a 25-μm core fiber.

fundamental mode of the doped core using standard mode matching techniques. In fact it was found that beam quality was easier to maintain in this fiber than in a fiber with a high 0.17-NA core, even though the core was substantially smaller (18 μm). Single-mode beam profiles and an example M^2 measurement are shown in Fig. 22.21 for a fiber length appropriate for making efficient amplifiers. Although good beam quality has been demonstrated in laboratory

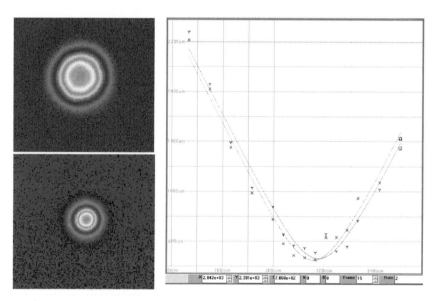

Figure 22.21 Near-field and far-field beam profile from large mode area (LMA) EYDF-25/300 under seeded amplifier conditions and M^2 measurement (used, with permission, from reference [51]).

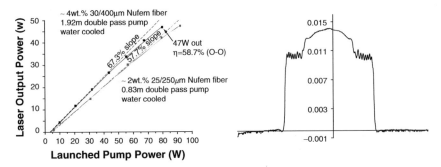

Figure 22.22 Slope efficiency measurements for compositionally optimized thulium-doped fiber [52] (left) and the fiber refractive index profile of a pedestal-type large mode area (LMA) thulium-doped fiber (right).

based experiments with high NA large core fibers [42], developing components and designing systems to guarantee a beam quality specification would be particularly challenging. It is believed that the adoption of a pedestal fiber design will make the challenges associated with delivering good quality beam profiles at eye-safe wavelengths manageable.

Compositional development of thulium-doped fiber has also been performed to optimize the thulium and alumina concentrations required to promote the two-for-one cross-relaxation process [51], and slope efficiencies as high as 68% (170% quantum efficiency) have been reported [52] (Fig. 22.22). An index profile of the LMA thulium-doped fiber (LMA-TDF-25/250) showing the refractive indices of the core and the pedestal relative to the cladding is also shown in Fig. 22.22. The absorption of the fiber at 790 nm was measured to be around 4.5 dB/m and the absorption spectrum is shown in Fig. 22.17. The fiber was demonstrated to have a near–diffraction-limited beam quality and an M^2 of less than 1.3 was reported [52].

22.7 CONCLUSIONS

Today both CO_2 and YAG lasers are routinely capable of producing CW output powers in the range of sub-Watt to multi-kiloWatt and as such have become the mainstay of the laser cutting and welding industry. Despite numerous advancements in their design, these lasers are still typified by poor wall-plug efficiencies (typically 1–10%) and/or relatively poor optical beam qualities. On the other hand, developments in laser diode technology, fiber design, and beam-combining techniques have meant that cladding pumped ytterbium-doped fiber lasers have attracted growing interest as a route to highly efficient (20–40%

wall-plug efficiencies), high output power, high beam quality (near–diffraction-limited) lasers for a vast array of material processing applications. More specifically, fiber lasers have a number of distinct advantages over their more conventional alternatives including size, reliability, wavelength selectivity, heat dissipation, wall-plug efficiency, and operational cost. Furthermore, they may be operated without the need for active cooling or optical alignment.

LMA fiber technology has become an established means of power scaling fiber lasers and amplifiers, and as the basic fiber design has become standardized, this has encouraged component manufacturers to design and build high-power LMA-compatible devices. Consequently, the ytterbium-doped fiber laser is now challenging more traditional bulk solid-state and gas laser systems in a range of both industrial and military, sensing and material processing applications. Similarly, the recent evolution of LMA erbium/ytterbium and LMA thulium-doped fibers may be expected to spur the development of high-power lasers and amplifiers in the eye-safe regions of 1.5 and 2.0 μm, respectively.

REFERENCES

[1] Snitzer, E. 1961. Optical maser action of Nd3+ in a barium crown glass. *Phys. Rev. Lett.* 7:444–446.

[2] Snitzer, E. 1963. Neodymium glass laser. In: *Proc. 3rd Int. Conference of Quantum Electronics,* pp. 999–1019, Paris, 1963.{Au: Pls provide full book title and the publisher name and place where published.}

[3] Koester, C. J. and E. Snitzer. 1964. Amplification in a fiber laser. *Applied Opt.* 3:1182–1186.

[4] Stone, J. and C. Burrus. 1974. Neodymium-doped fiber lasers: Room temperature CW operation with an injection laser pump. *IEEE J. Quant. Electr.* 10:794.

[5] Snitzer, E. et al. 1988. Double-clad offset core Nd fiber laser. In: *Conference Proc. of Optical Fiber Sensors,* PD5, New Orleans.

[6] Dominic, V. et al. 1999. 110 W fiber laser. In: *Conference Proceedings of OSA—Optical Soc. of Am. CLEO,* CPD-11, Washington, DC, 1999.

[7] Paschotta, R. et al. 1997. Ytterbium doped fiber amplifiers. *IEEE J. Quant. Electr.* 33(7):1049–1056.

[8] Limpert, J. et al. 2003. 500 W continuous-wave fiber laser with excellent beam quality. *Electr. Lett.* 39(8):645.

[9] Nilsson, J. and B. Jaskorzynska. 1993. Modeling and optimization of low-repetition-rate high-energy pulse amplification in CW-pumped erbium-doped fiber amplifiers. *Opt. Lett.* 18(24):2099.

[10] Gloge, D. 1971. Weakly guiding fibers. *Appl. Opt.* 10:2252–2258.

[11] Offerhaus, H. L. et al. 1998. High energy single-transverse-mode Q-switched fiber laser based on a multimode large-mode area erbium-doped fiber. *Opt. Lett.* 23(21):1683–1685.

[12] Nilsson, J. et al. 1997. Ytterbium^{3+}-ring-doped fiber for high-energy pulse amplification. *Opt. Lett.* 22(14):1092.

[13] Griebner, U. et al. 1996. Efficient laser operation with nearly diffraction limited output from a diode-pumped heavily Nd-doped multi-mode fiber. *Opt. Lett.* 21:266.

[14] Renaud, C. C. et al. 1999. Compact high-energy Q-switched cladding-pumped fiber laser with a tuning range over 40nm. *IEEE Photon. Technol. Lett.* 11(8):976–978.

[15] Fermann, M. E. 1998. Single-mode excitation of multimode fibers with ultrashort pulses. *Opt. Lett.* 23:52–54.

[16] Koplow, J. P. et al. 2000. Single-mode operation of a coiled multimode fiber amplifier. *Opt. Lett.* 25:442.

[17] Kliner, D. et al. 2004. Fiber laser technology reels in high power results. *OE Magazine* 4(1): 32–35.

[18] Liu, C.-H. et al. 2004. 810 W continuous-wave and single-transverse-mode fiber laser using 20 mm core Yb-doped double-clad fiber. *Electr. Lett.* 40:1471–1472.

[19] Jeong, Y. et al. 2004. Ytterbium doped large-core fiber laser with 1.36kW continuous-wave output power. *Opt. Exp.* 25:6088–6092.

[20] Gapontsev, V. P. 2005. Recent progress in beam quality improvement of multi-kW fiber lasers. *Photon. West* 5709-11, San Jose, January 2005. See also www.ipgphotonics.com.

[21] Tankala, K. et al. 2004. Large mode area double clad fibers for pulsed and CW lasers and amplifiers. *Photonics West,* 5335-22, San Jose.

[22] Ghatak, A. K. et al. 1998. Design of waveguide refractive index profile to obtain flat modal field. *SPIE Proc.* 3666:40.

[23] Dawson, J. W. et al. 2003. Large flattened mode optical fiber for high output energy pulsed fiber lasers. In: *Conference on Lasers and Electro-Optics,* CWD5, Washington, DC.

[24] Torruellas, W. et al. 2005. High peak power Ytterbium doped fiber amplifiers. In: *Tech. Digest Solid State and Diode Laser Technology Review* (SSDLTR), Fiber-7, Los Angeles.

[25] Ueda, K. et al. 2002. 1 kW CW output from fiber embedded lasers. In: *Proceedings Conference on Lasers and Electro-Optics,* CPDC4, Long Beach. OSA—Opt. Soc. Am. Washington, DC.

[26] Noda, J. et al. 1986. Polarization maintaining fibers and their applications. *J. Lightwave Technol.* 4(8):1071–1089.

[27] Koplow, J. P. et al. 2000. Polarization-maintaining, double-clad fiber amplifier employing externally applied stress-induced birefringence. *Opt. Lett.* 25(6):387–389.

[28] Duling, I. N., III, and R. D. Esman. 1992. Single-polarisation fiber amplifier. *Electr. Lett.* 28(12):1126–1128.

[29] Tajima, K. 1990. Er3+-doped single-polarisation optical fibers. *Electr. Lett.* 26(18): 1498–1499.

[30] Kliner, D. A. V. et al. 2001. Polarization-maintaining amplifier employing double-clad bow-tie fiber. *Opt. Lett.* 26(4):184–186.

[31] Tankala, K. et al. 2003. PM-double clad fibers for high-power lasers and amplifiers. Paper presented at Photonics West, 4974-40, San Jose.

[32] Machewirth, D. P. et al. 2003. Polarization-maintaining double-clad optical fibers for coherent beam combining. Presented at Solid State Laser Conference, New Mexico, 2003.

[33] Available at www.nufern.com.

[34] Khitrov, V. et al. 2005. Linearly polarized high power fiber laser with monolithic PM-LMA fiber and LMA-based grating cavities. Presented at Photonics West, 5709-10, San Jose.

[35] Liu, A. et al. 2005. 60-W green output by frequency doubling of a polarized Yb-doped fiber laser. *Opt. Lett.* 30:67–69.

[36] Khitrov, V. et al. 2006. 242W single-mode CW fiber laser operating at 1030nm lasing wavelength and with 0.35nm spectral width. Presented at ASSP, WD5, Lake Tahoe.

[37] Wickham, M. et al. 2003. High power fiber array coherent combination demonstration. In: *Proceedings 16th SSDLTR,* HPFib-5.

[38] Minden, M. et al. 2004. Self organized coherence in fiber laser arrays. *Proc. SPIE* 5335-14:89–97.

[39] Cheng, M.-Y. et al. 2005. High-energy and high-peak-power nanosecond pulse generation with beam quality control in 200 μm core highly multimode Yb-doped fiber amplifiers. *Opt. Lett.* 30(4):358–360.

[40] Koroshetz, J. et al. 2004. High power eye-safe fiber transmitter for free space optical communications. Presented at Advanced Solid-State Photonics, MA2, Santa Fe.

[41] Shen, D. Y. et al. 2005. Highly efficient Er,Yb-doped fiber laser with 188W free-running and >100W tunable output power. *Opt. Exp.* 13(13):4916–4921.

[42] Yusim, A. et al. 2005. 100 Watt single-mode CW linearly polarized all-fiber format 1.56-μm laser with suppression of parasitic lasing effects. Proceedings of SPIE Vol. 5709:69–77, 2005.

[43] Jackson, S. D. and S. Mossman. 2003. Efficiency dependence on the Tm^{3+} and Al^{3+} concentrations for Tm^{3+}-doped silica double-clad fiber lasers. *Appl. Opt.* 42(15):2702–2707.

[44] Frith, G. et al. 2005. 85W Tm^{3+}-doped silica fiber laser. *Electr. Lett.* 41(12):1207–1208.

[45] Valley, G. C. 2001. Modeling cladding-pumped Er/Yb fiber amplifiers. *Opt. Fiber Technol.* 7:21–44.

[46] Zhou, X. and H. Toratani. 1995. Evaluation of spectroscopic properties of Yb^{3+}-doped glasses. *Phys. Rev. B* 52(22):15889–15897.

[47] Grubb, S. G. et al. 1991. High-power sensitised erbium optical fibre amplifier. *Proc. OFC* PD7:31–33.

[48] Arai, K. et al. 1986. Aluminium or phosphorus co-doping effects on the fluorescence and structural properties of neodymium-doped silica glass. *J. Appl. Phys.* 59:3430–3436.

[49] Jackson, S. D. 2004. Cross relaxation and energy transfer upconversion processes relevant to the functioning of 2 μm Tm^{3+}-doped silica fibre lasers. *Opt. Commun.* 230:197–203.

[50] Shen, D. Y. et al. 2005. High-power and ultra-efficient operation of a Tm^{3+}-doped silica fiber laser. Presented at Advanced Solid State Photonics, MC-6.

[51] Tankala, K. et al. 2006. New developments in high power eye-safe LMA fibers. Presented at Photonics West, 6102-06, San Jose.

[52] Frith, G. P. and D. G. Lancaster. 2006. Power scalable and efficient 790 nm pumped Tm^{3+}-doped fibre lasers. Presented at Photonics West, 6102-08, San Jose.

Chapter 23

Optical Fibers for Biomedical Applications

Moshe Ben-David[1] and Israel Gannot[2]

[1]*Department of Biomedical Engineering, Tet-Aviv University, Tel-Aviv, Israel*
[2]*Department of Electrical and Computer Engineering, George Washington University, Washington, D.C.*

23.1 INTRODUCTION

Optical fibers have revolutionized medicine in many ways. They have advanced both diagnostics and treatments capabilities. Their major advantages lie in the fact that they are thin and flexible so they can be introduced into the body to remotely sense, image, and treat.

This capability enabled the introduction of minimally invasive procedures, which became the preferred choice for surgery. Such procedures minimize the postoperative pain and discomfort, as well as shorten—and sometimes eliminate—hospitalization time, thus saving on costs and reducing the number of missed workdays. They also reduce the risk of contamination from the hospital environment. And maybe, most importantly, they allow the patient return to the comfort of his or her home sooner.

People are taking it naturally to have endoscopes in the clinical practice. This instrument—and all its variations—is based on the existence of a coherent fiber bundle, which allows imaging of internal organs while working their way through the natural orifices of the body or in some cases through minimal incisions. Fiber bundles were patented in the 1920s by Baird in England and Hansell in the United States in parallel, but most important progress and realization were made by Van Heel and Capany in the 1950s (also two parallel unconnected works). Now imaging bundles can have 100,000 fibers that are small enough and flexible to be used in various types of endoscopes.

In this chapter, we describe new progress in this exciting field of fiber optics into the infrared (IR) regimen and other new fiber-based sensing methods in the visible, ultraviolet (UV), and IR.

We begin (Section 23.2) with the development of fibers and waveguides that are capable to transmit high energies for tissue ablation, thus replacing cumbersome medical laser arms based on tubes, joints, and reflecting mirrors. We continue (Section 23.3) with an example of state-of-the-art operation of calculi fragmentation in the salivary gland ducts. The next two sections deal with two sensing methods that make use of fibers: absorption spectroscopy, which uses the hollow core to make spectral measurements of different biological gases, and evanescent wave sensing, which uses the penetration of waves outside the borders of the delivery medium.

Section 23.6 deals with remote sensing of temperature, which can be applied through endoscopes as well.

The last section brings us into the imaging field again; however, it expends our capabilities into the mid-IR (MIDIR) range, thus allowing collection of additional important data that lie in this spectral range for temperature mapping for diagnostic and feedback mechanisms for treatments.

23.2 MEDICAL LASER ARMS

The most straightforward use of lasers in medicine is laser surgery such as benign lesion removal and tissue ablation. Although a variety of lasers are on the market, only a few might be suited for medical applications, and there is a need to optimize the laser for the required applications. The most common lasers used for medical applications are $CO_2(\lambda = 10.6\,\mu m)$, $Er/YAG(\lambda = 2.93\mu m)$, $Ho/YAG(\lambda = 2.07\,\mu m)$, $Nd/YAG(\lambda = 1.06\,\mu m)$, and some others in the visible range.

A CO_2 *laser* is often referred to as the "surgical laser" because its action most resembles traditional surgery. Unlike that of any other medical laser, its action on tissue is directly visible as it is used. The CO_2 laser was the first laser widely used by surgeons and is still the most used of all the medical lasers. Strongly absorbed by water, which constitutes more than 80% of soft tissue, this laser emits continuous wave (CW) or pulsed far-IR light at 10.6 μm, which can be focused into a thin beam and used to cut like a scalpel or defocused to vaporize, ablate, or shave soft tissue. The CO_2 laser may be operated in pulsed mode or used with scanning devices to precisely control the depth and area of ablation. Its uses include

- Removal of benign skin lesion, such as moles, warts, keratoses [1]
- As a "laser scalpel" in patients or body areas prone to bleeding [2–4]
- "No-touch" removal of tumors, especially of the brain [5, 6] and spinal cord [7–9]
- Laser surgery for snoring (LAUP) [10–12]

- Shaving, dermabrading, and resurfacing scars, rhinophyma, skin irregularities [13, 14]
- Cosmetic laser resurfacing for wrinkles and acne [15–17]

Er/YAG laser emits a MIDIR beam at 2.93, which coincides with the absorption peak for water. Its principal use is to ablate tissue for cosmetic laser resurfacing for wrinkles and skin irregularities [18–20]. The laser has been advertised to offer advantages of reduced redness, decreased side effects, and rapid healing compared to the pulsed or scanned CO_2 laser but does so by its limited penetration into tissue, which limits the results compared to the more versatile CO_2 laser. It has also been used as a dental drill substitute to prepare cavities for filling [21–26].

Ho/YAG Laser is relatively new to the medical/dental fields. It emits a MIDIR beam at 2.07 μm. Its principal use is to precisely ablate bone and cartilage, with many applications in orthopedics for arthroscopy [27–32], urology for lithotripsy [33–35] (removal of kidney stones), ENT for endoscopic sinus surgery [36, 37], and spine surgery for endoscopic disc removal. The Ho/YAG laser was approved for TURP (prostate removal) [38–41].

To facilitate laser surgery and other applications, one has to direct the laser light from the laser to the application point. For the visible and near-IR region up to 2 μm, this can be done by using standard silica fibers. However, once we choose to use lasers in the MIDIR ($>2\,\mu$m), other means are needed. There are two means to deliver MIDIR laser power to the application point: articulated arms and MIDIR optical fibers. Each of the solutions could be coupled to a hand-piece that shapes the laser beam to a desired energy distribution.

At the beginning of laser use in medicine, energy was transmitted by articulated arms, which was a set of tubes connected by joints and reflecting mirrors to three freedom ranks (Fig. 23.1). At first, this setup suffered from power drifts and required a lot of maintenance. In time, it was improved and the laser power drifts were reduced, but still the device was cumbersome (Fig. 23.2) and limited to external use.

Although many MIDIR lasers still use articulated arms, a better choice is to use an optical fiber. The main advantages of optical fibers are their size and flexibility. These features enable the operator more maneuverability and flexibility (Fig. 23.3) but most important may allow operating inside the body in minimally invasive procedures.

Whether the laser beam is guided from the laser using an articulated arm or an optical fiber, in most of the cases there is a hand-piece at their distal end (Fig. 23.4). A hand-piece is a device that serves a few purposes. First, it enables the operator to easily hold the distal end of the optical fiber or articulated arm and may force the operator to work at a predetermined working distance. Moreover, it serves as a beam shaper, which reshapes the laser beam to a

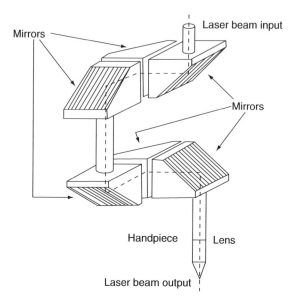

Figure 23.1   Articulated arm design.

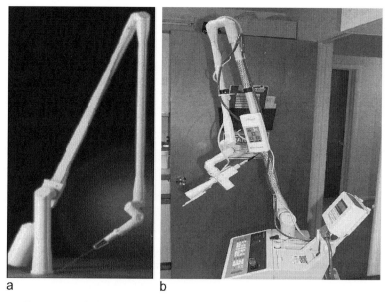

a   b

Figure 23.2   Articulated arm (a), laser with articulated arm (b) (ShoreLaser).

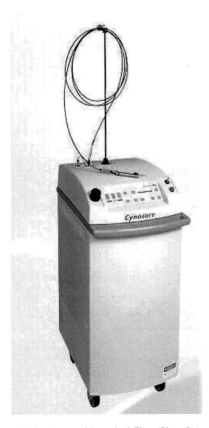

Figure 23.3 Laser with optical fiber (ShoreLaser™).

predefined spot size. Last, it may allow the use of cooling gas flow to the treated tissue in order to cool it quickly and avoid thermal damage.

23.3 TRANSENDOSCOPIC SURGICAL APPLICATION

Quite a few fiber-based applications have been developed and used for the past years. We have chosen to present one new application in which we took part. This demonstrates the advantages of this method in minimally invasive surgery.

Sialolithiasis (salivary gland stones) is the most frequently occurring of the diseases affecting the salivary glands. In postmortem studies, incidence was found to be 1.2%, while in hospital admissions, it is estimated to be 1 per

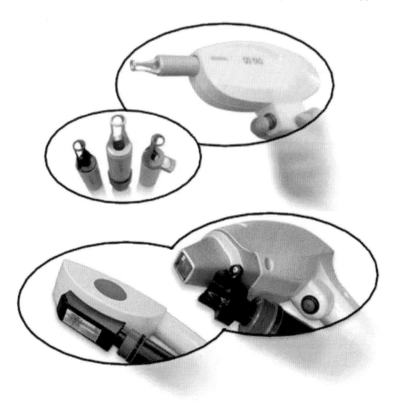

Figure 23.4   Hand-piece (MsQ® Israel).

15,000 [42]. Sialolithiasis results in a mechanical obstruction of the salivary duct causing repetitive swelling during meals, which can remain transitory or be complicated by bacterial infections [43]. The introduction of sialoendoscopy in the mid 1990s has significantly improved the diagnosis and treatment of salivary gland inflammatory diseases, yet the endoscopic removal of larger and impacted salivary gland stones remains challenging, often requiring sialoadenectomy, which is associated with the risk of facial paralysis [44–47] and requires general anesthesia and hospitalization. Although the patient can function without one of the glands, it is involved with decreased secretion of saliva, which creates dry mouth and potentially more dental caries. Extracorporeal shockwave lithotripsy is also used for treatment [48]. This technique requires several sessions at intervals of a few weeks. The remaining stone debris may function as the ideal nidus for further calcification and sialolithiasis recurrence. This technique creates potential risk to bones and teeth. The success rate is also rather low and does not pass 50% in the best cases [49]. It is much less effective for stones of 7-mm diameter. Interventional sialoendoscopy with a wire basket through a miniaturized endoscope was also carried out in the 1990s [50]. It has good success for the 1- to 3-mm diameter stones, but most stones are 3–7 mm.

Endoscopic laser lithotripsy can potentially treat most cases of salivary gland stones with minimal complications while preserving a functional salivary gland. In urology, the holmium laser has become the gold standard for endoscopic lithotripsy with delivery of the energy through highly reliable, low-OH silica fibers with core diameters as small as 200 microns [51]. Some successful attempts have been reported on the use of this laser for sialolithiasis, but the smaller anatomy of the salivary ducts increases the risk to surrounding soft tissues [52–54].

Metal hollow waveguides were used because they are extremely durable, inexpensive, and biocompatible and can withstand high energies and powers. Hollow metal waveguides can be produced with low attenuation (<0.5 dB/m) and can be bent to a radius of curvature as low as 5 cm. Hollow waveguides must be hermetically end-sealed for use in a water environment because even a small amount of water in the waveguide will completely attenuate the beam. Silver hollow waveguides of 0.9 mm outer diameter and 0.13 mm wall thickness were prepared with an AgI internal dielectric coating [55]. A total of 99.9% pure silver tubes with a 20 RMS inner surface finish were used. The tube's inner surface underwent chemical reaction with a solution of iodine dissolved in ethanol, resulting in a thin dielectric layer of AgI. Coating deposition time and iodine concentrations were adjusted to optimize transmission at around 3 microns. Polished sapphire rods of 0.63-mm diameter and 5-mm length were cemented at the tip of the waveguide using a biocompatible adhesive. A plano-convex ZnSe lens was designed to focus the 1-mm output aperture of the OpusDent fiber into the 0.63-mm waveguide, which was mounted in a Luer/SMA connector to facilitate quick connection to both the erbium laser and the sialoendoscope (Fig. 23.5).

23.3.1 Clinical Tests

Helsinki committee approval was obtained to conduct human clinical trials at the outpatient facility of the Oral & Maxillofacial Department, Barzilai Hospital, Ashkelon, Israel, and performed by Professor O. Nahlieli.

The trial was conducted on 17 patients (9 females, 8 males) aged 11–72 years, with salivary stones of 1–15 mm in diameter located in the posterior part of the salivary ducts. Altogether, 21 stones located in 18 glands, 16 submandibular and 2 parotid, were involved in the trial. Stone size and location were documented by plain radiographs, sialography, and high-resolution ultrasound

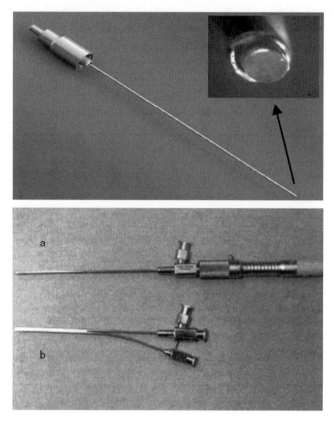

Figure 23.5   Waveguide with luer lock and sapphire tip (top). Sialoendoscope (bottom).

before the procedure, immediately after the procedure (radiography only), and at 12 weeks postoperative. Videoendoscopy of the procedure was recorded and included a postoperative view of the lumen in the area (Fig. 23.6).

Patients were followed up for lumen patency, symptoms of sialoadenitis, and any potential complications, immediately, at 4 weeks, and at 12 weeks after the procedure.

All procedures were performed under local anesthesia in an ambulatory environment, using standard video-sialoendoscopy protocol. Under this protocol, the orifice of Wharton or Stensen duct is first identified and a lacrimal probe gently inserted. Papillotomy is preferably performed with a CO_2 laser. Lacrimal dilators from 1 to 3 mm in diameter are then sequentially used for duct dilation. The rigid scope, connected to a videocamera and monitor, is introduced with the help of isotonic saline irrigation. The salivary ducts are diagnosed and the stone

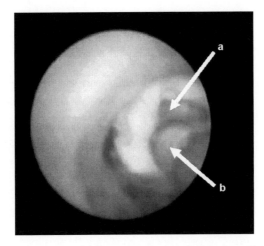

Figure 23.6 An endoscopic view of the fragmentation. (a) Cavities formed within the fragmented stone are clearly seen. (b) The sapphire tip must remain in contact with the calculus for fragmentation to occur.

is located. If no contraindications are detected, the fiber is inserted through the operating channel of the endoscope until its distal end is within the field of view of the endoscope. Actual laser irradiation is performed only under clear vision, when the fiber is seen to be in direct contact with the stone and tangential to the duct. Adequate irrigation is maintained throughout the procedure using an intravenous bag connected to the irrigation port on the sheath of the endoscope.

The fiber was connected to a standard Lumenis Opus 20, dental erbium laser. Fragmentation usually began at a setting of 150 mJ/pulse and a pulse rate of 10 Hz.

While the initial plan was to attempt full fragmentation on all stones, it was soon realized that because of anatomical and operational constraints, such as tortuous ducts or poor visibility, alternative techniques to assist in stone removal can be applied. Consequently, three alternative methods were used depending on each case: total fragmentation, mostly for stones under 5 mm; creation of a traction point for miniforceps; and separation of surrounding soft tissues in cases of impacted stones.

Of the 21 stones treated, 5 were fully fragmented, 7 stones of 5–7 mm were prepared for extraction by miniforceps, and 9 stones with diameters up to 15 mm were released from surrounding soft tissues for subsequent endoscopic or surgical removal. A total of 15 of the 18 treated glands returned to normal function without any symptoms, whereas 2 nonfunctional glands remained nonfunctional but were completely asymptomatic. No other complications have been noted during the 2–12 months of follow-up so far.

23.4 ABSORPTION SPECTROSCOPY

23.4.1 Introduction

Spectroscopy studies the way electromagnetic radiation interacts with matter. When radiation meets matter, the radiation is either scattered, emitted, or absorbed. This gives rise to three principal branches of spectroscopy. *Emission spectroscopy* observes light emitted by atoms excited by radiation–matter interactions. *Raman spectroscopy* monitors light scattered from molecules. *Absorption spectroscopy* studies radiation absorbed at various wavelengths.

In absorption spectroscopy, light illuminates a sample of material to be analyzed. The sample, which can be liquid, solid, or gas, is usually enclosed in an absorption cell. Each element or compound in the sample absorbs particular wavelengths of light, resulting in one or more dark lines on its spectrum. These lines are a "fingerprint" identifying which chemical substances are present in the sample and their quantities, as well as other information about detailed structure and activity.

When an atom or molecule absorbs energy, electrons are promoted from their ground state to an excited state. In a molecule, the atoms can rotate and vibrate with respect to each other. These vibrations and rotations also have discrete energy levels that can be considered as being packed on top of each electronic level.

Absorption spectroscopy is not limited to a specific wavelength region and can be observed at a wide range of wavelengths. However, the absorption in each wavelength range can be accounted to light interaction with different types of energy levels in the examined sample. The absorption in the UV and visible spectrum is due to electronic absorption, whereas that in the IR region is due to vibrational and rotational energy levels of the molecules (Fig. 23.7).

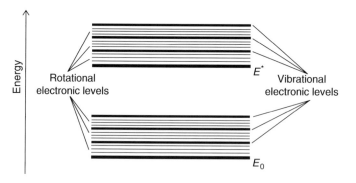

Figure 23.7   Energy levels in a molecule.

The absorption of UV or visible radiation corresponds to the excitation of outer electrons. The absorption in organic molecules is restricted to certain functional groups (chromophores) that contain valence electrons of low excitation energy. The spectrum of a molecule containing these chromophores is complex. This is because the superposition of rotational and vibrational transitions on the electronic transitions gives a combination of overlapping lines. This appears as a continuous absorption band. Many inorganic species (such as semiconductors) show charge–transfer absorption and are called *charge–transfer complexes*. For a complex to demonstrate charge–transfer behavior, one of its components must have electron-donating properties and another component must be able to accept electrons. Absorption of radiation then involves the transfer of an electron from the donor to an orbital associated with the acceptor.

IR radiation does not have enough energy to induce electronic transitions as seen with UV. Absorption of IR is restricted to compounds with small energy differences in the possible vibrational and rotational states. For a molecule to absorb IR radiation, the vibrations or rotations within a molecule must cause a net change in the dipole moment of the molecule. The alternating electrical field of the radiation interacts with fluctuations in the dipole moment of the molecule. If the frequency of the radiation matches the vibrational frequency of the molecule, then radiation will be absorbed, causing a change in the amplitude of molecular vibration.

23.4.2 Medical Applications of Absorption Spectroscopy

Absorption spectroscopy is difficult to perform *in vivo* because the human tissue scatters light considerably, so a quantitative analysis is difficult to obtain. However, it can be performed *in situ*. *In situ* studies had been conducted for several applications: glucose [56, 57] and other blood analytes analysis, NO and CO_2 measurements in breath [58] and molecular tissue mapping [59].

Quick and accurate measurements of blood analytes without sample preparation may enable a better treatment. Kim et al. [56] showed that it is possible to measure the blood glucose levels in whole blood without sample preparation. They used absorption spectroscopy in the MIDIR region 6.5–11 μm. In this wavelength region, there are distinct glucose absorption lines, along with several hemoglobin absorption lines. The hemoglobin absorption lines are used for calibration. The measured prediction error ranged between 10 and 50 mg/dl.

Gas analysis of human breath may provide valuable biomedical and clinical information. The detection of gas traces such as ammonia, NO (Table 23.1), may be used for noninvasive medical diagnosis and monitoring the success of a medical treatment. As for today, the most common method for exhaled gas

Table 23.1

Exhaled gas traces and the illness they indicate

Gas type	Indicated illness
Ammonia	Renal failure
Acetone	Diabetes
NO	Asthma
Ethanol	Glucose level
Isoprene	Biochemical stress

analysis is mass spectroscopy with different kinds of ionization. This method provides good results, but it cannot be realized as a turnkey instrument.

Spectroscopic analysis of exhaled gas has been investigated for several years. Most of the groups have been investigating exhaled gas in the near-IR (NIR) and MIDIR region. However, there is also research on spectroscopic response in the UV. Eckhardta et al. [60] investigated gas spectroscopy in the UV using hollow waveguides. The main advantage of conducting spectroscopic analysis in the UV is that the absorption of oxygen in that region is small, thus enabling measurement in air. The main advantage of using a fiber instead of a flow cell is the ability to achieve a longer path length and, therefore, better sensitivity. They showed theoretically that it is possible to measure several gas traces simultaneously and measured experimentally NO.

The use of MIDIR spectroscopy for exhaled gases has been investigated thoroughly by a number of research teams around the world [58, 61–63]. Roller et al. [58, 61, 63] used a tunable laser in the 5.2-μm region and an absorption cell to measure the amount of NO in human exhaled breath. They showed that measuring NO and CO_2 simultaneously eliminates the need for system calibration. Using their system design, NO levels of 44 parts per 10^9 to 20 parts per 10^9 were measured, which is the near-normal level.

Ganser et al. [62] developed an on-line sensing measurement of nitric oxide (NO) traces based on CW quantum cascade laser, operating near 5.2 m. They used a Faraday modulation technique that provides unique selectivity because it exploits the magnetic moment of NO molecules. The minimum detectable NO concentration was found to be 4 ppb (sampling time: 10 seconds).

Absorption spectroscopy can be used to detect abnormalities in tissues and organs. Tumors or other irregularities in tissue and organs might be detected directly by measuring their spectral response. Another method is monitoring the organ metabolism, blood flow in the tissue, and blood oxygenation.

Hynes et al. [59] investigated the use of microspectroscopy for dental applications. Chronic periodontitis is an inflammatory disease of the supporting structures of the teeth. IR microspectroscopy has the potential to simultaneously monitor multiple disease markers, including cellular infiltration and collagen catabolism and, hence, differentiate diseased and healthy tissues. They showed that connective tissue contains lower densities of DNA, protein, and lipids compared to higher densities in epithelial tissue. Collagen-specific tissue mapping by IR microspectroscopy revealed much higher levels of collagen deposition in the connective tissues compared to that in the epithelium. Thus, inflammatory events such as cellular infiltration and collagen deposition and catabolism can be identified by IR microspectroscopy.

Neurological complications during critical illness remain a common cause of morbidity and mortality. The current methods used to monitor cerebral function including electroencephalography, jugular bulb mixed venous oxygen saturation, and transcranial Doppler either require an invasive procedure or are not sensitive enough to effectively identify patients at risk for cerebral hypoxia. Tobias et al. [64] investigated NIR spectroscopy as a method to measure oxygen in tissue. This method is noninvasive, which works similar to pulse oximetry, allows for the penetration of living tissue, and provides an estimate of brain tissue oxygenation by measuring the absorption of IR light by tissue chromophores.

23.5 EVANESCENT WAVE SPECTROSCOPY

23.5.1 Introduction

NIR and MIDIR spectroscopy is a reliable method for obtaining the fingerprint of solids, liquids, and gases. It can detect small amounts of materials and quantify the amount of material inside a given sample. One of the methods of NIR and MIDIR spectroscopy is evanescent wave spectroscopy (EWS), which is also known as *attenuated total reflectance spectroscopy* (ATRS).

Evanescent waves are the attenuated waves that "leak" from an optical waveguide, guiding crystal or fiber core to its cladding or surroundings in the case of a waveguide, crystal, or optical fiber without a cladding (Fig. 23.8). Because the evanescent waves are highly attenuated, their interaction with the waveguide's surroundings is limited to a small distance from the guiding material. The amount of attenuation depends on the wavelength that is transmitted through the guiding material and the type of material in the surroundings. The penetration depth d_p is given by Eq. (23.1), where λ is the wavelength, θ is the angle of incidence, n_1 and n_2 are the index of refraction of the surrounding and the core, respectively. d_p is very small, usually in the range of 0.5 μm:

$$d_p = \frac{\lambda}{2\pi\sqrt{n_2^2 \sin^2 \theta - n_1^2}} \tag{23.1}$$

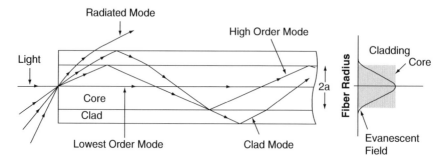

Figure 23.8   Guided mode and evanescent wave in an optical fiber.

23.5.2 Experimental Setups

There are several optical setups for performing EWS. Each setup contains a tunable light source (tunable laser, monochromator), a detector, and depending on the examined object, some kind of a sample chamber. One of the most used systems for EWS is an FTIR. Two typical EWS systems are shown in Fig. 23.9. The first system (Fig. 23.9a) is known as *fiber optic EWS* (FEWS) [65]. The experimental setup uses an FTIR, along with a partially uncladded optical fiber made of AgBrCl [66] or chalcogenide [67] glass (if the measurement is done in the MIDIR region) or silica (if the measurement is done in the UV, VIS, or NIR region). These fibers are inserted into a sample chamber or are placed in contact with the sample. The fiber's surface may be sensitized to bond a certain material to increase the sensor sensitivity [68].

The second method [69] (Fig. 23.9b) is to couple the light into a crystal that serves as a waveguide. The crystal is then placed in contact with the sample. Here, as in the previous setup, it is possible to use different light sources. The light source choice depends on the examined material.

To obtain good sensitivity, it is recommended that the contact length between the fiber–crystal be as long as possible [70, 71]. Each reflection from the fiber–crystal wall increases the interaction between the light and the sample. As the number of reflections increases, the light attenuation decreases and a clearer spectrum is being recorded. The number of reflections can be obtained using geometrical optical and is given by Eq. (26.2), where L is the interaction length, d is the fiber diameter, and θ is the angle of incidence:

$$N(\theta, d, L) = L \frac{\tan(90 - \theta)}{d} \tag{23.2}$$

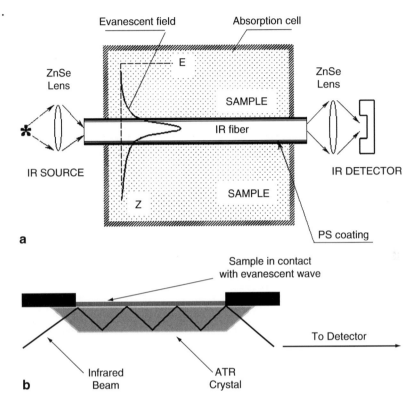

Figure 23.9 Evanescent wave spectroscopy setups. (a) Fiber optic evanescent wave spectroscopy setup. (b) Optical crystal evanescent wave spectroscopy setup (used, with permission, from reference [69]).

Equation (23.2) shows that it is beneficial to use long fibers/crystals. However, one has to take into account the size limitation of the sample itself. Although long fibers are usually applicable to liquid or gas samples, shorter fibers/crystals would be needed for solid samples and *in vivo* measurements.

Lately a new type of EWS sensors was suggested. These sensors are based on two dimensional photonic crystals [72–74]. It was shown that solutions and biological fluid that enters the air holes and gaps of the photonic crystal causes a shift in their photonic band-gap. The amount of shift in the photonic band-gap is proportional to the amount of biomolecule in the solution. EWS might be applicable to many spectroscopic applications. Remote analysis is required for many industrial applications in chemistry and biochemistry. A few examples are *in situ* monitoring of chemical reactions [75–77], on-site pollution control [78], *in vivo* monitoring of biological processes [79], and biomedical sensing such as noninvasive glucose measurements [80] and cancer diagnosis [81].

23.5.3 Chemical Sensing

23.5.3.1 Chemical Analysis and Monitoring

All organic species exhibit vibrational modes within the 2- to 16-μm wavelength region. This makes FEWS an excellent tool for remote detection of a wide variety of chemicals and biochemicals ranging from proteins and enzymes to simple molecules such as ethanol and benzene.

Chemical detection of butanone [82] by means of FEWS was reported during the late 1980s. Chemical analyses were then performed on acetone, ethanol, and sulfuric acid using chalcogenide fibers [83, 84]. It was shown that the detection versus concentration curve follows a linear Beer–Lambert law, allowing quantitative analysis of solute samples. A wider range of organic species, including carcinogens such as benzene, toluene, and trichloroethylene, were later detected [85–87]. Sensitivity of several parts per million levels was achieved.

Spectral monitoring performed continuously and in real time allows one to observe and follow changes in molecular structures throughout a chemical reaction. For example, conversion of fructose and glucose into ethanol can be monitored quantitatively during the fermentation process in cider production [88]. Similarly, variation in the primary amine signal can be used to follow the curing stage in epoxy resins [76, 82], and C–H vibrations can be used to characterize the polymerization of styrene films [89]. Finally, remote fiber sensing is particularly useful for remotely characterizing reactions that take place in hazardous environments such as microwave ovens. For example, the microwave-assisted synthesis of diethoxypenthane can be monitored using the shift in the C–O band [75].

Ouyang et al. [90] have shown that it is possible to detect large biomolecules using two-dimensional photonic crystals based on macroporous silicon microcavities. They were able to detect rabbit immunoglobulin G (IgG) (150 kDa, 1 Da = 1 gmol^{-1}). The sensing performance of the device was tested and it was shown that concentrations down to 1–2 μM could be detected with sensitivity of 1–2% of a protein monolayer. Another study by Jensen et al. [74] demonstrated highly efficient evanescent wave detection of fluorophore-labeled biomolecules in aqueous solutions positioned in the air holes of the microstructured part of a photonic crystal fiber. The air-suspended silica structures located between three neighboring air holes in the cladding crystal guide light with a large fraction of the optical field penetrating into the sample even at wavelengths in the visible range. An effective interaction length of several centimeters is obtained when a sample volume of less than 1 μl is used.

23.5.3.2 Pollution Control

A portable FTIR and long optical fibers may enable one to conduct environmental on-site pollution measurements in water reservoirs, soil, and marine environments with very high sensitivity. Fuel contamination was detected in soil [91, 92], and several researches were conducted on different liquids, especially in water [65, 68, 92–95]. It was shown that pollutants such as benzene could be detected in water. Lately real-time monitoring during a pilot scale test of a natural aquifer in Munich, Germany, took place [78, 96]. During that study, variations in pollutant concentration profiles were recorded following injection of test quantities of chlorinated hydrocarbons in the aquifer inlet. Control measurements performed simultaneously by chromatographic analysis were in good agreement with the optical detection. This study and other studies demonstrated the potential of FEWS for permanent pollution-monitoring devices in water wells.

23.5.4 Biochemical Sensing

23.5.4.1 Biochemical Analysis

Biological molecules, microorganisms, and tissues have an IR fingerprint, which permits differentiation among different kinds of bacteria and species. Moreover, it may enable the identification of changes in the metabolic process [97] such as structural changes in tissue and protein unfolding.

Several animal studies had been conducted on body fluids drawn from animals, especially blood serum. Keirsse et al. [98] showed that it is possible to differentiate between a fat mouse blood serum to the one from a healthy mouse. Shimony et al. [99] investigated the amount of bovine albumin in water suspension.

Kishen et al. [100] monitored mutans streptococci activity in human saliva. They used FEWS to monitor a bacterial-mediated biochemical reaction. To achieve this, a short length of the cladding is removed; the fiber core surface is treated and coated with a thin film of porous glass medium using the sol-gel technique. The mutants streptococci-mediated reaction with sucrose is monitored using a photosensitive indicator, which is immobilized within the porous glass coating. Spectroscopic analysis shows that the transmitted intensity at 597 nm increases conspicuously when monitored for 120 minutes. Two distinct phases are observed, one from 0 to 60 minutes and the other from 60 to 120 min. A negative correlation coefficient between the rate of increase in absorption peak intensity was seen ($R = -0.994$).

Cytron et al. [101] have measured the amount of salt in human urine in real time and without any sample preparation to quantitatively assess the urine composition for the diagnosis of urolithiasis in patients. Urine samples were obtained from two groups of patients: 24 patients with stone formation after shock-wave lithotripsy and 24 healthy subjects of similar age. IR absorption measurements were performed in real time, using IR-transmitting silver halide fibers. The absorption data were compared with the IR absorption spectra of aqueous solutions prepared in our laboratory with known concentrations of known urinary salts. The results were used for the study of the chemical composition of these salts in the urine samples and for a quantitative analysis of the concentration of the salts. They determined the composition of the stones in 20 of the 24 patients on the basis of the characteristic absorption peaks for the oxalates, carbonates, urates, and phosphates observed in their urinary samples. Using the method mentioned earlier, they found the concentration of different salts in urine with an average error of 20%.

23.5.4.2 *In Vivo* Measurements and Live Organisms Monitoring

One of the advantages of FEWS is the ability to collect the signal far from the working place. This feature allows one to perform measurements on live subjects and inside the patient body in minimally invasive procedures. Such measurements might replace conventional biopsies. Moreover, FEWS enables one to examine cell cultures, live microorganisms, and tissue explants directly within an experimental setup designed to provide appropriate environmental conditions.

In vivo spectroscopic analyses have been applied to medical diagnostics and to metabolic monitoring. As an example, noninvasive glucose measurements were performed through the mucous membrane of the lips of human patients [80]. Clinical tests on diabetic patients showed that the FEWS results closely matched the control measurements obtained by a glucose oxidase method. Glucose levels were monitored in real time after patients had been subjected to pulsatile intravenous glucose injections.

Similar *in vivo* measurements were applied to cancer diagnostics on anesthetized live animals [91]. Human breast tumors grown near the surface of mice skin showed noticeable spectral variations, and *in vivo* tests demonstrate the feasibility of quantitative measurement of dye clearance in the gastroesophageal tract [102]. Sukuta et al. [103] investigated skin cancer using FEWS using chemical factor analysis. They isolated the eigenspectra of biochemical species and some of the eigenspectra have been preliminarily identified as due to protein peptide bond and lipid carbonyl vibrations. Cluster analysis was used for classification, and good agreement with prior pathological classifications, specifically for normal skin tissue and melanoma tumors, has been found.

Live cell monitoring can be used to evaluate metabolic activities on various types of cell cultures. The dynamics of bacterial cell colonies were recorded during the swarming of bacterial biofilms [98]. The synthesis of lubricant slime during active migration of *Proteus mirabilis* bacteria was observed in the exopolysaccharide features of the second derivative spectra. Another study showed that it is possible to detect bacterial activity in human saliva [104]. The sensor determines the specific concentration of *Streptococcus mutans* in saliva, which is a major causative factor in dental caries.

Several groups had investigated human metabolism. Lucas et al. [105] studied the metabolic responses of human lung cells exposed to various toxicants. A monolayer of a lung epithelial cell line (A549) attached to a chalcogenide fiber surface was exposed to micromolar quantities of Triton X-100 surfactant. A rapid alteration of the cell membrane was observed within minutes of exposure. This study showed that because of shallow penetration depth of the evanescent wave, the FEWS technique collects strong signals from the cell itself while minimizing signal from the surrounding fluid [105]. Another study examined the water diffusion into the human skin [79].

23.6 FIBER OPTIC THERMAL SENSING

Changes in local temperature of blood, body organs, tissue, and skin in biological systems depend on the thermal energies received and lost, as well as on chemical reactions as metabolisms. Increase or decrease in temperature might indicate some kind of disease or an abnormal function of the body organ. Temperature measurements can be done using standard thermometers, thermocouples, thermistors, and radiometric methods and optical methods.

Although measurements of the body core temperature indicate some kind of illness, they do not indicate what the illness is and whether it is just a symptom of it. However, measurement of a local temperature of an organ might give more information about its condition. Moreover, a thermal image of an organ might reveal structural changes in the organ.

Optical fibers play a major role in temperature sensing. Optical fiber can be used as a temperature sensor; for example, it is possible to embed fluorescent material inside the fiber and monitor the fluorescence parameter changes caused by temperature changes [106]. Moreover, they can transmit thermal radiation to a radiometer that converts it to thermal readings. Finally, optical fiber bundles can be used to deliver, for example, thermal images for breast cancer imaging.

The use of optical fiber for temperature sensing and imaging has many advantages. They are very small (Fig. 23.10) and could be used during minimally invasive procedures. Furthermore, optical signals are not affected by EMI and RFI interferences, so these sensors can be used during magnetic resonance imaging (MRI) procedures.

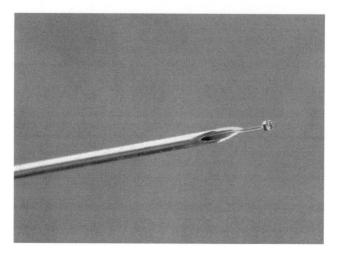

Figure 23.10 Optical fiber sensor by Fiso (Canada) (used, with permission, from reference [107]).

23.6.1 Fiber Optic Thermal Sensor

Fiber optic thermal sensors are based on two major technologies: changes in the relaxation time of fluorescent [106] material due to temperature changes, and changes in the optical path of a Fabry–Perot interferometer [107–110].

In fiber optic thermal sensors based on fluorescence, the fiber probe communicates optically with a temperature-sensitive fluorescence material such as phosphor [106] (Fig. 23.11). The fluorescence signal is activated by pulses laser diode and is transmitted back through the fiber to a detector.

The phosphor tip, at the far end, is either embedded in the medium to be measured or placed in contact with its surface. The probe is composed of silica fiber and various jacketing layers, all of which are stable over the full temperature measurement range of the instrument. The phosphor, which is typically stable to much higher temperatures than the glass itself, can respond to temperature in various ways: change of quantum efficiency, spectral shift of emission and/or excitation bands, and alteration of the fluorescence lifetime (decay time). Of these temperature-dependent mechanisms, change in decay time provides the most robust approach to measuring temperature [106]. The reason is that lifetime changes can be quantified in a manner independent of instrumental and environmental variables.

The Fabry–Perot fiber optic temperature sensor measures the changes in the interference pattern, which are caused by changes in the optical path of one of the interferometer legs. The changes of the optical path are caused by changes in

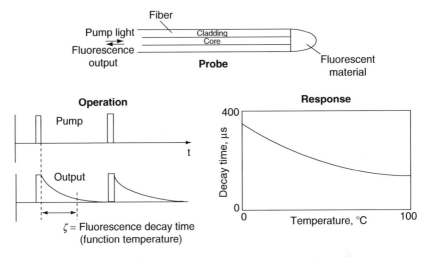

Figure 23.11   Fiber optic temperature sensor based on fluorescence.

the optical properties (index of refraction, Bragg grating reflectance, etc.) of the material when changes in temperature occur.

Two types of Fabry–Perot fiber optic temperature sensor could be found: intrinsic Fabry–Perot [108, 110] in which the perturbation of interest affects the fiber itself and extrinsic Fabry–Perot [109] in which the fiber is used only to guide the light. The structure of an extrinsic Fabry–Perot temperature sensor is shown in Fig. 23.12 [110]. The cavity can be formed by encapsulating two separate fibers inside a glass capillary tube (Fig. 23.12a) or by mounting a miniature cavity—formed by glass or silicon micromachining techniques—on the tip of a fiber (Fig. 23.12b), with a deformable membrane having a thin film reflector [108]. The power source could be either a laser (usually a laser diode) or an LED. When the sensor is exposed to the perturbation of interest (temperature, pressure, etc.) and the reference FPI is not, a difference in the optical path length is created. This change causes a change in the interference pattern at the interferometer output. This change is analyzed and is being converted into a temperature reading.

Fiber optic temperature sensors could be used for a wide range of medical applications. The main advantage of these sensors is that they are immune to EMI and RFI. Therefore, they could be used during other procedures such as MRI [111, 112] examinations. Other applications could be temperature measurements of internal organs during minimally invasive procedures.

In the field of minimally invasive surgery, radiofrequency (RF) waves are used locally via a probe or over a region of the body introduced in a tunnel. This

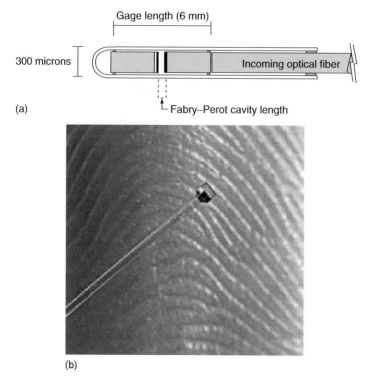

Figure 23.12   Extrinsic Fabry–Perot fiber optic sensors. (Photo courtesy of FISO).

method permits resection of masses or tumor treatment with minimal bleeding and minimal damage to surrounding tissue. However, it is also involved in close temperature monitoring. The small size and intrinsic immunity of optical sensors to electromagnetic fields permit continuous *in situ* temperature measurements without adding harm to the tissue. Integration of ultraminiature temperature sensor from Fiso, Inc. [112] into minimally invasive surgery involving RF waves could help the optimal control of these type of therapies.

23.6.2 Optical Fiber Radiometry

A warm body at room temperature ($\sim$300 °K) emits radiation according to Planck's law at the wavelength region of the mid-IR (3–20 μm). The radiation intensity is given by Stephan Boltzmann law:

$$I = \varepsilon \sigma T^4, \tag{23.3}$$

where I is the radiation intensity, ε is the emissivity, and σ is the Stephan Boltzmann constant. The emissivity of the human body is unity ($\varepsilon = 1$) and the radiation peak wavelength is around 10 μm.

As can be seen from Eq. (23.1), by measuring the radiation intensity, one can calculate the body's temperature. A device that measures the radiation intensity and converts it into a temperature reading is called *radiometer*. A typical radiometer usually consists of three parts: (1) optics that collects the IR radiation, (2) an IR detector that converts the photons into an electrical signal, and (3) an electronic system for signal processing. A fiber optic radiometer utilizes an optical fiber, which is suitable for MIDIR radiation transmission, as the collection optics. A typical fiber optic radiometer system is shown in Fig. 23.13 [113–115].

Fiber optic radiometers could be used for many medical applications. The most straightforward is measurement of body core temperature [116]. Another application is temperature measurements during MRI procedures. There are many cases when it is necessary to measure the temperature of a sample while it is being imaged in an MRI system. For example, there are surgical procedures in the brain, where the treatment of a tumor involves heating with a laser beam or cooling with a cold finger. Medical treatment depends on the accurate measurement of the temperature inside the tumor. In the space between the magnets poles, there are strong magnetic fields, strong gradients of magnetic fields and RF fields. The measurement of temperature in such a hostile environment with conventional instruments (e.g., thermocouples) is not easy.

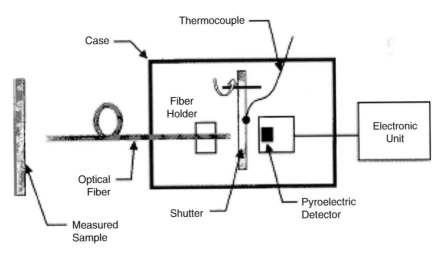

Figure 23.13 A typical fiber optic radiometric system (used, with permission, from reference [113]).

Thermal measurement of the skin during different procedures could be facilitated using a fiber optic radiometer. Temperature measurement of skin and tissue during laser heating may help prevent the doctor from damaging the surrounding tissue. This is important in applications such as skin resurfacing [117], laser tissue soldering [118], and burn generation [119]. Moreover, the temperature measurement could be used as a feedback [120] to the procedure and could automate some of the procedures by stopping the laser when the temperature reaches a certain value.

23.7 THERMAL IMAGING

The use of thermal imaging is increasing mainly because of the improvement in thermal cameras. *In vivo* studies [121] have increased the clinical value of this equipment in many disciplines such as breast cancer (risk assessment, detection, prognosis, and therapeutic monitoring), burn trauma (staging), diabetes, cardiology, neurology, urology, pulmonary, dermatology, ophthalmology, neonatology, pain management, and anxiety detection. Fewer clinical studies have reported the use of IR imaging of internal tissue surfaces intraoperatively. Neurosurgeons have exhibited the IR imaging capability to detect vascular occlusion and reperfusion, as well as its potential to explore cerebrovascular disease, epilepsy, functional cortical activation, and brain tumors.

Unlike direct IR measurements, utilizing IR imaging by minimally invasive means is a challenge as the IR images need to be transferred from the internal tissue surface to the IR camera trans-endoscopically. Because minimally invasive surgeries have been extensively replacing conventional surgeries during the last decade, resolving this problem has become essential.

Transmitting a thermal image through an endoscope could be done using a scanning fiber optic radiometer or IR fiber bundles coupled to an IR camera. The first method is quite complicated because scanning an image with a single fiber might take a long time and controlling the position of the fiber tip is challenging.

The second method is to use IR optical fiber bundles for transmitting the thermal radiation to an IR camera. Over the last 2 decades, several groups had been trying to manufacture IR bundles, unfortunately with limited success. The first ideas for IR imaging cones were reported by Kapany [122]. The first group who reported an IR-transmitting coherent bundle was Saito et al. [123] in 1985. They used As-S fibers, which transmit radiation in the 2- to 6-μm wavelength range. Their aim was to be able to map temperatures in hard-to-reach areas of engines, furnaces, and nuclear reactors. They have created 200–1000 fiber bundles with As-S core and FEP Teflon cladding. The bundle was coupled to an InSb camera (with 3.0–5.4 μm operating spectral range) to collect various thermal images. At first, a blackbody with uniform distribution of 773 °C was imaged, although different temperatures were measured at each spot in relation to each fiber in the bundle. This was caused by different transmittance of each fiber. After calibration for these differences, images of an iron and a candle flame were successfully taken.

In 1987, Klocek et al. [124] reported a more flexible bundle made from GeSeSb fibers that were fused together only at both ends to overcome the rigidity of Saito's bundle, which was fused along the entire length of the bundle. In 1991, Nishii et al. [125] reported using an As2S3-based imaging bundle. Nishii et al. [125] created three types of chalcogenide glasses. Using a method suggested by Kapany, 8400 fibers, 100 cm long, were stacked in a rectangle. The ends of the bundle were ground and polished for an optical flat surface. The bundle was inserted into a Teflon tube and coupled to an IR camera. Images of a human face were taken and different temperature areas were observed. The high NA of the fibers caused some resolution lowering. Anti-reflecting coatings on both surfaces improved the image as well, and temperatures as low as 25 °C could be captured by the camera through the bundle.

Ray Hilton, from Amorphous Materials in Garland, Texas, is also using chalcogenide glass to create thermal imaging bundles. The bundle created was 3600 elements of 68-μm diameter each with 52% active area. A USAF target with a heat source (human palm behind) was used to test resolution, which was found to be 7 lp/mm (Fig. 23.14). The dark spots in the picture are due to a few

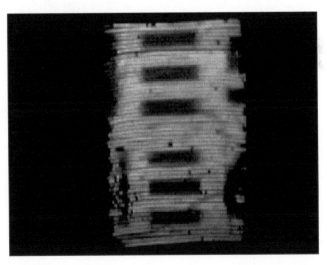

Figure 23.14 Thermal imaging through Hilton's chalcogenide bundle of the USAF target with the palm of the hand as heat source.

fibers that were damaged in the fabrication process. The human subject was also imaged by a 3- to 5-μm camera with reasonable results [126].

The chalcogenide glass was, in fact, the first used in endoscopic setting. Naghavi et al. [127] reported the first prototype of a flexible IR fiber optic catheter for the IR imaging of atherosclerotic plaques. The system had thermal resolution of 0.01 °C and spatial resolution of 100 μm. Temperature heterogeneity was successfully detected in a phantom model, simulating blood vessels and hot plaques, as well as in *in vivo* animal study. They concluded that the technique is feasible and can be used for thermal detection of vulnerable atherosclerotic plaques.

Katzir et al. [128–132] reported on an imaging bundle made of silver halide fibers. A 900-element bundle was used to transfer images of heated tungsten wire at 40 °C into a thermal camera with reasonable resolution. These types of bundles were used by Gannot's group as a feedback mechanism to optimize laser-based tissue ablation. Figure 23.15 shows an *in vivo* setup of an IR laser interaction on mouse tissue. Figure 23.16 shows the temperature mapping.

Gopal et al. [133] reported another type of imaging bundle based on internal coating of a polygon structure of 900 channels produced by collimated holes

Figure 23.15 *In vivo* ablation of mouse tissue through a hollow waveguide and temperature distribution imaging through a silver halide imaging bundle made by the Katzir group.

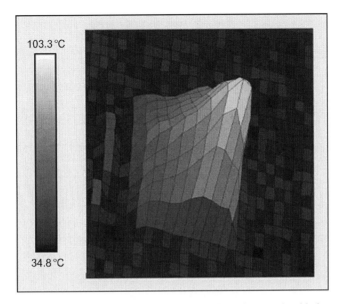

Figure 23.16 Three-dimensional temperature mapping of the tissue at the ablation area. Laser irradiation: 10.6 m, CW, 1 second elapsed, 0.4-W power ($50 \, W/cm^2 APD$), through 1-mm ID hollow waveguide; IRIFB: silver halide core-clad 100–230 μm; Phantom model: mouse's skin (abdomen).

(CA). The coating method used is the same as applied in the single-tube coating of fused silica or Teflon [134], but in this case, each element was 45 or 65 μm in diameter. The coated bundle is shown in Fig. 23.17. An image of a heated hot wire is shown in Fig. 23.18. The bundle created in this method is still not very flexible, and only about 50-cm long pieces are fabricated at this time. One of the great advantages of this bundle is the zero crosstalk between the elements.

23.7.1 Infrared Imaging and Tomography in Minimally Invasive Procedures

The investigation of IR imaging and/or thermocouple-based thermography using minimally invasive techniques has just begun. Several studies from various medical disciplines that have used experimental apparatuses are briefly summarized here.

Ogan et al. [135] have exploited trans-laparoscopic IR imaging to monitor surface renal temperatures during RF ablation. They claimed that IR imaging had enabled them to assess treatment adequacy and ablation margins. *In vivo*

Figure 23.17   Coherent bundle 900 channels internally coated with Ag/AgI layers.

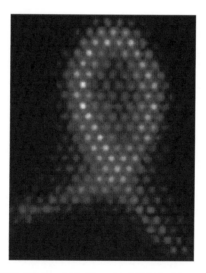

Figure 23.18   Hot wire OD: 0.2 mm; loop diameter: ~2 mm; 65 μm 900 HGW IRIFB; 2-in. imaging lens and 1-in. magnifying lens.

coronary bypass minimally invasive surgeries in beagles has been carried out by Nakagawa et al. [136] using IR imaging. In addition, a physiological saline of different temperatures was injected through the endoscope channel to see the changes in cortical images. They concluded that transendoscopic IR imaging (7–14 μm) may provide a noninvasive functional angiography. Stefanadis et al. [137] have developed a thermocouple-based thermography catheter and demonstrated in a human *in vivo* trial that thermal gradients between healthy and neoplastic tissue zones could be a useful criterion in the diagnosis of malignancy in tumors of the bladder. They found significant temperature differences (e.g., 1 °C) between patients with benign and malignant tumors. Endodontists McCullagh et al. [138] compared thermocouple-based thermography versus IR imaging of temperature rise on the *in vitro* root surface during the continuous wave of condensation technique. They found IR imaging a useful tool for mapping temperature change over a large area. In 2001, Cadeddu et al. [139] reported nine laparoscopic urologic procedures in patients, using IR imaging (3–5 μm) laparoscopically (i.e., via a rigid endoscope). They stated that IR imaging proved useful in differentiating between blood vessels and other anatomic structures and that, in contrast to conventional endoscopy, vessel identification, assessment of organ perfusion, and transperitoneal localization of the ureter were successful in all instances. They concluded that IR imaging is a potentially powerful adjunct to laparoscopic surgery.

Dayan et al. [140] have developed a method where thermal imaging can be used as a feedback mechanism to optimize laser ablation within body cavities. All these examples prove the necessity of a flexible imaging bundle in minimally invasive medical procedures for diagnostics and treatments.

REFERENCES

[1] Christenson, L. J. et al. 2000. Treatment of multiple cutaneous leiomyomas with CO_2 laser ablation. *Dermatol. Surg.* 26(4):319–322.
[2] Barzilay, B. et al. 1982. Comparative experimental study on the use of the carbon dioxide laser beam in partial nephrectomy. *Lasers Surg. Med.* 2(1):73–80.
[3] Chinali, G. et al. 1983. Liver regeneration after atypical hepatectomy in the rat. A comparison of CO_2 laser with scalpel and electrical diathermy. *Eur. Surg. Res.* 15(5): 284–288.
[4] Simonsen, E. et al. 1983. CO_2 laser for the treatment of invasive vulvar and vaginal cancer. *Neoplasma* 30(3):359–364.
[5] Black, P. M. et al. 1993. Present and future applications of lasers in neurosurgery. *Keio J. Med.* 42(4):169–170.
[6] Heppner, F. and P. W. Ascher. 1982. The CO_2 laser in neurosurgery. *Int. Adv. Surg. Oncol.* 5:385–396.
[7] Eltorai, I. et al. 1988. The use of the carbon dioxide laser beam in the surgery of pressure sores. *Int. Surg.* 73(1):54–56.

[8] Heppner, F. et al. 1987. CO_2 laser surgery of intramedullary spinal cord tumors. *Lasers Surg. Med.* 7(2):180–183.

[9] Herrmann, H. D. et al. 1988. Intramedullary spinal cord tumors resected with CO_2 laser microsurgical technique: Recent experience in fifteen patients. *Neurosurgery* 22(3):518–522.

[10] Kotecha, B. et al. 1998. Laser assisted uvulopalatoplasty: An objective evaluation of the technique and results. *Clin. Otolaryngol. Allied Sci.* 23(4):354–359.

[11] Krespi, Y. P. et al. 1994. Laser-assisted uvula-palatoplasty for snoring. *J. Otolaryngol.* 23(5):328–334.

[12] Remacle, M. et al. 2002. Laser-assisted surgery addressing snoring long-term outcome comparing CO_2 laser versus CO_2 laser combined with diode laser. *Acta Otorhinolaryngol. Belg.* 56(2):177–182.

[13] Lane, J. E. et al. 2002. Treatment of pearly penile papules with CO_2 laser. *Dermatol. Surg.* 28(7):617–618.

[14] Michel, J. L. et al. 2001. Resurfacing CO_2 laser treatment of linear verrucous epidermal nevus. *Eur. J. Dermatol.* 11(5):436–439.

[15] Song, M. G. et al. 1999. Resurfacing of facial angiofibromas in tuberous sclerosis patients using CO_2 laser with flash scanner. *Dermatol. Surg.* 25(12):970–973.

[16] Koo, S. H. et al. 2001. Laser punch-out for acne scars. *Aesthetic Plast. Surg.* 25(1):46–51.

[17] Fitzpatrick, R. E. et al. 1994. Laser ablation of facial cosmetic tattoos. *Aesthetic Plast. Surg.* 18(1):91–98.

[18] Kim, Y. J. et al. 2005. Analysis of hyperpigmentation and hypopigmentation after Er:YAG laser skin resurfacing. *Lasers Surg. Med.* 36(1):47–51.

[19] Drnovsek-Olup, B. et al. 2004. Repetitive Er:YAG laser irradiation of human skin: A histological evaluation. *Lasers Surg. Med.* 35(2):146–151.

[20] Airan, L. E. and G. Hruza. 2002. Current lasers in skin resurfacing. *Facial Plast. Surg. Clin. North Am.* 10(1):87–101.

[21] Crespi, R. et al. 2006. Er:YAG Laser Scaling of Diseased Root Surfaces: A Histologic Study. *J. Periodontol.* 77(2):218–222.

[22] Kim, J. H. et al. 2005. Effectiveness of an Er:YAG laser in etching the enamel surface for orthodontic bracket retention. *Dent. Mater. J.* 24(4):596–602.

[23] Karlovic, Z. et al. 2005. Erbium:YAG laser versus ultrasonic in preparation of root-end cavities. *J. Endod.* 31(11):821–823.

[24] Goncalves, M. et al. 2005. Influence of pulse frequency Er:YAG laser on the tensile bond strength of a composite to dentin. *Am. J. Dent.* 18(3):165–167.

[25] Chimello-Sousa, D. T. et al. 2006. Influence of Er:YAG laser irradiation distance on the bond strength of a restorative system to enamel. *J. Dent.* 34(3):245–251.

[26] Li, Z. Z. et al. 1992. Er:YAG laser ablation of enamel and dentin of human teeth: Determination of ablation rates at various fluences and pulse repetition rates. *Lasers Surg. Med.* 12(6):625–630.

[27] Hafez, M. I. et al. 2002. Ablation of bone, cartilage, and facet joint capsule using Ho:YAG laser. *J. Clin. Laser Med. Surg.* 20(5):251–255.

[28] Nagle, D. J. and M. A. Bernstein. 2002. Laser-assisted arthroscopic ulnar shortening. *Arthroscopy* 18(9):1046–1051.

[29] Pullin, J. G. et al. 1996. Effects of holmium:YAG laser energy on cartilage metabolism, healing, and biochemical properties of lesional and perilesional tissue in a weight-bearing model. *Arthroscopy* 12(1):15–25.

[30] Min, K. et al. 1996. Quantitative determination of ablation in weight of lumbar intervertebral discs with holmium:YAG laser. *Lasers Surg. Med.* 18(2):187–190.

[31] Vangsness, C. T., Jr. and C. F. Smith. 1995. Arthroscopic shoulder surgery with three different laser systems: An evaluation of laser applications. *Arthroscopy* 11(6):696–700.

[32] Holmium:YAG surgical lasers. *Health Devices* 1995; 24(3):92–122.

[33] Ramakrishnan, P. A. et al. 2005. Holmium laser cystolithotripsy in children: Initial experience. *Can. J. Urol.* 12(6):2880–2886.

[34] Lee, H. et al. 2006. Urinary calculus fragmentation during Ho:YAG and Er:YAG lithotripsy. *Lasers Surg. Med.* 38(1):39–51.

[35] Finley, D. S. et al. 2005. Effect of holmium:YAG laser pulse width on lithotripsy retropulsion *in vitro. J. Endourol.* 19(8):1041–1044.

[36] Gleich, L. L. et al. 1995. The holmium:YAG laser-assisted otolaryngologic procedures. *Arch. Otolaryngol. Head Neck Surg.* 121(10):1162–1166.

[37] Woog, J. J. et al. 1993. Holmium:YAG endonasal laser dacryocystorhinostomy. *Am. J. Ophthalmol.* 116(1):1–10.

[38] Bukala, B. and J. D. Denstedt. 1999. Holmium:YAG laser resection of the prostate. *J. Endourol.* 13(3):215–219.

[39] Kuntz, R. M. et al. 2004. Transurethral holmium laser enucleation of the prostate compared with transvesical open prostatectomy: 18-month follow-up of a randomized trial. *J. Endourol.* 18(2):189–191.

[40] Matsuoka, K. et al. 1998. Holmium laser resection of the prostate. *J. Endourol.* 12(3): 279–282.

[41] Yamanishi, T. et al. 2001. Transurethral holmium:YAG laser prostatectomy using a side-firing fiber for bladder outlet obstruction due to benign prostatic enlargement: Urodynamic evaluation of surgical outcome. *Eur. Urol.* 39(5):544–550.

[42] Escudier, M. P. and M. McGurk. 1999. Symptomatic sialoadenitis and sialolithiasis in the English population, an estimate of the cost of hospital treatment. *British Dental Journal* 186(9):463–466.

[43] Berry, R. 1995. Sialoenitis and sialolithiasis: Diagnosis and management. *Oral Maxillofac. Surg. Clin. North Am.*

[44] Nahlieli, O. and A. M. Baruchin. 1999. Endoscopic technique for the diagnosis and treatment of obstructive salivary gland diseases. *J. Oral Maxillofac. Surg.* 57(12):1394–1402.

[45] Nahlieli, O. and A. M. Baruchin. 2000. Long-term experience with endoscopic diagnosis and treatment of salivary gland inflammatory diseases. *Laryngoscope* 110(6):988–993.

[46] Nahlieli, O. et al. 2003. Endoscopic mechanical retrieval of sialoliths. *Oral Surg. Oral Med. Oral Pathol. Oral Radiol. Endod.* 95(4):396–402.

[47] Baurmash, H. D. 2004. Submandibular salivary stones: Current management modalities. *J. Oral Maxillofac. Surg.* 62(3):369–378.

[48] Iro, H. et al. 1989. Extracorporeal shockwave lithotripsy of salivary-gland stones. *Lancet.* 2:114–115.

[49] Capaccio, P. et al. 2004. Extracorporeal lithotripsy for salivary calculi: A long-term clinical experience. *Laryngoscope* 114(6):1069–1073.

[50] Nahlieli, O. and A. M. Baruchin. 1997. Sialoendoscopy: Three years' experience as a diagnostic and treatment modality. *J. Oral Maxillofac. Surg.* 55(9):912–918.

[51] Fried, N. M. 2006. Therapeutic applications of lasers in urology: An update. *Expert Rev. Med. Devices* 3(1):81–94.

[52] Katz, P. and M. H. Fritsch. 2003. Salivary stones: Innovative techniques in diagnosis and treatment. *Curr. Opin. Otolaryngol. Head Neck Surg.* 11(3):173–178.

[53] Kerr, P. D. et al. 2001. Endoscopic laser lithotripsy of a proximal parotid duct calculus. *J. Otolaryngol.* 30(2):129–130.

[54] Marchal, F. and P. Dulguerov. 2003. Sialolithiasis management: The state of the art. *Arch. Otolaryngol. Head Neck Surg.* 129(9):951–956.

[55] Croitoru, N. et al. 1990. Hollow fiber waveguide and method of making same. US Patent 5567471.

[56] Kim, Y. J. et al. 2003. Determination of glucose in whole blood samples by mid-infrared spectroscopy. *Applied Opt.* 42(4):745–749.

[57] Luo, Y. et al. 2005. Simultaneous analysis of twenty-one glucocorticoids in equine plasma by liquid chromatography/tandem mass spectrometry. *Rapid Commun. Mass Spectrom.* 19(10):1245–1256.

[58] Roller, C. et al. 2002. Simultaneous NO and CO_2 measurement in human breath with a single IV–VI mid-infrared laser. *Opt. Lett.* 27(2):107–109.

[59] Hynes, A. et al. 2005. Molecular mapping of periodontal tissues using infrared microspectroscopy. *BMC Med. Imaging.* 5:2.

[60] Eckhardta, H. S. et al. 2006. Fiber-optic based gas sensing in the UV region. Presented at BIOS 2006.

[61] Bakhirin, Y. A. et al. 2004. Mid infrared quantum cascade laser based off-axis integrated cavity output spectroscopy for biogenic nitric oxide detection. *Applied Opt.* 43(11): 2257–2266.

[62] Ganser, H. et al. 2002. Online monitoring of biogenic nitric oxide with a QC laser-based Faraday modulation technique. *Applied Phys. B.* 78(3):513–517.

[63] Roller, C. et al. 2002. Nitric oxide breath testing by tunable-diode laser absorption spectroscopy: Application in monitoring respiratory inflammation. *Applied Opt.* 41(28):6018–6029.

[64] Tobias, J. D. 2006. Cerebral oxygenation monitoring: Near-infrared spectroscopy. *Expert Rev. Med. Devices* 3(2):235–243.

[65] Sheshnev, A. et al. 2000. Mathematical simulation of water and benzene penetration in thin polystyrene films and experimental investigation of the mass transport processes by FEWS-FTIR spectroscopy procedure. Presented at the First International Conference on Mathematical Modeling and Simulation of Metal Technologies, Ariel, Israel.

[66] Raichlin, Y. et al. 2003. Evanescent-wave infrared spectroscopy with flattened fibers as sensing elements. *Opt. Lett.* 28(23):2297–2299.

[67] Lucas, P. et al. 2005. Advances in chalcogenide fiber evanescent wave biochemical sensing. *Anal. Biochem.* 351(1):1–10.

[68] Janotta, M. et al. 2003. Sol-gel-coated mid-infrared fiber-optic sensors. *Appl. Spectrosc.* 57(7):823–828.

[69] Perkin, E. FT-IR spectroscopy—attenuated total reflectance (ATR). Perkin Elmer. Available at www.perkinelmer.com.

[70] Hocde, S. et al. 2001. Chalcogenide based glasses for IR fiber chemical sensors. *Solid State Sci.* 3:279–284.

[71] Katz, A. et al. 1994. Qualitative evaluation of chalcogenide glass based fiber evanescent wave spectroscopy. *Appl. Opt.* 33:5888–5894.

[72] Haurylau, M. et al. 2005. Optical properties and tunability of macroporous silicon 2-D photonic bandgap structures. *Phys. Stat. Sol.* 202:1477–1481.

[73] Saarinen, J. J. et al. 2005. Optical sensor based on resonant porous silicon structures. *Opt. Exp.* 13(10):3754–3764.

[74] Jensen, J. B. et al. 2004. Photonic crystal fiber based evanescent-wave sensor for detection of biomolecules in aqueous solutions. *Opt. Lett.* 29(17):1974–1976.

[75] Boussard-Pledel, C. et al. 1999. Infrared glass fiber for evanescent wave spectroscopy. *Proc. SPIE* 3596:91–98.

[76] Dury, M. A. et al. 1993. Mid-IR tapered chalcogenide fiber optic attenuated total reflectance sensors for monitoring epoxy resin chemistry. *Proc. SPIE* 2069:113–120.

[77] Zhang, X. H. et al. 2003. Applications of chalcogenide glass bulks and fibers. *J. Ooptoelecron. Adv. Mat.* 2003:1327–1333.

[78] Steiner, H. et al. 2003. *In situ* sensing of volatile organic compounds in groundwater: First field tests of a mid-infrared fiber-optic sensing system. *Appl. Spectrosc.* 57(6):607–613.

[79] Raichlin, Y. et al. 2002. Infrared fiber optic evanescent wave spectroscopy for the study of diffusion in human skin. *Proc. SPIE* 4614:101–108.

[80] Uemura, T. et al. 1999. Noninvasive blood glucose measurement by Fourier transform infrared spectroscopic analysis through the mucous membrane of the lip: Application of a chalcogenide optical fiber system. *Front. Med. Biol. Eng.* 1999:137–153.

[81] Afasnasyeva, N. et al. 1999. Infrared fiber optic evanescent wave spectroscopy: Applications in biology and medicine. *Proc. SPIE* 3596:152–164.

[82] Compton, D. A. C. et al. 1988. *In situ* FT-IR analysis of a composite curing reaction using a mid-infrared transmitting optical fiber. *Appl. Spectrosc.* 42:972–979.

[83] Heo, J. et al. 1991. Remote fiber-optics chemical sensing using evanescent-wave interactions in chalcogenide glass fibers. *Appl. Opt.* 30:3944–3951.

[84] Rodrigues, M. and G. H. Siegel. 1991. Chalcogenide glass fibers for remote spectroscopic chemical sensing. *Proc. SPIE* 1591:225–236.

[85] Sangehra, J. S. et al. 1995. Infrared evanescent absorption spectroscopy of toxic chemicals using chalcogenide glass fibers. *J. Am. Ceram. Soc.* 78:2198–2202.

[86] Sanghera, J. S. et al. 1994. Infrared evanescent absorption spectroscopy with chalcogenide glass fibers. *Appl. Opt.* 33:6315.

[87] Charlton, C. et al. 2005. Infrared evanescent field sensing with quantum cascade lasers and planar silver halide waveguides. *Anal. Chem.* 77(14):4398–4403.

[88] Le Coq, D. et al. 2002. Infrared glass fibers for in-situ sensing, chemical and biochemical reactions. *Comptes Rendus Chimie* 5:907–913.

[89] Li, K. and J. Meichsner. 2001. *In situ* infrared fiber evanescent wave spectroscopy as a diagnostic tool for plasma polymerization in a gas discharge. *J. Phys. D Appl. Phys.* 34:1318–1325.

[90] Ouyang, H. et al. 2005. Macroporous silicon microcavities for macromolecules detection. *Adv. Func. Mater.* 15:1851–1859.

[91] Sanghera, J. S. et al. 2000. Development and infrared applications of chalcogenide glass optical fibers. *Fiber Integr. Opt.* 19:251–274.

[92] Shaviv, A. et al. 2003. Direct monitoring of soil and water nitrate by FTIR based FEWS or membrane systems. *Environ. Sci. Technol.* 37(12):2807–2812.

[93] Michel, K. et al. 2004. Monitoring of pollutant in wastewater by infrared spectroscopy using chalcogenide glass optical fibers. *Sens. Actuat. B* 2004(101):252.

[94] Raichlin, Y. et al. 2004. Investigations of the structure of water using mid-IR fiberoptic evanescent wave spectroscopy. *Phys. Rev. Lett.* 93(18):185703.

[95] Huang, G. G. and J. Yang. 2003. Selective detection of copper ions in aqueous solution based on an evanescent wave infrared absorption spectroscopic method. *Anal. Chem.* 75(10):2262–2269.

[96] Steiner, H. et al. 2003. Online sensing of volatile organic compounds in groundwater using mid-infrared fibre optic evanescent wave spectroscopy: A pilot scale test. *Water Sci. Technol.* 47(2):121–126.

[97] Hocde, S. et al. 2004. Metabolic imaging of tissues by infrared fiber-optic spectroscopy: An efficient tool for medical diagnosis. *J. Biomed. Opt.* 9(2):404–407.

[98] Keirsse, J. et al. 2004. Chalcogenide glass fibers for in-situ infrared spectroscopy in biology and medicine. *Proc. SPIE* 5459:61.

[99] Shimony, S. et al. 1987. Fourier transform infrared spectra of aqueous protein mixtures using a novel attenuated total internal reflection cell with infrared fibers. *Biochem. Biophys. Res. Commun.* 142:1059–1063.

[100] Kishen, A. et al. 2003. A fiber optic biosensor (FOBS) to monitor mutans streptococci in human saliva. *Biosens. Bioelectron.* 18(11):1371–1378.

[101] Cytron, S. et al. 2003. Fiberoptic infrared spectroscopy: A novel tool for the analysis of urine and urinary salts in situ and in real time. *Urology* 61:231–235.

[102] Poscio, P. et al. 1994. *In vivo* measurement of dye concentration using an evanescent-wave optical sensor. *Med. Biol. Eng. Comput.* 32(4):362–366.

[103] Sukuta, S. and R. Bruch. 1999. Factor analysis of cancer Fourier transform infrared evanescent wave fiberoptical (FTIR-FEW) spectra. *Lasers Surg. Med.* 24(5):382–388.

[104] John, M. S. et al. 2002. Determination of bacterial activity by use of an evanescent-wave fiber-optic sensor. *Appl. Opt.* 41(34):7334–7338.

[105] Lucas, P. et al. 2005. Evaluation of toxic agent effects on lung cells by fiber evanescent wave spectroscopy. *Appl. Spectrosc.* 59(1):1–9.

[106] Stokes, J. and G. Palmer. 2002. A fiber optic temperature sensor. *Sensors Magazine.*

[107] FISO Technologies, Inc. 2002. Product Literature. Aug. Issue. FISO Technologies. Quebec, Canada.

[108] Chen, Y. and H. F. Taylor. 2002. Multiplexed fiber Fabry–Perot temperature sensor system using white-light interferometry. *Opt. Lett.* 27:903–905.

[109] Easley, C. J. et al. 2005. Extrinsic Fabry–Perot interferometry for noncontact temperature control of nanoliter-volume enzymatic reactions in glass microchips. *Annal. Chem.* 77:1038–1045.

[110] Lee, C. E. and H. F. Taylor. 1991. Fiber-optic Fabry–Perot temperature sensor using a low-coherence light source. *J. Lightwave Technol.* 9:129–131.

[111] Beaulieu, J. 2002. Temperature measurement in an MRI scanner. Available at www.fiso.com.

[112] Hamel, C. and E. Pinet. 2006. Temperature and pressure fiber optic sensors applied to minimally invasive diagnostics and therapies. Presented at BIOS 2006.

[113] Sade, S. et al. 2002. Fiber-optic infrared radiometer for accurate temperature measurements. *Appl. Opt.* 41(10):1908–1914.

[114] Scharf, V. and A. Katzir. 2002. Four-band fiber optic radiometry for true temperature measurements during an exothermal process. *Opt. Eng.* 41(7):1502–1506.

[115] Uman, I. et al. 2004. All-fiber-optic infrared multispectral radiometer for measurements of temperature and emissivity of graybodies at near-room temperature. *Appl. Opt.* 43(10):2039–2045.

[116] Sade, S. and A. Katzir. 2001. Fiber optic infrared thermometer for accurate measurement of the core temperature of the human body. *Opt. Eng.* 40(6):1141–1143.

[117] Cohen, M. et al. 2001. Monitoring of the infrared radiation emitted from skin layers during CO_2 laser resurfacing: A possible basis for a depth navigation device. *Lasers Med. Sci.* 16:176–183.

[118] Shenfeld, O. et al. 1994. Silver halide fiber optic radiometric temperature measurement and control of CO_2 laser-irradiated tissues and application to tissue welding. *Lasers Surg. Med.* 14(4):323–328.

[119] Cohen, M. et al. 2003. Temperature controlled burn generation system based on a CO_2 laser and a silver halide fiber optic radiometer. *Lasers Surg. Med.* 32(5):413–416.

[120] Ilesiz, I. C. and A. Katzir. 2001. Thermal-feedback-controlled coagulation of egg white by the CO_2 laser. *Appl. Opt.* 40(19):3268–3277.

[121] Gannot, I. 2005. Thermal imaging bundle: A potential tool to enhance minimally invasive medical procedures. *IEEE Circuits Devices* 2005:28–33.

[122] Kapany, N. S. 1965. Recent developments in infrared fiber optics. *Infrared Phys.* 5(1):69–80.

[123] Saito, M. et al. 1985. Infrared image guide with bundled as-S glass-fibers. *Appl. Opt.* 24(15):2304–2308.

[124] Klocek, P. et al. 1987. Chalcogenide glass optical fibers and image bundles—properties and applications. *Opt. Eng.* 26(2):88–95.

[125] Nishii, J. et al. 1991. Coherent infrared fiber image bundle. 59(21):2639–2641.

[126] Hilton, A. 2001. Infrared imaging bundles with good image resolution. *SPIE Proc.* 4253:28–36.

[127] Naghavi, M. et al. 2003. Thermography basket catheter: *In vivo* measurement of the temperature of atherosclerotic plaques for detection of vulnerable plaques. *Cathet. Cardiovasc. Intervent.* 59(1):52–59.

[128] Paiss, I. and A. Katzir. 1992. Thermal imaging by ordered bundles of silverhalide crystalline fibers. *Appl. Phys. Lett.* 61(12):1384–1386.

[129] Paiss, I. et al. 1991. Properties of silver-halide core-clad fibers and the use of fiber bundle for thermal imaging. *Fiber Integr. Opt.* 10(3):275–290.

[130] Rave, E. et al. 2000. Ordered bundles of infrared-transmitting AgClBr fibers: Optical characterization of individual fibers. *Opt. Lett.* 25(17):1237–1239.

[131] Rave, E. et al. 2000. Thermal imaging through ordered bundles of infrared-transmitting silver-halide fibers. *Appl. Phys. Lett.* 76(14):1795–1797.

[132] Rave, E. and A. Katzir. 2002. Ordered bundles of infrared transmitting silver halide fibers: Attenuation, resolution and crosstalk in long and flexible bundles. *Opt. Eng.* 41(7):1467–1468.

[133] Gopal, V. et al. 2004. Coherent hollowcore waveguide bundles for infrared imaging. *Opt. Eng.* 43(5):1195–1199.

[134] Croitoru, N. et al. 1990. Characterization of hollow fibers for the transmission of infrared radiation. *Appl. Opt.* 29(12):1805–1809.

[135] Ogan, K. et al. 2003. Infrared thermography and thermocouple mapping of radiofrequency renal ablation to assess treatment adequacy and ablation margins. *Urology* 62(1):146–151.

[136] Nakagawa, A. H. et al. 2003. Intraoperative thermal artery imaging of an EC-IC bypass in beagles with infrared camera with detectable wavelength band of 7–14 micron: Possibilities as novel blood flow monitoring system. *Minimally Invas. Neurosurg.* 46(4):231–234.

[137] Stefanadis, C. T. et al. 2003. New balloon-thermography catheter for *in vivo* temperature measurements in human coronary atherosclerotic plaques: A novel approach for thermography? *Cathet. Cardiovasc. Interven.* 58(3):344–350.

[138] McCullagh, J. J. P. et al. 2000. A comparison of thermocouple and infrared thermographic analysis of temperature rise on the root surface during the continuous wave of condensation technique. *Int. Endodontic J.* 33(4):326–332.

[139] Cadeddu, J. J. and P. G. Schulam. 2001. Laparoscopic infrared imaging. *J. Endourol.* 15(1):111–116.

[140] Dayan, A. et al. 2004. Theoretical and experimental investigation of the thermal effects within body cavities during transendoscopical CO_2, laser-based surgery. *Lasers Surg. Med.* 35(1):18–27.

Chapter 24

Mechanical Strength and Reliability of Glass Fibers

Charles R. Kurkjian and M. John Matthewson

Photonic Components Reliability Group, Department of Materials Science and Engineering, Rutgers University, Piscataway, New Jersey

24.1 INTRODUCTION

Silica-based light-guide fibers have been produced and deployed with great success in spite of their well-known "brittleness." For instance, techniques have been developed that allow long lengths (tens of kilometers) of such fibers to be drawn and coated in-line. The preform and draw processes allow production fiber that has a flaw-free surface with, essentially, the theoretical strength under the conditions of use/test (~5.5 GPa) for most of its length. The few manufacturing defects that occur are eliminated by a proof-testing procedure that allows the remaining length of fiber to attain almost any desired guaranteed strength. While the commonly specified proof stress is approximately 700 MPa, by reducing the available continuous lengths and at some increase in cost, higher proof stresses are possible. Success of this sort has not been achieved with all glass fibers, however. Multicomponent silicate fibers, hollow waveguides, and photonic crystal guides have been studied much less, and while their capabilities can be predicted to some extent, very few data are available. The mechanical properties of other specialty glass fibers, such as heavy-metal fluoride (HMF), telluride, and chalcogenide glasses, have also been studied with varying degrees of success.

Intensive study and modification of the processes used for manufacturing the fiber preforms and the resulting fiber have resulted in the very high quality fiber described earlier. The use of very high quality raw materials for the core and cladding, control of the purity of the gases employed in the furnace, as well as control of the solid and gaseous impurities available to interact with the hot fiber in the draw environment proved very important in this regard. Because it was known that contact of the pristine fiber surface with any hard material would

735

easily cause surface damage and a precipitous reduction in strength, techniques were developed very early to enable a protective polymer coating to be applied in-line without itself causing damage to the fiber surface. Because of the purity of all of the materials employed, there are rarely any solid impurity particles in the body of the fiber, so in almost all cases, the strength of the fiber is the strength of its surface. Thus, the condition of the surface of the fiber determines its strength. Whatever the situation with regard to the "starting" strength of any given fiber or fiber type, several things must be kept in mind:

a. Fibers are normally coated in-line with a polymer, which provides protection from mechanical damage. The strength of bare fiber can be easily reduced precipitously by the development of such mechanical damage on the glass surface. The damage is the result of the contact of hard foreign particles on the glass surface, which produce strength reducing cracks or scratches. Such strengths show a very broad distribution and are unpredictable.
b. Delayed failure or fatigue causes the initial strength to decrease with time under load in glasses of most compositions.
c. In some cases, strength decreases are noted with time under "severe" conditions, even in the absence of an applied stress; this is termed "zero stress aging."
d. Glasses are essentially "frozen" liquids with no regular lattice, but with a random structure and fluctuations characteristic of a liquid. They are in a metastable state with respect to the crystal and heating to a critical temperature where crystallization may occur at a reasonable rate will certainly lead to strength degradation. Even if there is no crystallization, there may be relaxation of the glass structure and, therefore, changes in the glass properties that must be kept in mind when processing fibers for use in some applications.

24.2 REVIEW OF GLASS PROPERTIES

In this section, we review some of the important parameters of glass systems that can affect their mechanical behavior and mechanical reliability.

24.2.1 Noncrystallinity, the Glass Transition (T_g), and Relaxation Processes

Although there are several operational definitions of "glass," for our purpose, we define a glass as a material that has been cooled from the melt without crystallizing. In the region from T_m, the melting point and T_g, the glass transition temperature (the temperature at which the viscosity, $\eta \sim 10^{12.4}$ Pa/sec), the material is called a "supercooled" liquid (Fig. 24.1, [1]). In this state, it has the structure of the liquid, and although the mobility of the structure is reduced as the temperature is reduced, structural rearrangements can occur in experimental times. Because of this, the coefficient of thermal expansion is substantially higher ($\sim$2–3 times) than below T_g.

As seen in Fig. 24.1, changes in cooling rate from T_g will result in changes in room temperature properties such as density and refractive index. To a very rough approximation, we may consider these materials to behave as maxwellian solids, with $\eta = G\tau$, where G is the shear modulus and τ the shear relaxation time; at T_g, relaxation times of the order of minutes are found. Thus, metastable equilibrium is achieved in reasonable experimental times. Below T_g, relaxation increases with an activation energy of the order of 400–1000 kJ/mol, and thus, major structural changes cannot usually occur in experimental times. However, in the case of rather rapidly quenched materials, such as silica fibers, it has been found that substantial relaxations can occur below T_g.

Below T_g the mobility is so small that structural rearrangements do not occur in experimental times. This material is considered to be in the amorphous

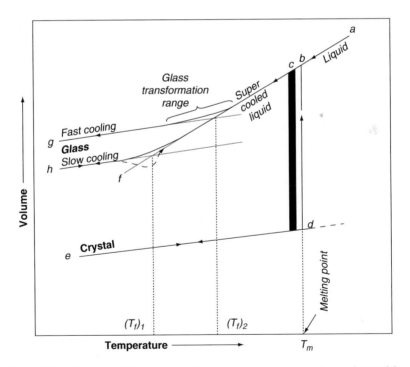

Figure 24.1 Schematic of the glass transition (used, with permission, from reference [1]).

or glassy state. On cooling from T_g to room temperature, the linear thermal expansion is the result of changes in bond lengths, whereas above T_g, a thermal expansion coefficient, which is at least twice as great, is due to structural rearrangements and to bond length changes. For materials such as silica, silicates, and some chalcogenides, upon reheating through T_g and up to the melting point, there is little probability of crystallization when heating at reasonable rates. Other glasses such as the HMFs may rather easily crystallize on either reheating or on the initial cooling. Thus, in such glasses, care must be taken to ensure that no crystallization occurs. In particular, surface-nucleated crystallization may be a problem both mechanically and optically.

Glasses differ from their crystalline counterparts in that their structure has no repeating lattice but is instead random. There may be both chemical and structural randomness, which may give rise to optical scattering. Regions of chemical or structural fluctuations may lead to fluctuations in solution rate as well. We postulate that this may lead to the zero stress aging described later.

24.2.2 Brittleness, Hardness, and Cracking

Most of the inorganic glasses with which we are concerned fail in a brittle manner. That is, they will fail by fast fracture (i.e., cracks that grow unstably to produce rapid and complete failure without any significant ductile deformation). Brittle materials are sensitive to the presence of cracks and other stress-concentrating defects, and if mishandled, it is possible to introduce significant damage with corresponding weakening. As a result, glasses are frequently mistakenly described as "brittle" if they are weak and easily break on handling. However, this is incorrect because "brittle" refers to crack propagation without ductility; whereas brittle materials are often weak, they can also be extremely strong if severe strength-degrading defects are avoided or eliminated. Pure silica and silicate glasses have high intrinsic strengths that are higher than most "strong" ductile metals. However, their brittleness means they are easily damaged. Contact of the newly drawn perfect surface of a silica light-guide fiber with a sharp hard object will result in an instantaneous and serious reduction in strength, by a factor up to 100 or more. The susceptibility to surface damage (cracking) is not easily understood or predicted. Although, as just stated, the glasses with which we are concerned are basically brittle, under the action of a complex state of applied stress such as found under a sharp indenter or scribe, they will flow at room temperature. This flow will result in the development of residual stresses and ultimately lead to the generation of cracks. Also, even blunt indenters, if subject to a high enough load, will result in the generation of cracks. Two investigations in the literature illustrate this nicely. Baikova et al. [2] studied the behavior of a glass plate when indented with a 2-mm diameter steel ball.

Their initial results showed a great deal of scatter in the load required to produce cracking and thus considerable variability in the strength reduction, certainly at loads in the range of 20–100 N, and occasionally at substantially lower loads. After cleaning the ball and the glass surface, they found the higher loads were required to cause fracture, and from this, they deduced that the presence of dust particles was the cause of the very much lower loads. With a clean ball on a clean surface, the load necessary to cause cracking and thus lower the strength of a silica plate was found to be approximately 100 N, in agreement with that predicted for this situation in which a ball is loaded on a flat substrate. In this case, the tensile stress that induces cracking is just outside the contact area, roughly the diameter of the ball. One of the present authors [3] has shown more quantitatively that the use of an indenter in the shape of the corner of a cube, rather than the much more blunt Vickers pyramid indenter with a 138-degree included angle, results in a reduction of the load to cause cracking by two orders of magnitude (<2 mN versus 2 N or more) (Fig. 24.2). In the case of the cube corner indenter, cracking is produced during loading and the appropriate equation to describe this effect is

$$P_r \sim P\left(\frac{E}{H}\right)^{1/2} \cot \psi^{2/3},\qquad(24.1)$$

where P_r is the crack opening force due to the applied load P, E is Young's modulus, H the hardness, and ψ is the indenter angle [4]. In the case of the

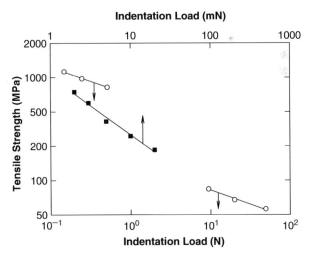

Figure 24.2 Comparison of the effect of indentation load on the resultant strength of silica fibers: Vickers (circles, bottom scale) and cube corner (squares, top scale) (data used, with permission, from reference [3]).

blunter Vickers pyramid indenter, cracking may occur on unloading and, therefore, is driven by the residual stresses developed during the plastic deformation associated with formation of the permanent impression.

24.2.3 Composition Effects

Table 24.1 lists a number of typical inorganic glasses that may be of use as light guides or in light-guide devices, together with some of their important properties. E is the Young's modulus, K_{IC} is the fracture toughness, H_V is the Vickers hardness, T_g is the glass transition temperature, and α is the coefficient of thermal expansion. Columns five through eight list some strength values and are described later. As indicated earlier, glasses are often referred to as *brittle,* whereas as can be seen here, they can have exceptionally high intrinsic strengths.

24.2.3.1 Silicates

The vast majority of commercial glasses are so-called multicomponent glasses (MCGs) that contain three to six oxide components. Early in the development of light-guide fibers, many workers attempted to employ these well-known and well-characterized glasses for long-length communications [5–7]. It was found very difficult to purify them to the extent necessary for low optical loss. In addition, it was found quite early on that by using vapor-phase processes and extreme purity liquid precursors for the preparation of pure glass-former glass fibers, high purity could be maintained throughout the fiber production. However, even so, it was not possible to produce MCG fibers because suitable liquid precursors were generally not available. In addition, simple two-component alkali silicates that were predicted to have ultimate scattering losses lower than silica could not be used because at the low alkali concentrations desired, phase separation was encountered [8]. Also, the addition of even small amounts of these oxides results in a glass with very different viscosity and thermal expansion, so processing of these with a silica cladding becomes difficult. Duncan et al. [9] looked at alkali and alkaline earth borosilicate glasses (silica and NBS glass), but as indicated earlier, had problems with purity. Figure 24.3 is a Weibull probability plot (see the later section on Weibull statistics) of the distributions of the strength of silica and an NBS glass and it is seen that reasonable strengths were obtained for such glasses, but their purity, and thus, optical transmission, was poor. In the future, it may be desirable to employ such glasses, if extremely low losses are not required, because their mechanical properties may be suitable.

In Fig. 24.3, we present failure strain rather than failure stress/strength. The reason for this is that a common technique used for testing these fibers

Table 24.1

Mechanical properties of various glasses

Glass	E GPa	K_{IC} MPa/m$^{1/2}$	H_V GPa	σ^a GPa	$E/5$ GPa	σ_K GPa	% theoretical[c]	n	T_g °C	$\alpha 10^{-6}/°C^{-1}$
Silica	72	0.75	14	14 (77 K)	14	14	100 (14/14)	20	1150	5
SLS	70	0.75	5.5	7.7 (77 K)	14	14	40 (5.5/14)	15	550	85
Chalcogenide	10–20	0.15–0.25	1.0–1.9	1.2 (300 K)	2–4	4.7	50–100 [2.4/(2–4)]	10–20	200–300	100–600
HMF	50–65	0.2–0.5	2.5–3.0	1.0–1.6 (300 K)	10–13	9.3	35 (3.2/9.3)	0–60	250–400	150–200
Tellurite	45	0.25–0.3	3.0–3.8	0.82 (300 K)	9	5.6	30 (1.6/5.6)	?	350	150

[a] Highest measured strengths at temperature given in parenthesis.
[b] Strengths are calculated using K_{glass}/K_{silica} from Eq. 24.3, assuming the initial sizes are equal.
[c] Strengths at 77 K are divided by the lower value from columns 5 or 7. The room temperature strengths in column 5 have been multiplied by 2 to approximate strengths at 77 K.

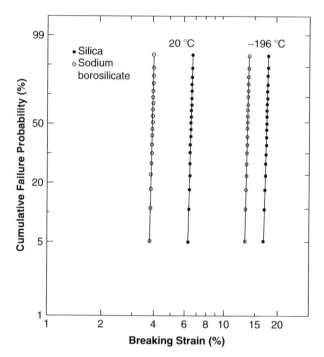

Figure 24.3   Failure strain (measured in two-point bend) of silica (●) and sodium borosilicate (○) glass fibers in liquid nitrogen and at room temperature (used, with permission, from reference [9]).

has been two-point bending rather than the more usual tension testing. Bending measurements directly determine the strain to failure and the elastic modulus is needed to convert to failure stress. Because the modulus is often not known, it is simpler to report failure strain, as has been discussed in the literature [10]. In a later section, the conversion of strain to stress is described in more detail.

24.2.3.2 Tellurite Glasses

Although tellurium dioxide does not form a glass in its pure state, oxide glasses based on TeO_2 have been found to have desirable light-emitting properties when doped with rare earth ions. In particular, the optical behavior of these glasses has been studied extensively for use as erbium-doped fiber amplifiers. To our knowledge, only one study of the mechanical strength of these glasses has been carried out [11].

24.2.3.3 Nonoxide Glasses

Because of their atomic sizes, bond types and bond strengths, oxide glasses show very strong optical absorptions in the near infrared (IR). Although the near-IR band edge may be moved somewhat further into the IR by using large heavy atoms in a silicate glass or by going to a germanium-based glass, to shift the band edge substantially, glasses based on anions other than oxygen are required. For this purpose, chalcogenide and HMF, non-silicate, nonoxide glasses have been studied intensively.

24.2.3.3.1 Chalcogenides

Although chalcogenide glasses were studied as early as 1912 [12], serious study of the individual chalcogens and chalcogenide glasses was not undertaken until the 1960s when interesting electronic properties were discovered [13, 14].

Chalcogenide glasses can be formed over a very wide range of composition. They contain one or more of the elements sulfur, selenium, or tellurium (chalcogens) with other elements from groups IV, V, and VI of the periodic table. For glasses having good transmission in the near IR (0.5–10 μm), most attention is now paid to glasses in the general composition ranges of As-S, As-Se, Ge-As-S, and Ge-Se-Te. Extrinsic absorption is most often caused by the presence of impurities of oxygen and hydrogen.

24.2.3.3.2 Heavy-Metal Fluorides

In 1975, Poulain et al. [15] unexpectedly discovered a fairly large range of glass-forming compositions based on HMFs; specifically zirconium fluoride. There followed a period of intense study because it had been shown theoretically that for highly pure glasses in these systems, there was the possibility of ultimate optical losses substantially lower than those for silica. It has subsequently been found impossible to produce either bulk glass or fibers with impurity levels sufficiently low for these theoretical losses to be realized. In addition, the difficulties with crystallization during fiber drawing restrict the compositions that are practically useful. However, they are still of some interest for certain applications.

Compositions most studied are ZBLAN (fluorides of zirconium, barium, lanthanum, and sodium) and ZBLAN with additions of AlF_3 [16].

24.2.3.4 Photonic Crystal Fibers

Photonic crystal fibers or "holey" fibers are just that. They are fibers having holes distributed over their cross-section [16]. Because they are normally a single material and normally silica, their material properties are the same as the materials listed in Table 24.1.

24.3 MECHANICAL PROPERTIES

In most of this section, we deal with general glass properties or with those of silica fibers. In the final portions of the section, we review nonsilica fiber properties.

In Table 24.1, columns five through eight, some data on the strengths of a variety of glasses are given. In column 5, the highest reported measured strengths are given, with the measurement temperature indicated in parenthesis. For silica and the soda-lime–silica glass, these are strengths measured at 77 K for what are thought to be flaw-free samples. The value for silica is the same as that for a silica-clad light-guide fiber because in this case the strength-controlling surface is composed of silica. We designate these *intrinsic inert strengths*—strengths measured on samples with no flaws under conditions where no delayed failure or fatigue can occur (77 K). Columns 6 and 7 give two approximations for the theoretical strengths. The first (column 6) is an approximation based on the fact that the strength is expected to be a substantial fraction of the elastic (Young's) modulus, in this case, we list $E/5$. It is seen that this gives the value measured for silica. Column 7 is an estimate based on the proportionality of strength (σ) and fracture toughness (K_{IC}) for a given flaw size, c, in Eq. (24.3). Here, it is assumed that the effective "crack" size in all glasses is the same and the data are normalized to the value of 14 GPa for silica. Column 8 is the ratio of the highest measured strength (estimated at 77 K as twice the room temperature value) to the lowest estimated theoretical value. As indicated, the value for silica is the theoretical value. For all other glasses except for As_2S_3, the percentage is 30–40%. For As_2S_3, it is possible that the fibers for which the measured strength at room temperature is 1.2 GPa may be very close to the inert, intrinsic, or theoretical "flaw-free" strength, approximately 2.4 GPa at 77 K.

24.3.1 Strength

24.3.1.1 Inert, Intrinsic Strength: Tension and Two-Point Bend

Figures 24.4 and 24.5—from the very important work of Proctor, Whitney, and Johnson in 1967 [17]—illustrate a number of significant points that we discuss later in this section. This work was carried out at Rolls Royce because of its interest in developing stronger aluminum aircraft parts by using silica fiber reinforcement. Figures 24.4 and 24.5 show the very significant temperature and

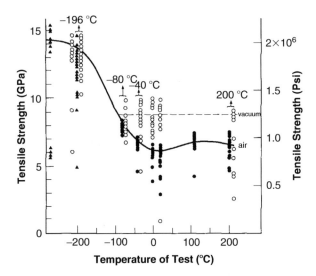

Figure 24.4 Effect of temperature on the tensile strength of silica fibers. Fibers were tested in the following environments: × liquid helium or nitrogen, ○ *in vacuo*, ● air, ▲ gaseous hydrogen (used, with permission, from reference [17]).

time dependence of the tensile strength of silica fibers. These figures are of great significance and are critical to the understanding of silica and silicate fiber strength. They essentially show the effect of water vapor on the "instantaneous" strength of a silica fiber. Two things should be noticed. The large increase in strength as the temperature is lowered below room temperature, and the time dependence of the strength at room temperature are both manifestations of

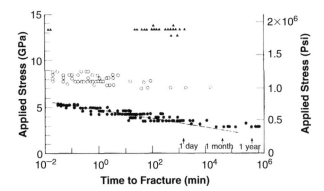

Figure 24.5 Effect of time on the tensile strength of silica fibers. × 77K, ○ *in vacuo*, ● ambient (used, with permission, from reference [17]).

"fatigue" or delayed failure. This is discussed in detail later. Here, we simply state that by testing under conditions in which the effects of moisture are eliminated (e.g., in a vacuum or at low temperature), there is no time dependence of strength; this is then termed the "inert" strength. Figure 24.3, from Duncan et al. [9] shows the inert failure strain measured at 77 K in two-point bending for fibers of silica and a sodium borosilicate glass (mentioned earlier). These latter data are shown to illustrate two important points. First, these data are plotted on a Weibull probability scale [18], as described later. The high value of the slope of the lines indicates that the strength distribution is very narrow. Second, many investigations of light-guide glass compositions, silicates, and the other glass types we discuss have employed the two-point bending technique for the study of the strength of these fibers. This technique has been described by a number of workers and has often been used in the study of light-guide fibers because of its ease and adaptability to many temperatures and environments [19]. However, this measurement method gives failure strain rather than failure stress/strength. It is known that the Young modulus (E) of silica and other silicate glasses are rather substantial functions of strain [20]. While little is known about the details of this nonlinear behavior, it is known that silica is anomalous in this regard. Its Young modulus increases with strain to at least approximately 10% strain, whereupon it decreases in a more normal way. In the evaluation of strength from failure strain at room temperature where the strain is less than 5–6%, the following equation is valid [21]:

$$\sigma = E\varepsilon; \; E = E_0(1 + \alpha\varepsilon), \tag{24.2}$$

where E_0 is the Young modulus in the limit of zero strain (72 GPa) and α corrects for the strain dependence of modulus. The accepted value of α for fused silica is 3 [21] for simple tensile stresses. However, asymmetry in the elastic modulus for tensile and compressive strains leads to a shift in the neutral axis of bent fiber such that the effective value of α that should be used for converting between bending stress and bending strain is 2.125 [22]. Thus, at room temperature where the initial failure strain in air is about 7%, the corresponding E is 83 GPa or about 15% higher for bending. While E at greater strains is not known exactly, it appears that at about 20% strain, the value at 77 K, the modulus has essentially returned to its zero-strain value of 70 GPa or so. This can be seen by dividing the 12.5 GPa strength measured by Proctor et al. [17] by the failure strain of 18% measured by France et al. [23] to give $E_{\varepsilon=18\%}$ approximately 69.4 GPa. Another example is that of a standard window or bottle glass composition, so-called soda-lime–silica glass. Ernsberger [24] has found a strength of approximately 7.7 GPa at 77 K, while Lower [25] has found a failure strain of approximately 17.7%. Thus, $E_{\varepsilon=17.7\%} \sim 43.5$ GPa whereas $E_0 \sim 72$ GPa. Very little is known about the strain dependence of E for other common silicate glasses.

In the preceding examples, we have shown data for strengths when there is no fatigue. In these cases, we have also chosen data for glasses that by our criterion [26] are flaw free. Thus, these strengths are inert, because they are free from atmospheric corrosion effects. Also, because they are flaw free, these strengths are considered intrinsic, that is, they are representative of the material and are not affected by the presence of flaws. A paper by one of the present authors has been concerned with a description and definition of intrinsic inert strength. It is suggested that the following criteria are necessary for intrinsic strength: The strength should be high with an essentially infinite Weibull modulus (i.e., the strength is single-valued), and therefore, there will be no size dependence of this strength (either fiber length or diameter). Results from Duncan et al. [9] show the room temperature and 77 K failure strains for silica, as well as for a borosilicate glass that had early been the subject of much study as a glass for possible light-guide use (Fig. 24.3). As can be seen, these failure strains are high and basically single valued. In addition, these measurements were made at 77 K. This has the effect of eliminating the effect of water in the ambient, so there is no delayed failure/fatigue. These measurements are thus "inert."

24.3.1.1.1 Flaws

As mentioned earlier, if there are no flaws, the strength is intrinsic. In practice there will be flaws of varying types, sizes, severity, distribution in length, and so on, as discussed later in the section on Weibull analysis. Flaws may be the result of scratching, adherence of solid foreign particles, or a change in the geometry of the fiber surface (see later discussion on aging). Reduction in strength may not be the result of a stress enhancement per se, but simply the result of a residual stress, perhaps because of a change in composition of the glass because of an impurity or an indentation too small to produce a crack. In the case of a residual compressive stress, it will simply be added to the applied stress in calculating failure stress.

In most cases, flaws produced as described earlier will initially have a sharp tip and the strength reduction resulting from this is one of the following:

Sharp flaws:

$$\sigma = \frac{K_{IC}}{Yc^{1/2}}, \tag{24.3}$$

where σ is the strength, K_{IC} is the critical stress intensity factor or toughness, and Y is the crack shape parameter (normally taken to be ~ 1.24), and c is the crack length.

Blunt flaws:

In the preceding discussion, we have been concerned with cracks that have sharp tips and can thus be described by Eq. (24.3). In a later section, we deal with flaws that have a rounded tip. In this case, assuming an elliptically shaped flaw:

$$\sigma_a = \frac{\sigma_u}{1 + 2(c/b)}, \tag{24.4}$$

where σ_u is the ultimate strength (i.e., the intrinsic strength of the material), σ_a is the applied stress, and the denominator represents the stress concentration factor at the "crack" tip, where c is the major and b the minor axis of the elliptical flaw. Such macroscopically blunt flaws were studied by Inniss et al. [27] and are described in a later section on aging.

24.3.1.2 Effect of Fiber Diameter

In spite of experimental studies showing no such effect, there is often still the perception that the strength of glass fibers increases as the diameter decreases in fiber diameter. This is primarily due to the classic paper of A. A. Griffith in 1920 [28]. In this work, he showed that the strength of a common soda-lime silicate glass (SLS) was a strong function of the fiber diameter. While this idea persisted for some time and is often still believed today, this effect was shown to be false by Otto in 1955 [29] for a similar glass, E-glass, and perhaps more completely and convincingly by Cameron in 1960 [30]. For these MCGs, it would appear that fibers must be drawn above T_l, its liquidus temperature (the temperature at which the last crystal melts), in order for a high strength single-mode distribution to be obtained. When drawn at lower temperatures, which would normally be the case when drawing thicker fibers, a lower strength mode appears. Thus, the mean or average strength of this fiber diameter will be lower than is expected for that diameter. Cameron showed clearly that the high strength mode was independent of diameter for these E-glass fibers.

While this diameter independence has not been demonstrated unequivocally in silica, the outstanding evidence is such as to be convincing. First, the essential constancy of the strength results on silica over the years, apparently independent of drawing conditions, show closely similar results. The work of Proctor et al. [17] shown in Figs. 24.4 and 24.5 was carried out on fibers of 20- to 80-μm diameter and they indicate no effect of diameter on strength. Most high-quality studies of the strength of silica for light guides have been made on 125-μm diameter fibers and have shown essentially the value found by Kurkjian and Paek [31], namely approximately 5.5 GPa at room temperature, which is very similar to the value found by Proctor et al. at room temperature [17]. There have been few other studies of the tensile strength of silica glass at 77 K. However, Smith and Michalske [32] measured 35-μm silica fibers in tension at UHV ($<6.6 \times 10^{-6}$Pa) at room temperature and found strengths approximately 11–14 GPa. Because it may be assumed that measurements at this high vacuum condition would be essentially inert, as were those of Proctor et al. at 77 °K, this agreement is expected. It is also interesting to note that Griffith himself showed that the strengths of silica rods of 2-mm diameter were about 7 GPa at room temperature. While this was a crude bending measurement, it was firm enough and to Griffith, essentially unexplainable, so that he exempted it from his consideration of the diameter effect.

The interest in nano materials has just produced some interesting results in this regard. The group of Mazur [33] at Harvard has been studying the optical guiding behavior of silica fibers of diameters less than the wavelength of light (see Chapter 11 for more details). They have made initial efforts to measure the strength of these fibers in both tension and two-point bending. While the diameter control is very good (better than 0.1%) due to the clever "self-modulating" technique used to draw these "nanowires," both of these testing techniques are difficult with fibers of this size, so the scatter in the data is quite large (Weibull modulus ~5). While their data seem to show a dependence of strength on diameter, within the statistics, there is no significant dependence and the mean value of the combined results of tension and bending is approximately 5 GPa, not too different for that normally found for 125-μm light-guide fibers. Additional work on these fibers would appear to be warranted. All of these data are in Fig. 24.6, together with results for optical fiber with diameters up to 1 mm [34]. We will consider there to be no effect of fiber diameter on strength.

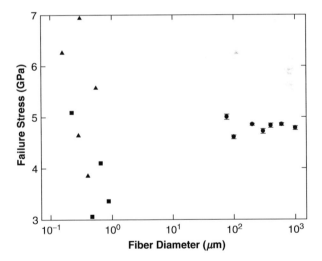

Figure 24.6 Effect of diameter on the strength of silica fibers measured in ambient on the time scale of approximately 10 seconds. Measurements: ● two-point bending, averaged for several measurements [34], ▲ two-point bending single measurements [33], ■ tension single measurements [33].

24.3.1.3 Effect of Ambient Chemistry

Later in this chapter, we discuss in detail the effects of ambient chemistry on the time dependence of strength. Here we simply show how changes in two important environmental parameters affect the short time strength (i.e., the strength recorded in seconds at room temperature). Figure 24.7 [35] shows the effect of ambient relative humidity, while Fig. 24.8 [36] shows the effect of the pH of the solution in which the measurements were made.

In the section on intrinsic strength, we showed the effect of water on the strength of silica fibers. In Fig. 24.7, we show in detail the effect of changes in water concentration or activity (relative humidity) on the strength measured on a time scale of approximately 10 seconds. As shown earlier, during the time of measurement, the strength has actually decreased from its inert value of approximately 14 GPa (or ~20% strain) to a value of the order of 5 GPa. As the relative humidity is increased, the strength decreases continuously, from about 6 to 5 GPa.

The behavior illustrated in Fig. 24.7 deals with the effect of relative humidity. In the case of aqueous solutions, there is also an effect of the pH of the solution and this is shown in Fig. 24.8. In aqueous environments, reaction with hydroxyl ion is rapid compared to reaction with molecular water, explaining the sensitivity to pH.

24.3.1.4 Tension

Light-guide fibers that are manufactured today show very high and very consistent tensile strengths. Figure 24.9 [37] shows a Weibull probability plot

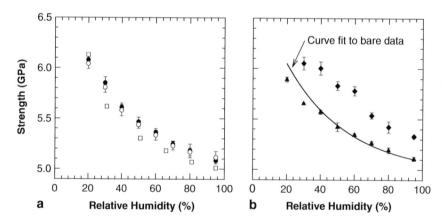

Figure 24.7 Effect of relative humidity on the strength of light-guide fibers. (a) data for ○ bare fibers, ● acrylate coated fibers, and □ silicone coated fiber. (b) data for ▲ polyimide and ◆ silicone coated fibers compared to the bare data. (Data used, with permission, from reference [35].)

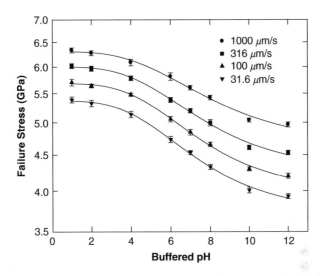

Figure 24.8 Effect of pH on the strength of light-guide fibers (data used, with permission, from reference [36]).

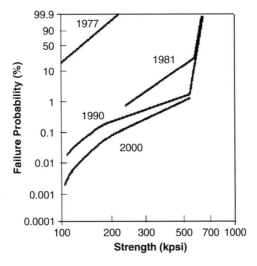

Figure 24.9 Weibull plot of strength of 20-m gage length of silica light-guide fibers showing how improvements have been made over the years [37]. (1 kpsi = 6.9 MPa.)

of the tensile strength of light-guide fibers tested in tension for a 20-m gage length representing the improvements in strength over the years. For simple uniaxial tension,

$$\sigma = \frac{L}{\pi r^2},\tag{24.5}$$

where σ is the tensile strength, L is the applied axial load, and r is the fiber radius. Therefore, in tension, the stress for a given load is higher when the fiber has a smaller diameter. Table 24.2 shows what axial loads are required to generate a range of values of stress for several fiber diameters in the range of practical interest.

In Fig. 24.9, it can be seen that the curve is bimodal. It has been shown that the very narrow high-strength mode is the "intrinsic" strength of the fiber under the test conditions. That is, the lengths of fibers showing this high uniform strength are flaw free. If such breaking is accomplished in the absence of water (say in vacuum or at 77 K), the strength of such flaw-free fibers is approximately 14 GPa. Under normal ambient conditions, the measured strength is time dependent and is approximately 5.5 GPa at room temperature and tested in 10 seconds (Fig. 24.9). There should be no slope to this curve. However, while very high values of the Weibull slope, m (which correspond to a narrow distribution, see later discussion), can be measured in short lengths, many experimental factors contribute to the more moderate value of $m \sim 50$. The lower strength broader distribution reflects a variety of "manufacturing" defects or flaws. These flaws act to reduce the strength of the fiber by acting as stress-concentrators according to Eq. (24.3).

The exact nature of the failure probability–strength curve obviously depends on a number of factors but is represented reasonably by the lowest curve in Fig. 24.9 for good current manufacturing practices. As is seen later, this probability curve depends on the gage length tested.

Table 24.2

Axial loads (for tensile loading) and bending radii (for flexural loading) that result in a given stress for fiber diameters of 125 and 80 μm

Stress (GPa)	Tensile load (N)		Bend radius (mm)	
	125-μm fiber	80-μm fiber	125-μm fiber	80-μm fiber
0.2	2.5	1.0	22.6	14.5
0.5	6.1	2.5	9.1	5.8
1	12.3	5.0	4.6	3.0
2	24.5	10.1	2.4	1.5

The bend radii are calculated using Eq. (24.2) with $E_0 = 72$ GPa and $\alpha = 2.125$ (i.e., fused silica)

24.3.1.4.1 Weibull Statistics

As we have discussed, the strength of brittle materials is controlled by the presence of stress concentrating defects. Such defects are distributed in size/ severity, nature, position, and orientation. As a result, the average measured strength of a set of specimens depends on the specimen size because larger specimens are more likely to contain larger defects and so are expected to be weaker on average. Weibull's weakest link theory for brittle fracture [18] assumes that failure is caused by the most severe defect and provides the following result for how the cumulative probability of failure by an applied stress σ, $F(\sigma)$, is related to σ and the length of the fiber, L,

$$F(\sigma) = 1 - \exp[-Lf(\sigma)] \tag{24.6}$$

or

$$\ln \ln \frac{1}{1-F} = \ln L + \ln f(\sigma) \tag{24.7}$$

where $f(\sigma)$ is a monotonically increasing function of σ. $f(\sigma)$ is not known *a priori*, but an empirical power law is ubiquitously assumed, which corresponds to the Weibull distribution:

$$f(\sigma) = (\sigma/\sigma_0)^m, \tag{24.8}$$

where m is the Weibull shape parameter (an inverse measure of distribution width) and σ_0 is the Weibull scale parameter, which is a measure of centrality of the strength distribution. This results in the well-known Weibull distribution. Equation (24.7) then becomes

$$\ln \ln \left(\frac{1}{1-F} \right) = \ln L + m \ln \sigma - m \ln \sigma_0. \tag{24.9}$$

Considering the failure of specimens with two lengths, L_1 and L_2, the applied stresses that give the same probability of failure, σ_1 and σ_2, are related by

$$\ln \frac{\sigma_1}{\sigma_2} = \frac{1}{m} \ln \frac{L_2}{L_1}. \tag{24.10}$$

Equation (24.10) is valid for any value of F, so it is valid for the mean strength, median strength, or indeed any other percentile. Therefore, as expected, longer specimens are weaker than short ones.

This formalism shows how the strength depends on specimen length. However, it results in σ_0 having the strange units of [stress] × [length]$^{1/m}$. σ_0 can be given the expected units of stress in several ways, but the simplest is to drop the length dependence in Eq. (24.9) to give

$$\ln \ln \left(\frac{1}{1-F} \right) = m \ln \sigma - m \ln \sigma_0. \tag{24.11}$$

For a given set of specimens, the length dependence of strength is ignored and the strength data are interpreted in terms of Eq. (24.11); when other specimen lengths are considered, Eq. (24.10) is used. Equation (24.11) may be visualized using a "Weibull plot," which is a graph of lnln $[1/(1-F)]$ versus lnσ, which, if the strength data are described by a Weibull distribution, is a straight line of slope m and intercept on the horizontal lnσ axis of lnσ_0. σ_0 is the 63rd percentile.

If the strengths are measured of specimens of length L_1, then the expected strengths for specimens of length L_2 can be calculated using Eq. (24.10). Graphically, Eq. (24.10) represents a sideways shift on the Weibull plot by an amount $\ln(L_2/L_1)/m$ or equivalently a vertical shift along the probability axis of $\ln(L_2/L_1)$. An example of a Weibull plot is shown in Fig. 24.10 [38]. The right-hand axis is linear in $\ln\ln[1/(1 - F)]$, while the left-hand axis is the actual failure probability and is nonlinear. The shift in the strengths with test length in Fig. 24.10 is illustrated in Fig. 24.11 where the mean strength is shown schematically as a function of gage/test length [38].

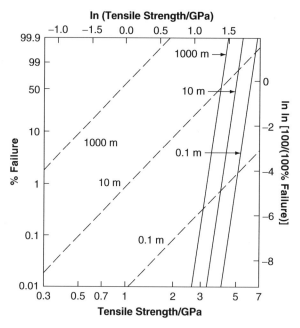

Figure 24.10 Schematic of Weibull plots for different fiber test lengths: Dashed lines are for a broad low strength distribution and solid lines are for a narrow high strength distribution. (Modified, with permission, from reference [38].)

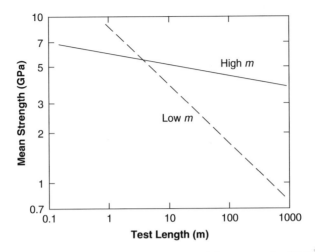

Figure 24.11 Schematic of mean strength versus test length for a narrow low Weibull modulus, *m* (solid line), and a broad high Weibull modulus (dashed line) strength distributions. (Modified, with permission, from reference [38].)

24.3.1.5 Bending

Light-guide fibers are often deployed in a bent configuration. The fundamental equation for the bending of a fiber into a circular loop is the following:

$$\varepsilon = \frac{r}{R}, \tag{24.12}$$

where ε is the maximum tensile strain that is on the outside of the bend, r is the radius of the fiber, and R is the bend radius or the radius of curvature. Therefore, for the same maximum strain (or stress), the minimum permitted bend radius is proportional to the fiber diameter. As a result, for fiber of a given strength subjected to primarily bending stresses, employing thinner fiber permits bending to tighter radii. In contrast, for fiber subjected primarily to axial tension, thicker fiber can withstand higher axial loads (Eq. [24.5]).

As already mentioned, testing of optical fibers is often carried out in two-point bending. This is because of the simplicity of the testing and because the very short length tested avoids manufacturing defects. The application of a bending stress will result in failure when the tensile stress on the outside of the bend reaches the failure stress. While Eq. (24.12) is appropriate for a circular loop of fiber, the strain in two-point bending is approximately 20% higher. Because the fiber will fail when the stress and not the strain reaches a critical value, the stress must be calculated from the bending strain. This is done by multiplying the failure strain by Young's modulus at this strain. The strain dependence of Young's modulus must be accounted for as shown in Eq. (24.2).

Table 24.2 compares the axial loads for tensile loading and bend radii for bending that result in a range of stresses and for two fiber diameters in the range of practical interest.

24.3.2 Fatigue

We have noted that many glasses are weaker when measured in ambient conditions compared to measurements in "inert" environments, such as liquid nitrogen (77 K). The phenomenon causing this effect is termed *fatigue*. Most inorganic glasses show a time-dependent strength or delayed failure. This effect is apparent either on testing dynamically, when the measured strength is lower for slower stressing rates, or statically by applying a fixed load (stress), which eventually leads to failure, with the time to failure decreasing with increasing stress. At least in oxide glasses, the mechanism is the combined effect of stress and the hydrolysis of strained siloxane bonds at the "crack" tip by ambient moisture:

$$\equiv Si - O - Si \equiv + H_2O \rightarrow \ \equiv Si - O - H + H - O - Si \equiv . \qquad (24.13)$$

In nonoxide glasses, the mechanism is often not completely determined. Figure 24.12 [39] shows results of fatigue measurements over a rather extended time range. This is the result of the combination of a number of studies using various techniques. The combined results form a single curve.

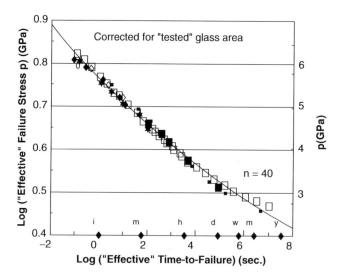

Figure 24.12 Compilation of fatigue data for silica light-guide fibers (used, with permission, from reference [39]).

A rough rule of thumb is that the long time strength is approximately one-third of the short time strength. A more complete analysis of a given fiber/device is, of course, ultimately required.

As indicated earlier, it is believed that the very strong short sections of light-guide fiber are flaw free. In spite of this, the generally applied model for fiber fatigue involves the slow propagation of cracks in the sample:

$$V = AK_I^n, \tag{24.14}$$

where V is the crack growth velocity, K_I is the stress intensity factor, and A and n are empirical fit parameters. This equation appears to be appropriate whether the high or low strength behavior is considered. n is the stress corrosion suscep-tibility parameter. A higher value provides greater fiber reliability. It has been shown that because of the very severe dependence of the velocity on stress, it is difficult to tell whether the empirical power law or a physically more meaningful exponential form are more appropriate and thus the simpler power law is commonly used, primarily because of its mathematical convenience. The power law does, however, provide more optimistic predictions of reliability than expo-nential models.

Service conditions are usually assumed to have constant applied stress and under such "static fatigue" conditions the time to failure, t_f, is given by

$$t_f = \frac{2}{(n-2)AY^2}\left(\frac{\sigma_i}{K_{IC}}\right)^{n-2}\frac{1}{\sigma^n} = B\frac{\sigma_i^{n-2}}{\sigma^n}, \tag{24.15}$$

where

$$B = \frac{2}{AY^2(n-2)}\frac{1}{K_{IC}^{n-2}}, \tag{24.16}$$

and where σ is the service stress and σ_i is the inert strength (i.e., the strength that would be measured when no fatigue occurs). This is effectively the strength of the fiber at 77 K. This is rarely known, but for silica it is approximately twice the strength measured at room temperature in times of the order of seconds (i.e., the normally measured strength). Equation (24.15) shows how the failure times for two different applied stresses are related:

$$(t_f\sigma^n)_1 = (t_f\sigma^n)_2. \tag{24.17}$$

For both high and low strength silica fibers, it is assumed with reasonable justification, that $n \approx 20$.

For strength measurement using a constant stress rate, $\dot{\sigma}$, the failure stress, σ_f, is given by

$$\sigma_{f,\dot{\sigma}}^{n+1} = \frac{2(n+1)\dot{\sigma}}{AY^2(n-2)}\left(\frac{\sigma_i}{K_{IC}}\right)^{n-2} = (n+1)B\dot{\sigma}\sigma_i^{n-2}. \tag{24.18}$$

Laboratory testing that has been accelerated by using aggressive environments (high temperature, high water activity) has shown an additional strength reduction mechanism in addition to fatigue. This "aging" phenomenon must be considered because of its potential negative impact on reliability.

24.3.3 Aging

In the study of the strength and reliability of silica light guides, testing is often carried out under extreme conditions to accelerate the degradation and, thus, hopefully allow prediction out to longer times/lifetimes. When this was done early in the study of silica light-guide fibers, an effect that had not really been known earlier was discovered in both aging and fatigue [40, 41]. If fiber is held for some time under no stress and subsequently strength tested, it is found that depending on the condition of holding, or aging, the strength may have degraded. This is now known as *zero stress aging,* or more simply *aging.* Figure 24.13 shows results of a rather long duration aging test of a polymer-coated silica fiber aged in distilled water at 25 °C and a shorter time test in 100 °C water [42]. As can be seen, for some prolonged period, no strength degradation occurs, but eventually rather precipitous strength degradation occurs. This effect has been shown to be present in fatigue as well and the time scale depends on the state of the water [43] (Fig. 24.14). The important fact to be learned from Fig. 24.13 [42, 44] is that the activation energy for the process responsible for this strength degradation is approximately 80 kJ/mol, the activation energy for the reaction of silica with water.

The model that has been used to describe this aging effect involves surface roughening. As we have discussed, glasses are essentially supercooled liquid glasses and, as such, contain both chemical and structural fluctuations. These fluctuations have different thermochemical properties, including water solubility and dissolution rate. These differences lead to roughening of the glass surface as shown in the atomic force microscopy (AFM) images in Fig. 24.15 when the surface is exposed to an aggressive environment [45]. This effect was studied in greater detail using a model system by Inniss et al. [27].

A more pronounced effect was shown in the borosilicates studied by France et al. [6] and is shown in Fig. 24.16. In this case, not only are these glasses less durable, but they show composition fluctuations that will contribute to the unequal solution rate effects. Such fluctuations have been shown in optical scattering studies.

The preceding discussion relates to the aging behavior of the pristine fiber strength, rather than the behavior of the occasional defects. Evidence suggests that aging weak defects in aggressive environments can lead to strength recovery

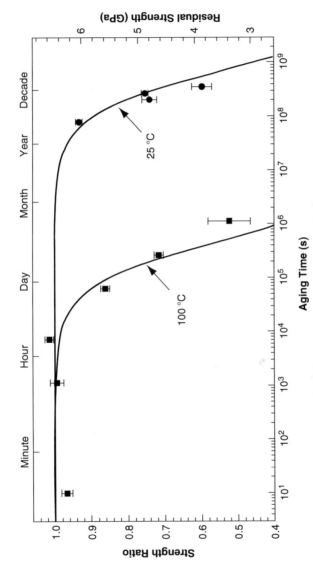

Figure 24.13 Effect of temperature on the aging of silica light-guide fibers (data used, with permission, from reference [42]). The ratio of the strength after aging to the strength before aging is shown as a function of the aging time in both 25 °C (■) and 100 °C (●) water.

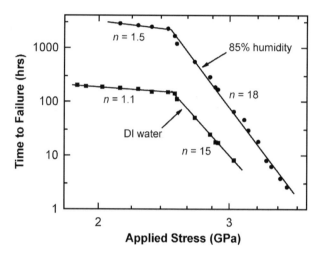

Figure 24.14 Effect of liquid water and water vapor on the static fatigue of silica light-guide fibers (data used, with permission, from reference [43]).

in much the same way as etching in hydrofluoric acid (a discussion of which is presented later). However, strength loss of the strong sections of fiber might be more important because the fiber might become too weak to handle along its entire length.

24.3.4 Nonsilicate Glasses

The measurement of the strength, fatigue, and aging of nonsilicate glasses has not been very extensive. The measurements have often been made in two-point bending and, therefore, usually not for long lengths of fiber. While this is valuable for the assessment of the measured strength relative to the theoretical strength, these are not necessarily practical or useful strengths. In this section, we discuss all of these properties in nonsilicate glasses.

24.3.4.1 Heavy-Metal Fluoride Glasses

The mechanical properties of HMF glasses are quite sensitive to the presence of liquid water, but not as sensitive to water vapor. The highest strengths for HMF glass fibers whose surface had been etched are approximately 1.4 GPa [46]; because the measurement was made in bending, the calculation of the failure stress depends on knowledge of the Young modulus. Carter [47] found

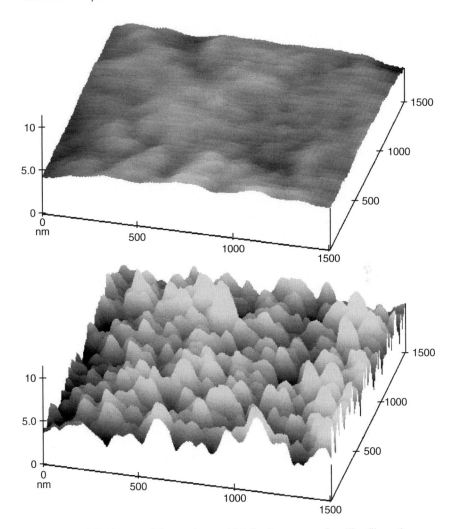

Figure 24.15 AFM images of the roughness which develops upon aging silica fibers–the upper image of the unaged surface while the lower image is from a fiber that had been aged for 168 hours in 90 °C pH 7 buffer solution (used, with permission, from reference [45]).

similar values and attributed the less than theoretical strength to the presence of bubbles and crystals. Her results (Fig. 24.17) illustrate the differences that may be seen in testing different fiber lengths.

Perhaps the most detailed study on fatigue and aging in HMF glasses is that of Colaizzi and Matthewson [48] who characterized glasses whose compositions included AlF_3 (Fig. 24.18).

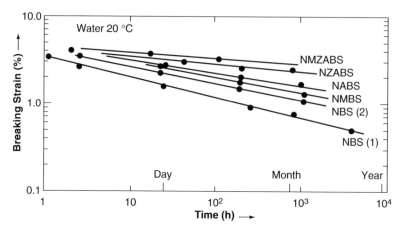

Figure 24.16 Aging of fibers of alkali borosilicate glasses of various compositions (used, with permission, from reference [6]).

24.3.4.1.1 Chalcogenide Glasses

Studies of the two-point bend strength of arsenic sulfide glass fibers as seen in Table 24.1 show the apparent approach to that theoretically expected strength, namely 1.5–2.5 GPa. Devyatkh et al. [49] published a value for strength under ambient conditions of approximately 1.2 GPa. Sanghera [50] (Fig. 24.19), measured about a half of that value (0.570 GPa) at room temperature, but a value of 0.854 GPa at 77 K (i.e., ~1.5 times greater than the room temperature strength). Applying this ratio to the data of Devyatkh et al. [49] would give an "inert" strength of approximately 2.4 GPa or $E/\sigma \sim 6$ for their fibers. In fact these data are presented as strength, although the measured values are actually failure strain. In the case of Devyatkh et al. [49], estimated strengths at 77 K amount to a failure strain of approximately 16%, very nearly that found for silica.

While the situation with respect to the fatigue and aging of silica and silica-based glasses is clearly due to the presence of water, the situation for chalcogenide glasses is not as clear; there is no clear-cut experimental evidence for the critical chemistry. Dianov et al. [51] give values of $n \sim 10$–20, but the scatter in the data (Weibull modulus, $m \sim 10$) leaves this in doubt. The data of Sanghera (Fig. 24.19) are interesting in this regard. Here, we see that the ratio of strengths at 77 K and room temperature is approximately 1.5, while the ratio of room temperature to metal-coated strength is approximately 5. One possibility we may suggest here is that the fatigue is due to the interaction of the glass with oxygen rather than water.

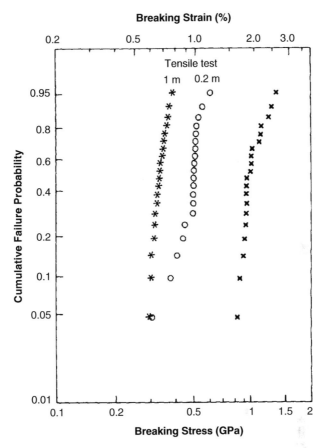

Figure 24.17 Weibull plot of the strength ZBLAN fibers measured in bending and in tension with test lengths of 0.2 and 1 m (used, with permission, from reference [47]).

24.3.5 Photonic Crystal or "Holey Fibers"

The only set of experimental results on photonic crystal (PC) fibers is that of Zhou et al. [52]. These workers have found that the two PC fibers they have studied, with respectively 90 and 6 holes, show breaking loads effectively 10 and 20% higher than that found for standard single mode fused silica clad fibers. When account is taken of the differences in cross-sectional area (3% and 0.4%, respectively), the strengths are 4.45, 5.0, and 5.4 GPa, respectively, for the standard single-mode fiber, 90-hole fiber, and 6-hole fibers. However, the experiments were not well controlled in that the three fibers had different

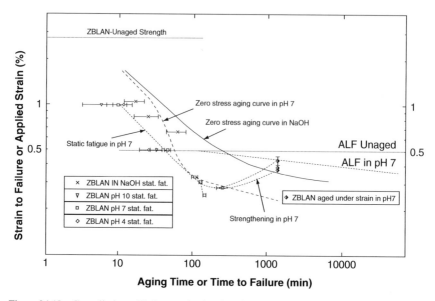

Figure 24.18 Compilation of fatigue and aging data for ZBLAN and aluminum fluoride-doped ZBLAN ("ALF") fibers (used, with permission, from reference [48]).

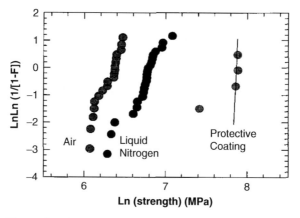

Figure 24.19 Effect of environment on the bending strength of As_2S_3 fibers (used, with permission, from reference [50]).

polymer coatings, so the difference in strengths might not be significant. Further, because the flaws causing failure are substantially smaller than the holes or the distance between the holes, the presence of the holes is not expected to have any effect on strength.

24.4 COATINGS

24.4.1 General Comments and Polymer Coatings

Very early in the development of the light-guide technology, it was clear that the pristine surface of the as-drawn fibers would have to be protected from mechanical damage by applying a thin polymer coating to isolate the fiber surface from any hard sharp particles in the environment. As their development progressed, dual coatings were developed; the inner compliant coating reduced the microbending losses, while the harder more abrasion resistant outer secondary coating made the assembly more robust. Contrary to much early literature, polymer coatings were found to be quite permeable to water [53]. It was found that penetration through typical coatings, which are approximately 60 μm thick is accomplished in times of the order of 1 hour or less. Therefore, one must still expect strength degradation from both fatigue and aging. The details of the coating chemistry are apparently of importance to the aging behavior of the fiber and perhaps less so to the fatigue behavior, but no detailed explanation has yet been given.

24.4.2 Metal Coatings

As indicated earlier, the polymer coatings that have been applied to light-guide fibers have been shown to be quite permeable to water, so once steady state has been attained, the fatigue and aging behavior of coated and bare fibers are similar. It is natural, therefore, that a fair amount of early work was done on coatings that were hermetic (i.e., those that would isolate the fiber surface from water). Figure 24.20 [54] shows that the use of a metal coating can result in a strength that is similar to that obtained at 77 K, that is, the inert strength. The results in Fig. 24.20 are given as failure strength and should probably be corrected because they were taken in two-point bending and Eq. (24.2) was used to calculate failure stress with $\alpha = 3$ instead of the value of $\alpha = 2.125$ appropriate for bending. However, Eq. (24.2) has only been verified for strains up to approximately 6% and is incorrect for the failure strains seen in Fig. 24.20.

24.4.3 Inorganic Coatings

Many inorganic coatings have been tried, but the best has been found to be carbon, even though the behavior is complicated and variable. Figure 24.21 [55] illustrates the effect of draw speed on the strength and fatigue parameter, *n*. This

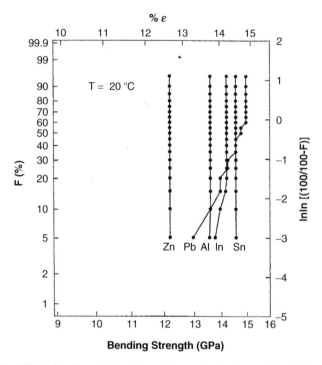

Figure 24.20 Effect of melt-applied metal coating on the bend strength of silica light-guide fibers (used, with permission, from reference [54]).

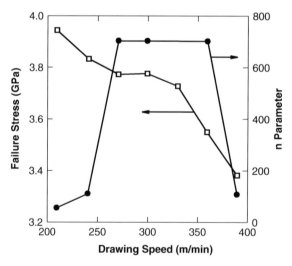

Figure 24.21 Effect of fiber draw speed on the resultant fiber strength and fatigue parameter, *n*, for carbon coated silica (data used, with permission, from reference [55]).

is an effect connected primarily to the thickness of the coating. However, it is known that the coating "composition," thickness, and roughness influence the strength, fatigue (water permeability), and hydrogen permeability [56].

24.5 HANDLING AND POST-DRAW PROCESSING

In our earlier discussions, we considered short time fiber strength, long time fiber strength, fatigue and aging, all things that must be considered in the design of the system under consideration if some reasonable assumptions about the strength of the starting fiber are made. On the other hand, some type of fiber handling and processing are very often required in the fabrication and assembly of any given device. In this section, we describe some specific operations that may introduce unpredictable damage. By keeping in mind certain factors that can affect the strength of the fiber as a result of such processing, it should be possible to minimize both these strength reductions and reduce unexpected reliability consequences.

24.5.1 Fiber Stripping

All-silica light-guide fibers have a coating (usually a polymer) applied in-line during drawing to protect the glass surface from mechanical damage. Use of fibers in practice very often necessitates the removal of this coating. Three general techniques can be used to accomplish this: mechanical, chemical, or thermal.

While stripping in the field most commonly uses a tool similar to diagonal cutting wire strippers where the stripping hole is somewhat larger than the fiber diameter, this will inevitably lead to a strength reduction due to the sort of mechanical damage illustrated in Fig. 24.22 [57]. While the strength reduction can be considerable in some cases, in general, there are still substantial lengths that have been left undamaged. Knowing this, it is possible to employ the very simple and quick mechanical stripping technique in conjunction with a proof-testing procedure to guarantee a desired strength.

Other techniques have been employed that remove the coating with less surface damage and, therefore, with less strength reduction. The technique that has been found to produce the cleanest surface with no damage due to the process itself is chemical stripping using hot ($\sim$200 °C) concentrated sulfuric acid [58]. This process effectively removes most polymer coatings in a few seconds, producing a "clean" surface. It has been shown that by careful handling and stripping in this way, no strength reduction is observed [59]. In certain cases, soaking in acetone or methylene chloride may be used. These chemicals

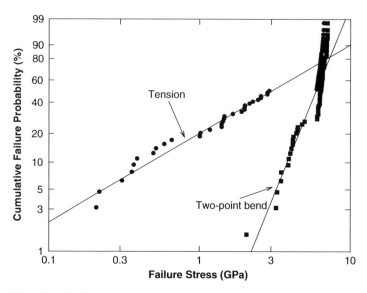

Figure 24.22 Weibull plot of the strength of mechanically stripped fiber measured in tension (●) and bending (■) (data used, with permission, from reference [57]).

cause swelling of the coating, which may then be stripped off with no strength reduction. There are also a number of thermal strippers available on the market, which can produce quite acceptable stripped fiber strengths.

Park et al. [60] described an interesting and quite different approach. Their stripping technique makes use of the fact that the primary and secondary polymer coatings on dual-coated fiber have distinctly different properties. By rapid local heating, the compliant primary coating will rapidly expand and effectively "explode" the more brittle primary coating with a resultant rather small decrease in fiber strength.

24.5.2 Fiber Cleaving

As discussed earlier, in the fracture of inorganic glasses, we are dealing with failure in a brittle manner. This results in a fracture surface that is generated by the propagation of a crack perpendicular to the axis of the applied stress. If the applied stress is appropriate and the crack is of the right size and orientation, the crack will propagate such that a smooth "mirror" fracture surface is generated perpendicular to the fiber axis. The details of this process are given in the section on fractography. Here, we simply illustrate the type of fracture surfaces that can be developed. Figure 24.23 [61] shows micrographs of fracture surfaces. In each case, the fracture was generated by a scratch at the bottom surface (i.e., the

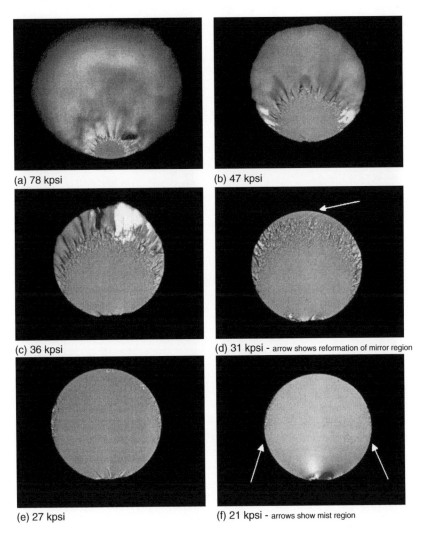

(a) 78 kpsi

(b) 47 kpsi

(c) 36 kpsi

(d) 31 kpsi - arrow shows reformation of mirror region

(e) 27 kpsi

(f) 21 kpsi - arrows show mist region

Figure 24.23 Fracture surfaces of fibers that broke at a range of failure stresses (a) 0.538 GPa, (b) 324 MPa, (c) 248 MPa, (d) 214 MPa, (e) 186 MPa, and (f) 145 MPa (used, with permission, from reference [61]).

6 o'clock position). Figure 24.23a–f shows the fracture surface for increasingly large scratches, and consequently the fracture stress, as indicted by each micrograph, becomes less. It is seen that the smooth region/mirror becomes larger as the strength decreases. The objective of a good cleave is to have this mirror be the size of the entire fiber so that good mating is achieved. The details of

fractography are described later. Care must be taken that a simple cut is made and that there is not extensive damage to the fiber surface.

24.5.3 Splicing

The joining of two sections of optical fiber can be done either mechanically or by fusion splicing. In mechanical splicing, the ends of the two fibers are joined by gluing them inside a sleeve or ferrule. While this technique is simple and rapid, it is bulky and not particularly robust. For high-quality, high reliability work, especially for undersea use, fusion splicing is often employed. Figure 24.24 illustrates some very early splicing work [62]. This shows the importance of avoiding damaging the fiber during stripping or subsequent handling. In the case of both splicing and tapering, some portion of the fiber is re-heated. In the center region of either a splice or a taper, the fiber is completely melted and, thus, when cooled will be expected to have original perfect fiber strength. At intermediate positions along the fiber, the maximum temperature will be intermediate between the melting temperature and room temperature. In some

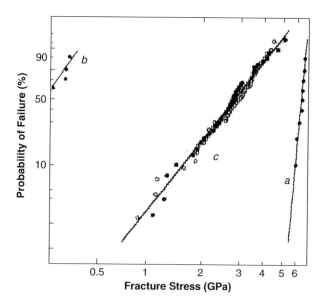

Figure 24.24 Early strength results for fusion splices: (a) original fiber strength, (b) coating mechanically stripped, and (c) coating chemically stripped (used, with permission, from reference [62]).

of these regions, the strength degradation due to the interaction of the fiber surface with water in the atmosphere will be rapid enough to cause strength degradation during the processing time. Figure 24.25 shows later work by Krause [63]. Here, it is shown that by careful stripping and by adhering to appropriately "clean" conditions, splices showing very little strength degradation can be obtained, while if in addition to "clean" conditions, the conditions are also "dry" (accomplished by the use of a H_2/Cl_2 torch), essentially no strength degradation is suffered. While this is often not feasible in practice, strengths approaching those obtained under these ideal condition can be achieved.

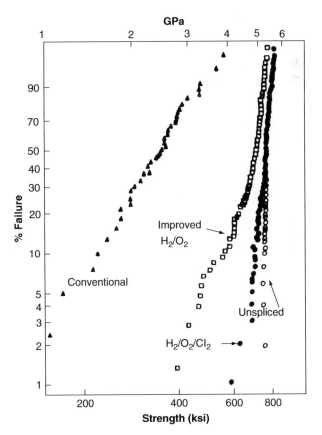

Figure 24.25 Weibull plot of the strengths of fusion splices made using various methods (1 ksi = 6.9 MPa) (used, with permission, from reference [63]).

It should be kept in mind that the temperature gradient, which necessarily accompanies any such heating of a portion of the fiber, such as in tapering or thermal lensing, may lead to substantial strength degradation by the interaction of the hot surface with moisture described earlier.

24.5.4 Polishing

Grinding and polishing of the ends of fibers involves many complicated and poorly understood processes. In the grinding process, sharp hard particles scratch the glass surface resulting in gouges and cracks. Some of these cracks cause chipping of the surface and thus removal of some of the glass. Grinding is used to rapidly approach the final desired size and shape. The final shape and surface quality are produced by polishing. Polishing is similar to grinding except that much finer abrasive particles are used. In each of these processes, a series of decreasing abrasive particle sizes is employed so that the damage created in one step is removed in the next step. It is important that the abrasive used in one step be removed before the next step. Polishing has been shown to be a combination of mechanical and chemical processes, so the appropriate choice of both polishing material and size is important.

24.5.5 Soldering/Pigtails

Fibers often need to be soldered into packages, many times to achieve hermeticity or to precisely maintain dimensional stability. It would seem that the use of the very strong melt-coated fibers described earlier would be the first choice. While this is sometimes the case, in most cases, more or less standard metallization processing is used using—for instance—reactive metals such as Ti or Cr, which are evaporated or sputtered onto the glass surface and covered with a solderable and oxidation-resistant final layer. While this process provides good adherence of metal to glass, the resulting assembly is usually rather weak, presumably due to the strength reduction accompanying the sputtering of the Ti or Cr layer. Little detailed work has been published regarding the mechanism of strength reduction is such systems; it would seem reasonable to attribute the weakening to the reaction of the sputtered metal with the perfect glass surface. In the case of a silica fiber coated with aluminum by the melt process, high fiber strength, good adherence of metal to glass, and good joint pull strength have been achieved (Fig. 24.26) [64]. While not completely understood, although there is apparently interaction between the silica surface and the aluminum coating, the strength is recovered upon chemically stripping the aluminum [65].

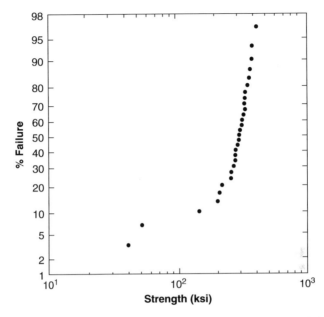

Figure 24.26 Strength of solder joints to aluminum-coated fiber (1 ksi = 6.9 MPa) (used, with permission, from reference [64]).

24.5.6 Recovery of Handling Damage: Etching

The use of hydrofluoric acid (HF) for the etching of glass surfaces has been practiced commercially for many years. Proctor et al. [66] found that etching of damaged commercial soda-lime–silica glass rods resulted in substantial recovery of strength. In the case of silica rods, the recovery was not nearly as great; the maximum room temperature strength recorded was approximately 2.5 GPa. Vitman et al. [67] partially recovered strength of an abraded fiber. Miyama et al. [68] etched silica light-guide splices and were able to increase the mean fiber strength from 0.8 GPa to approximately 2 GPa. Kurkjian et al. [69] showed that etching of high-strength silica light-guide fibers resulted in a decrease in normal room temperature two-point bending strength from 5.5 GPa to a steady-state plateau value of approximately 3.5 GPa (Fig. 24.27). It was suggested that the smooth perfect fiber becomes roughened by the acid etch and that the roughness acts as stress concentrators as described earlier in the section on aging. Thus, although additional work is needed, we suggest that by appropriate etching of fibers damaged by handling abrasion or other mechanisms, strength increases to at least 3.5 GPa should be possible. Of particular practical importance, Kurkjian et al. [69] found that the weakening of the perfect fiber was accomplished both

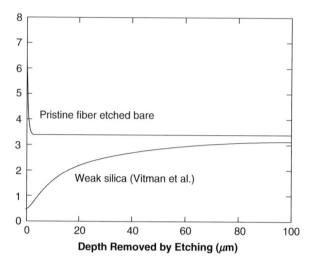

Figure 24.27 Strength increase in weak fiber and strength decrease in flawless fiber on exposure to HF.

with bare fibers and with coated fibers. Thus, it would be expected that strengthening could also be accomplished by etching through the coating, which has been found [70].

24.6 FRACTOGRAPHY

A very useful practical tool is fractography. This is the microscopic study of fracture surfaces. Such studies can give valuable information about the stress level at which a fiber failed, the distribution of failure stress (tension/bending/torsion), and can give an indication of the reason for failure. Figure 24.23 illustrates the features that may be seen on a typical fracture surface. At the early stages of propagation, a crack will grow slowly and during this time will produce fracture surface that is quite smooth; this region is the "mirror." The crack continuously increases in velocity, gradually approaching its limiting velocity (approximately one-third the longitudinal sound velocity) at which point the excess energy begins to be taken up by the creation of extra fracture surface; the surface roughens (the "mist" region) and at higher speeds produces steps in the surface (the "hackle" region).

It has been empirically found that the radius of the mirror region, r_m, is directly related to the stress at failure, σ, by

$$\sigma = \frac{A}{r_m^{1/2}}, \tag{24.19}$$

where A, the mirror constant, is a parameter found empirically. It has been found that for all glasses considered here, the ratio of the mirror radius, r_m and the original flaw size, c, is approximately 0.1. From Eq. (24.19), it can be seen that A has the units of stress intensity and is essentially proportional to K_{IC}. If $c \sim r_m/10$, then, $K_{IC} \sim 0.4A$.

24.7 PROOF-TESTING AND RELIABILITY

As we have discussed earlier, the majority of the length of fiber has an extremely high strength, but occasional weak defects are encountered, which are introduced during manufacturer. These weak defects control the practical reliability of fiber that is of useful length (i.e., more than a few millimeters). Review of Eq. (24.9) for the Weibull strength distribution shows that even for very low applied stress, σ, there is a small but finite probability of failure, F. Mechanical reliability can only be completely ensured if the applied stress is actually zero, which cannot be achieved in practice, especially for applications in which the fiber is bent. The situation is greatly improved by "proof-testing"— which involves applying a short pulse of stress to the entire fiber length—the weakest defects will fail under the proof stress and so are eliminated. If the fiber breaks, it can be discarded or spliced (followed by proof-testing of the splice, of course). The resulting lengths of fiber will have a certain minimum assured strength after proof testing. Choice of the proof stress is a compromise: A high proof stress gives fiber with a high ensured strength but increases the frequency of proof failures and hence decreases yield. Lowering the proof stress improves yield but reduces the minimum strength. For silica fiber in telecommunications applications, the proof stress is typically 700 MPa (100 kpsi).

Proof-testing of long lengths of fiber is normally performed using a continuous tester in which the fiber is fed around a pair of pulleys that apply a given tensile load to the section of fiber between them. While useful for long lengths of fiber, this technique is not suitable for short lengths. In that case, the fiber is usually proof-tested by passing it through a series of rollers that bend the fiber in several directions so the entire surface of the fiber experiences a stress close to the maximum bending stress.

While the idea of proof-testing is simple, in practice it is complicated by fatigue. Firstly, proof-testing is conducted in ambient environment, so the surviving fiber has lost some strength during proof cycle. This is a minor effect, and because the weakest fiber is removed, the average strength of the surviving fiber is increased by proof-testing. Secondly, proof-testing ensures a minimum strength, which is the strength immediately after proofing (i.e., the minimum inert strength, σ_i, is the proof stress, σ_P). As a result, if the strength of lengths of proof-tested fiber is measured, fatigue during the measurement means that some of the strengths are actually lower than the proof stress, although this occurrence is rare. Thirdly, the proof stress cannot be removed instantaneously so that fatigue occurs during the unloading phase. As a result, the minimum ensured strength is a little lower than σ_P. Crack growth during unloading is minor and is suppressed by unloading from the proof stress as quickly as possible. These three effects have been studied in detail and reliability models have been developed.

24.7.1 Minimum Strength Design

The most common and conservative approach to reliability modeling is known as the "minimum strength design." Here, the maximum applied stress is limited to some fraction of the minimum fiber strength, usually taken to be the proof stress. With this approach, one assumes that the length of fiber under stress is no weaker than the proof stress. This is appropriate in applications in which many kilometers of fiber are stressed or where the fiber has been weakened by handling events where the protective coating has been removed, for example, fusion splicing. For silica-clad optical fiber with a value of the stress corrosion parameter, $n \sim 20$, the allowable applied stress is shown in Fig. 24.28 for three common stress events. One is allowed to stress fiber to approximately one-fifth the proof stress for long-term events, one-third the proof stress for installation events lasting hours, and one-half the proof stress for short-term processing stresses. These guidelines are shown in terms of bend radius in Fig. 24.29. Assuming the typical proof stress of 700 MPa, the allowable bend radius for, say, fiber stranded about a central member in a cable is 32 mm. The maximum allowable tensile stress during installation of an aerial cable is 230 MPa for the same fiber. Pulleys used for guiding fiber during processing should be no greater than 25-mm diameter.

24.7.2 Failure Probability Design

In splice enclosures and many photonic device applications, relatively short lengths of fiber are permanently coiled and stored. As shown earlier, the probability of encountering a flaw in a short length of today's fiber is low. Thus, a failure probability design methodology is appropriate in these situations. A general rule of thumb is that one can consider a failure probability design for deployed lengths of less than a kilometer.

Failure probability is incorporated into Eq. (24.15) through the inert strength, σ_i,

$$t_f = B(F\{\sigma_i\})^{n-2}\,\sigma_a^{-n}. \tag{24.20}$$

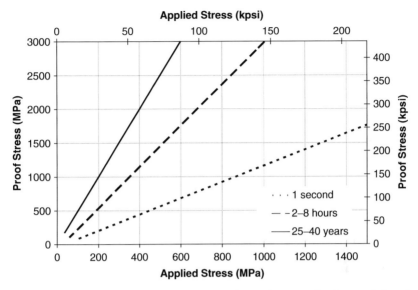

Figure 24.28 Allowable stress as a function of proof stress for processing (1 second), installation (hours), and in-service (years) stress events.

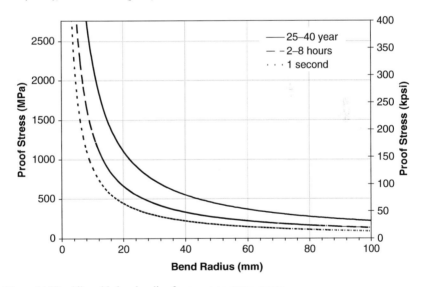

Figure 24.29 Allowable bend radius for common stress events.

The strength distributions, such as those shown in Fig. 24.9, can be directly substituted into Eq. (24.20), provided one accounts for the fatigue that naturally occurs when testing in air [71]. When such modeling was being developed, it was recognized that one should account for fatigue during processing events like proof-testing and in-service life. The Telecommunications Industry Association has captured these developments into a reliability guideline based on the power law crack velocity model [72]. One of the most notable contributions is that of Mitsunaga et al. [73] where field failure probability is estimated from knowledge of the proof-testing conditions and the proof test break rate, N_P,

$$F = 1 - \exp\left(-N_P L \frac{m}{n-2} \frac{\sigma_a^n t_s}{\sigma_P^n t_P} \right), \tag{24.21}$$

where L is the length of fiber, σ_a is the applied stress in service which is applied for a service life t_s, σ_P is the proof stress which is applied for a time t_P. The Weibull modulus, m, is that of the pre–proof-test strength distribution. This model assumes that the distribution of flaws surviving proof-testing can be estimated from those failing proof-testing.

ACKNOWLEDGMENTS

We thank G. Scott Glaesemann of Corning, Inc., Corning, New York, for helpful comments and especially for contributions to the section on proof-testing and reliability.

REFERENCES

[1] Varshneya, A. K. 1993. *Fundamentals of Inorganic Glasses*. Academic Press, New York.
[2] Baikova, L. G. et al. 1967. Nature of the damage of high-strength glass and possibilities of protecting glass. *Sov. Phys. Sol. State* 9(2).
[3] Semjonov, S. L. and C. R. Kurkjian. 2001. Strength of silica optical fibers with cube corner indents. *J. Non-Cryst. Solids* 283/1-3:220–224.
[4] Pharr, G. M. et al. 1993. Measurement of fracture toughness in thin films and small volumes using nanoindentation techniques. In: *Mechanical Properties and Deformation Behavior of Materials Having Ultra-fine Microstructure* (M. Natasi, ed.), pp. 449–461. Springer, Berlin, Germany.
[5] Koizumi, K. et al. 1974. New light focusing fibers made by a continuous process. *Appl. Optics* 13:255–260.
[6] France, P. W. et al. 1983. Strength and fatigue of multicomponent optical glass fibres. *J. Mat. Sci.* 18:785–792.
[7] French, W. G. et al. 1975. Glass fibers for optical communications. *Annu. Rev. Mat. Sci.* 5:373–394.
[8] Lines, M. E. 1994. Can the minimum attenuation of fused silica be significantly reduced by small compositional variation? 1. Alkali metal dopants. *J. Non-Cryst. Solids* 171:209–218.
[9] Duncan, W. J. et al. 1984. The effect of environment on the strength of optical fiber. In: *Strength of Inorganic Glasses* (C. R. Kurkjian, ed.), pp. 309–328. Plenum Press, New York.
[10] Matthewson, M. J. 1994. Optical fiber mechanical testing techniques. *Proc. Soc. Photo-Opt. Instrum. Eng. Crit. Rev.* CR50:32–59.

[11] Kraus, J. et al. 2002. Bending strength and fractographic analysis of zinc tellurite glass modified fibres. *Proc. Soc. Photo-Opt. Instrum. Eng.* 4639:30–39.

[12] Merwin H. E. and E. S. Larsen. 1912. Mixtures of amorphous sulphur and selenium as immersion media for the determination of high refractive indices with the microscope. *Am. J. Sci.* 34:42–47.

[13] Dewald, J. F. et al. 1962. *J. Electrochem. Soc.* 109:243C.

[14] Kolomiets, B. T. and E. A. Lebedev. 1963. *Radiotecknika i Radioelectronika* 8:2037.

[15] Poulain, M. et al. 1975. Verres fluores au tetrafluorure de zirconium proprietes optiques d'un verre dope au Nd^{3+}. *Mat. Res. Bull.* 10(4):243–246.

[16] Harrington, J. A. 2004. *Infrared Fibers and Their Applications.* SPIE Press, Bellingham, WA.

[17] Proctor, B. A. et al. 1967. The strength of fused silica. *Proc. Roy. Soc.* A297:534–557.

[18] Weibull, W. 1951. A statistical distribution of wide applicability. *J. Appl. Mech.* 18.

[19] Matthewson, M. J. et al. 1986. Strength measurement of optical fibers by bending. *J. Am. Ceram. Soc.* 69:815.

[20] Krause, J. T. et al. 1979. Deviations from linearity in the dependence of elongation upon force for fibres of simple glass formers and of glass optical lightguides. *Phys. Chem. Glasses* 20(6):135–139.

[21] Glaesemann, G. S. et al. 1988. Effect of strain and surface composition on Young's modulus of optical fibers. *OFC'88 Tech. Digest* 48.

[22] Griffioen, W. 1992. Effect of nonlinear elasticity on measured fatigue data and lifetime estimations of optical fibers. *J. Am. Ceram. Soc.* 75(10):2692–2696.

[23] France, P. W. et al. 1980. Liquid nitrogen strengths of coated optical glass fibres. *J. Mat. Sci.* 15:825–830.

[24] Ernsberger, F. M. 1969. Tensile and compressive strength of pristine glasses by an oblate bubble technique. *Phys. Chem Glasses* 10(6):240–245.

[25] Lower, N. P. 2004. Failure studies of glass fibers, PhD thesis. University of Missouri, Rolla, MO.

[26] Kurkjian, C. R. et al. 2003. The intrinsic strength and fatigue of oxide glasses. *J. Non-Cryst. Solids* 316:114–124.

[27] Inniss, D. et al. 1993. Chemically corroded pristine silica fibers: Sharp or blunt flaws? *J. Am. Ceram. Soc.* 76:3173.

[28] Griffith, A. A. 1920. The phenomena of rupture and flow in solids. *Phil. Trans. Roy. Soc.* A221:163.

[29] Otto, W. H. 1955. Relationship of tensile strength of glass fibers to diameter. *J. Am. Ceram. Soc.* 38:122.

[30] Cameron, N. M. 1968. The effect of environment and temperature on the strength of E-glass fibres, Pt. 1. High vacuum and low temperature. *Glass Tech.* 9(1):14.

[31] Kurkjian, C. R. and U. C. Paek. 1983. Single-valued strength of "perfect" silica fibers. *Appl. Phys. Lett.* 42:251.

[32] Smith, W. A. and T. M. Michalske. 1990. Report for DOE contract no. DE-AC04-70DPOO789.

[33] Mazur, E. 2004. Personal communication.

[34] Lin, B. L. and M. J. Matthewson. 1996. "Diameter Dependence of the Strength of Optical Fibers." Unpublished work.

[35] Armstrong, J. L. et al. 2000. Humidity dependence of the fatigue of high-strength fused silica optical fibers. *J. Am. Ceram. Soc.* 83(12):3100–3108.

[36] Taylor, A. T. and M. J. Matthewson. 1998. Effect of pH on the strength and fatigue behavior of fused silica optical fiber. In: *Proc. 47th Int. Wire Cable Symp,* pp. 874–879. INCS, Eatontown, NJ.

[37] Glaesemann, G. S. 2006. Personal communication.

[38] Kurkjian, C. R. 1977. Tensile strengths of polymer coated fibers for use in optical communications. In: *Proc. XI Int. Cong. Glass*, p. 469.

[39] Breuls, A. 1993. A COST comparison of *n* values obtained with different techniques. In: *COST Proc. Turin*.

[40] Krause, J. T. 1979. Transitions in the static fatigue of fused silica fiber lightguides. In: *Proc. 5th ECOC, Postdeadline paper* 19.1-1–19.1-4.

[41] Wang, T. T. and H. M. Zupko. 1978. Long-term mechanical behavior of optical fibres coated with a u.v.-curable epoxy. *J. Mat. Sci.* 13:2241–2248.

[42] Kurkjian, C. R. and M. J. Matthewson. 1996. Strength degradation of lightguide fibres in room temperature water. *Elec. Lett.* 32(8).

[43] Matthewson, M. J. and H. H. Yuce. 1994. Kinetics of degradation during fatigue and aging of fused silica optical fiber. *Proc. Soc. Photo-Opt. Instrum. Eng.* 2290:204–210.

[44] Krukjian, C. R. et al. 1996. Room temperature strength degradation of optical fibers. *SPIE, 2611* 34–37.

[45] Rondinella, V. V. et al. 1994. Coating additives for improved mechanical reliability of optical fiber. *J. Am. Ceram. Soc.* 77(1):73–80.

[46] Schneider, H. W. 1988. Strength in fluoride glass fibers. In: *Halide Glasses V.* (M. Yamane and C. T. Moynihan, eds.), pp. 561–570. Trans Tech Publications, Inc., Uetikon-Zurich.

[47] Carter, S. F. 1990. Mechanical properties. In: *Fluoride Glass Optical Fibres* (P. W. France et al., eds.), pp. 219–236. Blackie and CRC Press, Florida.

[48] Colaizzi, J. and M. J. Matthewson. 1994. Mechanical durability of ZBLAN and Aluminum fluoride-based optical fiber. *J. Lightwave Technol.* 12(8):1317–1324.

[49] Devyatkh, G. G. et al. 1999. Recent developments in As-S glass fibers. *J. Non-Cryst. Solids* 256/257:318–322.

[50] Sanghera, J. 2005. Strength of arsenic sulfide fibers. Personal communication.

[51] Dianov, E. M. et al. 1990. Mechanical properties of chalcogenide glass optical fibers. *Proc. SPIE* 1228:92–100.

[52] Zhou, J. et al. 2004. High tensile strength photonic crystal fibers. *OFC'04 Tech. Digest* WI2.

[53] Mrotek, J. L. et al. 2001. Diffusion of moisture through optical fiber coatings. *J. Lightwave Technol.* 19(7):988–993.

[54] Bogatyrjov, V. A. et al. 1988. High strength hermetically sealed optical fiber. *JETP Lett.* 14:343.

[55] Orcel, G. et al. 1995. Hermetic and polymeric coatings for military and commercial applications. *Proc. Int. Wire Cable Symp.* 330–334.

[56] Kurkjian, C. R. and H. Leidecker. 2001. Strength of carbon-coated fibers. *Proc. Soc. Photo-Opt. Instrum. Eng.* 4215:134–143.

[57] Kurkjian, C. R. 2006. Distribution of stripping-induced flaws. *Proc. Soc. Photo-Opt. Instrum. Eng.* 4215:150–157.

[58] Rondinella, V. V. and M. J. Matthewson. 1993. Effect of chemical stripping on the strength and surface morphology of fused silica optical fiber. *Proc. Soc. Photo-Opt. Instrum. Eng.* 2074:52–58.

[59] Matthewson, M. J. et al. 1997. Acid stripping of fused silica optical fibers without strength degradation. *J. Lightwave Technol.* 15(3):490–497.

[60] Park, H. S. et al. 1999. A novel method of removing optical fiber coating with hot air stream. *OFC/IOOC99 Tech. Digest* 2:371–373.

[61] Castilone, R. J. et al. 2002. Relationship between mirror dimensions and failure stress for optical fibers. *Proc. Soc. Photo-Opt. Instrum. Eng.* 4639:11–20.

[62] Krause, J. T. et al. 1981. Strength of fusion splices for fibre lightguides. *Elec. Lett.* 17:232.

[63] Krause, J. T. 1986. Ultrahigh strength fibre fusion splices by modified H_2/O_2 flame fusion. *Elec. Lett.* 22:1075–1076.

[64] Simpkins, P. G. and C. R. Kurkjian. 1995. Aluminum-coated silica fibres: Strength and solderability. *Elec. Lett.* 31(9):747–748.

[65] Bogatyrjov, V. A. et al. 1992. Heat resistant optical fibers hermetically sealed in aluminum. *Tech. Phys. Lett.* 8(698):699.

[66] Proctor, B. A. 1962. The effects of hydrofluoric acid etching on the strength of glass. *Phys. Chem Glasses* 3(1):7–27.

[67] Vitman, F. F. et al. 1996. Influence of temperature on the strength of etched fused quartz glass in its high strength state. *Sov. Phys. Sol. State* 8(5):1195.

[68] Miyama, Y. et al. 1983. Submarine optical-fiber cable containing high-strength splices. *J. Lightwave Tech.* 1(1):184–189.

[69] Kurkjian, C. R. et al. 2004. Effects of heat treatment and HF etching on the strength of silica lightguides. *Proc. Soc. Photo-Opt. Instrum. Eng.* 5465:223–229.

[70] Kurkjian, C. R. et al. 2006. Recovery of strength caused by HF etching. Unpublished work.

[71] Castilone, R. J. et al. 2000. Extrinsic strength measurements and associated mechanical reliability modeling of optical fiber. *Proc. 16th NFOEC.*

[72] Technical report on the power-law theory of optical fibre reliability. IEC SC86A/WG1 NO-17A, IEC (March, 1997).

[73] Mitsunaga, Y. et al. 1982. Failure prediction for long length optical fiber based on proof testing. *J. Appl. Phys.* 53(7):4847–4853.

Index

783

Vapor axial deposition (*Continued*)
rare earth-doped fiber fabrication,
202–209

W

Water peak loss, *see* Zero water peak fiber
Waveguide dispersion
definition, 38, 42
group delay equation, 42
multilayer index profiles, 44–45
wavelength dependence, 43–44
Weapon systems, development, 3
Weibull statistics, tensile strength,
753–754

X

XPM, *see* Cross-phase modulation

Y

Ytterbium-doped fiber
co-doped Er/Yb systems, 222–223,
689–690

large-mode-area fiber lasers
advantages, 673–675
bend loss, 676
diameter limits, 675–676
near-field spatial profile, 677
numerical aperture, 675–676, 678
stimulated Brillouin scattering,
678–679
photo-darkening, 231–234

Z

Zero water peak fiber
applications, 134
fabrication, 134–137
hydrogen aging loss, 139–141
performance maintenance, 137–138
strength paradoxes, 273–276
water peak loss and components, 134,
136
ZWP fiber, *see* Zero water peak fiber

Printed and bound by CPI Group (UK) Ltd, Croydon, CR0 4YY

10/05/2025

01866494-0001